The ARRL

Operating Manual

Seventh Edition

Editor
Dana George Reed, KD1CW

Contributing Editors
Chuck Hutchinson, K8CH
Joel Kleinman, N1BKE
R. Dean Straw, N6BV
Larry Wolfgang, WR1B

Cover Design
Sue Fagan

Production and Proofreading
Michelle Bloom, WB1ENT
Paul Lappen
Jayne Pratt Lovelace
Jodi Morin, KA1JPA
David Pingree, N1NAS
Joe Shea
Edward Vibert

About the Cover

Background photo: One of the Nashua (NH) Area Radio Club Field Day antennas. *Photo by Maureen Heedles.*

Top left: Two members of the Harvard (University) Wireless Club celebrate the club's 90th anniversary by firing up some vintage Heath gear at W1AF during a special event operation from the club station in Cambridge, Massachusetts. *Photo courtesy Harvard Gazette —Kris Snibbe.*

Center: Rich, KG2PU (left), and John, K2RXR, operate the W2GSA Marconi special event station from Twin Lights lighthouse, Atlantic Highlands, New Jersey.

Lower right: During a recent Jamboree on the Air, Jared Boswel of Troop 543 in Florida experiences the art of radio communication with another scout in Montana. Volunteer Chris Duvall, KF4RRM, lends a hand.

Published by
ARRL
The national association for Amateur

Printed in the USA

Seventh Edition
Second Printing, 2002

ISBN: 0-87259-793-8
ISBN: 0-87259-809-8 CD-ROM

Contents

Foreword

The ARRL Operating Manual belongs in every ham's shack. No matter if you've just received your first Amateur Radio license or if you're an experienced veteran on the bands, this volume contains help and information for you.

In these pages you'll find a wealth of reference material in an easy-to-use format. You'll also find explanations of activities and modes. All of this is designed to help you increase your enjoyment and skill.

Every chapter of this edition has been reviewed and updated by folks that are experienced operators. The manual has been reorganized to make it easier for you to use. The chapters on Operating Basics, Rules and Regulations, FM, Contesting, Image Communications and Online Resources have been completely rewritten.

The FCC rules Part 97 start with the basis for Amateur Radio, and in the first place state as a principle, "Recognition and enhancement of the value of the amateur service to the public as a voluntary noncommercial communication service, particularly with respect to providing emergency communications."

In the next year, I challenge you to engage in three activites. First, learn to handle formal message traffic. You don't have to join a net, just learn how. If you already know how, then sharpen your skills.

Second, renew your commitment to Public Service. Get involved and train with those who serve your community. Use your radio and your skills to make your community a better place to live.

Finally, *enjoy* Amateur Radio. Do something that's fun. Try a new band, mode or activity. The pages of this book are filled with ideas that are both fun and challenging.

As always, we want to hear from you. This operating manual is meant for you. We conducted a survey to help us plan this edition, and based on that survey we prepared this edition. We want to hear from you. Let us know what you think by filling out the feedback form at the back of this book. We want our publications to meet your needs.

David Sumner, K1ZZ
Executive Vice President
Newington, Connecticut
June 2000

About the ARRL

The national association for Amateur Radio

The seed for Amateur Radio was planted in the 1890s, when Guglielmo Marconi began his experiments in wireless telegraphy. Soon he was joined by dozens, then hundreds, of others who were enthusiastic about sending and receiving messages through the air—some with a commercial interest, but others solely out of a love for this new communications medium. The United States government began licensing Amateur Radio operators in 1912.

By 1914, there were thousands of Amateur Radio operators—hams—in the United States. Hiram Percy Maxim, a leading Hartford, Connecticut inventor and industrialist, saw the need for an organization to band together this fledgling group of radio experimenters. In May 1914 he founded the American Radio Relay League (ARRL) to meet that need.

Today ARRL, with approximately 170,000 members, is the largest organization of radio amateurs in the United States. The ARRL is a not-for-profit organization that:

- promotes interest in Amateur Radio communications and experimentation
- represents US radio amateurs in legislative matters, and
- maintains fraternalism and a high standard of conduct among Amateur Radio operators.

At ARRL headquarters in the Hartford suburb of Newington, the staff helps serve the needs of members. ARRL is also International Secretariat for the International Amateur Radio Union, which is made up of similar societies in 150 countries around the world.

ARRL publishes the monthly journal *QST*, as well as newsletters and many publications covering all aspects of Amateur Radio. Its headquarters station, W1AW, transmits bulletins of interest to radio amateurs and Morse code practice sessions. The ARRL also coordinates an extensive field organization, which includes volunteers who provide technical information and other support services for radio amateurs as well as communications for public-service activities. In addition, ARRL represents US amateurs with the Federal Communications Commission and other government agencies in the US and abroad.

Membership in ARRL means much more than receiving *QST* each month. In addition to the services already described, ARRL offers membership services on a personal level, such as the ARRL Volunteer Examiner Coordinator Program and a QSL bureau.

Full ARRL membership (available only to licensed radio amateurs) gives you a voice in how the affairs of the organization are governed. ARRL policy is set by a Board of Directors (one from each of 15 Divisions). Each year, one-third of the ARRL Board of Directors stands for election by the full members they represent. The day-to-day operation of ARRL HQ is managed by an Executive Vice President and his staff.

No matter what aspect of Amateur Radio attracts you, ARRL membership is relevant and important. There would be no Amateur Radio as we know it today were it not for the ARRL. We would be happy to welcome you as a member! (An Amateur Radio license is not required for Associate Membership.) For more information about ARRL and answers to any questions you may have about Amateur Radio, write or call:

ARRL—The national association for Amateur Radio
225 Main Street
Newington CT 06111-1494
Voice: 860-594-0200
Fax: 860-594-0259
E-mail: **hq@arrl.org**
Internet: **www.arrl.org/**

Prospective new amateurs call (toll-free):
800-32-NEW HAM (800-326-3942)
You can also contact us via e-mail at **newham@arrl.org**
or check out *ARRLWeb* at **http://www.arrl.org/**

The ARRL—At Your Service

ARRL Headquarters is open from 8 AM to 5 PM Eastern Time, Monday through Friday, except holidays. Call **toll free** to join the ARRL or order ARRL products: **1-888-277-5289** (US), M-F only, 8 AM to 8 PM Eastern Time.

If you have a question, try one of these Headquarters departments . . .

	Telephone	*Electronic Mail*
Joining ARRL	860-594-0338	**membership@arrl.org**
***QST* Delivery**	860-594-0338	**circulation@arrl.org**
Publication Orders	860-594-0355	**pubsales@arrl.org**
Regulatory Info	860-594-0236	**reginfo@arrl.org**
Exams	860-594-0300	**vec@arrl.org**
Educational Materials	860-594-0301	**ead@arrl.org**
Contests	860-594-0232	**contests@arrl.org**
Technical Questions	860-594-0214	**tis@arrl.org**
Awards	860-594-0288	**awards@arrl.org**
DXCC/VUCC	860-594-0234	**dxcc@arrl.org**
Advertising	860-594-0207	**ads@arrl.org**
Media Relations	860-594-0328	**newsmedia@arrl.org**
QSL Service	860-594-0274	**buro@arrl.org**
Scholarships	860-594-0230	**foundation@arrl.org**
Emergency Comm	860-594-0265	**emergency@arrl.org**
Clubs	860-594-0267	**clubs@arrl.org**
Hamfests	860-594-0262	**hamfests@arrl.org**

You can send e-mail to any ARRL Headquarters employee if you know his or her name or call sign. The second half of every Headquarters e-mail address is **@arrl.org**. To create the first half, simply use the person's call sign. If you don't know their call sign, use the first letter of their first name, followed by their complete last name. For example, to send a message to John Hennessee, N1KB, Regulatory Information Specialist, you could address it to **jhennessee@arrl.org** or **N1KB@arrl.org**.

If all else fails, send e-mail to **hq@arrl.org** and it will be routed to the right people or departments.

Technical Information Service

The ARRL answers questions of a technical nature for ARRL members and nonmembers alike through the Technical Information Service. Questions may be submitted via e-mail (**tis@arrl.org**); Phone (860-594-0214); Fax (860-594-0259); or mail (TIS at ARRL, 225 Main Street, Newington, CT 06111). The TIS also maintains a home page on *ARRLWeb*: **http://www.arrl.org/tis**. See the **References** chapter of this *Handbook* for a more detailed description of the numerous technical links available at this site. Also, please note that the "Technical Information Server" or *Info Server* service previously available via e-mail has been discontinued.

ARRL ON THE WORLD WIDE WEB

You'll find *ARRLWeb* at: **http://www.arrl.org/**

At the ARRL Web page you'll find the latest W1AW bulletins, a hamfest calendar, exam schedules, an on-line ARRL Publications Catalog and much more. We're always adding new features to *ARRLWeb,* so check it often!

Members-Only Web Features

As an ARRL member you enjoy exclusive access to our Members-Only Web features. Just point your browser to **http://www.arrl.org/members/** and you'll open the door to benefits that you won't find anywhere else.

• *QST* Product Review Archive. Get copies of *QST* product reviews from 1980 to the present.

• *QST/QEX* searchable index (find that article you were looking for!)

• Previews of contest results and product reviews. See them here before they appear in *QST!*

• Access to your information in the ARRL membership database. Enter corrections or updates on line!

Stopping by for a visit?

We offer tours of Headquarters and W1AW at 9, 10 and 11 AM, and at 1, 2 and 3 PM, Monday to Friday (except holidays). Special tour times may be arranged in advance. Bring your license and you can operate W1AW anytime between 10 AM and noon, and 1 to 3:45 PM!

Would you like to write for *QST*?

We're always looking for new material of interest to hams. Send a self-addressed, stamped envelope (2 units of postage) and ask for a copy of the *Author's Guide*. (It's also available via the ARRL Info Server, and via *ARRLWeb* at **http://www.arrl.org/qst/aguide/**)

Press Releases and New Products/Books

Send your press releases and new book announcements to the attention of the ***QST* Editor** (e-mail **qst@arrl.org**). New product announcements should be sent to the Product Review Editor (e-mail **reviews@arrl.org**).

ARRL Audio News

The best way to keep up with fast-moving events in the ham community is to listen to the ARRL Audio News. It's as close as your telephone at 860-594-0384, or on the Web at **http://www.arrl.org/arrlletter/audio/**

Interested in Becoming a Ham?

Just pick up the telephone and call toll free 1-800-326-3942, or send e-mail to **newham@arrl.org**. We'll provide helpful advice on obtaining your Amateur Radio license, and we'll be happy to send you our informative Prospective Ham Package.

The Amateur's Code

The Radio Amateur is:

CONSIDERATE...never knowingly operates in such a way as to lessen the pleasure of others.

LOYAL...offers loyalty, encouragement and support to other amateurs, local clubs, and the American Radio Relay League, through which Amateur Radio in the United States is represented nationally and internationally.

PROGRESSIVE...with knowledge abreast of science, a well-built and efficient station and operation above reproach.

FRIENDLY...slow and patient operating when requested; friendly advice and counsel to the beginner; kindly assistance, cooperation and consideration for the interests of others. These are the hallmarks of the amateur spirit.

BALANCED...radio is an avocation, never interfering with duties owed to family, job, school or community.

PATRIOTIC...station and skill always ready for service to country and community.

—The original Amateur's Code was written by Paul M. Segal, W9EEA, in 1928.

Chapter 1

Operating Amateur Radio

By ARRL Staff

Amateur Radio is a realm of magic that puts people in contact with other people—around their towns or cities, or around the world. Amateur Radio provides opportunities to learn about and experiment with technology. It's also a way to learn electronic communication skills. It's a means to overcome the limitations of physical handicaps and enter an open arena of communications. It's fun, and a great way to make friends. It's an opportunity to participate in public service activities. And it's an opportunity to serve your community and make your neighborhood a better place to live.

There is one basic idea linking the various opportunities in ham radio—that of communicating, whether by talking, sending pictures or digital data, or sending Morse code. You have the ability to communicate with other hams, as Amateur Radio operators are traditionally called. Sure, you could just pick up a telephone, or you could log onto the Internet to communicate with other people. But the real magic of communicating by radio is very often the sheer randomness of making contact. You never know who might come back to you when you call CQ ("calling all hams").

This *ARRL Operating Manual* covers the most popular modes used in ham radio. Whether you are a new ham or are rejoining the hobby after a long absence, this book can help you decide what to operate, where to operate and how to operate. Think of your participation in Amateur Radio as a never-ending journey—there's always something new to explore, always something new to do!

WHY AMATEUR RADIO EXISTS

Amateur Radio owes it existence to international and national regulations. These regulations reflect the value of Amateur Radio as perceived by national and international leaders. In the United States, the fundamental purpose of the Amateur Radio Service is expressed in the following principles outlined in Federal Communications Commission rule §97.1:

(a) Recognition and enhancement of the value of the amateur service to the public as a voluntary noncommercial communication service, particularly with respect to providing emergency communications.

(b) Continuation and extension of the amateur's proven ability to contribute to the advancement of the radio art.

(c) Encouragement and improvement of the amateur service through rules which provide for advancing skills in both the communications and technical phases of the art.

(d) Expansion of the existing reservoir within the amateur radio service of trained operators, technicians, and electronics experts.

(e) Continuation and extension of the amateur's unique ability to enhance international goodwill.

Amateur Radio is a Service

If you have read the Public Service column in *QST*, you've no doubt heard of the exploits of hams from all walks of life who have selflessly donated their time by providing

Fourteen hams in one family! It all began with Greg, KD6CPN, and Julie, KE6BKK, Sue of Ranchos Palos Verdes, CA. Their son Kevin, age 9, is KF6LYY. Like a row of toppling dominos the rest of the family followed suit. Several Hawaiian cousins passed their exams while visiting the rest of the clan. Only 4-year-old Christopher remains unlicensed—but he has to learn how to read first! Top row, from left to right: Amy, KH7EZ; Nancy, NH6OU; Julie, KE6BKK; Eleanor, KF6NAU; Julius, KF6NAQ; Kitty, KF6MRI; Laura, KF6MRL; Katherine, KF6MRC. Lower row, from left to right: Alan, KH7FA; Kevin, KF6LYY (front); Jeff, AH6IX; Greg, KD6CPN; Christopher; Darryl, KJ6UD; Natalie, KF6NFU.

Mark, KBØYRK (left), and Bill, KBØMGU (right), use a portable amateur television unit to send flood-damage survey video back to the Grand Forks, North Dakota, Emergency Operations Center. The extent of the terrible Grand Forks Flood of 1997 is readily apparent in the photo.

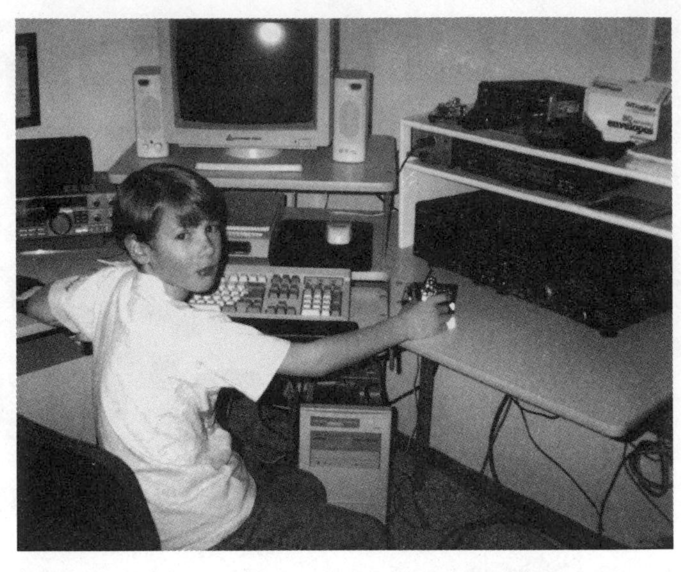

Jeff, NØRY, age 11, raced with his dad, Joe, NØLF, to the Amateur Extra level. Pop beat Jeff by passing the exam one day before Jeff did, but their Extra tickets were issued on the same day. Jeff was only 10 when he passed the exams, and he went through the ranks from Novice to Extra in just six months. Jeff's proud mom, Ruth, is NØNUI.

emergency communication. Many, many more hams provide this service than are given recognition by the media, but that's all part of being a ham—the intrinsic reward is the satisfaction of doing a job well.

You may or may not be called at some time to provide this service to your community. But by being prepared, honing your on-the-air operating skills to their sharpest, maximizing your equipment to obtain the best from it, and being prepared to pitch in should the need arise, you will be ready. In so doing, you will derive untold hours of satisfaction from the exciting hobby we know as Amateur Radio.

THE WIDE DIVERSITY OF HAMS

A retired California housewife became a ham so that she could serve her community as a volunteer. Her husband is a long-time ham who is well known around the world for his operating skills and exploits. He regularly talks with hams in all parts of the world, but that's not her goal. In fact, most of the folks she talks to are less than 20 miles away. She is an active leader in emergency communications and practices and trains others regularly. When the need arises, she and other members of her club provide emergency communications services to the agencies that protect her home and her community.

A boy asks his father how radios and computers work. They talk a bit and then begin to experiment. The talks and experiments lead to further study and the topics broaden. The boy's interest in technology continues to grow. He studies for and passes his first Amateur Radio license examination, and shortly after that he upgrades. He enjoys communicating, but the technology is what really attracts his interest. What does his future hold? No doubt he'll have a career in some field of electronics.

The young woman, every inch a competitor, shouted with joy as she read the results. Not only had she beaten everyone else in her entry category in the contest, but she had also set a world record. She felt great!

The middle-aged man turned in his chair when the computer beeped an alert. Yes! The station he had wanted to contact had been spotted on the air. That station was set up on an island in the Indian Ocean by a couple of hams, a husband-and-wife team who spend their vacations operating Amateur Radio from exotic locations. Yes, they have an antenna atop the apartment building where they live, but there's a big difference between operating from home and operating from an exotic location that great numbers of hams want to contact.

The young girl, having come home from school, went into her bedroom and turned on her radio. No music came from the radio. Instead, she picked up the microphone and called another Amateur Radio station. A familiar voice answered her call, "Hi, honey, how are you today? Over." She replied, "I'm fine, grandpa. Let me tell you what happened today..."

You might think you know these folks. It's possible, since they are real. But the descriptions fit many persons. All kinds of people enjoy communicating over the air. Kings and diplomats, homemakers and waiters, engineers and scientists, teachers and students, doctors and bookkeepers find fun, challenge and fulfillment in Amateur Radio. Day or night, hams are communicating with each other, helping others, doing technical experiments or just simply having fun by meeting new people.

THE WIDE VARIETY OF HAM ACTIVITIES

Hams are involved in all sorts of fun, challenging and fulfilling activities. That means not just from home, but on the move—in a car, or on a bike or hiking in the mountains. You might be on the sea or in the air—even in Earth orbit, where many ham astronauts have enjoyed using their radios.

The chapter titles of this book provide only partial insight into the scope of activities in which Amateur Radio operators are engaged. So what sorts of things do hams do?

Near, Far, Wherever You Are

Some enjoy talking to their local friends and neighbors using small, inexpensive, low-powered radios that operate on fixed channels. Repeaters extend the limited range of their radios by amplifying and retransmitting their signals. Depending on where the repeater is located, you can talk with other hams 50 or even 100 miles away. Distances up to 30 miles are common.

Others are enchanted with faraway places with strange-sounding names. The object of their quest is distance (DX). This usually translates to trying to contact as many distinct political and geographic areas as they can. For many of these the quest doesn't end until the contact is confirmed with a *QSL* card—a postcard confirming contact with another station—in the mail.

Some enjoy the thrill and adventure of travel to and operating from distant and exotic places. That doesn't necessarily mean the farthest corners of the world. It could be a vacation trip to a warm island in the Caribbean to escape the freezing winter weather, or on a business trip to the Middle East.

Out of This World!

And hams are not limited to communication on the Earth. Many enjoy talking with astronauts and cosmonauts in orbit. As this is written, hams around the world are awaiting their first radio contacts with other hams aboard the International Space Station.

Hams also communicate with each other through satellites that hams themselves have designed and built. Some even bounce their signals to each other off the moon. Others use an active auroral zone or an ephemeral meteor trail to reflect transmissions. Hams in fact have been pioneers in many modes of communication, paving the way for many different commercial uses.

This slow-scan TV picture sent from the Russian *Mir* space station shows cosmonauts Gennady Padalka (left) and Sergei Ardeyev looking over the operator's manual for the Kenwood TM-V7A FM transceiver.

Competitions

The thrill of competition calls some hams to enter contests for a weekend of intense activity. It's a way to test station capability and operator ability. For the busy ham, it's a great way to cram a lot of contacts into a short period of time. For others, a contest involves operating away from home—perhaps with friends at a "super station."

Competition takes on many forms. Hams who love to collect awards compete against themselves. Are they able to make contact with all the states? All the counties? All the prefectures of Japan or the provinces of Spain?

Some hams take to foot or vehicle with direction-finding equipment to see who's the best at finding a hidden transmitter. Their competition may be local or international. Contesting is to Amateur Radio what the Olympic Games are to worldwide amateur athletic competition: a showcase to display talent and learned skills, as well as a stimulus for further achievement through competition. Increased operating skills and greater efficiency may be the end result of Amateur Radio contesting, but the most common experience is *fun*.

The contest operator is also likely to have one of the better signals on the band—not necessarily the most elaborate station equipment, but a signal enhanced by constant experimentation and improvement. Contest operation encourages station and operator efficiency. Nearly every contest has competitors vying to see which one can work the most stations (depending on the rules of the particular contest) in a given time frame. In some contests, the top-scoring stations have consistently worked 100 or more stations per hour for the entire 48-hour contest period.

The ARRL contest program is so diverse that one contest or another appeals to almost every type of ham—the beginning contester and the old hand, the newest Technician and the most experienced Extra Class veteran, the "Topband" (160 meter) buff and the microwave enthusiast. Complete contest entry rules and results appear in *QST* a few months before the contest. There is at least one contest for every interest.

Working Cooperatively

Those involved in *public service* and *traffic handling* work together cooperatively. Traffic handling involves passing messages to others over the amateur bands. Hams handle *third-party traffic* (messages for nonhams) in both routine situations and in times of disaster. Public service communications make Amateur Radio a valuable public resource, one that has been recognized by Congress and a whole host of federal, state and local agencies that serve the public.

Nets are regular gatherings of hams who share a mutual interest and who use the net (short for "network") to further that interest. Their most common purpose is to pass traffic or participate in one of the many other ham activities, from awards chasing and DXing to just plain old talking among longtime friends.

County and State hunting nets are very popular since they provide a frequency to work that 49th and 50th state for the ARRL WAS (Worked All States) award. DX nets provide the same opportunity for those interested in working

DX stations without the competition of contests or by ceaselessly tuning around looking for "new ones."

There are nets dedicated to beginner or slow-speed CW operation. These can help a newcomer sharpen operating skills. The *ARRL Net Directory*, available from ARRL HQ for $5 as this is written, is a compilation of public service nets.

DXing

DX (ham shorthand for long distance) holds a special fascination for many hams. It can be correctly defined in different ways. To most amateurs, DX is the lure of seeing how far away you can establish a QSO—the greater the distance, the better. DX is a personal achievement, bettering some previous "best distance worked," involving a set of self-imposed rules.

DX can also be competing on a larger scale, trying to break the *pileups* of people calling a rare DX station. DXers often aim for one of the DX-oriented awards. DXing can be a full-time goal for some hams and a just-for-fun challenge for others. Regardless of whether you turn into a serious DXer or just have a little fun searching for entities you have yet to work, DXing is one of the most fascinating aspects of Amateur Radio.

Award Hunting

Some hams spend most of their time in a pursuit that never ends—earning awards. The ARRL offers the awards listed in **Table 1-1**. Most other national amateur societies, private clubs and contest groups sponsor awards and certificates for various operating accomplishments. Many of the awards are very handsome paper certificates or intricately designed plaques very much in demand by awards chasers.

Table 1-1
ARRL Operating Awards

Award	***Qualification***
Friendship Award	Contact 26 stations with calls ending A through Z.
Rag Chewers' Club	A single contact 1/2 hour or longer
Worked All States (WAS)	QSLs from all 50 US states
Worked All Continents (WAC)	QSLs from all six continents
DX Century Club (DXCC)	QSLs from at least 100 foreign entities
VHF/UHF Century Club (VUCC)	QSLs from many grid locators
A-1 Operator Club	Recommendation by two A-1 operators
Code Proficiency	One minute of perfect copy from W1AW qualifying run
Old Timers Club	Held an Amateur Radio license at least 20 years
ARRL Membership	ARRL membership for 25, 40, 50, 60 or 70 years

Rag Chewing

"Chewing the rag" refers to getting on the air and spending minutes (or hours!) in interesting conversation on virtually any and every topic imaginable. Without a doubt, the most popular operating activity is rag chewing. The rag chew may be something as simple as a brief chat on a 2-meter or 440-MHz FM repeater as you drive across town. It also may be a group of friends who have been meeting on 15-meter SSB every Saturday afternoon for 20 years. The essential element is the same—hams talking to each other on any subject that interests them. The ARRL awards a certificate that declares membership in the *Rag Chewers' Club*, earned by carrying on a conversation of half an hour or longer. Most hams will tell you that it is not difficult to earn.

Mary, TI3AMY, and her husband Olbert, TI3OMY, at their station in Turrialba, Costa Rica.

PSK31, Packet and Other Digital Modes

In the past, RTTY (Baudot teletype) was the only digital mode. As computers became popular, then almost essential in the ham shack, the number of digital modes expanded quickly. The fist computer-based mode was packet radio. Packet requires either a full computer or a terminal (display and keyboard) at each end, in addition to a packet radio controller and the radios themselves. The Packet Cluster has become an unparalleled tool for the DXer hunting for needed countries, while APRS, another offshoot of packet, is another popular application. For details, see the VHF Digital Communications and Networks and DXing chapters.

PSK31 is a relatively new digital mode that has become very popular. It is a narrow-band, real-time digital mode that exploits the ubiquitous sound card and Windows operating system. The software is even available free of charge over the Web. Other digital modes include PACTOR, Clover and G-TOR. See the HF Digital Communications chapter for details.

GETTING STARTED

Okay, so you're interested in Amateur Radio (you probably wouldn't be reading this book if you weren't!). We hope we've piqued your interest about some of the neat things hams do. So, how do you get the information you need about specific aspects of our fascinating hobby? How do you go about getting your first license? And how do you go about setting up a station so you can actually start communicating with other hams?

First, let us introduce you to an organization dedicated specifically to Amateur Radio. It is the ARRL—the national association for Amateur Radio. The ARRL is the only not-for-profit organization set up to serve the more than 600,000 Amateur Radio operators (hams) in the United States, and you should become a member.

ARRL produces an entire line of publications dedicated to the radio amateur, including the book you are now reading. These materials provide great ways to keep informed of news and technical developments. Moreover, through your ARRL membership you will be supporting ongoing efforts in the national and international arenas that will help ensure that ham frequencies remain ham frequencies in spite of pressures from commercial interests.

Just about anything you may need to know—whether it be from the technical or operating sides of the hobby—can be found in an ARRL publication or obtained directly from the HQ staff. The basic beginner's publication is *Now You're Talking!*, which is used as a textbook by many instructors at local radio clubs.

The world-renowned ARRL journal, *QST*, published since 1915, is also an excellent source of technical, operating, regulatory and feature articles on all aspects of Amateur Radio. Since it is published each month, *QST* is a timely source of information all hams can use, and a significant benefit of ARRL membership. Contact the ARRL for membership information at 1-888-277-5289, or send e-mail to **circulation@arrl.org**.

ARRL on the Internet—*ARRLWeb*

ARRL also has a strong Internet presence—*ARRLWeb at* **www.arrl.org**. The Internet-connected ham can find late-breaking Amateur Radio news, as well as columns, feature articles, and information on virtually anything related to ham radio. A search engine helps visitors to zero in on specific pages, and a US call sign server is available to all.

ARRL members may register to access special Web site features and services. One of the most popular is the *QST* "Product Review." Members can also search the on-line archive of *QST* "Product Review" columns from the past 20 years. Product reviews provide no-nonsense, often critical technical and operating reviews of equipment to help guide your purchasing decisions. Members can also search the indexes of *QST* and *QEX*.

In addition, ARRL members can use the Web site to keep abreast of DXCC news, share their knowledge and experience via "The Doctor is On-Line" site, check HF propagation from the US to significant DX locations around the world, or sign up to receive free ARRL e-mail products. These include weekly editions of *The ARRL Letter* as well as W1AW/ARRL bulletins (general, DX, propagation, satellite and Keplerian), ARRL Division and Section news, new ARRL product announcements, membership expiration notification, and amateur license expiration notification.

On-the-Air Bulletins

Up-to-the-minute information on everything of immediate interest to amateurs—from Federal Communications Commission policy-making decisions to propagation predictions, to news of DXpeditions—is also transmitted in W1AW bulletins. W1AW is the Amateur Radio station maintained at ARRL HQ in Newington, Connecticut. Bulletins are transmitted at regularly scheduled intervals on CW (Morse code), phone (voice) and several digital (teletype) modes on various frequencies. In addition there are Morse-code practice broadcasts. A schedule of W1AW transmissions appears regularly in *QST* and may also be obtained from ARRL HQ or on *ARRLWeb*.

LOCAL AMATEURS/CLUBS

When you seek answers to your questions about ham radio, don't overlook another great source of information—the experience of your fellow hams. There is nothing hams like to do better than share the vast wealth of experience they have in the hobby. Being human, most hams like to brag a little about their on-the-air exploits. A general question on a particular area of operation to a ham who has experience in that area is likely to bring you all kinds of data. A dedicated DXer will talk for hours on techniques for work-

Hey, ho, and up she rises! Members of the Northwest Amateur Radio Society near Houston, Texas, haul up a triband beam and a 40-foot tower at the W5NC Field Day site.

ing a rare DX station. Similarly, a traffic handler will be only too happy to give you hints on efficient net procedures.

For the new ham (or potential ham), the problem may not be in asking the proper questions, but in finding another amateur of whom to ask the questions. That's where Amateur Radio clubs play an important role.

As a potential ham or a new ham, you might not know of an amateur in your immediate area. A good choice would be to try to find the time and meeting place for a nearby Amateur Radio club. What better source of information for the not-yet-licensed and the newly licensed than a whole club full of experienced hams? If you are not yet licensed, your local Amateur Radio club is also the place to find Morse code and electronics theory courses that will prepare you for the FCC exams.

ARRL Field and Educational Services can refer you to an ARRL-affiliated Amateur Radio club near you. The ARRL is more than happy to supply this information. To find a club near you, check *ARRLWeb* at **www.arrl.org/FandES/field/club/clubsearch.phtml**.

There is also a toll-free telephone number for prospective hams: 800-32NEW-HAM (800-326-3942), or you can send e-mail to **newham@arrl.org**.

HOW TO GET ON THE AIR

Let's say you've done your homework. You've studied hard and you passed your license exam. Congratulations! Like many newcomers to Amateur Radio, you have a Technician class license, giving you privileges on the VHF and UHF amateur bands. Getting on the air for you may be as simple as finding a new (or used) VHF hand-held radio, charging the batteries and then talking with your new ham friends through a local FM repeater.

This is exactly what many hams are looking for—reliable, fun communication with a small group of friends in their own and nearby communities. With your Technician license in hand, you also can talk or send Morse code through the amateur satellites or to space shuttle astronauts. Or, with some additional equipment and some easy-to-find software, you can explore the VHF digital modes such as APRS. In addition, you can use SSB voice or CW to contact other stations on the VHF and UHF bands. These modes require transceivers and antennas that are different from those used for FM.

Other new hams will find it hard to resist the lure of the HF bands. You'll hear hams on the repeater or at your radio club talking enthusiastically about their adventures on the lower-frequency bands. Someone may regale you with tales of how she has made friends many thousands of miles away, across oceans and continents. She might talk about the magic of that time when a ham in Calcutta answered her CQ on 20 meters. She'll smile broadly, recalling that the Indian ham was as interested as she is in jazz and programming computers.

By now, you can't help wondering what it takes to earn HF-band operating privileges. Your first step is to learn to receive Morse code at 5 words per minute. As a Technician who has passed the code test, you'll have limited access to HF. To earn the right to transmit on *all* the HF bands, you'll need to pass another written exam, the General. Beyond that is the Amateur Extra license, which allows access to every part of every ham band.

It is a bit more challenging to put together an HF station than a VHF station. HF equipment generally covers more modes and has greater requirements for frequency stability than an FM-only radio. Thus, HF gear is generally more expensive, and the antennas at HF are bigger than the flexible rubber antennas used on hand-held transceivers or the VHF ground-plane antenna mounted on the roof of your car or house.

There are so many different possible HF station configurations that it may be hard to choose the very best station to start out with. Most hams face other constraints—such as having a limited budget to spend for ham gear, or having restrictions on the size and location of antennas. Some hams have a difficult time installing any sort of antenna outdoors and they must resort to indoor or perhaps easily hidden "stealth" wire antennas.

The first step in selecting your new station should be to make up a list. This list simply answers a few questions. Will you be operating HF or only VHF? How much room do you have—in other words, how big can your shack or operating position be? Do you have room for long antennas, such as HF dipoles, or high antennas such as HF verticals and VHF/UHF arrays? How much do you plan to spend, including rig, furniture, coax and wire?

Some of these questions may not apply to your situation, and you may need to list the answers to other questions not discussed here. The result of making your personal list, however, is to give you an idea of what you want, as well as your limitations.

The next step is to do some research. You have many sources of information on ham gear:

1) ***Hands-on experience***. Try to use as many different pieces of gear as possible before you decide. This applies to VHF/UHF as well as HF gear. Ask a nearby ham friend or one or more of your fellow radio club members if you can use their station. Try your club's station. Note what you like and what features you don't care for in each of the stations you tried.

2) ***Radio club members***. Ask members of your local radio club about their personal preferences in gear and antennas. Be prepared for a great volume of input. Every amateur has an opinion on the best equipment and antennas. Years of experimentation usually go into finding just the right station equipment to meet a particular amateur's needs. Listen and take note of each ham's choices and reasons for selecting a particular kind of gear. There's a lot of experience, time, effort and money behind each of those choices.

3) ***Advertisements***. *QST* is chock-full of ads for all the newest up-to-date equipment, as well as some premium used gear. Read the ads, and don't be afraid to contact the manufacturers of the gear for further information. Compare specifications and prices to get the best deal. If there is a dealer close to you, pay him a visit.

4) ***Product Reviews***. *QST* also contains detailed Product Reviews, written by ARRL staff, that include reliable mea-

surements made in the ARRL Laboratory. These definitive reviews pull no punches in describing the good and the less-desirable features of the equipment being evaluated.

IN THE SHACK—YOUR EQUIPMENT

After you've made your final choice (it's not really final, as most hams will trade station equipment often during their ham careers), consider the sources where you might get the best deal, whether it be a new or used transceiver. Remember to include shipping and handling in the cost, and inquire about warranty service. Several possible sources of equipment are:

1) ***Local amateurs***. Many hams will have spare used gear and may be willing to part with this gear at a reasonable price. Be sure you know what a particular rig is going for on the open market before settling on a final price. If you are new to the hobby and you buy a rig from a local club member, you may be able to talk him or her into "Elmering" you (helping you) with the rig's installation and operation.

2) ***Hamfests/flea markets***. Many radio clubs run conventions called hamfests. Usually one of the big attractions of these events is the flea market or equipment sales. Much used gear, usually in passable shape, can be found at reasonable prices. Local distributors and manufacturers of new ham gear sometimes show up at these events to sell equipment.

3) ***Local electronics dealers***. If you are lucky enough to live near an electronics distributor who handles a line or two of ham gear, so much the better. The dealer can usually answer any questions you may have, will usually have a demonstration unit and will be pleased to assist you in purchasing your new gear.

4) ***Mail order***. There are many mail-order ham equipment distributors to choose from. Some deal in new equipment, some deal in used equipment and some deal in both. Check the ads in the ham publications, such as those found in *QST*, for the equipment you want. The ads are also an excellent place to find out the going price of a particular piece of gear. Since prices sometimes change quickly, many dealers list a toll-free number to call to obtain a price. Don't be shy—call them.

A detailed treatment of used equipment and getting on the air quickly and effectively is in the ARRL's book *Now You're Talking!*. The ARRL *Radio Buyer's Sourcebook* contains dozens of in-depth Product Reviews from *QST*. Other *Sourcebook* articles will help you understand how the tests were conducted and what the results mean. If you're looking for a rig, you should have a copy of the *Sourcebook*.

Take your time deciding what gear to get. You may even consider building your gear. There's a lot of satisfaction to be had in telling the operator on the other end of your QSO, "The rig here is home brew." *The ARRL Handbook for Radio Amateurs* has extensive coverage of construction and basic-to-advanced electronics theory information.

How Much Power?

Some hams find the low initial cost of low power (*QRP*) HF equipment very attractive, especially since some are available in kit form. Nevertheless, QRP is not for most beginners. In the ham radio community, there are both *QRO* (power output up to the full legal limit of 1500 watts) and QRP (5 watts output—or less) enthusiasts. Most hams run 100-150 watts—the power level of typical transceivers. There are times when it is necessary to run the full legal power limit to establish and maintain solid communications or to compete effectively in a DX contest on certain bands. Most often, the 100-watt level is more than enough to provide excellent contacts.

On VHF the situation is slightly different. Unless you are trying to work several hundred miles in an opening or operate a weak-signal mode, 5 to 10 watts is enough power when coupled with a good antenna for excellent local communications.

VHF/UHF Gear

Again, if you're interested mainly in voice and maybe packet radio communication with local amateurs, all you need is an FM transceiver. FM transceivers are available for hand-held and mobile use; the mobile rig can be used at home if you have a 12-V battery or suitable power supply. **Fig 1-1** illustrates the choices available for a VHF or UHF station and how portable, mobile and home (fixed) rigs and antennas can be used.

1) ***Hand-held transceivers***. Hand-held transceivers put out from under a watt to 5 watts or more. If the hand-held you're considering is capable of high-power operation, you'll want to be able to switch to low power when possible, to conserve the battery. Most hand-helds offer a HIGH/LOW power switch. Good used hand-held transceivers sell from $150 and up. New transceivers start at about $250. See *The ARRL Radio Buyer's Sourcebook* for more information. Remember: With higher power levels you will want to use a remote antenna. Radiation of a lot of power in front of your eyes or near your head is not a good idea. This important topic is discussed in the Safety chapter of *The ARRL Handbook for Radio Amateurs*, 1995 and later editions.

2) ***Mobile transceivers*** have power outputs ranging from 10 to 50 watts. In populated areas with many repeaters, any more than 10 watts is probably unnecessary. Mobile

Canoe mobile in the wilderness. John, KL7JR, used a Yaesu FT-747GX transceiver and an Outbacker antenna to make more than 300 contacts last fall from Bridge Lake, British Columbia, as KL7USI/VE7, a special-event operation for the Canadian Islands Awards program.

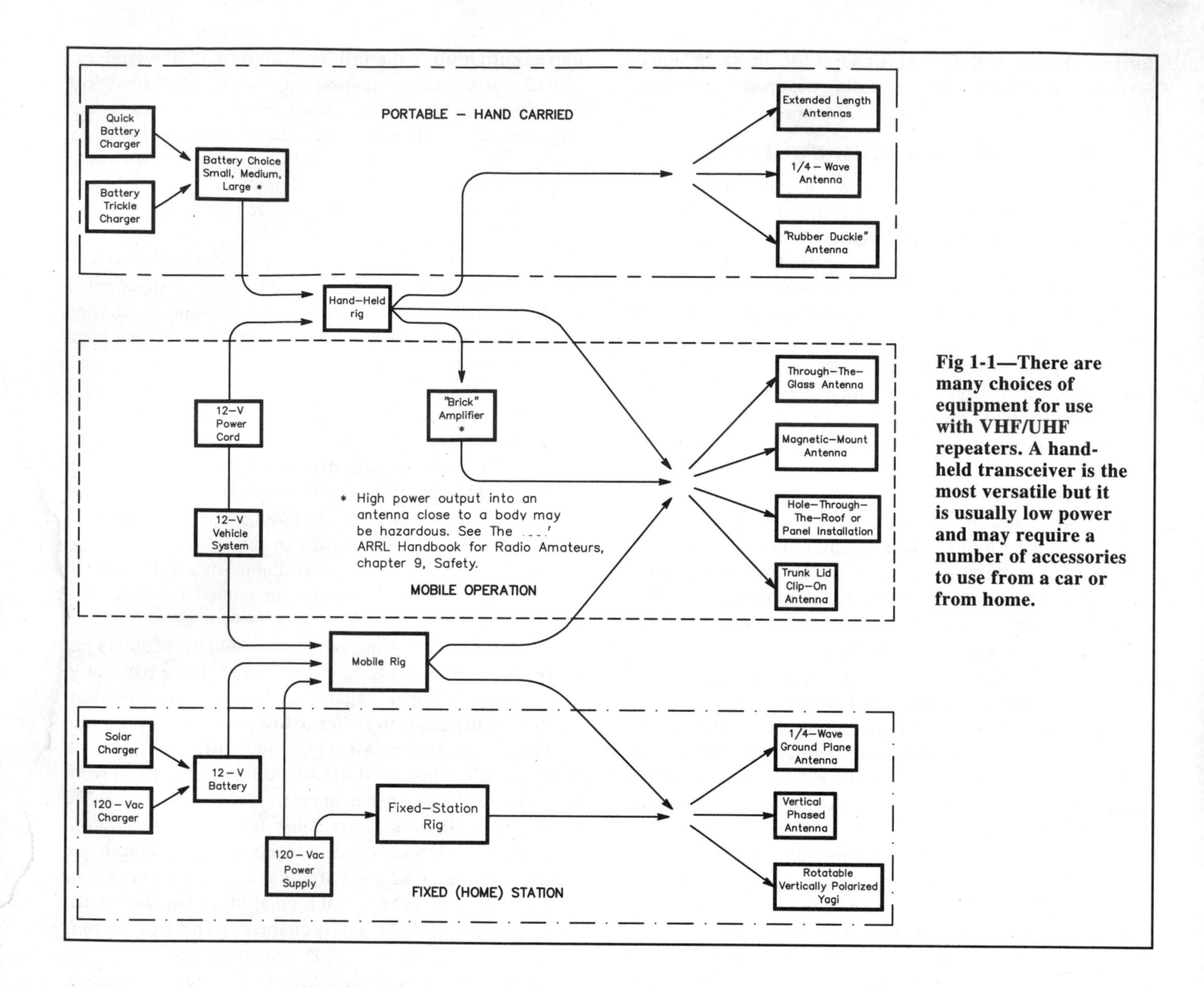

Fig 1-1—There are many choices of equipment for use with VHF/UHF repeaters. A hand-held transceiver is the most versatile but it is usually low power and may require a number of accessories to use from a car or from home.

transceivers are available used from $150 and up. New transceivers start at about $250 for a 10-watt unit. A mobile transceiver can be used indoors, too, if you have an adequate power supply. Selection of a mobile antenna is just as important as selection of the rig. Most hams are disappointed when they try to use a hand-held transceiver in a car with the rubber ducky antenna attached to the rig. An antenna outside the metal skin of the car will work much better.

3) ***Fixed rigs***. As shown in Fig 1-1, a mobile rig connected to a power supply or 12-V automotive battery makes a good fixed station. It also can be disconnected and moved into the car for mobile operation. Powering the rig from a battery has the added advantage of allowing emergency operation when the local power lines go down in a storm.

4) ***VHF packet radio***. Most VHF packet radio operation takes place on the 2-meter and 440-MHz bands. A mobile transceiver makes a good packet-radio station rig, too. You'll need a computer or terminal, and a terminal node controller (TNC). Used TNCs are available for $70 or so, but new ones are only slightly more expensive. Newer units may offer "mailbox" features, where other hams can leave messages for you when you aren't home.

HF Equipment

1) ***Used separate transmitter/receiver combinations***. At one time in the 1960s and 1970s all the major equipment manufacturers sold separate receivers and transmitters designed to transceive together. Examples are the Heath SB-401/SB-303, Drake T-4/R-4, Kenwood T-599/R-599 and Collins 32S-3/75S-3. Each of these rigs was a classic in its time, but they are outdated now. They lack many refinements common in modern transceivers, including the ability to transmit on the 30, 17 and 12-meter ham bands. All the transmitters mentioned, and most of the receivers, used vacuum tubes.

Years and high temperatures can take their toll on electronic equipment. Tubes are expensive and are becoming harder to find. Capacitors tend to degenerate, and finding replacements rated at the operating voltages of tubes can be a real chore. If you can find a good deal and know the equipment is in good working order (don't take the seller's word unless you know him or her well!), you may be able to pick up a pair for under $200. Don't be surprised, however, at some of the prices quoted. Some rigs, such as the Collins S-line, command high prices from collectors. If you're looking for a first rig, the cost of a used rig such as the S-line is such that you're better off considering a more modern, used transceiver.

2) ***Transceivers***. HF transceivers (a transmitter plus a receiver) have been common since the late 1960s. Like all older gear, older transceivers are likely to have maintenance problems. Mobile use subjects a radio to a great deal of vibration and wide ranges of temperature. A rig showing signs of having been used for mobiling may not be a good choice for your first station.

Transceivers manufactured in the 1970s were partly solid state. The transmitters usually had three tubes: two in the power amplifier and another serving as the driver stage. Some transceivers from this era had built-in ac power supplies. Mechanical parts, such as those used in the tuning assembly, may be impossible to obtain. Make sure everything works properly before you buy. Transceivers from this era probably won't operate on the 30, 17 and 12-meter bands.

Transceivers manufactured after about 1980 featured fully solid-state designs. Tuning mechanisms had fewer mechanical parts to wear, and all-band operation became common. Many feature general-coverage receivers that continuously tune from below 100 kHz to 30 MHz. With the general-coverage receiver you can listen to time and frequency-standard stations such as WWV and WWVH and enjoy the variety of shortwave broadcasting.

New transceivers don't usually offer many more features than the better-equipped used gear mentioned in the previous paragraph. Of course, new equipment comes with a factory warranty, and you know no one opened the case to make modifications.

3) ***Homemade equipment***. Although most hams will not want to design or build their first stations, if you have a background in electronics, you might want to "homebrew" your own. Kits are available from a variety of sources, and many proven circuits appear in *The ARRL Handbook for Radio Amateurs*.

Accessories

A few accessories found in almost all stations either help set up and test equipment or help operate it.

SWR Indicator

This device is handy for testing an antenna and feed line when the antenna is first erected, and later to make sure the antenna is still in good shape. If an antenna tuner is used with a multiband antenna system, such as a 135-foot or a G5RV dipole, an SWR (standing-wave ratio) indicator is essential for ensuring the tuner is adjusted for a reasonable SWR. Many modern HF transceivers include an SWR indicator. External SWR meters can cost as little as $30. If you plan to operate both VHF and HF you may have to buy one for HF and a second for VHF/UHF.

Antenna Tuner

There are many antenna tuners available on the market, costing between $75 and $1500. The circuits are pretty simple, though, and many first-time homebrew projects are antenna tuners. (*The ARRL Handbook for Radio Amateurs* contains plans for building them.) By carefully shopping at hamfest flea markets you can often assemble your own for less than $50 in parts.

Keys, Keyers and Paddles

If you're interested in CW, you will need some means of sending code. You may find it useful to start with a straight key until you feel you have the proper rhythm. Then you may wish to buy a keyer and paddles, or use your computer to send code. Some modern transceivers have keyers built in, so all you need is a paddle. Paddles may be standard or *iambic*. The iambic type requires less hand motion but takes a bit longer to master.

Keyers cost from $50 to $250, and good paddles can cost about the same. Code-transmitting programs are available for most computers; see the ads in *QST*.

Computers

Computers have become a very common—and for some people, a necessary—part of the ham shack. Some are used on the air; that is, to send and receive Morse code, fax, slow-scan TV or digital modes. Others are used for logging and record keeping. Hams use various types of computers, ranging from original IBM-PCs to UNIX-based work stations. The most popular type—having the most software available—is an IBM compatible unit.

When considering a computer for the shack remember that a computer uses various oscillators inside that can generate annoying spurious signals in your receiver. Some computers can be very sensitive to the presence of radio-frequency energy (RF), such as that generated by your nearby transceiver. For help with this topic see *The ARRL RFI Book*, which contains an entire chapter on solving computer problems in the shack.

Test Meter

An inexpensive digital multimeter capable of measuring voltage, current and resistance is very helpful around the shack. High accuracy is not needed for most projects. A $15 to $30 unit will pay for itself the first time you need to check the integrity of a coax connector you've just installed.

STATION SETUP

How you set up your station is determined by how much space you have, and how much equipment you have to squeeze into it. The table should be about 30 inches high and 30 inches deep. An old desk makes a good operating table. Build shelves for your equipment. Radios stacked on top of one another can't breathe and may overheat. It's also easier to change cables and move equipment around when you use shelves. Make sure you can get behind the gear to plug or unplug cables.

Ray Rising, K4LWJ/HK3BBR, is glad to be home and enjoying freedom and Amateur Radio. He was held captive by Colombian guerrillas for 810 days starting in March 1994. Ray is shown here with one of the few personal items his captors allowed him to keep: his SWL radio and an antenna made from a scouring pad. He says, "Did you know there is about 80 feet of wire in a scouring pad?" Ray is a missionary with a support arm of the Summer Institute of Linguistics, better known as the Wycliffe Bible Translators.

To prevent fatigue, place your key or paddle far enough from the edge of the table so your entire arm is supported. A microphone can be mounted on a stand placed on the table or on an extension that reaches in from the back or side. The best way to test the arrangement of your rig is to sit in the operating chair and operate the rig's controls. If the knobs are too low, try placing spacers or blocks under the front feet of the rig. Too high? Try placing the spacers under the rear legs.

Check that you can see the frequency dial. While it may be nice to place a keyboard right in front of your position, it may cause considerable strain on your back, shoulder and arms if you have to reach across the keyboard to operate the rig. In this case consider a T or U-shaped operating position.

Be sure to have an effective earth ground routed to your station. Use as many grounds as you can locate, bonding them together electrically and bringing the connection to the operating position. If you live on a higher floor and a ground is not available, use a cold-water pipe if you can.

Connections from the various ground points to the shack should be as short as possible. Since RF currents flow mainly on the surfaces of conductors, ground conductors such as shield braid from RG-8 coaxial cable or wide strips of flashing copper are best.

Safety should be a prime consideration. Many hams will not plug in their equipment for the first time until all the equipment cases have been tied to ground. Use a master switch and make sure other people in the house know where it is and how to use it. Consider running your shack through a *GFI* (ground fault interrupt) outlet. Unfortunately not all ham gear, especially older units, will operate with these devices. By detecting unbalanced line currents and shutting off the voltage when unbalance occurs, however, they could save your life!

ANTENNAS

Whatever antennas you select, ***install them safely***. Don't endanger your life or someone else's for your hobby! Antennas are important. The best (and biggest) transmitter in the world will not do any good if the signal is not radiated into the air. A good rule of thumb to follow is "always erect as much antenna as possible." The better your antenna array, the better will be your radiated signal. A good antenna system will make up for inadequacies or shortcomings in station equipment. A less-sensitive receiver hears better with a good antenna system and a bigger antenna system will make a QRP (low power) station sound a lot louder at the receiving end.

If you are thinking of a tower, talk to a few local hams before starting construction (or applying for a building permit). Local rules and ordinances may have a large impact on your plans.

Apply common sense to your antennas! Many hams act as though the world will end if they put up an antenna and measure an SWR greater than 1.5:1. For most purposes an SWR of 3:1 is perfectly acceptable at HF with good quality feed lines 100 feet or less in length.

First VHF/UHF Antennas

One of the nice things about VHF and UHF operation is that often the simplest antennas, if mounted high enough, will do an excellent job. Ground planes, J-poles and simple beams can either be purchased at a reasonable cost or often constructed in a few minutes. If you want to test a 2-element quad for 144-MHz, just take the design from the *ARRL*

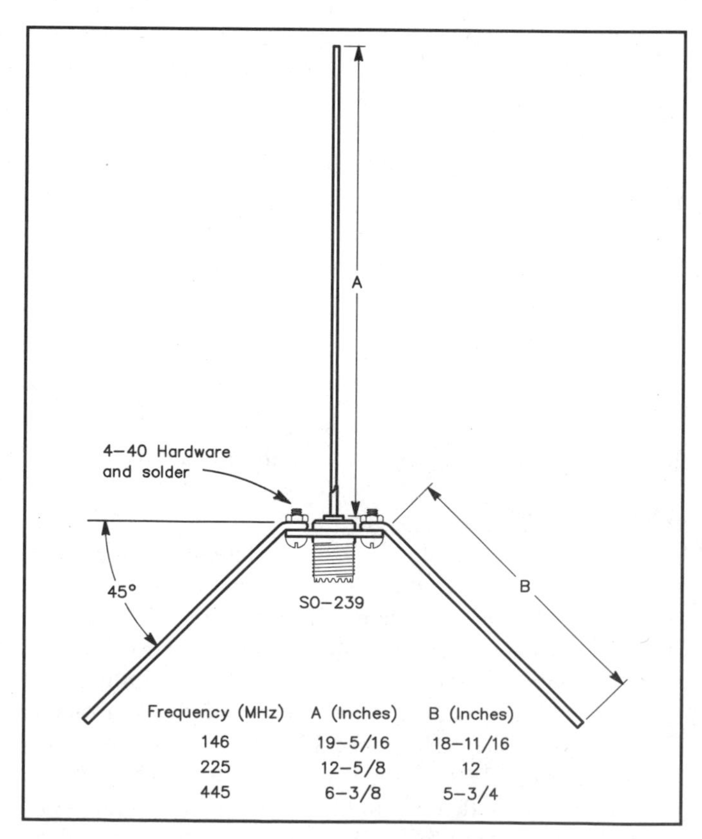

Fig 1-2—A simple ground-plane antenna for VHF. The elements are made from 3/32 or 1/16-inch brass welding rod or #10 or #12 bare copper wire.

Antenna Book, build it from scrap wood and heavy gauge copper or aluminum wire, and run a few tests. The unit you built may not stay up for a long time in bad weather, but it will be fine for determining if this is the sort of antenna you want to put up permanently.

A simple ground-plane antenna is shown in **Fig 1-2**. It can be mounted by taping the feed line and bottom connector to a pole so that the antenna extends over the top of the pole. It can also be suspended by a cord by lengthening the vertical element and bending the extra length into a loop. One end of the cord is fastened to the loop and the other end runs over a tree branch.

A Simple HF Antenna

The most popular first antenna is the half-wavelength dipole. It consists of a half wavelength of antenna wire with a feed line and an insulator at its center. See **Fig 1-3**. For a new ham's activities, a resonant dipole system is a good selection. It is very easy to erect and has a low SWR, a measure of how well an antenna is tuned to the transmitted frequency and its feed line.

A single-band dipole fed with low-loss feed line can also be used on other bands. In fact with an antenna tuner, a balun, and a random length center fed dipole you can actually operate on any HF band. The trick is to use a low-loss open-wire type of feed line and provide a match with an antenna tuner.

Where to Put the Antenna?

Think about your antenna location. Remember: Your antenna should be as high and as far away from surrounding trees and structures as possible. Never put an antenna near power lines! The dipole will require one support at each of its ends (perhaps trees, poles or even house or garage eaves), so survey your potential antenna site with this in mind. If you find space is so limited that you can't put up a straight-line dipole, don't give up. You can bend the dipole and still make plenty of contacts.

Some hams have been known to become very frustrated when they cannot put up a 135-foot long dipole that looks like it came out of an engineering design manual. Most dipoles are very close to the ground and still work well. Many hams successfully use a dipole suspended from the middle, called an inverted V. The performance of this antenna is so

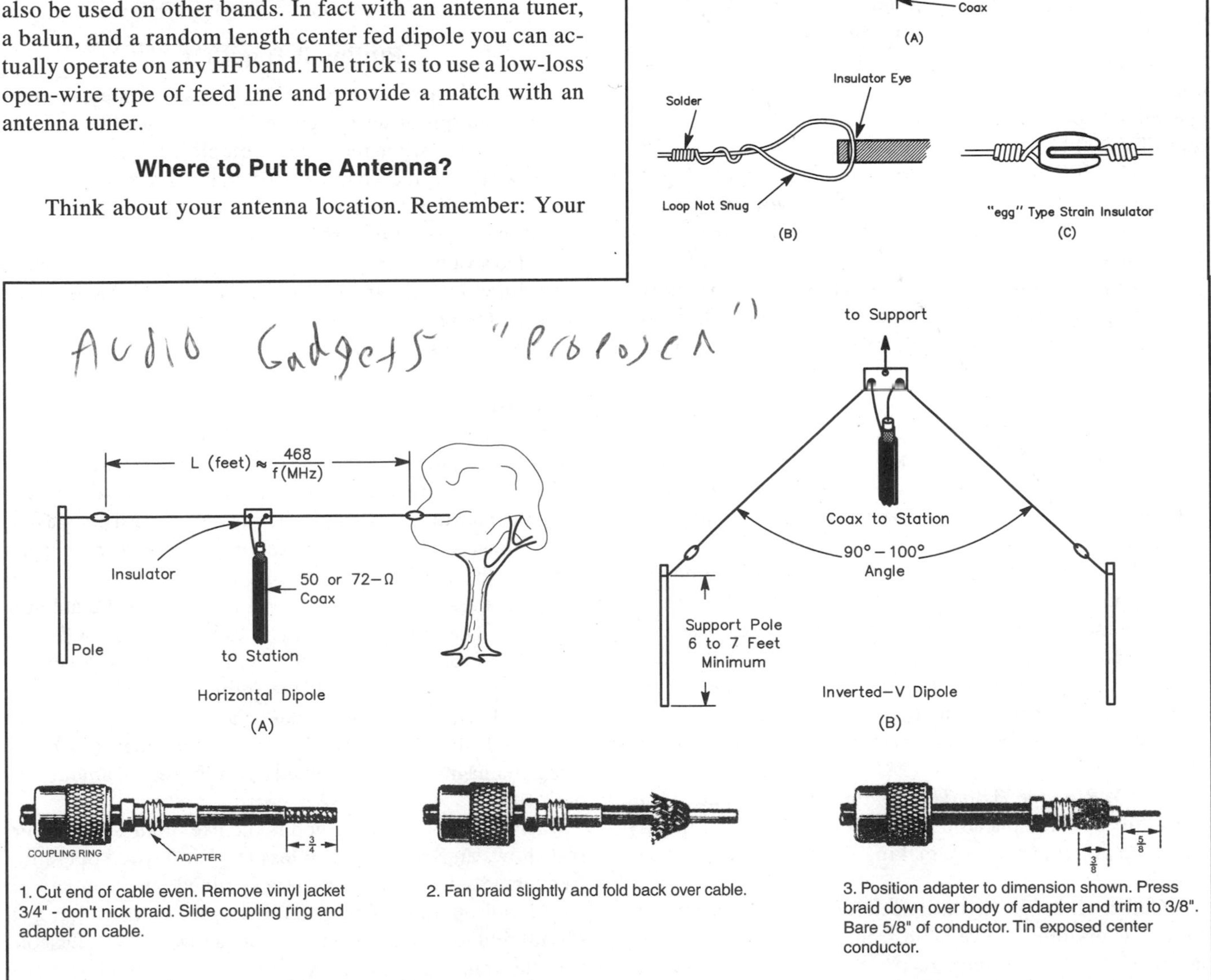

Fig 1-3—Building your own HF dipole antenna is a popular project.

satisfactory they are not even aware of the losses this configuration may have as compared to the ideal dipole. You can put up an antenna under almost any circumstances, but you may need to use your imagination.

Antenna Parts

If this is the first time you have tried to put up a dipole by yourself the following parts list will give you some guidance.

1) ***Antenna wire***. #12 or #14 hard-drawn Copperweld (copper-clad steel) is preferred, so the antenna won't stretch. It will be strong enough to support itself as well as the weight of the feed line connected at its center. Always buy plenty of wire. It never goes to waste!

2) ***Insulators***. You need one center and two end antenna insulators for a simple dipole.

3) ***Clamp***. Large enough to fit over two widths of your coaxial cable to provide mechanical support.

4) ***Coax***. Feed line made of a center conductor surrounded by an insulating dielectric. This in turn is surrounded by a braid called the shield and an outer insulating jacket. You need RG-58 or RG-8. Look for coax with heavy braid shield such as Belden or Times Wire and Cable. Stay away from cheap cable from unknown manufacturers, or too-good-to-be-true deals on surplus cable.

A good alternative is balanced open-wire line. This is constructed using two parallel pieces of wire connected and spaced with plastic rods or enclosed in a plastic jacket, similar to TV 300-Ω wire. Some open-wire line has pieces of the center plastic removed to form "windows," lightening the line and reducing the dielectric losses. If you are sure you want to build a single-band dipole, stick to the coax; otherwise consider the tuner, balun and open-wire configuration.

5) ***Connector***. Connects the feed line (coax) to your rig. This will probably be a PL-259 connector, standard on most rigs. If your radio needs another kind, check your radio's instruction manual for installation information. You also need connectors for the coax lines between your antenna tuner, SWR meter and your rig.

6) ***Electrical tape and coax sealant***. This is needed to cover the antenna ends and joints to make them waterproof. Otherwise water can get into coax, eventually ruining your antenna system.

7) ***Rope***. You need enough to tie the ends of the antenna to a supporting structure. Rope may be used. However, it does degenerate with time and weather. You can use conductive guy wire, with insulators. Conventional wisdom says you should make sure the length of the supporting wire is not resonant in a ham band. In practice, it's doubtful you'll see any difference in antenna performance with any length of guy wire.

8) ***SWR meter***. This is essential for your station and especially for antenna adjustment. SWR meters are readily available and inexpensive, making them easier to buy than to build.

Gather all the parts you'll need for your chosen antenna. Almost everything is available from your local electronics store or from suppliers advertising in *QST*. When this is done, the fun of actually putting together your antenna can begin.

Table 1-2
Antenna Lengths in Feet

	½ wavelength	***¼ wavelength***
80 m	126′ 6″	63′ 3″
40 m	65′ 8″	32′ 10″
15 m	22′ 2″	11′ 11″
10 m	16′ 7″	8′ 3″

Remember to add about 1 foot to each end of the dipole for tuning adjustment.

Assembly is quite simple. Your dipole consists of two lengths of wire, each approximately ¼ wavelength long at your chosen (or lowest) operating frequency. These two wires are connected in the center, at an insulator, to the feed line. In our antenna the feed line, which brings the signals to and from your radio, is coaxial cable. Calculate the length of the half-wave dipole by using this simple formula:

antenna length in feet = 468/frequency in MHz

(The information in **Table 1-2** has approximate lengths already calculated.) Now measure the antenna wire, keeping it as straight as possible.

Putting It Together

Carefully assemble your antenna, paying special attention to waterproofing the coax. Don't solder the antenna ends until later, when retuning is completed. Just twist them for now. Route the coax to your station, remembering to keep it as unobtrusive as possible. Cut the coax to a length that will leave some excess for strain relief so your rig won't be pulled around during strong winds! Install the connectors according to the diagrams. Now connect your SWR bridge between the feed line (coax) and the transmitter.

One trick used by old-timers is the addition of a 10,000-Ω resistor, soldered directly across the center insulator of the dipole. An ohmmeter connected from one side of the feed line to the other at the ground end should measure this value of 10,000 Ω. If it measures an open circuit, it means the feed line is disconnected or broken. A short circuit means the feed line or the connector is shorted. The high-value resistor has no effect on the antenna or SWR. Left in place it acts as a handy check on the antenna and feed line.

There are several ways to see if you built the antenna correctly. One very direct way is to buy or borrow an antenna noise bridge. The bridge will tell you the resonant frequency of the antenna. If the frequency is too low, you should shorten the length of the antenna (cutting equal amounts from both ends). Resonant frequency too high? The antenna must be lengthened by adding equal amounts to both ends.

Another common method is the use of an SWR meter. If it shows an SWR of 2:1 or less at your desired operating frequency, your antenna system is tuned, and you can go ahead and operate. If your SWR is greater than 2:1, you should retune your antenna to obtain a lower SWR. Disconnect the transceiver, and try shortening the antenna a few inches on each end. Keep notes! Reconnect the transceiver, and check the SWR. If the SWR goes lower, continue shortening the ends until you get the lowest possible SWR reading. Be careful not to shorten it too much! As the last step, when you are sure the length of the dipole is correct, solder the wire wrapped around the end insulators.

Problems and Cures

A high SWR that does not change when you change the antenna length by a few feet (HF only) probably means something is wrong beyond merely length misadjustments of your simple, one-band dipole. Check to see if your coax is open or shorted. Make sure your antenna isn't touching anything and that all your connections are sound.

When all systems are go, get on the air and operate. As you settle into that first QSO with your new antenna, enjoy those feelings of pride, accomplishment and fun that will naturally follow. After all, that's what Amateur Radio is all about!

EXPERIENCE? GET ON THE AIR!

Actual on-the-air operating experience is the best teacher. In this section, we will try to give you enough of the basics to be able to make it through your first QSO. If nothing else, read the following material for additional reference notations. Learn by doing and don't be afraid to ask questions of someone who might be able to help. After all, we're in this hobby together, and assistance is only as far away as the closest ham.

How do you develop good operating habits? This *ARRL Operating Manual* is an excellent place to start. An entire chapter is devoted to each major Amateur Radio activity, from working DX to space communications. Each chapter has been written by a ham with considerable experience in that area. Experience, even through the words of others, is a powerful teacher. Notice the table of contents; you'll be amazed at the diversity and amount of good solid reference material you have at your fingertips.

LISTEN, LISTEN, LISTEN...!

The most efficient way to learn to do something is with a coach or *Elmer*, a term dating back to the beginnings of Amateur Radio. An Elmer is someone who helps a newcomer become an established, competent ham. Since there are rules and an established way of doing things in the Amateur Radio Service, you'll learn most effectively by just plain, old-fashioned listening. Use a receiver (yours or one you borrow) and your ears. Tune around the amateur frequency bands, and just listen, listen, listen. Listen to as many QSOs as you can. Learn how the operators conduct themselves, see what works for other operators and what doesn't, and incorporate those good points into your own operating habits.

You'll be surprised how much operator savvy you can pick up just by listening. A little common sense goes a long way in helping you decide what is (and what isn't) good operating practice.

Lauren Ford, daughter of *QST* Managing Editor Steve Ford, WB8IMY, chats on 10 meters during the Kid's Day event in June 1999.

we didn't do this.

ON-THE-AIR EXPERIENCE FOR NEWCOMERS

When it is time for you to take the plunge, fire up the rig and make your first QSO. Go for it! Nothing can compare with actual on-the-air experience. Remember, everyone on the air today had to make their first QSO at some time or other. They had the same tentativeness you might have now, but they made it through fine. So will you.

It's the nature of the ham to be friendly, especially toward other members of this wonderful fraternity we all joined when we passed that examination. Trust the operator at the other end of your first QSO to understand your feelings and be as helpful as possible. After all, he or she was once in your situation!

Your First Contact

Are your palms sweaty, hands shaking and is there a queasy herd of butterflies (wearing spikes, perhaps) performing maneuvers in your stomach? Chances are you're facing your first Amateur Radio QSO. If so, take heart! Although some may deny it, the vast majority of hams felt the same way before firing up the rig for their first contact.

Although nervousness is natural, there are some preparations that can make things go a little more smoothly. Practicing QSOs face-to-face with a friend, or perhaps with several members of the local Amateur Radio club is a good way to ease the jitters.

Some find it useful to write down in advance information you will use during the QSO. One nervous newcomer made notes of her own QSO data on index cards. One card contained her location and another her name (no kidding!). But after a half dozen contacts, she gradually felt comfortable enough to forgo using her cards. So will you.

The ideal security blanket to have with you as you make your first few QSOs is your Elmer or another experienced ham. You'll find that after your initial nervousness wears off you will do just fine by yourself. And you will have honored a ham friend by allowing him or her the privilege of sharing your first on-the-air contacts.

Picking a Band

160 and 80 Meters

Eighty meters, and its phone neighbor, 75 meters, are favorites for ragchewing. I frequently check out the 80-meter Novice/Technician CW subband. There I find both newcomers as well as old-timers trying to work the rust out of their fists. Around 3600 kHz you'll find the digital modes, including RTTY, PSK31 and packet. The QRP frequency is 3560 kHz. If you hear a weak signal calling CQ near 3560, crank down your power and give a call. Another favorite frequency is 3579.5 kHz. If you live in the eastern half of North America, listen for W1AW on 3581 (CW), 3625 (digital) or 3999 kHz (SSB). W1AW runs 1000 watts to a modest antenna—an inverted V at 60 feet. If you can copy W1AW, you can probably work the East Coast, even with low power. AM operation is generally found between 3870 and 3890 kHz.

Even if you can't chase DX, you will find plenty to do on either band. Ionospheric absorption is greatest during the day, thus local contacts are common. At night, contacts over 200 miles away are more frequent, even with a poor antenna. Summer lightning storms make for noisy conditions in the summer, while winter is much quieter. You may also be troubled by electrical noise here. A horizontally polarized antenna, especially one as far from the building as possible, will pick up less electrical noise.

Topband, as 160 meters is often called, is similar to 80 meters. QSOs here tend to be a bit more relaxed with less QRM. DX is frequent at the bottom of the band. Don't let the length of a half-wave dipole for 160 keep you off the band; a 25 or 50-foot "long wire" can give you surprisingly good results. One favorite trick is to connect together the center conductor and shield of the coax feed line of a 40 or 80-meter dipole and load the resulting antenna as a "T," working it against ground.

40 and 30 Meters

I must confess to being biased in favor of these bands, especially 40 meters. If I could only have a receiver that covered one band, it would be 40. Running 10 watts from my East Coast apartment (indoor antenna) I can work European hams, ragchew up and down the coast, check into Saturday morning QRP nets, and listen to foreign broadcast stations besides. Yes, 40 is a little crowded. Look at the bright side: You won't be lonely. I think it's possible to work someone on 40 any time of the day or night.

From a modest station, daytime operation on 40 is easier than at night. The skip (distance you can work) during the day is shorter, but that helps reduce interference. Most other countries have SSB privileges down to 7050 kHz. During the day, they, and the broadcasters, won't bother you much. Outside the US, QRPers hang around 7030 kHz; in the US we use 7040 kHz. Digital operators work around 7060 to 7100 kHz. Don't miss the Novice/Technician CW subband from 7100 to 7150 kHz. Even when the foreign broadcasters are booming in, you usually can find someone to work there. Hams operating AM are typically around 7290 kHz. Forty is a good band for daytime mobile SSB operation, too. You'll find plenty of activity, and propagation conditions tend to be stable enough to allow you to ragchew as you roll along.

The 30-meter band has propagation similar to 40 meters. Skip distances tend to be a little longer on 30 meters, and it's not so crowded. At present, stations in the US are limited to 200 watts output on this band. DX stations seem to like the low end of the band, from 10100 to 10115 kHz. Ragchewers often congregate above 10115. We share 30 meters with other services, so be sure you don't interfere with them. SSB isn't allowed on 30, but you can use CW and the digital modes.

20 Meters

As much as I like 40 and 30 meters, I have many fond memories of 20 meters as well. When I upgraded my license to General in 1963, I made a beeline to 20 meters. To this day, I can't stay away for long. A 20-meter dipole is only 33 feet long, and that doesn't have to be in a straight line. Many US hams have worked their first European or Australian contacts with a dipole and 100 watts.

Many hams consider 20 meters the workhorse DX band. At the bottom of a solar cycle, 20 meters may be usable in a particular direction for only a few hours a day. Even then, 20 is usually open to somewhere in the world throughout the day and night. For example, from New England, 20 is open to some part of South America for 24 hours a day, whatever the level of sunspots might be. On the other hand, 20 meters can be open to the Far East for as much as 13 hours of the day (with very weak signals) when sunspot activity is low, while it can be open all day during periods of high solar activity.

There's plenty of room on the band. CW ragchewers hang out from 14025 to 14070 kHz, where you start hearing packet and other digital stations. The international QRP frequency is 14060 kHz. The sideband part of the band is sometimes pretty busy and then it may be difficult to make a contact with low power or a modest antenna. Look above 14250 for ragchewers. Impromptu discussion groups that sometimes spring up on 20 SSB make for interesting listening, even if you don't participate. If you like photographs, look around 14230 kHz for slow-scan TV. You'll need some extra equipment (as discussed in the Image Communications chapter of this book) to see the pictures.

17, 15 and 12 Meters

Except during years of high solar activity, you'll do most of your operating during daylight hours. Propagation is usually better during the winter months. Seventeen and 12 meters are relatively new amateur bands, so they aren't as crowded as 15 meters. Fifteen, though, is not nearly as crowded as 20. On 15, the QRP calling frequency is 21060 kHz. Don't overlook the Novice/Technician CW subband from 21100 to 21200 kHz. No special frequencies are used for QRP operation on 17 and 12. SSB operation is much easier on 17 and 12 because of lower activity. Low activity doesn't mean no activity—when those bands are open, you'll find plenty of stations to work. You only need one at a time, after all. Digital operation is found from 21070 to 21100 kHz, and around 18100 and 24920 kHz.

Some Russian RS-series satellites operate on 15 and 10 meters. Because of their low orbits, it isn't possible to work more than about 3000 miles on the RS satellites. However, their low orbits make it possible to use these satellites with low power and simple antennas.

Practical indoor, outdoor, mobile or portable antennas for these bands are simple to build and install. It's even possible to make indoor beam antennas for the range of 18 to 25 MHz.

10 Meters

The 10-meter band stretches from 28000 to 29700 kHz. During years of high solar activity, 10 to 25-watt transceivers will fetch plenty of contacts. When the sun is quiet, there are still occasional openings of thousands of miles. Ten meters also benefits from sporadic-E propagation. You'll find most sporadic-E openings in the summer, but they can happen anytime. Sporadic-E openings happen suddenly and end just as quickly. You may not be able to ragchew very long, but you'll be amazed at how many stations you can work.

SSB activity is heaviest in the Novice/Technician subband from 28300 to 28500 kHz. The Novice/Technician CW subband, 28100 to 28300 kHz, is also a good place to look for activity, as is the QRP calling frequency at 28060 kHz. You can operate 1200-baud packet radio on 10 meters, whereas we're limited to 300 baud on the lower bands. Digital operation takes place from 28070 to 28100 kHz.

Higher in the band, above 29000 kHz, you'll find amateur FM stations and repeaters, and the amateur satellite subband. AM operation is also popular between 29000 and 29200 kHz.

Operating on 50 MHz and Above

The VHF/UHF/microwave bands offer advantages to the low-power operator. The biggest plus is the relatively smaller antennas used. A good-sized 2-meter beam will easily fit in a closet when not in use. Portable and mobile operation on these bands is also easy and fun.

6 Meters

Six meters is perhaps the most interesting amateur band. When solar activity is high, worldwide QSOs are common. When solar activity is low, however, opportunities for long-distance communication decrease. Sporadic-E propagation, which I mentioned earlier, is the most reliable DX mode during periods of low solar activity.

With small antennas, like 3-element beams, it's possible to work 1000 miles on sporadic E. Three-element 6-meter beams don't fit well inside houses or apartments, but you might be able to put one in an attic or crawl space. Even if you can only use a dipole, you'll be able to work locals, and snag some more distant stations when the band opens.

Just about any mode found on the HF bands is used on 6 meters. CW and SSB operation take place on the lower part of the band. Higher up you'll find FM simplex and repeater stations. Another mode you'll sometimes find on 6 meters is radio control (RC) of model planes, boats and cars.

2 Meters

Simply stated, 2 meters is the most popular ham band in North America. From just about any point in the US, you can probably work someone on 2 meters, 24 hours a day. Most hams know about 2-meter FM and packet radio operation, but CW and SSB are used here too. There's even an amateur satellite subband on 2 meters.

CW and SSB operation is done mostly with horizontally polarized antennas. FM and packet operators use vertical polarization, while satellites can be worked with either. A popular 2-meter antenna called a *halo* is perfect for indoor or mobile use on CW or SSB. The omnidirectional *halo* has no gain, but you'll be able to work locals, and up to 100 miles during band openings.

FM and packet usually require only a simple vertical antenna. The *ARRL Repeater Directory* will tell you what repeaters are available in your area. This book lists all repeaters reported to ARRL for the bands from 29 to 10,000 MHz.

The 222 and 430-MHz Bands

Every mode used on 2 meters is found on 222 except satellite communication. The 430-MHz or 70-cm band is becoming more and more popular, too. Multiband hand-held and mobile transceivers are available at prices only slightly higher than single-band rigs. If you think you'd like to try these bands in addition to 2 meters, look into a multiband rig.

One mode you'll find on 70 cm that isn't allowed on the lower frequencies is fast scan amateur television (ATV). Assuming you already have a broadcast TV set, all you need is a receive converter, transmitter, antenna and camera. Inexpensive cameras designed for home-video use are fine for ATV. ATV repeaters may be found in larger metropolitan areas. They're listed in the *ARRL Repeater Directory*.

23 cm and Up

As you go higher in frequency, the size of antennas gets smaller. This fact allows you to use very high-gain antennas that aren't very big. Commercial equipment is available for the bands through 10,000 MHz (10 GHz). You'll also find easy-to-build kits (the tuned circuits are etched onto the circuit boards).

Because antennas are so small, it's possible to have 20 to 30-dB gain antennas that fit in your car's trunk. In comparison, a big 20-meter beam might offer only 10 dB of gain. Operating from the field with battery-powered equipment is very popular, especially during VHF/UHF/microwave contests. Thanks to high-gain antennas, contacts over several hundred miles are possible with equipment running 1 or 2 watts.—*Jim Kearman, KR1S, adapted from* Low Profile Amateur Radio, *published by the ARRL.*

OPERATING—WHAT, WHERE AND HOW

We're finally getting down to the real nitty-gritty. You have read a little about the basics of ham radio and have some ideas about selecting a rig and station equipment and erecting a decent antenna system. You may even have made your first on-the-air contact.

Now it's time to learn a little more about how to go about playing the ham radio game. For everyone to be able to communicate with everyone else effectively, we need to use accepted operating procedures. The next part of this chapter briefly describes the major modes of ham radio communication and a few of the procedures and conventions that hams use on the air. Other chapters discuss the operating procedures for many specialized modes of communication in greater detail. But first we are going to look at operating HF voice and CW (Morse code).

The sidebar *Picking a Band* gives the characteristics of the various ham bands, and **Fig 1-4** shows the Amateur Radio frequency allocations in the US.

CW Operating

CW (which stands for *continuous wave*) is a common bond among hams. Many take pride in their code proficiency. It takes practice to master the art of sending good

US Amateur Bands

ARRL *The national association for AMATEUR RADIO*

April 15, 2000

Novice, Advanced and Technician Plus Allocations

New Novice, Advanced and Technician Plus licenses will not be issued *after* April 15, 2000, but *existing* Novice, Technician Plus and Advanced class licenses are unchanged. Amateurs can continue to renew these licenses. Technicians who pass the 5 wpm Morse code exam *after* that date have Technician Plus privileges, although their license says Technician. They must retain the 5 wpm Certificate of Successful Completion of Examination (CSCE) as proof. The CSCE is valid indefinitely for operating authorization, but is valid only for 365 days for upgrade credit.

160 METERS

E,A,G

1800 1900 2000 kHz

Amateur stations operating at 1900-2000 kHz must not cause harmful interference to the radiolocation service and are afforded no protection from radiolocation operations.

80 METERS

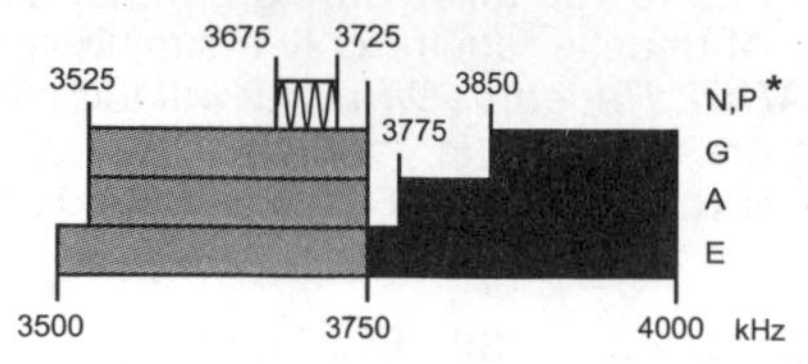

5167.5 kHz (SSB only): Alaska emergency use only.

40 METERS

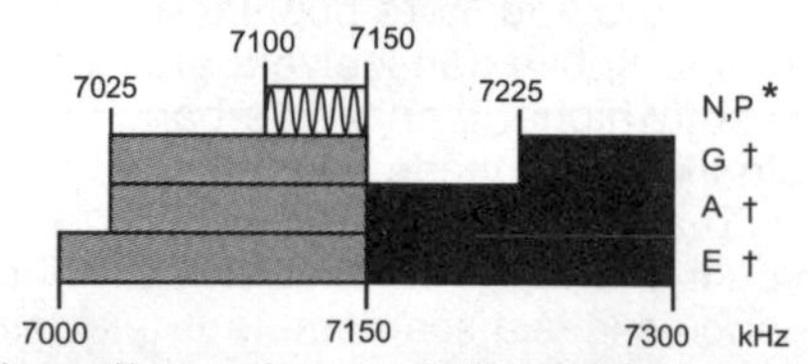

† Phone and Image modes are permitted between 7075 and 7100 kHz for FCC licensed stations in ITU Regions 1 and 3 and by FCC licensed stations in ITU Region 2 West of 130 degrees West longitude or South of 20 degrees North latitude. See Sections 97.305(c) and 97.307(f)(11). Novice and Technician Plus licensees outside ITU Region 2 may use CW only between 7050 and 7075 kHz. See Section 97.301(e). These exemptions do not apply to stations in the continental US.

30 METERS

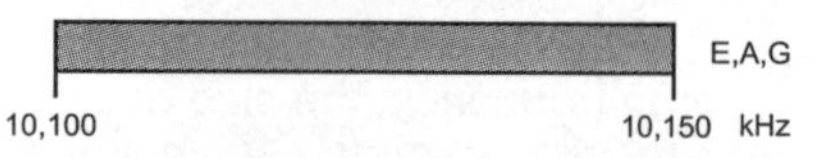

Maximum power on 30 meters is 200 watts PEP output. Amateurs must avoid interference to the fixed service outside the US.

20 METERS

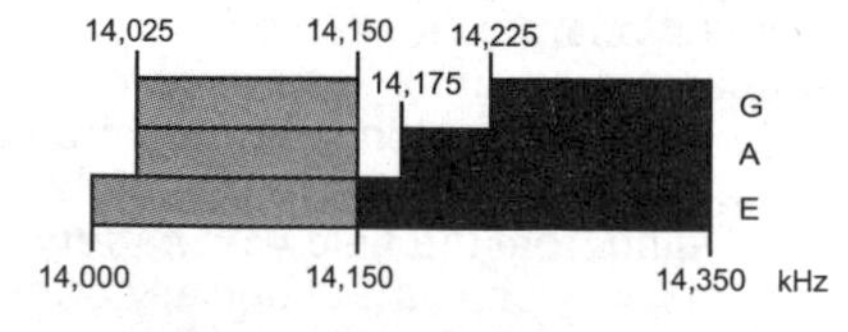

17 METERS

15 METERS

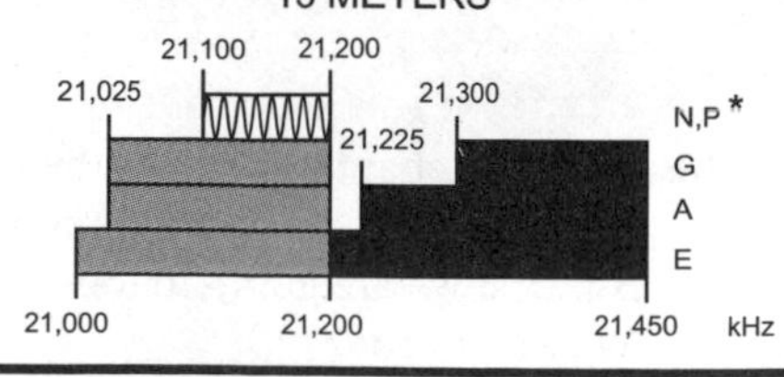

12 METERS

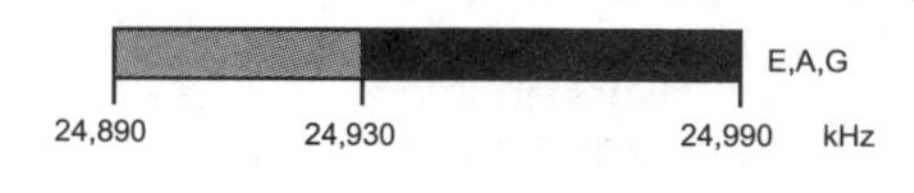

10 METERS

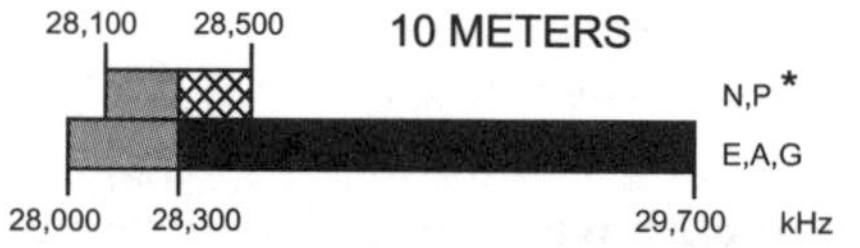

Novices and Technician Plus Licensees are limited to 200 watts PEP output on 10 meters.

6 METERS

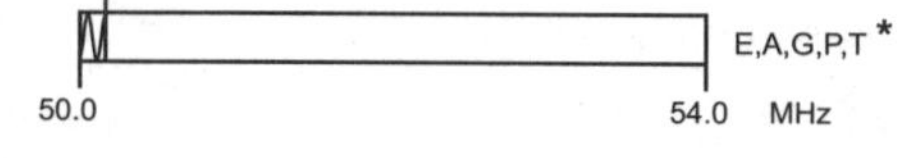

2 METERS

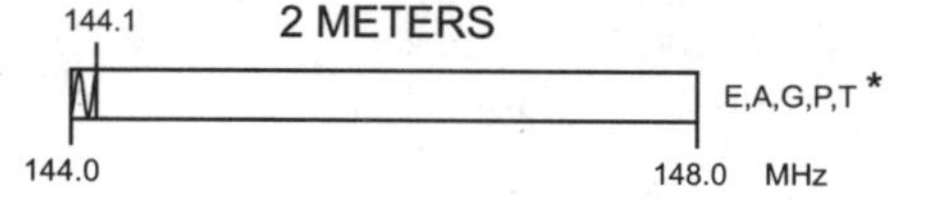

1.25 METERS

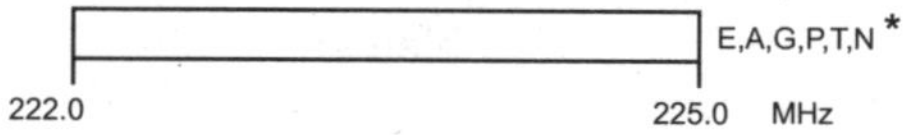

Novices are limited to 25 watts PEP output from 222 to 225 MHz.

70 CENTIMETERS **

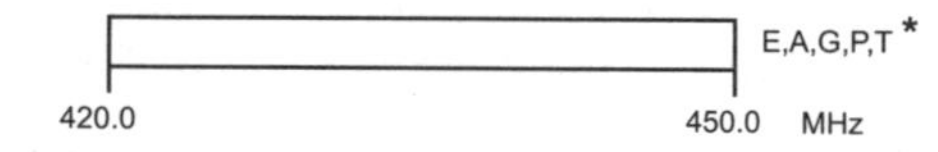

33 CENTIMETERS **

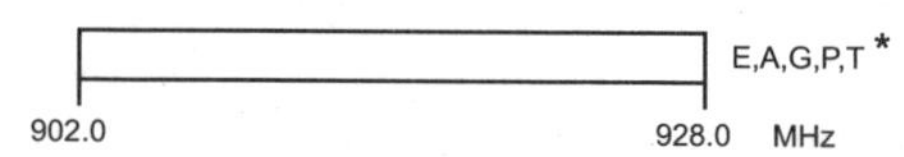

23 CENTIMETERS **

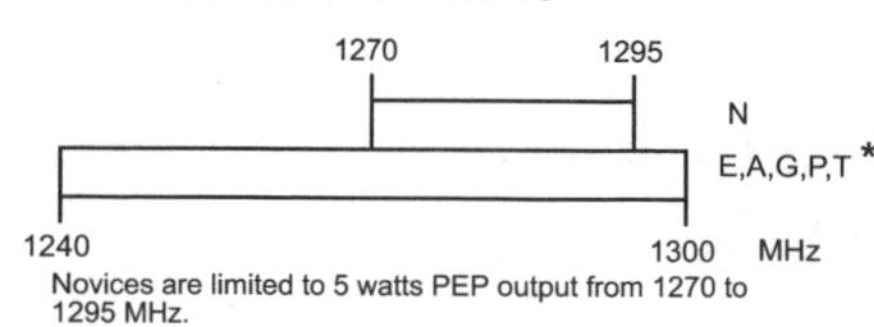

Novices are limited to 5 watts PEP output from 1270 to 1295 MHz.

US AMATEUR POWER LIMITS

At all times, transmitter power should be kept down to that necessary to carry out the desired communications. Power is rated in watts PEP output. Unless otherwise stated, the maximum power output is 1500 W. Power for all license classes is limited to 200 W in the 10,100-10,150 kHz band and in all Novice subbands below 28,100 kHz. Novices and Technicians are restricted to 200 W in the 28,100-28,500 kHz subbands. In addition, Novices are restricted to 25 W in the 222-225 MHz band and 5 W in the 1270-1295 MHz subband.

Operators with Technician class licenses and above may operate on all bands above 50 MHz. For more detailed information see *The FCC Rule Book.*

KEY

- = CW, RTTY and data
- = CW, RTTY, data, MCW, test, phone and image
- = CW, phone and image
- = CW and SSB phone
- = CW, RTTY, data, phone, and image
- = CW only

E = EXTRA CLASS
A = ADVANCED
G = GENERAL
P = TECHNICIAN PLUS
T = TECHNICIAN
N = NOVICE

*Technicians who have passed the 5 wpm Morse code exam are indicated as "P".

**Geographical and power restrictions apply to all bands with frequencies above 420 MHz. See *The FCC Rule Book* for more information about your area.

All licensees except Novices are authorized all modes on the following frequencies:

2300-2310 MHz
2390-2450 MHz
3300-3500 MHz
5650-5925 MHz
10.0-10.5 GHz
24.0-24.25 GHz
47.0-47.2 GHz
75.5-76.0, 77.0-81.0 GHz
119.98-120.02 GHz
142-149 GHz
241-250 GHz
All above 300 GHz

For band plans and sharing arrangements, see *The FCC Rule Book.*

ARRL ***We're At Your Service***

ARRL Headquarters	860-594-0200 (Fax 860-594-0259)	hq@arrl.org
Publication Orders	Toll-Free 1-888-277-5289 (860-594-0355)	orders@arrl.org
Membership/Circulation Desk	Toll-Free 1-888-277-5289 (860-594-0338)	membership@arrl.org
Getting Started in Amateur Radio	Toll-Free 1-800-326-3942 (860-594-0355)	newham@arrl.org
Exams	860-594-0300	vec@arrl.org
ARRL on the World Wide Web	www.arrl.org/	

Fig 1-4—Frequency Allocation Chart for US amateurs.

Table 1-3
Q Signals

These Q signals are the ones used most often on the air. (Q abbreviations take the form of questions only when they are sent followed by a question mark.)

QRG	Will you tell me my exact frequency (or that of ___)? Your exact frequency (or that of ___) is ___ kHz.
QRL	Are you busy? I am busy (or I am busy with ___). Please do not interfere.
QRM	Is my transmission being interfered with? Your transmission is being interfered with ___ (1. Nil; 2. Slightly; 3. Moderately; 4. Severely; 5. Extremely.)
QRN	Are you troubled by static? I am troubled by static ___. (1-5 as under QRM.)
QRO	Shall I increase power? Increase power. QRP Shall I decrease power? Decrease power. QRQ Shall I send faster? Send faster (___ WPM).
QRS	Shall I send more slowly? Send more slowly (___ WPM).
QRT	Shall I stop sending? Stop sending.
QRU	Have you anything for me? I have nothing for you.
QRV	Are you ready? I am ready.
QRX	When will you call me again? I will call you again at ___ hours (on ___ kHz).
QRZ	Who is calling me? You are being called by (on ___ kHz).
QSB	Are my signals fading? Your signals are fading.
QSK	Can you hear me between your signals and if so can I break in on your transmission? I can hear you between signals; break in on my transmission.
QSL	Can you acknowledge receipt (of a message or transmission)? I am acknowledging receipt.
QSN	Did you hear me (or) on ___ kHz? I did hear you (or ___) on ___ kHz.
QSO	Can you communicate with direct or by relay? I can communicate with ___ direct (or relay through).
QSP	Will you relay to ___? I will relay to.
QST	General call preceding a message addressed to all amateurs and ARRL members. This is in effect "CQ ARRL."
QSX	Will you listen to ___ on ___ kHz? I am listening to ___ on ___ kHz.
QSY	Shall I change to transmission on another frequency? Change to transmission on another frequency (or on ___ kHz).
QTB	Do you agree with my counting of words? I do not agree with your counting of words. I will repeat the first letter or digit of each word or group.
QTC	How many messages have you to send? I have ___ messages for you (or for ___).
QTH	What is your location? My location is ___ .
QTR	What is the correct time? The time is ___ .

code on a hand key, bug (semi-automatic key) or electronic keyer. It takes practice to get that smooth rhythm, practice to get that smooth spacing between words and characters, and practice to learn the sound of whole words and phrases, rather than just individual letters.

CW is an effective mode of communication. CW transceivers are simpler than their phone counterparts, and a CW signal can usually get through very heavy QRM (interference) much more effectively than a phone signal. Therefore, hams use shortcuts and abbreviations during a CW QSO. Many have developed within the ham fraternity, while some are borrowed from old-time telegraph operators. *Q signals* are among the most useful of these abbreviations. A list of the most popular Q signals is in **Table 1-3**.

You don't have to sit down and memorize this list. Copy the list and keep the copy on your operating table. You can also request Form FSD-218 from the ARRL or pick it up on *ARRLWeb*. It contains a handy list of Q signals and abbreviations for mounting at your operating position.

After using some of the Q signals and abbreviations a few times you will quickly learn the most popular of them without needing any reference. With time, you'll find that as your CW proficiency rises, you will be able to communicate almost as quickly on CW as you can on the voice modes. For more information on CW operating, see *Morse Code: The Essential Language*, published by ARRL. A list of common abbreviations is in **Table 1-4**.

It may not seem that way to you now, but your CW sending and receiving speed will rise very quickly with on-the-air practice. For your first few QSOs, carefully choose to answer the calls from stations that are sending at a speed you can copy (perhaps another first timer on the band?). Courtesy on the ham bands dictates that an operator will slow his or her code speed to accommodate another operator. Don't be afraid to call someone who is sending just a bit faster than you can copy comfortably. That operator will generally slow down to meet your CW speed. Helping each other is the name of the game in ham radio.

To increase your speed, you may wish to continue to copy the code practice sessions from W1AW, the ARRL HQ station. It might be a good idea to spend some time sending in step (on a code-practice oscillator—not on the air, of course) with W1AW transmissions; this approach will help develop your sending ability.

Correct CW Procedures

The best way to establish a contact, especially at first, is to listen until you hear someone calling CQ. CQ means, "I wish to contact any amateur station." Avoid the common operating pitfall of calling CQ endlessly; it clutters up the air and drives off potential new friends. The typical CQ would go like this: CQ CQ CQ DE K5RC K5RC K5RC K. The letter K is an invitation for any station to go ahead. If there is no answer, pause for 10 or 20 seconds and repeat the call.

If you hear a CQ, wait until the ham finishes transmitting (by ending with the letter K), then call him, thus: K5RC K5RC DE K3YL K3YL $\overline{\text{AR}}$. ($\overline{\text{AR}}$ is equivalent to *over*). In answer to your call, the called station will begin the reply by sending K3YL DE K5RC R. That R (*roger*) means that he has received your call correctly. Suppose K5RC heard someone calling him, but didn't quite catch the call because of interference (QRM) or static (QRN). Then he might come back with QRZ? DE K5RC K (Who is calling me?).

The QSO

During the contact, it is necessary to identify your station only once every 10 minutes and at the end of the communication. Keep the contact on a friendly and cordial level,

Table 1-4
Some Abbreviations for CW Work

Although abbreviations help to cut down unnecessary transmission, make it a rule not to abbreviate unnecessarily when working an operator of unknown experience.

AA	All after
AB	All before
ABT	About
ADR	Address
AGN	Again
ANT	Antenna
BCI	Broadcast interference
BCL	Broadcast listener
BK	Break; break me; break in
BN	All between; been
BUG	Semi-automatic key
B4	Before
C	Yes (correct)
CFM	Confirm; I confirm
CK	Check
CL	I am closing my station; call
CLD-CLG	Called; calling
CQ	Calling any station
CUD	Could
CUL	See you later
CW	Continuous wave (i.e., radiotelegraph)
DLD-DLVD	Delivered
DR	Dear
DX	Distance, foreign countries
ES	And, &
FB	Fine business, excellent
FM	Frequency modulation
GA	Go ahead (or resume sending)
GB	Good-by
GBA	Give better address
GE	Good evening
GG	Going
GM	Good morning
GN	Good night
GND	Ground
GUD	Good
HI	The telegraphic laugh; high
HR	Here, hear
HV	Have
HW	How
LID	A poor operator
MA, MILS	Milliamperes
MSG	Message; prefix to radiogram
N	No
NCS	Net control station
ND	Nothing doing
NIL	Nothing; I have nothing for you
NM	No more
NR	Number
NW	Now; I resume transmission
OB	Old boy
OC	Old chap
OM	Old man
OP-OPR	Operator
OT	Old timer; old top
PBL	Preamble
PSE	Please
PWR	Power
PX	Press
R	Received as transmitted; are
RCD	Received
RCVR (RX)	Receiver
REF	Refer to; referring to; reference
RFI	Radio frequency interference
RIG	Station equipment
RPT	Repeat; I repeat; report
RTTY	Radioteletype
RX	Receiver
SASE	Self-addressed, stamped envelope
SED	Said
SIG	Signature; signal
SINE	Operator's personal initials or nickname
SKED	Schedule
SRI	Sorry
SSB	Single sideband
SVC	Service; prefix to service message
T	Zero (number)
TFC	Traffic
TMW	Tomorrow
TNX-TKS	Thanks
TT	That
TU	Thank you
TVI	Television interference
TX	Transmitter
TXT	Text
UR-URS	Your; you're; yours
VFO	Variable-frequency oscillator
VY	Very
WA	Word after
WB	Word before
WD-WDS	Word; words
WKD-WKG	Worked; working
WL	Well; will
WUD	Would
WX	Weather
XCVR	Transceiver
XMTR (TX)	Transmitter
XTAL	Crystal
XYL (YF)	Wife
YL	Young lady
73	Best regards
88	Love and kisses

remembering that the conversation is not private and many others, including nonamateurs, may be listening. It may be helpful at the beginning to have a fully written-out script typical of the first couple of exchanges in front of you. A typical first transmission might sound like this: K3YL DE K5RC R TNX CALL. UR 599 599 QTH HOUSTON TX NAME TOM. HW? K3YL DE K5RC $\overline{\text{KN}}$. This is the basic exchange that begins most QSOs. Once these basics are exchanged the conversation can turn in almost any direction. Many people talk about their jobs, other hobbies, families, travel experiences, and so on.

Both on CW and phone, it is possible to be informal, friendly and conversational; this is what makes the Amateur Radio QSO enjoyable. During a CW contact, when you want the other station to take a turn, the recommended signal is $\overline{\text{KN}}$ (*go ahead, only*), meaning that you want *only* the contacted station to come back to you. If you don't mind someone else signing in, just K (*go ahead*) is sufficient.

Ending the QSO

When you decide to end the contact or the other ham expresses the desire to end it, don't keep talking. Briefly express your thanks for the contact: TNX QSO or TNX CHAT—and then sign out: 73 $\overline{\text{SK}}$ WA1WTB DE K5KG. If you are leaving the air, add CL (*closing*) to the end, right after your call sign.

These ending signals, which indicate to the casual

Table 1-5
ARRL Procedural Signals (Prosigns)

Situation	*CW*	*Voice*
check for a clear frequency	QRL?	Is the frequency in use?
seek contact with any station	CQ	CQ
after call to a specific named station or to indicate the end of a message	$\overline{\text{AR}}$	Over, end of message
invite any station to transmit	K	Go
invite a specific named station to transmit	$\overline{\text{KN}}$	Go only
invite receiving station to transmit	BK	Back to you
all received correctly	R	Received
please stand by	$\overline{\text{AS}}$	Wait, stand by
end of contact (sent before call sign)	$\overline{\text{SK}}$	Clear
going off the air	CL	Closing station

Additional RTTY prosigns

SK QRZ—Ending contact, but listening on frequency.
SK KN—Ending contact, but listening for one last transmission from the other station.
SK SZ—Signing off and listening on the frequency for any other calls.

Table 1-6
The Phonetic Alphabet

When operating phone, a standard alphabet is often used to ensure understanding of call letters and other spelled-out information. Thus Larry, WR1B, would announce his call as Whiskey Romeo One Bravo if he felt the station on the other end could misunderstand his call. Phonetics are not routinely used when operating VHF-FM.

A—Alfa (**AL** FAH)
B—Bravo (**BRAH** VOH)
C—Charlie (**CHAR** LEE)
D—Delta (**DELL** TAH)
E—Echo (**ECK** OH)
F—Foxtrot (**FOX** TROT)
G—Golf (GOLF)
H—Hotel (HOH **TELL**)
I—India (**IN** DEE AH)
J—Juliet (**JEW** LEE ETT)
K—Kilo (**KEY** LOH)
L—Lima (**LEE** MA)
M—Mike (MIKE)
N—November (NO **VEM** BERR)
O—Oscar (**OSS** CAR)
P—Papa (PAH **PAH**)
Q—Quebec (KEY **BECK**)
R—Romeo (**ROW** ME OH)
S—Sierra (SEE **AIR** AH)
T—Tango (**TANG** OH)
U—Uniform (**YOU** NEE FORM)
V—Victor (**VIK** TORE)
W—Whiskey (**WISS** KEY)
X—X-Ray (**EX** RAY)
Y—Yankee (**YANG** KEY)
Z—Zulu (**ZOO** LOU)

The **boldfaced** syllables are emphasized.

Table 1-7
The RST System

Readability
1—Unreadable
2—Barely readable, occasional words distinguishable
3—Readable with considerable difficulty
4—Readable with practically no difficulty
5—Perfectly readable

Signal Strength
1—Faint signals, barely perceptible
2—Very weak signals
3—Weak signals
4—Fair signals
5—Fairly good signals
6—Good signals
7—Moderately strong signals
8—Strong signals
9—Extremely strong signals

Tone
1—Sixty-cycle ac or less, very rough and broad
2—Very rough ac, very harsh and broad
3—Rough ac tone, rectified but not filtered
4—Rough note, some trace of filtering
5—Filtered rectified ac but strongly ripple-modulated
6—Filtered tone, definite trace of ripple modulation
7—Near pure tone, trace of ripple modulation
8—Near perfect tone, slight trace of ripple modulation
9—Perfect tone, no trace of ripple or modulation of any kind

If the signal has the characteristic steadiness of crystal control, add the letter X to the report. If there is a chirp, add the letter C. Similarly for a click, add K. (See FCC Regulations §97.307, Emissions Standards.) The above reporting system is used on both CW and voice; leave out the "tone" report on voice.

listener the status of the contact, establish Amateur Radio as a cordial and fraternal hobby. At the same time they foster orderliness and denote organization. These signals have no legal standing; FCC regulations say little about our internal procedures.

Table 1-5 contains a brief list of these procedural signals. You can also request Form FSD-220 from the ARRL. It contains this list plus other handy information for the operating position.

Phone Operating Procedures

These phone or voice operating procedures apply to operation on the HF bands as well as SSB on VHF/UHF. Procedures used on repeaters are different since the operation there is channelized—that is, anyone listening to the repeater will hear you as soon as you begin to transmit. Therefore there is no need to call CQ. Each repeater may use a slightly different procedure. There is a complete discussion of repeater operations in the FM chapter of this book.

Learning phone procedures for HF is exactly like learning CW operating procedure. Listen to what others are doing, and incorporate their good habits into your own operating style. Use common sense in your day-to-day phone QSOs, too:

(1) Listen before transmitting. Ask if the frequency is in use before making a call on any particular frequency.
(2) Give your call sign as needed, using the approved ITU (International Telecommunication Union) Phonetics. These phonetics are given in **Table 1-6**.

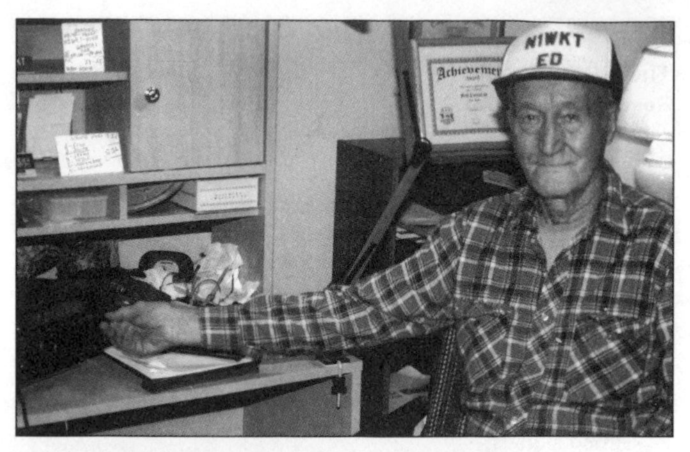

It's never too late to learn. Ed Linn, N1WKT, of Kensington, Connecticut, earned his Technician Plus license when he was 79 years old. Ed says he first became interested in Amateur Radio around 1930, but during the Great Depression he had no way to travel to New York City or Boston to take the FCC exam. More than 60 years later he finally realized his dream!

(3) Make sure your signal is clean. Do not turn your microphone gain up too high. If you have a speech processor, use it only when you are sure it is properly adjusted. Don't take the chance of transmitting spurious (out of band) signals.

(4) Keep your transmissions as short as possible to give as many operators as possible a chance to use the frequency spectrum.

(5) Give honest signal reports. **Table 1-7** lists what the various RST reports mean.

Whatever band, mode or type of operating you choose, there are three fundamental things to remember. The first is that courtesy costs very little and is often amply rewarded by bringing out the best in others. The second is that the aim of each radio contact should be 100% effective communication. The good operator is never satisfied with anything less. The third is that the "private" conversation with another station is actually *public*. Keep in mind that many amateurs are uncomfortable discussing so-called controversial subjects over the air. Also, never give any information on the air that might be of assistance to the criminally inclined. As an example, never state when you are going to be out of town!

Correct phone operation is more challenging than it first may appear, even though it does not require the use of code or special abbreviations and prosigns. This may be because operators may have acquired some imperfect habits in their pronunciation, intonation and phraseology even before entering Amateur Radio! To these might be added a whole new set of cliches and mannerisms derived from listening to below-par operators.

Using the proper procedure is very important. Voice operators say what they want to have understood, while CW operators have to spell it out or abbreviate. Since the speed of transmission on phone is generally between 150 and 200 words per minute, the matter of readability and understandability is critical to good communication. The good voice operator uses operating habits that are beyond reproach.

It is important to speak clearly and not too quickly. This is particularly important when talking to a DX station who does not fully understand our language.

Avoid using CW abbreviations (including HI) and Q signals on phone, although QRZ (for "who is calling?") has become accepted. Otherwise, plain language should be used. Keep jargon to a minimum. Some hams use "we" instead of "I," "handle" instead of "name" and "Roger" instead of "that's correct." These expressions are not necessary and do not contribute to better operating. No doubt you will hear many more.

Phone Procedure

There are three ways to initiate a voice contact: call CQ (a general call to any station), answer a CQ, or call at the end of the other person's QSO. If activity on a band seems low and you have a reasonable signal, a CQ call may be worthwhile.

Before calling CQ, it is important to find a frequency that appears unoccupied by any other station. This may not be easy, particularly in crowded band conditions. Listen carefully—perhaps a weak DX station is on frequency.

Always listen before transmitting. Make sure the frequency isn't being used *before* you come barging in. If, after a reasonable time, the frequency seems clear, ask if the frequency is in use, followed by your call: "Is the frequency in use? This is N2EEC." If, as far as you can determine, no one responds, you are ready to make your call.

As in CW operation, CQ calls should be kept short. Long calls are unnecessary. If no one answers, you can always call again. If you do transmit a long call, you may interfere with stations that were already on the frequency but whom you didn't hear in the initial check. In addition, any stations intending to reply to the call may become impatient and move to another frequency. If two or three calls produce no answer, there may be interference on the frequency. It's also possible that the band isn't open.

An example of a CQ call would be: "CQ CQ Calling CQ. This is N2EEC, November-Two-Echo-Echo-Charlie, Novem-

SAREX (the Space Amateur Radio Experiment) isn't just for American students. In 1997 several members of the Tsinghua University Amateur Radio Club (BY1QH) in Beijing, China, made contact with astronauts aboard the space shuttle *Columbia*. Also participating in the event were students from nearby Tsinghua High School.

Guest Operator

An FCC rules interpretation allows the person in physical control of an Amateur Radio station to use his or her own call sign when guest operating at another station. Of course, the guest operator is bound by the frequency privileges of his or her own operator's license, no matter what class of license the station licensee may hold. For example, Joan, KB6MOZ, a Technician, is visiting Stan, N6MP, an Extra Class licensee. Joan may use her own call sign at N6MP's station, but she must stay within the Technician subbands.

When a ham is operating from a club station, the club call is usually used. Again, the operator may never exceed the privileges of his or her own operator's license. The club station trustee and/or the club members may decide to allow individual amateurs to use their own call signs at the club station, but it is optional. In cases where it is desirable to retain the identity of the club station (W1AW at ARRL HQ, for example), the club may require amateurs to use the club call sign at all times.

In the rare instance where the guest operator at a club station holds a higher class of license than the club station trustee, the guest operator must use the club call sign and his/her own call sign. For example, Billy, KR1R, who holds an Amateur Extra Class license visits the Norfolk Technician Radio Club station, KA4CVX. Because he wants to operate the club station outside the Tech subbands, Billy would sign KA4CVX/KR1R on CW and "KA4CVX, KR1R controlling" on phone. (Of course, this situation would prevail only if the club requires that their club call sign be used at all times.)

In Pursuit of ... DX

"I'll never forget the thrill of my first DX contact. It was on CW in the Novice portion of 15 meters. I heard a G3 (a station in England) calling CQ, and no one answered right away so I decided to give it a try. Success! With a lump in my throat and sweat on my palms, I managed to complete my first DX QSO and become hopelessly hooked. DX is great!"

Many hams can tell a similar story. DX is Amateur Radio shorthand for *long distance*; furthermore, DX is universally understood by hams to be a station in a foreign country. Chasing DX is one of the most popular activities in our hobby, and—if you operate on the HF bands—you can get in on the fun!

Once you have passed a Morse code test, you can work DX on Technician frequency allocations almost as easily as can an Amateur Extra ham. You don't need a super kilowatt station and huge antennas. The beauty of DXing is that your operating skills can overcome deficiencies in your station equipment. Pick the right frequency band and the right time of day (or night) and the DX stations will be there—ready to talk to you.

Check the "How's DX?" column in *QST*. It contains tips on propagation and news of interest to DXers. And don't forget to listen to the weekly W1AW DX bulletin for the latest DX news. Good DX!

ber-Two-Echo-Echo-Charlie, calling CQ and standing by."

When replying to a CQ, both call signs should be given clearly. Use the ITU phonetic alphabet shown in Table 1-6 to make sure the other station gets your call correctly. Phonetics are necessary when calling into a DX pileup and initially in most HF contacts but not usually used when calling into an FM repeater.

When you are calling a specific station, it is good practice to keep calls short and to say the call sign of the station called once or twice only, followed by your call repeated twice. VOX (voice operated switch) operation is helpful because, if properly adjusted, it enables you to listen between words, so that you know what is happening on the frequency. "N2EEC N2EEC, this is W2GD, Whiskey-Two-Golf-Delta, Over."

Once contact has been established, it is no longer necessary to use the phonetic alphabet or sign the other station's call. According to FCC regulations, you need only sign your call every 10 minutes, or at the conclusion of the contact. (The exception is handling international third-party traffic; you must sign both calls in this instance.) A normal two-way conversation can thus be enjoyed, without the need for continual identification. The words "Over" or "Go Ahead" are used at the end of a transmission to show you are ready for a reply from the other station.

Signal reports on phone are two-digit numbers using the RS portion of the RST system (no tone report is required). The maximum signal report would be "59"; that is, readability 5, strength 9. On FM repeaters, RS reports are not appropriate. When a signal has fully captured the repeater, this is called "full quieting."

Conducting the Contact

Aside from signal strength, it is customary (as in a typical CW QSO described elsewhere in this chapter) to exchange location, name and usually a brief description of the rig and antenna. Keep in mind that many hams will not know the characteristics of your *Loundenboomer 27A3* and your *Signal Squirter 4*. Therefore, you may be better off just saying that your rig runs 100 watts output and the antenna is a trap dipole. Once these routine details are out of the way, you can discuss virtually anything the two of you find appropriate and interesting.

DX Contacts

DX can be worked on any HF band as well as occasionally on 6 meters. When the 11-year solar sunspot cycle favors 10 meters, worldwide communications on a daily basis are commonplace on this band. Ten meters is an outstanding DX band when conditions are right! A particular advantage of 10 for DX work is that effective beam-type antennas tend to be small and light, making for relatively easy installation.

Keep in mind that while many overseas amateurs have an exceptional command of English, they may not be familiar with many of our colloquialisms. Because of the language differences, some DX stations are more comfortable with the barebones type contact and you should be sensitive to their preferences. In unsettled propagation conditions, it may be necessary to keep the whole contact short. The good operator takes these factors into account when expanding on a basic contact.

When the time comes to end the contact, end it. Thank the other operator (once) for the pleasure of the contact and say good-bye: "73, N9ABC this is N2EEC, Clear." This is all that is required. Unless the other amateur is a good friend, there is no need to start sending best wishes to everyone in the household including the family dog! Nor is this the time to start digging up extra comments on the contact that will require a "final final" from the other station (there may be other stations waiting to call in). Please understand that during a band opening on 10 meters or on VHF, you should keep contacts brief so as many stations as possible can work the DX coming through.

Additional Recommendations

Listen with care. It is very natural to answer the loudest station that calls, but sometimes you will have to dig deep into the noise and interference to hear the other station. Not all amateurs can run a kilowatt.

Use VOX or push-to-talk (PTT). If you use VOX, don't defeat its purpose by saying "aaah" to keep the relay closed. If you use PTT, let go of the mike button every so often to make sure you are not "doubling" with the other station. Don't filibuster.

Talk at a constant level. Don't ride the mike gain. Try to keep the same distance from the microphone. Keep the mike gain down to eliminate room noise. Follow the manufacturer's instructions for use of the microphone. Some require close talking, while some need to be turned at an angle to the speaker's mouth. Speech processing (often built into contemporary transceivers) is often a mixed blessing. It can help you cut through the interference and static, but if too much is used, the audio quality suffers greatly. Tests should be made to determine the maximum level that can be used effectively, and this should be noted or marked on the control. Be ready to turn it down or off if it is not really required during a contact.

The speed of voice transmission (with perfect accuracy) depends almost entirely on the skill of the two operators concerned. Speak at a rate that allows perfect understanding as well as permitting the receiving operator to record the information.

RECORDKEEPING (LOGGING AND QSLING)

Although the FCC does not require logging except for certain specialized occurrences, you can still benefit by keeping an accurate log (see **Fig 1-5**). *The ARRL Logbook*, on sale at your local radio bookstore or directly from ARRL HQ, provides a good method for maintaining your log data at your fingertips.

The log entry should include:

1) The ca ll sign of the station worked.
2) The date and time of the QSO. Always use UTC (Universal Coordinated Time, sometimes also called GMT or Zulu time) when entering the date and time. Use UTC whenever you need a time or date in your ham activities. The use of UTC helps all hams avoid confusion through conversion to local time. **Fig 1-6** explains the UTC system.

See inside Cover — MODE
Output in Watts — POWER
UTC Recommended — TIME
RST. See back Inside Cover — REPORT
This column may also be used for contest-exchange info received. — TIME OFF

FIXED: DATE, FREQ., MODE, POWER
VARIABLE: TIME, STATION WORKED, REPORT, TIME OFF, COMMENTS, QSL

DATE	FREQ.	MODE	POWER	TIME	STATION WORKED	REPORT SENT	REPORT REC'D	TIME OFF	COMMENTS (QTH / NAME / QSL VIA)	QSL S	QSL R
16 Nov	3715	CW	100	1800	KA1EBV	589	479	1835	MANOMET, MA JOHN	✓	✓
	〃	CW	100	1840	WB7TPY	599	599	1855	TUFTS UNIVERSITY DAVE	✓	
17 Nov	28.125	CW	100	2002	KC4AAA	469	559	2003	AT SOUTH POLE! BIG PILE UP BURO	✓	
	28.380	SSB	100	2010	KAØHJD	59	58	2016	DES MOINES, IA KRISTEN	✓	
	〃	SSB	100	2010	WØSH	58	57	2016	PALM BAY, FL GARY COLLECTS OLD TELEGRAPH KEYS	✓	
	7.130	CW	100	2110	KA1GQT	579	589	2055	LINDA IS A NURSE	✓	✓
20 NOV	21.126	CW	100	1645	NA1L	599	599	1705	BRISTOL, CT DALE IS A LAWYER		
21 NOV	28.[illegible]80	RTTY	60	1018	WB8IMY	589	579	1052	CT STEVE WORKS AT ARRL		
	28.125	CW	100	1804	XZ1A	599	479	1804	MYANMAR! BILL Huge Pile up K5FUV	✓	✓

Fig 1-5—The ARRL Log Book is adaptable to all types of operating.

UTC Explained

Fig 1-6—Time is recorded both in log books and on QSL cards.

Ever hear of Greenwich Mean Time? How about Coordinated Universal Time? Do you know if it is light or dark at 0400 hours? If you answered no to any of these questions, you'd better read on!

Keeping track of time can be pretty confusing when you are talking to other hams around the world. Europe, for example, is anywhere from 4 to 11 hours ahead of us here in North America. Over the years, the time at Greenwich, England, has been universally recognized as the standard time in all international affairs, including ham radio. (We measure longitude on the surface of the Earth in degrees east or west of the Prime Meridian, which runs approximately through Greenwich, England, and which is halfway around the world from the International Date Line.) This means that wherever you are, you and the station you contact will be able to reference a common date and time easily. Mass confusion would occur if everyone used their own local time.

Coordinated Universal Time (abbreviated UTC) is the name for what used to be called Greenwich Mean Time. Twenty-four-hour time lets you avoid the equally confusing question about AM and PM. If you hear someone say he made a contact at 0400 hours UTC, you will know immediately that this was 4 hours past midnight, UTC, since the new day always starts just after midnight. Likewise, a contact made at 1500 hours UTC was 15 hours past midnight, or 3 PM (15 – 12 = 3).

Maybe you have begun to figure it out: Each day starts at midnight, 0000 hours. Noon is 1200 hours, and the afternoon hours merely go on from there. You can think of it as adding 12 hours to the normal PM time(3 PM is 1500 hours, 9:30 PM is 2130 hours, and so on. However you learn it, be sure to use the time everyone else does—UTC. See chart below.

The photo shows a specially made clock, with an hour hand that goes around only once every day, instead of twice a day like a normal clock. Clocks with a digital readout that show time in a 24-hour format are quite popular as a station accessory.

UTC	EDT/AST	CDT/EST	MDT/CST	PDT/MST	PST
0000*	2000	1900	1800	1700	1600
0100	2100	2000	1900	1800	1700
0200	2200	2100	2000	1900	1800
0300	2300	2200	2100	2000	1900
0400	0000*	2300	2200	2100	2000
0500	0100	0000*	2300	2200	2100
0600	0200	0100	0000*	2300	2200
0700	0300	0200	0100	0000*	2300
0800	0400	0300	0200	0100	0000*
0900	0500	0400	0300	0200	0100
1000	0600	0500	0400	0300	0200
1100	0700	0600	0500	0400	0300
1200	0800	0700	0600	0500	0400
1300	0900	0800	0700	0600	0500
1400	1000	0900	0800	0700	0600
1500	1100	1000	0900	0800	0700
1600	1200	1100	1000	0900	0800
1700	1300	1200	1100	1000	0900
1800	1400	1300	1200	1100	1000
1900	1500	1400	1300	1200	1100
2000	1600	1500	1400	1300	1200
2100	1700	1600	1500	1400	1300
2200	1800	1700	1600	1500	1400
2300	1900	1800	1700	1600	1500
2400	2000	1900	1800	1700	1600

Time changes one hour with each change of 15° in longitude. The five time zones in the US proper and Canada roughly follow these lines.

*0000 and 2400 are interchangeable. 2400 is associated with the date of the day ending, 0000 with the day just starting.

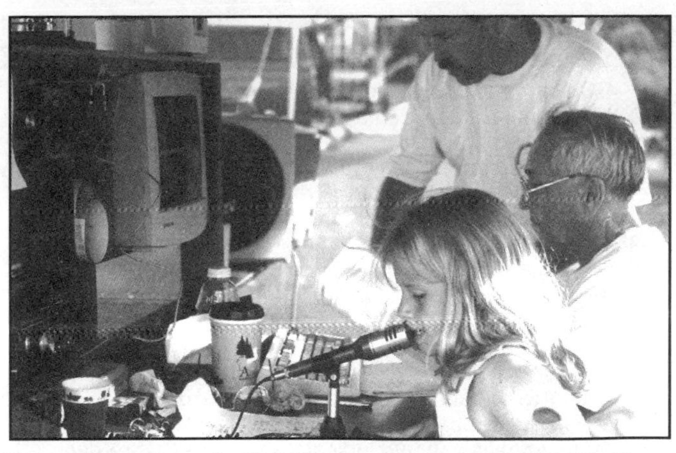

Three generations do Field Day together at the Woodbridge (VA) Wireless 80-meter station. Craig, KJ4LR, is standing behind his father, Vernon, KB4NZJ, while his daughter Catherine makes a phone contact.

Members of Woodbridge (Virginia) Wireless, W4IY, struggle in the oppressive heat to install a triband beam at their Field Day site.

3) The frequency or frequency band on which the QSO took place.
4) The emission mode of communication.
5) Signal reports sent and received.
6) Any miscellaneous data, such as the other operator's name or QTH that you care to record.

The FCC has promulgated a rather elaborate system of emission designators (see the Rules and Regulations chapter). For logging simplicity, the following common abbreviations are often used:

Abbreviation	*Explanation*
CW	telegraphy on pure continuous wave
MCW	tone-modulated telegraphy
SSB	single-sideband suppressed carrier
AM, DSB	double-sideband with full, reduced or suppressed carrier
FAX	facsimile
FM	frequency- or phase-modulated telephony
RTTY	radioteletype
AMTOR	time diversity radioteletype
P	pulse
TV or SSTV	television or slow-scan television

QSLing—The Final Courtesy

QSL cards are a tradition in ham radio. Exchanging QSLs is fun, and they can serve as needed confirmations for many operating awards. You'll probably want your own QSL cards, so look in *QST* for companies that sell them, or you may even want to make your own. Your QSL should be attractive, yet straightforward. All necessary QSO information on one side of the card will make answering a QSL a relatively simple matter.

A good QSL card should be the standard 3.5 × 5.5-inch size (standard post card size) and should contain the following information:

1) Your call. If you were portable or mobile during the contact, this should be indicated on the card.

2) The geographical location of your station. Again, portables/mobiles should indicate where they were during the contact.

3) The call of the station you worked. This isn't as simple as it sounds. Errors are very common here. Make sure it is clear if the call contains the numeral 1 (one), capital I ("eye") or lower case l ("el").

4) Date and time of the contact. Use UTC for both and be sure to convert the time properly from local time, if that's how you keep your log. It is best to write out the date in words to avoid ambiguity. Use May 10, 1995 or 10 May 1995, rather than 5/10/95. Most DX stations will use 8-2-95 to mean February 8, 1995. The day is written before the month in many parts of the world.

5) Frequency. The band in wavelength (meters) or approximate frequency in kHz or MHz is required.

6) Mode of operation. Use accepted abbreviations, but be specific. CW, SSB, RTTY and AM are clear and acceptable. FCC emission designations are not always understood by DX stations.

7) Signal report.

8) Leave no doubt the QSL is confirming a two-way contact by using language such as "confirming two-way QSO with" or "2 X" or "2-Way" before the other station's call. Other items, such as your rig, antenna and so on, are optional.

9) If you make any errors filling out the QSL, destroy the card and start over. Do not make corrections or mark-overs on the card, as such cards are not acceptable for awards purposes.

Now comes the problem of how to get your QSLs to the DX station. Sending them directly can be expensive, so many amateurs use the ARRL's Overseas QSL Service. This is an outgoing service for ARRL members to send DX QSL cards to foreign countries at a minimum of cost and effort.

To receive QSL cards from DX (overseas) stations, the ARRL sponsors incoming QSL Bureaus, provided free for all amateurs throughout the United States and Canada. Each call area has its own bureau staffed totally by volunteers. To expedite the handling of your QSL cards (both incoming and outgoing), be sure to follow the bureau's requirements at all times. See the DXing chapter for further details on the ARRL Overseas QSL Service and the ARRL QSL Bureau.

Table 1-8

The "Considerate Operator's Frequency Guide"

The following frequencies are generally recognized for certain modes or activities (all frequencies are in MHz).

Nothing in the rules recognizes a net's, group's or any individual's special privilege to any specific frequency. Section 97.101(b) of the Rules states that "Each station licensee and each control operator must cooperate in selecting transmitting channels and in making the most effective use of the amateur service frequencies. No frequency will be assigned for the exclusive use of any station." No one owns a frequency.

It's good practice—and plain old common sense—for any operator, regardless of mode, to check to see if the frequency is in use prior to engaging operation. If you are there first, other operators should make an effort to protect you from interference to the extent possible given that 100% interference-free operation is an unrealistic expectation in today's congested bands.

Frequency	Use
1.800-1.810	Digital modes
1.810	CW QRP
1.800-2.000	CW
1.843-2.000	SSB, SSTV and other wideband modes
1.910	SSB QRP
1.995-2.000	Experimental
1.999-2.000	Beacons
3.560	QRP CW calling frequency
3.590	RTTY DX
3.580-3.620	Data
3.620-3.635	Automatically controlled data stations
3.710	QRP Novice/Technician CW calling frequency
3.790-3.800	DX window
3.845	SSTV
3.885	AM calling frequency
3.985	QRP SSB calling frequency
7.040	RTTY DX QRP CW calling frequency
7.080-7.100	Data
7.100-7.105	Automatically controlled data stations
7.100	QRP Novice/Technician CW calling frequency
7.171	SSTV
7.285	QRP SSB calling frequency
7.290	AM calling frequency
10.106	QRP CW calling frequency
10.130-10.140	Data
10.140-10.150	Automatically controlled data stations
14.060	QRP CW calling frequency
14.070-14.095	Data
14.095-14.0995	Automatically controlled data stations
14.100	NCDXF/IARU beacons
14.1005-14.112	Automatically controlled data stations
14.230	SSTV
14.285	QRP SSB calling frequency
14.286	AM calling frequency
18.100-18.105	Data
18.105-18.110	Automatically controlled data stations
21.060	QRP CW calling frequency
21.070-21.100	Data
21.090-21.100	Automatically controlled data stations
21.340	SSTV
21.385	QRP SSB calling frequency
24.920-24.925	Data
24.925-24.930	Automatically controlled data stations
28.060	QRP CW calling frequency
28.070-28.120	Data
28.120-28.189	Automatically controlled data stations
28.190-28.225	Beacons
28.385	QRP SSB calling frequency
28.680	SSTV
29.000-29.200	AM
29.300-29.510	Satellite downlinks
29.520-29.580	Repeater inputs
29.600	FM simplex
29.620-29.680	Repeater outputs

Note

ARRL band plans for frequencies above 29.680 MHz are shown in *The ARRL Repeater Directory* and *The FCC Rule Book*. NCDXF/IARU beacons operate on 14.100, 18.110, 21.150, 24.930, and 28.200 MHz.

Computer Logging

If you are into personal computing, there are many log-keeping programs that will take your input, format your log file and even give you a printed output. In addition, computer programs for contest applications are discussed in the Contests chapter.

A well-kept log will help you preserve your fondest ham radio memories for years. It will also serve as a book-keeping system should you embark upon a quest for ham radio awards, or a complete collection of QSL cards.

QSLing

The QSL card is the final courtesy of a QSO. It confirms specific details about your two-way contact with another ham. Whether you want the other station's QSL as a memento of an enjoyable QSO, or for an operating award, it's wise to have your own QSL cards and know how to fill them out. That way, when you send your card to the other station, it will result in the desired outcome (his card sent to you).

Your QSL

Your QSL card makes a statement about you. It may also hang in ham radio shacks all over the world. So you will want to choose carefully the style of QSL that you have printed. There are many QSL vendors listed in the Ham Ads section of *QST* each month. A nominal fee will bring you many samples from which to choose or you may design and/or print your own style. The choice is up to you. See the accompanying sidebar for more information on QSLs.

THE BANDS

Operating on the VHF/UHF bands is relatively straightforward for most modes of operation commonly employed there. Operating on the HF bands is a little more variable, because the propagation depends on a number of factors that are quite literally in outer space. The main influence on the Earth's ionosphere is the Sun, but the magnetic field of the Earth gets into the act very much as well in determining how HF signals are propagated from one place to another.

The sidebar *Picking a Band* gives you a generalized idea of what to expect on both the HF and the VHF/UHF bands. Never forget, however, that part of the excitement and mystery of the HF bands lies in their unpredictability from day-to-day, or even from hour-to-hour.

Table 1-8 shows the Considerate Operator's Frequency Guide, and **Table 1-9** shows the North American VHF/UHF/EHF Calling Frequencies.

Propagation Beacons

By providing a steady signal on certain frequencies, HF beacons provide a valuable means of checking current propagation conditions—how well signals are traveling at the time you want to transmit.

Table 1-9
North American VHF/UHF/EHF Calling Frequencies

Band (MHz)	*Calling Frequency*
50	50.110 DX
	50.125 SSB US, local
	50.620 digital (packet)
	52.525 National FM simplex frequency
144	144.010 EME
	144.100, 144.110 CW
	144.200 SSB
	146.520 National FM simplex frequency
222	222.100 CW/SSB
	223.500 National FM simplex frequency
432	432.010 EME
	432.100 CW/SSB
	446.000 National FM simplex frequency
902	902.100
	903.100 East Coast
	906.500 National FM simplex frequency
1296	1294.500 National FM simplex frequency
	1296.100 CW/SSB
2304	2304.1 CW/SSB
10000	10368.1 CW/SSB
	10280.0 WBFM

VHF/UHF Activity Nights

Some areas do not have enough VHF/UHF activity to support contacts at all times. This schedule is intended to help VHF/UHF operators make contact. This is only a starting point; check with others in your area to see if local hams have a different schedule.

Band (MHz)	*Day*	*Local Time*
50	Sunday	6 PM
144	Monday	7 PM
222	Tuesday	8 PM
432	Wednesday	9 PM
902	Friday	9 PM
1296	Thursday	10 PM

NCDXF/IARU International Beacon Network

The Northern California DX Foundation, in cooperation with the International Amateur Radio Union (IARU), has established a widespread, multi-band beacon network. As this is written, this network operates on 14.100, 18.110, 21.150, 24.930 and 28.200 MHz. These beacons transmit at a sequence of power levels from 0.1 to 100 watts over a set period of time. A full description with the most up-to-date status is on their Web page at: **http://www.ncdxf.org/beacon.htm**.

That finishes our quick tour of operating procedures. Your knowledge will grow quickly with on-the-air experience. Enjoy the learning process. Ham radio is such a diverse activity that the learning process never stops—talk to most 80-year-old hams, and they'll tell you there is always something new to learn!

Chapter 2

Amateur Radio Rules and Regulations

Our national government grants us, as radio amateurs, certain privileges. In order to retain these privileges in the USA, we must conform to the rules and regulations of the Federal Communications Commission (FCC). These Telecommunications Rules are known officially as Title 47 CFR (Code of Federal Regulations), Part 97.

This chapter covers only the most obvious rules that pertain to amateurs' day-to-day operation. This is not a complete discussion of all the FCC rules and regulations. For complete coverage of FCC rules as they pertain to Amateur Radio, along with the complete text of Part 97, see *The ARRL's FCC Rule Book*, published by ARRL. (Locate a dealer near you or order online at **www.arrl.org/catalog/**.) You can also find the complete text of Part 97 on *ARRLWeb* at **www.arrl.org/FandES/regulations/news/part97/**.

Access to frequencies is essential to Amateur Radio. Without frequencies on which to operate, there would be no Amateur Radio. Since radio waves don't stop at international borders, the radio spectrum is a matter for international as well as national regulation. This chapter introduces you to some of the international organizations that affect Amateur Radio around the world.

YOUR LICENSE

You must have a license granted by the FCC to operate an Amateur Radio station in the United States, in any of its territories or possessions, or from any vessel or aircraft registered in the United States. Foreign amateurs must operate under the appropriate reciprocal operating authority to operate in FCC–regulated areas. The FCC sets no minimum or maximum age requirement for obtaining a license, nor does it require that an applicant be a US citizen. US citizens are required to pass an FCC license exam to obtain a US Amateur Radio license, though.

An Amateur Radio license incorporates two kinds of authorization. For the individual, the license grants operator privileges in one class. The class of license the operator has earned determines these privileges. The license also authorizes an Amateur Radio station to operate the physical transmitting equipment.

There are several ways that aliens, that is, amateurs who are citizens of another country, may obtain permission to operate an Amateur Radio station in places regulated by the FCC. There must be a multilateral or bilateral reciprocal operating arrangement in effect, to which the United States and the alien's government are parties, for amateur service operation on a reciprocal basis. For example, an English amateur—G3CCX—visiting friends in New Hampshire and operating would identify as W1/G3CCX.

Canadian citizens are permitted to operate an amateur station in the US with a copy of their Canadian license. Citizens of a country that is part of the European Conference of Postal and Telecommunications Administrations (CEPT) agreement may use their amateur licenses from that country to operate in the US. All they need is a copy of their CEPT license and proof of citizenship. No other paperwork or application form is required.

Citizens of certain countries in the Americas may use an International Amateur Radio Permit (IARP) to operate in the US. This document is issued according to the terms of the Inter-American Convention on an International Amateur Radio Permit. The IARP is not a license, but is a document issued either by the government or by the International Amateur Radio Union (IARU) member society in the amateur's home country. (In the US, ARRL is the IARU society.) The IARP document certifies that the person does have an Amateur Radio license from that country.

These agreements generally provide operator privileges that are similar to the privileges enjoyed by the alien operator in his or her own country. The maximum privileges are those of a US Amateur Extra class licensee.

In no case may a resident alien or US citizen operate under one of these agreements since all US citizens are required to have taken and passed the necessary US requirements. A person whose FCC-issued license was revoked or suspended can't use one of these reciprocal operating agreements to circumvent the FCC action, either! See the FCC Rules, Part 97 for the full details.

The United States effectively has six classes of operator licenses: Novice, Technician, Technician with 5 wpm Morse code exam credit, General, Advanced and Amateur Extra. Only three classes of *new* licenses will be issued by the FCC, however: Technician, General and Amateur Extra. Amateurs who hold a valid Novice or Advanced license may renew their license without penalty within 90 days before it expires, or within the 2-year grace period for renewal. The frequency privileges for these licensees remain unchanged. When Technician Plus class operators renew their licenses they will receive a new document showing a license class of Technician. Their old (expired) Technician Plus class license provides continued proof of their authorization to use the Novice HF bands, though.

After April 15, 2000, new Technician licensees must pass the 35 question Element 2 exam. The Technician license authorizes operation on all frequencies above 30 MHz. Upon passing the 5 wpm code test, a Technician is authorized to use all Novice HF privileges. General and Amateur Extra classes of license require knowledge that is more technical. The General class license requires a 35 question Element 3 exam and the Amateur Extra class license requires a 50 question Element 4 exam. In addition, the General and Extra class licenses require a 5 wpm Morse code exam, Element 1. In return for passing the more difficult technical exams, each higher-class license rewards the licensee with more privileges, including expanded frequency allocations and more modes of operation. See **Table 2-1**.

License Renewal/Modification

Once you have a license, you can keep it for your lifetime, but it must be renewed every 10 years. The old FCC Form 610 is no longer accepted. With the implementation of the Commission Registration System (CORES) on December 3, 2001, CORES is the primary FCC registration system though ULS, the Universal Licensing System, is still active. Everyone doing business with the FCC, including amateur licensees, must obtain and use a 10-digit FCC Registration Number (FRN) when filing. Amateur licensees currently registered in ULS have or should have been cross-registered in CORES and issued an FRN by mail. Amateurs can check to see if they have an FRN via a ULS license search at **www.fcc.gov/wtb/uls** or **www.arrl.org/fcc/fcclook.php3**. Many Internet call sign servers, including ARRL's, also can provide this information.

Now that CORES is mandatory, the FCC will "auto-register" all amateurs who seek to register in ULS and will issue them an FRN. Amateurs then should use their FRN in place of their Taxpayer Identification Number (TIN—typically an individual's Social Security Number) when filing applications with the FCC. New or upgrade license applicants not previously registered in ULS will be registered automatically in both CORES and ULS when they provide a TIN on a license application filed through a Volunteer Examiner Coordinator. Although both ULS and CORES will contain a licensee's FRN, updating information in one system will not update the other. CORES offers exemptions to amateur clubs and to foreign entities not holding a TIN/SSN. Club station applicants also

Table 2-1
Amateur Operator Licenses†

Class	*Code Test*	*Written Examination*	*Privileges*
Novice	5 wpm (Element 1)	(No *new* Novice licenses will be issued.)	Telegraphy on 3675-3725, 7100-7150 and 21,100-21,200 kHz with 200 W PEP output maximum; telegraphy RTTY and data on 28,100-28,300 kHz and telegraphy and SSB voice on 28,300-28,500 kHz with 200 W PEP max; all amateur modes authorized on 222.0-225.0 MHz, 25 W PEP max; all amateur modes authorized on 1270-1295 MHz, 5 W PEP max.
Technician		Technician-level theory and regulations. (Element 2)*	All amateur privileges above 50.0 MHz.
Technician with Morse Code credit	5 wpm (Element 1)	Technician-level theory and regulations. (Element 2)*	All Novice HF privileges in addition to all Technician privileges.
General	5 wpm (Element 1)	Technician and General theory and regulations. (Elements 2 and 3)	All amateur privileges except those reserved for Advanced and Amateur Extra.
Advanced		(No new Advanced licenses will be issued.)	All amateur privileges except those reserved for Amateur Extra class.
Amateur Extra	5 wpm (Element 1)	All lower exam elements, plus Extra-class theory (Elements 2, 3 and 4)	All amateur privileges.

†A licensed radio amateur will be required to pass only those elements that are not included in the examination for the amateur license currently held.

*If you hold an expired Technician license issued before March 21, 1987, you also have credit for Element 3. You must be able to prove to a VE team that your Technician license was issued before March 21, 1987 to claim this credit. If you hold an expired or unexpired Technician license issued before February 14, 1991, you also have credit for Element 1 if you provide proof of this to a VE team.

may use a trustee's TIN/SSN or a tax-exempt club's IRS-assigned EIN. Amateurs also may use FCC Form 160 to register in CORES, and those doing so will be mailed a CORES password for on-line access.

The preferred method of conducting business with the FCC is electronically at the FCC URL above. Amateurs conducting business with the FCC manually, that is, by paper form, should follow this procedure:

1) ARRL members may use the NCVEC Form 605, not an FCC form, but one accepted by the FCC if it goes through a VEC. The ARRL VEC offers this as a free membership service. This form can be found at **www.arrl.org/fcc/forms.html** and may be used for all uses except a request for a duplicate license and requests for new vanity call signs. This must be sent directly to the ARRL VEC in Newington, Connecticut, not the FCC. Amateurs may also use the FCC Form 605 which can be found at **www.fcc.gov/Forms/Form605/605main.pdf**.

2) ARRL members using the NCVEC Form 605 should send it to the ARRL VEC, 225 Main St., Newington, CT 06111. Amateurs using the FCC Form 605 must send it to the FCC, 1270 Fairfield Rd., Gettysburg, PA 17325. There is no fee for renewing an amateur license.

3) The FCC will not process renewal applications received more than 90 days prior to the expiration date of the license.

4) It's a good idea to make a copy of your application as proof of filing before your expiration date. If your application is processed before the expiration date, you may continue to operate until your new license arrives. Otherwise, you may not operate till your renewal is processed.

5) If your license has already expired, it may be still possible to renew since there is a two-year grace period.

6) If you have questions about modifying your license, contact the ARRL Regulatory Information Branch.

7) You are required to keep your mailing address up-to date. The FCC may cancel or suspend a license if their mail is returned as undeliverable.

Applications by clubs for club licenses must be made through a Club Station Call Sign Administrator (CSCSA) and ARRL is one. The FCC no longer accepts club applications. The NCVEC Form 605 must be used.

Information on the status of your application can be found at **www.arrl.org/fcc/fcclook.php3**, from the FCC at 1-888 225-5322 or from ARRL.

US AMATEUR CALL SIGNS

The International Telecommunication Union (ITU) Radio Regulations outline the basic principles used in forming amateur call signs. According to these regulations, an amateur call sign must consist of one or two letters (sometimes the first or second may be a number) as a prefix, followed by a number and then a suffix of not more than three letters. Refer to the References chapter for a complete listing of international call-sign prefixes.

Every US Amateur Radio station call sign is a combination of a 1 or 2-letter prefix, a number and a 1, 2 or 3-letter suffix. The first letter of every US Amateur Radio call sign is always an A, K, N or W. For example, in the call sign W1AW, the W is the prefix, 1 is the number, and AW is the suffix.

For many, but not all, Amateur Radio stations located within the continental United States, the number in the call sign designates the geographic area in which the station was originally licensed. The call-sign districts for the continental US are illustrated in **Fig 2-1**. The prefix/numeral designators for US Amateur Radio districts outside the continental US are shown in **Table 2-2**.

The FCC used to require the number (or in the case of US stations located outside the continental US, the prefix/number) to correspond to the call-sign district in which the station was located. Hams are now allowed to retain their present call sign even if the station location is moved permanently to another call-sign district.

The Vanity Call Sign Program, begun in 1995, allows individual amateurs and club stations to pick a particular call sign for a fee. (The exact amount is subject to change, and is $12 for a 10-year license at this writing, in December 2001. Contact the ARRL Regulatory Information Branch or check *ARRLWeb* for the latest details.) It's routine to hear station call signs on the air that do not correspond to the call-sign district in which the station is located. It is not necessary to identify a station as "portable," "mobile," "fixed" or "temporary."

ON-THE-AIR IDENTIFICATION REQUIREMENTS

The amateur rules specifically prohibit unidentified transmissions, but what is required for proper station identification? An amateur station must be identified at the end of a transmission or series of transmissions. You must also identify at intervals not to exceed 10 minutes during a single transmission or a series of transmissions of more than 10 minutes' duration.

The identification may always be given using the

Table 2-2
FCC-Allocated Prefixes for Areas Outside the Continental US

Prefix	*Location*
AH1, KH1, NH1, WH1	Baker, Howland Is.
AH2, KH2, NH2, WH2	Guam
AH3, KH3, NH3, WH3	Johnston Is.
AH4, KH4, NH4, WH4	Midway Is.
AH5K, KH5K, NH5K, WH5K	Kingman Reef
AH5, KH5, NH5, WH5 (except K suffix)	Palmyra, Jarvis Is.
AH6,7 KH6,7 NH6,7 WH6,7	Hawaii
AH7, KH7, NH7, WH7	Kure Is.
AH8, KH8, NH8, WH8	American Samoa
AH9, KH9, NH9, WH9	Wake Is.
AHØ, KHØ, NHØ, WHØ	Northern Mariana Is.
ALØ-9, KLØ-9, NLØ-9, WLØ-9	Alaska
KP1, NP1, WP1	Navassa Is.
KP2, NP2, WP2	Virgin Is.
KP3,4 NP3,4 WP3,4	Puerto Rico
KP5, NP5, WP5	Desecheo

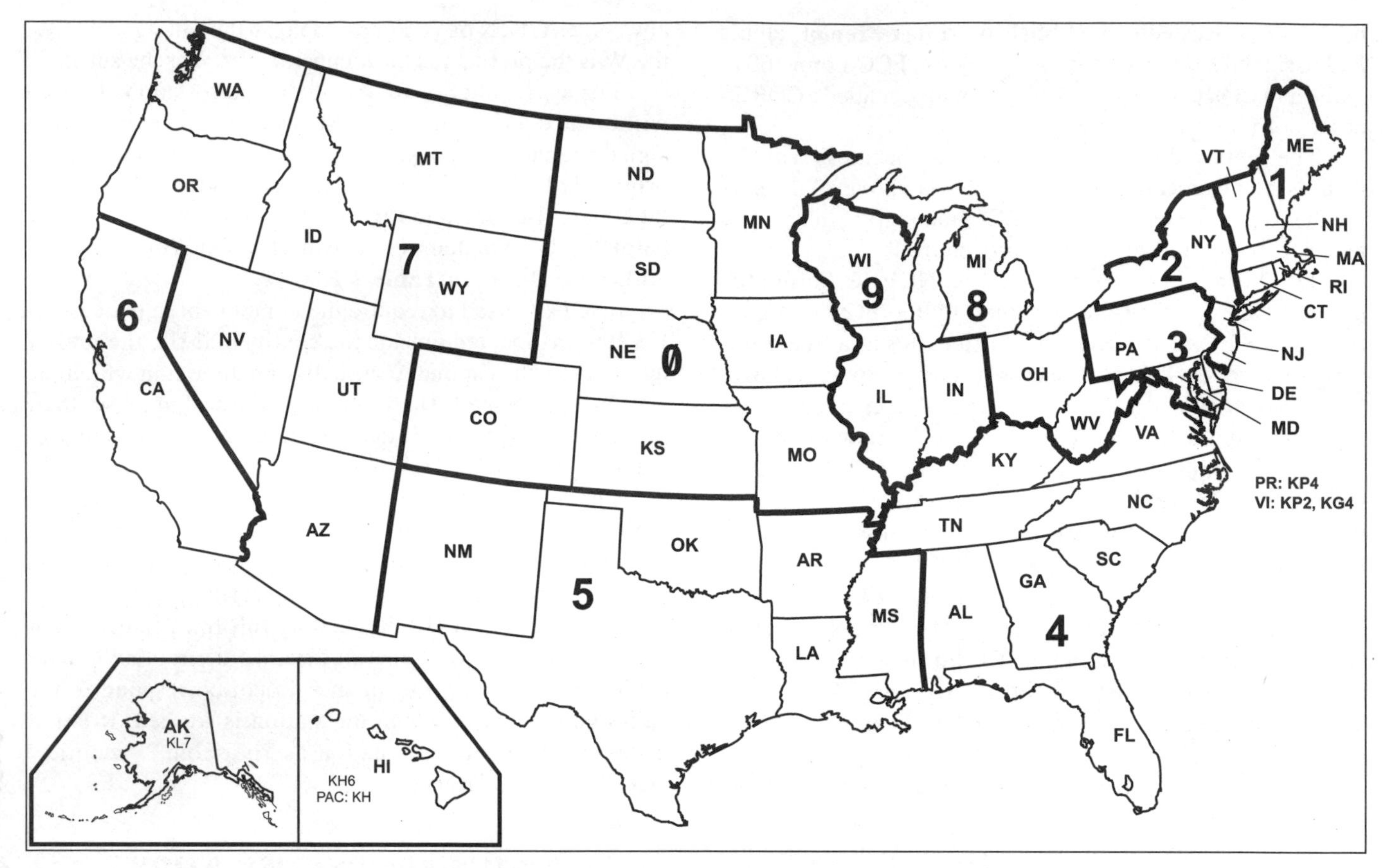

Fig 2-1—The 10 US Call Districts established by the FCC. In most cases, the number in a US amateur call sign indicates where that operator lived when the FCC first issued that call sign. Alaska is part of the seventh call district, but has its own set of prefixes: AL7, KL7, NL7 and WL7. Hawaii is part of the sixth call district but also has its own set of prefixes: AH6, KH6, NH6 and WH6.

International Morse code. If an automatic keying device is used for the CW identification, the speed must not exceed 20 wpm. Generally, the station identification can be given using the same mode as the communications. (The exception is Spread Spectrum communications, in which case the ID must be given using CW or phone emission.) When identifying on phone, you must use the English language.

At the end of an exchange of third-party communications (that is, on behalf of anyone other than the control operator) with a station located in a foreign country, you must also give the call sign of the station with which third-party traffic was exchanged.

If the station is operating under the authority of a reciprocal operating permit for alien amateur licensee, a special indicator is used. A letter and number indicating the location of the station is given, followed by the call sign issued by the operator's licensing country. (For Canadian amateurs, the letter and number follow the call sign.) At least once during each communications, the identification announcement must include the geographical location as nearly as possible. Give the city and state, commonwealth or possession along with the call sign.

Temporary Designator

Temporary designators must be used after a call sign under some circumstances. These designators are used for "instant upgrading," an FCC rule that permits holders of an amateur license who successfully pass an examination for a higher class of license to use the new privileges immediately after leaving the test session. The accredited Volunteer Examiners issue the amateur a certificate of successful completion of examination (CSCE), which authorizes the new privileges for one year or until the issuance of the new, upgraded, permanent license (whichever comes first). The temporary certificate also sets forth a two-letter indicator for the upgraded class of license. For example, "temporary AG" means the operator has upgraded to General.

When operating under the authority of a temporary amateur CSCE with privileges authorized by the certificate, but which exceed the privileges of the licensee's permanent station license, the station must be identified in the following manner:

1) On phone, by the transmission of the station call sign, followed by a word such as "temporary" followed by the special indicator shown on the certificate, appropriate to the newly earned class of license.

2) On CW and digital modes, by the transmission of the station call sign, followed by the slant mark, followed by the special indicator shown on the temporary amateur permit, for example, KB5NOW/AG.

Additional Identification Requirements

To meet the identification requirements of the amateur rules, the call sign must be transmitted on each frequency being used. If identification is made by an automatic device used only for identification by CW, the code speed must not exceed 20 words per minute. While the FCC does not require use of a specific list of words to aid in sending one's call sign on phone, it does encourage the use of a nationally or internationally recognized standard phonetic alphabet as an aid for correct phone identification.

PERMITTED COMMUNICATIONS

The US amateur rules clearly set forth the kinds of stations with which amateurs may communicate. These stations are:

1) Other amateur stations.

2) Stations in other FCC-licensed services during emergencies, and with RACES stations.

3) Any station that is authorized by the FCC to communicate with amateurs.

Additionally, certain "one-way" transmissions are allowed, including those for tuning up your rig, calling CQ, while in remote control operation, during emergency communications, for code practice and bulletins, and for retransmission of space-shuttle communications.

Every amateur should operate with one eye on the basis and purpose of the Amateur Radio Service, as set forth in Section 97.1 of the amateur rules. Adhere to these fundamental principles and you can't go wrong!

§97.1 Basis and Purpose

The rules and regulations in this part are designed to provide an amateur radio service having a fundamental purpose as expressed in the following principles:

(a) Recognition and enhancement of the value of the amateur service to the public as a voluntary noncommercial communication service, particularly with respect to providing emergency communications.

(b) Continuation and extension of the amateur's proven ability to contribute to the advancement of the radio art.

(c) Encouragement and improvement of the amateur radio service through rules which provide for advancing skills in both the communication and technical phases of the art.

(d) Expansion of the existing reservoir within the amateur radio service of trained operators, technicians, and electronics experts.

(e) Continuation and extension of the amateur's unique ability to enhance international goodwill.

THIRD-PARTY COMMUNICATIONS

Third-party traffic is defined in the US amateur rules as, "a message from the control operator (first party) of an amateur station to another amateur station control operator (second party) on behalf of another person (third party)." In other words, sending or receiving any communication on behalf of anyone other than yourself or the control operator of the other station. If you allow anyone to say "hello" into the microphone at your station, that is third-party traffic. Sending or receiving formal radiogram messages is third-party traffic (see the Traffic Handling chapter). Autopatching and phone patching (interconnecting your radio to the telephone system) are also third-party traffic. You, the control operator of the station, are the first party; the control operator at the other station is the second party; and anyone else participating in the two-way communication from either station is a third party.

Certain kinds of third-party traffic are strictly prohibited. All amateurs should know that they must not handle the following kinds of third-party traffic:

1) International third-party traffic, except with countries that allow it or except where the third party is a licensed amateur eligible to be a control operator of the station;

2) Third-party traffic involving material compensation, direct or indirect, paid or promised to any party.

Business Rules

Amateurs have quite a bit of flexibility concerning what is allowed in their public service and personal communications. Look at Section 97.113 when trying to decide if a particular use of the ham bands is allowed:

1) Is it expressly prohibited in the rules?

2) Is there compensation—is anyone being paid?

3) Does the control operator or his employer have a financial interest in the activity using the ham bands?

4) Are communications transmitted on a regular basis that could be reasonably furnished through other radio services?

Under the current business rules, an amateur may conduct his or her personal business over Amateur Radio, including such activities as ordering a pizza or making a dentist appointment. Although this activity is legal, some repeater owners/trustees may ask you not to use their repeater for some of these transactions.

Third-Party Traffic Outside the US

Third-party traffic is permitted only with countries that have entered into third-party traffic agreements with the US (see **Table 2-3**). This is not through reluctance on the part of the US Government, but because of prohibitions on the part of the other governments. The amateur at the other end may be forbidden to handle any type of third-party traffic. US amateurs must abide by this restriction and must have no participation in the handling of third-party traffic in such cases.

Participation by a Third Party

One of the best ways to get a nonham interested in Amateur Radio is to allow him or her to speak into a microphone and experience first-hand the excitement of amateur communication. While the US amateur rules allow active participation by third parties, the control operator must make certain that the communications abide by all the restrictions pertaining to third-party traffic. For example, the third party may not talk about business and may not communicate with any country that has not signed a third-party agreement with the US. Also, the control operator may not leave the controls of the station while the third party is participating

Table 2-3

International Third-Party Traffic (As of December 2001)

Occasionally, DX stations may ask you to pass a third-party message to a friend or relative in the States. This is all right as long as the US has signed an official third-party traffic agreement with that particular country, or the third party is a licensed amateur. The traffic must be noncommercial and of a personal, unimportant nature. During an emergency, the US State Department will often work out a special temporary agreement with the country involved. But in normal times, never handle traffic without first making sure it is legally permitted. Allowing an unlicensed person to talk over the radio is also third-party traffic. The same rules apply.

Prefix	*Country*
V2	Antigua/Barbuda
LU	Argentina
VK	Australia
V3	Belize
CP	Bolivia
T9	Bosnia-Herzegovina
PY	Brazil
VE	Canada
CE	Chile
HK	Colombia
D6	Comoros (Federal Islamic Republic of)
TI	Costa Rica
CO	Cuba
HI	Dominican Republic
J7	Dominica
HC	Ecuador
YS	El Salvador
C5	Gambia
9G	Ghana
J3	Grenada
TG	Guatemala
8R	Guyana
HH	Haiti
HR	Honduras
4X	Israel
6Y	Jamaica
JY	Jordan
EL	Liberia
V7	Marshall Islands
V6	Federated States of Micronesia, The
XE	Mexico
YN	Nicaragua
HP	Panama
ZP	Paraguay
OA	Peru
DU	Philippines
VR6	Pitcairn Island*
V4	St. Christopher/Nevis
J6	St. Lucia
J8	St. Vincent and the Grenadines
9L	Sierra Leone
ZS	South Africa
3DA	Swaziland
9Y	Trinidad/Tobago
TA	Turkey
GB	United Kingdom **
CX	Uruguay
YV	Venezuela
4U1ITU	ITU Geneva
4U1VIC	VIC Vienna

Notes:

* Since 1970, there has been an informal agreement between the United Kingdom and the US, permitting Pitcairn and US amateurs to exchange messages concerning medical emergencies, urgent need for equipment or supplies, and private or personal matters of island residents.

** Limited to special-event stations with callsign prefix GB (GB3 excluded).

Please note that the Region 2 Division of the International Amateur Radio Union (IARU) has recommended that international traffic on the 20 and 15-meter bands be conducted on the following frequencies:

14.100-14.150 MHz
14.250-14.350 MHz
21.150-21.200 MHz
21.300-21.450 MHz

The IARU is the alliance of Amateur Radio societies from around the world; Region 2 comprises member-societies in North, South and Central America, and the Caribbean.

Note: At the end of an exchange of third-party traffic with a station located in a foreign country, an FCC-licensed amateur must also transmit the call sign of the foreign station as well as his or her own call sign.

in amateur communications. The amateur rules allow third-party participation only if a control operator is present at the control point—and if he or she is continuously monitoring and supervising the third party's participation.

Third-party participation is not the same as designating another licensed amateur to be control operator of a station. Control operators designated by a station licensee may operate the station to the extent authorized by the control operator's class of license.

No person who has had their Amateur Radio license suspended or revoked by the FCC may participate in Amateur Radio communications as a third party.

LOGGING

Although there is no FCC requirement that amateurs keep logs of routine operating activities, amateurs may keep voluntary logs. An FCC representative may still mandate log-keeping in specific cases. An accurate station log can be useful in many situations. It can help prove when your station was on the air, and which bands and modes you may have been operating. In addition, it can be a lot of fun to look back through your log and remember the thrill of those special contacts.

AUTHORIZED POWER

Generally, an Amateur Radio transmitter may be operated with a PEP (peak envelope power) output of up to 1500 watts. There are circumstances, however, under which less power must be used. These are:

1) Under all circumstances, an amateur is required to use the minimum amount of transmitter power necessary to carry out the desired communications.

2) Novices are limited to a maximum of 5 watts PEP output on the 1270-MHz band, 25 watts PEP on the 222-MHz band, and 200 watts PEP on the 80, 40, 15 and 10-meter bands. The 200-watt limitation also applies to any licensed radio amateur who operates in the 80, 40 and 15-meter Novice subbands and to any amateur operating on the 30-meter band. Technician operators who have passed the 5 wpm Morse code exam are also limited to 200 watts PEP on the 10-meter band. The rules limit the maximum transmitter output power in the Amateur Radio Service to

1500-watts PEP. Higher-class licensees may use 1500-watts PEP in the Novice sections of 28 MHz, 222 MHz and 1270 MHz.

3) Certain power limitations apply to portions of some bands at VHF and UHF. See §97.313 of the FCC rules.

AMATEUR OPERATION IN A FOREIGN COUNTRY

Operation may not be conducted from within the jurisdiction of a foreign government unless that foreign government has granted permission for such operation. Some countries have allowed US amateurs to operate while visiting; however, not all countries permit foreigners this privilege. Every country has the right to determine who may and who may not operate an Amateur Radio station within its jurisdiction. There are severe penalties in some countries for operating transmitting equipment without the proper authority.

Canada-US Reciprocal Operating

No formal application or permit is required for US amateurs to operate amateur equipment in Canada, nor is any formal paperwork required for Canadian amateurs to operate in the United States. It's automatic. In all cases, visitors must stay within the band and mode restrictions of the host country. US radio amateurs may operate in Canada with the mode and frequency privileges authorized them in the United States. Canadian radio amateurs may operate in the US with the mode and frequency privileges authorized to them in Canada, but not exceeding the privileges of a US Extra Class operator.

US amateurs operating in Canada are required to give their station identification by using their US call sign, followed by the slant bar or equivalent words and then the Canadian prefix for their location.

Multilateral and Bilateral Operating Agreements

The CEPT Agreement

The European Conference of Postal and Telecommunications Administrations (CEPT) Recommendation T/R 61-01 benefits Amateur Radio operators from countries that are participants in the agreement. The US has been accepted as a "non-CEPT participant" in the agreement, that is, a non-European participant. A CEPT Amateur Radio license is the license issued to an amateur by the country of which the person is a citizen and licensee. A US amateur may operate in most European CEPT countries with the following items: proof of a US license (your original license — FCC Form 660), proof of US citizenship (usually a Passport or similar photo ID and birth certificate) and a copy of the FCC CEPT Public Notice.

The FCC Public Notice (DA 99-1098, released June 7, 1999) entitled "Amateur Service Operation in CEPT Countries" is written in three languages: English, French, and German. This document, written in all three languages, must be with you when you operate. The FCC Public Notice can be found on *ARRLWeb* at: **www.arrl.org/FandES/field/regulations/io/#cept**.

US amateurs may not operate in non-CEPT countries outside of Europe even though the country has also been accepted into the CEPT agreement. For example, a US amateur may not operate in Israel under this agreement even though both the US and Israel have been accepted into CEPT as "non-CEPT countries." In this case, the US amateur must obtain an Israeli reciprocal license unless their government says otherwise.

There are two classes of CEPT Amateur Radio license:

Class 1—This class permits use of all frequency bands allocated to the Amateur Service and Amateur-Satellite Service and authorized in the country where the amateur station is to be operated. It will be open only to those amateurs who have proven their competence in Morse code to their own administration. The European Radiocommunications Office (ERO) has determined that for the purposes of Recommendation T/R 61-01, the FCC Amateur Extra, Advanced, General and Technician with 5 wpm Morse code exam credit licenses are equivalent to CEPT Class 1.

Class 2—This class permits use of all frequency bands allocated to the Amateur Service and Amateur-Satellite Service above 30 MHz and authorized in the country where the amateur station is to be operated. The ERO has determined that for the purposes of Recommendation T/R 61-01, the FCC Technician license is equivalent to CEPT Class 2.

Notice that there is no CEPT equivalent to the FCC Novice license. Operation by Novices is not authorized under a CEPT Amateur Radio License.

Station Identification: When transmitting in the foreign CEPT country the license holder must use his or her national call sign preceded by the appropriate ITU designator, such as G/KB1EEE. The CEPT call sign prefix and the national call sign must be separated by the character "/" (telegraphy) or the word "stroke" (telephony). **Table 2-4** lists the participating CEPT countries (as of December 2001), along with the call sign identifier to be used with each CEPT license class.

The International Amateur Radio Permit (IARP)

The Inter-American Convention on an International Amateur Radio Permit (CITEL/Amateur Convention) allows visitors to operate stations temporarily in other countries of the Americas. Participation in the CITEL/Amateur Convention allows US citizens to operate amateur stations in seven countries within CITEL, a component of the Organization of American States. Under the CITEL/Amateur Convention, US amateur operators with an International Amateur Radio Permit (IARP) have reciprocal operating privileges for one year, or until their FCC license expires, whichever occurs first. A new permit must be obtained each year for IARP operations in certain countries in the Americas. **Table 2-5** lists the participating CITEL countries (in December 2001).

For a US citizen to operate an amateur station in a CITEL country, an International Amateur Radio Permit (IARP) is necessary. According to the CITEL Agreement, the IARP may be issued by a member-society of the Inter-

Table 2-4
CEPT Countries That Recognize the US Participation in the CEPT Radio Amateur License (As of December 2001)

Country	*Call Sign Prefix(es)*	
	CEPT Class 1	*CEPT Class 2*
Austria	OE	OE
Belgium	ON	ON
Bosnia and Herzegovina	T9	T9
Bulgaria	LZ	LZ
Croatia	9A	9A
Cyprus	5B	5B
Czech Republic	OK	OK
Denmark	OZ	OZ
Faroe Islands	OY	OY
Greenland	OX	OX
Estonia	ES*	ES*
Finland	OH	OH
France	F	F
Corsica	TK	TK
Guadeloupe	FG	FG
Guiana	FY	FY
Martinique	FM	FM
St-Bartholomew	FJ	FJ
St-Pierre/Miquelon	FP	FP
St-Martin	FS	FS
Reunion	FR	FR
Mayotte	FH	FH
French Antarctica	FT	FT
Local permission also required:		
Glorieuse	FR	FR
Jean de Nova	FR	FR
Tromelin	FR	FR
Crozet	FT	FT
Kerguelen	FT	FT
St. Paul & Amsterdam	FT	FT
Terre Adelie	FT	FT
French Polynesia	FO	FO
Clipperton	FO	FO
New Caledonia	FK	FK
Wallis & Futuna	FW	FW
Germany	DL	DC
Hungary	HA	HG
Iceland	TF	TF
Ireland	EI	EI
Italy	I	I
Latvia	YL	YL
Liechtenstein	HBØ	HBØ
Lithuania	LY	LY
Luxembourg	LX	LX
Monaco	3A	3A
Netherlands	PA	PA
Netherland Antilles	PJ	PJ
Norway	LA	LC
Portugal	CT	CT
Azores	CU	CU
Madeira	CT	CT
Romania	YO	YO
Slovak Republic	OM	OM
Slovenia	S5	S5
Spain	EA	EB
Sweden	SM	SM
Switzerland	HB9	HB9
Turkey	TA	TA
Ukraine	UT	UT
United Kingdom	G	G
Isle of Man	GD	GD
N. Ireland	GI	GI
Jersey	GJ	GJ
Scotland	GM	GM
Guernsey	GU	GU
Wales	GW	GW

* Must include numeral corresponding to Administrative District

For updates, check the Web site of the European Radiocommunications Office: **www.ero.dk**. Look for "**Implementation**." At "**Decision/Recommendation**" select "**T/R 61-01**."

national Amateur Radio Union (IARU)—for the US, the IARU member society is the American Radio Relay League (ARRL). The permit describes its authority in four different languages. The FCC presently recognizes the ARRL as the issuing body for such permits. The ARRL offers this service to US citizens for their use when they travel to CITEL countries. The ARRL provides this service on a non-discriminatory basis, at no expense to the United States Government.

There are two classes of International Amateur Radio Permits:

Class 1 requires knowledge of the international Morse code and proven proficiency in CW. It carries all operating privileges in the host country (Technician with Morse code, General, Advanced or Extra class US licensees qualify for Class 1).

Class 2 does not require knowledge of telegraphy and carries all operating privileges above 30 MHz. It is, therefore, equivalent to the US Technician Class operator license. There is no equivalent Class description for the US Novice license; therefore, the US Novice license is not eligible to operate under IARP.

Station Identification: When the station is transmitting under the authority of an IARP, an indicator consisting of the appropriate letter-numeral designating the station location must be included before, after, or both before and after the call sign issued to the station by the licensing country. At least once during each intercommunication, the identification announcement must include the geographical location as nearly as possible to the city and state, commonwealth or possession of the station operation. An example of station identification under IARP by a US amateur operator while traveling in or near Lima, Peru might be, "this is OA1/W1XYZ, near Lima, Peru." If on CW, a sample ID procedure would be "DE OA1/W1XYZ NR LIMA PERU."

Table 2-5
CITEL Countries That Allow Operation with the International Amateur Radio Permit (IARP) (As of December 2001)

Country	*Prefix*
Argentina	LU
Brazil	PY
Canada	VE
Peru	OA
Uruguay	CX
US	K, N, W
Venezuela	YV

Application: Credentials will be issued to US amateurs and citizens upon receipt of a completed and signed application, along with a photocopy of the applicant's US FCC amateur license, a copy of the applicant's legal photo-ID and a 1 inch by 1 inch (up to 1.5 by 1.5 inch) color or black and white photo of the applicant (to be affixed to credentials), and the application fee (payable by check or money order to ARRL, or by credit card). US amateurs who are US citizens can obtain an IARP application from ARRL HQ by sending an SASE. There is an IARP application on *ARRLWeb* at: **www.arrl.org/FandES/field/regulations/io/index.html#iarp**.

A processing fee ($10 US Dollars in December 2001) is charged by ARRL to cover IARP credentials/authority creation and delivery to applicants. For domestic rush/courier delivery, add $10 (street addresses only).

Other Reciprocal Operating Agreements

Except for the Canadian, CEPT and CITEL agreements, US amateurs must obtain a written permit to be allowed to operate in any foreign country. The administrations of some countries have signed reciprocal operating agreements with the US. Such agreements facilitate the application procedures for getting permission to operate. These agreements, because they are reciprocal, also allow foreign amateurs the opportunity to receive permission to operate while in the US. If you want to operate a station in a foreign country, ARRL HQ can help you obtain forms and information. Send an SASE to ARRL HQ's Regulatory Information Branch for details. Include one unit of postage for every country requested. A few countries require a US licensee to hold a General or higher-class license to qualify for a permit to operate. You may be required to pay a licensing fee in order to obtain a permit, which must be obtained before equipment is taken into a country. Plan as far ahead as possible. **Table 2-6** is a list of countries that have signed reciprocal operating agreements with the US. The latest information and updates to this list can be found on *ARRLWeb* at: **www.arrl.org/FandES/field/regulations/io/recip.html**.

RADIO FREQUENCY INTERFERENCE

The FCC authorizes radio amateurs to use up to 1500 watts PEP output and many different bands of the radio spectrum. These privileges do not come without some effort on the part of the license applicant, however. A prerequisite for getting on the air is passing an examination that tests the applicant's technical knowledge. One area of special importance is knowing about radio frequency interference (RFI), its causes and its cures. The FCC expects radio amateurs to identify and solve most RFI problems, but RFI is not always a simple matter.

Much has been written about RFI. For example, the material in *The ARRL RFI Book* examines this subject from both a legal and technical standpoint. Also, *The ARRL Handbook for Radio Amateurs* devotes an entire chapter to solving radio frequency interference problems. If you are the recipient of interference complaints, these two books will be of interest to you.

A growing number of amateurs are unjustly accused of causing RFI. This occurs when an amateur transmitter is operating properly, and the responsibility for the interference lies with the electronic device experiencing the interference. For example, some television sets receive signals they are not supposed to because the manufacturer decided that reducing the per-unit cost was more important than incorporating adequate shielding in the circuit design. In these cases, the transmitter is operating with a "clean" signal; it is the television set that must be brought up to today's engineering standards. This usually means that the only effective cure must be performed by adding a filter at the television set. Trying to make your neighbor understand that it is his equipment that is at fault is usually no easy task, however.

Public Law 97-259 allows the FCC to require standards for unwanted-signal-rejection filtering in electronic home-entertainment devices. Also, some manufacturers have made some progress in dealing with this problem by providing free assistance to owners of their products who experience RFI. The onus for preventing the interference, however, continues to fall on amateurs.

Today, the interference problem is of huge proportions. This can be attributed to the rapid growth of the consumer electronics industry and the failure of manufacturers to

Table 2-6
Countries That Share Reciprocal/Licensing Agreements (As of December 2001)

V2	Antigua/Barbuda	OH	Finland	ZL	New Zealand
LU	Argentina	F	France *	YN	Nicaragua
VK	Australia	DL	Germany	LA	Norway
OE	Austria	SV	Greece	HP	Panama
C6	Bahamas	J3	Grenada	P2	Papua New Guinea
8P	Barbados	TG	Guatemala	ZP	Paraguay
ON	Belgium	8R	Guyana	OA	Peru
V3	Belize	HH	Haiti	DU	Philippines
CP	Bolivia	HR	Honduras	CT	Portugal
T9	Bosnia-Herzegovina	TF	Iceland	J6	St. Lucia
A2	Botswana	VU	India	J8	St. Vincent and the Grenadines
PY	Brazil	YB	Indonesia	S7	Seychelles
VE	Canada	EI	Ireland	9L	Sierra Leone
CE	Chile	4X	Israel	H4	Solomon Is.
HK	Colombia	I	Italy	ZS	South Africa
TI	Costa Rica	6Y	Jamaica	EA	Spain
9A	Croatia	JA	Japan	PZ	Suriname
5B	Cyprus	JY	Jordan	SM	Sweden
OZ	Denmark (incl. Greenland)	T3	Kiribati	HB	Switzerland
HI	Dominican Rep.	9K	Kuwait	HS	Thailand
J7	Dominica	EL	Liberia	9Y	Trinidad/Tobago
HC	Ecuador	LX	Luxembourg	TA	Turkey
YS	El Salvador	Z3	Macedonia	T2	Tuvalu
V6	Federated States of Micronesia	V7	Marshall Is.	G	United Kingdom**
3D	Fiji	XE	Mexico	CX	Uruguay
		3A	Monaco	YV	Venezuela
		PA	Netherlands		
		PJ	Neth. Antilles		

Notes: An automatic reciprocal agreement exists between the US and Canada, so there is no need to apply for a permit. Simply sign your US call followed by a slantbar and the Canadian letter/number identifier.

* Including French Guiana, French Polynesia, Guadeloupe, Amsterdam Island, Saint-Paul Island, Crozet Island, Kerguelen Island, Martinique, New Caledonia, Reunion, St. Pierre and Miquelon, and Wallis and Futuna Islands

** Including Bermuda, British Virgin Islands, Cayman Islands, Channel Islands (including Guernsey and Jersey) Falkland Islands (including South Georgia Islands and South Sandwich Islands), Gibraltar, Isle of Man, Montserrat, St. Helena,(including Ascension Island, Gough Island, Tristan Da Cuhna Island), Northern Ireland, and the Turks and Caicos Islands include proper shielding or filtering in the home-entertainment equipment. If you are accused of causing RFI:

1) Check your log. Were you operating at that time? (A complete log, although no longer required by the FCC, is very useful in interference situations.)

2) Check with your own nonamateur equipment. If you are not interfering with your own TV set, chances are the problem lies with your neighbor's receiver and not with your transmitter.

3) Solicit the cooperation of your neighbor in testing to determine the exact cause of the interference.

4) Check with your local radio club for a TVI committee or other assistance or contact your ARRL Section Manager for a referral to an ARRL Technical Coordinator or Technical Specialist in your area (Section Managers are listed on page 12 in *QST*).

5) Request RFI assistance from the manufacturer of the home-entertainment device.

6) Read *The ARRL RFI Book* and the Interference chapter in *The ARRL Handbook for Radio Amateurs*.

7) Be prompt, courteous and helpful. Amateur Radio's reputation is at stake—as well as your own.

8) For further assistance, write to the Technical Information Service at ARRL HQ and on the *ARRLWeb* at **www.arrl.org/tis**. You can find regulatory assistance for RFI at **www.arrl.org/FandES/field/regulations/rfi-legal**. Include as much detail as you can about the symptoms and circumstances, but avoid emotional commentary.

ON THE HORIZON

The FCC proposes additional rules changes periodically; check *QST* and *ARRLWeb* for news of late-breaking events.

The rules are dynamic because Amateur Radio is dynamic—constantly changing to meet and create new communications technologies to better serve the public and society.

AMATEUR FREQUENCY ALLOCATIONS AND BAND PLANS

There are bands of frequencies allocated to the Amateur Service extending from 1800 kHz (73 kHz in the UK) to over 300 GHz. **Table 2-7** will give you an overview of the radio spectrum, some nomenclature and the Amateur Service bands allocated in the International Telecommunication Union (ITU) Radio Regulations.

There is a tendency for band plans to lag reality. This is due in part to the time needed to research, invite and digest comment from amateurs, arrive at a mix that will serve the diverse needs of the amateur community, and adopt a formal band plan. This is a process that can take a year or more on the national level and a similar period in the International Amateur Radio Union (IARU). Nevertheless, new communications modes or the popularity of existing ones can make a year-or-two-old band plan look obsolete. Such revolutionary change has taken place recently with the popularity of data communications, particularly in the 20 and 2-meter bands. Changes of this magnitude cause the new users to scramble for frequencies and some of the existing mode users to draw their wagons in a circle. The national societies, their staffs and committees, and the IARU have the job of sorting out the contention for various frequencies and preparing new band plans. Fortunately, we have not exhausted all possible ways of improving our management of the spectrum so that all Amateur Radio interests can be accommodated.

The 160-Meter Band

The 160-meter band provides some excellent DX opportunities. The basic problem with allocations in this band has been competition with the Radiolocation Service. New pressures are possible as a result of planned expansion of the medium-frequency broadcast band in the 1605-1705 kHz range.

The 80-Meter Band

While US amateurs enjoy the use of 3500-4000 kHz, not all countries allocate such a wide range of frequencies to the 80-meter band. There are fixed, mobile and broadcast operations, particularly in the upper part of the band.

The 40-Meter Band

The 40-meter band has a big problem: International broadcasting occupies the 7100-7300 kHz band in many parts of the world. During the daytime, particularly when sunspots are high, broadcasting does not cause much interference to US amateurs. At night, however, especially when sunspot activity is low, the broadcast interference is heavy. Some countries allocate only the 7000-7100 kHz band to amateurs. Others, particularly in Region 2, allocate 7100-7300 kHz as well, which at times is subject to interference. The result is that there is a great demand for frequencies in the 7000-7100 kHz segment. The effect is that there are two band plans overlaid on each other: ours, spread out over 7000-7300 kHz and another one that compresses everything into 7000-7100 kHz. The IARU is trying to solve this problem internationally.

The 30-Meter Band

The 30-meter band is excellent for CW and digital modes. The only problem is that US amateurs must not cause harmful interference to the fixed operations outside the US. This restricts transmitter power output to 200 watts and is one reason for not having contests on this band.

The 20-Meter Band

The workhorse of DX is undoubtedly the 20-meter band. It offers excellent propagation to all parts of the world throughout the sunspot cycle and is virtually clean of interference from other services.

The 17-Meter Band

The 18068-18168 kHz band was awarded to amateurs on an exclusive basis, worldwide, at WARC-79. It was made available for amateur use in the US in January 1989.

The 15 and 12-Meter Bands

The 21000-21450 and 24890-24990 kHz bands are ex-

Table 2-7
The Electromagnetic Spectrum with Amateur Service Frequency Bands by ITU Region

Wave-length	*Frequency*		*Nomen-clature*	*Metric Band*	*Amateur Radio Bands by ITU Region* *Region 1*	*Region 2*	*Region 3*
1 mm	300 GHz						
		Microwaves	EHF Milli-metric	1 mm	241-250	241-250	241-250
				2 mm	142-149	142-149	142-149
				2.5 mm	119.98-120.02	119.98-120.02	119.98-120.02
				4 mm	75.5-81	75.5-81	75.5-81
				6 mm	[illegible]	[illegible]	[illegible]
1 cm	30 GHz						
			SHF Centi-metric	1.2 cm	24-24.25	24-24.25	24-24.25
				3 cm	10-10.5	10-10.5	10-10.5
				5 cm	5.65-5.85	5.65-5.925	5.65-5.85
				9 cm		3.3-3.5	3.3-3.5
10 cm	3 GHz						
			UHF Deci-metric	13 cm	2.3-2.45	2.3-2.45	2.3-2.45
				23 cm	1240-1300	1240-1300	1240-1300
				33 cm		902-928	
				70 cm	430-440	430-440	430-440
1	300 MHz						
			VHF Metric	1.25 m		222-225	
				2 m	144-146	144-148	144-148
				6 m		50-54	50-54
10	30 MHz						
			HF Deca-metric	10 m	28-29.7	28-29.7	28-29.7
				12 m	24.89-24.99	24.89-24.99	24.89-24.99
				15 m	21-21.45	21-21.45	21-21.45
				17 m	18.068-18.168	18.068-18.168	18.068-18.168
				20 m	14-14.350	14-14.350	14-14.350
				30 m	10.1-10.150	10.1-10.150	10.1-10.150
				40 m	7-7.1	7-7.3	7-7.1
				80 m	3.5-3.8	3.5-4	3.5-3.9
100	3 MHz						
			MF Hecto-metric	160 m	1.81-1.85	1.8-2	1.8-2
1000	300 kHz						
			LF Kilo-metric				
10,000	[illegible]						
			VLF Myria-metric				
100,000	3 kHz						

Note: This table should be used only for a general overview of where ITU Amateur Service and Amateur-Satellite Service frequencies fall within the radio spectrum. They do not necessarily agree with FCC allocations; for example, the 70-cm band is 420-450 MHz in the United States.

cellent for DX during the high part of the sunspot cycle. They also offer some openings throughout the rest of the sunspot cycle.

The 10-Meter Band

This is an exclusive amateur band worldwide. Its popularity rises and falls with sunspot numbers and propagation.

The VHF and Higher Bands

The 6-meter band is not universal, but the trend seems to be toward allocating it to amateurs as TV broadcasting vacates the 50-54 MHz band. Recent new 6-meter privileges for certain European amateurs adds to the worldwide interest in this band. It is also excellent for amateur exploitation of meteor-scatter communications using various modes including packet radio.

Two meters is heavily used throughout the world for CW, EME, SSB, FM and packet radio. Satellites occupy the 145.8-146 MHz segment.

In 1991, the FCC reallocated the 220-222 MHz band to the land mobile service. In partial compensation for that loss of amateur spectrum, acting on petition of the ARRL, the Commission allocated 219-220 MHz to the Amateur Radio Service on a secondary basis. This band segment is only for stations participating in fixed, point-to-point digital messaging systems. There are special provisions to protect domestic waterways telephone systems using that band. US amateurs have a primary allocation at 222-225 MHz, which is largely used for repeaters.

The 70-cm band is prime UHF spectrum. The 430-440 MHz band is virtually worldwide, whereas the 420-430 MHz and 440-450 MHz bands are not. Frequencies around 432 are used for weak-signal work, including EME, and the 435-438 MHz band is for amateur satellites. The 70-cm band is the lowest frequency band that can be used for fast-scan television and spread spectrum emissions.

As a result of Congressional action, the 2305-2310 MHz band is subject to auction for other radio services, thus giving rise to concern about Amateur Radio retention and useful access to this band. The ARRL is pursuing primary status for the Amateur Service in the 2300-2305 MHz band and continued use of the 2305-2310 MHz band.

The 33-cm band (902-928 MHz) is widely shared with other services, including Location and Monitoring Service (LMS), which is primary, and ISM (industrial, scientific and medical) equipment applications. A number of low-power devices including spread spectrum local area networks operate in this band under Part 15 of the FCC's Rules.

The 1240-1300 MHz band is used by amateurs for essentially all modes, including FM and packet. By regulation, the 1260-1270 MHz segment may be used only in the Earth-to-space direction when communicating with amateur satellites.

While the Amateur Service has a secondary allocation in the 2300-2450 MHz band in the international tables, in the United States the allocation is 2390-2400 MHz and 2402-2417 MHz primary, and 2300-2310, 2400-2402 and 2417-2450 secondary. Most of the weak-signal work in the US takes place around 2304 MHz, while much of the satellite activity is in the 2400-2402 MHz segment.

The remaining microwave and millimeter bands are the territory of amateur experimenters. It is important that the Amateur Service and the Amateur-Satellite Service use these bands, and contribute to the state-of-the-art in order to retain them. There is growing interest on the part of the telecommunications industry and the space science community to fully exploit the 20-95 GHz spectrum.

FREQUENCY COORDINATION

From the earliest days of Amateur Radio, hams simply listened on a frequency to see if it was in use, and if not, went ahead and transmitted on that frequency. Amateurs operating on HF still use this simple listen-before-transmit procedure of randomly assigning themselves a frequency for use over a short period. This simple procedure has served us well over the years, but is no longer the only "rule of the road."

On HF, *The ARRL Net Directory* provides a mild spectrum-management technique that might be called "frequency registration." This lets hams know where and when nets normally operate.

On the VHF and UHF bands, FM repeaters introduce a more difficult problem. Repeater users can't just "slide" up or down in frequency to find a clear frequency. The repeater equipment is set to operate on specific frequencies. If multiple repeaters try to operate on the same frequency pair — or even on nearby frequencies — within the same geographical area, interference is sure to result.

Regional repeater frequency coordinators maintain a database of repeaters in their area. They attempt to find suitable frequencies for new repeaters, so they will not cause interference to existing ones. The National Frequency Coordinators' Council (NFCC) is an association of the regional coordinators.

THE AMATEUR RADIO SPECTRUM

The electromagnetic spectrum is a limited resource. Every kilohertz of the radio spectrum represents precious turf that is blood sport to those who lay claim to it. Fortunately, the spectrum is a resource that cannot be depleted— one that if misused can be restored to normal as soon as the misuse stops. Every minute of every hour of every day, we have a fresh chance to use the spectrum intelligently.

Although the radio spectrum has been used in a certain way, changes are possible. Needs of the various radio services evolve with technological innovation and growth. There can be changes both in the frequency-band allocations made to the Amateur Service and how we use them. Also, to our benefit, signal propagation determines where a signal may be received. Specific frequencies may be reused numerous times around the globe. Amateur Radio is richly endowed with a wide range of bands starting at 1.8 MHz and extending above 300 GHz. Thus, we enjoy a veritable smorgasbord of bands with propagational "delicacies" of every type—direct wave, ground wave, sky wave, tropospheric scatter, meteor scatter and more. In addition, we can supplement these with moon reflections and active artificial satellites.

International Regulation of the Spectrum

Amateur Radio frequency band allocations don't just happen. Band allocation proposals must first crawl through a maze of national agencies and the International Telecommunication Union (ITU) with more adroitness than a computer-controlled mouse. Simultaneously, the proposals have to run the gauntlet of the competing interests of other spectrum users.

Treaties and Agreements

To bring some order to international relationships of all sorts, nations sign treaties and agreements. Otherwise (with respect to international communications) chaos, anarchy and bedlam would vie for supremacy over the radio spectrum. Pessimists think we already have some of that, but they haven't any idea how bad it could be without international treaties and agreements.

The International Telecommunication Union

The origins of the International Telecommunication Union (ITU) trace back to the invention of the telegraph in the 19th century. To establish an international telegraph network, it was necessary to reach agreement on uniform message handling and technical compatibility. Bordering European countries worked out some bilateral agreements. This eventually led to creation of the ITU at Paris in 1865 by the first International Telegraph Convention, which yielded agreement on basic telegraph regulations.

Plenipotentiary Conferences

The ITU has Plenipotentiary Conferences every four years. "Plenipotentiary" is a conference that is fully empowered to do business. The conferences determine general policies, review the work of the Union, revise the Convention if necessary, elect the Members of the Union to serve on the ITU Council, and elect the Secretary-General, Deputy Secretary-General, the Directors of the Bureaus and members of the Radio Regulations Board.

World Radiocommunication Conferences

ITU World Radiocommunication Conferences (WRCs) are held every two years with agendas agreed at the previous WRC and confirmed by the ITU Council. Various issues related to Amateur Radio may come up at these conferences. For example, direct amateur frequency allocations may be made or modified. In addition, changes made to the primary allocations of another service may affect the secondary status of Amateur Radio allocations.

Inter-American Telecommunication Commission (CITEL)

CITEL is the regional telecommunications organization for the Americas. It is a permanent commission under the Organization of American States (OAS), with a secretariat in Washington, DC. The CITEL Assembly meets every four years, while its committees meet one or more times yearly. ARRL and IARU Region 2 are active participants in Permanent Consultative Committee no. 3 (PCC.111 - Radiocommunications).

TELECOMMUNICATIONS REGULATION IN THE UNITED STATES

The Communications Act of 1934, as amended, provides for the regulation of interstate and foreign commerce in communication by wire or radio. This Act is printed in Title 47 of the US Code, beginning with Section 151.

Federal Communications Commission

The FCC is responsible, under the Communications Act, for regulating all telecommunications except that of the Federal government. This includes the Amateur Radio Service and the Amateur-Satellite Service, which are regulated under Part 97 of the Commission's rules. (Part 97 is available in *The ARRL's FCC Rule Book*, published by the ARRL. You can also find the complete text of Part 97 on *ARRLWeb* at: **www.arrl.org/FandES/field/regulations/news/part97/**)

The internal structure of the FCC is shown in **Fig 2-2**. The Amateur Radio Service comes under the Public Safety and Private Wireless Division of the Wireless Telecommunications Bureau.

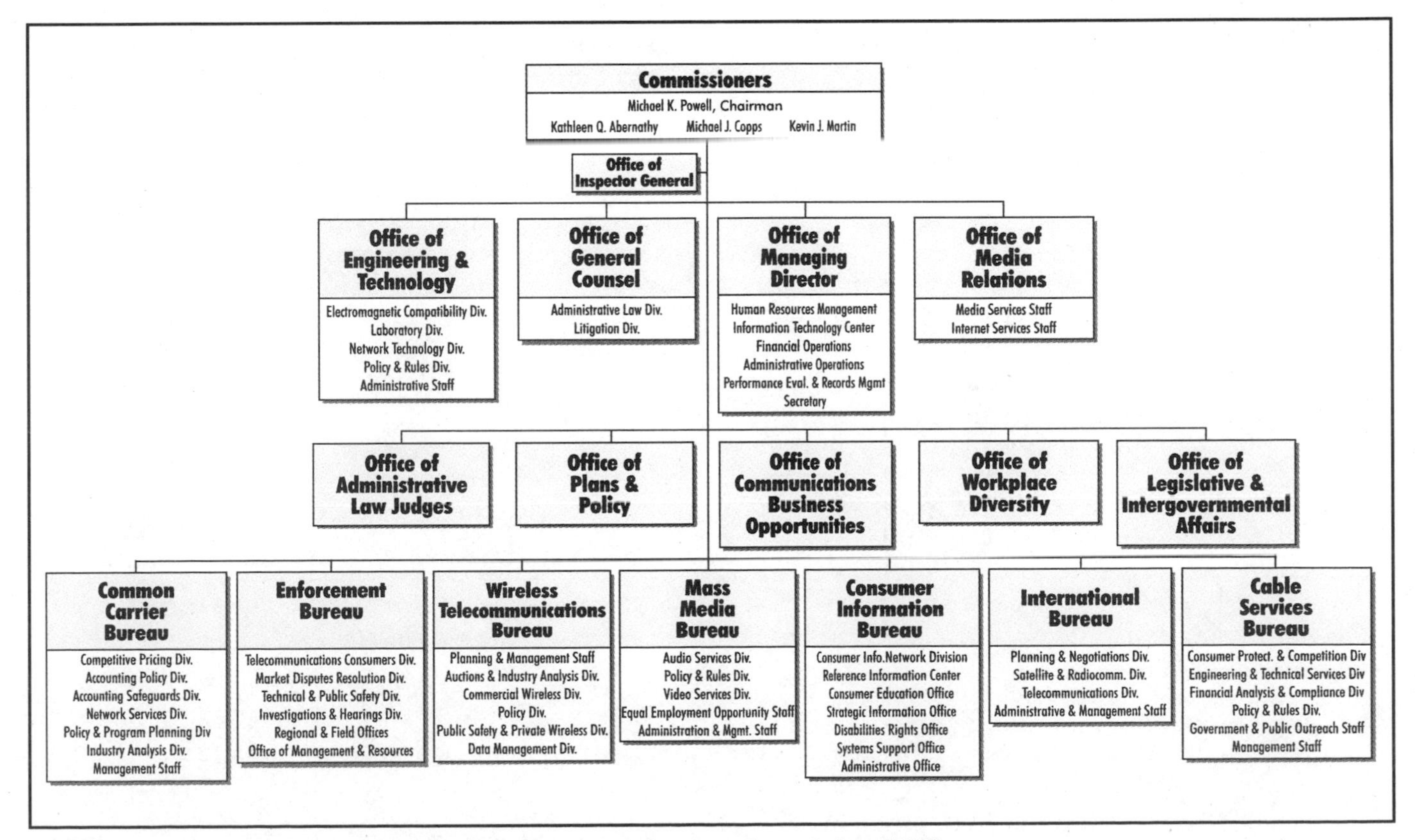

Fig 2-2 — An organization chart of the Federal Communications Commission (FCC).

Federal Government Telecommunications

The functions relating to assignment of frequencies to radio stations belonging to, and operated by, the United States Government are assigned to the Assistant Secretary of Commerce for Communications and Information (Administrator, National Telecommunications and Information Administration-NTIA). Among other things, NTIA:

- coordinates telecommunications activities of the Executive Branch.
- develops plans, policies and programs relating to international telecommunications issues.
- coordinates preparations for US participation in international telecommunications conferences and negotiations.
- develops, in coordination with the FCC, a long-range plan for improved management of all electromagnetic spectrum resources, including jointly determining the National Table of Frequency Allocations.
- conducts telecommunications research and development.

Obviously, the FCC and the NTIA work closely together on many issues relating to radio spectrum. Many Amateur Radio bands, especially the UHF and higher frequency bands are shared with a variety of government agencies. As long as we remain good sharing partners, these agencies can be powerful advocates for protecting Amateur Radio frequencies from other users.

Chapter 3

FM
The Friendly Mode

Jay Mabey, NUØX
PO Box 19022
Cedar Rapids, IA 52409
e-mail nu0x@arrl.net

If you were to ask a roomful of amateur operators if they operate on FM or repeaters, the answer would be a resounding "absolutely." There are solid reasons that more hams use VHF FM than any other mode. FM communications are clear, crisp and available nearly everywhere. In many urban areas there are simply no repeater frequency pairs left to assign.

FM communications are usually local in nature; you'll be talking to other hams who are 20 or 30 miles from your location. It's a more personal kind of communication, in that you can (and often do) meet the other amateur face to face.

Another advantage to repeaters is that they are *channelized*—you don't have to tune up and down the bands as you do with an HF radio. You simply talk and listen! Hearing your local friends on the air leads to a feeling of being a part of a community. It's good to know that like spirits are out there and ready to help, talk or have coffee at the push of a PTT button!

FM and the local repeater equate to the town meetinghouse or general store of Amateur Radio. What a great place to meet friends, old and new! The use of a repeater allows a low-power handheld radio or mobile rig to have great range, and with the signal strength of a repeater there's virtually no noise. The use of squelch circuits in FM transceivers means if there's no signal, there's no noisy static to annoy those around you. Those who have ever spent time on the road trying to tune in an HF signal know that your traveling companions tend to develop animosity toward your hobby, despite your best efforts to extol its virtues!

One of the most significant uses of FM and the local repeater is for serving the public. Many areas have active ARES groups on one or more local repeaters. See the Emergency Communications chapter for more about ARES and other public service groups and agencies. During times of severe weather or natural disasters, ARES groups mobilize and the local repeater becomes disaster central. The system can be utilized for handling emergency traffic, assisting law enforcement and other agencies with communications, situation reporting and the various other functions that Amateur Radio has been famous for during times of trouble.

SKYWARN-trained hams are seen on a daily basis out along the roads in tornado alley. These trained and skilled amateurs watch the sky and are constantly ready to communicate with the National Weather Service office (where there's also a ham in residence) as to the real world situation—adding eyes to the effectiveness of Doppler Radar systems.

In general, repeaters offer something for every interest within the amateur community. HF DXers and contest buffs can swap information. Traffic handlers can effectively pass traffic between local and section levels. Repeaters can send code practice for those studying for their licenses. Digital enthusiasts can communicate using repeaters dedicated to their specific modes. Adding to the versatility of repeater operation is the ability to communicate while walking, driving or simply relaxing in the comfort of your home.

WHAT IS A REPEATER?

A repeater is similar to any other Amateur Radio station, in that it uses a transmitter, a receiver and an antenna. The magic is in the fact that the receiver and transmitter in a repeater are on different frequencies and the output of the receiver is fed to the input of the transmitter. Thus, everything that the receiver hears is retransmitted simultaneously (repeated) by the transmitter. See **Fig 3-1**.

Of course, your radio's receiver and transmitter are also tuned to different frequencies (the opposite of those on the repeater). Your radio transmits on the repeater's *input* frequency and receives on the repeater's *output* frequency.

Sometimes, when it's not possible because of interference problems or other issues to place the repeater transmitter and receiver in the same location, a remote or auxiliary link is used. This link provides a relay from the receiver to the transmitter.

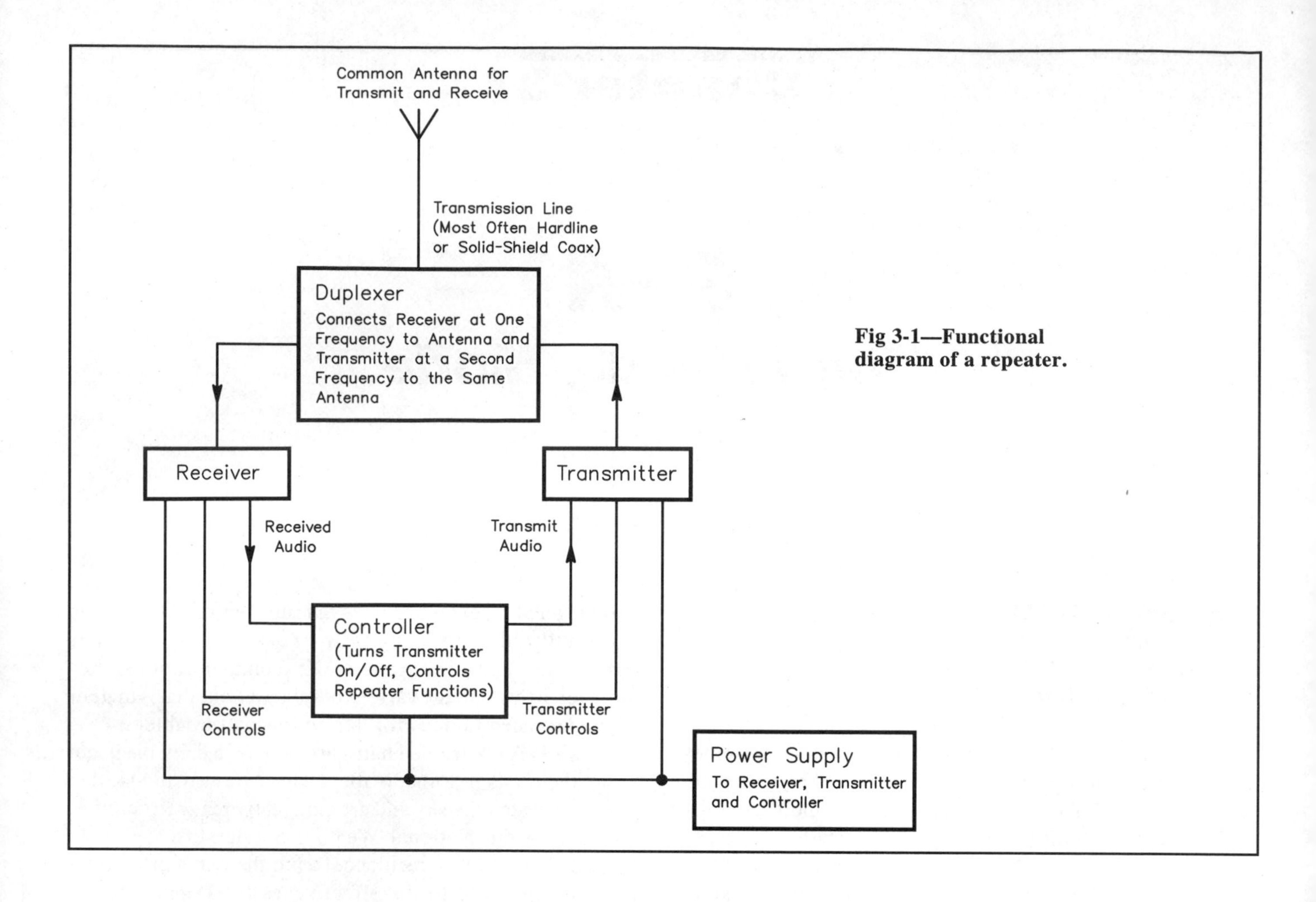

Fig 3-1—Functional diagram of a repeater.

Local control of a repeater in most cases is simply not possible, due mainly to their locations. Most operate under automatic control and have a control link (either radio or telephone line) that enables the system to be shut off if necessary by the system's control operator.

Frequency offset—the difference between the frequency on which the repeater hears and transmits—varies by repeater band. On 144 MHz the offset is 600 kHz; on 222 MHz it's 1.6 MHz and on 450 MHz it's 5 MHz. This frequency separation is generally built in to a modern transceiver's memory so it's transparent to the user. The radio chooses the offset depending upon which band is selected.

A repeater system may also include connections to receiver and transmitter combinations on other bands. For example, a 2-meter repeater linked to the 70-cm band may receive on 147.69 MHz and transmit on 147.09, while it also receives on 449.625 and transmits on 444.625 (see **Fig 3-2**). If a signal is present on 147.69 or on 449.625, it is retransmitted on both 147.09 and 444.625. These *crosslinked* repeaters may include coverage for several bands. Such systems are capable of operating on all bands at all times, or can be set up to have remote control selectability for the various links. Confused yet?

Many modern dual band VHF/UHF transceivers have the capability to do crossband linking. Before you set up your mini repeater in the mall parking lot, though, bear in mind that linking must be done above 222 MHz (the UHF side) in order to remain in accordance with FCC rules. There are also issues with the coordination of what is in effect a portable repeater, and your frequency coordinator may need to know of your plans in order to prevent needless interference to an existing coordinated repeater system in the area.

There are also more complex repeater systems that form integrated wide-coverage systems. Many of these systems cover entire states and in some cases (like the Evergreen Intertie and Cactus Systems) cover *several* states. These systems consist of linked repeaters located miles apart that are

440-MHz
Repeater
Controller
2-Meter
Repeater
Controller
Radio or Telephone Line Link
(2-Way)

Fig 3-2—Two repeaters using different bands can be crosslinked as shown.

interconnected by two-way VHF or UHF links. These systems allow operators to use a local repeater to make contacts with hams in distant cities or states. As an example, the Evergreen Intertie includes more than 23 repeater stations throughout California, Oregon, Washington, Idaho, Montana, British Columbia and Alberta. An amateur on Mt Shasta can chat with a ham in Edmonton, Alberta, while both are using low-power hand-held transceivers! Integrated wide-coverage repeater systems allow users to turn links on and off as needed and activate telephone autopatching and other features by using the dual-tone multifrequency (DTMF) keypads on their radios.

Licensing

With the exception of the Novice VHF/UHF subbands, you'll need a Technician class license (or higher) to operate in the VHF and UHF Amateur Radio spectrum (above 30 MHz). Technician class hams may also use VHF/UHF repeaters that transmit on frequencies not normally permitted for use by Technicians (for example, a 450-MHz repeater may have an output on 29.640 MHz—the 10 meter band). To own a repeater you must hold at least a Technician license.

Further clarification on licensing is available from the ARRL Regulatory Information Branch via *ARRLWeb* (**http://www.arrl.org/field/regulations**), e-mail (**reginfo**

FM and Repeater-Speak

access code: one or more numbers or symbols that are keyed into the repeater with a telephone tone pad to activate a repeater function, such as an autopatch.
autopatch: a device that interfaces a repeater to the telephone system to permit repeater users to make telephone calls. Often just called the *patch*.
break: the word used to interrupt a conversation on a repeater *only* to indicate that there is an emergency.
carrier-operated relay (COR): a device that causes the repeater to transmit in response to a received signal.
channel: the pair of frequencies (input and output) used by a repeater (example: the 94 machine would use the frequency pair 146.340 and 146.940).
closed repeater: a repeater whose access is limited to a select group. (see open repeater).
control operator: the Amateur Radio operator who is designated to control the operation of the repeater, as required by FCC regulations.
courtesy tone: an audible indication that a repeater user may go ahead and transmit.
coverage: the geographic area within which the repeater provides communications.
CTCSS: Continuous Tone Coded Squelch. A system of subaudible tones which operate the squelch (COR) of a repeater when the corresponding subaudible tone is present on a transmitted signal.
crossband: communications to another frequency band by means of a link interfaced with the repeater.
digipeater, digital repeater: a packet radio repeater.
duplex: a mode of communication in which you transmit on one frequency and receive on another frequency.
duplexer: a device that permits the use of one antenna for both transmitting and receiving with minimal degradation to either incoming or outgoing signals.
frequency coordinator: an individual or group responsible for assigning channels to new repeaters with a minimum chance of interference to existing repeaters.
full duplex: a mode of communication in which you transmit and receive simultaneously.
full quieting: a received signal that contains no noise.
half duplex: a mode of communication in which you transmit at one time and receive at another time.
hand-held transceiver: a portable transceiver small enough to fit in the palm of your hand, clipped to your belt or even in a shirt pocket.
input frequency: the frequency of the repeater's receiver (and your transceiver's transmitter).
intermod: interference caused by spurious signals generated by intermodulation distortion in a receiver front end or transmitter power amplifier stage.
key up: to turn on a repeater by transmitting on its input frequency.
LiTZ: Long Tone Zero (the i is added for pronunciation). An alert system that uses a DTMF tone pad-sent zero (0) keyed for at least three seconds to request emergency or urgent assistance.
machine: a repeater system (slang).
magnetic mount, mag-mount: an antenna with a magnetic base that permits quick installation and removal from a motor vehicle or other metal surface.
NiCd: a nickel-cadmium battery that may be recharged many times; often used to power portable transceivers. Pronounced *NYE-cad*.
offset: the spacing between a repeater's input and output frequencies.
open repeater: a repeater whose access is not limited.
output frequency: the frequency of the repeater's transmitter (and your transceiver's receiver).
over: a word used to indicate the end of a voice transmission.
polarization: the plane an antenna system operates in; most repeaters are vertically polarized.
radio direction finding (RDF): the art and science of locating a hidden transmitter.
Repeater Directory: an annual ARRL publication that lists repeaters in the US, Canada and other areas.
reverse autopatch: a device that interfaces the repeater with the telephone system and permits the users of the phone system to call the repeater and converse with on-the-air repeater users.
separation, split: the difference (in kHz) between a repeater's transmitter and receiver frequencies. Repeaters that use unusual separations, such as 1 MHz on 2 meters, are sometimes said to have odd splits.
simplex: a mode of communication in which you transmit and receive on the same frequency.
squelch tail: the noise burst heard in a receiver that follows the end of an FM transmission.
time-out: to cause the repeater or a repeater function to turn off because you have transmitted for too long.
timer: a device that measures the length of each transmission and causes the repeater or a repeater function to turn off after a transmission has exceeded a certain length.
tone pad: an array of 12 or 16 numbered keys that generate the standard telephone dual-tone multifrequency (DTMF) dialing signals; resembles a standard telephone keypad.

@arrl.org) or phone (860-594-0200).

The FCC has determined that, though shared by many users, the repeater is a licensed Amateur Radio station, and as such its use can be limited by its owner. This is the premise by which the closed repeater came into existence. Every area of the country has its share of open repeaters, but many also have closed repeaters that are limited to those hams who are members of the sponsoring group.

Coordination

Repeaters (though magical in nature) do not magically appear. Repeater systems are generally set up at significant personal expense and effort by an individual amateur operator or local club. The FCC has never mandated the choice of a repeater frequency. Rather, the regional frequency coordinator works with the owner to ensure the least possibility of interference to (or from) adjacent machines. The coordination process runs the gambit from one-man operations in sparsely populated areas to well-organized incorporated entities covering all or portions of many states. The process of coordination allows the repeater owner to place the repeater on the air on a selected frequency and ensures some measure of protection from someone else setting up a machine on the same frequency nearby.

This does not entail "ownership of the frequency." Instead, it reflects the fact that the FCC has historically sided with the *coordinated* repeater versus the noncoordinated system in interference disputes. Most coordinators have standards under which they work—generally distance between repeaters, power output, and geographical and terrain issues. Coordinators are all volunteers who have a very difficult job to do, considering that the number of repeater pairs is limited but the number of amateurs desiring a repeater is seemingly infinite! The NFCC (National Frequency Coordination Council) oversees the regional coordinators, sets basic standards for coordination and maintains a SPOC or Single Point of Contact between the coordination community and the FCC in areas where the groups must have a dialog.

Limiting Access

Most Amateur Radio repeaters are open to all users. There are no restrictions on the use of the repeater's functions. Limited-access repeaters do exist, however. Although some would argue that such operations go against the spirit of our hobby, a closed repeater is legal according to FCC regulations.

More often than not, especially in today's operating environment, you will find open repeaters that require the use of special codes or subaudible tones to gain access. The reason for tone encoding the access is to prevent interference, not to limit users of the system. In cases where extraneous transmissions often activate the repeater, the use of tone encoding is the only practical way to resolve the problem. How is access to these repeaters controlled? Most

What about Interference?

With an increase in the number of reports of repeater-to-repeater interference, the FCC is placing more emphasis on repeaters being coordinated. Repeater coordination is an example of voluntary self-regulation within the Amateur Service. Noncoordinated repeater operation may imply nonconformance with locally recognized band plans (e.g., an unusual frequency split) or simply that the repeater trustee has not yet applied for or received official recognition from the frequency coordinator.

In an effort to resolve repeater-to-repeater interference complaints, FCC has offered the following definition of Repeater Operation. *Where an amateur radio station in repeater or auxiliary operation causes harmful interference to the repeater or auxiliary operation of another amateur radio station, the two are equally and fully responsible for resolving the interference unless one station's operation is coordinated and the other's is not. In that case, the station engaged in the non-coordinated operation has primary responsibility to resolve the interference.* Harmful interference, by definition, is the kind that is helped to happen and not simply the sharing of frequencies by many repeater systems vying for very limited spectrum.

Table 3-1
CTCSS Tone Frequencies

The purpose of CTCSS (PL) is to reduce cochannel interference during band openings. CTCSS-equipped repeaters respond only to signals having the CTCSS tone required for that repeater. These repeaters do not respond to signals on their inputs that lacked the correct tone, and correspondingly do not transmit and repeat them, thus adding to the congestion.

The standard ANSI/EIA frequency codes with their Motorola alphanumeric designators are as follows:

67.0—XZ	118.8—2B	183.5
69.3—WZ	123.0—3Z	186.2—7Z
71.9—XA	127.3—3A	189.9
74.4—WA	131.8—3B	192.8—7A
77.0—XB	136.5—4Z	196.6
79.7—WB	141.3—4A	199.5
82.5—YZ	146.2—4B	203.5—M1
85.4—YA	151.4—5Z	206.5—8Z
88.5—YB	156.7—5A	210.7—M2
91.5—ZZ	159.8	218.1—M3
94.8—ZA	162.2—5B	225.7—M4
97.4—ZB	165.5	229.1—9Z
100.0—1Z	167.9—6Z	233.6—M5
103.5—1A	171.3	241.8—M6
107.2—1B	173.8—6A	250.3—M7
110.9—2Z	177.3	254.1—0Z
114.8—2A	179.9—6B	

In 1980, the ARRL Board of Directors adopted the 10-meter CTCSS tone controlled squelch frequencies listed below for voluntary incorporation into 10-meter repeater systems to provide a uniform system.

Call area	*Tone 1*	*Tone 2*
W1	131.8—3B	91.5—ZZ
W2	136.5—4Z	94.8—ZA
W3	141.3—4A	97.4—ZB
W4	146.2—4B	100.0—1Z
W5	151.4—5Z	103.5—1A
W6	156.7—5A	107.2—1B
W7	162.2—5B	110.9—2Z
W8	167.9—6Z	114.8—2A
W9	173.8—6A	118.8—2B
W0	179.9—6B	123.0—3Z
VE	127.3—3A	88.5—YB

often, via a technique called *continuous tone-controlled squelch system* (CTCSS). (Many hams refer to CTCSS as *PL*—a Motorola trademark that stands for *Private Line*.) When a transmitter is configured for CTCSS, it sends a subaudible tone along with the transmitted voice or other signals. The frequency of the CTCSS tone is below the lowest audio frequency other stations will pass to their speakers, but it's sensed by a suitably equipped repeater. The repeater is programmed to respond only to carriers that send the proper tone. This effectively locks out signals that don't carry the correct CTCSS tone. Modern VHF and UHF transceivers include the necessary circuitry to generate CTCSS tones, so if you know the one you need, you can simply program it on your rig.

Alphanumeric names are used to designate the tones, and the Electronic Industries Alliance (EIA) has developed 50 standard CTCSS tone frequencies. A list of current CTCSS tones is shown in **Table 3-1**.

OPERATING

FM repeaters provide the means to communicate efficiently. Before making your first transmission, however, you should be aware of some basic operating techniques. The following suggestions will assist you in operating a repeater as if you've been doing it for years.

1) *Monitor the repeater* to become familiar with any peculiarities in its operation. Each repeater is a different kingdom and many have specific and in some cases unwritten rules that apply—it's always best to *listen* for a while before jumping into the fray.

2) *To initiate a contact* simply indicate that you are on frequency. Various geographical areas have different practices on making yourself known, but generally "This is NUØX monitoring," will suffice. Please don't *kerchunk* (key up without identifying yourself) the repeater just to see if it's working. This procedure is annoying to other users. Taken a step too far, it may trigger the control operator to shut off the machine!

3) *Identify legally*. You must identify at the end of a transmission or series of transmissions and at least once each 10

PRESENT ARMS! The Northern Indiana VHF drill team consists of W9XD, WZ9M, KB9ATR, KC9XT, N9LVL and N9LBJ in the back row. Up front are N9IOX, Chris Kratzer and KB9GRP. ***(Photo courtesy of KB9GNU)***

Band Plans

The term *band plan* refers to an agreement among concerned VHF and UHF operators and users about how each Amateur Radio band should be arranged. The goal of a band plan is to reduce interference between all the modes sharing each band. Aside from FM repeater and simplex activity, CW, SSB, AM, satellite, amateur television (ATV) and radio control operations also use these bands. (For example, a powerful FM signal at 144.08 MHz could spoil someone else's long-distance CW contact.) The VHF and UHF bands offer a wide variety of amateur activities, so hams have agreed to set aside space for each type.

When considering frequencies for use in conjunction with a proposed repeater, be certain both the input and output fall within subbands authorized for repeater use, and do not extend beyond the subband edges. FCC Rules Part 97.205(b) defines frequencies currently available for repeater use.

The band plan for 2 meters, the most heavily used band for FM activity, follows.

Frequency (MHz)	*Operation*
144.00-144.05	EME (moonbounce) CW
144.05-144.10	General CW and weak signals
144.10-144.20	EME and weak-signal SSB
144.20	National SSB calling frequency
144.200-144.275	General SSB operation
144.275-144.300	Propagation beacons
144.30-144.50	OSCAR subband
144.50-144.60	Linear translator inputs
144.60-144.90	FM repeater inputs
144.90-145.10*	Weak signal and FM simplex
145.10-145.20	Linear translator outputs
145.20-145.50	FM repeater outputs
145.50-145.80	Miscellaneous and experimental modes
145.80-146.00	OSCAR subband
146.01-146.37	Repeater inputs
146.40-146.58	Simplex
146.52	National Simplex Calling Frequency
146.61-147.39	Repeater outputs
147.42-147.57	Simplex
147.60-147.99	Repeater inputs

*The following packet radio frequency recommendations were adopted by the ARRL Board of Directors in July 1987:

1) Automatic/unattended operations should be conducted on 145.01, 145.03, 145.05, 145.07 and 145.09 MHz.

a) 145.01 should be reserved for interLAN use.

b) Use of the remaining frequencies (above) should be determined by local user groups.

2) Additional frequencies in the 2-meter band may be designated for packet radio use by local coordinators.

Specific channels recommended above may not be applicable in all areas of the US. For example, amateurs in many areas conduct TCP/IP operations on 144.91 MHz and many PacketCluster nodes congregate on 144.95 MHz.

Prior to establishing regular packet radio use on any VHF/UHF channel, it is advisable to check with the local frequency coordinator. The decision on how the available channels are used should be based on coordination between local packet users.

Due to differences in regional coordination plans the simplex frequencies listed may be repeater inputs/outputs as well.

minutes during the communication. This is the FCC rule.

4) *Pause between your transmissions.* Sometimes the repeater will have a beep tone to indicate the appropriate pause for you. This allows other hams the opportunity to use the repeater (someone may have an emergency). On many repeaters a pause is necessary to reset the time-out timer.

5) *Keep transmissions short and thoughtful.* Your monologue may prevent someone with an emergency from using the repeater. If you talk long enough, you may actually time out the repeater. Something else to keep in mind is that your transmissions are being heard by non-hams who can pick them up on most VHF/UHF public service scanners. Remember that you are an ambassador for Amateur Radio operators and they will be judged on your behavior. Please don't give a bad impression of our service to others.

6) *Use simplex whenever possible.* If you can complete your QSO on a direct frequency, there is no need to tie up the repeater and prevent others from using it. In most areas the initial contact can be on the repeater. The conversation can then migrate to an agreed-upon simplex frequency.

7) *Use the minimum amount of power* necessary to maintain communications. This is required by FCC rules, and it reduces the possibility of accessing distant repeaters on the same frequency. In general, by the nature of the beast, you won't need a 60-W radio to access a local repeater.

8) *Don't break into a contact unless you have something to add.* Interrupting is no more polite on the air than it is in person.

9) Repeaters are intended primarily to facilitate mobile operation. During commuter rush hours, *home stations should relinquish the repeater to mobile stations*; some repeater groups have rules that specifically address this practice.

10) Many repeaters are equipped with *autopatch* facilities which, when properly accessed, connect the repeater to the telephone system to provide a public service. Although amateur operators may not accept compensation for the use of their radios, under certain conditions the radio/autopatch can be used to make appointments and order items. This is a decision made by the local repeater owner, however, so check with yours before making such calls. Autopatch facilities should never be used to avoid a toll call or where regular telephone service is available. Remember, autopatch privileges that are abused may be rescinded. Further guidelines and information on autopatch usage follow shortly.

11) All repeaters are assembled and maintained at considerable expense and effort. Usually an individual or group is responsible, and those who are regular users of a repeater should support (yes, probably financially . . .) the efforts of keeping the repeater on the air for all to use.

Autopatch Guidelines

Autopatch operation involves using a repeater as an interface to a local telephone exchange. Hams operating mobile or portable stations are able to use the autopatch to access the telephone system and place a call. Hams use autopatches to report traffic accidents, fires and other emergencies. There's no way to calculate the value of the lives and property saved by the intelligent use of autopatch facilities in emergencies. The public interest has been well served by amateurs with interconnect capabilities. As with any privilege, this one can be abused and the penalty for abuse could be the loss of the privilege for all amateurs. The suggested guidelines here are based on conventions that have been in use for years on a local or regional basis throughout the country. The ideas they represent have widespread support in the amateur community. Amateurs are urged to observe these standards carefully so our traditional freedom from government regulation may be preserved as much as possible.

1) Although it's not the intent of the FCC Rules to let Amateur Radio operation be used to conduct an individual's or an organization's commercial affairs, autopatching involving business affairs may be conducted on Amateur Radio. (The FCC has stated that it considers not-for-profit and noncommercial organizations to be businesses.) On the other hand, amateurs are strictly prohibited from accepting any form of payment for operating their ham transmitters they may not use Amateur Radio to conduct any form of business in which they have a financial interest and they may not use Amateur Radio in a way that benefits their employers economically.

Amateurs should generally avoid using Amateur Radio for any purpose that may be perceived as abuse of the privilege. The point of allowing hams to involve themselves in business communication is to make it more convenient and to remove obstacles from ham operations in support of public service activities. Before this rule was revised in 1993 it was often technically illegal for amateurs to participate in many charitable and community service events because the FCC regarded any organization commercial or noncommercial as a business with respect to the rules and prohibited hams from making any communications to in any way facilitate the business affairs of any party. That meant that operating a talk-in station for a local nonprofit radio club's hamfest constituted a violation!

So now it's legal to use ham frequencies, including autopatch facilities, to communicate in such a way as to facilitate a business transaction. The distinction is essentially whether the amateur operator or his employer has a financial stake in the communication. This means that a ham may use a patch to call someone about a club event or activity, to

Erik Johnson, KE3MB, with a K9 on his lap, frequents the local repeater.

make a dentist appointment, to order a pizza or to see if a load of dry cleaning is ready to be picked up. In such situations the ham isn't in it for the money. However, no one may use the ham bands to dispatch taxicabs or delivery vans, to send paid messages, to place a sales call to a customer or to cover news stories for the local media (except in emergencies if no other means of communication is available). If the ham is paid for, or will profit from, the communication it may not be conducted on an amateur frequency. That's why there are telephones and commercial business radio services available.

Use care in calling a business telephone via an amateur autopatch. Calls may be legally made to one's office to receive or to leave personal messages, although using Amateur Radio to avoid the cost of public telephones, commercial cellular telephones or two-way business-band radio isn't considered appropriate to the purpose of the amateur service. Calls made in the interests of highway safety, such as for the removal of injured persons from the scene of an accident or for the removal of a disabled vehicle from a hazardous location, are clearly permitted.

A final word on business communications: Just because the FCC says that a ham can place a call involving business matters on a repeater or autopatch doesn't mean that a repeater licensee or control operator must allow you to do so! If he or she prefers to restrict all such contacts, he or she has the right to terminate your access to the system. A club, for example, may decide that it would rather not have members order commercial goods over the repeater autopatch and may vote to forbid members from doing so. The radio station's licensee and control operator are responsible for what goes over the air and have the right to refuse anyone access to the station for any reason.

2) All interconnections must be made in accordance with telephone company rules and fee schedules (tariffs). If you have trouble obtaining information about them from telephone company representatives, the tariffs are available for public inspection at your telephone company office. Although some local telephone companies consider Amateur Radio organizations to be commercial entities and subject to business telephone rates, many repeater organizations as noncommercial volunteer public service groups have successfully arranged for telephone lines at repeater sites to be charged at the lower residential rate.

Some Popular Choices

The following is a list of the most popular 2-meter FM repeater frequency pairs in the US and Canada. With these channels programmed in your transceiver, the chances are good that you will quickly find an accessible repeater wherever your travels take you.

The top-10 frequency pairs (listen on the second one) are as follows:

146.34/146.94
146.22/146.82
146.16/146.76
146.07/146.67
146.25/146.85
146.37/146.97
146.13/146.73
146.31/146.91
146.28/146.88
146.40/147.00

3) Autopatches should not be made solely to avoid telephone toll charges. Autopatches should never be made when normal telephone service could be used just as easily. The primary purpose of an autopatch is to provide vital, convenient access to authorities during emergencies. Operators should exercise care, judgment and restraint in placing routine calls.

4) Third parties (non-hams) should not be put on the air until the responsible control operator has explained to them the nature of Amateur Radio. Control of the station must never be relinquished to an unlicensed person. Permitting a person you don't know well to conduct a patch in a language you don't understand amounts to relinquishing control because you don't know whether what they are discussing is permitted by FCC rules.

5) Autopatches must be terminated immediately in the event of any illegality or impropriety.

6) Station identification must be strictly observed.

7) Phone patches should be kept as brief as possible as a courtesy to other amateurs; the amateur bands are intended primarily for communication among radio amateurs, not to permit hams to communicate with non-hams who can only be reached by telephone.

8) If you have any doubt as to the legality or advisability of a patch, don't make it. Compliance with these guidelines will help ensure that amateur autopatch privileges will continue to be available in the future, which helps the Amateur Radio Service contribute to the public interest.

Finding a Repeater

To use a repeater, you must know one exists. There are various ways to find a repeater. Modern transceivers generally have a scan mode that searches for a frequency that is in use in the area where you are or local hams can provide information about repeater activity. There are also several very good listings (both written and software based) that can provide you with all the information available for repeaters in your area. Various clubs (especially Special Service Clubs or SSCs) sponsor repeaters. The ARRL publishes *The ARRL Repeater Directory*, an annual, comprehensive listing of repeaters throughout the US, Canada and other parts of the world. The ARRL also publishes *TravelPlus*, a map-based CD-ROM that allows you to trace your proposed route on a color map and print a list of repeaters along the way. In addition to simply identifying local repeater activity, these directories are perfectly suited for finding repeaters during vacations and business trips. You can find more information or place an order on *ARRLWeb*: **http://www.arrl.org/catalog/**. Once you find a repeater to use, take some time to listen and familiarize yourself with its operating procedures.

Your First Transmission

If the repeater is quiet, pick up your microphone, press the switch, and transmit your call sign. For example, "This is NUØX monitoring." This advises others on frequency that you have joined the system and are available to talk. After you stop transmitting, the repeater sends an unmodulated carrier for a couple of seconds to let you know it ís working. Chances are that if anyone wishes to make contact they will call you at this

An HT at 10,000 feet comes in handy for Dan Cui, N1CJD. He uses it as he pilots a "Paramotor," a type of powered paraglider. Not all hams will want to repeat Dan's operation.

time. Some repeaters have specific rules for making yourself heard, but usually your call sign is all you need.

It's not good repeater etiquette to call CQ. Efficient communication is the goal. You're not trying to attract the attention of someone who is casually tuning his receiver across the band. Except for scanner operation, there just isn't much tuning through the repeater bands—only listening to the machine.

If you want to join a conversation already in progress, transmit your call sign during a break between transmissions. The station that transmits next should acknowledge you. Don't use the word BREAK to join a conversation. BREAK generally suggests an emergency and indicates that all stations should stand by for the station with emergency traffic.

If you want to see if your buddy across town is on the air just call him like this: "N1ND this is NUØX." If the repeater is active, but the conversation in progress sounds as though it's about to end, be patient and wait until it's over before calling another station.

If the conversation sounds like it's going to continue for a while, transmit your call sign between transmissions. After

LiTZ Operation and Wilderness Protocol

Mutual Assistance Procedures for VHF/UHF FM

One of the great features of Amateur Radio is it gives hams the ability to provide mutual assistance to one another. There are two common procedures currently in place for mutual assistance on VHF/UHF FM frequencies. The first is LiTZ, a DTMF (touch-tone) based all-call priority alerting system. The second is the Wilderness Protocol.

LiTZ (the *i* is added to make it easier to pronounce) is a simple method to indicate to others on an amateur VHF/UHF FM radio channel that you have an immediate need to communicate with someone, anyone, regarding a priority situation or condition.

LiTZ stands for Long Tone Zero. The LiTZ signal consists of transmitting DTMF (touch-tone) zero for at least 3 seconds. After sending the LiTZ signal, the operator announces by voice the kind of assistance that is needed. For example:

(5-second-DTMF-zero) "This is KA7BCD."

"I'm on Interstate 5 between mileposts 154 and 155. There's a 3-car accident in the southbound lane. Traffic has been completely blocked. It looks like paramedics will be needed for victims. Please respond if you can contact authorities for help. This is KA7BCD."

If your situation does not involve safety of life or property, try giving a general voice call before using LiTZ. Use LiTZ only when your voice calls go unanswered or the people who respond can't help you.

When you see the notation LiTZ for a repeater in *The ARRL Repeater Directory*, that means it's highly likely that someone will receive and respond to LiTZ signals transmitted on the input frequency of the repeater. Please note, however, that if a CTCSS tone is needed to access that repeater you should transmit that CTCSS tone along with your LiTZ signal.

The type and nature of calls that justify the use of LiTZ may vary from repeater to repeater, just as other uses vary. Here are some general guidelines that may be suitable for most repeaters and simplex calling channels.

LiTZ Use Guidelines

Event/Situation	*Waking Hours (0700-2200 LT)*	*Sleeping Hours (2200-0700 LT)*
Calling CQ	no	no
Calling a buddy	no	no
Weekly Test of LiTZ	yes	no
Club Message	yes	no
Driving Directions	yes	no
Report Drunk Driver	yes	yes
Car Break Down	yes	yes
Safety Life or Prop	yes	yes

Wilderness Protocol

The Wilderness Protocol is a suggestion that those outside of repeater range should monitor standard simplex channels at specific times in case others have priority calls. The primary frequency is 146.52 MHz with 52.525, 223.5 446.0 and 1294.5 MHz serving as secondary frequencies.

This system was conceived to facilitate communications between hams that were hiking or backpacking in uninhabited areas, outside repeater range. However, the Wilderness Protocol should not be viewed as something just for hikers.

It can (and should) be used by everyone anywhere repeater coverage is unavailable. The protocol only becomes effective when many people use it.

The Wilderness Protocol recommends that those stations able to do so should monitor the primary (and secondary, if possible) frequency every three hours starting at 7 AM, local time, for 5 minutes (7:00-7:05 AM, 10:00-10:05 AM, . . . 10:00-10:05 PM, etc.).

Additionally, those stations that have sufficient power resources should monitor for 5 minutes starting at the top of every hour, or even continuously. Priority transmissions should begin with the LiTZ signal. CQ-like calls (to see who is out there) should not take place until four minutes after the hour.

one of the other hams acknowledges you, politely ask to make a quick call on the repeater. Usually, the other stations will allow you this brief interruption. Make your call short. If your friend responds to your call, ask him to move to a simplex frequency or another repeater, or to stand by until the present conversation is over. Thank the other users for letting you interrupt them to place your call.

Acknowledging Stations

If you're in the midst of a conversation and a station transmits its call sign between transmissions, the next station in queue to transmit should acknowledge that station and permit the newcomer to make a call or join the conversation. It's discourteous not to acknowledge him and it's impolite to acknowledge him but not let him speak. You never know; the calling station may need to use the repeater immediately. He may have an emergency on his hands, so let him make a transmission promptly. Always remember to pause briefly at the end of your transmission (or before you jump in and respond to someone). This allows others to make themselves known.

Brevity

Always try to keep transmissions as short as possible. Short transmissions permit more people to use the repeater. All repeaters promote this practice by having timers that shut down the repeater temporarily whenever the length of a transmission exceeds a preset time limit. With this in the back of their minds, most users keep their transmissions brief.

Learn the length of the repeater's timer and stay well within its limits. The length may vary with each repeater; some are as short as 15 seconds and others are as long as three minutes. Some repeaters vary their timer length depending on the amount of traffic on frequency: the more traffic, the shorter the timer. Another purpose of a repeater timer is to prevent extraneous signals (or someone accidentally sitting on the PTT button on their mobile microphone...) from holding the repeater on the air continuously. This could potentially cause damage to the repeater's transmitter.

Because of the nature of FM radio, if more than one signal is on the same frequency at one time, it creates a muffled buzz or an unnerving squawk. If two hams try to talk on a repeater at once, the resulting noise is known as a *double*. If you're in a roundtable conversation, it's easy to lose track of which station is next in line to talk. There's one simple solution to eradicate this problem forever: *Always pass off to another ham by name or call sign.* Saying, "What do you think, Jennifer?" or "Go ahead, N1TDY" will eliminate confusion and help avoid doubling. Try to hand off to whomever is next in the queue, although picking out anyone in the roundtable is better than just tossing the repeater up for grabs and inviting chaos!

The key to professional-sounding FM repeater operation is to be brisk and to the point, and to leave plenty of room for others. Keep it moving.

Identification

You must give your call sign at the end of each transmission or series of transmissions and at least every 10 minutes during the course of a contact. You don't have to transmit the call sign of any other station, including the one you're contacting. (Exception: You must transmit the other station's call sign when passing third-party traffic to a foreign country.)

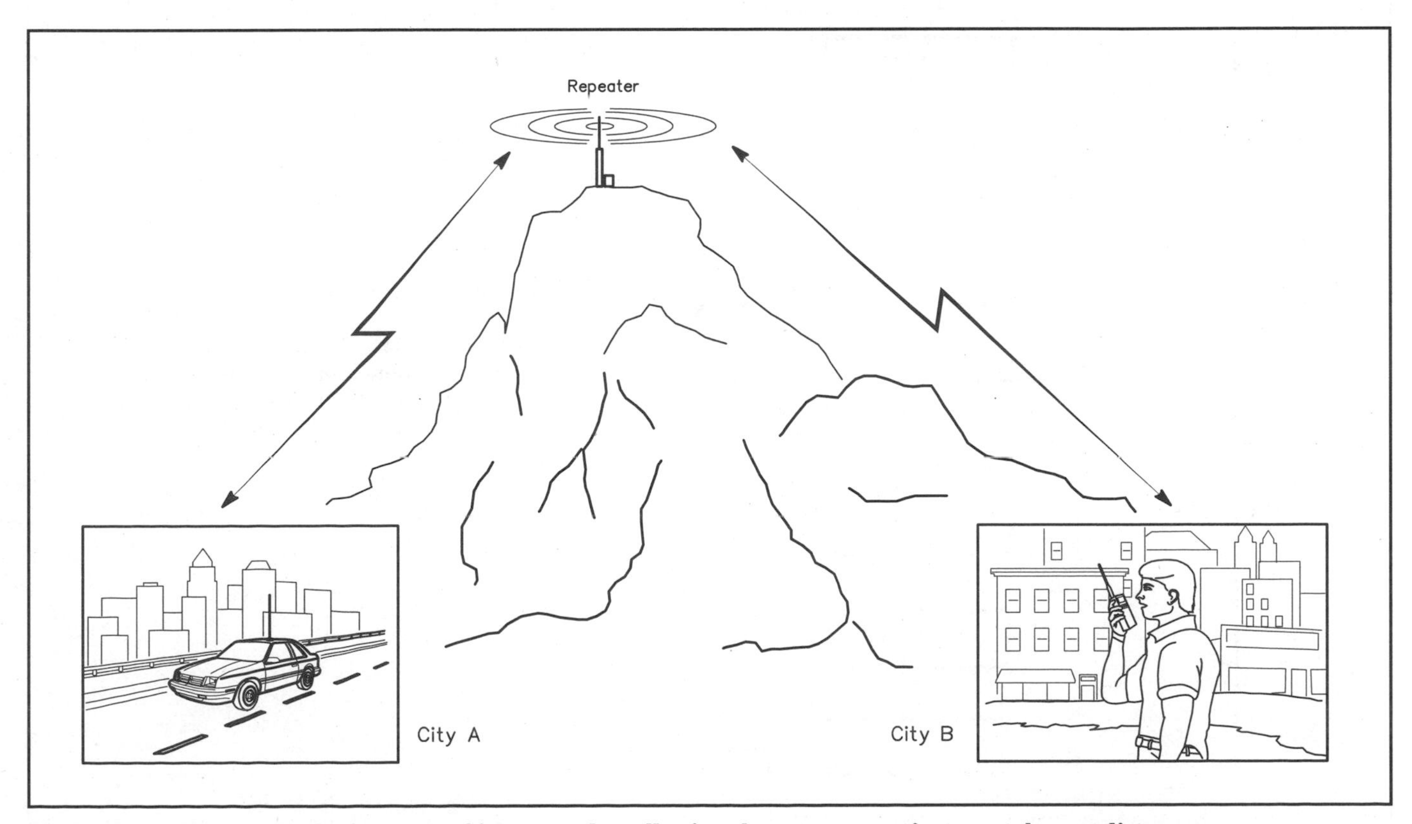

Fig 3-3—A repeater extends the range of its users, thus allowing them to communicate over longer distances.

As we are all familiar with FCC rules, we know it's illegal to transmit without station identification. Aside from breaking FCC rules, it's considered poor amateur practice to key your microphone to turn on a repeater without identifying your station. This is called *kerchunking* the machine. If you don't want to have a conversation, but simply want to check whether your radio works (or if you are able to access a particular repeater) just say something like "N1ND testing." This way you accomplish what you want and remain legal in doing so.

Go Simplex

Simplex is a fancy-sounding word for a direct contact on a single frequency. After you've made a contact on a repeater, move the conversation to a simplex frequency, if possible.

The function of a repeater is to provide communications between stations not able to communicate directly because of terrain or equipment limitations; see **Fig 3-3**. If stations *are* able to communicate without a repeater, they should. Always use simplex whenever possible so the repeater will be available for stations that need its facilities.

Simplex communication also offers a degree of privacy impossible to achieve on a repeater. There's also no time-out-timer to worry about or courtesy beep to wait for. The only caveat is to be aware, when selecting a frequency, to make sure it's designated for FM simplex operation. Band plans subdivide frequencies for specific modes of operation, such as satellite communications and weak-signal CW and SSB. If you select a simplex frequency indiscriminately, you may interfere with stations operating in other modes—and you may not even be aware of it.

How about some fun and challenge? VHF and UHF FM are also perfectly suited for sweepstakes, sprints, Field Day and other competitive events. In general, the only stipulation is that contacts must be made without using repeaters and conducted off the national simplex or calling frequencies. Watch for contest announcements in *QST*, bring that FM rig out to a mountaintop and call CQ contest sometime! VHF-UHF contesting is covered in greater detail in another chapter.

Fixed Stations and Prime Time

Repeaters are intended to *enhance* mobile and portable communications. During mobile operating prime time, fixed (home type) stations should yield to mobile stations. Some repeaters have an explicit policy dealing with this issue. When you're operating as a fixed station, however, don't abandon the repeater completely—monitor mobile activity. Your help may be needed in an emergency! Recommended operating priorities are shown in **Fig 3-4**.

Repeaters Cost Money

It takes time, a great deal of money, knowledge and energy to operate a reliable repeater system. How often do you stop to think of what goes into the machine you can conveniently key up 24/7. Obviously, nobody should feel compelled to join any repeater group, and you can certainly use thousands of open repeaters across the US without join-

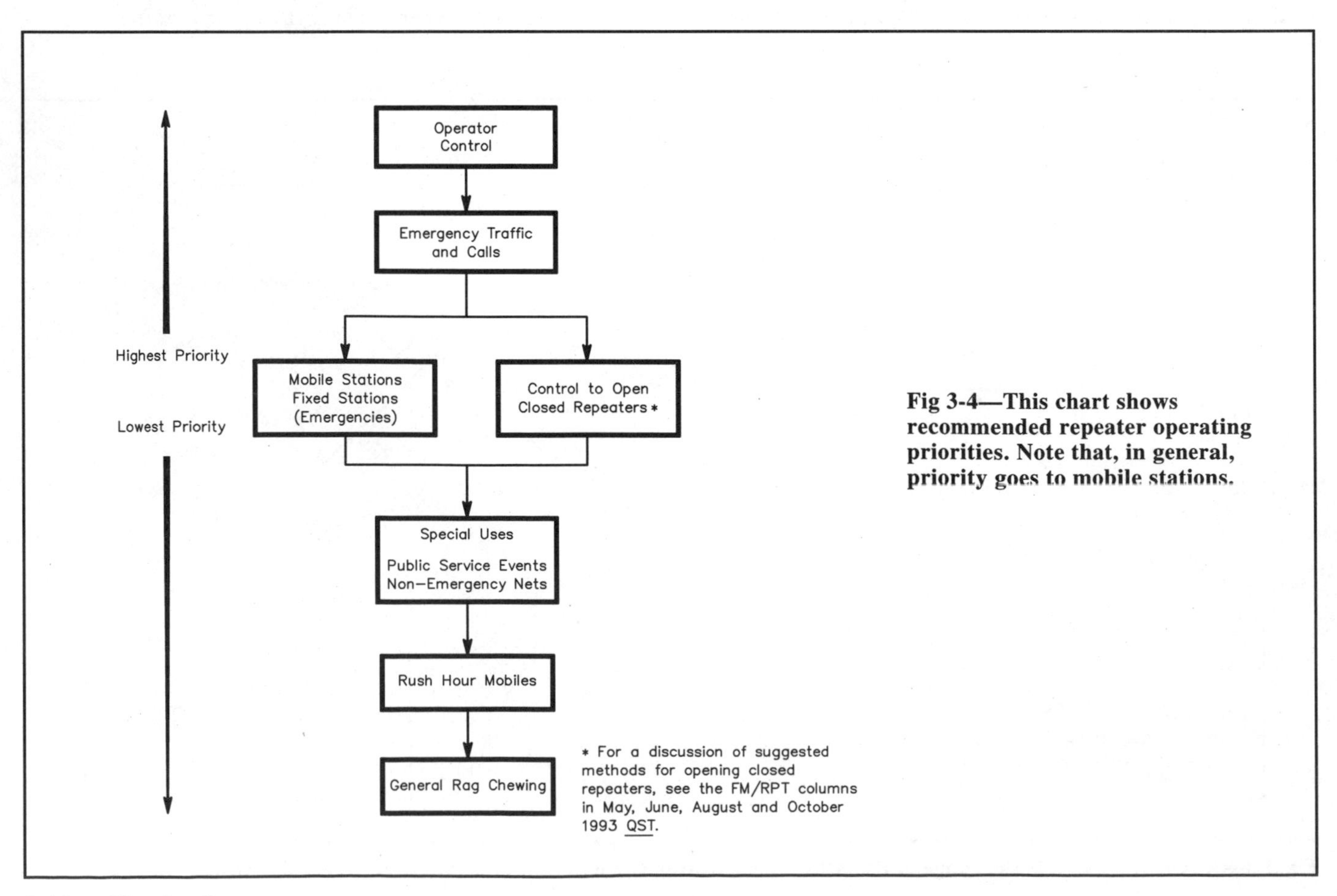

Fig 3-4—This chart shows recommended repeater operating priorities. Note that, in general, priority goes to mobile stations.

ing any clubs. On the other hand, if you frequent a system or just want to contribute to the cost of its upkeep, please feel free to render financial support, knowledge or muscle in its upkeep so it may be counted on in an emergency. Pooling of resources is the big plus in a sponsored repeater system—there is a vast pool of members to draw from.

An avid outdoorsman, Luigi, IC8GVV, always carries his hand-held radio when he climbs the cliffs near his home on the Isle of Capri.

Minimum Power

The VHF and UHF repeater bands can be a paradise for QRP enthusiasts. Make it a habit to run your transceiver on a low-power setting. There's usually no need to pump out heavy watts on VHF or UHF FM if you're within a reasonable range of a repeater or other station operating simplex.

Watch for Gremlins

Inspect your station regularly for loose connections, broken wires, antenna problems, intermittent grounds and other potential weak spots. Mobile installations are most prone to wear and damage. A well-designed antenna, quality feed line and properly installed connectors will improve your transmitted and received signals. This can also lead to fewer interference problems as a side effect! Never use a high-power external amplifier for a local contact. If a few watts won't bring up the repeater five miles away, 160 watts probably won't cut it, either.

When you're on a repeater frequency, use the minimum power necessary to maintain communications to avoid the possibility of accessing distant repeaters on the same frequency. Using minimum power is not only a courtesy to the distant repeaters, but also an FCC requirement.

Handling Traffic

Traffic handling is a natural for repeater operation. Where else can you find as many local outlets for the traffic coming down from the Section-level nets? Everyone who has a transceiver capable of repeater operation is a potential traffic handler. As a result, local traffic flourishes on FM repeaters.

The procedures for handling traffic on a repeater are about the same as handling traffic anywhere else. Handling traffic is also addressed elsewhere in this book. However,

Direction Finding

Radio direction finding (RDF), also known as foxhunting, rabbit hunting and hidden-transmitter hunting, has become a very popular type of VHF/UHF FM radiosport.

Here's how the scene plays out: You and the ham club folks are sitting around the local diner on a seasonable Sunday evening. After coffee, someone suggests a rabbit hunt might be fun. Excitement begins to grow, all the old hands dash off for their special DF antennas and black boxes as someone is elected to be the rabbit. The rabbit then hides a transmitter in the woods, a park or other inconspicuous place. For this purpose the transmitter can simply be a small self-contained unit set to transmit a signal at appropriate intervals, or it might be you in your car with the PTT button in your hot little hand.

After an appropriate time interval, the others in the hunt (equipped with their direction-finding equipment), go out to try to be the first to find the hidden transmitter. It can take minutes, or it can last until the rabbit is asked to identify the awesome place he or she located to hide! Foxhunting is great entertainment (and just as often keenly challenging team competition) at hamfests and club outings.

RDF also has a serious side. The FCC uses highly sophisticated DF equipment to track illegal signals to their source. Repeater operators can, and do, make use of the skilled DF and foxhunting folks on occasion to track down repeater jammers and unlicensed intruders. DF skills can also be handy for tracking down stolen transceivers that suddenly pop up on the air, obviously being operated by people who are not familiar with normal Amateur Radio operating procedures.

Locating a hidden transmitter is an art form and, like any other art, takes a great deal of practice to perfect. Homebrew DF antennas and equipment are both inexpensive and relatively easy to construct. So, next time you have the opportunity to participate in a foxhunt, take advantage of it to hone your direction-finding skills for the real thing.

During disasters, repeaters in the disaster area are used solely for emergency-related communication until the danger to life and property is past. (*Photo courtesy WA9TZL*)

HF Repeaters—Your DX Connection!

Imagine having a QSO with a station in Ecuador while you're on your way to work. Instead of a large HF rig under your dashboard, there is only a small transceiver that might easily be mistaken for a 2-meter FM unit at first glance. You're only running 10 or 20 watts, yet the Ecuadorian station is giving you a 59 report. How is this possible?

Well, all the FM repeater action isn't confined to the VHF and UHF bands. Narrowband voice FM is permitted from 28.3-29.7 MHz and wider-band FM voice is allowed from 29.0- 29.7 MHz. The upper end of the 29.0-29.7 MHz segment is populated with a number of FM repeaters. When propagation conditions are good, these repeaters can provide worldwide coverage!

It can be confusing to use a 10-meter FM repeater during good conditions. Your input signal may activate distant repeaters as easily as those close by. You may also hear multiple QSOs as several different transmitters battle for the repeaters (and your receiver's front end!). Sometimes band openings are as much a burden as they are a blessing. In 1980, the ARRL Board of Directors adopted a set of CTCSS access tones to be used voluntarily to provide a uniform nationwide system for 10-meter repeaters (see **Table 3-1**).

Where are They?

Most 10-meter FM repeaters operate between 29.520-29.680, with the FM simplex calling frequency at 29.600 MHz. (Avoid using 29.3-29.510 MHz for terrestrial contacts; ARRL recommends this segment be reserved exclusively for amateur satellite downlinks. Because these downlinks use CW or SSB, a powerful local FM signal would easily wipe out a weak-signal QSO, so respect this portion.)

There are four standard repeater frequency pairs, as follows (input/output MHz):

29.520/29.620
29.540/29.640
29.560/29.660
29.580/29.680

there are two things *unique* to repeater traffic handling: time-out timers and only occasional use of phonetics.

Repeaters have timers that shut down the repeater if a transmission is too long. Therefore, when you relay a message by repeater, release your push-to-talk (PTT) switch during natural breaks in the message to reset the timer. If you read the message in one long breath without resetting the timer, the repeater may shut down in the middle of your message.

Under optimal signal conditions, the audio quality of the FM mode of communications is excellent. It lacks the noise, static and interference common to other modes. Therefore, the use of phonetics and repetition is *not* necessary when relaying a message over a repeater. The only time phonetics are necessary is to spell out an unfamiliar word or words with similar-sounding letters. (An exception: The word *emergency* is always spelled out when it appears in the preamble of a formal radiogram.) On the rare occasion that a receiving station misses something, they can ask you to fill in the missing information.

The efficiency provided by repeaters has made repeater traffic nets very popular. There are probably one or more active repeater traffic nets in your area. A quick check of the *ARRL Net Directory* should provide you with one in your area. If there is no repeater traffic net in your vicinity, why not fill the need by starting one yourself? Contact your ARRL Section Manager (SM) or Section Traffic Manager (STM) for details. *QST* lists your local Section Manager near the front of every issue, or check *ARRLWeb*.

Satellite Gateway

A satellite gateway is a specially equipped station that links your local FM repeater to an Amateur Radio satellite. This makes it easier to try out a ham satellite without having to set up your own new equipment. It's your gateway to amateur satellite operation because it allows you to use the simplest ham radio station, a hand-held transceiver, to work stations around the world.

Operating through a gateway is easy. The Operations Controller (OC) is in charge of what goes into the repeater and keeps things in order. A list operation works well to ensure minimum confusion by those trying to use the gateway through the repeater. The OC instructs users to keep each contact within a maximum length, to speak clearly, and preferably, to say "over" to avoid doubling. Good discipline in gateway operation is essential. The signals from the gateway are heard over a large portion of the Earth, so efforts to minimize confusion are desirable.

It's permissible to call CQ through a repeater to work someone by satellite. More stations will respond to CQ from N1ND via gateway WB2NOM than QRZ OSCAR. Beyond these considerations, operating through a satellite gateway is straightforward. Common sense dictates the rest.

Chapter 4

VHF/UHF Operating
Beyond Repeaters

Michael Owen, W9IP and *QST* Staff

The radio spectrum between 30 and 3,000 MHz is one of the greatest resources available to the radio amateur. The VHF and UHF amateur bands are a haven for ragchewers and experimenters alike; new modes of emission, new antennas and state-of-the-art equipment are all developed in this territory. Plenty of commercial equipment is available for the more popular bands, and building your own gear is very popular as well. Propagation conditions may change rapidly and seemingly unpredictably, but the keen observer can take advantage of subtle clues to make the most of the bands. Most North American hams are already well acquainted with 2-meter (144 MHz) or 70-cm (440-MHz) FM. For many, channelized repeater operation is their first exposure to VHF or UHF. However, FM is only part of the VHF/UHF story! A great variety of SSB and CW activity congregates on the low ends of all the bands, from 6 meters all the way to the end of the radio spectrum—and even to light beyond.

HOW ARE THE BANDS ORGANIZED?

One of the keys to using this immense resource properly is knowing how the bands are organized. Each of the VHF and UHF bands is many megahertz wide, huge in comparison to any HF band. Different activities take place in separate parts of each. Even in the low ends of each VHF and UHF band, where SSB and CW activities congregate, there is a lot of space. Thus calling frequencies, unknown in the HF bands, help stations find each other.

By knowing the best frequencies and times to be on the air, you will have little trouble making plenty of contacts, working DX, and otherwise enjoying the world above 50 MHz. **Fig 4-1** shows suggested plans for using CW and SSB on the VHF and UHF bands.

By far the two most popular bands are 6 and 2 meters, followed by 70 cm. The 125 cm (222 MHz) and 33 cm (902 MHz) bands are not available worldwide. For that reason, commercial equipment is more difficult to find for these bands, and thus they are less popular.

The next two higher UHF bands at 23 cm (1296 MHz) and 13 cm (2304 MHz) are becoming more attractive, in part due to more commercial equipment being offered to meet the demands of amateur satellite operators.

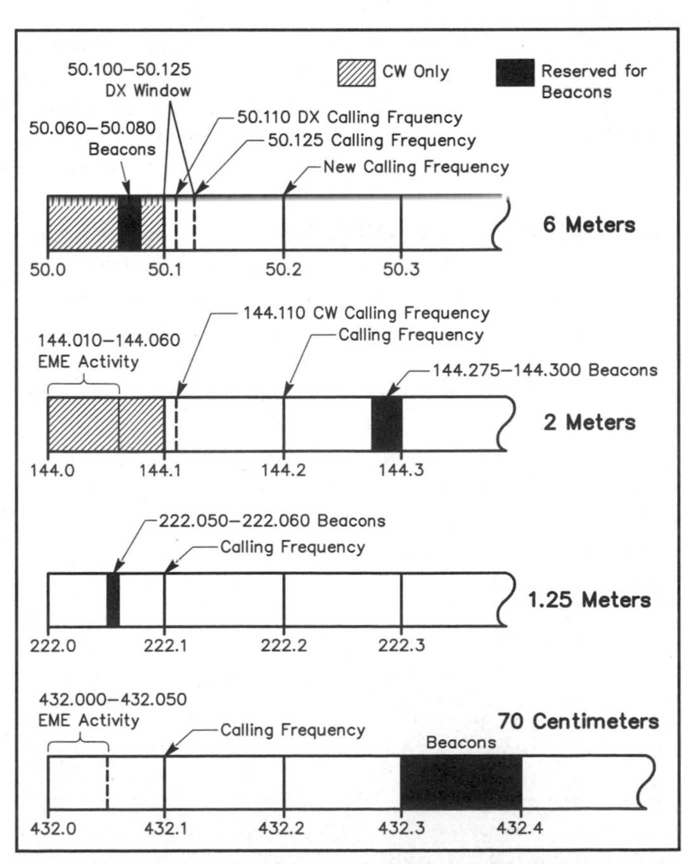

Fig 4-1—Suggested CW and SSB usage on the most popular VHF and UHF bands. Activity on 144 MHz and higher centers around the calling frequencies.

There is plenty of space on the VHF and higher bands for ragchewing, experimenting, working DX and many other activities unknown to the HF world. This means that whatever you like to do—chat with your friends across town, test amateur television or bounce signals off the moon—there is plenty of spectrum available. VHF/UHF is a great resource!

The key to enjoyable use of this resource is to know how everyone else is using it and to follow their lead. Basically, this means to listen first. Pay attention to the segments of the band already in use, and follow the operating practices that experienced operators are using. This way, you won't interfere with ongoing use of the band by others, and you'll fit in right away.

All of the bands between 50 and 1296 MHz have widely accepted calling frequencies (see **Table 4-1**). When there is little happening on the band, these are the frequencies operators will use to call CQ. For that reason, these are also the frequencies they are likely to monitor most of the time. Many VHF operators have gotten into a habit of tuning to one or more calling frequencies while doing something else around the shack. If someone wants a contact, you will already be on the right frequency to hear the call and make a contact.

The most important thing to remember about the calling frequencies is that they are not for ragchewing. After all, if a dozen other stations want to have a place to monitor for calls, it's really impolite to carry on a long-winded conversation on that frequency.

In most areas of the country, everyone uses the calling frequency to establish a contact, and then the two stations move up or down a few tens of kHz to chat. This way, everyone can share the calling frequency without having to listen to each other's QSOs. You can easily tell if the band is open by monitoring the call signs of the stations making contact on the calling frequency—you sure don't need to hear their whole QSO!

On 6 meters, a *DX window* has been established in order to reduce interference to DX stations. This window, which extends from 50.100-50.125 MHz, is intended for DX QSOs only. The DX calling frequency is 50.110 MHz. If you make a DX contact and expect to ragchew, you should move up a few kHz to clear the DX calling frequency. US and Canadian 6-meter operators should use the domestic calling frequency of 50.125 MHz for non-DX work. Because of increased crowding in the low end of the band in recent years, 50.200 MHz is becoming more popular as an alternative calling frequency. Once again, when contact is established, you should move off the calling frequency as quickly as possible.

Table 4-1
North American Calling Frequencies

Band (MHz)	*Calling Frequency*
50	50.090 general CW
	50.110 DX only
	50.125, 50.200 general SSB
144	144.110 CW
	144.200 SSB
222	222.100 CW, SSB
432	432.100 CW, SSB
902	902.100 CW, SSB
	903.100 CW, SSB (East Coast)
1296	1296.100 CW, SSB
2304	2304.1 CW, SSB
10,000	10,368.1 CW, SSB

ACTIVITY NIGHTS

Although it is possible to scare up a QSO on 50 or 144 MHz almost any evening (especially during the summer), in some areas of the country there is not always enough activity to make it easy to make a contact. Therefore, informal *activity nights* have been established so you will know when to expect some activity. Each band has its own night. There is some variation in activity nights from place to place, so check with someone in your area to find out about local activity nights.

Activity nights are particularly important for 222 MHz and above, where there are relatively few active stations on the air on a regular basis. If you have just finished a new transverter or antenna for one of these bands, you will have

Bernie Keiser, W4SW, with his dual-band rig (10 and 24 GHz).

Microwaves

The region above 1000 MHz is known as the microwave spectrum. Although European amateurs have made tremendous use of this rich resource, Americans lag behind somewhat. Commercial equipment from several European and US suppliers is becoming more widely available. The microwaves also provide great opportunities for building and experimenting, although entirely different construction techniques are required if amateurs are to build their own equipment. Nevertheless, the microwave bands are receiving a lot of attention from the experimentally minded among us. We can expect a considerable amount of progress as more amateurs try out the microwave bands.

The most popular microwave band in the USA is 10-GHz (3 cm). Using narrowband techniques (SSB/CW) and high-gain dish antennas, hams have made QSOs over 600 miles, but these are quite rare.

Most microwave activity is by prearranged schedules (at least in the US). This is partly because activity is so sparse and partly because antenna bandwidths are so narrow. With a 5° beamwidth, you would have to call CQ 72 times to cover a complete circle!

Common Activity Nights

Band (MHz)	Day	Local Time
50	Sunday	6:00 PM
144	Monday	7:00 PM
222	Tuesday	8:00 PM
432	Wednesday	9:00 PM
902	Friday	9:00 PM
1296	Thursday	10:00 PM

a much better chance to try them out during the band's weekly activity night. That doesn't mean there is no activity on other nights, especially if the band is open. It may just take longer to get someone's attention during other times.

Local VHF/UHF nets often meet during activity nights. Nets provide a regular meeting time for hams on 6 and 2 meters primarily, although there are regional nets at least as high as 1296 MHz. Several regional VHF clubs sponsor nets in various parts of the country, especially in urban areas. For those whose location is far away from the net control's location, the nets may provide a means of determining if your station is operating up to snuff, or if propagation is enhanced. Furthermore, you can sometimes catch a rare state or grid locator checking into the net. For information on the meeting times and frequencies of the nets, inquire locally.

WHERE AM I?

One of the first things you are bound to notice on the low end of any VHF band is that most QSOs include an exchange of grid locators. For example, instead of trying to tell a distant station, "I'm in Canton, New York," I say instead "My grid is FN24." It may sound strange, but FN24 is easier to locate on a grid locator map than my small town.

So what are these grid locators? They are 1° latitude by 2° longitude sections of the Earth. A grid locator in the center of the US is about 68 by 104 miles, but grids subtlely change size and shape, depending on their latitude.

Each locator has a unique two-letter/two-number identifier. The two letters identify one of 324 worldwide fields, which cover 10° latitude by 20° longitude each. There are 100 locators in each field, and these are identified by the two numbers. Exactly 32,400 grid locators cover the entire Earth. Two additional letters can be added for a more exact location, as in FN24kp. The extra two letters uniquely identifies a locale within a few miles.

There are several ways to find out your own grid square identifier. You can start by consulting **Tables 4-2** and **4-3**. By following the instructions shown in the tables, you will be able to locate your own grid. The hardest part is finding your location on a good map that has latitude and longitude on it; the rest is easy. Most high-quality road maps have this on the margins. Or, you could go a step further and purchase the topographic map of your immediate area from the US Geological Survey. For maps, write to US Geological Survey Information Services, Box 25286, Denver, CO 80225, tel 1-800-USA-MAPS. Once you have your latitude and longitude, the rest is a snap!

The ARRL publishes a grid-locator map (see **Fig 4-2**) of the continental United States and most populated areas of Canada. This map is available from ARRL for $1. If you are keeping track of grids for VUCC (the VHF/UHF Century Club award—see the Operating Awards chapter), you can color in each grid as you work it. Some operators use a light color when they work the grid for the first time, then color it more darkly when they receive a QSL. Others use color or pattern schemes to indicate different propagation modes. The ARRL also publishes a *World Grid Locator Atlas*, available for $5. On the Internet, point your browser to **http://www.arrl.org/locate/**

GRIDLOC is a simple computer program that can determine your grid locator from your latitude and longitude. The program also works in reverse, providing latitude and longitude for any grid. *GRIDLOC* is just one of 26 handy programs that are available on the *UHF/Microwave Experimenter's Software* diskette (3½-inch, for IBM-PCs and compatibles). The diskette is available from the ARRL for $10. The program is also available on *ARRLWeb* at **http://www.arrl.org/locate/gridinfo.html**

Table 4-2
How to Determine Your Grid Locator

1st and 2nd characters: Read directly from the map.

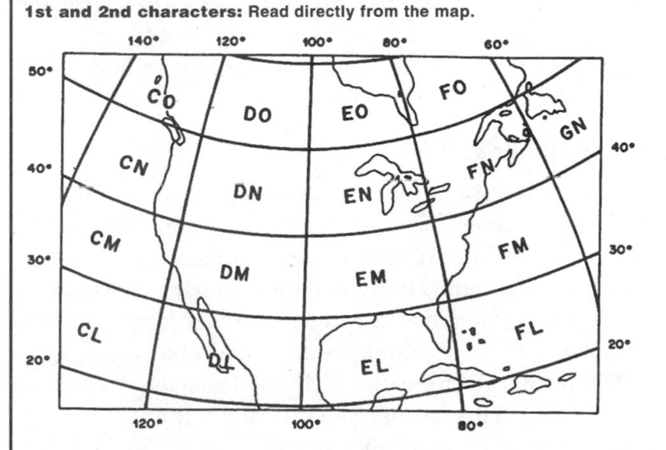

3rd character: Take the number of whole degrees west longitude, and consult the following chart.

Degrees West Longitude	Third Character	Degrees West Longitude	Third Character	Degrees West Longitude	Third Character
60-61	9	88- 89	5	114-115	2
62-63	8	90- 91	4	116-117	1
64-65	7	92- 93	3	118-119	0
66-67	6	94- 95	2	120-121	9
68-69	5	96- 97	1	122-123	8
70-71	4	98- 99	0	124-125	7
72-73	3	100-101	9	126-127	6
74-75	2	102-103	8	128-129	5
76-77	1	104-105	7	130-131	4
78-79	0	106-107	6	132-133	3
80-81	9	108-109	5	134-135	2
82-83	8	110-111	4	136-137	1
84-85	7	112-113	3	138-139	0
86-87	6				

4th character: This number is the same as the *2nd single digit* of your latitude. For example, if your latitude is 41° N, the 4th character is 1; for 29° N, it's 9, etc.

This four-character (2-letter, 2-number) designator indicates your 2° by 1° grid locator for VUCC award purposes.

Table 4-3
More Precise Locator

To indicate location more precisely, the addition of 5th and 6th characters will define the *sub-grid*, measuring about 4 × 3 miles in the central US. Longituide-latitude coordinates on maps, such as U.S. Department of the Interior Surveys, can be extrapolated to the nearest tenth of a minute, necessary for this level of locator precision. *This is not necessary in the VUCC awards program.*

5th character: If your number of degrees longitude is an *odd* number, see Fig A. If your number of degrees longitude is an *even* number, see Fig. B.

Odd Longitude* (Fig. A)

Minutes W. Longitude	*5th Character*
0- 5	L
5-10	K
10-15	J
15-20	I
20-25	H
25-30	G
30-35	F
35-40	E
40-45	D
45-50	C
50-55	B
55-60	A

Even Longitude* (Fig. B)

Minutes W. Longitude	*5th Character*
0- 5	X
5- 10	W
10-15	V
15-20	U
20-25	T
25-30	S
30-35	R
35-40	Q
40-45	P
45-50	O
50-55	N
55-60	M

6th character: Take the number of minutes *of latitude* (following the number of degrees) and consult the following chart.

Minutes N. Latitude	*6th Character*
0- 2.5	A
2.5- 5.0	B
5.0- 7.5	C
7.5- 10.0	D
10.0-12.5	E
12.5-15.0	F
15.0-17.5	G
17.5-20.0	H
20.0-22.5	I
22.5-25.0	J
25.0-27.5	K
27.5-30.0	L
30.0-32.5	M
32.5-35.0	N
35.0-37.5	O
37.5-40.0	P
40.0-42.5	Q
42.5-45.0	R
45.0-47.5	S
47.5-50.0	T
50.0-52.5	U
52.5-55.0	V
55.0-57.5	W
57.5-60.0	X

Propagation

Normal Conditions

What sort of range is considered normal in the world above 50 MHz? To a large extent, range on VHF is determined by location and the quality of the stations involved. After all, you can't expect the same performance from a 10-W rig and a small antenna on the roof as you might from a kilowatt and stacked beams at 100 feet.

On 2-meter SSB, a typical station probably consists of a low-powered multimode rig (SSB/CW/FM), followed by a 100-W amplifier. The antenna might be a single long Yagi at around 50 feet, fed with low-loss coax.

How far could this station cover on an average night using SSB? Location plays a big role, but it's probably safe to estimate that you could talk to a similarly equipped station about 200 miles away almost 100% of the time. Naturally, higher-power stations with tall antennas and low-noise receive preamps will have a greater range than this, up to a practical maximum of about 350-400 miles in the Midwest, less in the hilly West and East. It is almost always possible to extend your range significantly by switching from SSB to CW.

On 222 MHz, a similar station might expect to cover nearly the same distance, and perhaps 100 miles less on 432 MHz. This assumes normal propagation conditions and a reasonably unobstructed horizon. This range is a lot greater than you would get for noise-free communication on FM, and it represents the sort of capability the typical station should seek. Increase the height of the antenna to 80 feet and the range might increase to 250 miles, probably more, depending on your location. That's not bad for reliable communication!

Band Openings and DX

The main thrill of the VHF and UHF bands for most of us is the occasional band opening, when signals from far away are received as if they are next door. DX of well over 1000 miles on 6 meters is commonplace during the summer, and the same distance at least once or twice a year on 144, 222 and 432 MHz in all but mountainous areas.

DX propagation on the VHF/UHF bands is strongly influenced by the seasons. Summer and fall are definitely the most active times in the spectrum above 50 MHz, although band openings occur at other times as well.

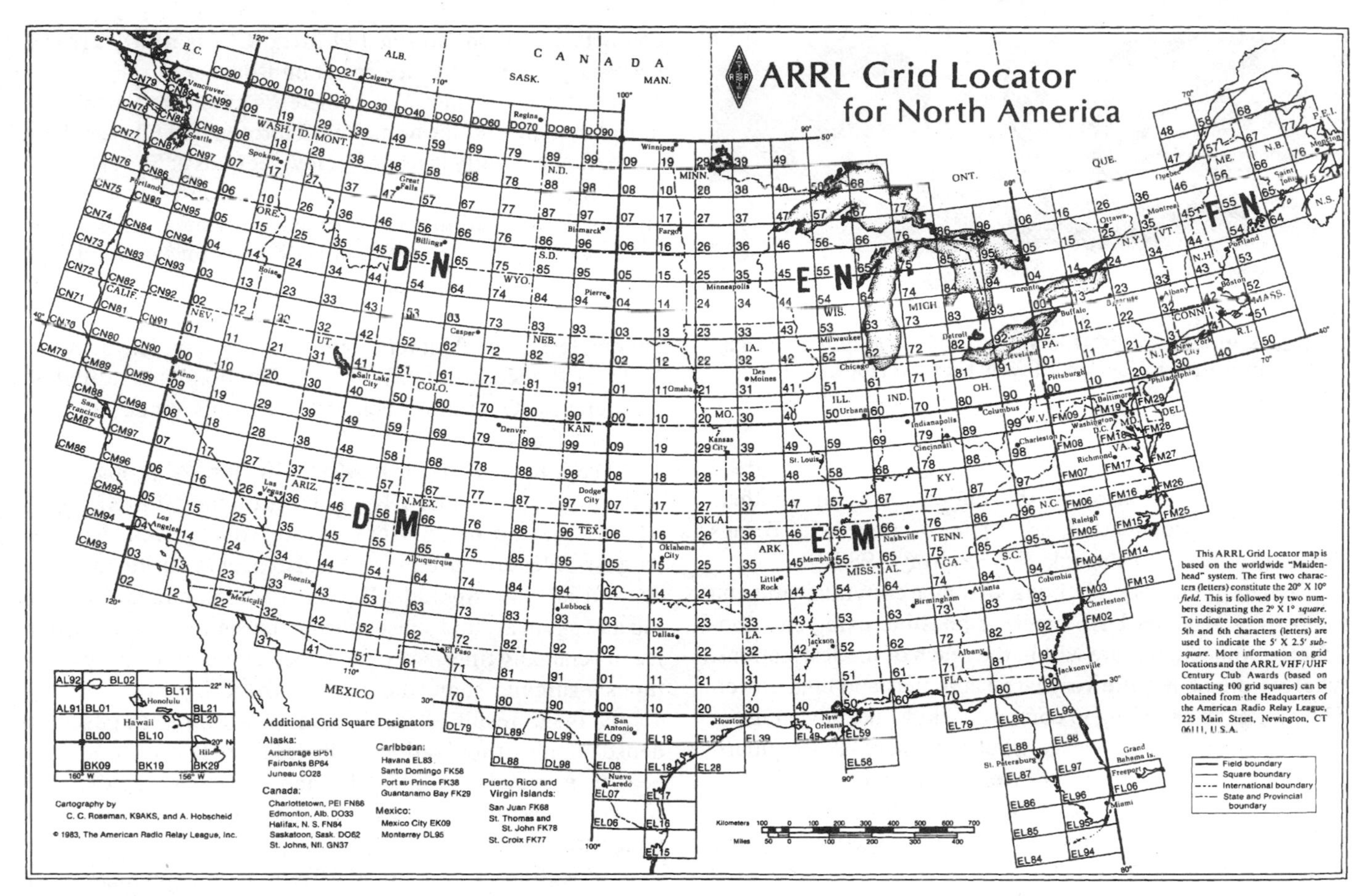

Fig 4-2—ARRL Grid Locator Map. A larger version is available from ARRL Headquarters.

There are many different ways VHF and UHF signals can be propagated over long distances. These can be divided between conditions that exist in the troposphere (the weather-producing lowest 10 miles of the atmosphere) and the ionosphere (between 50 and 400 miles high). The two atmospheric regions are quite distinct and have little effect on each other.

1) *Tropospheric ducting—or simply ducting.* Ducts are responsible for the most common form of DX-producing propagation on the bands above 144 MHz. Ducts are like natural waveguides that trap signals close to the Earth for hundreds of miles with little loss of signal strength. They come in several forms, depending on local and regional weather patterns. This is because ducts are caused by the weather. Ducts may cover only a few hundred miles, or they may include huge areas of the country at once.

a) *Radiation Inversion* (mostly summer). This common type of weak duct is caused by the Earth cooling off in the evening. The air just above the Earth's surface also cools, while the air a few hundred meters above remains warm. This creates an inversion that refracts VHF/UHF radio signals. If there is little or no wind, there may be a gradual improvement in the strength of signals out to a range of 100-200 miles (less in hilly areas) as evening passes into night. Radiation ducts mostly affect the bands above 144 MHz, and are seldom noticeable at 50 MHz.

b) *Broad, regional ducts* (late summer and early fall). These are the DXer's dream! In one of these openings, stations as far away as 1200 miles (and maybe more) are brought into range on VHF/UHF. The broad, regional type of ducting opening is caused by stagnation of a large, slow-moving high-pressure system. The stations that benefit the most are often on the south or western sides (the so-called back side) of the system. This sort of sluggish weather system may often be forecast just by looking at a weather map.

c) *Wave-cyclone tropo* (spring). These openings don't usually last long. They are brought about by an advancing cold front that interacts with the warm sector ahead of it. The resulting contrast in air temperatures may cause thunderstorms along the front. If conditions are just right, a band opening may result. These openings usually involve stations in the warm sector ahead of the cold front. You may feel that there's a pipeline between you and a DX station 1000 miles or more away. You may not be able to contact anybody else!

There are several other types of tropo openings, but space doesn't allow more than a brief discussion here. Coastal breezes sometimes cause long, narrow tropo openings along the East and Gulf Coasts. Rarely, cold fronts may cause brief openings as they slide under warmer air. US West Coast VHFers are always on the alert for the California-to-Hawaii duct that permits 2500-mile DX. For a very

Doug Shepard, VE1PZ, at his station near Pictou, Nova Scotia. Doug runs an IC-706 and a 1200-W amplifier into a 9-element 6-meter Yagi on a tower that overlooks the Northumberland Strait.

complete discussion of all the major forms of weather-related VHF propagation, see "The Weather That Brings VHF DX," by Emil Pocock, W3EP, May 1983 *QST*.

2) *The Scatter modes*. Long-distance communication on the VHF and lower UHF bands is possible using the ionosphere. Some modes are within the reach of modest stations, whereas others require very large antennas and high power. They all have one thing in common: They take advantage of the scattering of radio waves whenever there are irregularities in the atmosphere. Very briefly, the main types are:

a) *Tropospheric scatter*. Scattering in the troposphere is actually the most common form of propagation. You use it all the time without even knowing it. Tropo scatter is what creates ordinary beyond-line-of-sight contacts every day. It is the result of scattering from blobs of air of wavelength size that have slightly different temperature or humidity characteristics than surrounding air. Maximum tropo-scatter distance is the same as your normal working distance. There is a maximum theoretical tropo-scatter distance at VHF/UHF using amateur techniques of about 500 miles, regardless of frequency.

b) *Ionospheric forward scatter*. Scattering from the D and lower reaches of the E layers propagate VHF signals from 500 to a maximum of about 1000 miles. Signals at 50 MHz are apt to be very weak and fade in and out of the noise, even for the best equipped pairs of stations. Forward scatter contacts are much rarer at 144 MHz. CW is nearly always necessary. The best times are at mid-day.

c) *Meteor scatter*. When meteors enter the Earth's atmosphere, they ionize a small trail through the E layer. This ionization typically lasts only a few seconds at 50 MHz, and for even shorter periods at higher frequencies. Before it dissipates, the ionization can scatter, or sometimes reflect, VHF radio waves. Meteor scatter signals may not last long, but they can be surprisingly strong, popping suddenly out of the noise and then slowly fading away. It is quite common on 6 meters, less so on 2 meters, and rare indeed on 222 and 432 MHz. No successful QSOs have been completed at higher frequencies. Operating techniques for this mode are discussed later.

3) *Sporadic E (E_s or E-skip)*. This type of propagation is the most spectacular DX-producer on the 50-MHz band, where it may occur almost every day from late May to early August. A less intense E_s season also occurs during December and January. Sporadic E is most common in mid-morning and again around sunset during the summer months, but it can occur any time, any date. E_s occurs on 2 meters several times a summer somewhere across the US.

E_s is the result of the formation of thin but unusually dense clouds of ionization in the E layer. These clouds appear to move about, intensify, and disappear rapidly and without warning. The causes of sporadic E are not fully understood.

Reflections from sporadic-E clouds make single-hop contacts of 500 to 1400 miles possible on 50 MHz and much more rarely on 144 MHz. Only two sporadic-E like contacts have been reported above 220 MHz. Multi-hop E_s contacts commonly provide several coast-to-coast openings on 6 meters each summer and even opportunities to work Europe and Japan! The longest sporadic-E contacts are in excess of 6000 miles, but these are rare.

Sporadic-E signals are usually quite strong, allowing even the most modest station to make long-distance contacts. Openings may last only 15 minutes or go on all day. So far, there has been no satisfactory way to predict when these elusive openings will make their appearance, but they are exciting when they do happen.

4) *Aurora (Au)*. The aurora borealis, or northern lights, is a beautiful spectacle which is seen occasionally by those who live in Canada, the northern part of the USA and northern Europe. Similar southern lights are sometimes visible in the southernmost parts of South America, Africa and Australia. The aurora is caused by the Earth intercepting a massive number of charged particles thrown from the Sun during a solar storm. These particles are funneled into the polar regions of the Earth by its magnetic field. As the charged particles interact with the upper atmosphere, the air glows, which we see as the aurora. These particles also create an irregular, moving curtain of ionization which can propagate signals for many hundreds of miles.

Like sporadic E, aurora is more evident on 6 meters than on 2 meters. Nevertheless, 2-meter Au is far more common than 2-meter sporadic E, at least above 40° N latitude. Au is also possible on 222 and 432 MHz, and many tremendous DX

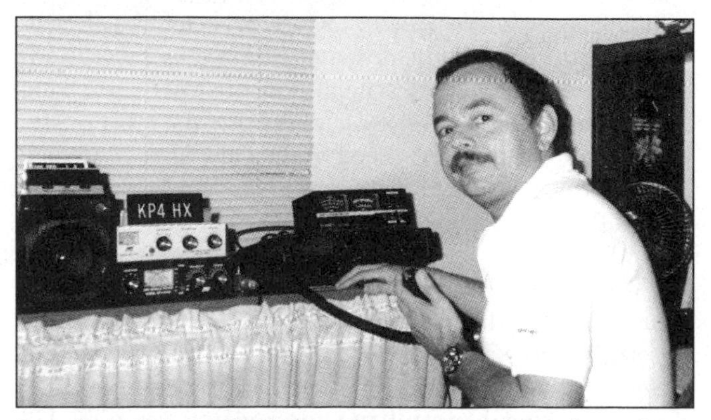

Braulio Feliciano, KP4HX, works stations throughout the US during summer band openings on 6 meters.

Doug Cooper, N7CNH, operating an all-mode transceiver and halo antenna on 2 meters from Spencer Butte south of Eugene, Oregon. Doug had to carry the equipment as he hiked to the top.

contacts have been made on these bands. Current record distances are over 1000 miles on 144, 222 and 432 MHz.

Aurora can be predicted to some extent from current reports of solar and geomagnetic activity, which can be found on most DX PacketClusters, on several WWW sites (such as those sponsored by the National Oceanic and Atmospheric Administration), and WWV broadcasts. At 18 minutes past each hour, WWV transmits a summary of the condition of the Earth's geomagnetic field. If the K index is 4 or above, you should watch for Au. Many VHFers have learned that a high K index is no guarantee of an aurora. Similarly, K indices of only 3 have occasionally produced spectacular radio auroras at middle latitudes. When in doubt, point the antenna north and listen!

Auroral DX signals are highly distorted. CW is the most practical mode, although SSB is sometimes used. Stations point their antennas generally northward and listen for the telltale hissing note that is characteristic of auroral signals. More about operating Au is presented later.

5) *Auroral E (AuE).* Signals propagated by auroral-induced sporadic E sound very much like ordinary E-skip, save that its causes and timing are quite different. Auroral E is induced by the same conditions that give rise to auroras, and like aurora, AuE is more common at northerly latitudes. Auroral E may accompany unusually strong auroras, but is more usually observed after midnight across the northern tier of states and Canada. Distances covered are similar to sporadic E.

6) *Transequatorial field-aligned irregularities (TE).* This unusual propagation mode creates paths of 2500 to 5000 miles on 50 MHz, and less commonly on 144 and 222 MHz. It involves some strict requirements. TE works only for stations equally distant from the geomagnetic equator. Common TE paths are from the Mediterranean to southern Africa, southern Japan to Australia, and the Caribbean and Venezuela to Argentina. The geomagnetic equator is displaced considerably to the south of the geographic equator in the Americas, so that only US stations from south Florida to southern California can normally make TE contacts into Argentina and Uruguay. TE-to E_s hook-ups on 50 MHz sometimes extend the possible coverage much further north.

TE appears almost exclusively in late afternoon and is more common around the March and September equinoxes, especially in years of high solar activity or when there is a geomagnetic storm. Signals have an unmistakable fluttery quality. TE is caused by two unusually dense regions of F-layer ionization that appear just north and south of the geomagnetic equator. Neither region is capable of propagating VHF signals over such long distances separately, but when linked together at the proper angles, some long north-south paths can result.

7) *Earth-Moon-Earth (EME or moonbounce).* This is the ultimate VHF/UHF DX medium! Moonbouncers use the Moon as a passive reflector for their signals, and QSO distance is limited only by the diameter of the Earth. Any two stations who can simultaneously see the Moon may be able to work each other via EME. QSOs between the USA and Europe or Japan are commonplace on VHF and UHF by using this mode. That's DX!

Previously the territory of only the biggest and most serious VHFers, moonbounce has now become more widely popular. Thanks to the efforts of pioneer moonbouncers such as Bob Sutherland, W6PO, and Al Katz, K2UYH, hundreds of stations are active, mainly on 144 and 432 MHz. This huge increase in activity especially by a handful of stations with gigantic antenna arrays, has encouraged many others to try making contacts via the moon.

Improvements in technology—low-noise amplifiers and better designs—have made it easier to get started. Also, several individuals have assembled gigantic antenna arrays, which make up for the inadequacy of smaller antennas. The result is that even modestly equipped VHF stations (150 W and one or two Yagis) are capable of making a few moonbounce contacts. Activity is constantly increasing. There is even an EME contest in which moonbouncers compete on an international scale.

Moonbounce requires larger antennas than most terrestrial VHF/UHF work. In addition, you must have a high-power transmitting amplifier and a low-noise receiving preamplifier to work more than the biggest guns. A modest EME station on 144 MHz consists of four long-boom Yagi antennas on an azimuth-elevation mount (for pointing at the Moon), a kilowatt amplifier and a low-noise preamplifier mounted at the antenna. On 432 MHz, the average antenna is eight long Yagis. You can make contacts with a smaller antenna, but they will be with only larger stations on the other end. Yagi antennas aren't the only type available; collinears and quagis are also widely used. Several UHFers have also built large parabolic dish antennas.

How Do I Operate on VHF/UHF?

Normal Conditions

The most important rule to follow on VHF/UHF, like all other amateur bands, is to listen first. Even on the relatively uncrowded VHF bands, interference is common near the calling frequencies. The first thing to do when you switch on the radio is to tune around, listening for activity. Of course, the calling frequencies are the best place to start

Jack Nyiri, AB4CR, setting up portable antennas for his rover entry in the ARRL September VHF QSO Party.

listening. If you listen for a few minutes, you'll probably hear someone make a call, even if the band isn't open. If you don't hear anyone, then it's time to make some noise yourself!

Stirring up activity on the lower VHF bands is usually just a matter of pointing the antenna and calling CQ. Because most VHF beams are rather narrow, you might have to call CQ in several directions before you find someone. Several short CQs are always more productive than one long-winded CQ. But don't make CQs too short; you have to give the other station time to turn the antenna toward you.

Give your rotator lots of exercise; don't point the antenna at the same place all the time. You never know if a new station or some DX might be available at some odd beam heading. VHFers in out-of-the-way locations, far from major cities, monitor the bands in the hope of hearing you.

Band Openings

How about DX? What is the best way to work DX when the band is open? There's no simple answer. Each main type of band opening or propagation mode requires its own techniques. This is natural, because the strength and duration of openings vary considerably. For example, you wouldn't expect to operate the same way during a 10-second meteor-burst QSO as during a three-day tropo opening.

The following is a review of the different main types of propagation and descriptions of the ways that most VHFers take advantage of them.

Tropospheric Ducting

Tropospheric ducts, whatever their causes, affect the entire VHF through microwave range, although true ducting is rarely observed at 50 MHz. Ducts often persist for hours at a time, sometimes for several days, so there is usually no panic about making contacts. There is time to listen carefully and determine the extent of the opening and its likely evolution.

Ducts are most common in the Mississippi Valley during late summer and early fall, and they may expand over much of the country from the Rocky Mountains eastward. Sprawling high-pressure systems that slowly drift southeastward often create strong ducts. The best conditions usually appear in the southwestern quadrant of massive highs.

Along the East Coast, ducting is more common along coastal paths of up to 1000 miles and sometimes longer. Stations in New England have worked as far as Cuba on 2 meters this way. The mountainous west rarely experiences long-distance ducting.

This void is partially made up by one of the most famous of all ducting paths, which creates paths from the West Coast to Hawaii. The famous trans-Pacific duct opens up several times a year in summer, supporting often incredibly strong signals over 2500 miles on 144 MHz through at least 5.6 GHz. Several world distance records have been made over this path. Other common over-water ducts appear across the Gulf of Mexico, mostly in early spring.

Ducts can be anticipated by studying weather maps and forecasts. Many VHFers also check television stations, especially in the UHF range, for early warnings of enhanced conditions. Check beacons on 144 MHz and higher, especially those you cannot ordinarily hear. Keep in mind that most forms of ducting intensify after sunset and peak just after sunrise. All of these techniques can enhance your chances of catching a ducting opening.

Ionospheric Forward Scatter

This mode is not used to its full potential, probably because forward scatter signals are often weak and easily overlooked. The best chances are for cooperating stations using CW in a quiet portion of the 50 or 144 MHz bands. Schedules may enhance the chances of success, but in any case, patience and a willingness to deal with weak fluttery signals is required. Contacts are sometimes completed via meteor-scatter enhancement that by chance occurs at the same time.

Meteor Scatter

Meteor scatter is very widely used on 50 and 144 MHz, and it has been used on 222 and 432 as well. Operation with this exciting mode of DX comes under two main headings: prearranged schedules and random contacts. Either SSB or CW may be used, although SSB is more popular in North America. European-style high-speed CW meteor-scatter techniques, which use a computer to send and help receive

Morse code sent at several hundred to several thousand characters per minute, has caught on in the US.

Most meteor-scatter work is done during major meteor showers, and many stations arrange schedules with stations in needed states or grid locators. In a sked, 15-second transmit-receive sequences are the norm for North America (Europeans use longer sequences). One station, almost always the westernmost, will take the first and third 15-second periods of each minute and the other station takes second and fourth. This is a very simple procedure that ensures that only one station is transmitting when a meteor falls. See accompanying sidebar.

A specific frequency, far removed from local activity centers, is chosen when the sked is set up. It is important that both stations have accurate frequency readout and synchronized clocks, but with today's technology this is not the big problem that it once was. Schedules normally run for $^1/_2$ hour or 1 hour, especially on 222 and 432 MHz where meteor-scatter QSOs are well earned!

The best way to get the feel for the meteor-scatter QSO format is to listen to a couple of skeds between experienced operators. Then, look in *QST* or ask around for the call sign of veteran meteor-scatter operators in the 800-1000 mile range from you (this is the easiest distance for meteor scatter). Call them on the telephone or catch them on one of the national VHF nets and arrange a sked. After you cut your teeth on easy skeds, you'll be ready for more difficult DX.

A lot of stations make plenty of meteor-scatter QSOs without the help of skeds. Especially during major meteor showers, VHFers congregate near the calling frequency of each band. There they wait for meteor bursts like hunters waiting for ducks. Energetic operators make repeated and brief CQs, hoping to catch an elusive meteor. When meteors blast in, the band comes alive with dozens of quick QSOs. For a brief time, normally five seconds to perhaps 30 seconds, 2 meters may sound like 20 meters! Then the band is quiet again. . . until the next meteor burst!

The quality of shower-related meteor-scatter DX depends on three factors. These three factors are well known or can be predicted. The most important is the radiant effect. The radiant is the spot in the sky from which the meteors appear to fall. If the radiant is below the horizon, or too high in the sky, you will hear very few meteors. The most productive spot for the radiant is at an elevation of about 45° and an azimuth of 90° from the path you're trying to work. The second important factor is the velocity of the meteors. Slow meteors cannot ionize sufficiently to propagate 144 or 222-MHz signals, no matter how many meteors there are. For 144 MHz, meteors slower than 50 km/s are usually inadequate (see accompanying sidebar detailing major meteor showers). Third, the shower will have a peak in the number of meteors that the Earth intercepts. However, because the peak of many meteor showers is more than a day in length, the exact time of the peak is not as important as most people think. Two interesting references are "Improving Meteor Scatter Communications," by Joe Reisert, W1JR, in June 1984 *Ham Radio*, and "VHF Meteor Scatter: an Astronomical Perspective," by M. R. Owen, W9IP, in June 1986 *QST*.

It takes a lot of persistence and a good station to be successful with random meteor scatter. This is mainly because you must overcome tremendous interference in addition to the fluctuations of meteor propagation. At least 100 W is necessary for much success in meteor scatter, and a full kilowatt will help a lot. One or two Yagis, stacked vertically, is a good antenna system. Antennas with too much gain have narrow beamwidths, and so often cannot pick up many usable meteor-scatter bursts.

In populated areas, it can be difficult to hear incoming meteor-scatter DX if many local stations are calling CQ. Therefore, many areas observe 15-second sequencing for random meteor-scatter QSOs, just as for skeds. Those who want to call CQ do so at the same time so everyone can listen for responses between transmissions. Sometimes a bit of peer pressure is necessary to keep everyone together, but it pays off in more QSOs for all. The same QSO format is used for scheduled and random meteor-scatter QSOs.

Sporadic E

Sporadic-E signals are generally so loud and openings last long enough that no special operating techniques are necessary to enjoy this mode. On 6 meters, 10-W stations with simple antennas can easily make contacts out to 1000 miles or so. The band may open for hours on any summer day, with signals constantly shifting, disappearing and re-

Ernst Willert, DK3FF is bouncing his 10-GHZ SSB signals off rain showers in the vicinity of Niederkassel-Rheidt, Germany. Ernst has used this unusual propagation technique to contact numerous central European stations.

appearing. A sporadic-E opening on 2 meters is much less common and typically lasts for less than an hour. Signals out to 1000 miles or so can be unbelievably strong, yet there is reason to be more alert. E-sip openings on 2 meters are rare and do not usually last long.

The main question for those hoping for sporadic E is when the band will open. Aside from knowing sporadic E is more common in summer mornings and early evenings than any other times, there is no satisfactory way to predict E-skip. It can appear any time.

Sporadic E affects lower frequencies first, so you can get some warning by listening to 10 meters. When E-skip shortens to 500 miles or so, it is almost certain that there is propagation somewhere on 6 meters. Some avid E-skip fans monitor TV channel 2 or 3 (if there is no local station) for signs of sporadic E.

Aurora and Auroral E

Aurora favors stations at high latitudes. It is a wonderful blessing for those who must suffer through long, cold winters because other forms of propagation are rare during the winter. Aurora can come at almost any time of the year. New England stations get Au on 2 meters about five to 10 times a year, whereas stations in Tennessee get it once a year if they're lucky. Central and Southern California rarely hear aurora.

The MUF of the aurora seems to rise quickly, so don't wait for the lower-frequency VHF bands to get exhausted before moving up in frequency. Check the higher bands right away.

You'll notice aurora by its characteristic hiss. Signals are distorted by reflection and scattering off the rapidly

Meteor-Scatter Procedure

In a meteor-scatter QSO, neither station can hear the other except when a meteor trail exists to scatter or reflect their signals. The two stations take turns transmitting so that they can be sure of hearing the other if a meteor happens to fall. They agree beforehand on the sequence of transmission. One station agrees to transmit the 1st and 3rd 15 seconds of each minute, and the other station takes the 2nd and 4th. It is standard procedure for the western-most station to transmit during the 1st and 3rd.

It's important to have a format for transmissions so you know what the other station has heard. This format is used by most US stations;

Transmitting	***Means you have copied;***
Call signs	nothing, or only partial calls
Call plus signal report (or grid or state)	full calls-both sets
ROGER plus signal report(or grid or state)	full calls, plus signal report (or grid or state)
ROGER	ROGER from other station

Remember, for a valid QSO to take place, you must exchange full call signs, some piece of information, and acknowledgment. Too many meteor QSOs have not been completed for lack of ROGERs. Don't quit too soon; be sure the other station has received your acknowledgment. Often, stations will add 73 when they want to indicate that they have heard the other station's ROGER.

Until a few years ago, it was universal practice to give a signal report which indicated the length of the meteor burst. S1 meant that you were just hearing pings, S2 meant 1-5 second bursts, and so on. Unfortunately, virtually everyone was sending S2, so there was no mystery at all, and no significant information was being exchanged.

Grid locators have become popular as the piece of information in meteor-scatter QSOs. More and more stations are sending their grid instead of S2. This is especially true on random meteor-scatter QSOs, where you might not know in advance where the station is located. Other stations prefer to give their state instead. For an excellent summary of meteor-scatter procedure, see "Meteor Scatter Communications" by Clarke Greene, K1JX, in January 1986 *QST* (pp 14-17). Also see "Hooked on Meteors!" by Tom Hammond, WD8BKM, in May 1995 *QST* (p 74).

Meteor Showers

Every day, the Earth is bombarded by billions of tiny grains of interplanetary debris, called meteors. They create short-lived trails of E-layer ionization which can be used as reflectors for VHF radio waves. On a normal morning, careful listeners can hear about 3-5 meteor pings (short bursts of meteor-reflected signal) per hour on 2 meters.

At several times during the year, the Earth passes through huge clouds of concentrated meteoric debris, and VHFers enjoy a meteor shower. During meteor showers, 2-meter operators may hear 50 or more pings and bursts per hour. Here are some data on the major meteor showers of the year. Other showers also occur, but they are very minor.

Major Meteor Showers

Shower	*Date range*	*Peak date*	*Time above quarter max*	*Approximate visual rate*	*Speed km/s*	*Best Paths and Times (local)*
Quadrantids	Jan 1-6	Jan 3/4	14 hours	40-150	41	NE-SW (1300-1500), SE-NW (0500-0700)
Eta Aquarids	Apr 21-May 12	May 4/5	3 days	10-40	65	NE-SW (0500-0700), E-W (0600-0900), SE-NW (0900-1100)
Arietids	May 29-Jun 19	Jun 7	?	60	37	N-S (0600-0700 and 1300-1400)
Perseids	Jul 23-Aug 20	Aug 12	4.6 days	50-100	59	NE-SW (0900-1100), SE-NW (0100-0300)
Orionids	Oct 2-Nov 7	Oct 22	2 days	10-70	66	NE-SW (0100-0300), N-S (0100-0200 and 0700-0900), NW-SE (0700-0800)
Geminids	Dec 4-16	Dec 13/14	2.6 days	50-80	34	N-S (2200-2400 and 0500-0700)

moving curtain of ionization. They sound like they are being transmitted by a leaking high-pressure steam vent rather than radio. SSB voice signals are so badly distorted that often you cannot understand them unless the speaker talks very slowly and clearly. The amount of distortion increases with frequency. Most 50-MHz Au contacts are made on SSB, where distortion is the least. On 144, 222 and 432 MHz, CW is the only really useful means of communicating via Au.

If you suspect Au, tune to the CW calling frequency (144.110) or the SSB calling frequency (144.200) and listen with the antenna to the north. Maybe you'll hear some signals. Try swinging the antenna as much as 45° either side of due north to peak signals. In general, the longest-distance DX stations peak the farthest away from due north. Also, it is possible to work stations far south of you by using the aurora; in that case your antenna is often pointed north.

High power isn't necessary for aurora, but it helps. Ten-watt stations have made Au QSOs but it takes a lot of perseverance. Increasing your power to 100 W will greatly improve your chances of making Au QSOs. As with most short-lived DX openings, it pays to keep transmissions brief.

Aurora openings may last only a few minutes or they may last many hours, and the opening may return the next night, too. If WWV indicates a geomagnetic storm, begin listening on 2 meters in the late afternoon. Many spectacular Au openings begin before sunset and continue all evening. If you get the feeling that the Au has faded away, don't give up too soon. Aurora has a habit of dying and then returning several times, often around midnight. If you experience a terrific Au opening, look for an encore performance about 27 to 28 days later, because of the rotation of the sun.

Auroral-E propagation is nearly a nightly occurrence in the auroral zone, but it may accompany auroras at lower latitudes. Do not be surprised to hear rough-sounding auroral signals on 6 meters slowly turn clear and strong! Auroral E is probably more common after local midnight after any evening when there has been an aurora. Auroral E propagation can last until nearly dawn. Six-meter contacts up to 2500 miles across northern latitudes (such as Maine to Yukon Territory) are not uncommon. Two-meter auroral E may also be commonplace in the arctic, but it is very rarely observed south of Canada.

Propagation Indicators

Active VHFers keep a careful eye on various propagation indicators to tell if the VHF bands will be open. The kind of indicator you monitor is related to the expected propagation. For example, during the summer it pays to watch closely for sporadic E because openings may be very brief. If your area only gets one or two per year on 2 meters, you sure don't want to miss them!

Many forms of VHF/UHF propagation develop first at low frequency and then move upward to include the higher bands. Aurora is a good example. Usually it is heard first on 10 meters, then 6 meters, then 2 meters. Depending on your location and the intensity of the aurora, the time delay between hearing it on 6 meters and its appearance on 2 meters may be only a few minutes, to as much as an hour, later. Still, it shows up first at low frequency. The same is true of sporadic E; it will be noticed first on 10 meters, then 6, then 2.

Don Twombly, WB1FKF, setting up his 10-GHz station on the top of Mt Wachusett (Massachusetts).

Tropospheric propagation, particularly tropo ducting, acts in just the reverse manner. Inversions and ducts form at higher frequencies first. As the inversion layer grows in thickness, it refracts lower and lower frequencies. Even the most avid VHF operator may not notice this sequence because of lower activity levels on the higher bands. It may be true that ducting affects the higher bands before the lower ones, but it is more likely to be noticed first on 2 meters, where many more stations are normally on the air. Six-meter tropo openings are very rare because few inversion layers ever develop sufficient thickness to enhance such long-wavelength signals.

How can you monitor for band openings? The best way is to take advantage of commercial TV and FM broadcast stations, which serve admirably as propagation beacons. Television Channels 2 through 6 (54-88 MHz) are great for catching sporadic E. As the E opening develops, you will see one or more stations appear on each channel. First, you may see Channel 2 get cluttered with stations 1000 miles or more away, then the higher channels may follow. If you see strong DX stations on Channel 6, better get the 2-meter rig warmed up. Channel 7 is at 175 MHz, so you rarely see any sporadic E there or on any higher channels. If you do, however, it means that a major 2-meter opening is in progress!

The gap between TV Channel 6 and 7 is occupied partly by the FM broadcast band (88-108 MHz). Monitoring that spectrum will give you a similar feel for propagation conditions.

Several amateurs have built converters to monitor TV video carrier frequencies. A system like this can be used to keep track of meteor showers. It's a way of checking to tell if meteors are plentiful or not, even if there appears to be little activity on the amateur VHF bands. It also can alert you to aurora and sporadic E. A variety of systems for monitoring propagation have been discussed by Joe Reisert, W1JR, in the June 1984 issue of *Ham Radio*.

EME: Earth-Moon-Earth

EME, or moonbounce, is available any time the Moon is in a favorable position. Fortunately, the Moon's position may be easily calculated in advance, so you always know when this form of DX will be ready. It's not actually that simple, of course, because the Earth's geomagnetic field can play havoc with EME signals as they leave the Earth's atmosphere and as they return. Not only can absorption (path loss) vary, but the polarization of the radio waves can rotate, causing abnormally high path losses at some times. Still, most EME activity is predictable.

You will find many more signals off the Moon during times when the Moon is nearest the Earth (perigee), when it is overhead at relatively high northern latitudes (positive declination), and when the Moon is nearly at full phase. The one weekend per month which has the best combination of these three factors is informally designated the activity or skeds weekend, and most EMEers will be on the air then. This is particularly true for 432 EME; 144-MHz EME is active during the week and on non-skeds weekends as well. See accompanying sidebar on EME operating practices.

Hilltopping and Portable Operation

One of the nice things about the VHF/UHF bands is that antennas are relatively small, and station equipment can be packed up and easily transported. Portable operation, commonly called hilltopping or mountaintopping, is a favorite activity for many amateurs. This is especially true during VHF and UHF contests, where a station can be very popular by being located in a rare grid. If you are on a hilltop or mountaintop as well, you will have a very competitive signal. See accompanying sidebar for further information.

Dick Bremer, WB6DNX, sets up a demonstration of his 2304-MHz gear.

Hilltopping is fun and exciting because hills elevate your antenna far above surrounding terrain and therefore your VHF/UHF range is greatly extended. If you live in a low-lying area such as a valley, a drive up to the top of a nearby hill or mountain will have the same effect as buying a new tower and antenna, a high-power amplifier and a preamplifier, all in one!

The popularity of hilltopping has grown as equipment has become more portable. The box and brick (compact multimode VHF rig and solid-state 100 to 200-W amplifier) combination is ideally suited for mobile and portable operation. You need no other power source than a car battery, and even with a simple antenna your signals will be outstanding.

Many VHF/UHFers drive to the top of hills or mountains and set up their station. A hilltop park, rest area or farmer's field are equally good sites, so long as they are clear of trees and obstructions. You should watch out for high-power FM or TV broadcasters who may also be taking advantage of the hill's good location; their powerful signals may cause inter-modulation problems in your receiver.

Antennas, on a couple of 10-foot mast sections, may be turned by hand as the operator sits in the passenger seat of the car. A few hours of operating like this can be wonderfully enjoyable and can net you a lot of good VHF/UHF DX. In fact, some VHF enthusiasts have very modest home stations but rather elaborate hilltopping stations. When they notice that band conditions are improving, they hop in the car and head for the hills. There, they have a really excellent site and can make many more QSOs.

In some places, there are no roads to the tops of hills or mountains where you might wish to operate. In this case, it is a simple (but sometimes strenuous) affair to hike to the top, carrying the car battery, rig and antenna. Many hilltoppers have had great fun by setting up on the top of a fire tower on a hilltop, relying on a battery for power.

Some VHF/UHFers, especially contesters, like to take the entire station, high power amplifiers and all, to hilltops for extended operation. They may stay there for several days, camping out and DXing. Probably the most outstanding example of this kind of operation is put on regularly by the group at W2SZ/1. Dozens of operators and helpers assemble this multiband station, often with moonbounce capability, and operate major VHF and UHF contests from Mount Greylock in Western Massachusetts. Their winning scores in virtually every VHF contest they have entered testifies to their skill and the effectiveness of hilltopping!

Contests

The greatest amount of activity on the VHF/UHF bands occurs during contests. VHF/UHF contests are scheduled for some of the best propagation dates during the year. Not only are propagation conditions generally good, but activity is always very high. Many stations come out of the woodwork just for the contest, and many individuals and groups go hilltopping to rare states or grids.

A VHF/UHF contest is a challenging but friendly battle between you, your station, other contesters, propagation and Murphy's Law. Your score is determined by a combination of skill and luck. It is not always the biggest or loudest station that scores well. The ability to listen, switch bands quickly and to take advantage of rapidly changing propagation conditions is more important than brute strength.

There are quite a few contests for the world above 50 MHz. Some are for all the VHF and UHF bands, while others are for one band only. Some run for entire weekends and others for only a few hours. Despite their differences, all contests share a basic similarity. In most North American contests, your score is determined by the number of contacts (or more precisely, the number of QSO points) you make, times some multiplier, which may be ARRL DXCC countries, grid locators, or some other factor. In most of the current major contests, you keep track of QSOs and multipliers

Moonbounce Operating Practices

After traveling 400,000 km, bouncing off a poorly reflective Moon, and returning 400,000 km, EME signals are quite weak. A large antenna, high transmitter power, low-noise preamplifier and very careful listening are all essential for EME. Nevertheless, hundreds of amateurs have made EME contacts, and their numbers are growing.

The most popular band for EME is 144 MHz, followed by 432 MHz. Other bands with regular EME activity are 1296 and 2304/2320 MHz.

Moonbounce QSO procedure is different on 144 and 432 MHz. On 144 MHz, schedule transmissions are 2 minutes long, whereas on 432 they are 2½ minutes long. In addition, the meaning of signal reports is different on the two bands. The difference in procedure is somewhat confusing to newcomers. Most EMEers operate on only one band, so they grow accustomed to the procedure on that band pretty quickly.

Signal Report and Meaning

	144 MHz	*432 MHz*
T	Signal just detectable	Portions of calls copyable
M	Portions of calls copyable	Complete calls copied
O	Both calls fully copied	Good signal, easily copied
R	Both calls, and O signal report copied	Calls and report copied

What does this difference in reporting mean? Well, the biggest difference is that M reports aren't good enough for a valid QSO on 2 meters but they are good enough on 432. So long as everyone understands the system, then there is no confusion. On both bands, if signals are really good, then normal RST reports are exchanged.

The majority of EME QSOs are made without any prearranged schedules. Almost always, there is no rigid transmitting time-slot sequencing. Just as in CW QSOs on HF, you transmit when the other station turns it over to you. This is particularly true during EME contests, where time-slot transmissions slow down the exchange of information. Why take 10-15 minutes when two or three will do?

On the other hand, many QSOs on EME are made with the assistance of skeds. This is especially the case for newcomers and rare DX stations.

On 144 MHz, each station transmits for 2 minutes, then listens as the other station transmits. Which 2-minute sequence you transmit in is agreed to in advance. The common terminology is even and odd. The even station transmits the 2nd, 4th, 6th, 8th. . . 2-minute segment of the hour, while the odd station transmits the 1st, 3rd, 5th. . . Therefore, 0030-0032 is an even slot, even if the sked begins on the half-hour. In most cases, the westernmost station takes the odd sequence. An operating chart, similar to many which have been published, will help you keep track.

Transmitting Sequence

Odd (eastern)	*Even (western)*
00-02	02-04
04-06	06-08
08-10	10-12
12-14	14-16
16-18	18-20
20-22	22-24
24-26	26-28
28-30	30-32
32-34	34-36
36-38	38-40
40-42	42-44
44-46	46-48
48-50	50-52
52-54	54-56
56-58	58-00

During the schedule, at the point when you've copied portions of both calls, the last 30 seconds of your 2-minute sequence is reserved for signal reports; otherwise, call sets are transmitted for the full 2 minutes.

On 432 MHz, sequences are longer. Each transmitting slot is 2½ minutes. You either transmit first or second. Naturally, first means that you transmit for the first 2½ minutes of each 5 minutes.

144-MHz Procedure—2-Min Sequence

Period	*1½ minutes*	*30 seconds*
1	Calls (W6XXX DE W1XXX)	Continue calls
2	W1XXXDE W6XXX	T T T T
3	W6XXX DE W1XXX	O O O O
4	RO RO RO RO	DE W1XXX K
5	R R R R R	DE W6XXX K
6	QRZ? EME	DE W1XXX K

432-MHz Procedure—2½-Min Sequence

Period	*2 minutes*	*30 seconds*
1	VE7BBG DE K2UYH	Continue calls
2	K2UYH DE VE7BBG	Continue calls
3	VE7BBG DE K2UYH	T T T
4	K2UYH DE VE7BBG	M M M
5	RM RM RM RM	DE K2UYH K
6	R R R R R	DE VE7BBG SK

On 222 MHz, some stations use the 144-MHz procedure and some use the 432-MHz procedure. Which one is used is determined in advance. On 1296 and above, the 432 procedure is always used.

EME operation generally takes place in the lowest parts of the VHF bands: 144.000-144.070; 222.000-222.025; 432.000-432.070; 1296.000-1296.050. Terrestrial QSOs are strongly discouraged in these portions of the bands. Activity on 10 GHz EME, which requires only a few watts when used with dish antennas 10 feet or more in diameter, is becoming increasingly popular.

VHF Mountaintopping

Mountaintopping and hilltopping (not quite so ambitious) have been popular activities as old as VHF operating itself. Few are blessed with a home operating location that facilitates a total command of the frequency. Consequently, ardent VHFers construct bigger and bigger arrays and amplifiers for the home station in order to produce that booming signal. Some, however, utilize the great equalizer to compete on an equal or superior footing with the big home stations-namely, a mountaintop location. Here, perched high above all the home stations, a small portable rig with a single Yagi antenna only a few feet off the ground suddenly sounds like a kilowatt feeding a killer antenna installed at home. Simple equipment performs amazingly well from a mountaintop QTH on VHF.

A mountaintop expedition can vary from a spur-of-the-moment Sunday afternoon picnic to a full-fledged weekend contest. Quick trips can also be conducted during band openings. Since a contest optimizes the opportunity to work a lot of stations on VHF, this sidebar is mostly a how-to for weekend contest operation conducted by one or two people. But this can be scaled down to a mountaintop stay of shorter duration.

The Times Are a Changin'

Old-timers will remember the drudgery of lugging boat anchors up rocky crevices. Dragging equipment and generators weighing hundreds of pounds up steep mountainsides was no picnic. Those who suffered sprained backs soon gave it up. The advent of solid-state equipment has made mountaintopping a far less strenuous activity. Even some of the highly competitive HF types have found new worlds to conquer above 50 MHz. A key factor in this revival was the introduction of a worldwide grid locator system in 1983. The use of grid locators in the major VHF contests has tickled the innermost secret desire of every radio amateur—to be on the receiving end of a DX pileup. Now instead of going on safari to some distant DX land, you can head for the mountains—some nearby mountaintop located in a rare grid.

Choosing a Site

Choosing a mountaintop site involves considering how far you want to travel to get there, accessibility to the top of the mountain and its all-important grid locator. Ideally, your mountain is only a short driving distance away, towers into the cirrusphere, has a six-lane interstate to the top and rises within a grid that has never before been on the air!

Obviously, some of these considerations may have to be compromised. Your first step to finding VHF heaven involves extensive study of a road atlas. How far do you want to travel? Where are the mountains? How high are they? Can you drive to the top? Draw in the grid line boundaries so you can tell which square it is in. If you're not sure, ask some active VHFers which are the difficult grids to work. When you start zeroing in on a potential site, you may want to get a topographic survey map of the area to determine access roads and direction of drop-off from the summit.

Never operate from a mountaintop without first scoping it out in person. Access is most important. Thus far, I've operated from sites which I have reached by car, ferry, gondola, 4-wheel drive, motor home and hiking. Unless you are going with mini-radios and gel-cells, you want to get there without backpacking it. A passable road to the top is ideal. When checking out a 2-meter site, bring a compass and 2-meter FM hand-held. A call on 146.52-MHz simplex should tell you how good the location is. Are you blocked in any direction? Is it already RF-city with commercial installations—a potential source of interference? Will you be able to clear any trees with a lightweight mast? Then the prime requisite: Is there a picnic table permanently at the site? If not, plan on bringing an operating table and chair, even though they add considerable bulk and weight to transport.

Once you've selected an operating site, be sure you have secured the necessary permission to use the site. This may simply require verbal permission from some authority or the owner, or, it could involve a lengthy exchange of correspondence with a state environmental agency of forests and parks and the signing of a liability release. But be sure you have permission. The last thing you want is the local sheriff shining a flashlight in your eyes at 3 AM while you are enjoying a fantastic tropo opening on 2 meters. You'll find rangers on fire watch most helpful in pointing out how to obtain necessary permission.

Power Source

Unless you have the extreme luck to find a location that will permit you to just plug in, make plans for providing your own power. With a single-band operation from a car (with antenna mast mounted just outside the car window), the car battery will probably suffice. Run a set of heavy-duty jumper cables directly to the car battery. Even a solid-state brick amplifier can be run off the car battery without ill effects. Just in case, park the car facing downhill!

For a more serious effort involving several VHF and UHF bands, a generator is recommended. Unless you are running high-power amplifiers, a small generator is all you should need. Attractive generators in the 500 to 1000-W category that look more like American Tourister luggage are now available. Mine is a 650-W beauty that weighs in at 43 pounds, and runs for four hours on a half gallon of petrol. And quiet? You can hear the wings of a Monarch butterfly flutter at 50 paces. A 5-gallon gasoline can provides more than enough flammable juice for a contest weekend.

Equipment

A mountaintop location effectively places your antenna atop a natural tower of hundreds, or perhaps, thousands of feet. With this height advantage, compact, lightweight, low-powered radios that can be boosted up to the 50- to 100-W range with solid-state amplifiers will perform nicely. Low-powered portable transceivers are manufactured for just this purpose. The popular 10 and 25-W multimode rigs are also quite adequate. Many discontinued models can be obtained at a substantial savings through the Ham-Ads section of *QST*. Use of transverters operating with mobile-type HF radios should also be considered.

If you don't have any sizable trees to get over, you can use simple mast sections that fit together. I use 5-foot sections available at the popular shack of radios. They are easily transportable. Important too is the method of antenna rotation. If you can install the antenna mast right next to the operating position, do it. It will save all the hassle of installing motorized rotators. Nothing beats the Armstrong method for speed and simplicity. I use a cross-piece of aluminum tubing mounted with U-bolts to the mast at arm level. See **Fig 4-3**. This provides instantaneous antenna-peaking capability—a necessity on VHF/UHF. While home stations are twirling their antennas in every direction trying to peak a weak signal, I've already worked him!

Installing antennas for several bands on the same mast is recommended. They should be oriented in the same direction. Many contacts on UHF are the result of moving stations over from other bands. For example, in a contest if you move a multiplier to 432 MHz after first working on 2 meters, and both antennas are on the same mast, you will first want to peak the signal on 144 MHz. Then, when you move to 432 MHz where the antennas are probably a bit more sharp and propagation perhaps marginal, both antennas will be pointing at each other for maximum signal. This can make the difference in whether the contact is made.

If your mountaintop operation involves staying overnight, additional attention must be paid to having the proper survival equipment. The most luxurious way to go is a van or RV. Otherwise, a tent may be required. For the rugged outdoors type, this can be as appealing as the

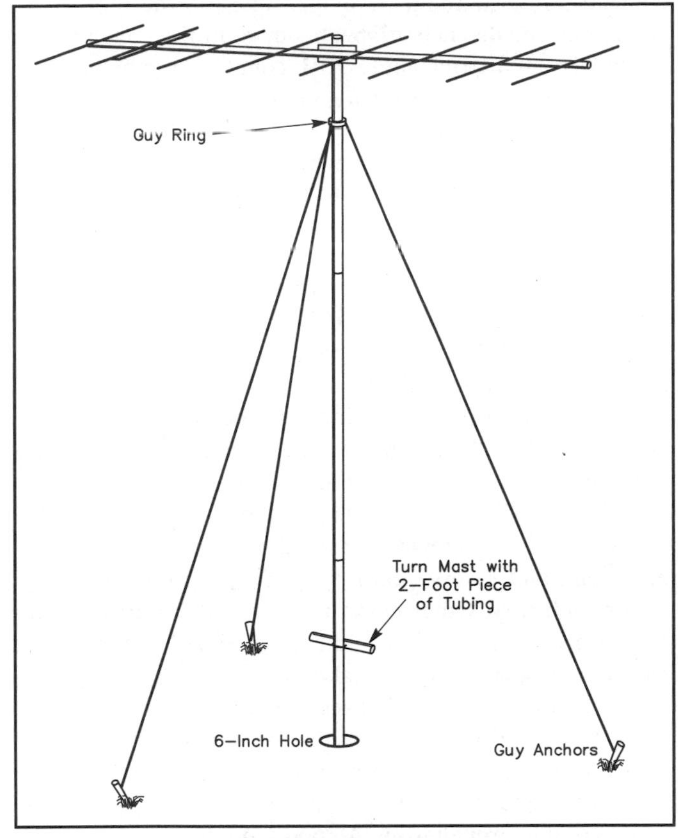

Fig 4-3—A portable mast, assembled from interlocking sections of aluminum tubing or commercial mast, can raise one or two small Yagis to a height of 25 feet. The mast is rotated by hand.

radio part. I find that cooking steaks over a campfire with a canopy of stars overhead (while a programmable keyer is calling CQ) is half the fun. But keep this aspect of the operation also simple as possible. Champagne and caviar can be held for another time. I've also found out the hard way that one can expect heavy winds on mountaintops. Large tents blow down easily in such weather.

Just because you're hilltopping in July or August, don't expect it will always be T-shirt and shorts weather. No matter what the season, expect to need a heavy jacket after dark. I always bring a heavy flannel shirt and ski jacket for night, and shorts in the daytime. And bring lightweight raingear, just in case. In the western mountains or during the winter anywhere, it can get cold and snow is always a possibility. And depending on the habitat, don't be surprised to be introduced to a critter or two, especially after dark!

Getting Started

Okay, you've read this far and are beginning to say to yourself: "Self, I think I'd like to try that." But there is a little voice of caution in you that says: "Don't go bonkers until you've sampled a little first." Good advice!

Start out by setting up on an easily accessible mountain for an afternoon during a contest period on a single band. For the first effort, I recommend 2 meters. With so many 10-W multimode rigs out there in radioland, 2 meters is your bread and butter band. Using a multi-element Yagi a few feet off the ground of a strategically located mountain or hill can whet your appetite. I first got hooked by operating from the side of a highway on Hogback Mountain, Vermont, with a 3-W portable 2-meter radio to a 30-W brick and 11-element Yagi. I was astounded by the results, with contacts hundreds of miles away. This launched my interest in acquiring more equipment for portable mountaintop use, each operation adding a new band or better antenna. The basic formula of keeping it lightweight and simple has prevailed, however.

Now what's keeping you from operating from atop that mountain?—*John F. Lindholm, W1XX*

by band. In other words, you can work the same station on each band for separate QSO and multiplier credit.

Listed below are the major VHF through microwave contests in North America. Other contests are popular in Europe and Asia, but these are not listed here because information about them is not usually available to most of us. All the listed contests are open to all licensed amateurs, regardless of their affiliation with any club or organization. Detailed rules for each contest are published in *QST*, *CQ* and the newsletters of the sponsoring organizations. For more information about contesting, see the Contesting chapter of this book.

1) *ARRL VHF Sweepstakes*: This one-weekend contest occurs in January. It is favored by clubs because it permits club members to pool their individual scores for the club's total. In addition, individuals and multioperator groups compete. Scores are determined by total QSO points per band multiplied by the number of grid squares worked per band. Activity is quite high during this midwinter contest, even though propagation conditions are often poor.

2) *ARRL June VHF QSO Party*: This contest is the highlight of the contest season for most VHFers. Scheduled for the second full weekend of June, the June Contest is often the most exciting of the major contests. Conditions on all the VHF bands are usually good, with 6 meters leading the way. This contest covers all the VHF and UHF bands, and QSOs and multipliers are accumulated for each band. Score is determined by multiplying QSO points per band by grid locators per band. Single band and multiband awards are given to high-scoring individuals in each ARRL section. In addition, multioperator groups compete against each other, and the competition can be fierce!

3) *CQ Worldwide VHF Contest*: This contest is international in scope and is usually held over a July weekend. Score is the number of different grid locators worked per band times the number of QSO points per band. A multitude of awards are available.

4) *ARRL UHF Contest*: As its name suggests, this contest is restricted to the UHF bands (plus 222 MHz). It takes place over a full weekend in August. All UHF bands are permitted, and grids are the multiplier. Less equipment is involved for this contest than for contests which cover all VHF/UHF bands, so many groups go hilltopping for the UHF Contest.

Branimir Antolic, 9A9B, puts Croatia on the moon—actually on the air via EME—with this lofty antenna array.

5) *ARRL September VHF QSO Party*: The rules for this contest are identical to those for the June VHF QSO Party. This contest is also very popular, and many multioperator groups travel to rare states and grids for it. By the second weekend of September, most 6-meter sporadic E has disappeared, but tropo conditions are often extremely good. Therefore, the bands above 144 MHz are the scene of tremendous activity during the September contest. Some of the best tropo openings of recent decades have taken place during this contest, and they are made even better by the high level of activity.

6) *ARRL 10-GHz and Up Cumulative Contest*: Yes, there is a contest for those hardy souls who inhabit the microwave world of 10 GHz and higher. It takes place over two late summer weekends. Operators who change their locations during the contest may work the same stations over again. Scoring is the cumulative total of QSO points (at 100 points per contact) and distance of each contact in kilometers, for each band used. This is great fun, as nearly all participants head for their favorite hilltops and mountain-tops for maximum range.

7) *ARRL EME Contest*: This contest is devoted to moonbounce. It takes place over two full weekends spaced almost a month apart, usually in the fall. The date of the contest is different each year because of the variable phase of the Moon. The dates are usually chosen by active moonbouncers to coincide with the best combination of high lunar declination, perigee and the full phase of the Moon. This contest is international in scope. Score is the number of QSO points made via EME per band, multiplied by the number of US call districts and ARRL countries per band. Hundreds of EMEers participate in this challenging test of moonbounce capability.

Contests are lots of fun, whether you're actively competing or not. You don't have to be a full-time competitor to participate or to enjoy yourself! Most participants are not really competing in the contest, but they get on the air to pass out points, to have fun for awhile, and to listen for rare DX. Others try their hardest for the entire contest, keep track of their score and send their logs in for awards. Either way, the contest is a fun challenge.

If you don't plan on being a serious competitor, then you just need to know the exchange. The exchange is the minimum information that must be passed between each station to validly count the QSO in the contest. In all of the terrestrial VHF contests sponsored by the ARRL you need only send and receive calls and grids, with acknowledgments each way. Other contests require some different information, such as serial numbers or signal reports, so it pays to check the rules to make sure.

A serious contest effort requires dedication and effort, as well as a station that can withstand a real workout. Contests are a challenge to operators and equipment alike. A good contest score is the result of hard work, a good station and favorable propagation. A good score is something to be proud of, especially if there is lots of stiff competition. And with the popularity of VHF and UHF contests these days, competition is always stiff!

Serious contesting requires using a computer logging program. Several are available through vendors who advertise in *QST*, *NCJ* and other journals. Laptop computers are favored because they tend to produce less interference to the VHF bands and they are easier to take on portable operations. You can submit your log electronically via e-mail or on a disk. Check the current contest rules for exact requirements.

Mark Mandelkern, K5AM, operates on the VHF bands with an entirely home-brew station from Las Cruces, New Mexico. The amplifiers run 1500 W, and the transceiver has special built-in features for weak-signal VHF reception.

Many operators don't feel that their stations are competitive on all the VHF and UHF bands. This is no problem. You can contest on one, two, or as many bands as you like. Indeed, some contesters specialize. Most contest results recognize achievements on each band separately. This has the advantage of concentrating your efforts where your station is the strongest, allowing you to devote full time to just one band. You don't need to be high powered to compete as a single-band entry. Location makes a lot of difference, and hilltopping single-banders have had tremendous success, particularly if they go to a rare grid square.

Other operators like to try for all-band competition. In this case, it's a real advantage to be able to hop from one band to another. You can quickly check 6 meters for activity while also tuning 2 meters. Or, if you work a rare grid on one band, you can take advantage of the opportunity by asking the other station to switch bands right then. Some contesters work one station on all possible bands within two minutes by band hopping! This is a speedy way to increase your grids and QSOs.

There are several other entry classes for the major VHF contests and variations on these classes in other events. (Check the rules for each contest, as entry classes vary somewhat.) Multioperator entries involve more than one operator. Many large multioperator efforts have operating positions for a dozen or more separate bands. Now that is an effort to put together! If your group is not ready for the big time, limited multioperator entries submit scores for four bands only.

Cowles Andrus, KB4CNI, sets his sights on the moon with this antenna array.

Chris Hazlitt, KL7FB, is the proud owner of this impressive moonbounce antenna array near Palmer, Alaska. Look closely and you'll see Chris standing on the tower.

In either case, successful multiop stations have learned to work cooperatively. The ability to operate simultaneously on several bands without mutual interference is important. Multiop efforts also have a foolproof system for passing messages from one operating position to another, especially referrals of a station from one band to another. If the operator of the 2-meter station in a multioperator entry works WØVD, for example, that operator may suggest immediately moving to 222 MHz and then 432 MHz. A message has to be sent to the operators at the 222 and 432 MHz positions to expect a call from WØVD at a particular frequency and time. Handwritten notes are simple and convenient if the operating positions are close together. Several logging programs have sophisticated features to pass messages among several computers that are networked together.

The rover class allows a station with one or two operators to move among two or more grid locators during the contest. Rovers may contact the same stations all over again from a different grid. Most rovers operate from a car, van or truck specially equipped for convenience and efficiency. Hilltops in rare grids are favored rover sites. Rovers must move the entire station, antennas and power supply to qualify. Special scoring rules apply to rovers.

Other popular entry classes appeal to single operators who contest with modest stations. The limited single operator does not limit the number of bands you may operate, but does place maximum power limits on each band. The single-operator QRP class limits power output on all bands to a maximum of 10 W. Stations must run on a portable power source and may not be located at fixed or home locations.

SOURCES OF INFORMATION

Many VHF/UHF operators like to keep abreast of the latest happenings on the bands such as new DX records, band openings or new designs for equipment. There are many excellent sources for current information about VHF/UHF. In addition, they provide a way to share ideas and ask questions, and for newcomers to become familiar with operation above 50 MHz.

The main sources are nets, newsletters and published

Bernhard Dobler, DJ5MN, obviously doesn't have to deal with deed restrictions or zoning ordinances at his home.

columns. Each has its own use and appeal; active VHFers usually seek out at least one of them.

Nets

Several nets meet regularly on the HF and VHF bands so that VHF/UHFers can chat with each other. These are listed below. It's a good idea to listen first, before checking in the first time, so you'll know the format of the net. Some nets like to get urgent news, scheduling information and other hot topics out of the way early, and save questions and discussion until later. Others are more free-form. You can learn quite a bit just by tuning in to these nets for a few weeks. Regular participants in the nets are often very knowledgeable and experienced. The technical discussions which sometimes take place can be very informative.

1) *Nets on 75 meters*. The Central States VHF Society sponsors a net on 3.818 MHz each Sunday evening (0230 UTC). It is open to all those interested in VHF and higher. The net provides a terrific source of technical information and is a good place to arrange for schedules and experimental tests.

A similar informal VHF net meets on Monday evenings (0200 UTC Tuesday, 0100 UTC during the summers) on 3.843 MHz. Both frequencies are also used for spontaneous nets, especially during contests, meteor showers and other unusual VHF propagation events.

2) *VHF Nets*. There are dozens of weekly nets that operate on 144 MHz around the country. Many 2-meter nets are sponsored by the Sidewinders on Two (SWOT) Amateur Radio Club. For current information on these nets, contact Howard Hallman, WD5DJT (wd5djt@swbell.net), or check his Web page at: **http://www.home.swbell.net/wd5djt**.

Other nets on 50 through 1296 MHz are sponsored by regional clubs, including the Northeast Weak Signal Society (New England), Mt Airy VHF Society (eastern Pennsylvania) and Western States Weak Signal Society (California). Other nets are run by local groups. Check locally for dates, times and frequencies.

3) *EME Nets*. International EME operators meet on 14.345 MHz each Sunday to arrange EME schedules and pass information. At 1600 UTC, the 432 MHz and higher EME crowd occupies the frequency. Starting at 1700 UTC, the more numerous 144-MHz EME operators gather. The 2-meter EME net often goes for more than an hour.

4) *International 6-Meter Liaison*. Avid 6-meter DX operators around the world monitor 28.885 MHz whenever there is a chance for intercontinental DX. There is no formal net, but it is a place to exchange information (especially on current propagation conditions), arrange schedules, and pass other information related to 6-meter operating.

Awards

Several awards sponsored by ARRL help stir activity on the VHF and higher bands. (See the Operating Awards chapter.) Many of those originally designed for HF operators, including Worked All States (WAS), Worked All Continents (WAC) and DX Century Club (DXCC) are also coveted by operators on 6 meters and higher. The VHF/UHF Century Club (VUCC) is designed only for the world above 50 MHz.

WAS requires contacts with all 50 United States. Like other ARRL awards, QSL cards are required. The quest for this award has probably been responsible for much of the technical advancement of VHF/UHFers during the past three decades. Moonbounce activity has benefited most directly because WAS on any band above 144 MHz requires EME capability. Even so, more than a dozen stations in the central part of the US have worked 48 states without resorting to moonbounce.

More than 1000 6-meter operators have earned WAS. During the peak of the most recent sunspot cycle, F_2 openings made transcontinental contacts common for many months. Amateurs in the continental US worked Alaska, Hawaii and tons of DX during that time. The solar cycle peaks at the turn of the century, so many more operators may get their chance to work elusive states. In quiet years, multi-hop E_s openings occur each summer to provide the slim chance of WAS on 6 meters.

Using portable moonbounce stations, several enterprising groups have mounted EME-DXpeditions to rare states. This has allowed quite a few hard working VHF/UHFers to complete WAS, even when there was no resident EME activity in some states. At last count, over 100 stations had received WAS on 2 meters and more than 10 on 70 cm.

Many 6-meter operators have also worked all continents with the assistance of F_2 propagation and many more will achieve WAC during solar cycle 23. WAC is not impossible on the higher bands, as many 144 and 432 EME stations have accomplished this feat. WAC on the higher bands depends on EME activity from under-represented continents.

Several top-scoring stations in the annual EME contests have made QSOs on all continents during a single weekend. The rarest continent is probably South America, where only a small handful of EMEers are active. WAC is not possible on 222 or 902 MHz because these bands are not authorized for amateur use outside of ITU Region 2 (North and South America).

Who says portable operators have to rough it? John Hawley, N2PBY, and Al Tencza, NX2Q, show us a better way.

DXCC is also within the capability of the VHF crowd. More than 200 DXCC awards have already been earned by 6-meter operators from around the world—unthinkable just 20 years ago. Even more astonishing are the dozen 2-meter EME operators with DXCC certificates hanging on their walls! DXCC has yet to be claimed on any higher band, but there is no technical reason why it cannot be done. It simply requires more EME activity from other countries.

ARRL breathed new life into the VHF and UHF bands in January 1983 when it announced the VUCC awards. Somewhat later, the required exchange in all terrestrial ARRL-sponsored VHF and UHF contests was changed to include grid locators, and activity in those contests increased as well. You can do a lot of grid-hunting during a good contest!

The VUCC award is based on working and confirming a certain number of grids on the VHF/UHF bands. For 50 and 144 MHz, the number is 100. On 222 and 432 MHz, the number is 50, and on 902 and 1296 MHz the minimum number is 25. 2.3-GHz operators need to work 10 grids, and five grids are required on the higher microwave bands. This is quite a challenge, and qualifying for the VUCC award is a real accomplishment!

VUCC endorsement stickers are available for those who work specified numbers of additional grids above the minimum required for the basic award. Quite a few stations have exceeded 600 grids on 6 meters and 200 on 2 meters, and a few have worked over 100 grids on 432. Impressive!

Problems

With the ever-increasing sharing of the VHF and UHF spectrum by commercial, industrial and private radio services, a certain amount of interference with amateur operation is almost inevitable. Amateurs are well acquainted with interference, and so we normally solve interference problems by ourselves.

Several main types of interference are common. The first is our old friend, television interference (TVI). Fundamental overload, particularly of TV Channels 2 and 3 from 6-meter transmitters, is still common. Some Channel 12 and 13 viewing is bothered by 222-MHz transmissions. Fortunately, as modern TV manufacturers have slowly improved the quality of their sets, the amount of TVI is beginning to decline.

A more common form of TVI is cable television interference (CATVI). It results from many cable systems distributing their signals, inside shielded cables, on frequencies which are allocated to amateurs. As long as the cable remains a closed system, in which all of their signal stays inside the cable and our signals stay out, then everything is usually okay. Despite the design intentions, cable systems are deteriorating because of age and have begun to leak. When a cable system leaks, your perfectly clean and proper VHF signal can get into the cable and cause enormous amounts of mischief. By the same token, the cable company's signals can leak out and interfere with legitimate amateur reception.

Other forms of radio frequency interference (RFI) is a common complaint of owners of unshielded or poorly designed electronic entertainment equipment. Amateur transmissions,

Ron Hranac, NØIVN, making one of his 266 QSOs in a recent ARRL 10-GHz Cumulative Contest. All of Ron's QSOs were on 10 GHz.

especially high power, may be picked up and rectified, causing very annoying problems. RFI may include stereos, videocassette recorders and telephones. The symptoms of RFI usually include muffled noises which coincide with keying or SSB voice peaks, or partial to total disruption of VCR pictures.

A few general principles may help in beginning your search for a solution to TVI/RFI problems:

1) With ordinary TVI, be sure your transmitter is clean, all coax connectors are tightened, and a good dc and RF ground is provided in your shack—before you look elsewhere for the cause of the problem. Then, find out if TVI affects all televisions or just one. If it's just one, then the problem is probably in the set and not your station.

2) With CATVI, remember that it is the cable company's responsibility to keep its system closed and in compliance with FCC rules. Unfortunately, the FCC's limits are loose enough that in some cases there will be interference-causing leakage from the system which is still within FCC limits. In that case, there is no easy solution to the problem. You may be pleasantly surprised by the cooperative attitude of some cable TV operators. (Cable company technicians often have worked overtime trying to solve CATVI complaints, but not everyone is so fortunate.)

3) In dealing with RFI, the main goal is to keep your RF out of the entertainment system. This is often solved rather simply by bypassing the speaker, microphone and power leads with disc ceramic capacitors. In other cases, particularly some telephones, you must also employ RF chokes. For further information on RFI and CATVI, consult *The ARRL RFI Book*, published by ARRL.

Several other types of interference plague VHF/UHF amateurs. These are the receive-only kinds of interference, which just affect your receiving capability. One form has already been mentioned: CATV leakage. For example, cable channel E is distributed on a frequency in the middle of the amateur 2-meter band. If the cable system leaks, you may experience very disruptive interference. Reducing leakage or perhaps eliminating the use of channel E may be satisfactory. This problem has vexed many stalwart VHFers already, and no end is in sight.

Scanner birdies may be a problem in your area. All scanners are superheterodyne-type receivers, which generate local oscillator signals, just as your receiving system does. Unfortunately, many scanners have inadequate shielding between the local oscillator and the scanner's antenna. The result is radiation of the oscillator's signal each time its channel is scanned—a very annoying chirp, swoosh or buzz sound every second or so. If the scanner's local oscillator frequency happens to fall near a frequency you're listening to, you'll hear the scanner instead. Amateurs can easily pick up 10-15 scanner birdies within 5 kHz of the 2-meter calling frequency (for example, a New York State Police frequency is about 10.7 MHz above 144.200!). Very little can be done about this problem aside from installing tuned traps in everyone's scanners—an unattractive prospect to most scanner owners.

Many amateurs who operate 432 MHz have lived with radar interference for years. Radar interference is identified by a very rapid burst of noise which sounds vaguely like ignition noise, repeated on a regular basis. Although some radars are being phased out, many remain. Amateurs are secondary users of the 420-450 MHz band, so we must accept this interference. The only solution may be directional

Jud Snyder, K2CBA, braved the blackflies to operate from Mt Greylock in Western Massachusetts.

antennas which may null out the interference, not a very satisfying alternative in some cases. 432-MHz EMEers sometimes hear radar interference off the Moon!

NEW FRONTIERS

The VHF/UHF spectrum offers amateurs a real opportunity to contribute to their own knowledge as well as the advancement of science in general. We as amateurs are capable of several aspects of personal research which are not possible for the limited resources of most scientific research organizations. Although these organizations are able to conduct costly experiments, they are limited in the time available, geographic coverage and, sometimes, by the *It's impossible* syndrome. Hams, on the air 24 hours a day, scattered across the globe, don't know that some aspects of propagation are impossible, so they occasionally happen for us.

For example, professionals suggest that VHF meteor scatter is limited to frequencies below about 150 MHz, yet hams have made dozens of contacts at 222 and 432 MHz. A spirit of curiosity and a willingness to learn is all that is required to turn VHF/UHF operating into an interesting scientific investigation.

Can amateurs make real contributions to understanding the limits of radio propagation? You bet! The following list includes just a few of the possibilities:

1) *Aurora on 902 or 1296*. Many aurora contacts have been made on 432 MHz; is aurora practical on 902 and 1296 MHz? Is aurora polarization sensitive?

2) *Unusual forms of F-layer ionospheric propagation*—FAI, TE and what else? Little is known about their characteristics, how they relate to overall geomagnetic activity and frequency. Amateurs can discover a lot here.

3) *Multiple-hop sporadic E on 50 MHz*. Single-hop propagation on this band is a daily occurrence during the summer. In recent years, multiple-hop openings have not been uncommon. Now that many European countries are beginning to allow 6-meter amateur operation, we can find out how many hops are actually possible! Several contacts over 6000 miles are well documented. Is there a limit to multiple-hop E_s?

4) *Meteor scatter*. Astronomers are very interested in the orbits of comets and their swarms of debris which give rise to meteor showers. It is often difficult for astronomers to observe meteors (because of clouds, moonlight and so on). Amateurs can make very substantial contributions to the study of meteors by keeping accurate records of meteor-scatter contacts.

5) *Polar-region propagation*. VHF radio propagation across the arctic regions is not well documented. Auroral E is probably a nightly occurrence in the auroral zone and has not been fully exploited. Some interesting paths at 50 and 144 MHz may yet be discovered.

6) *How effective is diversity reception* in different types of VHF/UHF reception? Stations with two or more antennas could investigate.

7) *Newer modes of data transfer*. Packet, AMTOR, PSK31 and the like—what are some of the limits to their use? Who knows how many amateurs may attempt to send packet information via EME.

Chapter 5

HF Digital Communications

Steve Ford, WB8IMY
ARRL Headquarters

Hams have been swapping data on the bands below 30 MHz for longer than you might imagine. From the end of World War II until the early '80s, radio telegraphy, better known as RTTY, was a thriving HF digital mode. In fact, it was the only HF digital mode available to hams at the time.

In the old days of RTTY, setting up a station wasn't a trivial exercise. You had to make room for a bulky mechanical teletype machine, cobble together an interface to your radio and install an oscilloscope to help you tune the signal. You sent text by typing on the teletype's awkward "green keys," and you read the other station's replies on sheets of yellow paper.

Then, in 1983, AMTOR made its debut, coinciding with the rising popularity of personal computers. (Personal computers revolutionized RTTY as well, banishing the clattering teletypes to the pages of history.) AMTOR was the first amateur digital communication mode to offer error-free text transmission.

From the early '80s the pace of change quickened. Packet radio emerged by the middle '80s and, for a time, reigned as the most popular form of amateur digital communication. As microprocessor technology became more sophisticated, we saw the rise of modes such as Clover and PACTOR that were capable of error-free exchanges under marginal band conditions (weak signals, interference and so on). In the late '90s computer soundcards were harnessed to create a new digital mode for casual operating: PSK31.

Let's take a brief look at the state of the amateur HF digital world today...

RTTY

RTTY is still the popular granddaddy of HF digital, although its popularity has been seriously undercut by PSK31. RTTY is primarily used for casual conversation, and it's still the mode of choice for digital contesting and DXing.

This mode uses the five-bit Baudot code. These five bits allow only 32 unique combinations. That's not enough to cover the letters of the alphabet and the digits 0 through 9. You also need punctuation and some control characters. Two of the combinations are used to shift between character sets called "letters" and "figures." This allows the other 30 combinations to have one of two meanings—much like it is on a keyboard, but it does not allow for upper and lower case letters.

PACKET

Although packet technology had been in existence since the early '70s, hams embraced it with gusto in the middle '80s. (Personal computers, again, were the driving force.) Packet is an error-correcting mode, which means that it is capable of communicating error-free information, including binary data (for images, software applications, etc). The problem with packet, as far as HF communication is concerned, is that it requires strong, "quiet" signals at both ends of the path to function efficiently. Packet doesn't tolerate signal fading, noise or interference, which makes it a poor choice for the chaotic world of HF. Despite its meager performance, HF packet remains stubbornly alive, though most of the activity is concentrated in overseas operations in Third

Table 5-1
Popular HF Digital Frequencies

Band (Meters)	*Frequencies (MHz)*
10	28.070–28.130
12	24.920–24.930
15	21.060–21.099
17	18.100–18.110
20	14.060–14.099[1]
30	10.120–10.150
40	7.060–7.099[2]
80	3.580–3.640

Notes
[1]This is the most active HF digital band
[2]Digital operators on this band should avoid interfering with operators in ITU Regions 1 and 3 who have phone privileges in this portion of 40 meters.

World nations where the affordability of packet equipment is still a strong drawing card.

Packet and the other modes that follow in this list make use of the (7-bit) ASCII character set. That means 128 different letters, symbols and control characters.

PACTOR

PACTOR strolled onto the telecommunications stage in 1991. It combined the best aspects of packet (the ability to pass binary data, for example) and the robust error-free nature of AMTOR. It was eagerly embraced by HF digital equipment manufacturers and became the number-one HF digital communication mode in a remarkably short period of time. PACTOR was popular for mailbox operations and other forms of message handling. As with packet, the Internet caused a serious decline in PACTOR activity, but it remains the most popular of the error-free modes.

PACTOR II debuted in the mid '90s as a rival to Clover, and the two have been doing battle for the hearts, minds and pocketbooks of HF communicators (commercial and amateur) ever since. Like Clover, PACTOR II uses DSP techniques and complex data coding to achieve extraordinary performance. Also like Clover, the necessary equipment can be quite expensive, which has slowed PACTOR II's acceptance in the ham community. In 1999 the creators of PACTOR II unveiled a pared-down processor that offered the same performance, but at a more attainable cost.

CLOVER

Clover was unveiled in 1993 by the HAL Communications Corporation. It was one of the first HF digital modes to use sophisticated data coding, coupled with complex modulation schemes and digital processing technology, in an effort to overcome the vagaries of HF. Clover promised and delivered impressive performance even in the face of weak signals and terrible band conditions. This performance initially came at a stiff price—one that few hams could afford. As you'd expect, the high cost of Clover technology dampened enthusiasm in the beginning. Price reductions later in the decade and the introduction of Clover II helped the mode retain a dedicated following.

PSK31

PSK31 could be viewed as a high-octane cousin of RTTY. It is not an error-free digital mode, but it offers excellent weak-signal performance. Peter Martinez, G3PLX, the same person who created AMTOR, invented PSK31. For a few years PSK31 languished in obscurity because special DSP hardware was necessary to use it. But in 1999 Peter designed a version of PSK31 that needed nothing more than a common computer soundcard. It was a simple piece of software that ran under *Windows* and used the soundcard as its interface to the transceiver. Peter made the software available at no cost on the Internet, and that was like tossing a lighted match into a can of gasoline. Within a few months PSK31 exploded in the HF digital community. As this chapter was being written, PSK31 was emerging as a possible successor to RTTY for casual ragchewing, contesting and DXing.

HELLSCHREIBER

Hellschreiber is not a new mode (having been pioneered in the 1920s and '30s by Rudolf Hell), but a number of hams are just beginning to discover its possibilities. Unlike all of the other modes we've discussed so far, Hellschreiber is visual. That is to say, the signals "paint" text on your screen much in the same sense that a television or fax signal paints an image.

One variation of Hellschreiber known as Feld-Hell works its magic by keying a CW transmitter ON for every black portion in a text character, and OFF for every white space. Feld-Hell has drawn some interest among low power (QRP) operators because you can operate with simple (but stable) CW transmitters. Feld-Hell is the most popular of the Hellschreiber modes.

You also can send text imagery by using different frequencies (tones) to represent the black and white areas. This

Don't Overdrive Your Transceiver!

When you're setting up your rig for your first PSK31 transmission, the temptation is to adjust the output settings for "maximum smoke." This can be a serious mistake because overdriving your transceiver in PSK31 can result in a horrendous amount of splatter, which will suddenly make your PSK31 signal much wider than 31 Hz—and make you highly unpopular with operators on adjacent frequencies.

As you increase the transmit audio output from your soundcard or multimode processor, watch the ALC indicator on your transceiver. The ALC is the automatic level control that governs the audio drive level. When you see your ALC display indicating that audio limiting is taking place, you are feeding too much audio to the transceiver. The goal is to achieve the desired RF output with little or no activation of the ALC.

Unfortunately, monitoring the ALC by itself is not always a sure bet. Many radios can be driven to full output without budging the ALC meter. You'd think that it would be smooth sailing from there, but a number of rigs become decidedly nonlinear when asked to provide SSB output beyond a certain level. (sometimes this nonlinearity can begin at the 50% output level.) We can ignore the linearity issue to a certain extent with an SSB voice signal, but not with PSK31 because the immediate result, once again, is splatter.

So how can you tell if your PSK31 signal is really clean? Unless you have the means to monitor your RF output with an oscilloscope, the only way to check your signal is to ask someone to give you an evaluation on the air. The PSK31 programs that use a waterfall audio spectrum display can easily detect "dirty" signals. The splatter appears as rows of lines extending to the right and left of your primary signal. (Overdriven PSK31 signals may also have a harsh, clicking sound.)

If you are told that you are splattering, ask the other station to observe your signal as you slowly decrease the audio level from the soundcard or processor. When you reach the point where the splatter disappears, you're all set. Don't worry if you discover that you can only generate a clean signal at, say, 50 W output. With PSK31 the performance differential between 50 W and 100 W is inconsequential.

version of Hellschreiber is called Multi-Tone Hell, or simply MT-Hell. Hellschreiber has a small but loyal following on the HF bands today.

MFSK16

MFSK16 appeared on the world stage in 1999 thanks to an innovative sound-card program developed by Nino Porcino, IZ8BLY, known as *Stream*. MFSK16 uses multiple tones in an effort to overcome the problems of HF propagation. The signals are easy to recognize by their musical sound.

Like PSK31 and Hellschreiber, MFSK16 is not an error-correcting mode. Even so, MFSK16 boasts of remarkably good performance under marginal conditions. As of 2001, MFSK16 was available only for *Windows* PCs.

COMPUTERS AND TRANSCEIVERS

Thanks to personal computers and microprocessor technology it is relatively easy to assemble an HF digital station these days. You have several choices available to you, depending on the resources you already own and how much you want to spend.

Obviously, you'll need a personal computer. A reasonably fast PC or Mac will do the job. Laptops also make fine HF digital computers, and even some palmtops can be pressed into service.

The next important component is an SSB transceiver. It does not have to be a fancy, feature-filled radio—especially if you are talking about casual HF digital operating. For burst modes such as PACTOR and Clover, however, you'll need a transceiver that can switch from transmit to receive very quickly. (See the sidebar "The Need for Speed.")

DIGITAL TO ANALOG—AND BACK AGAIN

Data exists in your computer in the form of changing voltages. Five volts might represent a binary "1" while zero volts may represent a binary "0." But a radio can't transmit changing voltages—at least not without a little help.

For digital communication our "helper" is a modulator/demodulator, otherwise known as a *modem*. A modem takes the data from your computer and translates it into shifting audio tones. One tone represents a binary "1" and another represents a binary "0." The difference in their frequencies is their *shift*. The HF RTTY standard is a 170-Hz shift, although this has "migrated" to a 200-Hz shift thanks to accommodations made for packet and PACTOR modems. Regardless of the frequency, the rest is easy. Feed the tones to an SSB transceiver and you're in business!

This basic setup is called *AFSK*, or audio frequency-shift keying. Your transceiver manual may refer to it as *FSK*, and this can be a little confusing. True FSK involves changing the frequency of your rig's master oscillator in sync with the data. However, few modern transceivers actually do this. Most simply take the audio tones from your modem and transmit them in the same manner as your voice. The result on the receiving end is the same.

The Need for Speed

When you're operating the burst modes such as PACTOR and Clover, your success will be largely measured by how fast your transceiver can switch from transmit to receive.

For example, a PACTOR station transmits a 960-ms long data block, then waits up to 170 ms for a 120-ms ACK or NAK signal from the receiving station. If the ACK or NAK fails to arrive before this 170-ms window closes, the data block will be repeated. PACTOR's patience is not endless though. If the ACKs or NAKs still fail to arrive after a certain number of repeated data blocks, your processor will declare the link "broken."

With this critical timing in mind, it's important to understand that the time it takes for your SSB transceiver to switch from transmit to receive is deducted from the 170-ms window. This has a substantial effect on how far you can communicate with PACTOR.

Let's say that your rig is a bit on the slow side. It switches from transmit to receive at a lazy 90 ms. Deduct 90 from 170 and you'll see that your window has suddenly shrunk to 80 ms.

Now deduct the delays at the receiving station. Let's be generous and say that his rig can switch within 40 ms. Now your window has collapsed to 40 ms. A radio signal can travel approximately 3700 miles—roundtrip—in 40 ms. This means that the maximum range of your PACTOR station would be roughly 4000 miles.

It is best to shop for an SSB transceiver that has the fastest transmit/receive turnaround. *QST* product reviews often publish this information in their data tables. Look for it and let it guide you to the best choice. When in doubt, the fewer milliseconds, the better!

And what about reception? The modem waiting patiently at the other end of the path is equipped with tone decoders and very sharp audio filters. It will respond only to the proper tones transmitted with the expected shift. Off-frequency tones are ignored, and tones separated by incorrect shifts never make it past the filters. But when the tones hit the targets, the modem instantly converts them into data pulses—which soon wind up in your computer.

What I've just described is a simple, two-tone system. If you're talking about the more advanced digital modes, things can get more complicated. More than one tone may be used, or there may be some legerdemain with phasing or frequency. The essential principle remains the same: Digital to analog (audio), then back again.

Basic Modems

You can do RTTY on the cheap with a modem consisting of a handful of parts. With this type of simple modem you can use shareware programs such as HamComm (See the sidebar "HF Digital Software on the Web") and get on the air in no time.

You can build this modem yourself, or buy it ready to go. There are simple stand-alone modems such as the MFJ-1214 and others that are built directly into DB-25 shells that plug into your computer's COM or LPT (printer port). The Tigertronics BP-2M is a good example of a small

HF Digital Software on the Web

A few words about Web addresses: Every attempt was made to ensure that the following Web addresses were active at the time this edition was published. Web addresses change with astonishing frequency, however, so don't be surprised if a particular address is suddenly "broken." The person who manages the site may have removed it from the Web, or the site may have been moved to another address. When you encounter a faulty address, try using a Web search engine such as Google, AltaVista, Yahoo or Lycos to relocate the site.

BlasterTeletype: RTTY with your sound card. **www.geocities.com/SiliconValley/Heights/4477/**

DigiPan: PSK31 software with easy point-and-click tuning. **members.home.com/hteller/digipan/**

DSP-CW: CW and RTTY with your sound card. **www.zicom.se/dsp/index.html**

HamComm: Allows you to communicate in RTTY, AMTOR and a few other digital modes when used with a simple modem. **www.pervisell.com/ham/hc1.htm**

Hellschreiber: *IZ8BLY Hellschreiber,* **iz8bly.sysonline.it/**

Mix32W: *Windows*-based RTTY and PSK31 with your soundcard. **tav.kiev.ua/~nick/my_ham_soft.htm**

Multimode: RTTY and PSK31 for Macintosh computers. PowerPCs recommended. **www.blackcatsystems.com/software/multimode.php3**

PSK31: For Linux, **aintel.bi.ehu.es/psk31.html**

MFSK16: *Stream,* **iz8bly.sysonline.it/**

PSK31: *WinPSK,* **www.qsl.net/ae4jy/**

multimode modem that offers impressive performance for less than $100. MFJ Enterprises makes a device known as the MFJ-1213 that will put you on RTTY for less than $50.

Soundcards as HF Modems

You may already be the proud owner of an HF modem—your PC sound card! Sound cards do exactly what modems do: Convert data to audio and audio to data. With the proper software running in the background, your PC can become a high-performance digital communication machine.

In recent years PCs have become increasingly powerful and soundcards have become as ubiquitous as hard drives. As a result, there appears to be a growing movement in the ham community toward using soundcards as radio "modems."

Brian Beezley, K6STI, was among the first to use the soundcard as a high-performance modem with his *RITTY* software. Others quickly followed. When PSK31 for *Windows* exploded onto the scene in the late '90s, it was implemented entirely in software using soundcards. The same is now true for the Hellschreiber and MFSK16 modes. The only external hardware you need is a simple switching circuit to key your radio (see **Fig 5-1**).

Of course, there is a catch or two. The first is that your computer must be powerful enough to run the processor-intensive software. The second catch is the soundcard. HF digital software is written to be compatible with a wide variety of soundcards, but not with all soundcards. If you're in doubt, stick with the Creative Labs "SoundBlaster" series. Creative Labs set the soundcard standard years ago and it remains the standard today.

Multimode Processors

Until soundcard software appeared on the market, the most common HF digital interface was the multimode processor, or what some refer to as a multimode TNC. This device offers several HF digital communication modes in a single unit that usually sits alongside your radio. The exception is the HAL Communications P38 that comes in the form of a circuit "card" that fits inside your PC.

Multimode processors offer several HF digital modes, plus a cornucopia of features such as message mailboxes, remote control and much more. The processors are relatively easy to hook up between your computer and your transceiver. The computer uses terminal software to "talk" to the processor. The processor manufacturer usually provides this software, but there are third-party sources as well. Between the processor and the radio you need to run cables for receive audio, transmit/receive switching and transmit audio.

But with so many soundcard-based modes available, what are the advantages of purchasing an outboard processor?

• You can use almost any computer with a multimode processor. The processor merely asks the computer to send and receive information. This is a easy task for even the oldest machines. As long as the computer has terminal software that can communicate with the processor—or if the computer can run the terminal software provided by the manufacturer—you'll be on the air in no time. You could use an "ancient" 286 or 386 PC, a used $200 laptop or whatever.

• Many multimode processors offer features that are not presently available with soundcard-based software. Mailboxes and automatic, stand-alone operations come to mind. Your processor can be set up to receive and process messages without the need for supervision by your computer. You can ignore the processor completely and use your PC to do other work.

• Clover and PACTOR II are available only with

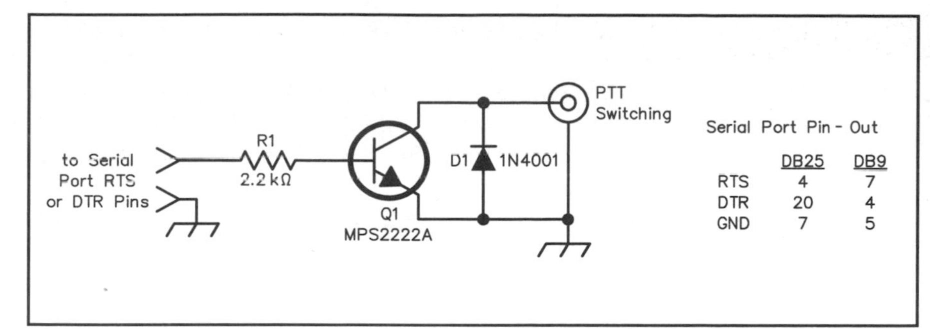

Fig 5-1—This simple circuit will allow your computer's COM port to key your transceiver when you're using soundcard software. You can choose to connect to either the RTS or DTR pins at the COM port. Some software packages key the transceiver by sending a logic "high" to one pin or the other. RadioShack part numbers are shown in parentheses.
D1—1N4001 diode (276-1101).
R1—2.2 kΩ resistor, 1/4 W (271-1325).
Q1—MPS-2222A transistor (276-2009).

multimode processors. If you wish to try these modes, you must purchase a HAL Communications unit (for Clover) or an SCS PTC-IIe (for PACTOR II). All of these products will also operate RTTY, PACTOR and, in the case of the SCS PTC-IIe, PSK31.

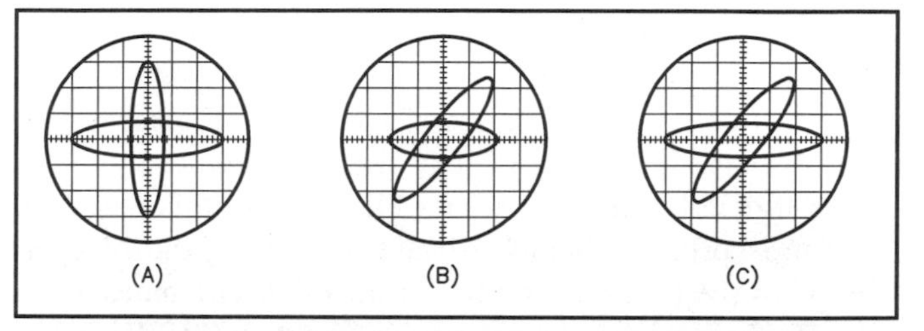

Fig 5-2—This is the classic crossed bananas RTTY tuning display. Pattern A indicates that the signal has been tuned corrected. At B the receiver is slightly off frequency, while C indicates that the transmitting station is using a shift that differs from your processor or modem setting. Although few RTTY operators use oscilloscope tuning today, modern tuning indicators still rely on the same principle. Some even attempt to emulate the crossed bananas display.

CONVERSING WITH RTTY

As with so many aspects of Amateur Radio, I highly recommend that you begin by listening. Tune between 14.080 and 14.099 MHz and listen for the long, continuous *blee-blee-blee-blee* signals of RTTY. (If you hear chirping, it isn't RTTY!)

Make sure your transceiver is set for lower sideband (LSB). That is the RTTY convention. The exception is when you are operating your rig in the "FSK" mode (sometimes labeled "DATA" or "RTTY"). Many hams prefer operating this way because it allows them to use narrow IF filters to screen out interference. When you operate in this fashion, your modem, soundcard or processor is not generating the mark and space tones. They're merely sending data pulses to the radio and the radio is creating its own mark and space signals. (This often requires a special connection to the transceiver. Consult your manual.)

As you tune in the signal, watch your tuning indicator. Tune slowly until you see that the mark and space tones are being decoded.

The old method of tuning was to use an oscilloscope. You tweaked the signal until two ellipses representing the mark and space tones crossed in the middle of the screen. This was known as the crossed bananas display (see **Fig 5-2**). Some modern software and hardware tuning indicators pay homage to the crossed bananas by providing a display in the shape of a plus (+) sign. In software you tune the radio until you see the +, which means that the mark and space tones have been correctly tuned, detected and decoded. In hardware such as the HAL Communications P38DX multimode processor you tune until the + display glows in sync with the RTTY signal. The KAM98, MFJ and Timewave processors use a "bouncing" LED bargraph, but the idea is fundamentally the same.

As you tune in the signal, you should see letters marching across your screen. Notice how the conversation flows just like a voice or CW ragchew.

KF6I DE WB8IMY . . . YES, I HEARD FROM SAM JUST YESTERDAY. HE SAID THAT HIS TOWER PROJECT WAS ALMOST FINISHED. KF6I DE WB8IMY K

The operators simply type back and forth, ending their transmissions with "K" to indicate that it is the other person's turn to transmit.

If you want to call CQ, the procedure is simple. Some programs have "canned" CQ messages that you can customize with your call sign. Others allow you to type off the air, filling a buffer with your CQ and storing it temporarily until you're ready to go.

Let's assume that you have a CQ stored in your buffer right now. Tap the key that puts your rig in the transmit mode. Now tap the key that spills the contents of your buffer into the modem. If you're monitoring your own signal, you'll hear the delightful chatter of RTTY and see something like this on your screen:

CQ CQ CQ CQ CQ CQ WB8IMY WB8IMY WB8IMY
CQ CQ CQ CQ CQ CQ WB8IMY WB8IMY WB8IMY K K

Now jump back to receive. That's all there is to it!

Notice how my CQ is short and to the point. Did you also notice that I repeated everything several times? *Remember that RTTY lacks error detection.* If you want to make certain that the other station copied what you sent, it helps to repeat it. You'll see this often in contest exchanges.

For example:

N6ATQ DE N1RL . . . UR 549 549. STATE IS CT CT. DE N1RL K

On the other hand, if you know that the other station is copying you well, there is no need to repeat information.

The Mystery Signal

What happens if you stumble across a RTTY signal that you cannot copy? The signal is strong enough, and you seem to be doing all the right things, but you see gibberish on your screen or nothing at all. Why?

It looks like you need to do a little detective work. Check the following:

• Is the signal "upside down"? The RF frequency of the mark signal is usually higher than the RF frequency of the space signal, but there is no law that dictates this standard. With most processors and software you can flip the relationship with the push of a button or the click of a mouse.

• Is the station running in upper sideband (USB) rather than lower sideband (LSB)?

• Are the operators really using signals that have 170 or 200-Hz shifts at 45 baud? Some RTTY operators might choose to run at 75 baud when sending lengthy text. Others may choose to use an oddball shift.

CATCH THE "WARBLE" WITH PSK31

The first step in setting up a PSK31 station is to jump onto the Web and download the software you need, according to the type of computer system you are using. As this book went to press there were PSK31 programs available for *Windows* (98, 95 and 3.1), *Linux*, *DOS* and the Mac.

Once you have the software safely tucked away on your hard drive, install it and read the "Help" files. Every PSK31 program has different features—too many to cover in this chapter. Besides, the features will no doubt change substantially in the months and years that follow the publication of this book. The documentation that comes with your software is the best reference.

The Panoramic Approach

All of the PSK31 programs are good; there isn't a bad apple in the bunch. But when I was introduced to *DigiPan*, it was love at first sight.

One of the early bugaboos with PSK31 had to do with tuning. Most PSK31 programs required you to tune your radio carefully, preferably in 1-Hz increments. In the case of the original G3PLX software, for example, the narrow PSK31 signal would appear as a white trace on a thin *waterfall* display. Your goal was to bring the white trace directly into the center of the display, then tweak a bit more until the phase indicator in the circle above the waterfall was more-or-less vertical (or in the shape of a flashing cross if you were tuning at QPSK signal). Regardless of the software, PSK31 tuning required practice. You had to learn to recognize the sight and sound of your target signal. With the weak warbling of PSK31, that wasn't always easy to do. And if your radio didn't tune in 1-Hz increments, the receiving task became even more difficult.

Nick Fedoseev, UT2UZ and Skip Teller, KH6TY, designed a solution and called it *DigiPan*. The "pan" in *DigiPan* stands for "panoramic"—a complete departure from the way most PSK31 programs work. With *DigiPan* the idea is to eliminate tedious tuning by detecting and displaying not just one signal, but entire *groups* of signals.

If you are operating your transceiver in SSB without using narrow IF or audio-frequency filtering, the bandwidth of the receive audio that you're dumping to your sound card is about 2000 to 3000 Hz. With a bandwidth of only about 31 Hz, a lot of PSK31 signals can squeeze into that spectrum. *DigiPan* acts like an audio spectrum analyzer, sweeping through the received audio from 100 to 3000 Hz and showing you the results in a large waterfall display that continuously scrolls from top to bottom. What you see on

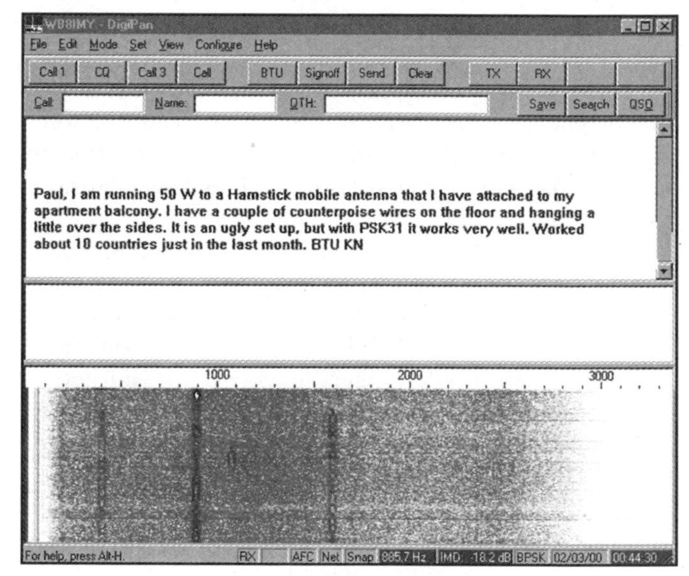

DigiPan **software panoramically displays many PSK31 signals at once. Tuning is as easy as clicking your mouse.**

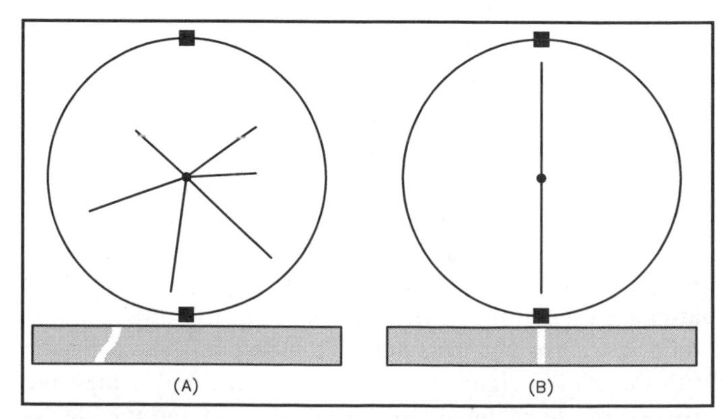

Fig 5-3—Tuning in a PSK31 BPSK signal using the original *Windows* software developed by Peter Martinez, G3PLX. As you tune across the signal, a white trace will appear (A). Just make a slight adjustment and the PSK31 signal is properly tuned (B).

your monitor are vertical lines of various colors that indicate every signal that *DigiPan* can detect. Bright yellow lines represent strong signals while blue lines indicate weaker signals.

The beauty of *DigiPan* is that you do not have to tune your radio to monitor any of the signals you see in the waterfall. You simply move your mouse cursor to the signal of your choice and click. A black diamond appears on the trace and *DigiPan* begins displaying text. You can hop from one signal to another in less than a second merely by clicking your mouse! If you discover someone calling CQ and you want to answer, click on the transmit button and away you go—no radio adjustments necessary. (Like the original PSK31 software, *DigiPan* also automatically corrects for frequency drift.)

This panoramic approach to digital signal communication, in my opinion, is one of the most important developments in the history of PSK31. It makes this exciting mode more "user friendly" and accessible to a larger audience and a larger assortment of radios. In recent years another popular program was introduced that used panoramic tuning: *WinPSK*. You can download both *DigiPan* and *WinPSK* on the Web free.

On the Air

Most of the PSK31 signals on 20 meters are clustered around 14.070 MHz, start by parking your radio in the vicinity and booting up *DigiPan*. **Do not touch your rig's VFO again.** Just place your mouse cursor on one of the vertical signal lines and right click. That's all there is to it!

PSK31 signals have a distinctive sound unlike any digital mode you've heard on the ham bands. You won't find PSK31 by listening for the *deedle-deedle* of a RTTY signal, and PSK31 doesn't "chirp" like the TOR modes. PSK31 signals *warble*—that's the best way I can describe them.

One remarkable aspect of *DigiPan* is that it allows you to see (and often copy) PSK31 signals that you otherwise cannot hear. It is not at all uncommon to see several strong signals (the audible ones) interspersed with wispy blue ghosts of very weak "silent" signals. I've clicked on a few of these ghosts and have been rewarded with text (not error free, but good enough to understand what is being discussed).

Using *DigiPan* reminds me of the sonar operators in the movie *The Hunt for Red October*. There is an eerie excitement in finding one of those ghostly traces and muttering to yourself, "Hmmm...what do we have here? An enemy submarine rigged for silent running? A distant pod of killer whales? Or Charlie in Sacramento running 5 W to his attic dipole?"

Spend some time tracking down PSK31 signals and watching the conversations. With a little practice you'll discover that tuning becomes much easier. You'll also find that you develop an "ear" for the distinctive PSK31 signal.

A PSK31 Conversation

Conversing with PSK31 is identical to RTTY. For example:

Yes, John, I'm seeing perfect text on my screen, but I can barely hear your signal. PSK31 is amazing! KF6I DE WB8IMY K

I know what you mean, Steve. You are also weak on my end, but 100% copy. WB8IMY DE KF6I K

Some PSK31 programs and processor software offer type-ahead buffers, which allow you to compose your response "off line" while you are reading the incoming text from the other station. The original PSK31 software for *Windows* lacked this feature, although it did allow you to send "canned" pre-typed text blocks known as *brag files*. (Brag files are usually descriptions of your station set up.)

PSK31 conversations flow casually, just like RTTY. The primary difference is that you will usually experience perfect or near-perfect copy under conditions that would probably render RTTY useless.

PACTOR

Unlike RTTY and PSK31, PACTOR is a burst mode. That is, it sends and receives data in segments or bursts rather than in a continuous stream. When a burst of data arrives, the receiving station quickly checks for errors caused by noise, interference or fading. If the data is corrupted, the receiving station transmits a brief signal known as a NAK (nonackowledgment) and the data burst is repeated. If the data arrives intact, or if the receiving station has enough information to "repair" any errors, an ACK (acknowledgment) is sent and the next block of data is on the way. This system of rapid-fire data bursts, ACKs and NAKs guarantees a 100% error-free information exchange between stations.

Burst-mode transmissions such as PACTOR are relatively easy to find on the air. Just listen for the characteristic chirping as the stations toss data back and forth. You will not encounter many live PACTOR conversations, but they do exist. The easiest way to start a live PACTOR chat is by calling CQ.

You call CQ on PACTOR using forward error correction, or FEC. The FEC signal sounds like very fast RTTY, but in reality it is a stream of data in which each character is repeated twice for redundancy—there are no ACKs or NAKs. Obviously, this means that an FEC transmission is not error-free, but the copy is good enough to pull out the call sign of the sending station. When you're sending a PACTOR CQ, that's really all that matters.

If you hear a signal that you suspect is a PACTOR FEC CQ, tune carefully until your processor or software indicates that it has locked (synchronized) with the FEC signal. Within a short time you should begin to see text on your screen.

CQ CQ CQ DE WB8IMY WB8IMY WB8IMY
CQ CQ CQ DE WB8IMY WB8IMY WB8IMY
CQ CQ CQ DE WB8IMY WB8IMY WB8IMY K K

Notice that this CQ uses several short lines of text rather than a few long lines. This helps stations synchronize more easily.

Starting a Conversation

Answering a CQ in PACTOR is straightforward. Depending on the software you're using, it may be as simple as entering:

CALL N1BKE
or, CONNECT N1BKE

. . . at the cmd: prompt. Some types of software streamline the process even further. There may be pop-up boxes where you simply enter a call sign.

Once you make contact, the conversation proceeds in turns. This means that one station "talks" (the ISS or information sending station) while the other "listens" (the IRS or information receiving station). When the ISS has had his say, he sends a special control signal known as the *over* command that immediately reverses the roles—suddenly you are the ISS and he is the IRS. Introduce yourself and ask a question about where he lives, or what he does for a living. Use the over command to flip the link again. A conversation is underway!

WB8ISZ DE WB8IMY . . . Hello, Dave. My name is Steve and my location is Wallingford, Connecticut. How do you copy? BTU. K <send the over command>

The HF E-Mail Connection

The Internet has become the e-mail medium of choice for most hams, but there is a sizeable group of amateurs who often travel beyond the reach of the Internet. This group includes hams at sea, travelers in recreational vehicles (RVs), missionaries, scientists and explorers.

The Winlink 2000 Network

More than 60 HF digital stations worldwide have formed a remarkably efficient e-mail network. Running *WinLink* software and using PACTOR or PACTOR II, these facilities transfer e-mail between HF stations. WinLink 2000 stations scan a variety of HF digital frequencies on a regular basis, listening on each frequency for about two seconds. By scanning through frequencies on several bands, the WinLink 2000 stations can be accessed on whichever band is appropriate according to your location and the propagation conditions at the time.

Accessing a WinLink 2000 Station

If you're already set up to operate PACTOR or PACTOR II, you can connect to a WinLink 2000 station right away. Just choose a station, dial in the frequency and transmit the connect request, along with the proper call sign. You'll find stations and frequency lists on the Web at **www.winlink2000.com**. WinLink 2000 stations usually scan through several frequencies. If you can't seem to connect, the WinLink 2000 station may already be busy with another user, or propagation conditions may not be favorable on the frequency you've chosen. Either try again later, or try to connect on another band.

To exchange e-mail with nonham friends and family on the Internet, you'll need the latest version of the AirLink software. You'll find it free for downloading at **www.airlink2000.com**.

Depending on the software you are using, sending the over command is usually as easy as tapping a single key. The software usually provides some sort of visual indicator to show which mode you are in—IRS or ISS—in case you become confused!

The Evolution of the PACTOR BBS

Many PACTOR BBSs in the early '90s acted as HF e-mail gateways to the VHF amateur packet radio network. You could connect to a PACTOR BBS on, say, 20 meters, and post a message that would eventually reach a distant packet BBS on 2 meters.

With the advent of affordable Internet access, amateur packet networks declined substantially to the point where they could no longer handle messages reliably. To adapt to the new rules of the game, many PACTOR BBSs evolved into Internet e-mail gateways. These HF e-mail outlets have integrated into a global system known as WinLink 2000. This network has proven popular among amateurs who find themselves in distant locations without Internet access: sailing enthusiasts, missionaries, adventurers, etc. By connecting to a PACTOR BBS on HF, they can send and receive Internet e-mail with ease. (See the sidebar "The HF E-mail Connection".)

Not all PACTOR BBSs are gateways. Some are simply clearinghouses for information. It's fun to check into a BBS and explore what it has to offer. When you connect to a BBS, you usually see a command line similar to the one shown below.

Welcome to WB8SVN's PACTOR BBS in Placentia, CA. Enter command: A,B,C,D,G,H,I,J,K,L,M,N,P,R, S,T,U,V,W,X,?,* >

After you enter a command you *do not* need to send the over code. The BBS will flip the link automatically after it receives the command from you.

You can use PACTOR BBSs to exchange messages with other PACTOR operators. And with PACTOR's ability to handle binary data, you can even download small programs from the BBS and run them on your computer!

PACTOR II

The state of the art never remains fixed. This is as true of PACTOR as it is of any communication mode. At the 1993 Dayton HamVention the inventors of PACTOR—now incorporated as Special Communications Systems (SCS) of Hanau, Germany—unveiled their newest creation: *PACTOR II*.

You can think of PACTOR II in terms of being a supercharged version of PACTOR (or perhaps we should refer to it as PACTOR I!). In good conditions PACTOR II boasts a data transfer rate up to six times faster than PACTOR. At the same time, PACTOR II is capable of maintaining links in conditions where the signal-to-noise ratio is at -18 dB. This means that PACTOR II can carry on an exchange even when the signals are virtually inaudible. Such remarkable performance is also conservative when it comes to bandwidth; a PACTOR II signal only occupies about 500 Hz of spectrum.

The complex *pi/4-DQPSK* modulation system used by

PACTOR II requires DSP technology and fast microprocessors. With DSP you let the *software* decide the composition of the signal, not the hardware. DSP is much more flexible, allowing you to create the signal you want without having to worry about hardware filters and so on. The tradeoff is that you must use high-speed microprocessors to handle all the incoming and outgoing information at a decent rate.

PACTOR II is also *backward compatible* with PACTOR. That is, a PACTOR II operator can communicate with a PACTOR operator and vice versa. Many HF digital BBSs, including Internet e-mail gateways, include PACTOR II among their operating modes.

PACTOR II is only available in the PTC-II series multimode processors manufactured (or licensed) by SCS. In addition to PACTOR II, the PTC-II offers other HF modes such as RTTY, SSTV, packet and even PSK31.

PACTOR II also operates in nearly the same fashion as PACTOR. In fact, the link is initially established using the PACTOR protocol, then automatically switched to PACTOR II if both stations are using PACTOR II processors. And as with PACTOR, you must take turns during the conversation, sending an "over" command to allow the other operator to send data.

CLOVER

Clover is an advanced HF digital communication system that the late Ray Petit, W7GHM, developed in a joint venture with HAL Communications of Urbana, Illinois. Clover uses a four-tone modulation scheme. An enhanced version using improved DSP technology came along later and was officially christened Clover II, but I'll refer to the mode simply as "Clover" for the sake of clarity.

Depending on signal conditions, several different modulation formats can be selected manually or automatically. Each tone is phase- and/or amplitude-modulated as a separate, narrow-bandwidth data channel. As you might guess, the resulting Clover signal is very complex!

For example, when the tone pulses are modulated using quadrature phase-shift modulation (QPSM), the differential phase of each tone shifts in 90° increments. Two bits of data are carried by each tone for a total of eight bits in each 32-ms frame. The resulting block data rate is about 250 bits per second. Clover is capable of even higher data rates when using 16-phase, four-amplitude modulation (16P4A). In this format, Clover perks along at 750 bit/s.

The complex, higher-speed modulation systems are used when conditions are favorable. When the going gets rough, Clover automatically brings several slower (but more robust) modes into play.

Even with these ingenious adaptive modulation systems, errors are bound to occur. That's where Clover's Reed-Solomon coding fills the gaps. Reed-Solomon coding is used in all Clover modes. Errors are detected at the receiving station by comparing check bytes that are inserted in each block of transmitted text. When operating in the ARQ (automatic repeat request) mode, Clover's damaged data can often be reconstructed without the need to request repeat transmissions. This is a major departure from the techniques used by PACTOR. Of course, Clover can't always repair data; repeat transmissions—which Clover handles automatically—are sometimes required to get everything right.

With the combination of adaptive modulation systems and Reed-Solomon coding, Clover boasts remarkable performance—even under the worst HF conditions. The only Amateur Radio digital mode with the potential to match Clover's performance is PACTOR II.

Clover Handshaking

As you may recall, PACTOR uses an "over" command to switch the link so that one station can send while the other receives. Clover links must be switched as well, but the switching takes place *without* using *over* commands.

When two Clover stations make contact, they can send limited amounts of data to each other (up to 30 characters in each block) in what is known as the *chat mode*. If the amount of data waiting for transmission at one station exceeds 30 characters, Clover automatically switches to the *block data mode*. The transmitted blocks immediately become larger and are sent much faster. The other station, however, remains in the chat mode. Because of precise frame timing, all of this takes place without the need for either operator to change settings or send *over* commands. The Clover controllers at both stations "know" when to switch from transmit to receive and vice versa. And what if both stations have large amounts of data to send at the same time? Then they *both* switch to the data block mode. This high degree of efficiency is transparent to you, the operator. All you have to do is type your comments or select the file you want to send—Clover takes care of everything else!

Clover features an FEC mode similar to that used by PACTOR. You use the Clover FEC to call CQ, or to send transmissions that can be received by several stations at once. (In the Clover ARQ mode, only two stations can communicate at a time.)

Clover on the Air

Most Clover-equipped stations are dedicated to relaying high-volume message traffic, often functioning as HF/Internet e-mail gateways. This makes casual Clover operating the exception rather than the rule, although you will find occasional keyboard-to-keyboard "live" chats.

Clover signals are relatively easy to recognize. The data bursts vary in length. Some are short, while others can last several seconds. The signals make a staccato brrrrrr sound rather than the chirping rhythms of PACTOR.

When it comes to on-the-air operating, Clover is different from any of the modes we've discussed so far. To call CQ, for example, the Clover controller sends a CW identification followed by a raucous stream of data. Unlike other digital modes, you do not see "CQ CQ CQ" flowing across your screen. In fact, you see nothing at all.

The controller sends CQ in the form of data signals that appear as CQ "flags" to other Clover stations. When another Clover operator tunes in your signal, all he sees is a statement on his screen announcing that you are calling CQ. At that point he can ignore you or press a single key to establish a Clover connection.

WE'LL "SEE" YOU IN HELL!

Are the Hellschreiber modes really HF digital? Or, are they a hybrid of the analog and digital worlds?

Some argue that the Hellschreiber modes are more closely related to facsimile since they display text on your computer screen in the form of images (not unlike the product of a fax machine). On the other hand, the elements of the Hellschreiber "image text" are transmitting using a strictly defined digital format rather than the various analog signals of true HF fax or SSTV.

The Hellschreiber concept itself is quite old, developed in the 1920s by Rudolf Hell. Hellschreiber was, in fact, the first successful direct printing text transmission system. The German Army used Hellschreiber for field communications in World War II, and the mode was in use for commercial landline service until about 1980.

As personal computers became ubiquitous tools in ham shacks throughout the world, interest in Hellschreiber as an HF mode increased. By the end of the 20th century amateurs had developed several sophisticated pieces of Hellschreiber software and had also expanded and improved the Hellschreiber system itself.

Like RTTY and PSK31, the Hellschreiber modes are intended for live conversations. Most of the activity is found on 20 meters, typically between 14.061 and 14.070 MHz. As with RTTY and PSK31, you simply type and send your text. The main difference involves what is actually seen on the receiving end.

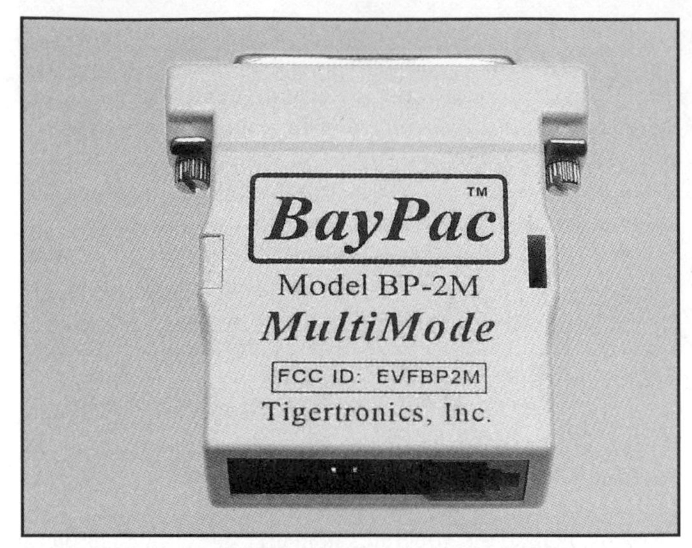

Among the easiest and most affordable ways to get started with RTTY are the Tigertronics BP-2M or MFJ-1213B modems.

Feld-Hell

Feld-Hell is the most popular Hellschreiber mode among HF digital experimenters. It has its roots in the original Hellschreiber format, adapted slightly for ham use.

Each character of a Feld-Hell transmission is communicated as a series of dots, with the result looking a bit like the output from a dot-matrix printer. A key-down state is used to indicate the black area of text, and the key up state is used to indicate blank or white spaces. One hundred and fifty characters are transmitted every minute. Each character takes 400 ms to complete. Because there are 49 pixels per character, each pixel is 8.163 ms long.

Feld-Hell characters can be sent by using a simple CW transmitter, but most operators prefer to achieve the same effect by feeding 900 or 980-Hz tones to an SSB transmitter. Whichever method is used, the timing requirements are precise. Fortunately, Rudolf Hell developed a simple technique of printing the text *twice* to deal with the effects of phase shifts and small timing errors. (The text is not transmitted twice, however.) In this sense you could say that Feld-Hell is a *quasi-synchronous* mode. The font was designed so that the top and bottom of each line of text could be matched, if necessary, to create readable words, no matter what phase relationship existed between transmitting and receiving equipment.

MT-Hell

MT-Hell or *Multi-Tone* Hell transmissions are similar in concept to Feld-Hell, but rather than using on/off keying to represent the black and white pixels that make up the text, MT-Hell uses *frequency variations*. As with Feld-Hell, all you need is a reasonably stable SSB transceiver to operate MT-Hell.

Shifting audio tones are used to designate the black/white pixel elements, and each row of pixels is sent using a different frequency. The DSP software at the receiving end only detects the *presence* of the MT-Hell signaling frequencies—it doesn't care about their shapes or amplitudes. Since noise and interference are amplitude-modulated in nature, the software can effectively "ignore" the garbage. The result is excellent performance under marginal conditions.

The data rate of MT-Hell is not fixed, because there is no attempt to achieve any sort of synchronization. The only timing constraints are those that ensure that the transmitted characters are displayed in the proper order and with minimal distortion. Using slow transmission rates and DSP detection software, the weak-signal performance can be improved even further.

There are at least four variations of MT-Hell in use today:

C/MT-Hell, or *Concurrent MT-Hell*, uses many tones (seven or more), usually transmitted at the same time. It is the most popular of the MT-Hell modes.

S/MT-Hell or *Sequential MT-Hell* was invented in 1998 by ZL1BPU. It uses only a few tones, typically five or seven, but never more than one at a time. *This is the only MT-Hell version that is suitable for use with CW QRP rigs.*

FSK-Hell is a two-tone system commonly using 980 Hz for black and 1225 Hz for white with a 245-Hz shift. You'll find it included in *Windows*-based software developed by IZ8BLY. G3PPT's *FELDNEW8* also offers FSK-Hell. Both programs operate at 122.5 baud.

Duplo-Hell is a relatively new variant invented by IZ8BLY. The font and format are identical to Feld-Hell, except that two columns are transmitted at the same time, using two on-off keyed tones. The tones and shift are the same as FSK-Hell.

MFSK16

To begin your exploration of MFSK16, you must download IZ8BLY's *Stream* software from the Web. (See the sidebar "HF Digital Software on the Web".)

So what does this new mode consist of? Well, there are 16 tones, sent one at a time at 15.625 baud, and they are spaced only 15.625 Hz apart. Each tone represents four binary bits of data. With a bandwidth of 316 Hz, the signal easily fits through a narrow CW filter.

Stream offers a generous collection of tools along the top, separate transmit and receive windows, a good collection of definable "macro" buttons, and an excellent "waterfall" tuning display. Along the bottom is a list of settings and parameters, plus the date and time. There is also a drop-down log window, for automatic logging and insertion of QSO information, and a very useful "QSP" window for relaying incoming text.

Nino's software actually includes three modes! The default mode is MFSK16 (16 tone 16 baud MFSK with FEC), and there is also slower but more sensitive MFSK8 (32 tones 8 baud with FEC). Both modes have the same bandwidth, just over 300 Hz, but sound quite different. The other new mode is one of Nino's own, 63PSKF, which is a 63 baud PSK mode, like PSK31 but faster, and with full-time FEC. PSK63F is about 100 Hz bandwidth. The MFSK and PSK modes are complementary, as Nino's new mode is great for short path DX and local QSOs. You'll have no trouble telling them apart, and no trouble telling Nino's PSK63F from PSK31, because it is twice as wide. As a standard of comparison, the software includes PSK31 as well.

The software is very simple to use—start typing and it transmits, and press F12 to end the transmission. The trouble comes in tuning in the MFSK signals. It takes some skill and a certain patience learning to tune in MFSK, but the results are worth the effort.

Because the tones are closely spaced and the filters very narrow, you must have a very stable transceiver; and you must use the tuning provided with the software, not the transceiver tuning, and certainly not the RIT. The software allows you to tune up and down in 1 Hz steps, or click on the waterfall for exact tuning. The waterfall has a zoom function, and zoom X3 is best.

The AFC is good, but you need to be within about 5 Hz of the right place to start with. The AFC works on the idle tone, which appears often enough the AFC to catch it. You can see this happen as the Phase Scope comes alive. You can also manually tune by clicking on the waterfall display in just the right spot, or use the Up/Down Frequency buttons to tweak the tuning.

Tuning is done using the waterfall display. Under the

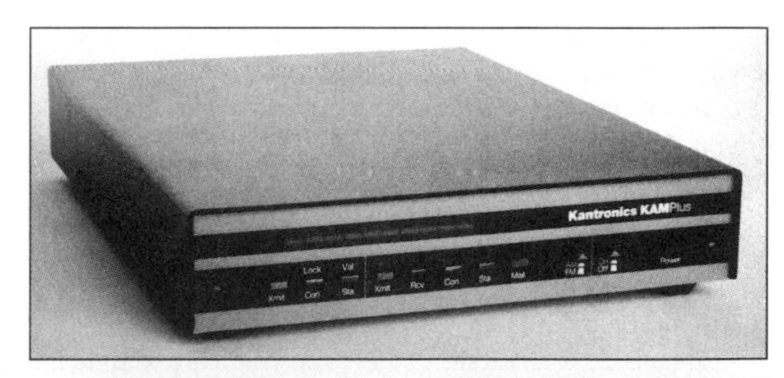

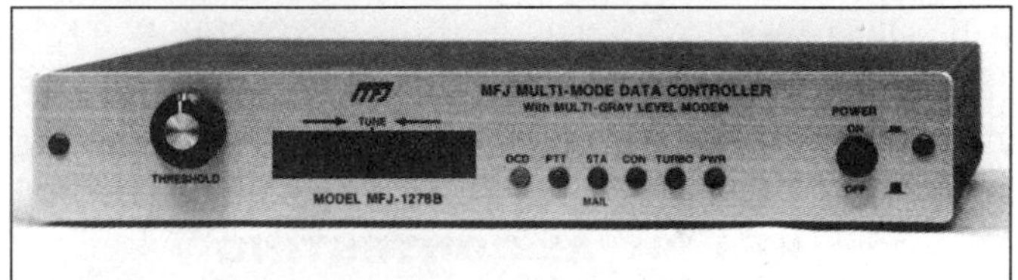

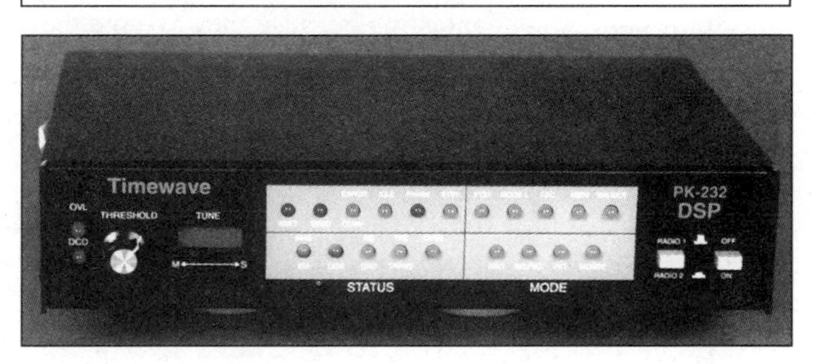

Popular multimode communication processors include the Kantronics KAM Plus, the MFJ-1278B and the Timewave PK-232/DSP.

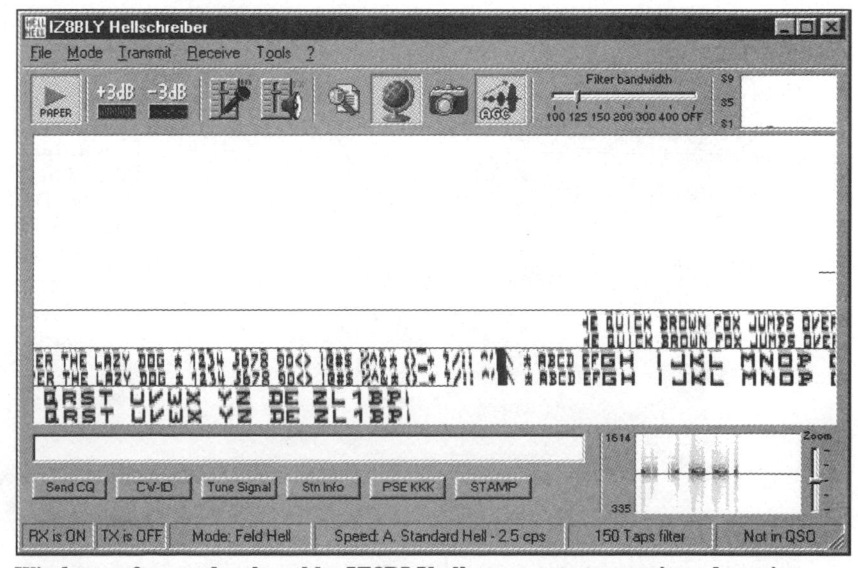

***Windows* software developed by IZ8BLY allows you to transmit and receive Feld-Hell, several MT-Hell modes and Morse using your computer soundcard.**

lower horizontal line (red on the screen) you'll see a broad band towards the left. This is the idle carrier, the lowest of the 16 tones. This carrier is transmitted briefly at the start of each over, and returns at the end, or whenever the operator stops to think. All you need to do is center the red line on this carrier, and the AFC will keep it there. During the over, you'll see little black vertical stripes all over the waterfall, with gray "side-lobes" above and below. These are the transmitted symbols, and once again, you can adjust the software tuning so the red line centers the lowest of these symbols. Unfortunately, while this is easy when the signal is already tuned, finding the correct spot on a weak signal during an over is not so simple and takes a little practice.

Once you've found the right spot, almost perfect text will start to appear on the screen, although it is delayed by some 3-4 seconds as the data trickles through the error correction system, and appears one or two words at a time.

The mode is a delight to use once you learn to tune in. The typing speed is fast, and while changeover from transmit to receive is not as fast as RTTY or Hellschreiber, it is quite good enough for rag-chewing and nets.

HF DIGITAL CONTESTING

Some hams shun contesting because they assume they don't have the time or hardware necessary to win—and they are probably right. But winning is not the objective for most contesters. You enter a contest to do the best you can, to push yourself and your station to whatever limits you wish. The satisfaction at the end of a contest comes from the knowledge that you were part of the glorious frenzy, and that you gave it your best shot!

Contesting also has a practical benefit. If you're an award chaser, you can work many desirable stations during an active contest. During the ARRL RTTY Roundup, for example, some hams have worked enough international stations to earn a RTTY DXCC. Jump into the North American RTTY QSO Party and you stand a good chance of earning your Worked All States in a single weekend.

Always remember that contesting is ultimately about enjoyment. The thrill of the contest chase gets your heart pumping. The little triumphs, like working that distant station even though he was deep in the noise, will bring smiles to your face.

You'll make a number of friends along the way, too. Why do you think they call it the "brotherhood of contesting"?

Digital Contesting Tips for the "Little Pistol"

If you are like most amateurs, your station falls into the Little Pistol category. Mine does, too. The Little Pistol station usually includes a low- to medium-priced 100-W transceiver feeding an omnidirectional antenna such as a dipole, vertical, end-fed wire and so on. Even if you are blessed with a beam antenna on a tower or rooftop, in all likelihood you are still a Little Pistol.

At the opposite end of the spectrum are the Big Gun stations. They have multiple beams on tall towers, phased array antennas on the lower bands, incredibly long Beverage receiving antennas and more. They own high-end (read: expensive) transceivers that boast razor-sharp selectivity, and they usually couple them to 1.5 kW linear amplifiers.

So how does a Little Pistol go about beating a Big Gun in a digital contest? With rare exceptions, he can't. Banging heads with a Big Gun station is usually an exercise in futility. He is going to hear more stations, and more stations are going to hear him. Focus, instead, on using your wits and equipment to get the best score possible. You probably won't beat a Big Gun, but you could make him sweat!

Check Your Antenna

How long has it been since you've inspected your antenna? Is everything clean and tight? And have you considered trying a different antenna? Contests are terrific testing

If you want to try Clover as well as RTTY and PACTOR, the solutions are HAL Communications P38 (left) or DXP38 (below). The P38 plugs into your PC buss slot while the P38DX is an outboard unit. Both products sell for less than $400.

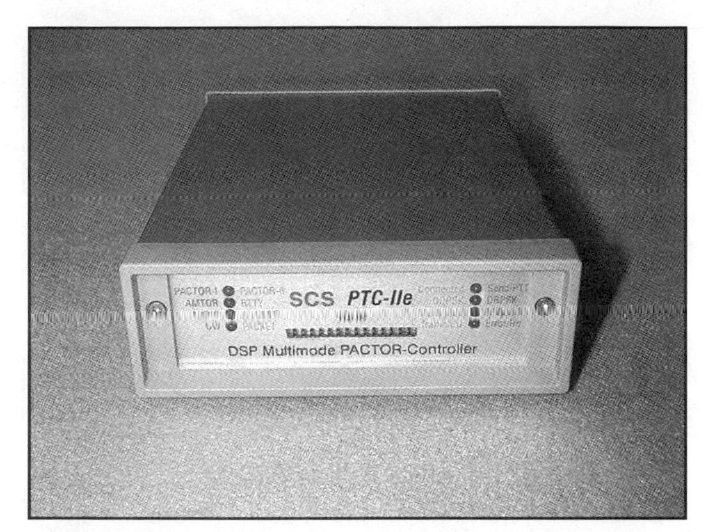

The SCS PTC-IIe is the only multimode communication processor that offers PACTOR II. The original PTC-II was priced at nearly $1000, which made it slow to catch on in the ham community. The PTC-IIe, however, sells for less than $700.

environments for antenna designs. For example, if you've been using a dipole for a while, why not try a loop? Just set up the antenna temporarily for the duration of the contest and see what happens.

Check Your Equipment

If you are having problems with RF getting into your processor or soundcard, fix these glitches before you find yourself in a contest! Also, consider spending $100 or so to install a 500-Hz IF filter in your radio if it doesn't have one already. Outboard audio DSP filters are helpful to cut down the signal clutter, but a good IF filter will make a world of difference in a contest environment. Finally, check your computer and software. Make sure you understand the program completely. Set up any "canned" messages/responses and test them before the contest begins.

Understand Propagation

If you are participating in a DX contest, is it a good idea to prowl the 40-meter band in the middle of the day? No, it isn't. Forty meters is only good for DX contacts after sundown. If you are hunting Europeans or Africans on 10 or 15 meters, when is the best time to look for them? The answer is late morning to early afternoon.

In other words, let propagation conditions guide your contest strategy. Your goal is to squeeze the most out of every band at the right time of day. For example, I might start the ARRL RTTY Roundup on 10 meters and then bounce between 10, 15 and 20 meters until sunset. After sundown, I'll concentrate on 20 and 40 meters. In the late night and the wee hours of the morning, I'll probably limit myself to 40 and 80 meters.

In some contests you can enter as a single-band station. Which band should you choose? That depends on the contest and your equipment. If you have a 15-meter beam and you are entering a DX RTTY contest, it makes sense to concentrate on 15 meters, even though you'll probably find yourself with nothing to do in the late evening after the band shuts down. Twenty meters makes sense as the ultimate around-the-clock band, but it is often very crowded during contests and populated with Big Gun stations. In some instances you may find it more profitable to concentrate your efforts on another band without so many signals.

Hunting and Pouncing vs "Running"

The common sense rule of thumb is that a Little Pistol station should only hunt and pounce. That means that you patrol the bands, watching for "CQ CONTEST" on your monitor, and pouncing on any signals you find. The Big Guns, on the other hand, often set up shop on clear frequencies and start blasting CQs. If conditions are favorable, they'll start hauling in contacts like a commercial fishing boat! This is known as *running*.

In many cases Little Pistols are probably wasting valuable time by attempting to run. There are situations, however, where running *does* make sense. If you've pounced on every signal you can find on a particular band, try sending a number of CQs yourself to catch some of the other pouncers. If you send five CQs in a row and no one responds, don't bother to continue. Move to another band and resume pouncing.

Seek Ye the Multipliers

Every contest has multipliers. These are US states, DXCC countries, ARRL sections, grid squares and so on, depending on the rules of the contest. A multiplier is valuable because it multiplies your total score.

Let's say that DXCC countries are multipliers for our hypothetical contest. You've amassed a total of 200 points so far and in doing so you made contacts with 50 different DXCC countries.

200 × 50 = 10,000 points

Those 50 multipliers made a huge difference in your score! Imagine what the score would have been if you had only worked 10 multipliers?

If given a choice between chasing a station that won't give me a new multiplier and pursuing one that will, I'll spend much more time trying to bag the new multiplier.

Choose the Right Mode

Many contests allow the use of all HF digital modes, but RTTY is by far the most popular. Yes, you can probably make some contacts during the ARRL RTTY Roundup using PACTOR, for example, but you'll miss RTTY contact points while doing so. As the saying goes, "When in Rome do as the Romans do."

There are some contests that are devoted to one mode in particular. With the increasing popularity of PSK31, for example, PSK31-*only* contests are springing up. Watch for the TARA PSK31 Rumbles in April and October, and the CCCC PSK31 contest in September. In October you'll find the DARC Hellschreiber contests. (See the "Digital Contest Calendar" sidebar).

Digital Contest Calendar

See *QST* or the *National Contest Journal* for complete rules.

Month	Date	Contest
January	1st	SARTG New Year's RTTY Contest
January	First full weekend	ARRL RTTY Roundup
January	Fourth full weekend	BARTG RTTY Sprint
February	First full weekend	NW QRP Club Digital Contest
February	Second weekend	WorldWide RTTY WPX Contest
March	Third weekend	BARTG Spring RTTY Contest
April	First full weekend	EA WW RTTY Contest
April	Third weekend	TARA PSK31 Rumble
April	Fourth full weekend	SP DX RTTY Contest
May	First full weekend	ARI International DX Contest
May	Second weekend	VOLTA WW RTTY Contest
June	Second weekend	ANARTS WW RTTY Contest
July	Third weekend	North American RTTY QSO Party
July	Fourth full weekend	Russian WW RTTY Contest
August	Third weekend	SARTG WW RTTY Contest
August	Last full weekend	SCC RTTY Championship
September	First full weekend	CCCC PSK31 Contest
September	Last full weekend	CQ/RJ WW RTTY Contest
October	First full weekend	TARA PSK31 Rumble
October	First full weekend	DARC International Hellschreiber Contest
October	Second Thursday	Internet RTTY Sprints
October	Third weekend	JARTS WW RTTY Contest
November	Second weekend	WAE RTTY Contest
December	First full weekend	TARA RTTY Sprints
December	Second weekend	OK RTTY DX Contest

Contest Software

No one says you have to use software to keep track of your contest contacts, but it certainly makes life easier! One of the fundamental elements of any contest program is the ability to check for duplicate contacts or dupes. Working the same station that you just worked an hour ago is not only embarrassing, it is a waste of time. The better contest programs feature immediate dupe checking. When you enter the call sign in the logging window, the software instantly checks your log and warns you if the contact qualifies as a dupe. The more sophisticated programs "know" the rules of all the popular digital contests and they can quickly determine whether a contact is truly a dupe under the rules of the contest in question. Some contests, for example, allow you to work stations only once, regardless of the band. Other contests will allow you to work stations once per band.

A good software package will also help you track multipliers. It will display a list of multipliers you've worked, or show the ones you still need to find.

Any of the contest software packages advertised in *QST* and the *National Contest Journal* will function well. To minimize headaches, however, my suggestion is to stick with software written specifically for digital contesting. There are three popular digital contesting software packages on the market today:

- ***RTTY*** by WF1B is an excellent DOS package that is compatible with almost any multimode communication processor. *RTTY* is not a mere logging program; it is a terminal program that sends and receives data from processor or modem. *RTTY* checks for dupes and "flags" them in color on your screen. It prepares your log for electronic submission. *RTTY* can also control your radio, changing bands and frequencies on the fly. WF1B will run on 486-level PCs and will also operate in a DOS window under *Windows 95* and *98*. You'll find complete information about *RTTY* on the Web at **http://www.wf1b.com/**.

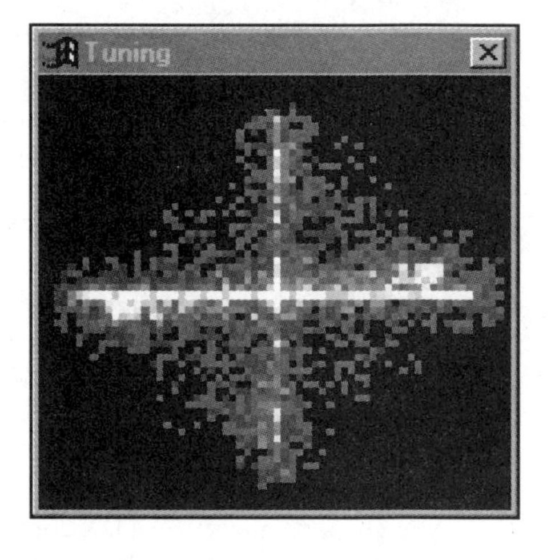

The RTTY software tuning indicator provided by *WriteLog*, an RTTY contest software package.

- OH2GI ***Ham System*** is another DOS package similar to *RTTY* in many respects. *Ham System* is particularly popular among European contesters and is compatible with almost every known modem and processor. See the *Ham System* Web site at **http://www.kolumbus.fi/jukka.kallio/**.
- ***WriteLog*** by Ron Stailey, K5DJ, is a full-featured *Windows*-based software package with an interesting twist. While *WriteLog* can be used with any processor or modem, it also has the *built-in* ability to send and receive RTTY and PSK31 using your soundcard. Like *RTTY*, *WriteLog* offers automatic dupe checking and flagging, multiplier displays, radio control, canned messages and more. See the *WriteLog* site on the Web at **http://www.contesting.com/writelog/**.

Chapter 6

VHF Digital Communications

Stan Horzepa, WA1LOU
One Glen Avenue
Wolcott, CT 06716-1442
Email wa1lou@arrl.net
http://www.tapr.org/~wa1lou

The Amateur Radio frequency allocation in the VHF spectrum consists of three bands: 6 meters (50-54 MHz), 2 meters (144-148 MHz) and 1.25 meters (219-220 an 222-225 MHz). Within these three bands, the primary digital mode of operation is packet radio.

Over the years, the operation of packet radio in the VHF spectrum has changed dramatically. In the 1980s, an international network of packet radio bulletin boards (PBBS) was the mode *du jour*. The PBBS network with its automatic mail forwarding capability permitted hams to originate e-mail at their local 2 meter PBBS and have it delivered anywhere in the world where there was a PBBS. And there were PBBSs everywhere!

The PBBS network was so popular that it gradually became overloaded with the amount of e-mail it had to handle. With the growth of the Internet, hams abandoned the PBBS network in favor of the Internet and its seemingly inexhaustible e-mail capabilities.

While use of the PBBS network dwindled, packet radio was being used for other applications. During the 1990s, these applications supplanted the PBBS network as the VHF digital mode of choice. As a result, today you will find such packet radio applications as gateways, DX PacketCluster, Automatic Position Reporting System (APRS), Propagation Network (PropNET) and satellite packet communications, in addition to PBBSs.

The purpose of this chapter is to describe how to use the VHF digital applications at the turn of the century. (The Satellite chapter in this book includes a discussion of VHF packet radio satellite applications.)

ASSEMBLING A VHF DIGITAL STATION

VHF digital communications requires three components:

- Terminal Node Controller (TNC)
- Computer
- VHF radio equipment

The following describes each of these components as they relate to VHF digital communications.

Terminal Node Controller (TNC)

Packet radio is the communication mode in which the output of a computer or computer terminal is assembled into bundles or packets according to a set of rules called a protocol. Each packet is transmitted to a remote radio station where it's checked for its integrity, then disassembled and its information fed to a terminal for you to read.

Packets are assembled and disassembled by a packet assembler/disassembler or PAD. Since the input and output of a PAD is digital, there is a digital-to-analog/analog-to-digital conversion between output/input of the PAD and radio transmitter/receiver. A modulator-demodulator (modem) performs this conversion. Typically, a PAD and its associated modem are packaged into one unit called a terminal node controller (TNC), which is connected between the computer or computer terminal and the radio equipment as illustrated in **Fig 6-1**.

Basically, a TNC is a very intelligent modem. Whereas a telephone modem permits a computer to communicate by means of the telephone and computer networks like the Internet, a TNC permits a computer to communicate by means of a radio network. Just as an intelligent telephone modem augments its functions by including a wide range of built-in commands to facilitate computer-telephone communications, the built-in intelligence of a TNC facilitates packet radio communications.

Both a telephone modem and a TNC are connected to a communications medium: the telephone modem to the telephone line and the TNC to the radio. Telephone modems and TNCs are both connected to the serial interface of a computer.

To initiate computer-telephone communication, you command an intelligent modem to address (dial the tele-

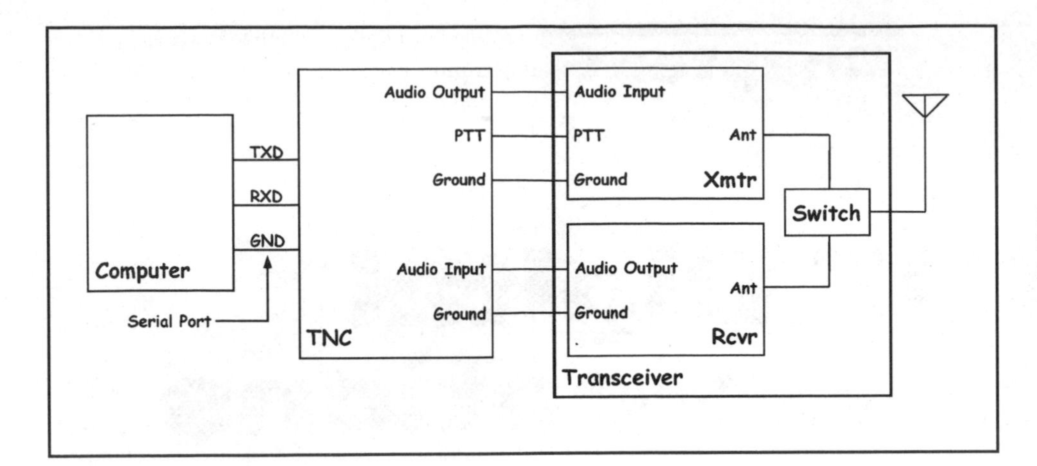

Fig 6-1-The components and their interconnection in a typical packet radio station.

phone number of) the other computer. Similarly, to initiate computer packet radio communications, you command a TNC to address (make a connection with the call sign of) another Amateur Radio station.

The TNC is the heart and soul of a VHF digital radio station. Without a TNC, you cannot operate packet radio. Choosing a TNC for VHF communications is relatively painless because most TNCs are designed for VHF communications; that is, they are ready to roll at 1200 bit/s, which is the most common data rate used for VHF packet communications.

Instead of using a TNC to perform the TNC function of a VHF digital station, you can connect a packet modem to a computer and run software on the computer that emulates a TNC. Packet modems in kit and assembled form are available from a number of sources for approximately $50. Typically, the modems are bundled with TNC-emulation software.

Computer

When the TNC was invented, its user interface was intended to be a computer terminal. By means of the terminal's keyboard, the user entered commands that controlled the TNC. Once communications are established, the user entered information on that same keyboard for transmission to the other station. Via its display, the terminal allowed the user to read the TNC's responses to commands and to read the information that was sent by the other station.

Software

Today, with the proliferation of personal computers, terminals are used only rarely. Instead, most Amateur Radio operators run software on their computers to communicate with their TNCs.

Some hams use terminal emulation software for TNC communications. Basically, terminal emulation causes your computer to act as a computer terminal and provides a direct interface to your TNC. You enter TNC commands and messages to other stations at the computer keyboard, and the computer monitor displays the responses to your commands and messages just as a computer terminal display does.

A variety of terminal software is available, from programs used to access mainframe computers over wire links to software that provides access to bulletin board systems (BBSs) via telephone lines. One of the most common terminal emulation programs is *Hyperterminal*, which is bundled with the *Windows* operating system. (You can find it in the Start > Programs > Accessories submenu of *Windows*.)

Other hams use software that is specifically designed for TNC communications. Such software offers features that are optimized for packet communications. For example, if you communicate with more than one station simultaneously (a multiconnect situation), some packet terminal programs allow you to open separate communication windows for each multiple connection. No terminal-emulation program package offers that feature!

There are a variety of sources for packet software. For example, TNC manufacturers Kantronics and MFJ have packet radio software that is optimized for their hardware products. Other providers offer a variety of packet radio software. Browse through the pages of *QST* and other ham radio publications to view advertisements for the current offerings. Meanwhile, online sources for freeware and shareware packet radio software include **ftp.ucsd.edu/hamradio/packet/termprogs/**, **ftp.tapr.org/software_lib/terminal/** and (for Mac users) **http://www.g0oanint.demon.co.uk/packet.html**.

Serial Port

Aside from software for communications with a TNC, the computer must have a serial port interface for the physical connection to the TNC. Almost all TNCs use the EIA/TIA-232-E (EIA-232 for short) interface with its ubiquitous 25-pin connector to provide a serial connection to a computer. Most computers have an EIA-232 interface, and connecting them to a TNC is simple. On the other hand, some computers have no interface at all and EIA-232 may be an option that's available through the addition of a PC board.

There are computers that use other interfaces. For example, TIA/EIA-422-B (EIA-422 for short) is found on some computers, including the Macintosh. EIA-422 is close enough to EIA-232 that it can be made to work by properly wiring the interfaces. (Cables that interface a Macintosh computer to a telephone line modem will interface a Mac to a TNC, too.)

Some newer computers now come with a USB interface that is not compatible with the TNC's EIA-232 interface. If your computer uses USB, you will have to use a converter in order to make the connection to the TNC.

Although EIA-232 supports 25 signals, any TNC you use will need only five of them (the signals on pins 2-5 and pin 7) and many TNCs can get by using even fewer signals. Check the TNC's manual and see what you can do to economize your computer-to-TNC cabling.

Radio Equipment

Obviously, you need radio equipment that operates on the VHF band of your choice (6, 2 or 1.25 Meters) for VHF digital communications. There can be a variety of components that compose the *equipment* at the RF end of a packet station. Some of it is of little concern to us. For example, as long as the antennas and feed line are capable of putting a signal on the desired packet frequency, it satisfies our requirements. Other RF hardware needs closer inspection, however.

Our primary concern is the speed at which the radio equipment can switch back and forth between the transmitting and receiving modes; that is, the radio equipment's receive-to-transmit and transmit-to-receive turnaround time. A TNC can switch between the transmitting and receiving modes very quickly—so quickly, in fact, that it must wait for the RF equipment to switch before it can continue to communicate.

One factor in the turnaround time is the physical switching of an antenna, internally in a transceiver and externally with an optional power amplifier and/or receiver preamplifier. The older the equipment, the more likely that switching is performed mechanically (and relatively slowly) by a relay.

The more modern the equipment, the more likely it is that the switching is accomplished electronically (and relatively quickly). However, this improvement in antenna switching may be compromised by the frequency synthesizer circuitry that is also found in modern radio equipment. After switching between the transmitting and receiving modes, synthesizers require some time to lock on frequency before they are ready.

In addition to the turnaround time, another problem is that the modem-to-radio interface of older radios depends on audio-response filters and audio levels intended for microphones and speakers. This can lead to incorrect deviation of the transmitted signal, noise and hum on the audio and other problems. Splatter filters and deviation limiters distort frequency response and further reduce the performance of the packet radio system. You are stuck in this environment unless you want to modify the radio. Performing surgery on your typical modern VHF FM voice transceiver may be difficult because of the use of LSI devices, surface-mounted components and miniaturization.

On the other hand, many modern transceivers now provide a modem interface that bypasses the microphone and speaker support circuitry. In addition to this improved modem interface, some modern transceivers also claim to support higher data rates such as 9600 bit/s. Such plug-and-play transceivers should be your choice when you are gearing up for VHF digital communications. Check the pages of *QST* for news, reviews and advertisements for these radios that are designed with digital modes in mind.

Hardware Connections

Installing the packet radio hardware is straightforward. Fig 6-1 illustrates a typical installation.

Radio Connection

The TNC has two ports: a radio port and a computer/serial port. Typically, the radio port requires four connections to the radio equipment: audio input, audio output, press-to-talk (PTT) and ground.

Connect the audio input of the TNC to the audio output of your radio. Typically, the audio output of your radio is a speaker or headphone connector, but some radios have optional audio outputs (sometimes labeled AFSK OUT). Again, connection to such an optional audio output avoids TNC disconnection when you switch to voice and may bypass circuits intended for processing voice and/or insert circuits intended for processing data. If your radio does not have separate AFSK jacks, the phone patch input and output jacks often provide an acceptable alternative.

Connect the audio output of the TNC to the audio input of your radio. Typically, the audio input of your radio is the microphone input (MIC) connection, but some radios have separate audio inputs for AFSK tones (sometimes labeled AFSK IN). If this connection is available, it is better to use it rather than the microphone input because you will not have to disconnect the TNC from the microphone connector whenever you want to use the radio in the voice mode. Also, the AFSK input may bypass circuits in the radio that are intended for voice and/or may insert circuits intended for data. Voice circuits are not necessarily beneficial to data transmission, so bypassing them is a good thing. On the other hand, circuits intended to improve data transmission should be used whenever possible.

Connect the PTT line of the TNC to a PTT connection on your radio. Usually, PTT is available at the microphone connector, but the PTT line is sometimes brought out to another connector as well. Again, connection to the optional PTT jack is preferable; this avoids cable changes when you switch modes.

Finally, connect the TNC ground to the ground connection that accompanies the other connections to your radio; that is, the ground that accompanies the radio's MIC, PTT, speaker or AFSK In/Out connections.

Computer Connection

The computer serial ports of most TNCs are compatible with interface standard EIA-232, which defines 25 signals that may be transferred via the interface. However, the TNC needs only three of those signals for connection to a computer: Transmitted Data, Received Data and Signal Ground.

The TNC typically uses a female 25-pin D-type (DB-25) connector for the computer/serial port. This requires using a male DB-25 connector with pins 2, 3 and 7 cabled to a connector that mates with the serial port of the computer.

The computer typically uses a male DB-25 or male 9-pin D-type (DB-9) connector for its serial port. This necessitates using a female DB-25 or DB-9 connector at the computer end of the computer-to-TNC cable. For DB-25-to-DB-25 cabling, pins 2, 3 and 7 (Transmitted Data, Received Data and Signal Ground) are connected between each DB-25 connector. For DB-25-to-DB-9 cabling, pins 2 and 3 (Transmitted Data and Received Data) are connected between each connector, and pin 7 (Signal Ground) of the DB-25 is connected to pin 5 of the DB-9.

To avoid the time and expense of building a cable, use the cable connecting your computer serial port to your external telephone line modem, if you have one.

The TNC's EIA-232 serial port is compatible with the serial ports of most computers with the exception of Macintosh computers. Until recently, Macs used an EIA-422 interface, which does not pose a big problem and only requires a male DB-25-to-Macintosh serial port cable to make the connection. However, current and recently obsolesced Macs use a Universal Serial Bus (USB) instead of EIA-422. So, you must obtain an adapter that permits you to connect a serial port device (such as a TNC) to the Mac's USB.

CONFIGURING A TNC

Before you can go on the air, you must set some of the parameters in your TNC. Some packet radio programs configure the TNC for you when you start the software. You simply enter some information when you start the software. The software then transfers that information to your TNC. If you are using packet radio software, refer to the documentation that accompanied the software to find out if the software performs the configuration for you.

If your packet radio software does not configure the TNC or if you are using terminal emulation software, then you have to configure the TNC manually. To configure and control a TNC, you enter commands into the keyboard of your computer. Commands may be entered only when the TNC's command prompt (cmd:) is displayed by your computer display. When this prompt is displayed, type the desired command, then press the Enter key. For example, to command the TNC to break a connection to another station, you type DISCONNE at the command prompt, then press Enter. In this chapter, entering commands will be represented as:

cmd: **DISCONNE <Enter>**

Many TNC commands may be abbreviated to save keystrokes. For example, instead of typing DISCONNE for the disconnect command, you can simply type the letter D. In this chapter, each command will be printed partially in uppercase characters and partially in lowercase characters. The uppercase and lowercase characters together represent the full name of the command, whereas the uppercase characters alone represent the abbreviated version of the command. For example, the disconnect command will be represented as:

Disconne

where Disconne is the full name of the command and D is the abbreviated version.

All TNCs today are compatible with the AX.25, TNC 2 command set. Therefore, this chapter uses TNC 2 commands throughout its description of packet radio operation.

Many TNCs have unique commands that are variations of the AX.25 TNC 2 commands or are new commands that add some functionality to the TNC. Space does not permit listing all these unique commands here; refer to the documentation that accompanied your TNC for information.

Terminal Emulation Parameters

Your TNC must be compatible with the operating parameters of your terminal emulation software. The following describes the parameters that must be compatible with the software.

Data Rate—Set the data rate of your TNC's serial port to the bit/s rate of your terminal emulation software. Many TNC 2 compatibles set their data rate using *autobaud* (the TNC automatically adapts to the data rate of the terminal emulation software). Typically, this is accomplished by pressing the Enter key a number of times after powering the TNC. After receiving some of these keyboard entries, the TNC calculates the correct data rate, sets itself to that rate and sends its sign-on preamble to the terminal emulation software at the correct data rate.

Note that on older TNC 2 compatibles, the data rate is set by means of a rear panel DIP switch (positions 1 through 5) while the TNC is powered off. Toggling on DIP switch position 1 through 5 selects 300, 1200, 2400, 4800 or 9600 bit/s, respectively.

Echo—This function causes the TNC to print on your computer's display each character that you type into your keyboard. This TNC function should be turned on or off depending on whether your terminal emulation software does or does not provide this function. (Echo on is the TNC 2 default value.)

If two characters are displayed for each one that you type into the keyboard (you type HELLO and HHEELLLLOO is displayed), then both your terminal emulation software and TNC are echoing, so turn off the TNC's echo by typing:

cmd: **Echo OFF <Enter>**

If nothing is displayed when you type, then the echo function is turned off in both your terminal emulation software and TNC, so turn on the TNC's echo function by typing:

cmd: **Echo ON <Enter>**

Character Length—Set the TNC to the character length used by your terminal emulation software. (7 bits per character is the TNC default value.)

To select 7 bits per character, type:

cmd: **AWlen 7 <Enter>**

To select 8 bits per character, type:

cmd: **AWlen 8 <Enter>**

Parity—Set the TNC to the parity used by your terminal emulation software. (Even parity is the TNC default.)

To select *even* parity, type:

cmd: **PARity 3 <Enter>**

To select *odd* parity, type:

cmd: **PARity 1 <Enter>**

To select *no* parity, type:

cmd: **PARity 0 <Enter>**

or

cmd: **PARity 2 <Enter>**

Screen Length—Set the TNC to the maximum number of characters displayed on each line of your computer display (80 characters per line is the TNC default), by typing:

cmd: **SCREENLN *n* <Enter>**

where ***n*** is number 0 to 255, representing the number of characters per line displayed by your computer.

Line Feeds—This function causes the TNC to insert a line feed after each carriage return that is received in incoming packets. This TNC function should be turned on or off depending on whether your terminal emulation software does or does not provide this function. (Automatic line feed on is the TNC default.)

If a blank line follows each line of received text, then both your terminal emulation software and TNC are adding line feeds. Turn off the TNC's automatic line feed function by typing:

cmd: **AUtolf OFF <Enter>**

If each line of received text is printed over the previously received line of text, then neither your terminal emulation software nor TNC are adding line feeds, so turn on the TNC's automatic line feed function by typing:

cmd: **AUtolf ON <Enter>**

Those are the most important parameters that you need to set to configure your TNC for compatibility with your terminal emulation software.

By setting these parameters correctly, your computer display should now display text from your TNC in a clear, legible manner. If not, you should recheck these parameters and also check the physical connection between your TNC and computer.

Radio Parameters

Your TNC must also be compatible with your radio equipment. Use the following TNC commands to customize it to the needs of your radio equipment.

Data Rate—Set the radio-port data rate of the TNC to the data rate you will use on the air. Typically, this parameter is set by means of a command unique to your TNC, so refer to the manual that accompanies your TNC. By default, the data rate of most TNCs is 1200 bit/s because 1200 bit/s is the data rate that is used on the most popular packet radio subband, 2-meter FM.

On older TNC 2 compatibles, the on-the-air data rate is set by means of a rear-panel DIP switch (positions 6 through 8) while the TNC is powered off. Toggling on DIP switch position 6 through 9 selects 300, 1200 or 9600 bit/s, respectively.

Turnaround Time—As discussed earlier, when a transceiver switches from receive to transmit, there is a short delay after the switching begins and before intelligence can be sent over the air. Although this delay is only measured in milliseconds (ms), it is critical in packet because a TNC is capable of enabling a transmitter and sending intelligence almost instantaneously. Therefore, some delay must be programmed into the TNC to be compatible with the receive-to-transmit turnaround time of your transceiver. This delay may be varied from 0 to 1200 ms by typing:

cmd: **TXdelay *n* <Enter>**

where ***n*** is number 0 to 120 representing a delay in 10-ms increments [30 (300 ms) is the TNC default]. For example, to set the delay to 400 ms, set n to 40 (400 ms / 10 ms = 40).

Maybe your transceiver's turnaround time is specified in your transceiver's manual. Typically, it is not, so you will have to experiment to find a delay that is suitable. In general, if your transceiver uses solid-state switching, you may set the delay relatively low. If your transceiver uses a mechanical relay for switching, you should set the delay higher. Also, synthesized transceivers require a greater delay than crystal controlled transceivers. If you are using an external amplifier, you must take its switching delay into account, too.

Maximum Number of Unacknowledged Packets and Packet Length—The MAXframe and Paclen parameters

Joe Carcia, NJ1Q, station manager at W1AW, handles traffic on the local packet BBS.

are critical TNC parameters that you should adjust depending on the operating conditions (propagation and channel activity).

The MAXframe parameter controls the maximum number of outstanding unacknowledged packets the TNC will allow at any time. For example, if MAXframe is set to 4, the TNC may send as many as four packets without receiving acknowledgments for any of them. Once the MAXframe limit is reached, however, the TNC will not send a new packet until one of the outstanding packets is acknowledged.

The maximum number of bytes of data in each packet is controlled by the Paclen parameter. The TNC will never send a packet longer than the selected Paclen value. As data is entered from the computer to the TNC, the TNC counts each byte of data. When the Paclen value is attained, the TNC makes up a packet containing the data, sends it over the air and begins counting the number of bytes of data for the next packet. The TNC will send packets shorter than the selected Paclen value when it is specifically commanded to do so (whenever the SEndpac control character <Enter> is entered).

The default values for the MAXframe and Paclen parameters (4 outstanding packets each 128 bytes long) are selected for good VHF operating conditions. When VHF operating conditions are less than optimal (there is a high level of channel activity), the MAXframe and Paclen values should be reduced.

To change the value of the MAXframe parameter, type:

cmd: **MAXframe *n* <Enter>**

where ***n*** is a number from 1 to 7 representing the maximum number of outstanding unacknowledged packets.

To change the value of the Paclen parameter, type:

cmd: **Paclen *n* <Enter>**

where ***n*** is a number from 0 to 255 representing the maximum number of bytes of data in each packet. Note that 0 actually represents 256 bytes.

Number of Packet Retries—The TNC waits a preset time for an acknowledgment that each packet it transmits was received without error. If the time limit is exceeded, the TNC resends the packet and again waits for an acknowledgment from the receiving TNC. When the maximum number of retries is reached, the sending TNC enters the disconnected state. The number of allowable retransmission attempts is controlled by the REtry parameter and the amount of time between retries is controlled by the FRack parameter.

The REtry and FRack parameters should be adjusted upward or downward, depending on the operating conditions. If conditions are good (a low level of activity on the channel), the REtry and FRack parameters may be adjusted downward. If a packet cannot get through after one or two attempts under good conditions, there is probably an insurmountable problem with the link. For example, the intended receiving station may have gone off the air, so you might as well abandon the effort immediately.

If conditions are marginal (a medium level of activity on the channel), it may only take a little longer to get the packet through to the intended receiving station, so the REtry and FRack parameters may be adjusted upward.

If conditions are poor (a high level of activity on the channel), wait until conditions improve. Your packets are likely to collide with other packets because of the crowded channel.

By default, the REtry parameter value is set to 10 retries, but that may be changed by typing:

cmd: **REtry *n* <Enter>**

where ***n*** is a number from 0 to 15. Zero represents an infinite number of retries and 1 to 15 represents the maximum number of retries that will be attempted before the TNC stops repeating a packet. Note that the REtry parameter does not include the initial transmission of the packet; it only represents the number of retries after the initial packet transmission.

By default, the FRack parameter value is set to 3 seconds, but this may be changed by typing:

cmd: **FRack *n* <Enter>**

where ***n*** is a number from 1 to 15, representing the number of seconds that the TNC will wait for a packet acknowledgment before it again tries to obtain an acknowledgment. Note that the TNC automatically adjusts this value higher depending on the number of digipeaters used in the selected path of the packet according to the formula:

FRack × (2 × dr + 1) = adjusted FRack

where dr is the number of digipeaters in the selected path. For example, if FRack is set to 4 seconds and 2 digipeaters are in the selected path, the adjusted FRack value is 20 seconds (4 sec × ((2 × 2 digipeaters) + 1) = 20 sec).

Prioritized Acknowledgment—Most TNCs are capable of using the channel-sharing protocol called *prioritized acknowledgment*. This protocol gives priority to packet acknowledgments (ACKs) on a channel.

ACKs indicate that a packet has been received correctly. By giving an ACK priority, the station that sent the packet receives the ACK more quickly and, as a result, is less likely to resend the packet. (On channels where ACKs are not given priority, ACKs may be delayed long enough to cause the waiting station to give up and resend the packet, thus wasting precious time on the channel.)

Configuring a TNC for prioritized acknowledgment (for 1200-bit/s VHF FM) involves setting the following parameters as indicated:

cmd: Ackprior ON
cmd: ACKTime 14
cmd: DEAdtime 33
cmd: Dwait 33
cmd: Frack 8
cmd: RESptime 0
cmd: SLots 3

These settings are recommended as a starting point. Once you are on the air, you may adjust these settings as conditions warrant. Also, check your TNC' s manual in case it recommends different settings.

Station Identification—Perhaps the most important TNC Command is the one that inserts your call sign in the TNC. To program your TNC with your call sign, type:

cmd: **MYcall *WA1LOU* <Enter>**

where ***WA1LOU*** is the call sign of *your* station.

Optionally, a secondary station identification (SSID), number 1 through 15, may be appended to your call sign by typing a hyphen and the desired number immediately after the call sign in the MYcall command (for example, WA1LOU-4). SSIDs permit the same call sign to be used for different uses simultaneously. For example, the author's APRS tracker station is identified as WA1LOU-8, whereas, his APRS digipeater is identified as WA1LOU-15 (to differentiate it from WA1LOU-8). If no SSID is specified, then the SSID is assumed to be 0 (WA1LOU-0).

Monitoring Parameters

"Listen before you transmit" is a rule of thumb in Amateur Radio, and your TNC provides an array of commands to augment your monitoring mode.

The Monitor command causes the TNC to display the contents of each packet received by your station while in the Command Mode. The display also includes the call sign of the station originating each packet and the call sign of the station that is the intended recipient of each packet. For example, a monitored packet may appear as:

N1BKE>K8CH: Data

where N1BKE is the call sign of the station originating the packet, K8CH is the call sign of the intended destination of the packet and Data is the contents of the packet. (The Monitor function is on by default.) To enable (or disable) this function, type:

cmd: **Monitor ON** (or **OFF**) **<Enter>**

To reveal more information in the monitoring mode, you can enable the MRpt command, which results in the display of the call sign(s) of the station(s) that repeat a monitored packet, in addition to the call signs of the originating and destination stations. With MRpt on, a monitored packet is displayed as:

N1BKE>K8CH,K1RO-9,WB1ENT-5*: Data

where N1BKE is the station originating the packet, K8CH is the intended destination of the packet, K1RO-9 and WB1ENT-5 are the stations repeating the packet and Data is the packet's contents. The asterisk indicates that WB1ENT-5 is actually transmitting the packet you are receiving. (The MRpt function is on by default.) To enable (or disable) this function, type:

cmd: **MRpt ON** (or **OFF**) **<Enter>**

Other monitoring commands that may be useful are:

MAll (default: on) allows you to monitor both connected and unconnected packets;

MCOM (default: off) allows you to monitor connect requests, disconnect requests, connect/disconnect acknowledgments and non-connect/disconnect acknowledgments, as well, as packets containing data;

MCon (default: on) display packets from stations on frequency, while your TNC is connected to another station;

MHeard lists the last 18 stations received on frequency by typing:

cmd: **MHeard <Enter>**

MHClear clears the list of received stations that are logged for MHeard command recall by typing:

cmd: **MHClear <Enter>**

MStamp (default: off) stamps the date and time of each monitored packet. To use this function, the date and time must be entered into the TNC using the DAytime command, by typing:

cmd: **DAytime *yymmddhhmm* <Enter>**

where *yy* is the last two digits of the year (00-99), ***mm*** is two digits representing the month (00-12), ***dd*** is the day of the month (00-31), ***hh*** is the hour (00-23) and ***mm*** is the minute of the hour (00-59). (Note that your TNC has a 24-hour, rather than a 12-hour clock.) For example, to enter March 8, 1951, 1:30 PM, type:

cmd: **DAytime 5103081330 <Enter>**

where 51 is the last two digits of 1951, 03 represents the third month (March), 08 represents the eighth day of the month, 13 represents the thirteenth hour of the day and 30 represents the thirtieth minute of the hour.

To limit the display of packets from certain stations, use the LCalls command to list a maximum of eight stations that you do or do not wish to monitor by typing:

cmd: **LCAlls *aaaaaa,bbbbbb, . . . hhhhhh* <Enter>**

where ***aaaaaa,bbbbbb, . . . hhhhhh*** are the call signs of stations you do or do not wish to monitor.

Next, use the BUDlist command to limit your monitoring according to the LCAlls list of stations. Your computer monitor will only display packets originating from stations in the LCAlls list if you type:

cmd: **BUDlist ON <Enter>**

Your computer monitor will display packets of all stations except the stations in the LCalls list (the TNC default). if you type:

cmd: **BUDlist OFF <Enter>**

Most of your monitoring needs can be met with the wide range of TNC monitoring commands at your disposal.

General Operating Parameters

The majority of amateur packet activity occurs at VHF,

VHF Digital Channels

Channels	*Activity*
50.620	Digital and packet calling frequency
51.12-51.18	Digital repeater inputs
51.62-51.68	Digital repeater outputs
53.530	Propagation Network (PropNET)
144.39	APRS
144.91-145.090	Packet
147.585	Propagation Network (PropNET)
219-220	Packet network*
223.52-223.64	Digital and packet
223.71-223.85	Packet (local coordinator's option)

*FCC requires written notification to the ARRL 30 days prior to operating a network relay in this band segment.

on 2 meters to be specific. The most commonly used data rate is 1200 bit/s using frequency modulated AFSK. Most TNCs are optimized for VHF FM operation, so getting on the air is a simple matter of turning on your packet equipment and tuning to your favorite packet frequency.

Most activity occurs on the following 2-meter frequencies: 144.390 and the 144.910-145.090 MHz subband. The former is the APRS channel for the U.S. and Canada, while the latter represents 10 non-APRS packet channels spaced in 20-kHz increments with 145.010 MHz usually being the most active channel of the 10. In other words, to find non-APRS packet activity, start at 145.010, then look up and down the other nine channels in the subband.

There is other packet activity on 2 meters as well as on 6 and 1.25 meters. Refer to the sidebar entitled "VHF Digital Channels" for information concerning all VHF packet activity.

APPLICATIONS

Now that you have this high-speed, error-free mode of communications up and running, what can we do with it? The answer is "plenty." The following paragraphs describe some of the more popular pursuits that have been applied to the packet mode.

Basic Packeting

After your TNC is tailored to your station and your mode of operation, you can send your first packet. The following paragraphs describe how to initiate and conduct packet communications from the keyboard of your computer.

To contact another station you must make a connection with that station. To do so, type:

cmd: ***Connect W1AW*** **<Enter>**

where ***W1AW*** is the call sign of the station to contact.

If W1AW's packet station is on the air and receives your connect request, your stations will exchange packets to set up a connection between stations. When the connection is completed, your computer will display:

*** CONNECTED to W1AW

and your TNC automatically switches to the Converse Mode.

Everything you type into the computer keyboard is now *packeted*. It is sent to the other station whenever you press the <Enter> key (the TNC default for the Sendpac character) or whenever the byte length of your keyboard input equals the number of bytes selected with the Paclen command (the default is 128 bytes).

If you are finished conversing with the other station, return to the Command Mode by typing **<CTRL-C>** (which is the TNC default for the Command character). When the command prompt (cmd:) is displayed, type:

cmd: **Disconne <Enter>**

and your station will exchange packets with the other station to break the connection. When the connection is broken, your computer displays:

*** DISCONNECTED

If, in mid-contact, you wish to enter a command and then continue with the contact, enter the Command Mode by typing **<CTRL-C>**. When the command prompt (cmd:) is displayed, enter the desired command. When you are ready to return to your contact, type:

cmd: **CONVers <Enter>**

and you are back in the Converse Mode. As before, anything you type now is packeted and sent to the other station.

If the other station does not respond to your initial connect request, your TNC resends the request. It will continue to do so until the number of additional attempts equals the number selected with the Retry command (the default is 10 attempts). If the number of additional attempts exceeds the Retry command selection, your TNC stops sending connect requests and your computer displays:

*** retry count exceeded

* * * DISCONNECTED

(Retry not only sets the maximum number of times that a connect request is resent, it also sets the number of times any packet is resent without acknowledgment.)

Packet Networking

The packet network is a complex evolutionary system of packet stations that have been loosely organized to transfer packets between local or distant points. Depending on the locations of those points, the packets being transferred between them may travel through a simple network consisting of one station operating on one frequency. They may also travel through a variety of networks consisting of tens of stations operating on HF, VHF, UHF and the Internet.

In the past, if you wanted to connect to another packet station, you had to know the exact path through the network to make the connection. If one or more stations were in that path, you had to know (and list) each station's call sign and use them when you invoked the Connect command.

Can you imagine what would happen if you had to know the name of each intermediary post office for every letter that you mailed? If you accidentally omitted a single name your letter would not be delivered!

Luckily, the post office does not require that you be familiar with its vast network to mail a letter. Similarly, in many parts of the packet network, you do not have to be familiar with the network to make a connection with another station.

The PBBS automatic mail-forwarding system is part of the packet network. Sending a message via the PBBS automatic mail-forwarding system is similar to sending a letter through the postal system. In either case, you do not have to know who will be handling your mail before it is delivered. All you need to know is the identity of the intended recipient and the address where the intended recipient picks up his or her mail.

For example, to send mail via the postal system, you address the envelope with the intended recipient's identification (the recipient's name) followed by the address where the recipient picks up mail. To send mail via the PBBS automatic mail-forwarding system, you address the mail with the intended recipient's Amateur Radio identification (the recipient's call sign) followed by the at-sign (@) and the address where the recipient picks up packet mail. Packet operators usually receive their mail at their local PBBS, so the address for a packet message is simply the call sign of the destination PBBS. Once you have properly addressed your mail, the system (either postal or PBBS) does the rest, automatically forwarding your mail to its destination.

Digipeater Networking

Digipeating is a function built into every AX.25-compatible TNC that permits the TNC to receive, temporarily store and retransmit (repeat) the packet transmissions of other stations. A digipeater repeats only transmissions that are specifically addressed for routing through that digipeater, as opposed to a typical voice repeater that retransmits everything it hears.

In light of the other packet networking options that are available today, digipeating is a very basic form of networking. Only eight digipeater stations can be used between any two points that are attempting to transfer packets. In order to use the digipeaters, the call sign of each must be known and specified when invoking the initial Connect command. Even though digipeating is rudimentary in comparison to what is available today, it still serves a purpose when an intermediary station or two is needed in a pinch to complete a connection.

When you are unable to make a connection with another station, the digipeater function provides the ability to circumvent the lack of connectability. You only need to determine which on-the-air packet stations can send and receive signals between your station and the station you are trying to reach. Once the existence of an intermediary station is known, type:

cmd: **Connect *WB8IMY* Via *K1ZZ-1* <Enter>**

where ***WB8IMY*** is the call sign of the station to connect to and ***K1ZZ-1*** is the call sign of the station that will act as the digipeater.

When K1ZZ-1 receives your connect request, the TNC automatically enters the digipeater mode and stores your request in memory until the frequency is quiet. It then retransmits your request to WB8IMY on the same frequency. If WB8IMY 's packet station is on the air and receives your connect request, your station and his will exchange packets through K1ZZ-1 to set up a connection between stations. Once the connection is established, your computer displays:

*** CONNECTED to WB8IMY VIA K1ZZ-1

Your stations continue to use the facilities of the digipeater until the connection is broken.

Digital and voice repeaters both repeat, but the similarity ends there. A digipeater receives and transmits on the same frequency (whereas a voice repeater receives and transmits on different frequencies). As a result, a digipeater operates half duplex; that is, it does not receive and transmit at the same time (as compared to a voice repeater, which transmits whatever it receives simultaneously).

If one station in the digipeater mode is insufficient to establish a connection, as many as eight stations can be called upon for digipeater operation in your connect request. Additional digipeaters are appended to the Connect command separated by commas. (For example, the command

Rick Lindquist, N1RL, is a regular on the local PacketCluster.

Connect WB8IMY Via K1ZZ-1,KU7G-3 <Enter> causes your TNC to send the WB8IMY connect request to K1ZZ-1, which relays it to KU7G-3. Then, KU7G-3 relays it to WB8IMY.)

Any packet station can act as a digipeater. This occurs automatically without any intervention by the operator of the station being used as a digipeater. You do not need his permission, only his cooperation because he can disable his station's digipeater function by means of the Digipeat command. (In the spirit of Amateur Radio, most packet operators leave the digipeater function enabled and disable it only under special circumstances.)

Similar to voice repeaters, some stations are set up as dedicated digipeaters. They are usually installed in good radio locations by packet clubs. Aside from location, the other advantage of a dedicated digipeater is that it is always there (barring a calamity). Stations do not have to depend on the whims of other packet stations, which may or may not be on the air when their digipeater functions are needed most.

Beyond Digipeaters

A network node is a relay station that allows you to connect to stations that are not connectable directly (a function similar to that of a digipeater). A node differs from the digipeater function because it is more intelligent.

Say you want to connect to station X, who is five digipeater hops away (digipeaters A, B, C, D and E). To connect to X, you use the command **Connect X via A, B, C, D, E <Enter>**. If A, B, C, D and E were network nodes rather than digipeaters, you would first connect with node A, then you would command node A to connect with node E. Once connected to node E, you would command node E to connect with station X.

If you use digipeaters, you have to know the path between you and station X (via A, B, C, D and E). If you use nodes, the nodes already *know* the path to other nodes. All you have to know is which node is local to station X. If a path changes (a node disappears or a new one appears), the node is aware of the change because nodes exchange information with each other regularly.

Another advantage of nodes is that their throughput is theoretically better than digipeaters. With digipeaters, only the last station in the circuit acknowledges received packets. With nodes, each node acknowledges each packet it receives. If a packet is lost in a chain of digipeaters, your TNC is not aware of that fact until it fails to receive an acknowledgment from the station at the opposite end of the circuit. If a packet is lost in a node sequence, the node that fails to receive an acknowledgment retransmits the lost packet until the next node in the circuit acknowledges it. As you might guess, the recovery of lost packets is lightning fast.

Packet nodes were the brainchild of Ron Raikes, WA8DED, who released NET/ROM to the packet community in 1987. NET/ROM became very popular because it was inexpensive and easy to install (just replace one of your TNC's EPROMs with a NET/ROM EPROM).

After NET/ROM, other packet radio network schemes were developed and introduced, for example, FlexNet, KA-Node, RATS Open System Environment (ROSE), TCP/IP for packet radio in a variety of flavors, TexNet, TheNet and X1J. They all are different in the way they work, but they all have the same goal: to simplify network connections, while improving network integrity.

All have achieved this goal with regard to simplifying network connections especially when compared with digipeater networking. You simply inform the network node that you want to connect to a specified station that uses a specified node, and the local node knows the network path to that remote node and station.

Some nodes use a mnemonic identifier instead of their FCC-given call signs. Typically, the identifier is a three-to-six character acronym that identifies the node's location. The reason for using identifiers is to simplify what the user has to remember. For example, instead of having to remember that WB1SOX is the node located at Fenway Park, all you have to remember is the node's mnemonic identifier, FENWAY.

To use a node, you first connect to your local node by entering a Connect command using your local node's call sign or mnemonic identifier. For example, to connect with the Fenway Park node, type either:

cmd: **Connect WB1SOX <Enter>**

or

cmd: **Connect FENWAY <Enter>**

After you receive the connected message from your local node, you may connect to another station that is also local to your node, or you may connect to another node.

To connect to another station, enter the Connect command followed by the call sign of the other station. To connect to another node, enter the Connect command followed by the call sign or mnemonic identifier of the other node. After you connect to the other node, use the Connect command again to connect to a station that is local to the other node. When you disconnect from another station, the node(s) disconnect automatically.

The Connect command used to connect to a node and the Connect command used to ask a node to connect to another station are different. When you connect to a node, you are using a TNC command. When you ask a node to connect to another station, you are using a node command. In order for a node to receive a node command, it must pass through your TNC as data; your TNC must not interpret a node command as if it were a TNC command. To ensure that this does not occur, do not send a node command until you connect to a node (after you receive the node connection message). If you send the node command too soon, the TNC will try to interpret it and, when it can't, will send you an error message. Note that after you connect to your local node, you will not receive another command prompt until you disconnect from the node, so don't wait for a command prompt before sending commands to a node.

TCP/IP

TCP/IP is different. It is based on the networking scheme known as the Defense Advanced Research Project Agency's (DARPA) Transmission Control Protocol (TCP) and Internet Protocol (IP). Despite what that name implies, TCP and IP are only two parts of a collection of protocols that comprise the complete DARPA protocol used for data communications networking between mainframe computers worldwide.

Phil Karn, KA9Q, wrote the amateur packet implementation of TCP/IP (for DOS-based computers) and called it *NET*. Later, he rewrote and refined *NET* and called it *NOS*. Other software writers produced new versions of *NOS*. And other software writers produced versions of *NET* and *NOS* for other platforms like the *Mac* and *Linux* operating systems.

One function of TCP/IP is to allow each TCP/IP station to be an intelligent network node. These nodes are *intelligent* because they automatically route packets to their intended destination without any operator intervention or direction. TCP/IP frees the operator from figuring out which string of digipeaters or remote network nodes must be used to get packets delivered to their intended destinations. The operator simply commands the software to communicate with station X and the network does the work, automatically determining the route to get to station X.

Besides intelligent networking, TCP/IP has a packet radio terminal function that allows users to communicate keyboard-to-keyboard with another user. A file transfer function, FTP, allows a user to send and receive (upload and download) ASCII (text) and binary (executables, graphics, etc) files to and from any other TCP/IP station. TCP/IP includes a built-in mailbox that automatically sends and delivers mail between TCP/IP stations and regular PBBSs. What's more, TCP/IP is multitasking on any machine. All of the aforementioned functions and more can be used simultaneously.

To get started with TCP/IP, you need:

- An IP address, which is a unique number assigned to the computer used at your packet station for communications over the TCP/IP network. To get an IP address, contact your local IP address coordinator. The source for the current list of IP address coordinators is **ftp.ucsd.edu/hamradio/amprnets**.
- Since TCP/IP emulates many of the functions of a TNC, you would think that it would allow you to get on packet

without it. Well, you still need a TNC and it must be one that supports the KISS mode. Virtually all current TNCs support KISS and many of the antique TNCs that don't support KISS can be made *KISSable* by updating the ROM containing the TNC software. (KISS is the acronym for Keep It Simple, Stupid.)

- TCP/IP software. There are many versions of the software and there are many sources. Some of the sources are **ftp.ucsd.edu/hamradio/packet/tcpip/** and **ftp.tapr.org/software_lib/tcpip/**.

Once you obtain these three items, you set up and use the software as described by its accompanying documentation. (A very useful book that details the setup and operation of NOS is *NOSintro: TCP/IP over Packet Radio* by Ian Wade, G3NRW. The book is available from the ARRL.)

After you have everything set, let's see if it works!

Before you run TCP/IP, you must place your TNC in the KISS mode. To do so, at the TNC's normal command prompt, type:

cmd: **KISS ON <Enter>**

After your TNC is in the KISS mode, power off the TNC, then power it back on. As your TNC initializes itself, its STA and CON front panel LEDs will blink three times to indicate that it is in the KISS mode. Some TNCs only require you to enter RESTART or RESET. You don't need to turn them on and off. Check your TNC manual to be sure.

Now, run TCP/IP and, after the system prompt appears (net>), type:

net> **FINGER (insert your call sign) @ (insert your call sign) <enter>**

The system reads and displays the file that you created in your subdirectory. (The FINGER file is a short description of you and your TCP/IP station.) Now it's time to perform a test to see if it works on the air. (*FINGERing* yourself is an operation that occurs within the confines of your computer; nothing is sent over the air.)

Try to get a complete DOMAIN.TXT or HOSTS.NET file from another station—N1FB in this example. (DOMAIN.TXT or HOSTS.NET files are look-up tables of the TCP/IP stations that, at a minimum, list each station's call sign and IP address. If you cannot get one locally, try **ftp.ucsd.edu/hamradio/amprhosts**) You can achieve this by using the File Transfer Protocol (or FTP). At the system prompt, type:

ftp n1fb <enter>

If all goes well, your computer indicates that a session has been *Established* and that n1fb.ampr.org FTP is *ready* for you to log on. After entering "anonymous," N1FB asks that you "Enter PASS command," so enter **pass <your call sign>** in response. (Pass is short for password.)

The *Logged in* message should appear shortly indicating that you are in N1FB's public subdirectory. Send **dir** to get a list of the contents (a directory) of the public subdirectory. If the directory contains a DOMAIN.TXT or HOSTS.NET file, you can get a copy by entering **get domain.txt** or **get hosts.net**.

After entering the get command, your computer indicates that a data connection for retrieval of the file is opening, then displays nothing while the file is actually transferred. The file transfer can go on for quite a while. The keying of your transmitter is the only indication that anything is happening. When the transfer is completed, your computer displays "Get complete" followed by the number of bytes transferred and the "File sent ok" message. At the system prompt, enter **close** to end the FTP session and log off K1ZZ.

There are misconceptions about TCP/IP that have caused some potential TCP/IP users to steer clear of it. Let me assure you that the following aren't true:

- You must keep your radio equipment and computer on 24 hours a day.

For the viability of the TCP/IP network (not to mention your local electric company), around-the-clock operation is preferable, but it's not a necessity as far as receiving mail is concerned. If your TCP/IP station is on the air all the time, it is always ready to receive traffic heading its way. If it is not active around the clock, other TCP/IP stations can intercept and hold your station's traffic for automatic delivery to you when your station is on.

- You must use a DOS-based computer.

Although most computers using TCP/IP are DOS-based, versions of TCP/IP for other computers do exist. I use an Apple Macintosh and there are versions for other computers and operating systems, too.

- You must use a fast computer with a lot of RAM and hard disk storage.

Relatively slow computers are used successfully in the amateur TCP/ IP world. Fast clocks and oodles of memory and disk storage are nice, but you can TCP/IP without them.

- TCP/IP cuts you off from the rest of the packet world.

Many AX.25-to-TCP/IP gateway stations and PBBSs forward your mail between the networks, so you are not cut off from everyone else.

PACKET BULLETIN BOARD SYSTEM (PBBS)

A packet bulletin board system (PBBS) is essentially an electronic mailbox that you access via packet radio. It allows you to post mail on the system for later retrieval by the addressee. You can also retrieve mail that's addressed to you. The mailbox does not limit your message posting to stations that frequent the local PBBS. If you know a station that checks into a distant PBBS, you can post a message to that station and it will be routed automatically to its destination. Besides storing individual messages, files of interest to the general packet population may be stored as well.

Hank Oredson, WØRLI, became the father of PBBS when he decided to put a $50 surplus Xerox 820-1 *CP/M* computer to good use by writing software that permitted it to function as a BBS; not a telephone BBS, but a packet BBS. In short order, WØRLI PBBSs began showing up on the air throughout packet country and became so popular that other hams wrote PBBS software for their favorite computers.

As tastes and technology changed over the years, various PBBSs grew in and out of favor with the packet radio population. Today, the most popular PBBS is probably the Roy Engehausen, AA4RE, PBBS known as *BB*. The sidebar

(AA4RE PBBS *BB* Command Set) describes the commands of this system and the PBBS procedures described below are based on this command set as well. Even if your local PBBS is not a *BB*, the sidebar and following description will still be applicable because most PBBSs use similar, if not identical, commands for their basic operations.

Finding a PBBS

There are fewer PBBSs on the air than there used to be, but there is likely to be at least one or two active locally on the 2-meter subband of 144.910-145.090 MHz.

Logging On and Off

Once you locate a PBBS, you must "log on" in order to use it. Logging onto a PBBS is as simple as initiating a contact with any other packet station. Just use the Connect command. For example:

cmd: **Connect *AA1GW-4* <Enter>**

where ***AA1GW-4*** is the call sign and SSID of the PBBS. After you are connected to the PBBS, its preamble is displayed on your computer followed by a request for commands from you.

When you are finished using a PBBS, you log off the system by using the B (for bye) command. When the PBBS receives the B command, it logs you off the system and disconnects.

Reading Mail

To read messages that are posted on the PBBS, you must know what messages are there. If there is a message stored on the PBBS addressed to you, the PBBS informs you of that fact when you log onto the system. By using the RM (for *read mine*) command, you can read whatever mail is waiting for you. In response to the RM command, the PBBS retrieves each message and sends it to your computer. After you have read your messages, use the KM (for *kill mine*)

AA4RE PBBS *BB* Command Set

The following is a synopsis of the commands that are available with the AA4RE Packet Bulletin Board System, *BB*, software.

General Commands

Command	Description
B	Log off PBBS.
NE	Toggle between short and extended command menu.
NH *x*	Enter the call sign (*x*) of the PBBS where you normally send and receive mail.
NN *x*	Enter your name (*x*) in system.
NZ *n*	Enter your ZIP Code (*n*).
T	Ring bell at the SYSOP's computer for one minute in order to converse with SYSOP.

Information Commands

Command	Description
DU *x*	Display information concerning station whose call sign is *x*.
J	Display list of TNC ports used by the PBBS.
JL	Display call signs of stations recently connected to PBBSs.
JN	Display call signs of stations currently connected to PBBSs.
J *x*	Display call signs of stations heard on port *x*.
H	Display summary of PBBS commands.
H *x*	Display description of command *x*.
I	Display information about PBBS.
V	Display PBBS software version.

Message Commands

Command	Description
KM	Kill all messages addressed to you that you have read.
K *n*	Kill message numbered *n*.
KT *n*	Kill NTS traffic numbered *n*.
L	List all new messages since you last logged on PBBS.
L$	List all messages with BIDS containing a specific pattern.
L@ *x*	List all messages addressed for forwarding to PBBS whose call sign is *x*.
L< *x*	List all messages from station whose call sign is *x*.
L> *x*	List all messages addressed to station whose call sign is *x*.
LA	List all type A bulletin messages.
LB	List all type B bulletin messages.
LD<> *yymmdd*	List all messages older than year *yy* month *mm* day *dd*.
LD> *yymmdd*	List all messages newer than year *yy* month *mm* day *dd*.
LF	List all forwarded messages.
LL *n*	List the last *n* messages stored on PBBS.
LM	List all messages addressed to you or from you.
L *n*	List all messages numbered higher than *n*.
LN	List all unread messages.
LP	List all private messages.
LS	List all messages with subjects containing a specific pattern.
LT	List all NTS traffic.
LU	List all unread messages from you.
LY	List all read messages.
R$	Read all messages with BIDS containing a specific pattern.
R@ *x*	Read all messages addressed for forwarding to PBBS whose call sign is *x*.
R< *x*	Read all messages from station whose call sign is *x*.
R> *x*	Read all messages addressed to station whose call sign is *x*.
RA	Read all type A bulletin messages.
RB	Read all type B bulletin messages.
RD< *yymmdd*	Read all messages older than year *yy* month *mm* day *dd*.
RD> *yymmdd*	Read all messages newer than year *yy* month *mm* day *dd*.

command to delete all of the messages you have read.

To read other messages that are on the PBBS, use the L (for *list*) command to view a list of all the messages that have been stored on the PBBS since the last time you logged onto the system. The PBBS lists each message by its message number, so if you wish to read a particular message use the R (for *read*) command by typing:

R *n* <Enter>

where ***n*** is the number of the message to be read.

Sending Mail

To send a message via a PBBS, use the S (for *send*) command. In response to the S command, the PBBS asks you for the call sign of the addressee. To post the message for retrieval on the local PBBS, simply type the call sign of the station to receive the message, then press **<Enter>**. If you want to have your message forwarded to a station at another PBBS, type the call sign of the station to receive the message, followed by the at-sign (@) and the call sign of the destination PBBS, then press **<Enter>**. For example:

***WA1STO@W9JJ* <Enter>**

where ***WA1STO*** is the call sign of the station that is to receive the message and ***W9JJ*** is the call sign of the destination PBBS where the message will be sent for retrieval. To save a step, you can address the message at the same time you invoke the S command (for example, S ***WA1STO @W9JJ***).

Next, the PBBS prompts you for the title of the message. Type a short two- or three-word title that represents the contents of the message, then press **<Enter>**. Next, type the contents of the message and end the last line of the message with **<CTRL-Z>** or **/EX**, then press **<Enter>** to indicate the end of the message. When the PBBS receives the <CTRL-Z>, it stores the message for later retrieval or forwarding.

To assist mail-forwarding, you can append geographic information to the message' s address. For example, to give the message addressed to WA1STO@W9JJ a little push, you can append .CT.USA.NA to W9JJ, which results in an address of WA1STO@W9JJ.CT.USA.NA. (CT, USA and NA are the abbreviations for the state, country and continent, respectively in which W9JJ is located.) This is known as hierarchical addressing.

REPLY *n*	Respond to the message numbered *n*.
RH *n*	Read message(s) numbered *n* with full message header displayed.
RH *n1-n2*	Read message(s) numbered *n1* through *n2* with full message header displayed.
RM	Read all messages addressed to you.
R *n*	Read all message(s) numbered *n*.
R *n1-n2*	Read all message(s) numbered *n*1 through *n*2.
RS	Read all messages with subjects containing a specific pattern.
RT	Read all NTS traffic.
SB ALL	Send a bulletin message addressed to everyone.
SP *x*	Send a private message to station whose call sign is *x*.
SR *n*	Respond to the message numbered *n*.
ST *xxxxx@NTSyy*	Send NTS message to ZIP Code xxxxx in state *yy*.
S *x*	Send message to station whose call sign is *x*.
S *x@y*	Send a message to station whose call sign is *x* at PBBS whose call sign is *y*.
File Commands	
DB *xyz*	Using protocol *x*, download from directory named *y*, the file named *z*.
D *xy*	Download from directory named *x*, file named *y*.
UB *xyz*	Using protocol *x*, download to directory named *y*, the file named *z*.
U *xy*	Upload to directory *x*, the file named *y*.
W	List what directories are available.
WD *x*	List what files are available in directory named *x* along with their date and time stamp.
WX *x*	List what files are available in directory named *x* along with their date and time stamp and size.

Once Is Enough!

If the PBBS does not respond to a command immediately, be patient and do not resend the command. One unique feature of packet is that whatever you send is received perfectly at the other end. Sometimes to achieve this result takes a number of attempts, especially if there is a lot of activity on frequency. If you send the same command twice, the PBBS eventually receives it twice and responds to it twice! If the response is a long one, your repeated command wastes valuable time.

DX PACKETCLUSTER

How do DXers find out about DX? They used to depend on the print media to spread the DX word. The problem is that the print media news is not always current. DXpedition plans may change at the last minute or a new country may pop on the air without warning.

DX spotting alleviates this problem. When someone spots new DX, he or she spreads the word by making an announcement on the local DX spotting frequency. During the 1970s and 1980s, the DX spotting frequency was typically a 2-meter FM voice repeater. The problem is that if you turned down the volume of your 2-meter radio to dig out a weak signal on 20 or if you were out of the shack or fast asleep, you missed the spot. Also, if you were not in range of that 2-meter repeater, you had no chance of hearing the spot. Packet and *DX PacketCluster* software has changed that forever, and DX spotting has become one of the most popular applications of packet communication.

DX PacketCluster consists of specialized software running on a packet radio station computer, which allows multiple stations to connect to that station or node. The software also allows each node to connect to other nodes to form a DX PacketCluster network. So, when a station connects to a node and makes an announcement, all stations connected to that node and all the nodes in the network receive the same announcement.

For example, let's say that K1CE announces that Abu Dabi is active on 7005.0 kHz. If K1CE connects to a *DX PacketCluster* node that is part of your local *PacketCluster* network, you will be able to copy his announcement even if you cannot copy K1CE directly.

If the spot was announced while you were digging out a weak one on 20, you would not miss it as long as your packet radio station was running silently in the background. And even if you missed the spot for some reason, you could get it later by asking the *DX PacketCluster* for old spots.

If you wanted to talk with K1CE, the *DX PacketCluster* permits you to do that, too. What if you wanted to know K1CE's location? You can use the *PacketCluster* to ask K1CE exactly where he is located. Besides conversing with any station on the *PacketCluster* network, users may conduct conferences (discussions involving more than two people).

Another function of the *DX PacketCluster* is its bulletin board system that operates in a fashion similar to a standard PBBS. Like other PBBSs, the *PacketCluster* system tells you when you have new mail. To read mail, or any message or bulletin, you use the R (for *read*) command. To post a message or bulletin of your own, you use the S (for *send*) command. To obtain a list of the messages and bulletins that are on the board, you use the DI (for *directory*) command and to remove a message or bulletin from the board, you use the K (for *kill*) command. The sidebar (*DX PacketCluster* Command Set) provides a synopsis of the *PacketCluster* commands.)

Show Me DX

Perhaps the most powerful command in the *DX PacketCluster* command set is the SH (for *show*) command. Invoke the SH/DX (for *show DX*) command, and the *PacketCluster* lists the last five DX announcements including the DX station's operating frequency, call sign, the time and date of the announcement and other pertinent information concerning the DX station (long path, LOUD, weak and so on). If you wish to limit the SH/DX command, you may do so. For example, SH/DX 10 causes the *PacketCluster* to recount the last five DX announcements for 10 meters only.

When XK1DDO is at the bottom of a pileup on 20 meters, you can use the *DX PacketCluster* to get some ammunition before you try to work the station. Assuming you have already entered your longitude and latitude in the *PacketCluster*'s database (using the SET/LO for *set location* command), invoke the SH/H (for *show heading*) command for Jabru (SH/H XK1). In response, the *PacketCluster* performs some calculations and sends results that will look something like this:

XK1 Jabru: 107 degs
Q dist: 7238 mi, 11649 km
Reciprocal heading: 309 degs

Now you can use the SH/SU (for *show sun*) command (SH/SU XK1) and the *PacketCluster* responds with:

XK1 Jabru Sunrise: 0455Z Sunset: 1703Z

Finally, try using the SH/M (for *show MUF*) command (SH/M XK1) and the *PacketCluster* responds:

Jabru propagation: MUF: 22.7 MHz LUF: 2.3 MHz

Other variations of the Show command include the SH/W (for *show WWV*) command to obtain WWV propagation information and the SH/U (for *show users*) command to obtain a list of all the other stations connected to the *DX PacketCluster* node or network.

As you can see, the *DX PacketCluster* is a powerful tool. Although similar tools existed in the past, only packet could make it as powerful as it is today.

DX PacketCluster Command Set

The following is a synopsis of the commands that are available with the *DX PacketCluster* software.

General Commands

B	Disconnect from cluster.
Q	Disconnect from cluster.
SET/H	Indicate that you are in your radio shack.
SET/LO *abcdef*	Set your station's latitude as *a* degrees *b* minutes *c* north or south and longitude *d* degrees *e* minutes *f* east or west, e.g., SET/LO 41 33 N 73 0 W.
SET/NA *x*	Set your name as *x*.
SET/NE *x*	Store in database that you need country(s) whose prefix(s) is *x*.
SET/NOH	Indicate that you are not in your radio shack.
SET/NON *x*	Delete from database country(s) you no longer need whose prefix(s) is *x*.
SET/QTH *x*	Set your QTH as location *x*.
T *x*	Talk to station whose call sign is *x*. Send <CTRL-Z>to terminate talk mode.
T *xy*	Send to station whose call sign is *x*, a one-line message *y*.

Information Commands

?	Display a summary of commands.
DX *xyz*	Announce DX station whose call sign is *x* on frequency *y* with optional comment *z*.
?H	Display a summary of commands.
H	Display a summary of commands.
SH/AN	Display last five announcements addressed to all.
SH/BU	Display list of bulletin files.
SH/BULLA	Display list of messages addressed to all.
SH/C	Display physical configuration of cluster.
SH/C/N	Display list of nodes connected to cluster.
SH/CL	Display names of nodes and number of users in cluster.

AUTOMATIC POSITION REPORTING SYSTEM (APRS)

APRS is the acronym for Automatic Position Reporting System, which is software that uses the unconnected packet radio mode to graphically indicate the position of moving and stationary objects on maps displayed on a computer monitor. Unconnected packets are used to permit all stations to receive each transmitted APRS packet on a one-to-all basis rather than the one-to-one basis afforded by connected packets.

Virtually all VHF APRS activity occurs on 2 meters, specifically on 144.39 MHz, which is recognized as *the* APRS operating channel in the United States and Canada. Like most other 2-meter packet operations, APRS operates at 1200 bit/s.

APRS software is available in a variety of flavors.

- *APRS*, the original flavor, is intended for *DOS*
- *MacAPRS* is intended for *Mac OS*
- *WinAPRS* is intended for *Windows 95, 98* or *NT*
- *APRS+SA* is intended for *Windows 95, 98* or *NT* while running DeLorme *Street Atlas USA*
- TH-D7 and TM-D700 are Kenwood's handheld and mobile transceivers with built-in TNCs and built-in versions of APRS.
- *pocketAPRS* is intended for *Palm OS*

To avoid confusion throughout this chapter, I will refer to the *DOS* version of APRS as *APRS (DOS)*.

All versions of APRS software are shareware. The primary source for APRS software and maps is the Internet. Two excellent sites are **http://www.tapr.org** and **http://www.tawg.org**

SH/COM	Display available database commands.
SH/D *n*	Display the last five DX announcements for frequency band *n*.
SH/D *n1-n2*	Display the last five DX announcements for the frequency range of *n1-n2*.
SH/D *x*	Display the last five DX announcements for station whose call sign is *x*.
SH/D	Display the last five DX announcements.
SH/FI	Display names of files in general files area.
SH/H *x*	Display heading and distance to country whose prefix is *x*.
SH/LOC *x*	Display the longitude, latitude and distance of station whose call sign is *x*.
SH/LOG *x*	Display cluster's log entries for station whose call sign is *x*.
SH/LOG	Display last five entries in cluster's log.
SH/M *x*	Display MUF for country whose prefix is *x*.
SH/NE *x*	Display call signs of stations needing country whose prefix is x.
SH/NE *x*	Display needed countries for station whose call sign is x.
SH/NO	Display system notice.
SH/PR *x*	Display country(s) and zone(s) assigned to prefix *x*.
SH/ST *x*	Display information about the station whose call sign is *x*.
SH/SU *x*	Display the sunrise and sunset times in country whose prefix is *x*.
SH/U *x*	Display name and location of station whose call sign is *x*.
SH/U	Display call signs of stations connected to the local node.
SH/U/F	Display call signs of stations connected to the cluster.
SH/W	Display last five WWV solar flux announcements.
W SF=*xxx*, A=*yy*, K=*zz,a*	Announce and log WWV propagation information where *xxx* is the solar flux, *yy* is the A-index, *zz* is the K-index and *a* is the forecast.
	Message Commands
A *x*	Send message *x* to all stations connected to the local node.
A/F *x*	Send message *x* to all stations connected to the cluster.
A/*x y*	Send to stations connected to node *x*, message *y*.
DE *n*	Delete message numbered *n*.
DI	List all new messages since you last logged on to cluster.
DI/A	List all messages on local node.
DI/N	List messages added since you last invoked the DI command.
DI/O	List messages addressed from or to you.
K *n*	Delete message numbered *n*.
L	List all new messages since you last logged on to cluster.
R *n*	Read message numbered *n*.
R	Read message(s) addressed to you.
REP	Reply to the last message read by you.
REP/D	Reply to and delete the last message read by you.
S *x*	Send a message to station whose call sign is *x*.
S/P	Send a private message.
S/RR *x*	Send a message with return receipt to station whose call sign is *x*.
	File Commands
R/*x y*	Read from file directory *x*, file named *y*.
TY/FI *x*	Display contents of file named *x*.
UP/B *x*	Upload a bulletin file named *x*.
UP/F *x*	Upload a file named *x*.
UPD/*x*	Update the database file named *x*.

Assembling an APRS Station

The standard configuration for packet radio hardware (radio-to-TNC-to-computer) also applies to APRS until you add a GPS receiver to the mix. Note that you do not need a GPS receiver for a stationary APRS installation, nor do you need a computer for a mobile APRS (tracker) installation. In these installations, the extra port or special cable is not an issue. It only becomes an issue when you desire both a computer and GPS receiver in the same installation.

One way of accomplishing this is by using a TNC or computer that has an extra serial port for a GPS receiver connection. Lacking the extra serial port, you can use a hardware single port switch (HSP) cable to connect a TNC and GPS receiver to the same serial port of your computer. The HSP cable is available from a number of sources including TNC manufacturers Kantronics, MFJ and PacComm.

Whichever GPS connection you use, make sure that you configure the APRS software so that it is aware that a GPS receiver is part of the hardware configuration and how the GPS receiver connection is accomplished.

Configuring APRS Software

You can not simply install the APRS software and run it because configuring the software is critical to its proper operation.

There are certain parameters that are important for the correct operation of the software. Many others are not critical, and those can be fine-tuned later after you get on the air and familiarize yourself with the operation of APRS. The important parameters are as follows:

- Call sign and Secondary Station Identifier (SSID) that you will use for APRS.
- Map icon—This setting determines the icon or symbol used to represent your station on the APRS maps.
- Location (latitude and longitude)—If you will be using a GPS receiver with APRS, the receiver will provide your latitude and longitude to the software automatically. If you are not using a GPS receiver, you will have to provide that information. If you know your latitude and longitude, you can configure the software with that information. If you do not know your latitude or longitude, you can get an approximation of that information after you start the APRS software by moving the computer cursor to the your location on the APRS map. The software will display the coordinates of the cursor on the map and you can then transfer the coordinates to the software.
- TNC type—You must indicate the type of TNC you are using, e.g., one-port or dual-port. In some cases, the software may also need to know the brand and model of TNC you are using.
- TNC computer port—You must indicate which computer port is used for the TNC connection and you must configure that port so that its parameters (data rate, character length, parity, etc.) are compatible with the TNC.
- Enable the GPS receiver—If you are using a GPS receiver with APRS, you must configure the software so it is aware that a GPS receiver is part of the APRS set-up and how it is connected in the set-up.
- GPS receiver computer port—If the GPS receiver, if any, is connected via a computer port, you must indicate which computer port is used and you must configure that port so that its parameters (data rate, character length, parity, etc.) are compatible with the GPS receiver.
- Unproto path—This is the digipeater path used for the retransmission of your packets. Until you become familiar with the APRS network topology in your area, set this parameter to RELAY,WIDE or leave it set to its default selection.
- Position report rate—This setting determines how often your station automatically sends packets containing your current location. At one extreme, this setting should be set to transmit often for fast moving objects (like planes

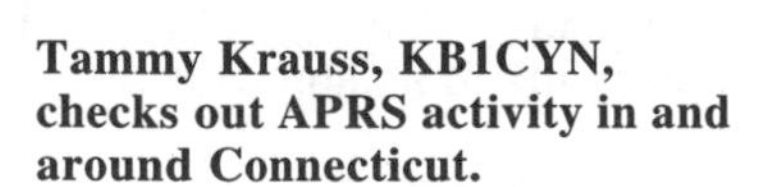

Tammy Krauss, KB1CYN, checks out APRS activity in and around Connecticut.

and trains). At the other extreme, this setting should be set to transmit least often for stationary objects (like a building).

- Local time at your station.
- Registration number—You will receive a registration number when you pay the shareware fee for the software. As soon as you receive this number, enter it into the software because without this number, you have to reconfigure the software each time you start it.

As you can imagine, different versions of APRS have different ways of performing the configuration. Here are some general guidelines for configuring each version of APRS.

- *APRS* (*DOS*) is the easiest version of APRS to configure. When you start the program for the first time, the program prompts you for the pertinent parameters for configuration.
- *MacAPRS* is configured by means of its Settings menu, particularly the Station Settings, GPS Settings, TNC Settings, MacAPRS Settings and Communications submenus under the Settings top menu. After selecting the parameters, you must toggle on the Open Port parameter under the Settings menu.
- *WinAPRS* is configured by means of its Settings menu, particularly the Station, APRS, Serial Port and TNC submenus under the Settings top menu. After selecting the parameters, you must toggle on the Open Port parameter under the Settings menu.
- *APRS+SA* is configured by means of its Setting menu, particularly the Main Parameters and Profile tabs under the Settings menu.
- *TH-D7* is configured by pressing its MENU button, selecting 2, then setting the parameters under 2, that is, 2-1 through 2-C.
- *pocketAPRS* is configured by means of the Settings button that appears when you start the software for the first time. Tapping the Settings button transfers you to the *pocketAPRS* Settings and Station Settings windows. When you are finished configuring those windows, select Transmit Control from the Settings menu to complete the configuration.

The preceding paragraphs describe the basics for configuring the APRS software. To fine-tune the configuration, refer to the ARRL publication entitled *APRS: Tracks, Maps and Mobiles* by Stan Horzepa, WA1LOU, which devotes a whole chapter to installing and configuring the various versions of APRS.

Using Maps

The most important part of using APRS involves using the APRS maps. And the most important part of using APRS maps is focusing a map on a particular location and magnifying the map in order to get a detailed view of the map and the APRS activity at that location. In all versions of APRS, centering the map on a location involves either moving the cursor to that location or moving the map so that the location is centered in the map window.

To magnify a map, you press the **<Page Down>** key in all versions of APRS except *pocketAPRS*, which requires that you select the plus (+) magnifying glass icon at the bottom of the map window, then tap the portion of the map you wish to magnify. To demagnify a map, you press the **<Page Up>** key in all versions of APRS except *pocketAPRS*, which requires you to select the minus (–) magnifying glass icon, then tap the map. All versions of APRS also have menu commands to zoom in and out of a map.

Tracking

Tracking moving objects on maps is the primary function of APRS. This makes APRS a great tool for public service communications. The ability to replay tracks just adds to this tool's functionality.

In order to track a moving object, that object must transmit its position as it moves. The moving object may do this automatically with a GPS receiver configured to a transmitter or manually by an operator providing position information to a computer running APRS aboard the moving object or remotely by an operator, placing an object like a storm system on an APRS map. In either case, each time a new position is received, the icon of that object moves accordingly on the APRS maps.

The following paragraphs describe how to track and replay tracked stations (**Figs 6-2** and **6-3** illustrate tracking and replaying).

APRS (*DOS*)–To replay a track: press **O**, press **R**, then type the call sign of the station whose track you wish to replay or press **<Enter>** to replay all the tracks.

APRS+SA–Select the **Track** tab, then click on the **Enable Tracking** box if it does not contain a check mark. Type the call sign of the station to track in the **<Callsign>** field, then press **Enter**. Click on the **Map All Reports** or **Map Most Recent** button to view the track of the selected station on the *Street Atlas* map.

MacAPRS–This version of APRS tracks all moving objects automatically. To replay a track: select the most recent map icon (the last position received) of the station to replay, then press **Command-R**.

WinAPRS–This version of APRS also tracks all moving

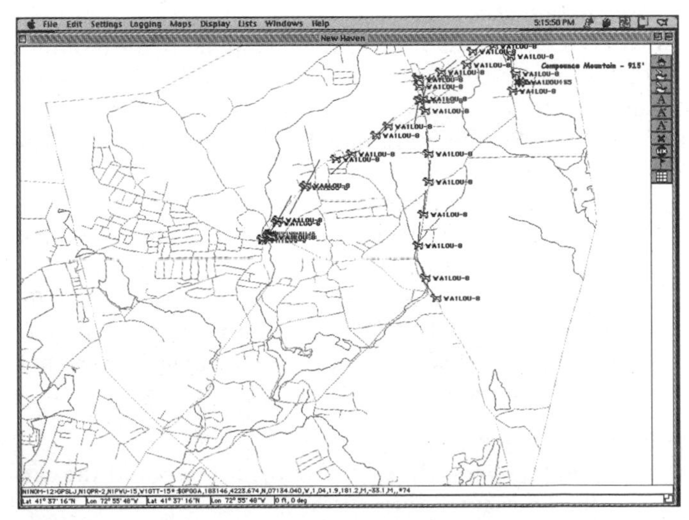

Fig 6-2—MacAPRS tracks WA1LOU-8 roving around downtown Wolcott, CT.

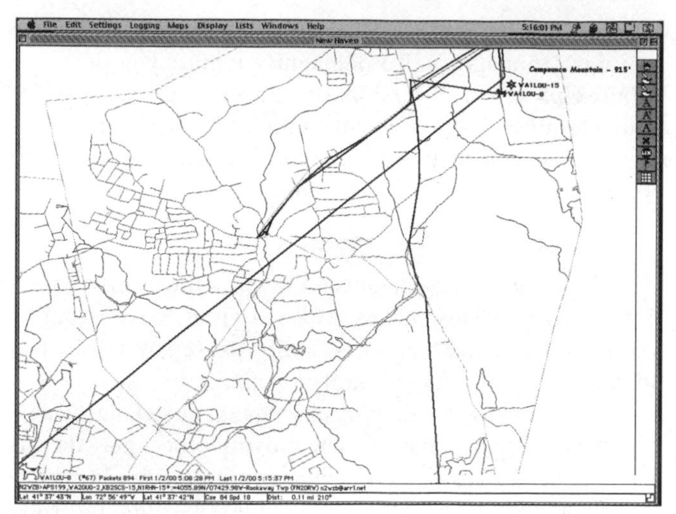

Fig 6-3-MacAPRS replays the track of WA1LOU-8's tour of downtown Wolcott, CT.

objects automatically. To replay a track: select the most recent map icon (the last position received) of the station to replay, then press **R**.

Adding Objects

Adding other things, i.e., objects, to the APRS map permits you to broadcast the location of those objects to all the APRS maps in your network. For example, you can display the location of a storm, a portable emergency operation center, or a lead runner in a marathon. These are all objects whose positions are critical but do not have the APRS equipment aboard to broadcast their positions themselves.

As the position, speed and course of an object change, you can enter updated information in order to adjust the icon of the object on the APRS map. And, when the storm, emergency or public service event is over, you can remove the object from the APRS map.

The following describes how to add or delete an object on an APRS map. It also describes how to change the location of an object.

• To add objects:

APRS (DOS)—Press **I**, then **A** and respond to the prompts.

APRS+SA—Select **Create Object . . .** from the **Objects** menu, fill in the blanks, then click on the **OK** button.

MacAPRS—Press **Command-E**, fill in the blanks, then click on the **OK** button.

pocketAPRS—Select the cross (+) icon, then tap the map where you want the object to appear. Fill in the blanks, then select the **Done** button.

WinAPRS—Select **Edit/Add Weather Object . . .** for weather objects or **Edit/Add Station/Object . . .** for other objects from the **Edit** menu. Fill in the blanks, then click on the **OK** button.

• To delete objects:

APRS (DOS)—Press **P**, select the object's icon with the cursor, then press **<Enter>**. Press **D**, then **Y**.

APRS+SA—Select the **Objects** tab, then select the object with the right mouse button. Select **State**, then **Kill**.

MacAPRS—Select **Station List** from the **List** menu, select the object, then press **<Delete>**.

pocketAPRS—Select the arrow icon, tap the object on the map, then select **Delete** from the drop-down menu.

WinAPRS—Select **Station List** from the **List** menu, select the object, then press **<Backspace>**.

• To change the location of objects:

APRS (DOS)—Select the object's icon with the cursor, then press **<Enter>**. Move the cursor to the new location, then press **I**.

MacAPRS—Select the object's icon, then press **Command-E**. Change the necessary parameters, then click on the **OK** button.

WinAPRS—Select the object's icon, then select **Edit/Add Weather Object . . .** for weather objects or **Edit/Add Station/Object . . .** for other objects from the **Edit** menu. Change the necessary parameters, then click on the **OK** button.

Messages and Bulletins

In addition to using maps, APRS permits keyboard-to-keyboard communications. You can send and receive one-line messages to other stations and transmit bulletins to all stations containing information that may be critical to an ongoing event or to announce an approaching event.

The following describes how to send and receive messages and bulletins with APRS.

• To send a message:

APRS (DOS)—Press **S** and respond to the prompts.

APRS+SA—Press **<Ctrl-P>**, fill in the blanks, then click on the **Send** button.

MacAPRS—Press **Command-M**, fill in the blanks, then click on the **OK** button.

pocketAPRS—Select **Send Message** from the **Main** menu, fill in the blanks, then click on the **Send** button.

TH-D7—Press the **MSG** button, then select **INPUT**. Fill in the blanks, then press the **OK** button twice.

WinAPRS—Press **F7**, fill in the blanks, then click on the **OK** button.

• To read a message:

APRS (DOS)—Press **R**.

APRS+SA—Select the **Traffic** tab.

MacAPRS—Press **Command-4**.

pocketAPRS—Select **Messages** from the **Views** menu.

TH-D7—Press the **MSG** button, select **LIST**, then select a message to read.

WinAPRS—Select **Message List** from the **Lists** menu.

• To send a bulletin:

APRS (DOS)—Press **S**. At the **To:** prompt, type **BLN#**, where # represents line number 1 through 9 of the bulletin, then enter one line of the bulletin at the **Entr MsgText:** prompt.

APRS+SA—Press **<Ctrl-B>**, type the contents of the bulletin, then click on the **Send** button.

MacAPRS—Press **Command-M**, click on the **Bulletin** button, type the contents of the bulletin, then click on the **OK** button.

pocketAPRS—Select **Send Message** from the **Main** menu, then select **Bulletin** from the pick list. Type the contents of the bulletin, then click on the **Send** button.

TH-D7—Press the **MSG** button, then select **INPUT**. At the **To:** prompt, type **BLN#**, where # represents line number 1 through 9 of the bulletin, then press the **OK** button twice.

WinAPRS—Press **F7**, then click on the **Bulletin** button. Type the contents of the bulletin, then click on the **OK** button.

• To read a bulletin:

APRS (DOS)—Press **B**.

APRS+SA—Select the **Bulletins** tab.

MacAPRS—Press **Command-4**.

pocketAPRS—Select **Messages** from the **Views** menu, then select **Bulletins** from the pick list.

TH-D7—Press the **MSG** button, select **LIST**, then select a message to read.

WinAPRS-Select **Message List** from the **Lists** menu.

APRS on the Internet

APRS stations called IGates are connected to the Internet to relay local APRS data to central Internet sites called APRServers. APRServers delete bad and duplicate packets, then allow other stations to access that data to view APRS activity worldwide.

If your APRS computer is connected to the Internet and your Web browser is Java-capable, you can connect to an APRServer (such as **http://www.aprs.net/usa1.html**) and display international APRS activity on your APRS maps.

To view nationwide and worldwide APRS activity:

- *APRS (DOS)*—With the serial port configured for a modem connection, press **F**, then **G**. APRS dials a server and downloads live backup files for each area of the country. When the call is over, press **F**, then **L** to load the desired live backup files.
- *APRS+SA*—Click the **Setup** menu, then the **Internet** tab. Configure the settings, then click on the **Connect** button and select **Close** from the **File** menu.
- *MacAPRS*—Select **TCP/IP Connections** from the **Settings** menu, then select **Conn to APRServe Network**.
- *WinAPRS*—Select **TCP/IP Connections** from the **Settings** menu, then select **Connect to APRServe Network**.

IGates also permit you to send one-line email messages to Internet email addresses. After addressing the message and transmitting it on the air, it is relayed to the Internet by any IGate that receives. (You receive an acknowledgment from the IGate after it relays your message.)

To send a message, address it to EMAIL and insert the Internet address as the first item in the message. For example, to send email to wd5ivd@tapr.org, the message would look something like this:

wd5ivd@tapr.org Where is the DCC next year?

PROPAGATION NETWORK (PROPNET)

Propagation Network (PropNET) is a continent-wide digital network trying to unlock the mysteries of VHF propagation. The network operates on 53.530 and 147.585 MHz by transmitting Unnumbered Information (UI) packets during meteor showers and the late summer tropospheric ducting season.

Participating stations transmit short packets frequently and monitor the channel for packets from stations beyond the range of normal propagation. APRS software is used by many participants because of the software's ability to plot the origination of each received packet on a map automatically.

If you do not have a powerful transmitter and/or high-gain antenna, you still can participate as a receive-only station. Receive-only stations are just as important as transmitting stations. Since receive-only stations do not transmit and are receiving continuously, they will never miss anything interesting during a transmission. For example, in 1999 WA1LOU participated as a receive-only station during a meteor shower and was one of the first stations to receive anything unusual (two packets from a station over 500 miles away on 2 meters).

Since PropNET is an ongoing experiment, its operating parameters are constantly being refined, especially with regard to the optimal TNC settings. So, it is recommended that you check the PropNET Web site for the current *modus operandi* (**http://go.to/BEACONet**).

To follow the activity of PropNET, you can subscribe to its email list (at **http://www.tapr.org**). By subscribing to the list, you will know when PropNET is active and be able to exchange information concerning your experiences during that activity.

THE 21ST CENTURY

At the dawn of the 21st Century, it is a good time to reflect on what it will bring with regard to VHF digital applications. TCP/IP and APRS are likely to become more popular as the 21st Century opens up. *DX PacketCluster* is also likely to have another growth spurt as sunspot conditions improve and the higher HF bands open up more frequently. You can expect the Internet to become a bigger player in the VHF digital applications as well.

The VHF digital world has been an interesting place at the end of one century and is sure to be even more interesting in the future.

Chapter 7

Traffic Handling

Getting the Message Through

Maria Evans, KT5Y, ARRL Staff and ARRL National Traffic System Officials

For just pennies a day, you can protect you and your family from all sorts of catastrophic illnesses with the Mutual of Podunk health care policy. . . .

Yes, at the amazing low price of $9.95, you can turn boring old potatoes, carrots and okra into culinary masterpieces with the Super Veggie-Whatchamadoodler—a modest investment for your family's mealtime happiness. . . .

Tired of all that ugly fat on your otherwise-beautiful body? Our Exer-Torture home gym will trim those thunder thighs in 30 days or your money back. . . .

Ah, those pitches. Everybody is trying to sell something—even the participants in the specialty modes of Amateur Radio. Our eyes light up thinking about all those wondrous gizmos that digitize, packetize and equalize. Those kinds of specialty modes are easily remembered and can quickly gain popularity. Most people, though, forget about the oldest specialty mode in Amateur Radio—traffic handling.

Admittedly, it's hard for traffic handlers to compete with other specialty modes because it just doesn't look exciting. After all, tossing messages from Great Uncle Levi and Grandma Strauss sure doesn't put you on the leading edge of our high-tech hobby, does it? Yet, over 500 nets are members of the ARRL National Traffic System (NTS), probably one of the most highly organized special interests of Amateur Radio. Today's traffic handlers are at this very moment setting new standards for traffic handling via PACTOR, AMTOR and packet—as well as using traditional modes. If you enjoy emergency-communications preparation, traffic handling is for you. Sure, it's true that over 90% of all messages handled via Amateur Radio are of the "at the state fair, wish you were here" or the "happy holidays" variety—certainly not life-and-death stuff. But consider this: Your local fire department often conducts drills without ever putting out a real fire, your civil defense simulates emergencies regularly, and many department stores hire people to come in and pretend to be shoplifters to check the alertness of their employees. Similarly, when real emergencies rear their ugly heads, traffic handlers just take it all in stride and churn messages out like they always do.

Traffic handling is also an excellent way to paint a friendly picture of Amateur Radio to the nonamateur public. What sometimes seems to be unimportant to us is seldom unimportant to that person in the address block of a message. Almost every traffic handler can relate stories of delivering a Christmas message from some long-lost friend or relative that touched the heart of the recipient such that they could hear the tears welling up at the other end of the phone. Those happy recipients will always mentally connect Amateur Radio traffic handling with good and happy things, and often this is the most satisfying part of this hobby within a hobby.

Next time you go to a hamfest, see if you can spot the traffic handlers. They are almost always reveling in a big social cluster, sharing stories and enjoying a unique camaraderie. Traffickers always enthusiastically look forward to the next hamfest, because it means another chance to spend time with their special friends of the airwaves. These friendships often last a lifetime, transcending barriers of age, geographical distance, background, and gender or physical ability.

Young or old, rural or urban, there's a place reserved for you on the traffic nets. Young people often can gain respect among a much older peer group and obtain high levels of responsibility through the traffic nets, good preparation for job opportunities and scholarships. "Nine to fivers" on a tight schedule can still manage to get a regular dose of Amateur Radio in just 15 to 30 minutes of net operation—a lot of hamming in a little time. Retired people can stay active in an important activity and provide a service to the general public.

Even if you live in the sticks, where you rarely get a delivery, you can still perform a vital function in NTS as a net control station (NCS) or as a representative to the upper echelons of NTS. You don't have to check in every night (a popular myth about traffic handling—if you can donate time just once a week you are certainly welcome on NTS). You don't need fancy antennas or huge amplifiers, and you don't even need those ARRL message pads. For the cost of a pad or paper and a pencil, you can interface with a system that covers thousands of miles and consists of tens of thousands of users. A world of fun and friendship is waiting for you in traffic handling. You only have to check into a net to become part of it.

MAKING THE BIG STEP: CHOOSING A NET AND CHECKING INTO IT

Checking into a net for the first time is a lot like making your first dive off the high board. It's not usually very pretty, but it's a start. Once you've gotten over the initial shock of hitting the water that hard, the next one comes a lot easier. But, like a beginning dive you can do a little advance preparation that will get you emotionally prepared for your first plunge.

Before you attempt to interface with the world of NTS, you need the right "software." Get a copy of the *ARRL Net Directory*, available from ARRL ($4) and two free ARRL operating aids—FSD-3, the list of ARRL numbered radiograms, and FSD-218, the "Q-signals for nets and message form" card. *The Public Service Communications Manual* is also recommended reading. A number of these operating aids and references are available via the Amateur Radio Public Service page of *ARRLWeb* (**http://www.arrl.org/field/pubservice.html**). After you have a basic understanding of the materials, you're ready to pick out a net in your ARRL section/state that suits you.

Go through the *Net Directory* and match up your time schedule with the nets in your section/state. Your section/state slow-speed CW net is a good choice (or your neighboring section, if yours doesn't have one). But if you still don't have the confidence to try the "big" CW section net, don't feel embarrassed about checking into a slow-speed net. Slow-speed nets are chockfull of veterans that help with NCS and NTS duties, and they're willing to help you, too. Perhaps you'd like to try the section phone net or weather net or a local 2-meter net. At any rate, you are the sole judge of what you want to try first.

Once you've chosen a net, it's a good idea to listen to it for a few days before you check in. Although this chapter will deal with a generalized format for net operation, each net is "fingerprinted" with its own special style of operation, and it's best to become acquainted with it before you jump in. If you have a friend who checks into the net, let your friend "take you by the hand" and tell you about the ins and outs of the net.

When the big day arrives, keep in mind that everybody on that net had to check into the net for the first time once. You aren't doing anything different than the rest of them, and this, like your first QSO, is just another "rite of passage" in the ham world. You will discover that it doesn't hurt, and that many other folks will be pleased, even happy, that you checked in with them. (See accompanying sidebar for general recommendations on how to make your NCS love you!)

How to be the Kind of Net Operator the Net Control Station (NCS) Loves

As a net operator, you have a duty to be self-disciplined. A net is only as good as its worst operator. You can be an exemplary net operator by following a few easy guidelines.

1) Zero beat the NCS. The NCS doesn't have time to chase all over the band for you. Make sure you're on frequency, and you will never be known at the annual net picnic as "old so-and-so who's always off frequency." Double-check that RIT control if your radio has one.

2) Don't be late. There's no such thing as "fashionably late" on a net. Liaison stations are on a tight timetable. Don't hold them up by checking in 10 minutes late with traffic.

3) Speak only when spoken to by the NCS. Unless it is a bona fide emergency situation, you don't need to "help" the NCS unless specifically asked. If you need to contact the NCS, make it brief. Resist the urge to help clear the frequency for the NCS or to "advise" the NCS. The NCS, not you, is boss.

4) Unless otherwise instructed by the NCS, transmit only to the NCS. Side comments to another station in the net are out of order.

5) Stay until you are excused. If the NCS calls you and you don't respond because you're getting a "cold one" from the fridge, the NCS may assume you've left the net, and net business may be stymied. If you need to leave the net prematurely, contact the NCS and simply ask to be excused (QNX PSE on CW).

6) Be brief when transmitting to the NCS. A simple "yes" (C) or "no" (N) will usually suffice. Shaggy dog tales only waste valuable net time.

7) Know how the net runs. The NCS doesn't have time to explain procedure to you while the net is in session. After you have been on the net for a while, you should already know these things.

8) Before the net begins, get yourself organized. Have all the materials you will need to receive traffic at hand. If you have messages to send, have them grouped by common destination according to the procedure of the net you're participating in. Nothing is more frustrating to the other operators—especially the one waiting to take your traffic—than being told "Wait a minute, I've got it here somewhere."

9) When receiving traffic and you have a question about the accuracy of anything passed in the message, don't tie the net up with discussions about ZIP codes, telephone area codes, etc. Just tell the NCS that you would like discuss message number 123 with the sending station after the net, or request to be allowed to move off frequency to clear up the matter. Remember, only the originating station can change the message—with the exception of the word count—between the number and the signature. Any suggested changes can be added as an "Op Note" if necessary.

10) Have a current copy of the *Public Service Communications Manual* and read it.

11) Don't "freelance" your traffic. Wait your turn to pass your messages as directed by the NCS.

ON CW

First, we'll pay a visit to a session of the Missouri CW Net, not so long ago, on an 80-meter frequency not so far away. KØSI is calling tonight's session, using that peculiar CW net shorthand that we aren't too used to yet.

KØSI: MCWN MCWN MCWN DE KØSI KØSI KØSI QND QNZ QTC? K

(Translation: Calling the Missouri CW Net, calling the Missouri CW Net. This is KØSI. This is a directed net, zero beat me. Any traffic? Over)

In the meantime, NDØN, K2ONP and WØOUD are waiting in the wings to check in. Since each is an experienced traffic handler, each listens carefully before jumping in, so as to not step on anyone.

NDØN: N
KØSI: N

(Notice that NDØN just sent the first letter of his suffix, and the NCS acknowledged it. This is common practice, but if you have the letter E or T or K as the first letter of your suffix, or if it is the same as another net operator's first letter, you might use another letter. It's not a hard-and-fast rule.)

NDØN: DE NDØN GE PETE QTC SOUTH CORNER 1 AIØO 1 K

(NDØN has one piece of traffic for South Corner, and one for AIØO.)

KØSI: NDØN DE KØSI GE JOHN R $\overline{AS}$

(Good evening, John, roger (I acknowledge) your traffic list. Wait/standby)

K2ONP: M
KØSI: M
K2ONP: DE K2ONP GE PETE TEN REP QRU K

(K2ONP is representative to the NTS Tenth Region Net tonight, and he has no traffic.)

KØSI: K2ONP DE KØSI HI GEO TU $\overline{AS}$
WØOUD: BK
KØSI: BK
WØOUD: DE WØOUD GE PETE QRU K
KØSI: WØOUD DE KØSI GE LETHA QNU TU $\overline{AS}$

(Good evening, Letha, the net has traffic for you. Please stand by.)

ARRL QN Signals for CW Net Use

QNA* Answer in prearranged order.
QNB* Act as relay Between______ and ______.
QNC All net stations Copy.
I have a message for all net stations.
QND* Net is Directed (controlled by net control station.)
QNE* Entire net stand by.
QNF Net is Free (not controlled.)
QNG Take over as net control station.
QNH Your net frequency is High.
QNI Net stations report In.
I am reporting into the net. (Follow with a list of traffic or QRU.)
QNJ Can you copy me?
QNK* Transmit messages for______ to______.
QNL Your net frequency is Low.
QNM* You are QRMing the net. Stand by.
QNN Net control station is ______.
What station has net control?
QNO Station is leaving the net.
QNP Unable to copy you.
Unable to copy ______.
QNQ* Move frequency to _____ and wait for ______ to finish handling traffic. Then send him traffic for ______.
QNR* Answer_____ and Receive traffic.
QNS Following Stations are in the net.* (Follow with list)
Request list of stations in the net.
QNT I request permission to leave the net for ______ minutes.
QNU The net has traffic for you. Stand by.
QNV* Establish contact with ______ on this frequency. If successful, move to _____ and send him traffic for ______.
QNW How do I route messages for ___?
QNX You are excused from the net.*
Request to be excused from the net.
QNY* Shift to another frequency (or to __ kHz) to clear traffic with ______.
QNZ Zero beat your signal with mine.
*For use only by the Net Control Station.

Notes on use of QN Signals

The QN signals listed above are special ARRL signals for use in amateur CW nets only. They are not for use in casual amateur conversation. Other meanings that may be used in other services do not apply. Do not use QN signals on phone nets. Say it with words. QN signals need not be followed by a question mark, even though the meaning may be interrogatory.

These "Special QN Signals for New Use" originated in the late 1940s in the Michigan QMN Net, and were first known to Headquarters through the then head traffic honcho W8FX. Ev Battey, W1UE, then ARRL assistant communications manager, thought enough of them to print them in *QST* and later to make them standard for ARRL nets, with a few modifications. The original list was designed to make them easy to remember by association. For example, QNA meant "Answer in Alphabetical order," QNB meant "Act as relay Between ...," QNC meant "All Net Copy," QND meant "Net is Directed," etc. Subsequent modifications have tended away from this very principle, however, in order that some of the less-used signals could be changed to another, more needed, use.

Since the QN signals started being used by amateurs, international QN signals having entirely different meanings have been adopted. Concerned that this might make our use of QN signals with our own meanings at best obsolete, at worst illegal, ARRL informally queried FCC's legal branch. The opinion then was that no difficulty was foreseen as long as we continued to use them only in amateur nets.

Occasionally, a purist will insist that our use of QN signals for net purposes is illegal. Should anyone feel so strongly about it as to make a legal test, we might find ourselves deprived of their use. Until or unless we reach such an eventuality, however, let's continue to use them, and use them right. After all, we were using them first.

(Since Letha lives in South Corner, the NCS is going to move WØOUD and NDØN off frequency to pass the South Corner traffic.)

KØSI: ØN?

NDØN: HR

(HR [here], or C [yes] are both acceptable ways to answer the NCS, who will usually use only your suffix from here on out to address questions to you.)

KØSI: OUD?

WØOUD: C

KØSI: NDØN ES WØOUD QNY UP 4 SOUTH CORNER K

(Go up 4 kHz and pass the South Corner traffic.)

NDØN: GG (going)

WØOUD: GG

(When two stations go off frequency, the receiving station always calls the transmitting station. If the NCS had said WØOUD QNV NDØN UP 4 GET SOUTH CORNER, WØOUD would have called NDØN first on frequency to see if she copied him. This is done often when conditions are bad. If they don't make connection, they will return to net frequency. If they do make connection, and pass the traffic, they will return as they are done.)

AIØO: O

KØSI: O

AIØO: DE AIØO GE PETE QRU K

KØSI: AIØO DE KØST GE ROB QNU UP 4 AIØO WID NDØN AFTER WØOUD THEN BOTH QNX 73 K

(Good evening, Rob, the net has traffic for you. Please go up 4 and get one for you from NDØN [WID means "with"] after he finishes with WØOUD. Then, when you are both finished, you and NDØN are both excused from the net. 73!)

AIØO: 73 GG

As you can see, it doesn't take much to say a lot on a CW net. Now that all the net business is taken care of, the NCS will start excusing other stations. Since K2ONP has a schedule to make with the Tenth Region Net, he will be excused first.

KØSI: K2ONP DE KØSI TU GEO FER QNI NW QRU QNX TU 73 K

K2ONP: TU PETE CUL 73 DE K2ONP SK

WØOUD: OUD

(Letha is back from receiving her traffic.)

KØSI: OUD TU LETHA NW QRU QNX 88 K

WØOUD: GN PETE CUL 88 DE WØOUD SK

Now, the NCS will close the net.

KØSI: MCWN QNF [the net is free] GN DE KØSI CL

When you check into a CW net for the first time, don't worry about speed. The NCS will answer you at about the speed you check into the net. You will discover that everyone on a CW net checks in with a different speed, just as everyone has a different voice on SSB. It's nothing to be self-conscious about. As the saying goes, "We're all in this together." The goal is to pass the traffic correctly, with 100 percent accuracy, not burn up the ether with our spiffy fists. Likewise, don't hesitate to slow down for someone else. Remember, when handling traffic, 100 percent accuracy is the minimum acceptable performance level!

ON SSB

Now, let's tune in a session of the Missouri Single Sideband Net.

As we look in on KØPCK calling tonight's session, keep in mind these few pointers:

1) The net preamble, given at the beginning of each session, will usually give you the information you need to survive on the net. Method of checking in varies greatly from net to net. For instance, some section nets have a prearranged net roll, some take checking by alphabetical order, or some even take checkins by geographical area. Don't feel intimidated by a prearranged net roll. Those nets will always stand by near the end of the session to take stations not on the net roster.
2) As you listen to a net, you will find that on phone, formality also varies. Some SSB nets are strictly business, while others are "chattier." However, don't always confuse lack of formality with looseness on a net. There is still a definite net procedure to adhere to.
3) Once, on a close play, the catcher asked umpire Bill Klem, "Well, what is it?"

"It ain't nothin' till I call it," he growled.

By the same token, you need to keep in mind that the NCS is the absolute boss when the net is in session. On CW nets, this doesn't seem to be much of a problem to the tightness of operation, but on phone nets, sometimes a group of "well-meaners" can really slow down the net. So don't "help" unless NCS tells you to.

Who Owns the Frequency?

Traffic nets sometimes have difficulties when it comes time for the call-up, and a ragchew is taking place on the published net frequency. What to do? Well, you could break in on the ragchew and ask politely if the participants would mind relinquishing the frequency. This usually works, but what if it doesn't? The net has no more right to the frequency than the stations occupying it at net time, and the ragchew stations would be perfectly within their rights to decline to relinquish it.

The best thing to do in such a case is to call the net near, but not directly on, the normal frequency—far enough (hopefully) to avoid causing QRM, but not so far that net stations can't find the net. Usually, the ragchewers will hear the net and move a bit farther away—or even if they don't, the net can usually live with the situation until the ragchew is over.

It is possible to conceive of a situation, especially on 75-meter phone, in which the net frequency is occupied and the entire segment from 3850 to 4000 kHz is wall-to-wall stations. In any case, it is not productive to argue about who has the most right to a certain frequency. Common courtesy says that the first occupants have, but there are many extenuating circumstances. Avoid such controversies, especially on the air.

Accordingly, net frequencies should be considered "approximate," inasmuch as it may be necessary for nets to vary their frequencies according to band conditions at the time. Further, no amateur or organization has any preemptory right to any specific amateur frequency.

Handling Instructions

HXA—(Followed by number.) Collect landline delivery authorized by addressee within ____ miles. (If no number, authorization is unlimited.)
HXB—(Followed by number.) Cancel message if not delivered within ______ hours of filing time; service originating station.
HXG—Report date and time of delivery (TOD) to originating station.
HXD—Report to originating station the identity of station from which received, plus date and time. Report identity of station to which relayed, plus date and time, or if delivered report date, time and method of delivery.
HXE—Delivering station get reply from addressee, originate message back.
HXF—(Followed by number.) Hold delivery until______ (date).
HXG—Delivery by mail or landline toll call not required. If toll or other expense involved, cancel message and service originating station.

An HX prosign (when used) will be inserted in the message preamble before the station of origin, thus: NR 207 R HXA50 W1AW 12 . . . (etc). If more than one HX prosign is used, they can be combined if no numbers are to be inserted; otherwise the HX should be repeated, thus: NR 207 R HXAC W1AW . . . (etc), but: NR 207 R HXA50 HXC W1AW . . . (etc). On phone, use phonetics for the letter or letters following the HX, to insure accuracy.

Since net time is upon us, let's get back to the beginning of the Missouri Single Sideband Net.

"Calling the Missouri Single Sideband Net, calling the Missouri Single Sideband Net. This is KØPCK, net control. The Missouri Single Sideband Net meets on 3988 kHz nightly at 6 PM for the purpose of handling traffic in Missouri and to provide a link for out-of-section traffic through the ARRL National Traffic System. My name is Ben, Bravo Echo November, located in Toad Lick. When I call for the letter corresponding to the first letter of the suffix of your call, please give your call sign only."

"Any low-power, mobile or portable stations wishing to check in?"

(wait 5 seconds)

"Any relays?"

(wait 5 seconds)

"This is KØPCK for the Missouri Sideband Net. Do we have any traffic?"

"KØORB traffic"

"KØORB, Good evening Bill. List your traffic, over."

"Good evening, Ben. Two out-of-state."

"Very good. Who is our Tenth Region Rep tonight?"

"Good evening, Ben. This is NIØR, Ten Rep."

"NIØR, this is KØPCK. Hi, Roger. Please call KØORB, move him to 3973 and get Bill's out-of-state traffic."

"KØORB, this NIØR. See you on 73."

"Going. KØORB."

Now, let's sit back and analyze this. As you can see, the format is pretty much identical except that it takes more words. Oh, yes, one other difference . . . you may have noticed that not one single Q-signal was used! Q-signals should not be used on phone nets. Work hard at avoiding them, and you will reduce the "lingo barrier," making it a little less intimidating for a potential new check in.

TWO SPECIAL CASES

HF digital operations are almost identical to CW nets, except for the method of transmission, but you need to remember this: It is always important to zero beat, but here it is crucial!

The other "exception to the rule" is the local 2-meter FM net. Many 2-meter nets are designed for ragchewing or weather spotting, so often if you bring traffic to the net, be prepared to coach someone in the nuances of traffic handling.

A few other things to remember:

1) Unlike a "double" on CW or SSB, where the NCS might even still get both calls, a "double" on an FM repeater either captures only one station, or makes an ear-splitting squeally heterodyne. Drag your feet a little before you check in, so you are less likely to double.
2) Be especially aware to wait for the squelch tail or courtesy beep. More people seem to time out the repeater on a net than at any other time!
3) Remember that a lot of people have scanners, many with the local repeater programmed in on one of the channels. Design your behavior in such a way that it attracts nonhams to Amateur Radio. In other words, don't do anything you wouldn't do in front of the whole town!

MAKING IT, TAKING IT AND GIVING IT AWAY: MESSAGE HANDLING AND MESSAGE FORM

By this stage, you have probably been checking into the net for a while, and things have started to move along quite smoothly on your journey as a traffic handler. But in the life of any new traffic op, the fateful day comes along when the NCS points RF at you and says those words that strike fear in almost every newcomer: "Go up four, get Cornshuck Hollow."

Now what? You could suddenly have "rig trouble" or a "power outage" or a "telephone call," or just bump your dial and disappear. After all, it has been done before, and everyone that didn't do it sure thought of it the first time they were asked to take traffic! Of course, there is a more honorable route—go ahead and take it! Chances are you will be no worse than anyone else your first time out.

To ease the shock of your first piece of traffic, maybe it would be a good idea to go over message form "by the book"—ARRL message form, that is.

ARRL MESSAGE FORM—THE RIGHT WAY, AND THE RIGHT WAY

A common line of non-traffickers is, "Aw, why do they have to go through that ARRL message-form stuff? It just confuses people and besides, my message is just a few words or so. It's silly to go through all that rigamarole."

Well, then, let's imagine that you are going to write a

Checking Your Message

Traffic handlers don't have to dine out to fight over the check! Even good ops find much confusion when counting up the text of a message. You can eliminate some of this confusion by remembering these basic rules:

1) Punctuation ("X-rays," "Queries") count separately as a word.

2) Mixed letter-number groups (1700Z, for instance) count as one word.

3) Initial or number groups count as one word if sent together, two if sent separately.

4) The signature does not count as part of the text, but any closing lines, such as "Love" or "Best wishes" do.

Here are some examples:

- Charles J McClain—3 words
- W B Stewart—3 words
- St Louis—2 words
- 3 PM—2 words
- SASE—1 word
- ARL FORTY SIX—3 words
- 2N1601—1 word
- Seventy three—2 words
- 73—1 word

Telephone numbers count as 3 words (area code, prefix, number), and ZIP codes count as one. ZIP + 4 codes count as two words. Canadian postal codes count as two words (first three characters, last three characters.)

Although it is improper to change the text of a message, you may change the check. Always do this by following the original check with a slash bar, then the corrected check. On phone, use the words "corrected to."

Book Messages

When sending book traffic, always send the common parts first, followed by the "uncommon" parts. For example:

R N0FQW ARL 7 BETHEL MO SEP 7 BT

ARL FIFTY ONE BETHEL SHEEP FESTIVAL LOVE BT
PHIL AND JANE BT

NR 107 TONY AND LYN CALHOUN AA
160 NORTH DOUGLAS AA SPRINGFIELD IL 62702
BT

NR 108 JOE WOOD AJ0X AA
84 MAIN STREET AA
LAUREL MS 39440 BT

NR 109 JEAN WILCOX AA
1243 EDGEWOOD DRIVE AA
LODI CA 95240 N

Before sending the book traffic to another operator, announce beforehand that it is book traffic. Say "Follows book traffic." Then use the above format. On CW, a simple HR BUK TFC will do.

THE AMERICAN RADIO RELAY LEAGUE
RADIOGRAM
VIA AMATEUR RADIO

NUMBER	PRECEDENCE	HX	STATION OF ORIGIN	CHECK	PLACE OF ORIGIN	TIME FILED	DATE
133	R	HXG	WØMME	ARL 7	MT. PLEASANT, IA	17ØØ	SEP 1

TO

MR AND MRS JEFF HOLTZCLAW
RT 1 BOX 127
TONGANOXIE, KS 66Ø86

THIS RADIO MESSAGE WAS RECEIVED AT
AMATEUR STATION ____ PHONE ____
NAME ____
STREET ADDRESS ____
CITY AND STATE ____

TELEPHONE NUMBER 913 555 1212

ARL	FIFTY	ONE	OLD	THRESHERS
REUNION	LOVE			

UNCLE CHUCKIE

REC'D	FROM	DATE	TIME	SENT	TO	DATE	TIME
	WØSS	SEP 1			WØFRC	SEP 2	

THIS MESSAGE WAS HANDLED FREE OF CHARGE BY A LICENSED AMATEUR RADIO OPERATOR, WHOSE ADDRESS IS SHOWN IN THE BOX AT RIGHT ABOVE. AS SUCH MESSAGES ARE HANDLED SOLELY FOR THE PLEASURE OF OPERATING, NO COMPENSATION CAN BE ACCEPTED BY A "HAM" OPERATOR. A RETURN MESSAGE MAY BE FILED WITH THE "HAM" DELIVERING THIS MESSAGE TO YOU. FURTHER INFORMATION ON AMATEUR RADIO MAY BE OBTAINED FROM A.R.R.L. HEADQUARTERS, 225 MAIN STREET, NEWINGTON, CONN. 06111

THE AMERICAN RADIO RELAY LEAGUE, INC., IS THE NATIONAL MEMBERSHIP SOCIETY OF LICENSED RADIO AMATEURS AND THE PUBLISHER OF QST MAGAZINE. ONE OF ITS FUNCTIONS IS PROMOTION OF PUBLIC SERVICE COMMUNICATIONS AMONG AMATEUR OPERATORS. TO THAT END, THE LEAGUE HAS ORGANIZED THE NATIONAL TRAFFIC SYSTEM FOR DAILY NATION-WIDE MESSAGE HANDLING.

PRINTED IN U.S.A.

Fig 7-1—Example message properly entered on the ARRL message form.

letter to your best friend. What do you think would happen to your letter if you decided that the standardized method the post office used was "silly," so you signed the front, put the addressee's address where the return address is supposed to go, and stamped the inside of the letter? It would probably end up in the Dead Letter Office.

Amateur message form is standardized so it will reach its destination speedily and correctly. It is very important for every amateur to understand correct message form, because you never know when you will be called upon in an emergency. Most nonhams think all hams know how to handle messages, and it's troublesome to discover how few do. You can completely change the meaning of a piece of traffic by accident if you don't know the ARRL message form, and as you will see later, this can be a real "disaster." Learn it the right way, and this will never happen.

If you will examine the sample message in **Fig 7-1**, you will notice that the message is essentially broken into four parts: the preamble, the address block, the text and the signature. The preamble is analogous to the return address in a letter and contains the following:

1) The number denotes the message number of the originating station. Most traffic handlers begin with number 1 on January 1, but some stations with heavy volumes of traffic begin the numbering sequence every quarter or every month.
2) The precedence indicates the relative importance of the message. Most messages are Routine (R) precedence—in fact, about 99 out of 100 are in this category. You might ask, then why use any precedence on routine messages? The reason is that operators should get used to having a precedence on messages so they will be accustomed to it and be alerted in case a message shows up with a different precedence. A Routine message is one that has no urgency aspect of any kind, such as a greeting. And that's what most amateur messages are—just greetings.

The Welfare (W) precedence refers to either an inquiry as to the health and welfare of an individual in the disaster area or an advisory from the disaster area that indicates all is well. Welfare traffic is handled only after all emergency and priority

MARS

Most modern-day traffic handlers don't realize that our standard message preamble is largely fashioned after the form used in the Army Amateur Radio System (AARS), which had its heyday in the 1930s. The ARRL standard preamble prior to that time was quite different, but the AARS form was adopted because it had advantages and was so widely used. AARS nets were numerous in the 1930s, using WL calls on two frequencies (3497.5 and 6990 kc.) outside the amateur bands and many frequencies, using amateur calls and amateur participants, inside the bands.

MARS, the post-World War II successor to AARS, encompasses all three of the US armed services. Although it operates numerous nets, all of which are outside the amateur bands on military frequencies, thousands of amateurs participate in MARS. This service performs some traffic coverage that we amateurs are not permitted to perform. MARS, which stands for Military Affiliate Radio System, is conducted in three different organizations under the direction of the Army, Navy and Air Force. Therefore, MARS is not in the strictest sense Amateur Radio.

Nevertheless, amateur messages find their way into MARS circuits, and MARS messages find their way into amateur nets. In fact, NTS has a semiformal liaison with MARS to handle the many messages from families in the States to their sons and daughters serving overseas.

Traffic for some points overseas at which US military personnel are stationed can be handled via MARS, provided a complete military address is given, even though some of these points cannot be covered by Amateur Radio or NTS. The traffic is originated in standard ARRL form and refiled into MARS form (now quite a bit different from ours) when it is introduced into a MARS circuit for transmission overseas. In this manner, traffic may be exchanged with military personnel in Germany, Japan and a few other countries that do not otherwise permit the handling of international third-party traffic.

Traffic coming from MARS circuits into amateur nets for delivery by Amateur Radio are converted from MARS to amateur form and handled as any other amateur message. The exception occurs when traffic originates overseas. In this case, the name of the country in which it originates, followed by "via MARS," should appear as the place of origin, so it does not appear that such messages were handled illegally by Amateur Radio. There are places where US military personnel are stationed that even MARS cannot handle traffic with, presumably because of objections by the host country. This information is in the hands of MARS "gateway" stations and changes from time to time.

The amount of MARS traffic appearing on amateur nets is not great, since MARS has a pretty good system of handling it on MARS frequencies, but it is important that we maintain liaison as closely as possible since all civilian MARS members are US licensed amateurs.

Information concerning MARS may be obtained directly from the individual branches at these addresses:

Air Force MARS
HQ AFC4A/SYXR (MARS)
203 Losey Street
Room 3065
Scott AFB, IL 62225-5234
Web: **http://public.afca.scott.af.mil/public/mars1.htm**

Navy-Marine Corps MARS
Chief U.S. Navy-Marine Corps
Military Affiliate Radio System (MARS) – BLDG 13
NAVCOMMU WASHINGTON
Washington, DC 20397-5161
Web: **http://www2.acan.net/~navymars/**

Army MARS
HQ US ARMY SIGNAL COMMAND
ATTN: AFSC-OPE-MA (ARMY MARS)
Ft Huachuca, AZ 85613-5000
Web: **http://www.asc.army.mil/mars/**

Are You a Type-NCS Personality?

As net control station (NCS), it pays to remember that the net regulars are the net. Your function is to preside over the net in the most efficient, businesslike way possible so that the net participants can promptly finish their duties and go on to other ones. You must be tolerant and calm, yet confident and quick in your decisions. An ability to "take things as they come" is a must. Remember that you were appointed NCS because your Net Manager believes in you and your abilities.

1) Be the boss, but don't be bossy. It's your job to teach net discipline and train new net operators (and retrain some old ones!). You are the absolute boss when the net is in session, even over your Net Manager. However, you must be a "benevolent monarch" rather than a tyrant. Nets lose participation quickly one night a week when it's Captain Bligh's turn to call the net. If the net has a good turnout every night but one, that tells something about its NCS.

2) Be punctual. Many of the net participants have other commitments or nets to attend to; liaison stations are often on a tight schedule to make the NTS region or area net. If you, as NCS, don't care when the net starts, others will think it's okay for them to be late, too. Then traffic doesn't get passed in time, and someone may miss his NTS liaison. In short, the system is close to breaking down.

3) Know your territory. Your members have names—use them. They also live somewhere—by knowing their locations you can quickly ascertain who needs to get the traffic. As NCS, it's your responsibility to know the geography of your net. You also need to understand where your net fits into the scheme of NTS.

4) Take extra care to keep your antennas in good shape because an NCS can't run a net with a "wimpy" signal. Although you don't have to be the loudest one on the net, you do have to be heard. You will discover that the best way to do this is to have a good antenna system. A linear amplifier alone won't help you hear those weak check ins!

5) The NCS establishes the net frequency. Just because the *Net Directory* lists a certain frequency doesn't give you squatters rights to it if a QSO is already in progress there. Move to a nearby clear frequency, close enough for the net to find you. QRM is a fact of life on HF, especially on 75/80 meters, so live with it.

6) Keep a log of every net session. Just because the FCC dropped the logging requirements doesn't mean that you have to drop it. It's a personal decision. The Net Manager may need information about a check in or a piece of traffic, and your log details can be helpful to him in determining what happened on a particular night.

7) Don't hamstring the net by waiting to move the traffic. Your duty is to get traffic moving as quickly as possible. As soon as you can get two stations moving, send them off to clear the traffic. If you have more than one station holding traffic for the same city, let the "singles" (stations with only one piece for that city) go before the ones with more than one piece for that city. The quicker the net gets the traffic moved, the sooner the net can be finished and the net operators can be free to do whatever they want.

---

traffic is cleared. The Red Cross equivalent to an incoming Welfare message is DWI (Disaster Welfare Inquiry).

The Priority (P) precedence is getting into the category of high importance and is applicable in a number of circumstances: (1) important messages having a specific time limit, (2) official messages not covered in the emergency category, (3) press dispatches and emergency-related traffic not of the utmost urgency, and (4) notice of death or injury in a disaster area, personal or official.

The highest order of precedence is EMERGENCY (always spelled out, regardless of mode). This indicates any message having life-and-death urgency to any person or group of persons, which is transmitted by Amateur Radio in the absence of regular commercial facilities. This includes official messages of welfare agencies during emergencies requesting supplies, materials or instructions vital to relief of stricken populace in emergency areas. During normal times, it will be very rare.

3) Handling Instructions are optional cues to handle a message in a specific way. For instance, HXG tells us to cancel delivery if it requires a toll call or mail delivery, and to service it back instead. Most messages will not contain handling instructions.

4) Although the station of origin block seems self-explanatory, many new traffic handlers make the common mistake of exchanging their call sign for the station of origin after handling it. The station of origin never changes. That call serves as the return route should the message encounter trouble, and replacing it with your call will eliminate that route. A good rule of thumb is never to change any part of a message.

5) The check is merely the word count of the text of the message. The signature is not counted in the check. If you discover that the check is wrong, you may not change it, but you may amend it by putting a slash bar and the amended count after the original count. [See the "Checking Your Message" sidebar for additional information on the message check.]

Another common mistake of new traffickers involves "ARL" checks. A check of ARL 8 merely means the text has an ARL numbered radiogram message text in it, and a word count of 8. It does not mean ARRL numbered message no. 8. This confusion has happened before, with unpleasant results. For instance, an amateur with limited traffic experience once received a message with a check of ARL thirteen. The message itself was an innocuous little greeting from some sort of fair, but the amateur receiving it thought the message was ARRL numbered message thirteen—"Medical emergency situation exists here." Consequently, he unknowingly put a family through a great deal of unnecessary stress. When the smoke cleared, the family was on the verge of bringing legal action against the ham, who himself developed an intense hatred for traffic of any sort and refused to ever handle another message. These kinds of episodes certainly don't help the "white hat" image of Amateur Radio!

6) The place of origin can either be the location (City/State

Tips on Handling Message Traffic by Packet Radio

Listing Messages

After logging on to your local NTS-supported bulletin board, type the command LT, meaning List Traffic. The BBS will sort and display an index of all NTSXX traffic awaiting delivery. The index will contain information that looks something like this:

MSG#	TR	SIZE	TO	FROM	@BBS DATE	Title
200	TIN	282	NTSCT	KYIT	000225	QTC1 Hartford
198	TN	302	NTSRI	K1CE	000224	QTC1 Cranston
192	TN	215	NTSIN	WF4R	000224	QTC1 Indianapolis
190	TN	200	NTSCT	AJ6F	000224	QTC1 Waterbury
188	TN	315	NTSCT	KH6WZ	000224	QTC1 Newington
187	TN	320	NTSCT	K6TP	000224	QTC1 New Haven
186	TN	300	NTSAL	WA4STO	000224	QTC1 Birmingham
184	TN	295	NTSCA	WB4F	000224	QTC1 Fresno

Receiving Messages

To take a message off the Bulletin Board for telephone delivery to the third party, or for relay to a NTS Local or Section Net, type the R or RT command, meaning Read Traffic, and the message number. R 188 will cause the BBS to find and send the message text file containing the RADIOGRAM for the third party in Newington. The RADIOGRAM will look like any other, with preamble, address, text and signature, only some additional packet-related message header information is added. This information includes the routing path of the message for auditing purposes; e.g., to discern any excessive delays in the system.

After the message is saved to the printer or disk, the message should be KILLED by using the KT command, meaning Kill Traffic, and the message number. In the above case, at the BBS prompt, type KT 188. This prevents the message from being delivered twice.

At the time the message is killed, many BBSs will automatically send a message back to the station in the FROM field with information on who took the traffic, and when it was taken!

Delivering or Relaying a Message

A downloaded RADIOGRAM should, of course, be handled expeditiously in the traditional way: telephone delivery, or relay to a phone or CW net.

Sending Messages

To send a RADIOGRAM, use the ST command, meaning Send Traffic. The BBS will prompt you for the NTSXX address (NTSOH, for example), the message title which should contain the city in the address of the RADIOGRAM (QTC1 Dayton), and the text of the message in RADIOGRAM format. The BBS, usually within the hour, will check its outgoing mailpouch, find the NTSOH message and automatically forward it to the next station for relay. Note: Some states have more than one ARRL Section. If you do not know the destination ARRL Section ("Is San Angelo in the ARRL Northern, Southern or West Texas Section?"), then simply use the state designator NTSTX.

Unbundle your messages please: one NTS message per BBS message. Please remember that traffic eventually will have to be broken down to the individual addressee somewhere down the line for ultimate delivery. When you place two or more NTS messages destined for different addressees within one packet message, eventually the routing will require the messages to be broken up by either the BBS SYSOP or the relay station, placing an additional, unreasonable burden on them. Therefore it is good practice for the originator to expend the extra word processing in the first place and create individual messages per city regardless if there are common parts of other messages. This means that book messages are not suitable in packet at this time unless they are going to the same city. Bottom line: Messages should be sent unbundled. (Tnx NI6A for this tip)

We Want You!

Local and Section BBSs need to be checked daily for message traffic. SYSOPs and STMs can't do it alone. They need your help to clear RADIOGRAMS every day, seven days a week, for delivery and relay. If you are a traffic handler or are interested in learning about this important aspect of Amateur Radio, contact your Section Traffic Manager or Section Manager for information in your Section.

or City/Province) of the originating station or the location of the third party wishing to initiate a message through the originating station. Use standard abbreviations for state or province. ZIP or postal codes are not necessary. For messages from outside the US and Canada, city and country is usually used. If a message came from the MARS (Military Affiliate Radio System) system, place of origin may read something like "Korea via Mars."

7) The filing time is another option, usually used if speed of delivery is of significant importance. Filing times should be in UTC or "Zulu" time.
8) The final part of the preamble, the date, is the month and day the message was filed—year isn't necessary.

Next in the message is the address. Although things like ZIP code and phone number aren't entirely necessary, the more items included in the address, the better its chances of reaching its destination. To experienced traffic handlers, ZIP codes and telephone area codes can be tip-offs to what area of the state the traffic goes, and can serve as a method of verification in case of garbling. For example, all ZIP codes in Minnesota start with a 5. Therefore, if a piece of traffic sent as St Joseph, MO, with a ZIP of 56374 has been garbled along the way, it conceivably can be rerouted. So, when it comes to addresses, the adage "the more, the better" applies. The text, of course, is the message itself. You can expedite the counting of the check by following this simple rule—when copying by hand, write five words to a line. When copying with a typewriter, or when sending a

The Net Control Sheet

Net controlling is no easy task, requiring much talent on the part of the operator. A useful prop is a set system for keeping track of net operation so that you don't get mixed up and start to lose control. This can best be effected by a sheet of paper on which you record who has what traffic, who covers what and who is on what side frequency. Trying to keep this information in your head is a losing battle unless you have a remarkable memory.

There are many methods for doing this, depending to a great extent on what net is being controlled and the exact procedure used. In general, however, the best method is to list the calls of stations reporting in vertically down the page, followed by their coverage, if known. The coverage may be unnecessary if the NCS knows his stations, but it is a good idea to leave space for it in case an unfamiliar station reports in and you have to ask his location.

Next, list horizontally across the page the traffic reported by each station coming into the net, using destination (abbreviated) followed by the number of messages. From this you can see at a glance when traffic flow can start. In most nets, it is best to start it right away, and not wait until all stations have reported in. As traffic is passed, it can be crossed off. Whenever you get a station who has no traffic and for whom there is none, that station can be excused (QNX) and crossed off the list. As stations clear traffic and there is none for them, they also can be excused.

If side frequency (QNY) procedure is used, net controlling is a bit more complicated, but the use of such frequencies vastly speeds up the process. In this case, you will need to keep track of who is on which side frequency clearing which traffic, and you will be kept busy dispatching them, sending stations up or down to meet stations already on side frequencies, checking stations back into the net as they return, etc. Both your fingers and your key or mike button will be kept going, and it can be a nightmare if not handled properly.

Probably the best method is to keep the side frequencies on a separate sheet, each side frequency utilizing a separate column, labeled up 5, up 10, down 5, down 10, or whatever spacing intervals you find practical. As two stations are dispatched to a side frequency, enter the suffixes of their calls in the appropriate column, at the same time crossing out the traffic they are sent there to handle. When they return, cross them out of the side-frequency column; this side frequency can then accommodate two other stations. Of course, if you dispatch a third station to the side frequency to wait to clear traffic to one of the stations already there, enter him in the column also, and then only one of the two originally dispatched will return, so just cross off that one. When all your listed traffic is crossed off, the net is ready to secure (QNX QNF).

There are a number of refinements to this method, but the above is basic and a good way to start. Experience will soon indicate better ways to do it. For instance, K2KIR's novel approach appears in the sidebar "Net Controlling EAN."

message via a digital mode, type the message 10 words to a line. You will discover that this is a quick way to see if your message count agrees with the check. If you don't agree, nine times out of 10 you have dropped or added an "X-ray," so copy carefully. Another important thing to remember is that you never end a text with an "X-ray"—it just wastes space and makes the word count longer.

When counting messages, don't forget that each "X-ray" (instead of period), "Query" and initial group counts as a word. Ten-digit telephone numbers count as three words; the ARRL-recommended procedure for counting the telephone number in the text of a radiogram message is to separate the telephone number into groups, with the area code (if any) counting as one word, the three-digit exchange counting as one word, and the last four digits counting as one word. Separating the telephone number into separate groups also helps to minimize garbling. Also remember that closings such as "love" or "sincerely" (that would be in the signature of a letter) are considered part of the text in a piece of amateur traffic.

Finally, the signature. Remember, complimentary closing words like "sincerely" belong in the text, not the signature. In addition, signatures like "Dody, Vanessa, Jeremy, Ashleigh, and Uncle Porter," no matter how long, go entirely on the signature line.

At the bottom of our sample message you will see call signs next to the blanks marked "sent" and "received." These are not sent as the message, but are just bookkeeping notes for your own files. If necessary, you could help the originating station trace the path of the message.

KEEPING IT LEGAL

In the FCC rules under the "Prohibited transmissions" heading (§97.113), it states that "no amateur station shall transmit any communication the purpose of which is to facilitate the business or commercial affairs of any party ... except as necessary to providing emergency communications." Under the same heading, it further states that "no station shall transmit messages for hire or for material compensation, direct or indirect, paid or promised."

The FCC rules also have a section directly addressing traffic handling—§97.115, "Third party communications," which reads as follows:

(a) an amateur station may transmit messages for a third party to:

(1) Any station within the jurisdiction of the United States.

(2) Any station within the jurisdiction of any foreign government whose administration has made arrangements with the United States to allow amateur stations to be used for transmitting international communications on behalf of third parties. No station shall transmit messages for a third party to any station within the jurisdiction of any foreign government whose administration has not made such an arrangement. This prohibition does not apply to a message for any third party who is eligible to be a control operator of the station.

(b) The third party may participate in stating the message where:

(1) The control operator is present at the control point and is continuously monitoring and supervising the third party's participation; and

(2) The third party is not a prior amateur service licensee whose license was revoked; suspended for less than the balance of the license term and the suspension is still in effect; suspended for the balance of the license term and

Net Controlling EAN

At the NTS area and region levels, the complexity of the net control task often suggests the desirability of using a matrix form of log sheet. Many of the Eastern Area Net (EAN) Net Control Stations use the form shown here. A similar form for region nets can be made by replacing the RN column headings with the section net designations and the CAN/PAN columns with single column titled THRU (or EAN, CAN or PAN). Although this form can be used with no other accessories than a pencil, the use of moveable objects such as 6-32 hex nuts, buttons or push pins materially aids the visualization of which stations are on the net frequency and hence which stations are available for pairing off and passing traffic.

As stations are sent off frequency, their hex nuts are moved from the net column to the appropriate side frequency and a single diagonal line is drawn through the traffic being cleared. If, for any reason, the traffic is not cleared, a circle is drawn around the traffic total as a reminder to the NCS that it still must be cleared. (In other words, the circle overrides the slash.) Assuming the traffic is cleared on the second attempt, an opposite diagonal is drawn, thus totally crossing out the traffic quantity.

For the example shown, the following notes should explain how the various features of this NCS sheet are utilized:

Halfway through the Eastern Area Net session:

• W1EFW and VE3GOL are both clear and have been QNXed at 0055 and 0057, respectively.

• W8PMJ has cleared his CAN and PAN traffic, as well as 8RN traffic from W1TN. He has yet to clear his 313N, plus receive 8RN from W3YQ. He is presently DOWN TEN with W2CS.

• W2CS was previously sent off to clear his 4RN with W4NTO, but they were unable to complete the pairing. The NCS will try again later, perhaps by using a relay station.

• WB4PNY has a net report for K2KIR, to be sent direct to him if he QNIs.

• N2AKZ and KW1U are UP TEN, clearing 2RN traffic.

• KQ3T, W4NTO and K4ZK are standing by on the net frequency, waiting for new assignments.

There are many techniques for determining in what order to clear the traffic listed, but a couple of fairly common ones are worth mentioning.

1) Assign highest priorities to stations having the largest totals. Thus, W1NJM (PAN RX), W8PMJ (8RN TX and RX) and W3YQ (3RN TX) should be kept waiting as little as possible. If conditions are good, QNQing other stations to these three will speed things up immensely.

2) Tackle the smaller individual destination totals first. This allows early excusing of stations having small traffic totals to clear, and avoids a last-minute panic near the end of the net.

3) Clear short-haul pairings first. In the winter near a sunspot minimum, short-haul communications are most likely to be successful early in the net session before the skip has gone out. At all other times, especially in the summer with nets on Daylight Saving Time, it delays long-haul pairings until band conditions have improved for those paths.

Obviously, there are times when these (or any other algorithms) will conflict with each other. In the final analysis, a good NCS bases his decisions on the specifics of a given net session: the band conditions, the operators, the traffic distribution and the time remaining.—*Bud Hippisley, K2KIR*

						30			EASTERN AREA NET													
+30	+25	+20	+15	+10	+5	IN	STATION	OUT	1RN	2RN	3RN	4RN	8RN	ECN	CAN	PAN	OTR	-5	-10	-15	-20	-25
						35	W1EFW	55	▓													
				⬡		31	N2AKZ			▓												
						⬡30	KQ3T				▓											
						⬡30	W4NTO					▓										
						33	W8PMJ				2		▓		2/	1/			⬡			
						36	VE3GOL	57		1/				▓	1/							
					⬡	34	WB4PNY								▓		KIR					
						30	W1NJM									▓		⬡				
					⬡	30	(TX) W3YQ		3/	2		1/	4	1/	3/	2/						
						31	(TX) W1TN			1/	1/		3/			1/		⬡				
						32	(TX) W2CS					(2)/	3/			3			⬡			
						⬡30	K4ZK				1/					7						
				⬡		34	(TCC) KW1U		1/	1/		1/										

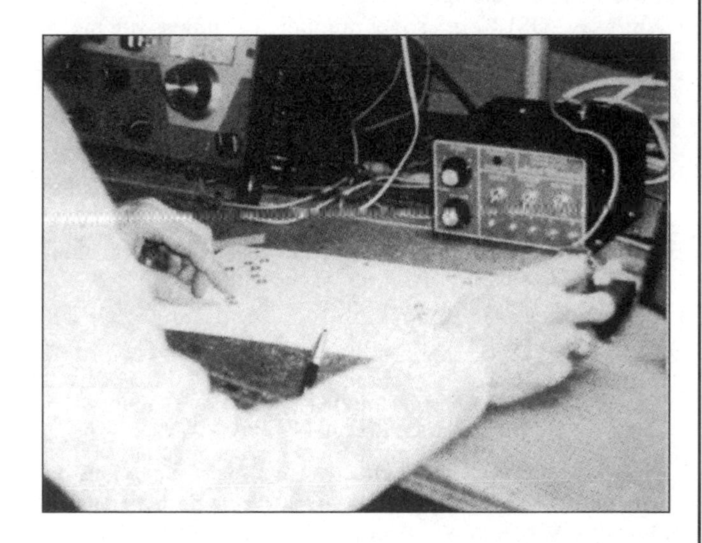

relicensing has not taken place; or surrendered for cancellation following notice of revocation suspension or monetary forfeiture proceedings. The third party may not be the subject of a cease and desist order which relates to amateur service operation and which is still in effect.

(c) At the end of an exchange of international third party communications, the station must also transmit in the station identification procedure the call sign of the station with which a third party message was exchanged.

Note that emergency communications is defined in §97.403 as providing "essential communication needs in connection with the immediate safety of human life and immediate protection of property when normal communications systems are not available."

It's self-explanatory. Every amateur should be familiar with these rules. Also, while third-party traffic is permitted in the US and Canada, this is not so for most other nations. A special legal agreement is required in each country to make such traffic permissible, both internally and externally (except if the message is addressed to another amateur). A list of third-party traffic agreements may be found at this ARRL Web Page address: **http://www.arrl.org/field/regulations/io/3rdparty.html**. Check this list before agreeing to handle any kind of international traffic.

Such third-party agreements specify that only unimportant, personal, nonbusiness communications be handled—things that ordinarily would not utilize commercial facilities. (In an emergency situation, amateurs generally handle traffic first and face the possible consequences later. It is not unusual for a special limited-duration third-party agreement to be instituted by the affected country during an overseas disaster.) The key point here is, particularly under routine day-to-day nonemergency conditions, if we value our privileges, we must take care not to abuse any regulations, whether it be on the national or international level.

SOME HELPFUL HINTS ABOUT RECEIVING TRAFFIC

1) Once you have committed the format of ARRL message form to memory, there's no need to use the "official" message pads from ARRL except for deliveries. Traffic handlers have many varied materials on hand for message handling. Some just use scrap paper. Many buy inexpensive 200-sheet 6 × 9-inch plain tablets available at stationery stores. Those who like to copy with a typewriter often use roll paper or fan-fold computer paper to provide a continuous stream of paper, separating each message as needed. As long as you can keep track of it, anything goes in the way of writing material.
2) Don't say "QSL" or "I roger number . . . " unless you mean it! It's not "Roger" unless you've received the contents of the message 100 percent. It's no shame to ask for fills (repeats of parts of the message). Make sure you have received the traffic correctly before going on to the next one.
3) Full (QSK) or semi-breakin can be very useful in handling traffic on CW. If you get behind, saying "break" or sending a string of dits will alert the other op that you need a fill.
4) You can get a fill by asking for "word before" (WB), "word after" (WA), "all before" (AB), "all after" (AA) or "between" (BN).

SENDING THE TRAFFIC

Just because you've taken a few messages, don't get the notion that being good at receiving traffic makes you a good sender, too. When one had to journey to the FCC office to take exams, one not only had to receive code in the code test, but send it as well. It used to be amazing how many folks could copy perfect 20-WPM code only to be inadequate on the sending portion. A lot of good ops became quite amazed that they could be so ham-fisted on a straight key!

Good traffic operators know they have to learn the nuances of sending messages as well as getting them. Your ability to send can "make or break" the other operator's ability to receive traffic in poor conditions. Imagine yourself and the other operator as a pair of computers interfaced over the telephone. Your computer (your brain) must successfully transfer data through your modem (your rig) over the lines of communication (the amateur frequencies) to the other modem, and ultimately, to the other computer.

Of course, for this transfer to be successful, two major items must be just right. The modems must operate at the same baud rate, and they must operate on matching protocol. Likewise, you must be careful to send your traffic at a comfortable speed for the receiving op, and use standardized protocol (standard ARRL message form). As you will see, these are slightly different for phone and CW, with even a couple of other deviations for HF digital or packet operations.

SENDING THE TRAFFIC BY CW

Someone once remarked, "The nice thing about CW traffic handling is that you have to spell it as you go along, so you don't usually have to spell words over." Also, the other main difference in CW traffic handling is that you tell someone when to go to the next address line or message section by use of the prosigns AA or BT. Keeping this in mind, let's show how our sample message would be sent:

NR 133 R HXG WØMME ARL 7 MOUNT PLEASANT IA 1700Z SEP 1
MR MRS JEFF HOLTZCLAW
ROUTE 1 BOX 127 AA
TONGANOXIE KS 66086
TEL 913 555 1212 BT
ARL FIFTY ONE OLD THRESHERS REUNION LOVE BT
UNCLE CHUCKIE AR N (if you have no more messages) or AR B (if you have further messages)

Now, let's examine a few points of interest:

1) You don't need to send "preamble words" such as precedence and check. The other operator is probably as familiar with standard ARRL form as you are (maybe more!).
2) The first three letters of the month are sufficient when sending the date.
3) In the address, always spell out words like "route" and "street."
4) Do not send dashes in the body of telephone numbers; it just wastes time.
5) Always, always, always spell out each word in the text! For example, "ur" for "your" could be misconstrued as the first two letters of the next word. Abbreviations are great for ragchewing, but not for the text of a radiogram.
6) Sometimes, if you have sent a number of messages, when you get to the next-to-last message, it's a good idea to send AR 1 instead of AR B to alert the other station that you have just one more.
7) If the other operator "breaks" you with a string of dits, stop sending and wait for the last word received by the other side. Then, when you resume sending, start up with that word, and continue through the message.

Becoming a proficient CW traffic sender is tough at first, but once you've mastered the basics, it will become second nature—no kidding!

SENDING THE TRAFFIC BY PHONE

Phone traffic handling is a lot like the infield fly rule in baseball—everyone thinks they know the rule, but in truth few really do. Correct message handling via phone can be just as efficient as via CW if and only if the two operators follow these basic rules:

1) If it's not an actual part of the message, don't say it.
2) Unless it's a very weird spelling, don't spell it.
3) Don't spell it phonetically unless it's a letter group or mixed group, or the receiving station didn't get it when you spelled it alphabetically.

Keeping these key points in mind, let's waltz through our sample message. This is how an efficient phone traffic handler would send the message:

"Number one hundred thirty three, routine, Hotel X-ray Golf, WØ Mike Mike Echo, ARL SEVEN, Mount Pleasant, Iowa, seventeen hundred Zulu, September one."

"Mr and Mrs Jeff Holtzclaw, route one, box twenty seven, Tonganoxie, T-O-N, G-A-N, O-X, I-E, Kansas, six six zero eight six. Telephone Nine one three, five five five, one two, one two. Break." You would then let up on the PTT switch and give the operator any fills needed in the first half of the message.

"ARL FIFTY ONE Old Threshers Reunion Love

Break Uncle Chuckie. End, no more" (if you have no more messages), "more" (if you have more messages).

Notice that in phone traffic handling, a pause—or with difficult addresses, "next line"—is the counterpart for AA. Also notice the lack of extraneous words. You don't need to say, "check," or "signature," or "Jones, common spelling." (If it's common spelling, why tell someone?) You only spell the uncommon. Most importantly, you speak at about half reading speed to give the other person time to write. If the receiving operator types, or if you have worked with the other op a long time and know his capabilities, you can speak faster. Always remember that any fill slows down the message more than if you had sent the message slowly to begin with!

OH, YES . . . THOSE EXCEPTIONS

Once again, HF digital and FM provide the exceptions to the rules. Digital traffic is very much like CW traffic. Use three or four lines between messages. This allows you to get four or five average messages on a standard $8^1/_2 \times 11$-inch sheet of paper.

When sending a message over your local repeater, remember that you often will be working with someone who isn't a traffic handler. It may be necessary to break more often (between the preamble and the address, for instance). Also, always make sure they understand about ARL numbered radiogram texts, and if they don't have a list, tell them what the message means. [The complete list of ARRL numbered radiogram texts appears in the References chapter and at the following ARRL Web Page address: **http://www.arrl.org/field/forms/fsd3.html**.] FM is a quiet mode, so you can get away with less spelling than you do on SSB.

If you yourself are already into traffic, don't try to force-feed correct traffic procedure in the case of someone just starting out; ease the person into it a little at a time. You will give a nontrafficker a more positive impression of traffic handling and may even make someone more receptive to joining a net. After all, our goal as traffic handlers is to have fun while being trained in accurately passing message traffic and enjoying it.

DELIVERING A MESSAGE

Up to now, all our traffic work has been carried out on the air. All of this changes, though, when we get a piece of traffic for delivery. Now we're tasked with contacting the general public with this message and expected to give it to someone. Unfortunately, many hams don't realize the importance of this action and miss an opportunity to engrave a favorable impression of Amateur Radio on nonhams. It's ironic that many hams can chat for hours on the air, but can't pick up the telephone and deliver a 15-word message without mumbling, stuttering or acting embarrassed. Delivering messages should be a treat, not a chore.

Let's go through a few guidelines for delivery, and if you keep these tips in mind, you and the party on the other end of the phone will enjoy the delivery.

1) Introduce yourself. Don't you hate phone calls from people you don't know and don't bother to give a name? Chances are they're trying to sell you something, and you brush them off. Most people have no idea what Amateur Radio is about, and it's up to you to make a good first impression.
2) Ask for the person named in the message. If he or she is not home, ask the person on the phone if they would take a message for that person.
3) Tell who the message is from before you give the message. Since the signature appears at the end of the message, most hams give it last, but you will hold the deliveree's attention longer if you give it first. When you get letters in the mail, you check out the return addresses first, don't you? Then you open them in some sort of order of importance. Likewise, the party on the phone will want to know the sender of the message first.

A good way to start off a delivery is to say something like, "Hi, Mr/Mrs/Miss so-and-so, my name is whatever, and I received a greeting message via Amateur Radio for you from wherever from such-and-such person." This usually gives you some credibility with your listener, because you mentioned someone they know. They will usually respond by telling such-and-such is their relative, college friend, etc. At that point, you have become less of a stranger in their eyes, and now they don't have to worry about you trying to sell them some aluminum siding or a lake lot at Casa Burrito Estates. Make sure you say it's a greeting message, too, to allay any fears of the addressee that some bad news is imminent.

4) When delivering the message, skip the preamble and just give the text, avoiding ARL text abbreviations. Chances are, Grandma Ollie doesn't give two hoots about the check of a message, and thinks ARL FORTY SIX is an all-purpose cleaner. Always give the "translation" of an ARL numbered text, even if the message is going to another ham.
5) Ask the party if they would like to send a return message. Explain that it's absolutely free, and that you would be happy to send a reply if they wish. Experienced traffickers can vouch that it's easy to get a lot of return and repeat business once you've opened the door to someone. It's not uncommon for strangers to ask for your name or phone number once they discover Amateur Radio is a handy way

to communicate with friends and relatives.

6) Notices of death and/or serious injury should only be handled as communication between emergency preparedness, Red Cross, Salvation Army or other relief agency officials and only in the absence of alternate commercial facilities. Radio amateurs should never, repeat never, be responsible for notifying individuals, third-party or otherwise, of death or serious injury. These should always be handled through the appropriate relief agencies.

TO MAIL OR NOT TO MAIL

Suppose you get a message that doesn't have a phone number, or the message would require a toll call. Then what? If you don't know anyone on 2 meters that could deliver it, or Directory Assistance is of no help, you are faced with the decision whether to mail it or not. There is no hard and fast rule on this (unless, of course, the message has an HXG attached). Always remember that since this is a free service, you are under no obligation to shell out a stamp or call telephone directory assistance just because you accepted the message.

Many factors may influence your decision. If you live in a large urban area, you probably have more deliveries than most folks, and mail delivery could be a big out-of-pocket expense that you're not willing to accept. If you live out in the wide open spaces, you may be the only ham for miles around, and probably consider mail delivery more often than most. Are you a big softie on Christmas or Mother's Day? If so, you may be willing to brunt the expense of a few stamps during those times of the year when you wouldn't otherwise. At any rate, the decision is entirely up to you.

Although you may be absolved from the responsibility of mailing a message, you don't just chuck the message in the trash. You do have a duty to inform the originating station that the message could not be delivered. A simple ARL SIXTY SEVEN followed by a brief reason (no listing, no one home for three days, mail returned by post office, and so forth) will suffice. This message always goes to the station of origin, not the person in the signature. The originating station will appreciate your courtesy.

NOW THAT YOU'RE MOVING UP IN THE WORLD

By now, you are starting to get a grasp of the traffic world. You've been checking in to a net on a regular basis, and you're pretty good at message form. Maybe you've even delivered a few messages. Now you are ready to graduate from Basic Traffic 101 and enroll in Intermediate Traffic 102. Good for you! You have now surpassed 80 percent of your peers in a skillful specialty area of Amateur Radio. However, there's still a lot to learn, so let's move on.

BOOK MESSAGES

Over the years, book messages have caused a lot of needless headaches and consternation among even the best traffic handlers. Many hams avoid booking anything just because they think it's too confusing. Truthfully, book messages are fairly simple to understand, but folks tend to make them harder than they actually are.

So, just what are book messages? Book messages are merely messages with the same text and different addresses. They come in two categories—ones with different signatures, and ones with the same signatures. Elsewhere in the chapter are examples. Often you will see book messages around holiday times and during fairs or other public events.

Oh, yes . . . one other thing about book messages. When you check into a net with a bunch of book messages, give the regular message count only. Don't say, "I have a book of seven for Outer Baldonia." Say instead, "Seven Outer Baldonia." Then, when you and the station from Outer Baldonia go off frequency to pass the traffic, tell him that it is book traffic. When he tells you to begin sending, give common parts first, then the "uncommon" parts (addresses and possibly signatures.) By following this procedure, you will avoid a lot of confusion.

Suppose you get a book of traffic on the NTS Region net bound for your state, but to different towns. When you take them to your section net, you will not be able to send them as a book, since they must be sent to different stations. Now what? Simply "unbook" them, and send them as individual messages. For instance, let's say you get a book of three messages for the Missouri section from the region net. Two are for Missouri City, and one is for Swan Valley. Simply list your traffic as Missouri City 2 and Swan Valley 1, for a total count of 3. Books aren't ironclad chunks of traffic, but a stepsaver that can be used to your advantage. They can be unbooked at any time. Use them whenever you can, and don't be afraid of them.

BECOMING NCS AND LIVING TO TELL ABOUT IT

Some momentous evening in your traffic career, you may be called upon to take the net. Perhaps the NCS had a power failure, or is on vacation, or perhaps a vacancy occurred in the daily NCS rotation on your favorite net. Should this be the case, consider yourself lucky. Net Managers entrust few members with net-control duties.

Of course, you probably won't be thinking how lucky you are when the Net Manager says "QNG" and sticks your call after it. Once again, just like your first check in or your first piece of traffic, you will just have to grit your teeth and live through it. However, you can make the jump easier by following these hints long before you are asked to be a net control:

1) Become familiar with the other stations on the net. Even if you never become NCS, it pays to know who you work with and where they live.

2) Pay close attention to the stations that go off frequency to pass traffic. What frequencies does the net use to move traffic? Which stations are off frequency at the moment? You will gain a feel for the net control job just by keeping track of the action.

3) Try to guess what the NCS will do next. You will discover many dilemmas when you try to second-guess the NCS. Often different amounts of traffic with equal precedence appear on the net, and a skillful NCS must rank them in order of importance. For instance, if you follow the NCS closely, you will discover that traffic for the NTS rep, such as out-of-state traffic, gets higher priority than one

Handy Hints for Handling Traffic at Fairs or Other Public Events

1) Although you may only be there a day or two, don't compromise your station too much. Try to put up the most you can for an antenna system because band conditions on traffic nets in the summer can really be the pits! Usually, you will be surrounded by electrical lines at fairs, so a line filter is a must. An inboard SSB or CW filter in your rig is a definite plus, too, and may save you many headaches.

2) Don't huddle around the rigs or seat yourself in the back of the booth. Get up front and meet the people. After all, your purpose is to "show off" Amateur Radio to perk the interest of nonhams.

3) Most people will not volunteer to send a piece of traffic, nor will they believe a message is really "free." It's up to you to solicit "business." Be cheerful.

4) Always use "layman's language" when explaining Amateur Radio to nonhams. Say "message," not "traffic." Don't ramble about the workings of NTS or repeaters; your listener just wants to know how Aunt Patty will get the message. "We take the message and send it via Amateur Radio to Aunt Patty's town, and the ham there will call her on the phone and deliver it to her," will do.

5) Make sure your pencils or pens are attached to the booth with a long string, or you will be out of writing utensils in the first hour!

6) Make sure there are plenty of instructions around for hams not familiar with traffic handling to help them "get the hang of" the situation.

7) Make sure your booth is colorful and attractive. You will catch the public's eye better if you give them something to notice, such as this suggested poster idea.

for the NTS rep's city. Situations like these are fun to second-guess when you are standing by on the net and will better prepare you for the day you might get to run the net.

Should that day arrive, just keep your cool and try to implement the techniques used by your favorite net-control stations. After a few rounds of NCS duties, you will develop your own style, and who knows? Perhaps some new hopeful for NCS will try to emulate you some day! See accompanying sidebars for further hints on developing a "type NCS" personality and on proper net-control methods.

HANDLING TRAFFIC AT FAIRS OR OTHER PUBLIC EVENTS

A very special and important aspect of message handling is that of how to handle traffic at public events. If the event is of any size, like a state fair, it doesn't take long to swamp a group of operators with traffic. Only by efficient, tight organization can a handful of amateurs keep a lid on the backlog.

No matter what size your public event, the following points need to be considered for any traffic station accessible to the public:

1) Often, more nontraffic handlers than traffic handlers will be working in the booth. This means your group will have to lay out a standard operating procedure to help the nontraffickers assist the experienced ops.

Jobs such as meeting the public, filling out the message blanks, sorting the "in" and "out" piles, and keeping the booth tidied up, can all be performed by people with little or no traffic skill, and is a good way to introduce those people into the world of traffic.

2) If you plan to handle fairly large amounts of traffic, the incoming traffic needs to be sorted. A good system is to have an "in-state" pile, an "in-region" pile, and an area pile for the three levels of NTS. After the traffic has been sent, it needs to be stacked in numerical order in the "out" pile. Keeping it in numerical order makes it easier to find should it need to be referred to.

Since your station will be on a number of hours, plan to check into your region and area, as well as your section, net. Another good idea is to have "helpers" on 2 meters who can also take some of your traffic to the region and/or area net. These arrangements need to be worked out in advance.

3) Make up your radiogram blanks so that most of the preamble is already on them, and all you need to fill in is the number, check and date. In the message portion, put only about 20-word lines to discourage lengthy messages. Try to convince your "customer" to use a standard ARL text so you can book your messages.

A real time- and headache-saver in this department is to fill out the message blank for the sender. This way, you can write in the X-rays and other jargon that the sender is unaware of.

4) Most importantly, realize that you sometimes have to work at getting "customers" as much as if you were selling something! Most people have no concept of Amateur Radio at all, and don't understand how message handling, works. ("How can they get it? They don't have a radio like that" is a very common question!)

Use posters to make your booth appealing to the eye. Make sure one of the posters is of the "How your message gets to its destination" variety, such as the one shown in this section. Don't go over someone's head when answering a question—explain it simply and succinctly.

Finally, don't be afraid to "solicit" business. Get up in front of the booth and say "hi" to folks. If they say "hi" back, ask them if they would like to send a free greeting to a friend or relative anywhere in the US (grandparents and grandchildren are the easiest to convince!). Even if they decide not to send a message, your friendliness will help keep our image of "good guys in white hats" viable among the general public, which is every bit as great a service as message handling. For further hints, see accompanying sidebar.

THE ARRL NATIONAL TRAFFIC SYSTEM—MESSAGE HANDLING'S "ROAD MAP"

Although you probably never think about it, when you check into your local net or section net, you are participating in one of the most cleverly designed game plans ever written—the National Traffic System (NTS). Even though the ARRL conceived NTS way back in 1949, and it has grown from one regular cycle to two or more, NTS hasn't outgrown itself and remains the most streamlined method of traffic handling in the world. (During this discussion, please refer to the accompanying Section/Region/Area map in **Fig 7-2**, the NTS Routing Guide in **Table 7-1** and the NTS Flow Chart in **Fig 7-3**.)

Actually, the National Traffic System can trace its roots to the railroad's adoption of Standard Time back in 1883, when radio was still only a wild dream. Three of the Standard Time Zones are the basis for the three NTS areas—Eastern Area (Eastern Time Zone), Central Area (Central Time Zone) and Pacific Area (Mountain and Pacific Time Zones). Within these areas are a total of 13 regions. The Atlantic Region Net is considered the Thirteenth Region since it handles international third party messages. You may wonder why NTS has 13 regions—why not just break it up into 10 regions, one for each US call-sign district? Ah, but check the map. You will discover that NTS not only covers the US, but our Canadian neighbors as well. Then, of course, the region nets are linked to section/local nets.

The interconnecting lines between the boxes on the flow chart represent liaison stations to and from each level of NTS. The liaisons from area net to area net have a special name, the Transcontinental Corps (TCC). In addition to the functions shown, TCC stations also link the various cycles of NTS to each other.

The clever part about the NTS setup, though, is that in any given cycle of NTS, all nets in the same level commence at approximately the same local time. This allows time for liaisons to the next level to pick up any outgoing traffic and meet the next net. In addition, this gives the TCC stations at least an hour before their duties commence on another area net or their schedule begins with another TCC station.

The original NTS plan calls for four cycles of traffic nets, but usually two cycles are sufficient to handle a normal load of traffic on the system. However, during the holiday season, or in times of emergency, many more messages are dumped into the system, forcing NTS to expand to four cycles temporarily. The cycles of normal operation are Cycle Two, the daytime cycle, which consists primarily of phone nets, and Cycle Four, the nighttime cycle, made up mostly of CW nets. In addition, Cycle One has been imple-

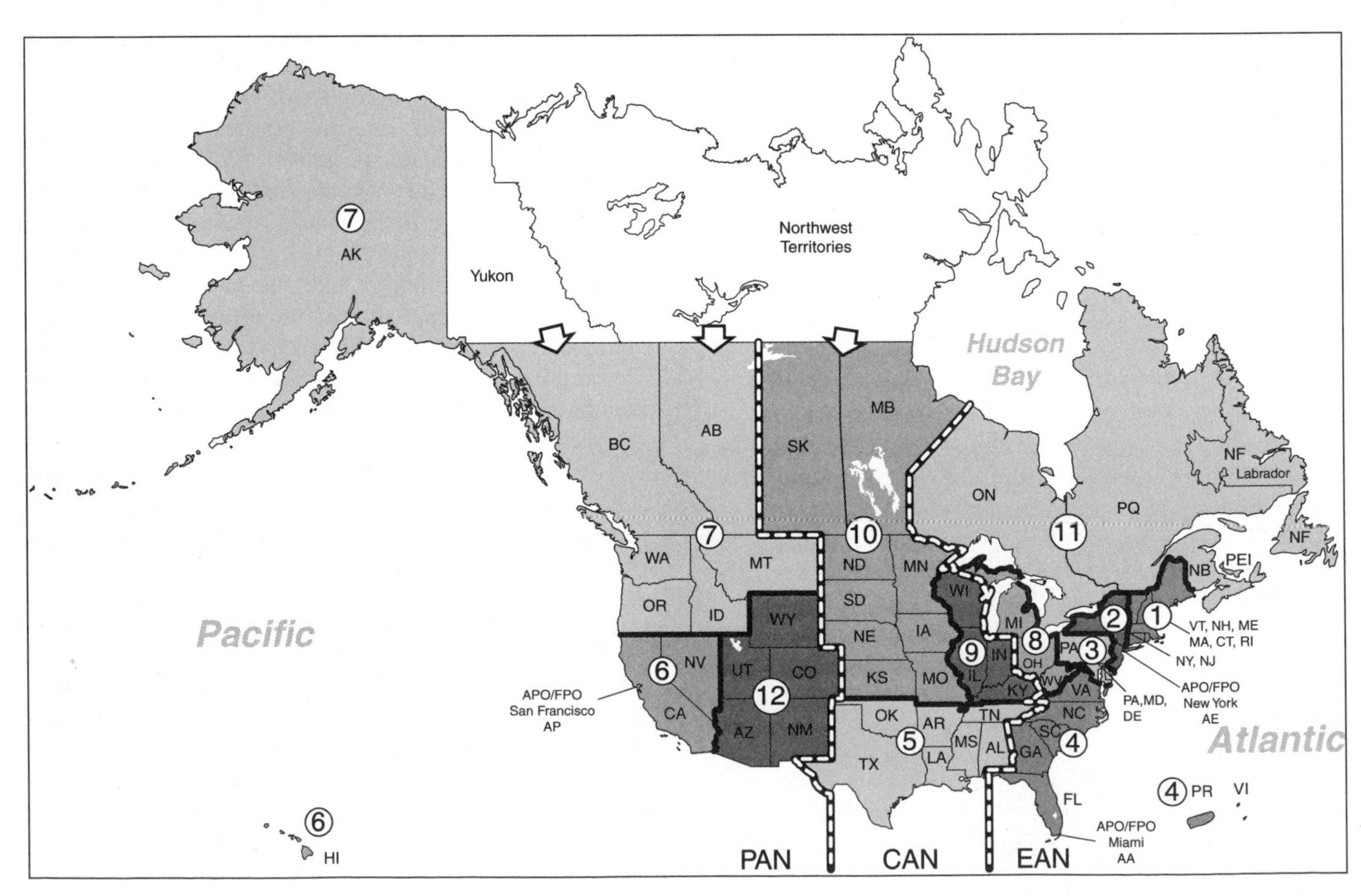

Fig 7-2—ARRL National Traffic System Area and Region map.

Table 7-1
National Traffic System Routing Guide

State/Province	*Abbrev.*	*Region*	*Area*
Alaska	AK	7	PAN
Alabama	AL	5	CAN
Alberta	AB	7	PAN
Arizona	AZ	12	PAN
Arkansas	AR	5	CAN
British Columbia	BC	7	PAN
California	CA	6	PAN
Colorado	CO	12	PAN
Connecticut	CT	1	EAN
Delaware	DE	3	EAN
Dist. of Columbia	DC	3	EAN
Florida	FL	4	EAN
Georgia	GA	4	EAN
Guam	GU	6	PAN
Hawaii	HI	6	PAN
Idaho	ID	7	PAN
Illinois	IL	9	CAN
Indiana	IN	9	CAN
Iowa	IA	10	CAN
Kansas	KS	10	CAN
Kentucky	KY	9	CAN
Labrador	LB	11	EAN
Louisiana	LA	5	CAN
Maine	ME	1	EAN
Manitoba	MB	10	CAN
Maryland	MD	3	EAN
Massachusetts	MA	1	EAN
Michigan	MI	8	EAN
Minnesota	MN	10	CAN
Mississippi	MS	5	CAN
Missouri	MO	10	CAN
Montana	MT	7	PAN
Nebraska	NE	10	CAN
Nevada	NV	6	PAN
New Brunswick	NB	11	EAN
New Hampshire	NH	1	EAN
New Jersey	NJ	2	EAN
New Mexico	NM	12	PAN
New York	NY	2	EAN
Newfoundland	NF	11	EAN
North Carolina	NC	4	EAN
North Dakota	ND	10	CAN
Nova Scotia	NS	11	EAN
Ohio	OH	8	EAN
Oklahoma	OK	5	CAN
Ontario	ON	11	EAN
Oregon	OR	7	PAN
Pennsylvania	PA	3	EAN
Prince Edward Is.	PEI	11	EAN
Puerto Rico	PR	4	EAN
Quebec	QC	11	EAN
Rhode Island	RI	1	EAN
Saskatchewan	SK	10	CAN
South Carolina	SC	4	EAN
South Dakota	SD	10	CAN
Tennessee	TN	5	CAN
Texas	TX	5	CAN
Utah	UT	12	PAN
Vermont	VT	1	EAN
Virginia	VA	4	EAN
Washington	WA	7	PAN
West Virginia	WV	8	EAN
Wisconsin	WI	9	CAN
Wyoming	WY	12	PAN
Virgin Islands	VI	4	EAN
APO/FPO AE		2	EAN
APO/FPO AA		4	EAN
APO/FPO AP		6	PAN
International		13	EAN

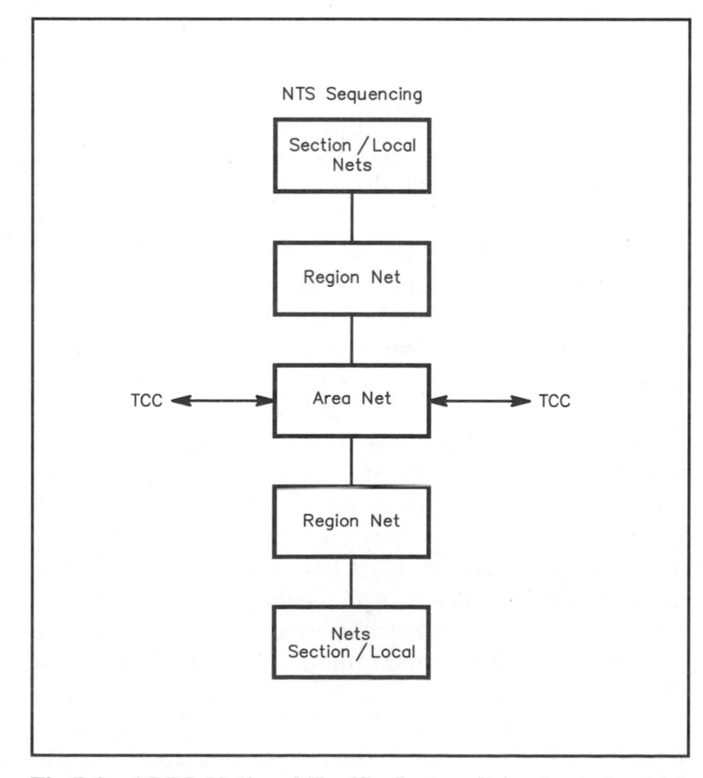

Fig 7-3—ARRL National Traffic System flow chart. See text.

mented in the Pacific Area and Cycle Three in the Eastern Area. Now that the rudiments of NTS have been covered, let's see where you fit in.

NTS AND YOU

Before the adoption of NTS, upper-level traffic handlers worked a system called the "trunk line" system, where a handful of stations carried the burden of cross-country traffic, day in day out. Nowadays, no one has to be an "iron man" or "iron woman" within NTS if they choose not to. If each liaison slot and TCC slot were filled by one person, one day a week, this would allow over 1000 hams to participate in NTS! Unfortunately, many hams have to double- and triple-up duties, so there is plenty of room for any interested amateur.

An NTS liaison spot one day (or night) a week is a great way to stay active in the traffic circuit. Many hams who don't have time to make the section nets get satisfaction in the traffic world by holding a TCC slot or area liaison once a week. If you would enjoy such a post, drop a note to your STM, Net Manager or TCC Director. They will be happy to add another to their fold.

However, remember that the area and region net are very different from your section or local net in one aspect. The function of the section or local net is to "saturate" its jurisdiction, so the more check ins, the better. On the upper-level nets,

Fame and Glory: Your Traffic Total, PSHR Report, and Traffic Awards and Appointments

Even if you handle only one message in a month's time, you should send a message to your ARRL Section Traffic Manager (STM) or ARRL Section Manager (SM) reporting your activity. Your report should include your total originations, messages received, messages sent and deliveries.

An origination is any message obtained from a third party for sending from your station. If you send a message to Uncle Filbert on his birthday, you don't get an origination. However, if your mom or your neighbor wants you to send him one with her signature, it qualifies (it counts as one originated and one sent). The origination category is essentially an "extra" credit for an off-the-air function. This is because of the critical value of contact with the general public and to motivate traffickers to be somewhat more aggressive in making their message-handling services known to the general public.

Any formal piece of traffic you get via Amateur Radio counts as a message received. Any message you send via Amateur Radio, even if you originated it, counts as a message sent. Therefore, any time you relay a message, you get two points: one received and one sent.

Any time you take a message and give it to the party it's addressed to, on a mode other than Amateur Radio, you are credited with a delivery. (It's okay if the addressee is a ham.) As long as you deliver it off the air (eg, telephone, mail, Internet/e-mail, in person), you get a delivery point.

Your monthly report to your STM, if sent in radiogram format, should look something like this:

NR 111 R NIØR 14 ST JOSEPH MO NOV 2
BLAIR CARMICHAEL WBØPLY
MISSOURI STM
FULTON MO 65251

OCTOBER TRAFFIC ORIG 2 RCVD 5 SENT 6 DLVD 1
TOTAL 14 × 73
ROGER NIØR

If you have a traffic total of 500 or more in any month, or have over 100 originations-plus-deliveries in a month, you are eligible for the Brass Pounders League (BPL) certificate, even if you did all that traffic on SSB! Make sure you send your traffic totals to your STM or SM. If you make BPL three times, you qualify for a one-time, call-sign engraved medallion for your shack.

Another mark of distinction is the Public Service Honor Roll (PSHR). You don't have to handle a single message to get PSHR, so it's a favorite among traffickers in rural areas. The categories have traditionally been listed in the Public Service column of *QST*, and they include checking into public service nets, acting as net control, handling an emergency message, or participating in a public service event. Stations that qualify for PSHR for 12 consecutive months, or 18 out of a 24-month period, will qualify for this one-time certificate. Again, candidates for PSHR need to report their monthly PSHR point total to their STM or SM. See the Public Service column for particulars.

Almost any station, with regular participation, can get an ARRL Net Certificate through the Net Manager. Once you have participated regularly in a net for three months, you are eligible. If you are an ARRL member, you can also become eligible for an Official Relay Station appointment in the ARRL Field Organization (for details concerning the ARRL Field Organization, see the References chapter). Either certificate makes a handsome addition to your shack.

Recognizing the needs of Amateur Radio operators new to the public service arena, ARRL Headquarters and the ARRL National Traffic System sponsor the National Traffic System Message Origination Award. To qualify for this one-time award, you must show proof of originating at least four properly formatted NTS radiograms in a one-month period to your ARRL Section Manager (SM) or Section Traffic Manager (STM) to receive your award certificate.

The local radio amateur community and your section leadership officials are ready to help you get involved in traffic handling. To contact your section officials, see page 12 of *QST*. Also, the ARRL Web Page has this referral information and links to some individual section Web Pages. Log onto **http://www.arrl.org/field/org/smlist.html**.

though, the name of the game is to move the traffic as quickly and efficiently as possible. Therefore, additional check ins—other than specified liaisons and stations holding traffic—only slow down the net. [However, if you are a station holding traffic to be moved, you can enter NTS at any level to pass your traffic, even if you've never been on an upper-level net before. Entering the system at the section or local level is preferred.]

If you are interested in finding out more about the "brass tacks" of the workings to NTS, get a copy of the *Public Service Communications Manual* (FSD-235), available for $1 (postpaid) from ARRL HQ. It's also available on the ARRL Web Page at this address: **http://www.arrl.org/field/pscm**. Every aspect of NTS is explained, as well as information about local net operating procedure, RACES and ARES operation. The PSCM will also orient you with net procedure of region and area NTS nets. It takes a little more skill and savvy to become a regular part of NTS, but the rewards are worth the effort. If you have the chance, go for it!

SO THERE YOU HAVE IT

Although this chapter is by no means a complete guide to traffic handling, it should serve as a good reference for veterans and newcomers alike. If you've never been involved in message handling, perhaps your interest has been piqued. Should that be the case, don't put it off. Find a net that's custom-made for you and check into it! You'll find plenty of fine folks that will soon become close friends as you begin to work with them, learn from them, and yes, even chat with them when the net is over.

The roots of traffic handling run deep into the history of Amateur Radio, yet its branches reach out toward many tomorrows. Our future lies in proving our worth to the nonham public, and what better way to ensure the continuance of our hobby than by uniting family and friends via Amateur Radio? Sure, it takes some effort, but a trafficker will tell you he stays with it because of the satisfaction he gets from hearing those voices on the other end of the phone say, "Oh, isn't that nice!" We've plenty of room for you—come grow with us!

The following ARRL NTS Officials helped with the revision of this chapter: Bill Thompson, W2MTA, Jim Leist, KB5W, Robert Griffin, K6YR, and Nick Zorn, N4SS.

Chapter 8

Emergency Communications

Serving Others

Richard R. Regent, K9GDF
5003 South 26th St
Milwaukee, WI 53221

People in desperate need of help can depend on Amateur Radio to serve them. Since 1913, ham communicators have been dedicated volunteers for the public interest, convenience and necessity by handling free and reliable communications for people in disaster-stricken areas, until normal communications are restored.

Emergency communication is an Amateur Radio communication directly relating to the immediate safety of human life or the immediate protection of property, and usually concerns disasters, severe weather or vehicular accidents. The ability of amateurs to respond effectively to these situations with emergency communications depends on practical plans, formalized procedures and trained operators.

PLANS

More than any other facet of Amateur Radio, emergency communication requires a plan—an orderly arrangement of time, talent and activities that ensures that performance is smooth and objectives are met. Basically, a plan is a method of achieving a goal. Lack of an emergency-communications plan could hamper urgent operations, defer crucial decisions or delay critical supplies. Be sure to analyze what emergencies are likely to occur in your area, develop guidelines for providing communications after a disaster, know the proper contact people and inform local authorities of your group's capabilities.

Start with a small plan, such as developing a community-awareness program for severe-weather emergencies. Next, test the plan a piece at a time, but redefine the plan if it is unsatisfactory. Testing with simulated-emergency drills teaches communicators what to do in a real emergency, without a great deal of risk. Finally, prepare a few contingency plans just in case the original plan fails.

One of the best ways to learn how to plan for emergencies is to join your local Amateur Radio Emergency Service (ARES) group, where members train and prepare constantly in an organized way. Before an emergency occurs, register with your ARRL Emergency Coordinator (EC). The EC will explain to civic and relief agencies in your community what the Amateur Service can offer in time of disaster. ARES and the EC are explained later in this chapter.

Communications for city or rural emergencies each require a particular response and careful planning. Large cities usually have capable relief efforts handled by paid professionals, and there always seems to be some equipment and facilities that remain operable. Even though damage may be more concentrated outside a city, it can be remote from fire-fighting or public-works equipment and law-enforcement authorities. The rural public then, with few volunteers spread over a wide area, may be isolated, unable to call for help or incapable of reporting all of the damage.

It's futile to look back regretfully at past emergencies and wish you had been better prepared. Prepare yourself now for emergency communications by maintaining a dependable transmitter-receiver setup and an emergency-power source. Have a plan ready and learn proper procedures.

PROCEDURES

Aside from having plans, it is also necessary to have procedures—the best methods or ways to do a job. Procedures become habits, independent of a plan, when everyone knows what happens next and can tell others what to do. Actually, the size of a disaster affects the size of the response, but not the procedures.

Procedures that should be widely known before disasters occur include how to coordinate or deploy people, equipment and supplies. There are procedures to use a repeater and an autopatch, to check into a net and to format or handle traffic. Because it takes time to learn activities that are not normally used every day, excessively detailed procedures will confuse people and should be avoided.

Specific Procedures

Your EC will have developed a procedure to activate

Distress Calling

An amateur who needs immediate emergency assistance (at sea, in a remote location, etc.) should call MAYDAY on whatever frequency seems to offer the best chance of getting a useful answer. MAYDAY is from the French *m'aider* (help me). On CW, use SOS to call for help. QRRR has been discarded by ARRL. The distress call should be repeated over and over again for as long as possible until answered.

The amateur involved should be prepared to supply the following information to the stations who respond to an SOS or MAYDAY:

- The location of the emergency, with enough detail to permit rescuers to locate it without difficulty.
- The nature of the distress.
- The type of assistance required (medical aid, evacuation, food, clothing, etc.).
- Any other information that might be helpful in the emergency area or in sending assistance.

the ARES group, but will need your help to make it work. A telephone alerting "tree" callup, even if based on a current list of phone numbers, might fail if there are gaps in the calling sequence, members are not near a phone or there is no phone service.

Consider alternative procedures. Use alerting tones and frequent announcements on a well monitored repeater to round up many operators at once. An unused 2-meter simplex frequency can function for alerting; instead of turning radios off, your group would monitor this frequency for alerts without the need of any equipment modifications. Since this channel is normally quiet, any activity on it would probably be an alert announcement.

During an emergency, report to the EC so that up-to-the-minute data on operators will be available. Don't rely on one leader; everyone should keep an emergency reference list of relief-agency officials, police, sheriff and fire departments, ambulance service and NTS nets. Be ready to help, but stay off the air unless there is a specific job to be done that you can handle efficiently. Always listen before you transmit. Work and cooperate with the local civic and relief agencies as the EC suggests; offer these agencies your services directly in the absence of an EC. During a major emergency, copy special W1AW bulletins for the latest developments.

Afterward, let your EC know about your activities, so that a timely report can be submitted to ARRL HQ. Amateur Radio has won glowing public tribute in emergencies. Help maintain this record.

For a comprehensive study of emergency communications procedures, see the *ARRL Emergency Coordinator's Manual* (FSD-9). Contact ARRL HQ Field and Educational Services for details on how to obtain a copy. "Experience is the worst teacher: it gives the test before presenting the lesson" (Vernon's Law). Train and drill now so you can be prepared for an emergency.

TRAINING

Amateurs need training in operating procedures and communications skills. In an emergency, radios don't communicate, but people do. Because amateurs with all sorts of varied interests participate, many of those who offer to help may not have experience in public service activities. Rarely are there enough trained operators, especially if a crisis persists for a long time.

Proper disaster training replaces chaotic pleas with smooth organized communications. The ARRL recognizes the need for emergency preparedness and emergency-communications training through sponsorship of the ARES. Well trained communicators respond during drills or actual emergencies with quick, effective and efficient communications. Each understands his or her role in the plan and sets a good example by knowing the proper procedures to use.

Whether training takes place with programs at club meetings, on the air or with a personal approach, the basic subjects should cover emergency communications, traffic handling, net or repeater operation and technical knowledge. Get as many people involved as possible to learn how emergency communications should be handled. Explain what's going on and assign each participant a useful role.

Practical on-the-air activities, such as the ARRL's Field Day and Simulated Emergency Test offer training opportunities on a nationwide basis for individuals and groups. Participation in such events reveals weak areas where discussions and more training are needed. In addition, drills and tests can be designed specifically to check dependability of emergency equipment, or to rate training in the local area.

One of the best sources for organized training activities is the Emergency Manager in your county Emergency Management Office. This person has access to many Federal Emergency Management Agency (FEMA) training programs, some of which deal with communications. Other FEMA training programs may help hams understand how to interface with local and state governments under conditions that develop during real emergencies. Indeed, in some states where ARES and RACES are integrated to some degree, Amateur Radio provides emergency communications directly for governmental bodies as well as for private institutions.

Field Day

The ARRL Field Day (FD) gets more amateurs out of their cozy shacks and into tents on hilltops than any other event. You may not be operating from a tent after a disaster but the training you will get from FD is invaluable.

In the ARRL Field Day, a premium is placed on sharp operating skills, adapting equipment that can meet challenges of emergency preparedness and flexible logistics. Amateurs assemble portable stations capable of long-range communications at almost any place and under varying conditions. Alternatives to commercial power in the form of generators, car batteries, windmills or solar power are used to power equipment to make as many contacts as possible. FD is held on the fourth full weekend of June, but enthusiasts get the most out of their training by keeping preparedness programs alive during the rest of the year.

Simulated Emergency Test

The ARRL Simulated Emergency Test (SET) builds emergency-communications character.

The purposes of SET are to:

- Help amateurs gain experience in communicating, using standard procedures under simulated emergency conditions, and to experiment with some new concepts.
- Determine strong points, capabilities and limitations in providing emergency communications to improve the response to a real emergency.
- Provide a demonstration, to served agencies and the public through the news media, of the value of Amateur Radio, particularly in time of need.

The goals of SET are to:

- Strengthen VHF-to-HF links at the local level, ensuring that ARES and NTS work in concert.
- Encourage greater use of digital modes for handling high-volume traffic and point-to-point welfare messages of the affected simulated-disaster area.
- Implement the Memoranda of Understanding between the ARRL, the users and cooperative agencies.
- Focus energies on ARES communications at the local level. Increase use and recognition of tactical communication on behalf of served agencies; using less amateur-to-amateur formal radiogram traffic.

Help promote the SET on nets and repeaters with announcements or bulletins, or at club meetings and publicize it in club newsletters. SET is conducted on the first full weekend of October. However, some groups have their SETs any time during the period of September 1 through November 30, especially if an alternate date coincides more favorably with a planned communications activity and provides greater publicity. Specific SET rules are published in *QST*.

Drills and Tests

A drill or test that includes interest and practical value makes a group glad to participate because it seems worthy of their efforts. Formulate training around a simulated disaster such as a tornado or a vehicle accident. Elaborate on the situation to develop a realistic scenario or have the drill in conjunction with a local event. Many ARRL Section Emergency Coordinators (SECs) have developed training activities that are specifically designed for your state, section or local area. County Emergency Managers are often well practiced in setting up exercises that can help you sharpen your communications and general emergency reaction skills.

During a drill:

1) Announce the simulated emergency situation, activate the emergency net and dispatch mobiles to served agencies.
2) Originate messages and requests for supplies on behalf of served agencies by using tactical communications. (Don't forget to label each message with a "this is a drill only" header, no matter what mode is used to transmit it.)
3) Use emergency-powered repeaters and employ digital modes.
4) As warranted by traffic loads, assign liaison stations to receive traffic on the local net and relay to your section net. Be sure there is a representative on each session of the section nets to receive traffic coming to your area.

After a drill:

1) Determine the results of the emergency communications.
2) Critique the drill.
3) Report your efforts, including any photos, clippings and other items of interest, to your SEC or ARRL HQ.

ARRL Emergency Coordinator and Certification Course

The ARRL has a certification program to provide training and formal recognition of amateur achievement in the field of emergency communications. This course is administered to ECs (or potential ECs) by the Section Emergency Coordinator for each section. The ARRL also offers the public service-oriented awards program of certificates for Public Service Honor Roll, Emergency Communications Commendation and Public Service Commendation.

Net Operator Training

Network discipline and message-handling procedures are fundamental emergency preparedness concepts. Training should involve as many different operators as possible in Net Control Station and liaison functions; don't have the same operator performing the same functions repeatedly or you will lose valuable training experience for the other members of the group. There should be plenty of work for everyone. Good liaison and cooperation at all levels of NTS requires versatile operators who can operate either phone or CW. Even though phone operators may not feel comfortable on CW and vice versa, encourage net operators to gain familiarity on both modes by giving them proper training. They can learn by logging for a regular operator in that mode.

The liaison duties to serve between different NTS region net cycles as well as between section nets are examples of the need for versatile operators.

Ask your ARRL Section Traffic Manager (STM) to visit your club to conduct a training seminar and to provide any operating tips pertaining to traffic nets in your section. If no local traffic net exists, your club should consider initiating a net on an available 2-meter repeater. Coordinate these efforts with the STM and the trustee(s) of the repeater you'll use. Encourage club members to participate in traffic-handling activities, either from their home stations or as a group activity from a message center.

Ask your ARRL Section Emergency Coordinator, county Emergency Manager or Emergency Coordinator to conduct a seminar for your group on communicating with first responder personnel such as police, firefighters or emergency medical technicians. They use procedures that are different from those used in Amateur Radio. It is likely that your ARES/RACES group will need to communicate with them in real situations.

For more information on NTS and traffic handling, see Chapter 7.

METHODS OF HANDLING INFORMATION

Emergency Operations Center

Amateur Radio emergency communications frequently use the Incident Command System (ICS). See **Fig 8-1**. The ICS is a way to control initial and subsequent activities in emergency and disaster situations.

Consider an automobile accident where a citizen or an amateur, first on the scene, becomes a temporary Incident Commander (IC) when he or she calls for or radios for help. A law-enforcement officer is dispatched to the accident scene in a squad car and, upon arriving, takes over the IC tasks. Relief efforts, like those in this simple example of an automobile accident, begin when someone takes charge, makes a decision and directs the efforts of others.

The Emergency Operations Center (EOC) responds to the IC by dispatching equipment and helpers, anticipating needs to supply support and assistance. It may send more equipment to a staging area to be stored where it can be available almost instantly or send more people to react quickly to changing situations.

If the status of an accident changes (a car hits a utility pole, which later causes a fire), the IC gives the EOC an updated report then keeps control even after the support agencies arrive and take over their specific responsibilities: Injuries—medical; fires—fire department; disabled vehicles—law enforcement or tow truck; and utility poles—utility company. By being outside the perimeter of dangerous activities, the EOC can use the proper type of radio communications, concentrate on gathering data from other agencies and then provide the right response.

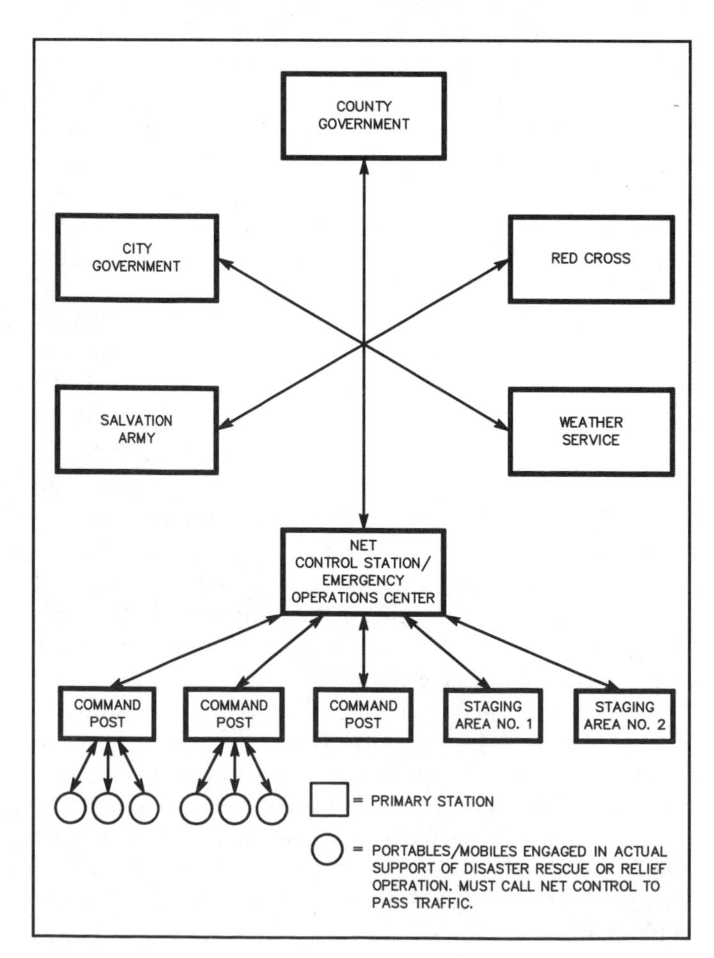

Fig 8-1—The interaction between the EOC/NCS and the command post(s) in a local emergency.

As an analogy, think of the ICs, who request action and provide information, being similar to net participants checking into an amateur net with emergency or priority traffic. The EOC, who coordinates relief efforts, then functions as a Net Control Station.

Whether there is a minor vehicle accident or a major disaster operation, the effectiveness of the amateur effort in an emergency depends mainly on handling information.

Incident Command System

The Incident Command System (ICS) is a management tool that provides a coordinated system of command, communications, organization and accountability in managing emergency events and is rapidly being adopted by professional emergency responders throughout the country. Amateurs should become familiar with ICS to work with agencies in a variety of multiple jurisdictions and political boundaries.

Incident Command Systems use:

- ***Clear text and common terms***. Ten codes are avoided (though hams working with first responders should be familiar with them just in case). All ICS participants including hams are expected to be familiar with ICS terminology. When the Incident Commander orders "a strike team of Type 2 trucks," everyone affiliated with filling the order knows exactly what is being requested.
- ***Unified Command***. The Incident Commander is the only boss and is responsible for the overall operation.
- ***Flexibility***. Functions such as planning, logistics, operations, finance and working with the press are described in detail so the organization size can change to match the particular incident's requirements. The IC can consist of only a single individual for a small incident or it can expand to a Command Staff for a large incident.
- ***Concise Span of Control***. Since management works well with a small number of people, the ICS typically is designed so that throughout the system, no leader has more than about five people reporting to them.

In some areas the ICS evaluates and determines what resources will be needed to start recovery. Amateur communicators are typically within the Logistics Section, Service Branch and Communications Unit of an ICS (**Fig 8-2**).

Tactical Traffic

Whether traffic is tactical, by formal message, packet radio or amateur television, success depends on knowing which to use.

Tactical traffic is first-response communications in an emergency situation involving a few operators in a small area. It may be urgent instructions or inquiries such as "send an ambulance" or "who has the medical supplies?" Tactical traffic, even though unformatted and seldom written, is particularly important in localized communications when

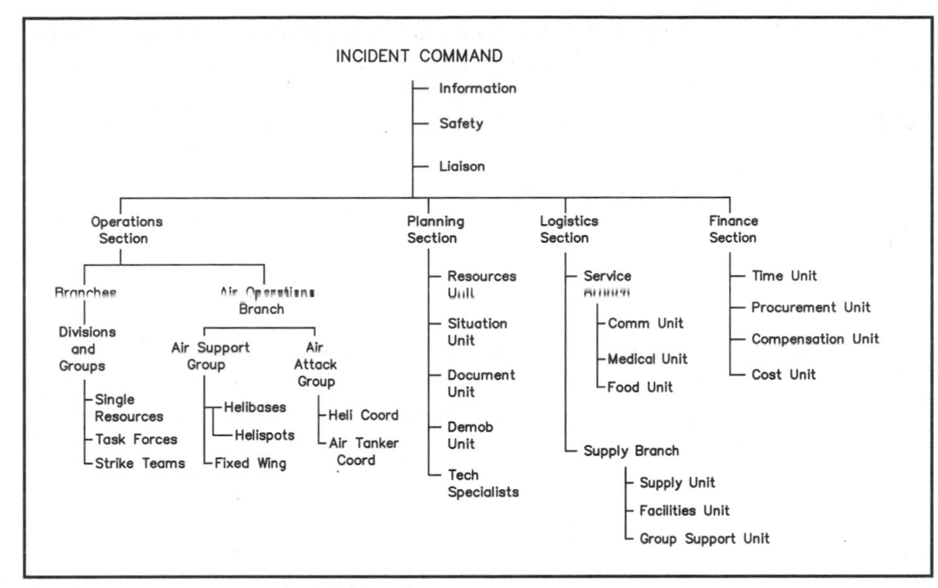

Fig 8-2—The Incident Command System structure.

working with government and law-enforcement agencies. Note, however, that logs should be kept by hams passing tactical traffic. A log may be relevant later for law enforcement or other legal actions, and can even serve to protect the Amateur Radio operator in some situations.

The 146.52 MHz FM calling frequency—or VHF and UHF repeaters and net frequencies (see **Fig 8-3**)—are typically used for tactical communications. This is a natural choice because FM mobile, portable and fixed-station equipment is so plentiful and popular. However, the 222 and 440 MHz UHF bands provide the best communications from steel or concrete stuctures, have less interference and are more secure for sensitive transmissions.

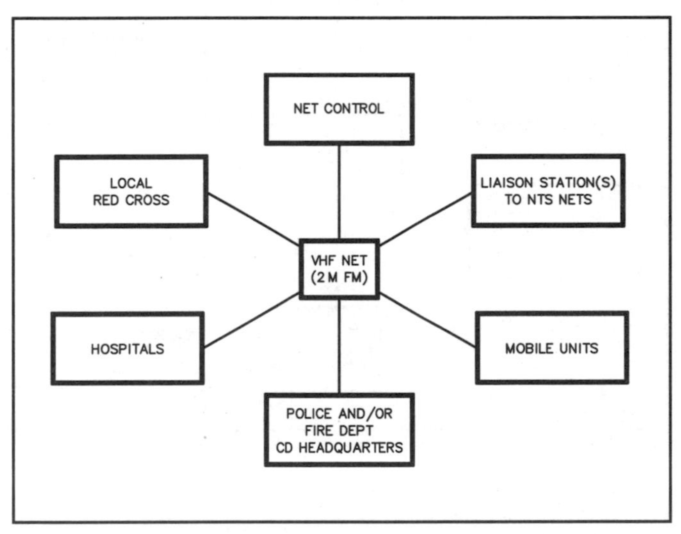

Fig 8-3—Typical station deployment for local ARES net coverage in an emergency.

One way to make tactical net operation clear is to use tactical call signs—words that describe a function, location or agency. Their use prevents confusing listeners or agencies who are monitoring. When operators change shifts or locations, the set of tactical calls remains the same; that is, the tactical call remains with the position even if the operators switch. Amateurs may use tactical call signs like "parade headquarters," "finish line," "Red Cross," "Net Control" or "Weather Center" to promote efficiency and coordination in public service communication activities. However, amateurs must identify with their FCC-assigned call sign at the end of a transmission or series of transmissions and at intervals not to exceed 10 minutes.

Another tip is to use the 12-hour local-time system for time and dates when working with relief agencies, unless they understand the 24-hour or UTC systems.

Taking part in a tactical net as an ARES team member requires some discipline and following a few rules:

1) Report to the Net Control Station (NCS) as soon as you

Stress Management

Emergency responders should understand and practice stress management. A little stress helps you to perform your job with more enthusiasm and focus, but too much stress can drive you to exhaustion or death.

Watch for these physiological symptoms:

- Increased pulse, respiration or blood pressure
- Trouble breathing, increase in allergies, skin conditions or asthma
- Nausea, upset stomach or diarrhea
- Muffled hearing
- Headaches
- Increased perspiration, chills, cold hands or feet or clammy skin
- Feeling weakness, numbness or tingling in part of body
- Feeling uncoordinated
- Lump in throat
- Chest pains

Cognitive reactions may next occur in acute stress situations; many of these signs are difficult to self-diagnose.

- Short term memory loss
- Disorientation or mental confusion
- Difficulty naming objects or calculating
- Poor judgment or difficulty making decisions
- Lack of concentration and attention span
- Loss of logic or objectivity to solve problems

Perhaps the best thing to do as you start a shift is to find someone that you trust and ask them to let you know if you are acting a bit off. If at some time they tell you they've noticed you're having difficulty, then perhaps it's time to ask for some relief. Another idea is to have some sort of stress management training for your group before a disaster occurs.

arrive at your assigned position.
2) Ask the NCS for permission before you use the frequency.
3) Use the frequency for traffic, not chit-chat.
4) Answer promptly when called by the NCS.
5) Use tactical call signs.
6) Follow the net protocol established by the NCS.
7) Always inform the NCS when you leave service, even for a short time.

In some relief activities, tactical nets become resource or command nets. A resource net is used for an event that goes beyond the boundaries of a single jurisdiction and when mutual aid is needed. A command net is used for communications between EOCs and ARES leaders. Yet with all the variety of nets, sometimes the act of simply putting the parties directly on the radio—instead of trying to interpret their words—is the best approach.

Formal Message Traffic

Formal message traffic is long-term communications that involve many people over a large area. It's generally cast in standard ARRL message format and handled on well established National Traffic System (NTS) nets, primarily on 75-meter SSB, 80-meter CW or 2-meter FM (see **Fig 8-4**). [In addition, there is a regular liaison to the International Assistance and Traffic Net, IATN, now officially designated the NTS Atlantic Region Net (ARN). The net meets on 14.303 MHz daily at 1130 UTC (1100 during the summer), to provide international traffic outlets, as suggested in **Fig 8-5**.]

Formal messages can be used for severe weather and disaster reports. These radiograms, already familiar to many agency officials and to the public, avoid message duplication while ensuring accuracy. Messages should be read to the originators before sending them, since the originators are responsible for their content. When accuracy is more important than speed, getting the message on paper before it is transmitted is an inherent advantage of formal traffic.

Packet Radio

Packet radio is a powerful tool for traffic handling, especially with detailed or lengthy text (see Chapter 6) or messages that need to be more secure than those transmitted by voice. Prepare and edit messages off line as text files. These can then be sent error free in just seconds, an important timesaver for busy traffic channels. Public service agencies are impressed by fast and accurate printed messages. Packet radio stations can even be mobile or portable. Relaying might be supplemented by AMTOR-Packet Link (APLink), a system equipped to handle messages between AMTOR HF and packet radio VHF stations.

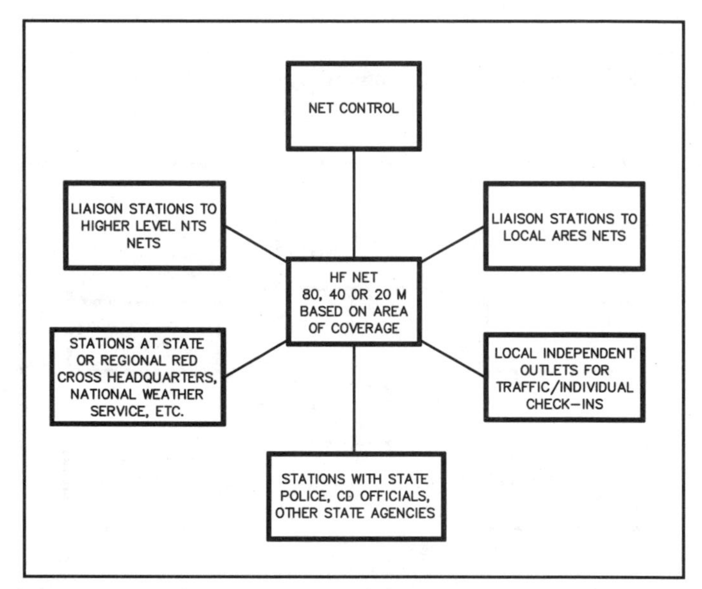

Fig 8-4—Typical structure of an HF network for emergency communications.

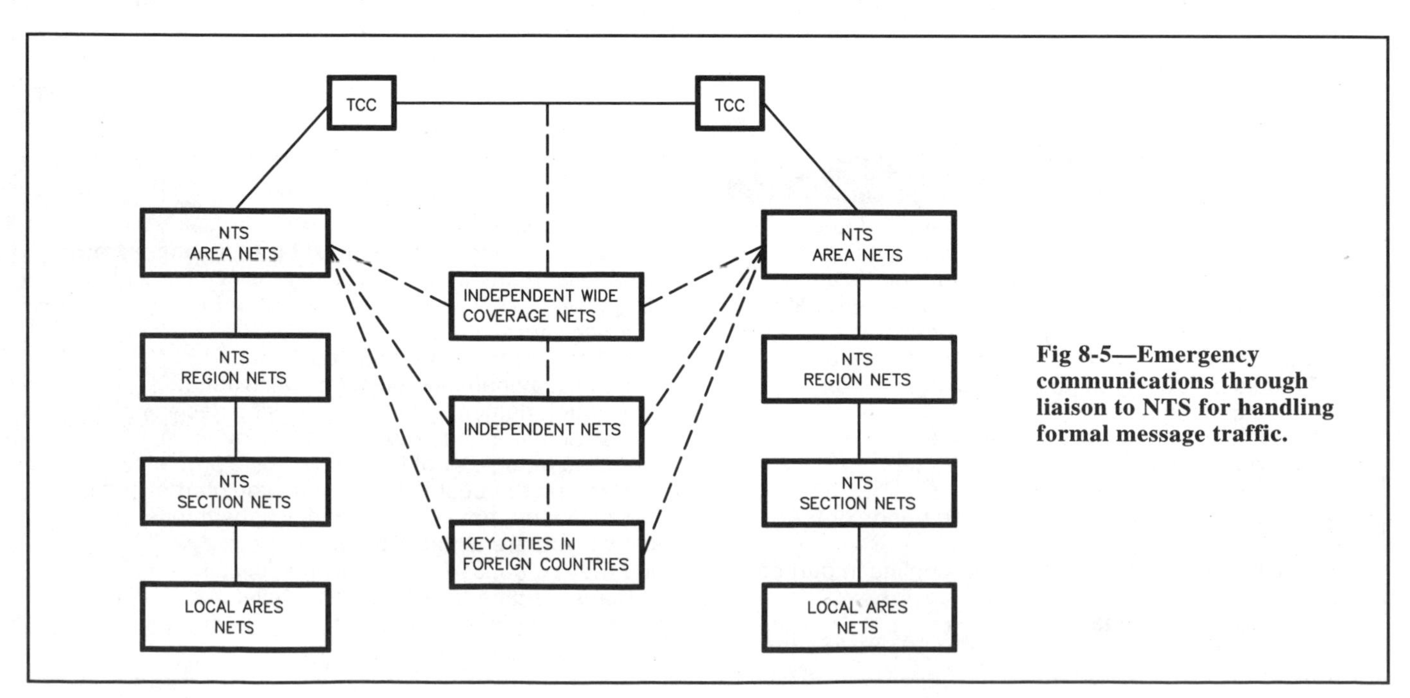

Fig 8-5—Emergency communications through liaison to NTS for handling formal message traffic.

Image Communications

Image communication offers live pictures of an area to allow, for example, damage assessment by authorities. Amateur Television (ATV) in its public service role usually employs portable Fast Scan Television (FSTV), which displays full motion, has excellent detail, can be in color and has a simultaneous sound channel. Although a picture is worth a thousand words, an ATV system requires more equipment, operating skill and preparation than using a simple hand-held radio.

Video cameras and 420-430 or 1240-1294 MHz radios can transmit public service images from a helicopter to a ground base station equipped with video monitors and a VCR for taping. Image communication works well on the ground, too. Video coverage of parades, severe weather, and even operation Santa Claus in hospitals adds another dimension to your ability to serve the public.

Slow Scan TV (SSTV) is also popular for damage assessment. Portable SSTV operations can use a digital still image camera, laptop computer and handheld or mobile transceiver. Signals may be relayed through a repeater to increase their range.

Automatic Packet Reporting System

APRS is an Amateur Radio technology that incorporates Global Positioning System (GPS) receiver tracking, Weather Instrumentation stations, digital cartography mapping, radio direction finding equipment and comprehensive messaging in one package. APRS has been adopted by some SKYWARN and ARES organizations as frontline technology for severe weather operations. Such a system provides accurate, real-time weather telemetry that SKYWARN operators and National Weather Service meteorologists use to issue severe weather warnings and advisories.

A combination of APRS and Emergency Management Weather Information Network (EMWIN) can be used to transmit warnings to field spotters to help track a storm's movement as well as report tornadoes or hail. A mobile APRS with GPS capability can be an essential tool during and following a large scale disaster to pinpoint critical locations in an area void of landmarks, such as forest fires or search and rescue activities. New APRS-friendly radios put tactical messaging capability in the palm of your hand.

Internet

The Internet provides a fluent, high speed conduit to communicate locally or over great distances. During nonemergency conditions, websites and e-mail are essential tools to help keep ECs in touch with many served agencies. Amateur Radio operators can contact their local organization or publicize emergency preparedness activites. Some Emergency Operation Centers have installed satellite-based Internet facilities with backup generator electricity.

AMATEUR RADIO GROUPS

The Amateur Radio Emergency Service

In 1935, the ARRL developed what is now called the Amateur Radio Emergency Service (ARES), an organization of radio amateurs who have voluntarily registered their capabilities and equipment for emergency communications (see **Fig 8-6**). They are groups of trained operators ready to serve the public when disaster strikes and regular communications fail. ARES often recruits members from existing clubs, and includes amateurs outside the club area, since emergencies do not recognize boundary lines.

Are you interested in public service activities, or preparing for emergency communications? Join ARES in your area and exchange ideas, ask questions and help with message centers, sports events or weather spotting. Any licensed amateur with a sincere desire to serve is eligible for ARES membership. The possession of emergency-powered equipment is a plus, but it is not a requirement. An ARES group needs to refresh its training with meetings, scheduled nets, drills or real emergencies. An effective ARES group is to our benefit, as well as to the benefit of the entire community. Information about ARES may be obtained from your ARRL Section Manager (listed on page 12 of *QST*) or ARRL Headquarters.

The Amateur Radio Emergency Service has responded countless times to communications emergencies. ARES also introduces Amateur Radio to the ever-changing stream of agency officials. Experience has proven that radio amateurs react and work together more capably in time of emergency when practice has been conducted in an organized group. There is no substitute for experience gained—before the need arises.

Official Emergency Station

After you get some ARES training and practice, you might want to refine your skills in emergency communica-

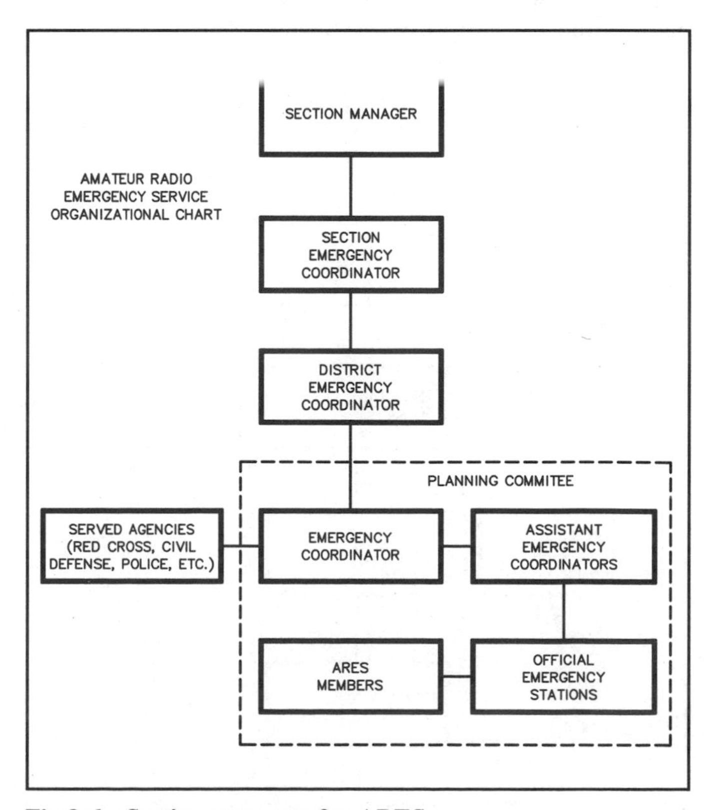

Fig 8-6—Section structure for ARES.

tions. If you possess a full ARRL membership, there are several opportunities available. The first is the OES appointment, which requires regular participation in ARES including drills, emergency nets and, possibly, real emergency situations. An OES aims for high standards of activity, emergency-preparedness and operating skills.

Emergency Coordinator

Next, when you feel qualified enough to become a team leader of your local ARES group, consider the EC appointment, if that position is vacant in your area. An EC, usually responsible for a county, is the person who can plan, organize, maintain response-readiness and coordinate for emergency communications (see **Fig 8-7**).

Much of the work involves promoting a working relationship with local government and agencies. The busy EC can hold meetings, train members, keep records, encourage newcomers, determine equipment availability, lead others in drills or be first on the scene in an actual disaster. Some highly populated or emergency-prone areas may also need one or more Assistant Emergency Coordinators (AEC) to help the EC. The AEC is an appointment made by an EC. The AEC position can be held by a ham with any class of license; they need not be ARRL members.

District Emergency Coordinator

If there are many ARES groups in an area, a DEC may be appointed. Usually responsible for several counties, the DEC coordinates emergency plans between local ARES groups, encourages activity on ARES nets, directs the overall communication needs of a large area or can be a backup for an EC. As a model emergency communicator, the DEC trains clubs in tactical traffic, formal traffic, disaster communications and operating skills.

Section Emergency Coordinator

Finally, there is one rare individual who can qualify as top leader of each ARRL Section emergency structure, the SEC. Only the Section Manager can appoint a candidate to become the SEC. The SEC does some fairly hefty work on a section-wide level: making policies and plans and establishing goals, selecting the DECs and ECs, promoting ARES membership and keeping tabs on emergency preparedness. During an actual emergency, the SEC follows activities from behind the scenes, making sure that plans work and section communications are effective.

Section Manager

The overall leader of the ARRL Field Organization in each section is the SM, who is elected by the ARRL membership in that section. The SM not only appoints the SEC to handle details of ARES and emergency-preparedness activities, but also appoints a Section Traffic Manager (STM) to handle details of the NTS and formal message traffic operation. Response to emergency and public service needs combines the ARES and the NTS. [For further details on the ARRL Field Organization, see Chapter 17.]

RADIO AMATEUR CIVIL EMERGENCY SERVICE

The Radio Amateur Civil Emergency Service (RACES) was set up in 1952 as a special phase of the Amateur Radio Service conducted by volunteer licensed amateurs. It is currently designed to provide emergency communications to local or state governmental agencies during times when normal communications are down or overloaded.

RACES operation is authorized during periods of local, regional or national civil emergencies by the FCC upon request of a state or federal official. While RACES was

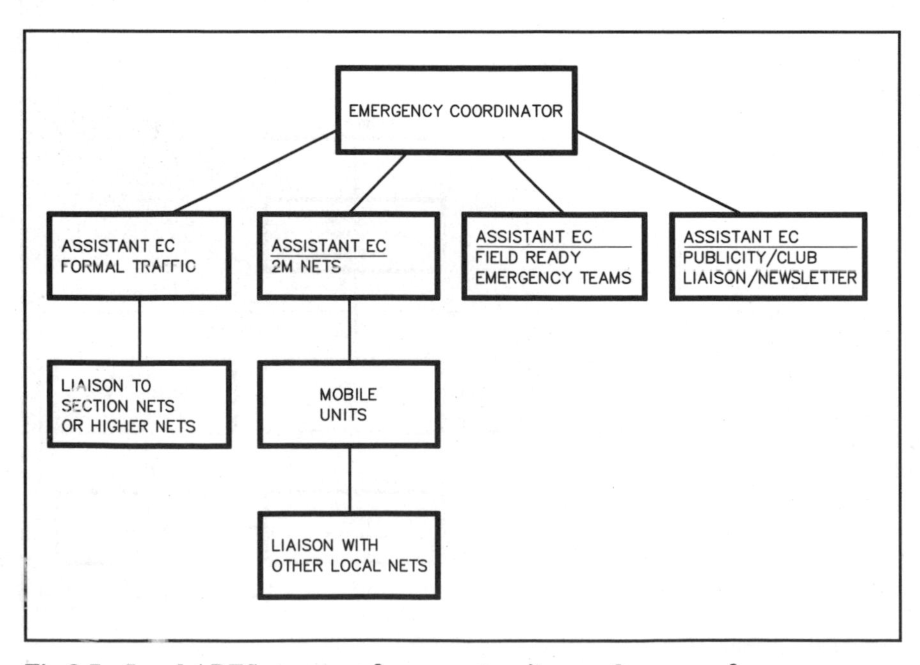

Fig 8-7—Local ARES structure for a county, city or other area of coverage.

Safety

Sometimes motivation to help in an emergency obscures clear thinking and following safe procedures.

Before you accept an assignment make sure that you are:

- adequately trained for the situation you are being asked to handle
- aware of dangers around you (situational awareness)
- practice common sense

If you are accompanying professional responders, make sure they are aware of your level of training so they don't unintentionally take you into a situation for which you aren't prepared. Be responsible for your own safety.

originally based on potential use during wartime, it has evolved over the years (as has the meaning of "civil defense" which is now called "civil preparedness"), to encompass all types of emergencies and natural disasters. RACES is sponsored by Federal Emergency Management Agency, but is usually administered by a state-level Office of Emergency Management.

Amateurs operating in a local RACES organization must be officially enrolled (registered). RACES operation is conducted by amateurs using their own primary station licenses, and by existing RACES stations. The FCC no longer issues new RACES (WC prefix) station call signs. Operator privileges in RACES are dependent upon, and identical to, those for the class of license held in the Amateur Radio Service. All of the authorized frequencies and emissions allocated to the Amateur Radio Service are also available to RACES on a shared basis. But in the event that the President invokes his War Emergency Powers, amateurs involved with RACES would be limited to certain frequencies, while all other amateur operation would be silenced.

When operating in a RACES capacity, RACES stations and amateurs registered in the local RACES organization may not communicate with amateurs not operating in a similar capacity. See FCC regulations for further information.

Although RACES and ARES are separate entities, the ARRL advocates dual membership and cooperative efforts between both groups whenever possible. For example, in Wisconsin, all members of ARES must be RACES-registered operators, and vice versa, and the current ARRL SEC is also the state RACES Chief Radio Officer. The RACES regulations make it simple for an ARES group whose members are all enrolled in and certified by RACES to operate in an emergency with great flexibility. Using the same operators and the same frequencies, an ARES group also enrolled as RACES can "switch hats" from ARES to RACES or RACES to ARES to meet the requirements of the situation as it develops. For example, during a "non-declared emergency," ARES can operate under ARES, but when an emergency or disaster is officially declared by a state or federal authority, the operation can become RACES with no change in personnel or frequencies.

Where there currently is no RACES, it would be a simple matter for an ARES group to enroll in that capacity, after a presentation to the civil-preparedness authorities. For more information on RACES, contact your State Emergency Management, Civil-Preparedness Office, or FEMA.

THE NATIONAL TRAFFIC SYSTEM

In 1949 the ARRL created the National Traffic System to handle medium and long-haul formal message traffic through networks whose operations can be expedited to meet the needs of an emergency situation.

The main function of NTS in an emergency is to link various local activities and to allow traffic destined outside of a local area to be systematically relayed to the addressee. In a few rare cases, a message can be handled by taking it directly to a net in the state where the addressee lives for rapid delivery by an amateur there within toll-free calling distance. However, NTS is set up on the basis of being able to relay large amounts of traffic systematically, efficiently and according to an established flow pattern. This proven and dependable scheme is what makes NTS so vital to emergency communications. See Chapter 7 for more information on NTS.

Additional details on ARES and NTS can be found in the *Public Service Communications Manual*, the *Emergency Coordinator's Manual* and *The ARRL Net Directory*, all published by the ARRL (online ordering available at: **http://www.arrl.org/catalog/**). Information on emergency communications and traffic handling also appears regularly in *QST*.

GOVERNMENT AND RELIEF AGENCIES

Government and relief agencies provide effective emergency management to help communities in disasters. They often rely on normal communications channels, but even reliable communications systems may fail, be unavailable or become overloaded in an emergency. Furthermore, in

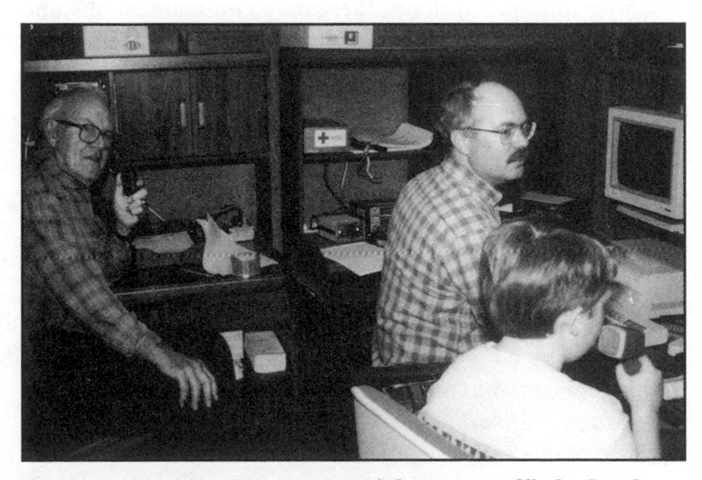

When the Ohio River flooded in 1997 the water level in Cincinnati rose to the point that commercial water traffic had to be suspended. Amateurs at the W8VVL QCEN radio room in the Red Cross building provided emergency communications. Operators shown on duty include (l-r) Vern Barhorst, WD4EEB; QCEN Communications Manager Mark Reising, WB9BVV; and Matthew Reising, KB8TVN.

disaster situations, agency-to-agency radio systems may be incompatible.

Fortunately, Amateur Radio communicators can serve and support in these situations. We can bridge communications gaps with mobile, portable and fixed stations. We also supply trained volunteers needed by the agencies for collection and exchange of critical emergency information.

As government and relief agencies are restrained by budget cuts, the general public is becoming more dependent on volunteer programs. They quickly recognize the value of efforts of radio amateurs who serve the public interest. This public recognition is important support for the continued existence and justification of Amateur Radio.

By using Amateur Radio operators in the amateur frequency bands, the ARRL has been and continues to be in the forefront of supplying emergency communications, either directly to the general public or through various agencies. In fact, there are several organizations that have signed official Memorandas of Understanding with the ARRL:

- The American National Red Cross
- The Salvation Army
- The Federal Emergency Management Agency
- The National Communications System
- The Association of Public Safety Communications Officials International
- The National Weather Service

The American National Red Cross and the Salvation Army extend assistance to individuals and families in times of disaster. Red Cross chapters, for example, establish, coordinate and maintain continuity of communications during disaster-relief operations (both national and international). These agencies have long recognized that the Amateur Radio Service, because of its excellent geographical station coverage, can render valuable aid.

The Federal Emergency Management Agency (FEMA) is a federal agency that supports state and local civil-preparedness and emergency-management agencies. With headquarters in Washington and 10 regional offices throughout the country, it can provide technical assistance, guidance and financial aid to state and local governments wishing to upgrade their emergency communications and warning systems. Since FEMA sponsors the RACES program, its recognition of ARRL-sponsored emergency preparedness programs can be a powerful tool in selling ARES capability to local emergency-management officials.

The National Communication System (NCS) is a confederation of government agencies and departments established by a Presidential Memorandum to ensure that the critical telecommunication needs of the Federal Government can be met in an emergency. The ARRL Field Organization continues to participate in communications tests sponsored by NCS to study telecommunications readiness in any conceivable national emergency. Through this participation, radio amateurs have received recognition at the highest levels of government.

The Association of Public Safety Communications Officials—International (APCO) and amateurs share common bonds of communications in the public interest. APCO is made up of law-enforcement, fire and public-safety communications personnel. These officials have primary responsibility for the management, maintenance and operation of communications facilities in the public domain. They also establish international standards for public-safety communications, professionalism and continuity of communications through education, standardization and the exchange of information.

Not many ARES groups have a "navy," but in Ozaukee County (Wisconsin) they do! OZARES (Ozaukee County ARES) is attached to Emergency Management (EM) of the county, and EM has a 40-foot Rescue Boat that responds to emergencies on adjoining Lake Michigan during the boating season. On the dock at the right is Captain Jack Morrison (N9SFG). Most of his crew are also hams. During emergency runs in inclement weather, OZARES members staff the Emergency Communications Center (ECC) in the county's Justice Center and keep in constant contact with the boat via ham VHF, marine and 800 MHz public service band radios. Weather radar in the ECC allows hams to give the rescue boat early warning of impending weather changes. This permits the crew to adjust their rescue activities for maximum life safety. The boat makes 50 to 60 emergency runs in a usual season. It has saved many lives and has prevented the loss of many thousands of dollars in property. (*Photo by WB9RQR*)

The National Weather Service (NWS) consists of a national headquarters in Washington, DC, six regional offices and over 200 local offices throughout the US. The ARRL Field Organization cooperates with NWS in establishing SKYWARN networks for weather spotting and communications. SKYWARN is a plan sponsored by NWS to report and track destructive storms or other severe weather conditions.

An increased awareness of radio amateur capabilities has also been fostered by ARRL's active participation as a member of the National Volunteer Organizations Active in Disaster (NVOAD). NVOAD coordinates the volunteer efforts of its member-agencies (the Red Cross and Salvation Army are among its members).

ARRL and radio amateurs continue to accelerate their presence with these agencies. This enhanced image and recognition of Amateur Radio attracts more "customers" for amateurs and our communications skills.

PUBLIC SERVICE EVENTS

Amateur Radio has a unique responsibility to render emergency communications in times of disaster when normal communications are not available. Progressive experience can be gained by helping with public service events such as message centers, parades and sports events. Besides serving the community, participation teaches skills that can be applied during an emergency. Public service events are ideal training grounds for amateurs interested in providing emergency communications because they can realistically practice tactical communications.

According to FCC regulations, amateurs are responsible for providing aid during emergencies, but cannot receive compensation for this aid. This requirement does not prohibit hams from assisting at public events sponsored by for-profit organizations.

Many Amateur Radio emergency groups use ham fax software to receive early severe weather warnings—direct from NOAA satellites or HF broadcasts. Hurricane Henrietta set off a few alerts in 1995, but moved out to sea.

Message Centers

Message centers are Amateur Radio showcase stations set up and operated in conspicuous places to send free radiogram messages for the public. Many clubs provide such radiogram message services at shopping malls, fairs, information booths, festivals, exhibits, conferences and at special local events to demonstrate Amateur Radio while handling traffic. Message centers also afford opportunities to train operators, to practice handling messages that are similar to disaster Welfare traffic, and to show the public that amateurs are capable, serious and responsible communicators. A plan for a message center should include a display, equipment, public relations and operators.

Successful centers include eye-catching organized displays and attractive stations that promote Amateur Radio and lure newcomers to the hobby. Cover a table with brightly colored cloth. Put out a few radio magazines and hang up some QSLs, maps, posters and operating aids. ARRL Headquarters can supply literature, pamphlets, press handouts and various informative videotapes to help make your club display a success.

Equipment should be simple, safe and uncomplicated. Try only a few of the following: HF stations with CW, SSB, or digital capability and VHF stations with FM, packet radio, satellite or ATV activities. Keep expensive radios secure from possible theft, out of reach from visitors or stored under a protected table. Decide on whether to use power from batteries, commercial power or generators. Check prior to the event for receiver hash or public-address system interference. To instill interest, have an easy-to-operate shortwave receiver accessible for spectators to tune and listen. To promote a good public-relations image, consider the visitors; use an extension loudspeaker, desk microphone or computer readout of code. Clear, local contacts impress people more than long-distance, scratchy QSOs.

It may be necessary to form a barrier to separate visitors and busy operators. Even so, have friendly and knowledgeable club members available. They should clearly describe to the passerby what Amateur Radio can do, answer questions, explain the message process, assist in preparing messages and so on. Avoid ham jargon. Keep explanations simple and informative; because of the public's varying degrees of interest, only a few will be interested in specific details. Discuss the variety of Amateur Radio, even if it is not all demonstrated. Most visitors will want to know about the role of Amateur Radio in the community, about FCC exams and the license process.

Well-planned message centers can handle thousands of messages at one event. Some visitors bring their address books and excitedly write out bunches of messages. Others inquisitively watch or wait for their turns. Handling messages for the local community is a valuable way amateurs can enhance public recognition. It also keeps you in practice and your equipment ready for emergencies. Before getting involved in message centers, you should become familiar with the traffic-handling procedures described in Chapter 7. (This chapter also contains additional hints on exhibit-station operation.)

Parades

From a communications viewpoint, parades simply relocate people and equipment, with a few requests for supplies or medical aid along the way. Therefore, parades simulate evacuations.

The first step in handling parade communications is to select a group representative who will contact the parade organizers and other officials during the planning stages. This representative can then instruct the ham group and describe what must be done.

As for most public service events, arrive early. Wear proper identification, whether it is a special jacket, cap or badge. The basic order for the day is to provide communications. Leave parade decisions to officials and first-aid or medical care to Red Cross personnel.

Certain operators should be assigned to find parade officials and help them exchange information. Others work in pairs or teams with the parade marshall and parade specialists like the band, float, vehicle and marcher officials. These officials will be busy once the bus loads of uniformed marchers arrive at the staging area. Set up net controls at parade headquarters where a telephone is available, at an announcement or information booth or at the staging area near the starting line. Fill vacant positions with standby operators. Establish a procedure for helping lost children.

Next, after initial parade staging is underway, many other operators have specific on-route assignments. Position hand-held-radio operators at intervals along the parade route where assignments can be carried out immediately. Operators can walk, march or ride along with the parade to report any specific problems. If possible, it's a good idea to place one amateur on top of a high building to spot general problems. Communicators near loud marching bands will need earphones. Hot weather can keep first-aid volunteers busy with victims of dehydration, sunstroke and heat exhaustion. In some cases, a request for an ambulance, paramedic or water will be needed. Parade floats may need repairs, fuel or towing.

Then, at the destaging area, more officials rely on communications to prevent clogging the area and stopping the parade. Large parades can require over 75 amateurs to handle communications, and they must plan and practice many months before the event.

As a result, amateurs provide reliable communications to resolve unanticipated problems and handle last-minute changes with immediate action. For more information on public service events, see the *ARRL Special Events Communications Manual*. Contact ARRL HQ Field and Educational Services for details on how to obtain a copy.

Sports Events

Sports events include all types of outdoor athletic activities where there is competition or movement, or when participants are timed. Again, it is important to assess the communication needs and plan with event officials, law-enforcement officers, medical and first-aid personnel. Know the route and preview the course. Everyone should understand their functions.

Like parades, there are strategic locations where sport-event amateur communicators could be positioned to be most useful:

- With or shadowing event officials and organizers or law-enforcement officers.
- At check-in points, staging areas or starting lines.
- At water stops, aid stations and checkpoints around the course or route.
- Near road intersections or sharp turns where safety of crowds or participants is crucial.
- In the course preview car, pace car, follow-up vehicle or roving sheriff's squad.
- At the net control station.
- Alongside first aid or Red Cross stations, medical personnel or ambulance.
- At message centers.
- Near a telephone or at a home station with a telephone.
- At the destaging area or finish line.

There are many types of sports events in which amateurs serve. The list is nearly endless: air shows, balloon races, soapbox derbies, road rallies, solar car races, Boy Scout hikes, bicycle tours, regattas, golf tournaments, kayak and river-raft races, football games, mini-olympics, Special Olympics, World Olympics, ski-lift operations, slalom courses, cross-country meets, and so on.

A popular event is the *A-Thon*: the walkathon, marathon, sportathon, bikeathon or even a triathlon (usually running, bicycling and swimming or canoeing). Long-distance marathons require a large, disciplined force of communicators. This is the case with the New York City Marathon, which is served by well over 200 ham radio operators. Good coverage allows operators to radio the identifying numbers and locations of marathon participants who drop out of the course. This information is posted at a base station to help friends and relatives locate their favorite runners, or to help account for all participants.

Physical endurance and fitness is not only for the participants; hams use roller skates, bicycles, golf carts, motorcycles or cars around the routes of sport events. Some communicators cover a race at a particular point until all participants have passed, then they become mobile and leapfrog up the route to repeat the procedure. Roving mobiles, sometimes called *roamers*, can keep track of everyone's location, determine race miles remaining, warn of potential traffic hazards, find lost people or items, report harassing spectators or post signs for riders or runners. Yet, even with this variety, keep in mind that an amateur is not to be used as a parking-lot attendant or crowd-control police. Take only those tasks you're capable of handling, such as being a communicator.

Portable operators, those not on wheels, can report medical emergencies, give weather reports and handle messages for participants or spectators. Separate UHF channels for administrative purposes are useful for getting supplies like drinking cups, water or bandages, and to report malfunctioning facilities or request more volunteers. Telephones may be crowded in the area, so first-aid stations or checkpoints with bulletin boards can double as NTS message centers. It is then easy for the NCS to give assignments and tell the officials that the route is ready and participants are set to go.

During major sports events, the Red Cross can dispatch their own first-aid vehicles and keep them near crowds or hazardous areas for fast accident response. A specific medical-radio frequency is often assigned. Be prepared to report anything from minor medical situations to serious problems where an ambulance might be called to take the victim to a hospital for examination. If you do call for the ambulance, assign someone to flag it down and give the driver final directions to the victim's location.

Public service events are great teachers. Once you gain skill in working with the public in hectic and near-emergency conditions, you're well prepared to tackle communications for more difficult natural-disaster assignments.

NATURAL DISASTERS AND CALAMITIES

Nature relentlessly concocts severe weather and natural calamities that can cause human suffering and create needs which the victims cannot alleviate without assistance. Despite the spectrum of requirements desperately needed to help people in a disaster, it is generally understood and agreed that amateurs will neither seek nor accept any duties other than Amateur Radio communications. Volunteer communicators do not, for example, enforce local laws, make major decisions, work as common laborers, rent generators, tents or lights to the public and so on. Instead, amateurs simply handle *radio communications*.

Here are several typical categories of amateur disaster and calamity communications:
- Severe-weather spotting and reporting
- Supporting evacuation of people to safe areas
- Shelter operations
- Assisting government groups and agencies
- Victim rescue operations
- Medical help requests
- Critical supplies requests
- Health-and-Welfare traffic
- Property damage surveys and cleanup

Severe Weather Spotting and Reporting

Nasty weather hits somewhere every day. Long ago, amateurs exchanged simple information among themselves about the approach and progress of storms. Next, concerned hams phoned the weather offices to share a few reports they thought might be of particular interest to the public. National Weather Service forecasters were relying on spotters: police, sheriff, highway patrol, emergency government and trained individuals who reported weather information by telephone. But when severe weather strikes, professional spotters may be burdened with law-enforcement tasks, phone lines may become overloaded, special communication circuits might go out of service or, worse yet, there can be a loss of electrical power. Because of these uncertainties, weather-center officials welcomed amateur operators and encouraged them to install their battery-powered radio equipment on site so forecasters could monitor the weather

ARES Personal Checklist

The following represents recommendations of equipment and supplies ARES members should consider having available for use during an emergency or public service activity.

Forms of Identification
- ARES Identification Card
- FCC Amateur Radio license
- driver's license

Radio Gear
- VHF transceiver
- microphone
- headphones
- power supply/extra batteries
- antennas with mounts
- spare fuses
- patch cords/adapters (BNC to PL-259/RCA phono to PL-259)
- SWR meter
- extra coax

Writing Gear
- pen/pencil/eraser
- clipboard
- message forms
- logbook
- note paper

Personal Gear (short duration)
- snacks
- liquid refreshments
- throat lozenges
- personal medicine
- aspirin
- extra pair of prescription glasses
- sweater/jacket

Personal Gear (72-hour duration)
- foul-weather gear
- three-day supply of drinking water
- cooler with three-day supply of food
- mess kit with cleaning supplies
- first-aid kit
- personal medicine
- aspirin
- throat lozenges
- sleeping bag
- toilet articles
- mechanical or battery powered alarm clock
- flashlight with batteries/lantern
- candles
- waterproof matches
- extra pair of prescription glasses

Tool Box (72-hour duration)
- screwdrivers
- pliers
- socket wrenches
- electrical tape
- 12/120-V soldering iron
- solder
- volt-ohm meter

Other (72-hour duration)
- HF transceiver
- hatchet/ax
- saw
- pick
- shovel
- siphon
- jumper cables
- generator, spare plugs and oil
- kerosene lights, camping lantern or candles
- 3/8-inch hemp rope
- highway flares
- extra gasoline and oil

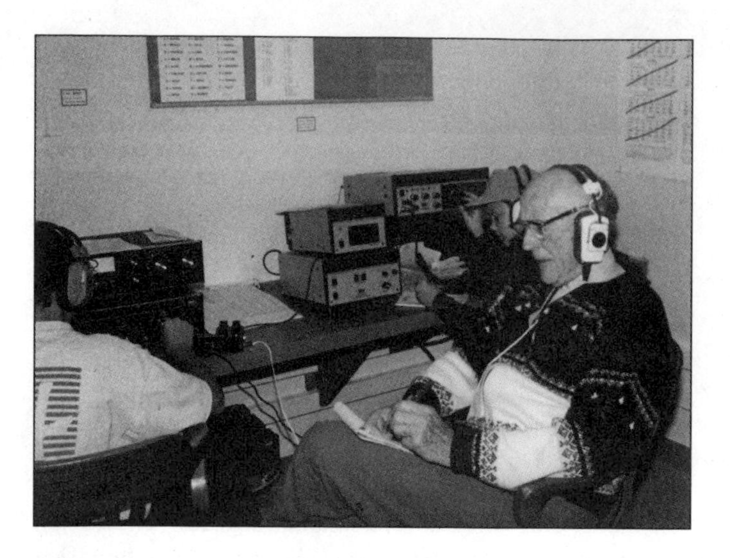

An OZARES training session finds Ray Meyer (N9PBY, left) operating 2 meters while 9-year old Jason Raasch (N9NNB, center) operates HF and octogenarian Gordon Hepburn (N9NLY, right) logs. The group trains in person at least monthly, and on-the-air weekly. OZARES members take their training seriously, because they know what they learn will be needed. They have participated in many emergencies including floods, tornadoes, severe thunderstorms, a search for a missing fugitive and even a boat sinking (see above). They are trained in some non-traditional areas for hams, such as vehicular traffic control, the same hazardous materials (HazMat) training required of police officers and fire fighters, marine communications and communications with the 800 MHz public service trunking system in use in Ozaukee County. Two days after their first training session in vehicular traffic control, a severe thunderstorm brought down many power lines in the county. Several members dispatched to the field found themselves reporting downed, sparking power lines with their HT in one hand while using the other hand to prevent vehicular traffic from running over the dangerous wires. (*Photo by WB9RQR*)

nets, request specific area observations and maintain communications in a serious emergency.

Those first, informal weather nets had great potential to access perhaps hundreds of observers in a wide area. Many Meteorologists-in-Charge eagerly began to instruct hams in the types of information needed during severe-weather emergencies, including radar interpreting. Eventually, Amateur Radio SKYWARN operations developed as an important part of community disaster preparedness programs. Accurate observations and rapid communications during extreme weather situations now proves to be fundamental to the NWS. Amateur Radio operators nationwide are a first-response group invaluable to the success of an early storm-warning effort. Weather spotting is popular because the procedures are easy to learn and reports can be given from the relative safety and convenience of a home or an auto.

For example, during a severe-storm episode in Wisconsin in 1994, there was danger of local flooding. A quick check of conditions at the homes of hams operating in an ARES net rapidly revealed trouble spots over a 40 square mile area. This information proved invaluable to emergency government and weather officials, yet no ham had to leave home to participate.

Weather reports on a severe-weather net are limited to drastic weather data, unless specifically requested by the nct-control operator. So, most amateurs monitor net operations and transmit only when they can help.

Weather forecasters, depending on their geographical location, need certain information.

During the summer or thunderstorm season report:

- Tornadoes, funnels or wall clouds.
- Hail.
- Damaging winds, usually 50 miles per hour or greater.
- Flash flooding.
- Heavy rains, rate of 2 inches per hour or more.

During the winter or snow season report:

- High winds.
- Heavy, drifting snow.
- Freezing precipitation.
- Sleet.
- New snow accumulation of 2 or more inches.

Here's a four-step method to describe the weather you spot:

1) *What*: Tornadoes, funnels, heavy rain and so on.
2) *Where*: Direction and distance from a known location; for example, 3 miles south of Newington.
3) *When*: Time of observation.
4) *How*: Storm's direction, speed of travel, size, intensity and destructiveness. Include uncertainty as needed. ("Funnel cloud, but too far away to be certain it is on the ground.")

Alerting the Weather Net

The Net Control Station, using a VHF repeater, directs and maintains control over traffic being passed on the Weather Net. The station also collates reports, relates pertinent material to the Weather Service and organizes liaison with other area repeaters. Priority Stations, those that are assigned tactical call signs, may call any other station without going through net control. The NCS might start the net upon hearing a National Oceanic and Atmospheric Administration (NOAA) radio alert, or upon request by NWS or the EC. The NCS should keep in mind that the general public or government officials might be listening to net operations with scanners.

Here are some guidelines an NCS might use to initiate and handle a severe-weather net on a repeater:

1) Activate alert tone on repeater.
2) Read weather-net activation format.
3) Appoint a backup NCS to copy and log all traffic—and to take over in the event the NCS goes off-the-air or needs relief.
4) Ask NWS for the current weather status.
5) Check in all available operators.
6) Assign operators to priority stations and liaisons.
7) Give severe-weather report outline and updates.
8) Be apprised of situations and assignments by EC.
9) Periodically read instructions on net procedures and types

of severe weather to report.
10) Acknowledge and respond to all calls immediately.
11) Require that net stations request permission to leave the net.
12) During periods of inactivity and to keep the frequency open, make periodic announcements that a net is in progress.
13) Close the net after operations conclude.

At the Weather Service

The NCS position at the Weather Service, when practical, can be handled by the EC and other ARES personnel. Operators assigned there must have 2-meter hand-held radios with fully charged batteries. The station located at the NWS office may also be connected to other positions with an off-the-air intercom system. This allows some traffic handling without loading up the repeater. Designate a supplementary radio channel in anticipation of an overload or loss of primary communications circuits.

If traffic is flowing faster than you can easily copy and relay, NWS personnel may request that a hand-held radio be placed at the severe-weather desk. This arrangement allows them to monitor incoming traffic directly. Nevertheless, all traffic should be written on report forms. If a disaster should occur during a severe-weather net, shift to disaster-relief operations.

Repeater Liaisons

Assign properly equipped and located stations to act as liaisons with other repeaters. Two stations should be appointed to each liaison assignment. One monitors the weather repeater at all times and switches to the assigned repeater just long enough to pass traffic. The second monitors the assigned repeater and switches to the weather repeater just long enough to pass traffic. If there aren't enough qualified liaison stations, one station can be given both assignments.

Weather Warnings

NWS policy is to issue warnings only when there is absolute certainty, for fear of the "cry wolf" syndrome (premature warnings cause the public to ignore later warnings). Public confidence increases with reliable weather warnings. When NWS calls a weather alert, it will contact the local EC by phone or voice-message pager, or the EC may call NWS to check on a weather situation.

Hurricanes

A hurricane is declared when a storm's winds reach 75 miles per hour or more. These strong winds may cause storm-surge waves along shores and flooding inland.

A Hurricane *Watch* means a hurricane may threaten coastal and inland areas. Storm landfall is a possibility, but it is not necessarily imminent. Listen for further advisories and be prepared to act promptly if a warning is issued.

A Hurricane *Warning* is issued when a hurricane is expected to strike within 24 hours. It may include an assessment of flood danger, small-craft or gale warnings, estimated storm effects and recommended emergency procedures.

Amateurs, in the 4th and 5th call areas in particular, can spot and report the approach of hurricanes well ahead of any news service. In fact, their information is sometimes edited and then broadcast on the local radio or TV to keep citizens informed.

The Hurricane Watch Net on 14.325 MHz, for example, serves either the Atlantic or Pacific during a watch or warning period and keeps in touch with the National Hurricane Center. Frequent, detailed information is issued on nets when storms pose a threat to the US mainland. In addition to hurricane spotting, local communicators may announce that residents have evacuated from low-lying flood areas and coastal shore. Other amateurs across the country can help by relaying information, keeping the net frequency clear and by listening.

Tornadoes

A tornado is an intensely destructive whirlwind formed from strongly rising air currents. With winds of up to 300 miles per hour, tornadoes appear as rotating, funnel-shaped clouds from gray to black in color. They extend toward the ground from the base of a thundercloud. Tornadoes may sound like the roaring of an airplane or locomotive. Even though they are short-lived over a small area, tornadoes are the most violent of all atmospheric phenomena. Tornadoes that don't touch the ground are called *funnels*.

A Tornado *Watch* is issued when a tornado may occur

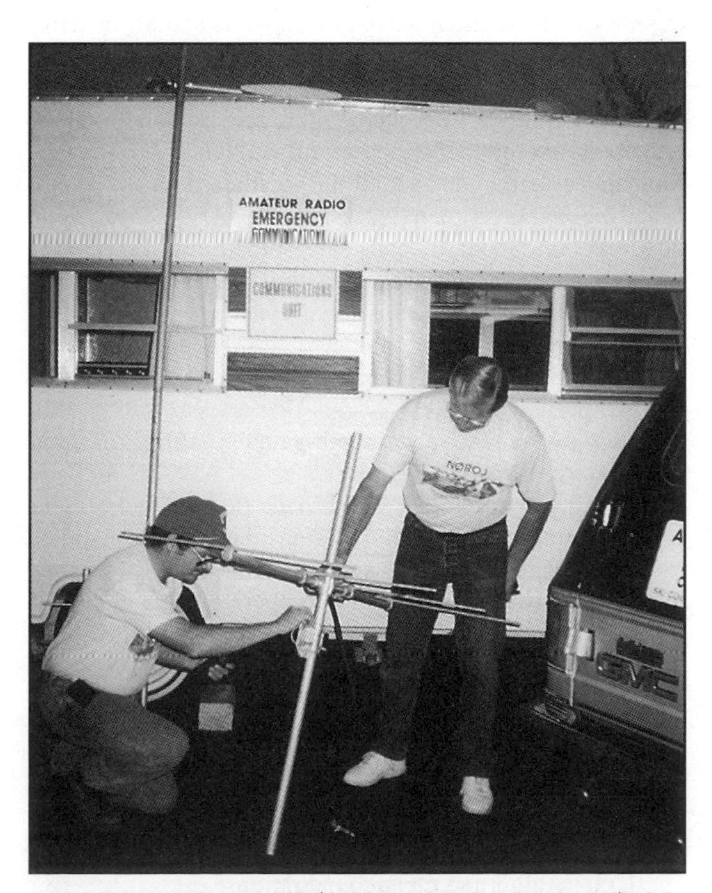

ARES EC Dick Doyal, KBØAHR, and Dale Morrie, NØROJ, prepare a 2-meter antenna for the communication van in Glenwood Springs, Colorado, where fires ravaged a large area. *(Photo by John Radloff, NØMOR)*

near your area. Carefully watch the sky.

A Tornado *Warning* means take shelter immediately, a tornado has actually been sighted or indicated by radar. Protect yourself from being blown away, struck by falling objects or injured by flying debris.

"Tornado alley" runs in the 5th, 9th and 10th US call areas. Amateurs in these areas often receive Tornado Spotter's Training and refresher courses presented by NWS personnel.

Amateur Radio's quick-response capability has reduced injuries and fatalities by giving early warnings to residents. Veteran operators know exactly how serious a tornado can be. They've seen tornadoes knock out telephone and electrical services just as quickly as they flip over trucks or destroy homes. Traffic lights and gas pumps won't work without electricity, creating problems for motorists and fuel shortages for electric generators.

After a tornado strikes, amateurs provide communications in cooperation with local government and relief agencies. Welfare messages are sent from shelters where survivors receive assistance. Teams of amateurs and officials also survey and report property damage.

Floods, Mud Slides and Tidal Waves

Floods occur when excessive rainfall causes rivers to overflow their banks, when heavy rains and warmer-than-usual temperatures melt excessive quantities of snow, or when dams break. Floods can be minor, moderate or major. Floods or volcanic eruptions may melt mountain snow, causing mud slides. A tidal wave or tsunami is actually a series of waves caused by a disturbance that may be associated with earthquakes, volcanoes or sometimes hurricanes.

Don't wait for the water level to rise or for officials to ask for help; sound the alarm and activate a weather net immediately. Besides handling weather data for NWS, enact the response plans necessary to relay tactical flood information to local officials. Assist their decision making by answering the following questions:

1) Which rivers and streams are affected and what are their conditions?
2) When will flooding probably begin, and where are the flood plain areas?
3) What are river-level or depth-gauge readings for comparison to flood levels?

Mobile operators may find roads flooded and bridges washed out. Flood-rescue operations then may be handled by marine police boats with an amateur aboard. If power and telephones are out, portable radio operators can help with relief operations to evacuate families to care facilities where a fixed station should be set up. The officials will need to know the number and location of evacuees. As the river recedes, the water level drops in some areas, but it may rise elsewhere to threaten residents. Liaisons to repeaters downstream can warn others, possibly through the federal Emergency Alert System, of impending flooding. Property-damage reports and welfare traffic will usually be followed by disaster relief and cleanup operations.

Winter Storms

A Winter Storm *Watch* indicates there is a threat of severe winter weather in a particular area. A Winter Storm *Warning* is issued when heavy snow (6 inches or more in a 12-hour period, 8 inches or more in a 24-hour period), freezing rain, sleet or a substantial layer of ice is expected to accumulate.

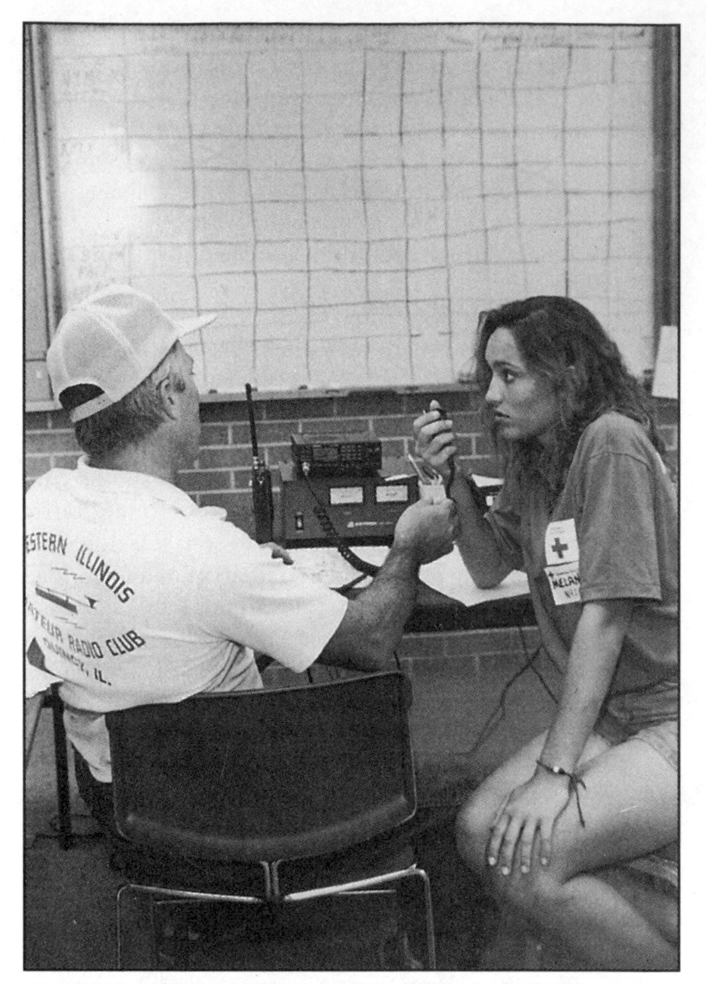

Jim Funk, N9JF, and his daughter Melanie, N9IQV, operate from the Adams County American Red Cross Chapter House in Quincy, Illinois, during the Great Flood of 1993. ***(Photo by Sandy Martin)***

Freezing rain or freezing drizzle is forecast when expected rain is likely to freeze when it strikes the ground. Freezing rain or ice storms can bring down wires, causing telephone and power outages.

Sleet consists of small particles of ice, usually mixed with rain. If enough sleet accumulates on the ground, it will make roads slippery.

Blizzards are the most dangerous of all winter storms. They are a combination of cold air, heavy snow and strong winds. Blizzards can isolate communities. A Blizzard *Warning* is issued when there is considerable snow and winds of 35 miles per hour or more. A *Severe Blizzard Warning* means that a heavy snow is expected, with winds of at least 45 miles per hour and temperatures 10° F or lower.

Travelers advisories are issued when ice and snow are expected to hinder travel, but not seriously enough to require warnings. Blizzards and snowstorms can create vehicle-traffic problems by making roads impassable, stranding motorists or drastically delaying their progress.

In some areas, local snowmobile club members or 4-wheel-drive-vehicle enthusiasts cooperate with hams to coordinate and transport key medical personnel when snowdrifts render roads almost impassable. They may also assist search teams looking for motorists.

Brush and Forest Fires

A prolonged period of hot, dry weather parches shrubs, brush and trees. This dry vegetation, ignited by lightning, arson or even a helicopter crash, can start a forest fire. The fires quickly become worse when winds spread the burning material.

An amateur who spots a blaze that has the potential of growing into a forest fire can radio the Park Service and ask them to dispatch a district ranger and firefighting equipment. When fires are out of control, hams help with communications to evacuate people, report the fires' movement or radio requests for supplies and volunteers. ARES groups over a wide geographical area can set up portable repeaters, digital links or ATV stations.

Amateurs, in the 6th and 7th US call areas in particular, should get adequate safety training, including fire line safety and fire shelter deployment. Even then, travel to a fire operation only upon receiving clear dispatch instructions from a competent authority. If you don't have proper training, inform whoever is in charge of your lack of training to prevent being given an inappropriate or dangerous assignment.

Fire safety rules are of special importance in an emergency, but also should be observed every day to prevent disaster. Since most fires occur in the home, even an alert urban amateur can spot and report a building on fire. Fires are extinguished by taking away the fuel or air (smother it), or by cooling it with water and fire-extinguishing chemicals. A radio call to the fire department will bring the needed control. The Red Cross generally helps find shelter for the homeless after fires make large buildings, such as apartments, uninhabitable. So even in cities, communications for routing people during and after fires may be needed.

Earthquakes and Volcanic Eruptions

Earthquakes are caused when underground forces break and shift rock beneath the surface, causing the Earth's crust to shake or tremble. The actual movement of the earth is seldom a direct cause of death or injury, but can cause buildings and other structures to collapse and may knock out communications, telephone service and electrical power. Most casualties result from falling objects and debris, splintering glass and fires.

Earthquakes strike without warning, but everyone knows that the quake happened. Amateurs should first ensure that the immediate surroundings and loved ones are safe, and then begin monitoring the ARES frequencies. Amateurs are often the first to alert communities immediately after earthquakes and volcanic eruptions occur. Their warnings are definitely credited with lessening personal injuries in the area.

Amateurs may assist with rescue operations, getting medical help or critical supplies and helping with damage appraisals. There will be communications, usually on VHF or UHF, for Red Cross logistics and government agencies. After vital communications is handled, the rest of the world will be trying to get information through Health-and-Welfare traffic.

Shelter Operations

A shelter or relief center is a temporary place of protection where rescuers can bring disaster victims and where supplies can be dispensed. Many displaced people can stay at the homes of friends or relatives, but those searching for family members or in need are housed in shelters.

Whether a shelter is for a few stranded motorists during a snowstorm, or a whole community of homeless residents after a disaster, it is an ideal location to set up an Amateur Radio communication base station. An alternate station location would be an ARES mobile communications van, if available, near the shelter.

Once officials determine the locations for shelters, radio operators can be assigned to set up equipment at the sites. In fact, the amateur station operators could share a table with shelter registration workers. Make sure you obtain permission for access to the shelter to assist and do not upset the evacuees. Use of repeaters and autopatches allows Welfare phone calls for those inside the shelters to inform and reassure friends, families and relatives. Use of a computer may help to control and distribute information. *ARESdata* is a packet radio database program that works well with emergency and public service communications (see December 1990 *QST*, page 75).

Health-and-Welfare Traffic

There can be a tremendous amount of radio traffic to handle during a disaster. This will free phone lines that remain in working order for emergency use by those people in peril.

Shortly after a major disaster, Emergency messages within the disaster area often have life-and-death urgency. Of course, they receive primary emphasis. Much of their local traffic will be on VHF or UHF. Next, Priority traffic, messages of an emergency-related nature but not of the utmost urgency, are handled. Then, Welfare traffic is originated by evacuees at shelters or by the injured at hospitals and relayed by Amateur Radio. It flows one way and results in timely advisories to those waiting outside the disaster area.

Incoming Health-and-Welfare traffic should be handled only after all emergency and priority traffic is cleared. Don't solicit traffic going to an emergency because it can severely overload an already busy system. Welfare inquiries can take time to discover hard-to-find answers. An advisory to the inquirer uses even more time. Meanwhile, some questions might have already been answered through restored circuits.

Shelter stations, acting as net control stations, can exchange information on the HF bands directly with destination areas as propagation permits. Or, they can handle formal traffic through a few outside operators on VHF who, in turn, can link to NTS stations. By having many NTS-trained amateurs, it's easy to adapt to whatever communications are required.

Property Damage Surveys

Damage caused by natural disasters can be sudden and extensive. Responsible officials near the disaster area, paralyzed without communications, will need help to contact appropriate officials outside to give damage reports. Such data will be used to initiate and coordinate disaster relief. Amateur Radio operators offer to help but often are unable to cross roadblocks established to limit access by sightseers and potential looters. Proper emergency responder identification will be required to gain access into these areas. In some instances, call-letter license plates on the front of the car or placards inside windshields may help. It's important for amateurs to keep complete and accurate logs for use by officials to survey damage, or to use as a guide for replacement operators.

Accidents and Hazards

The most difficult scenarios to prepare for are accidents and hazardous situations. They are unpredictable and can happen anywhere. Generally, an emergency *autopatch* is used only to report incidents that pose threats to life or personal safety, such as vehicle accidents, disabled vehicles or debris in traffic, injured persons, criminal activities and fires.

Using the keypad featured on most modern VHF handheld and mobile radios, the operator activates a repeater autopatch by sending a particular code. The repeater connects to a telephone line and routes the incoming and outgoing audio accordingly. By dialing 9-1-1 (or another emergency number), the operator has direct access to law-enforcement agencies.

Vehicle Accidents

Vehicle-accident reports, by far the most common public service activity on repeaters, can involve anything from bikes, motorcycles and automobiles, to buses, trucks, trains and airplanes. Law-enforcement offices usually accept reports of such incidents anywhere in their county and will relay information to the proper agency when it pertains to adjacent areas.

Here's a typical autopatch procedure:

1) Give your call and say "emergency patch."
2) Drop your carrier momentarily.
3) Key in the access code.
4) Dial the emergency number (usually 911).
5) Wait for police or fire operator.
6) Answer the questions that the operator asks.
7) After operator acknowledges, dump the patch by keying in the dump code.
8) Give your call and say "patch clear."
9) Don't stay keydown! Talk in short sentences, releasing your push-to-talk switch after each one, so the operator can ask you questions.

When you report a vehicle accident, remain calm and get as much information as you can. This is one time you certainly have the right to break into a conversation on a repeater. Use plain language, say exactly what you mean, and be brief and to the point. Do not guess about injuries; if you don't know, say so. Some accidents may look worse than they really are; requesting an ambulance to be sent needlessly could divert it away from a bona-fide accident injury occurring at the same time elsewhere. And besides, police cruisers are generally only minutes away in an urban area.

Here's what you should report for a vehicle accident:

1) Highway number (eg, I-43, SR-94, US-45).
2) Direction of travel (North, South, East, West).
3) Address or street intersection, if on city streets, or closest exit on highway.
4) Traffic blocked, or if accident is out of traffic.
5) Apparent injuries, number and extent.
6) Vehicles on fire, smoking or a fuel spill.

Example: "This is WB8IMY, reporting a two-car accident, I-94 at Edgerton, northbound, blocking lane number two, property damage only."

The first activities handled by experts at a vehicle-accident scene are keyed to rescue, stabilize and transport the victims. Then they ensure security, develop a perimeter, handle vehicle traffic and control or prevent fires from gasoline spills. Finally, routine operations restore the area with towing, wrecking and salvage.

The ability to call the police or for an ambulance, without depending on another amateur to monitor the frequency, saves precious minutes. Quick reaction and minimum delay is what makes an autopatch useful in emergencies. The autopatch, when used responsibly, is a valuable asset to the community.

Bill, KBØMGU, Ed, NØZAT, and Gerry, NØNGW, staff the ARES station at the Grand Forks Emergency Operations Center (EOC) during the April 1997 flood.

Freeway Warning

Many public-safety agencies recommend that you do not stop on freeways or expressways to render assistance at an accident scene unless you are involved, are a witness or have sufficient medical training. Freeways are extremely dangerous because of the heavy flow of high-speed traffic. Even under ideal conditions, driver fatigue or inattentiveness, high speeds and short distances between vehicles often make it impossible to stop a vehicle from striking stationary objects. If you must stop on a freeway, pull out of traffic and onto distress lanes. Exercise extreme caution to protect yourself. Don't add to the traffic problem. Instead, radio for help.

WORKING WITH PUBLIC SAFETY AGENCIES

Amateur Radio affords public-safety agencies, such as local police and fire officials, with an extremely valuable resource in times of emergency. Once initial acceptance by the authorities is achieved, an ongoing working relationship between amateurs and safety agencies is based on the efficiency of our performance. Officials tend to be very cautious and skeptical about those who are not members of the public-safety professions. At times, officials may have trouble separating problem solvers from problem makers, but being understaffed and on a limited budget, they often accept communications help if it is offered in the proper spirit.

Here are several image-building rules for working with safety agencies:

- Maintain group unity. Work within ARES, RACES or local club groups. Position your EC as the direct link with the agencies.
- Be honest. If your group cannot handle a request, say so and explain why. Safety personnel often risk their lives based on a fellow disaster-worker's promise to perform.
- Equip conservatively. Do not have more flashing lights, signs, decals and antennas than used by the average police or fire vehicle. Safety professionals are trained against overkill and use the minimum resources necessary to get the job done.
- Look professional. In the field, wear a simple jump suit or jacket with an ARES patch to give a professional image and to help officials identify radio operators.
- Respect authority. Only assume the level of authority and responsibility that has been given to you.
- Publicize Amateur Radio. If contacted by members of the press, restrict comments solely to the amateurs' role in the situation. Emergency status and names of victims should only come from a press information officer, or the government agency concerned.

The public service lifeline provided by Amateur Radio must be understood by public-safety agencies before the next disaster occurs. Have an amateur representative meet with public-safety officials in advance of major emergencies so each group will know the capabilities of the other. The representative should appear professional with a calm, businesslike manner and wear conservative attire. It is up to you to invite the local agencies to observe or cooperatively participate with your group.

Assisting the Police

Act as a communicator and radio for the police when you spot criminal activities. Memorize a description of the suspects for apprehension, or information about a vehicle for later recovery. Use caution—uniformed personnel arriving on the scene do not know who you are. Be an observer; let the officers act. For instance, if you see a vehicle traveling at high speed at night without lights, report the location and direction of travel to the law enforcement agency. Don't follow the vehicle.

Some Amateur Radio groups provide police communication assistance involving free taxi service to citizens who do not wish to drive home on New Year's Eve. They also act as lookouts for vandals on city streets and on freeway overpasses on Halloween. Hams have even reported sighting a person who is about to make a life-threatening jump into a river. Assisting the police in seemingly small ways can make a big difference in the community.

Search and Rescue

Amateurs helping search for an injured climber use repeaters to coordinate the rescue. A small airplane crashes, and amateurs direct the search by tracking from its Emergency Locator Transmitter. No matter what the situation, it's reassuring to team up with local search-and-rescue organizations who have familiarity with the area. Once a victim is found, the hams can radio the status, autopatch for medical information, guide further help to the area and plan for a return transportation. If the victim is found in good condition, Amateur Radio can bolster the hopes of base-camp personnel and the family of the victim with direct communications.

Even in cities, searches are occasionally necessary. An elderly person out for a walk gets lost and doesn't return home. After a reasonable time, a local search team plans and coordinates a search. Amateurs take part by providing communications, a valuable part of any search. When the missing person is discovered, there may be a need to radio for an ambulance for transportation to a nearby hospital.

Hospital Communications

Hospital phones can fail. For example, when a construction crew using a backhoe may accidentally cut though the main trunk line supplying telephone service to several hundred users, including the hospital. Such major hospital telephone outages can block incoming emergency phone calls. In addition, the hospital staff cannot telephone to discuss medical treatment with outside specialists.

Several hand-held-radio equipped amateurs can first handle emergency calls from nursing homes, fire departments and police stations. They can also provide communications to temporarily replace a defective hospital paging system. Next, they can help restore critical interdepartmental hospital communications and, finally, communications with nearby hospitals.

Preparation for hospital communications begins by cooperating with administrators and public-relations personnel. You'll need their permission to perform inside signal checks, install outside antennas or set up net control stations in the hospital.

Working with local hospitals doesn't always involve extreme situations. You may simply be asked to relay information from the poison control center to a campsite victim. Or you may participate in an emergency exercise where reports of the "victims'" conditions are sent to the hospital from a disaster site. One typical drill involved a simulated crop-duster plane crash on an elementary school playground during recess. The doctors and officials depended on communications to find casualties who were contaminated by crop-dusting chemicals. Again, never give the names of victims or fatalities over the air; this information must be handled only by the proper agency.

Toxic-Chemical Spills and Hazardous Materials

A toxic-chemical spill suddenly appears when gasoline pours from a ruptured bulk-storage tank. A water supply is unexpectedly contaminated, or a fire causes chlorine gas to escape at an apartment swimming pool. On a highway, a faulty shut-off valve lets chemicals leak from a truck, or drums of chemicals fall onto the highway and rupture. Amateur communications have helped in all these situations.

Caution: don't rush into a Hazardous Materials (HAZMAT) incident area without knowing what's involved, or you may well become a victim yourself. Vehicles carrying 1000 pounds or more of a HAZMAT are required by Federal regulations to display a placard bearing a four-digit identification number. Radio the placard number to the authorities, and they will decide whether to send HAZMAT experts to contain the spills.

Follow directions from those in command. Provide communications to help them evacuate residents in the immediate area and coordinate between the spill site and the shelter buildings. Hams also assist public service agencies by setting flares for traffic control, helping reroute motorists and so on. We're occasionally asked to make autopatch calls for police or fire-department workers on the scene as they try to determine the nature of the chemicals.

The National Transportation Safety Board, the Environmental Protection Agency and many local police, fire, and emergency government departments continue to praise ARES volunteers in their assistance with toxic spills.

EMERGENCY COMMUNICATION: SHIFTING NEEDS OF SERVED AGENCIES

The remainder of this chapter written by Rick Palm, K1CE, speaks to the question: "Is telecommunication technology rendering Amateur Radio obsolete in the disaster relief community?"

The use of Amateur Radio for public service, especially emergency communications, is as old as the ARRL itself. In the *first* issue of *QST* (December 1915), the membership is informed of a letter sent to the Secretary of the Navy by ARRL President Hiram Percy Maxim in response to the country's preparation for war. In that letter offering the ARRL's services to the military, Mr. Maxim states that:

> The League is "purely an amateur organization." Further, "the exchange of and delivery of messages is absolutely complimentary, and no consideration for transmission of a message is allowed under any circumstances. Regular radio telegraphic methods are employed. A sample of our official message blank is enclosed herewith."

Although he was a relatively new ham, Leo Sutherby, VE2PUP, was involved with communications immediately following the September 1998 crash of SwissAir flight 111. Leo is shown operating as net control from the shack of Tom Caithness, VE1GTC.

In a separate but similar letter to the Secretary of War, there is this direct reference to public service communications:

> In times of peace we also have confronting us sudden disasters, such as flood, fire or strike. Dayton, Ohio, was an example of a disastrous flood, which destroyed telegraphic and telephonic communications, and made it possible for the amateur wireless operator to render invaluable help. A fire which destroyed the central station of the telegraph and telephone Companies in a city, would also place that city in a very dangerous situation. The amateur wireless station would be the first place looked to in such an emergency.

Thus, an incredible record of public service and emergency communications is born, and it continues to this day, more than 80 years later. However, here is one more excerpt from Mr. Maxim's letter that is particularly worthy of note here:

> "Many of our stations have had no expense spared upon them, and are equipped *better than most commercial stations*." (emphasis added)

Can we say the same today, that amateur stations are better "equipped" than "commercial" stations? Joel Kandel, KI4T, a telecommunications professional and former chairman of the ARRL National Emergency Response Committee (ANERCOM) sounded an alarm in a *QST* Public Service column editorial: "More than at any time in the history of technological development, the world is witnessing the emergence and proliferation of new communication modes and media. The Amateur Radio Service can consider itself threatened by this development for two reasons: Many of the proponents of the new media are after our spectrum, and . . . much of the technology is fairly user friendly and very efficient."

Commercial systems have now outpaced Amateur Radio in technical capability. In the face of this development, is Amateur Radio still useful as an emergency response resource? With the expected efficiency of message handling rising as new systems come on line, will amateurs be able to step up and meet these expectations?

Will the FCC continue to support our service and our spectrum when a global Internet can handle greater volumes of messages with system redundancy to allow rerouting when segments are destroyed by disaster? New satellite services are arriving, providing voice, fax, data (Internet), video, and radio determination to portable handheld phones and palm-top terminals—with little or no terrestrial infrastructure to be damaged in a disaster. The Iridium system is accessible for worldwide communication by portable cellphone-like instruments and pagers.

Where do we go from here?

In 1998, ARRL sponsored four public service conferences around the country to identify future telecommunication needs of served agencies and reconcile them with corresponding present and future capabilities of the ARES program. The question we wanted to try to answer is, how will Amateur Radio continue to play an important role in providing emergency telecommunications in the future, in the face of new technology and the greater capabilities of such systems as the global mobile-satellite services. And what could our served agency representatives tell us to help us adapt to their evolving needs.

Each conference drew roughly 50 conferees, who were handed many solid ideas by served agency representatives, steering us in a good direction for the future. Here's a look at a few of the salient points.

- *More cross-training in served agency functions will add value to our contributions.*

This was a recurring theme in all four conferences. Especially in the case of the American Red Cross, amateurs can increase their value to the agency by taking advantage of training courses offered for several disaster relief functions: damage assessment, shelter management, mass care and feeding, for examples.

Traditionally, amateurs have declined to perform functions unrelated to their primary radio communication interest and training. Indeed, ARRL literature has cautioned amateurs against providing unrelated services, encouraging them to concentrate on their role as radio communicators only. The caution was founded on a healthy concern that amateurs performing unrelated functions for which they were not trained would become liabilities rather than assets, and cause Amateur Radio to lose credibility. However, if we do not broaden our perspective on Amateur Radio's traditionally limited role, we minimize our utility and risk finding ourselves outside looking in when it comes to serving agencies engaged in future disaster relief. Proper training and certification in the various functions are the keys to a successful bid for greater utility and corresponding perpetuation of our public service tradition.

- *Integration of amateur systems with non-amateur systems will increase utility and value.*

Kandel and others who have worked on ARRL planning and advisory committees have proposed greater integration of amateur radio systems with other telecommunication systems, despite concerns that we would dilute the "purity" of the amateur service and give other interests a foot in our door. Kandel says this is happening anyway: APRS, an amateur radio developed technology, now incorporates computers, the Internet, weather stations and the GPS. As another example, WinLink developers have interfaced Internet e-mail and HF radio using PACTOR. Hybrid systems such as these may represent an opportunity to enhance our contributions to served agencies for the future. Emergency managers and NWS personnel at the conferences all appreciated the value of APRS and similar systems in gathering information, and supporting communications from locations not serviced by their own.

- *More cross-training with other radio systems will make amateur operators more valuable in the EOC.*

A Collier County, Florida, emergency manager said at the Florida conference that he has a need for operators who are capable of operating *all* communication systems in the EOC. "When hams come in, they should be able to operate anything."

Thus, we may not only need to integrate amateur systems with other systems, but we may also want to promote programs to train our ARES volunteers to develop proficiency with new systems that do not directly integrate with amateur radio. Over the last few years, Florida has seen members of Amateur Radio response teams carrying cell phones and pagers. During Hurricane Andrew, municipal agencies often tasked amateurs with operating government radios. In 1992, most municipal communication employed conventional two-way radio. Today, it's 800 MHz trunking. Do amateurs today know what a trunked radio system is and how it operates? Could our volunteers operate a satellite telephone terminal to relay messages? And move it from one location to another and aim it correctly at the satellite?

The Telecommunications Committee of the Florida Emergency Planners Association (FEPA) has been asked to extend training to ARES and RACES operators in new technologies adopted by county and state emergency management agencies. ARRL should consider publishing a comprehensive training manual featuring information on *all* modern communication systems, rendering our amateurs more valuable in the emergency operations center. When the dust settles in a disaster, we will be remembered for how useful we were to the overall response, not whether we were using our own equipment or someone else's, concludes Kandel.

- *We should continue to emphasize our role in providing interagency communications during multi-agency responses.*

The provision of interagency communication has always been a solid, traditional role for Amateur Radio, and it appears that the need will be perpetuated for the future. Another recurring theme across the four conferences was that, despite the institution of new telecommunication technologies in the public safety sector, there is a continuing inability

Steve Lewis, N8TFD, at the WARN net control console, for the April 9, 1999 tornado disaster that hit Cincinnati. Later, the Queen City Emergency Net assisted the Cincinnati Chapter of the American Red Cross during relief operations.

of responding agencies to communicate with one another at a disaster site. As emergency responses become more sophisticated and the proliferation of new disaster response agencies continues, the need for interagency communications at a disaster site will become even more profound.

- *Amateurs need to present a unified front to the agencies they serve.*

This observation came from a high-level state official in Florida's Department of Emergency Management, and sent up a big red flag with our conferees. Infighting and turf wars, especially between ARES and RACES, are as old as the RACES program itself. Toss in the mix of clubs who claim monopolies over served agencies and emergency responses in their communities, and the result is confusion and disorganization as seen by the agency officials who don't have the time to deal with it.

An educational campaign with our volunteers and clubs should be undertaken to ensure that Amateur Radio presents an orderly, professional face to the agencies we serve.

- *Amateurs should emphasize the unique "decentralization" characteristic of the amateur service.*

Radio amateurs are already geographically dispersed throughout the areas to be affected by disaster. We are found in just about every community across the country. We're already everywhere that relief agencies would like to be, but can't, because of the obvious limitations. We need to reaffirm this unique characteristic in selling ourselves to the served agency community.

The value of decentralized radio services such as Amateur Radio has been recognized internationally by entities such as the United Nations' Working Group on Emergency Telecommunications. Not everyone will own a portable satellite telephone, at least not in the near future, owing to its expense, and the fact that they are not yet proven, reliable devices. We still hold a relative monopoly over low-cost, decentralized communications resources.

- *Expanding our client base with new agencies to serve will maximize opportunities for Amateur Radio.*

A good suggestion came from ARRL Emergency Coordinator June Jeffers, KBØWEQ, at the Midwest Regional Public Service Conference. If we lose the opportunity of serving some agencies as a result of our being displaced by new technology, we can hedge our bets for the future by expanding our client base; i.e. by drafting new and possibly nontraditional entities to serve. Looking for new clients to serve, Jeffers has been working with district schools in establishing an emergency communication network, with a permanent station at district HQ.

Jerry Boyd, K6BZ, a well-respected public safety official, ARRL Section Emergency Coordinator, and frequent *QST* Public Service column contributor identified other sectors for expanding our client base: public works departments, utility companies, transportation companies, hospitals, convalescent centers, senior citizen homes, and child care centers. See his foresighted editorial in the Public Service column, February 1995 *QST*, p 80.

Summary

Is there cause for alarm that Amateur Radio may not have a future in public service communications? In many, especially rural, parts of the country, it will be a long time before sophisticated telecommunication systems become available to public safety and related agencies. In others, new technology is here now and our role has already been diminished as a result. It is clear that new telecommunication tools will ultimately affect the needs of all of our served agencies in the not-too-distant future. We cannot become complacent. We must adapt to meet their evolving needs or we face reductions in opportunities to serve. That translates to less relevance, and a weakened position when it comes time to defend our spectrum needs in the face of increasing pressure from other interests.

If we work to adapt to our served agency's new needs, and keep up our public service record established more than 80 years ago intact, we can continue to count on the valuable returns: as examples, consider that both APCO and FEMA filed comments opposing the powerful Land Mobile Communications Council's petition to reallocate the 430-440 MHz segment of the amateur 70-cm band. In APCO's case, the organization had broken ranks as an LMCC member to do so. Such support from our served agencies is priceless and we must do everything we can to safeguard it.

Chapter 9

DXing

Contacting Those Faraway Places

Bill Kennamer, K5NX
ARRL Headquarters

DX IS . . . the essence of Amateur Radio. If it were not for the desire to send a signal over the hill, then over the water, then across the oceans, and around the world, Amateur Radio, or even wireless itself would probably have been dismissed as a useless laboratory phenomenon. But that first DXer, Guglielmo Marconi, put the signals over the hill, across the oceans, and around the world, and the world hasn't been the same since. It isn't an overstatement to say that worldwide communication owes everything to DXing, and to Marconi, the original DXer and DXpeditioner.

The history of DXing is long and varied. Of course it starts with those early efforts of Marconi, who very early turned to commercial development. It continued with amateurs whose experimentation allowed them to work stations farther and farther away. But the real dream was to bridge the oceans. The ARRL had a part in that dream. At the Board of Directors meeting at the first National Convention in Chicago, 1921, then Traffic Manager Fred Schnell presented a plan that would give the best possible chance for radio signals being heard across the Atlantic. For the transatlantic receiving test scheduled in the late fall, Paul Godley, 2XE, considered to be the foremost receiving expert in the United States at the time, and a member of the ARRL Technical Committee, was dispatched across the Atlantic with his best receiving equipment. Setting up in a tent on the coast of Scotland, Godley began his tests. By December 7, 1921, he was ready. Tuning across the bands, he began to hear a 270-meter spark. The operator's call sign, 1AAW, was clearly heard! (But 1AAW turned out to be a pirate! Even at the beginning pirates were one of the hazards of DXing.) The signals of more than 30 American amateurs were heard during this series of tests, waiting only for a two-way crossing to be completed.

Even then, the experimenting and modernization spurred by a desire for better DX performance was apparent. Godley reported that of the signals heard, over 60% had used CW rather than spark, and with less power. The death knell for spark had been sounded, and the future of tube transmission was ensured. All because of the desire for DX.

In late 1921 and early 1922, Clifford Dow, 6ZAC, who was located in Hawaii, heard signals from the western United States. He announced this in a letter to *QST*, and said if he could get some transmitting equipment, he believed he could make it across the Pacific. A group from the West Coast sent him the needed transmitter, and two-way contact was established on April 13, 1922, between Dow and 6ZQ and 6ZAF in California. Thus was yet another tradition of DXing established, that of providing equipment to activate a "new one."

Léon Deloy, 8AB, of France, had participated in the transatlantic tests of early 1923. His signal had been one of several heard by US amateurs, but no two way communication resulted. During the summer, he studied American receiving methods, and even came to Chicago for the National Convention. He returned to France with American receiving equipment, determined to be the first to make the transatlantic crossing.

After setting up and testing, by November Deloy was ready. He cabled ARRL Traffic Manager Schnell that he would transmit on 100 meters from 9 to 10 PM, beginning November 25. He was easily heard. By November 27, Schnell had secured permission for use of the 100-meter wavelength at 1MO and the station of John Reinartz, 1XAM. The two of them waited for Deloy.

For an hour Deloy called and sent messages. Then he signed. The first DX pileup began as Schnell and Reinartz both called. Deloy asked Reinartz to stand by, and worked first Schnell and then Reinartz for the first transatlantic QSOs. The age of DXing had finally truly begun.

From these small beginnings, to the later exploits of such as Bill Huntoon, Danny Wiel, Gus Browning, and the latter day exploits of Martti Laine, and such as the VKØIR group, DXing has grown. While the early days were marked by exploration and discovery, we now recognize that we have the equipment, power and antennas to hear a pin drop on the other side of the world and respond to it. While some discovery is still involved, and that mostly on the VHF/UHF bands, for most of us DXing takes the form of Radiosport at its zenith, a ritual of great importance to some, and a source of friendship and occasional enjoyment to others. DX IS. . .no doubt about it, the essence of Amateur Radio.

WHAT IS THIS THING CALLED DX?

In the early days of radio, DX became the acronym for "distance." At the time, DX could well have been determined by a new town, a new county, or even a new state. With the passage of time, DXers became more proficient, and DX is now generally accepted, at least at HF, as being contacts outside your own country. VHFers will consider it to be a new and distant grid square, while the microwave DXer will consider his DX in terms of miles. Our discussion will be primarily concerned with HF DXing.

A DXer most often pursues his hobby by chasing DX contacts that will bring him new credits for one of the popular award programs that he might be chasing. This may be an award such as the ARRL DX Century Club (DXCC), *CQ Magazine*'s Worked All Zones Award (WAZ), or the Radio Society of Great Britain's Islands on the Air program (IOTA). Each of these programs has a different objective. All, however, have in common the pursuit of DX, no matter the form it may take. A further discussion of US awards is in the Operating Awards chapter of this book.

Almost every ham has a bit of DXer ingrained in his system. It's only human nature to want to see how far away your equipment will work. The VHF operator may want to see how far away he can be from the repeater and still make it back with an HT. Listen to a repeater pileup on a tropo opening someday. You'll quickly see that the DX spirit exists in all of us. However, those who feel a stronger pull, a desire to achieve more, will soon hear the siren's call of DXing, and all manner of new transceivers, amplifiers and antennas will appear. In some, the bug has lain dormant for many years, only to break though with a sudden, passionate surge that takes hold of life's direction itself, and guides him to new altitudes in pursuit of his own particular dream.

In this chapter, we will look at the various stages a DXer may go through, and see how his operating techniques, equipment and need for knowledge varies. We will see what resources are needed at each stage, and will explain several of the different facets of the DX hobby.

The birthplace of radio and DXing, Sasso Marconi, the home of Guglielmo Marconi, is now the home of the Guglielmo Marconi Foundation. IY4FGM, the Foundation's station still provides DX from the original DX QTH.

DXING 101: FROM THE BEGINNING

Few individuals start out in Amateur Radio just to DX. Many were interested in just getting on the air, and suddenly bumped into a DX station. The fascination began at that point, and grew over time. Yet it is likely to be the beginning DXer who enjoys the hobby more. There's a lot to be said for just beginning, not knowing all the tricks, not knowing what to expect. The simple fact is that an experienced DXer is what he is because of one thing. . .experience! Experience doesn't come overnight, but through both the passage of time and by practicing. The beginning DXer certainly can get some practice, because everything is new. By the time he's completed his DXCC, he's learned a lot, and will be ready to go forward. And, while an experienced operator with a good station might work DXCC in a weekend, it's likely to take even a highly motivated beginner a year or more. So, let's examine a few things that should be developed over that first year.

EQUIPMENT

What to start with? Actually the beginning DXer can start with almost anything. Having said that, there are a few things the beginning DXer *shouldn't* do. One of them is QRP. Although some may have had some success with it in the past, the fact is that the beginning DXer doesn't need the additional frustrations that being inexperienced and operating with 5 watts will bring. So the first worthwhile piece of equipment for the DXer is a rig with at least 100 watts output. Rig selection is a complex subject, best left for a much longer discussion. In general, here are some suggested features for a starting point:

- 100 watts minimum output.
- Adequate receiver design. The better you can afford, the better you can perform. The old adage "You can't work 'em if you can't hear 'em" applies almost as much to rig selection as to hearing ability.

- Adequate filters. Narrow filters are needed for CW and SSB—a minimum of 500 Hz for CW, while 2.4 kHz is a desirable figure for SSB. These should be mechanical or crystal filters. Generally, outboard audio filters are not adequate for DXing on a crowded band.
- A good set of headphones. Usually this means headphones with a response between 200 to 3500 cycles. Remember, we're going DXing, not listening to hi-fi.
- A second VFO or separate receiver. These will be needed if you should encounter any split frequency pileups.

With this type of equipment, worldwide DX can be worked. In fact, many of the DX stations the DXer will work over his career are using nothing more than this minimum list themselves. It is how the equipment is set up and operated that will make the difference.

There are some tools the DXer should have in his possession before beginning. One is *The ARRL DXCC List* publication. Most of the information found there is also contained in the back of this book. Another important tool is a good world map, such as the *ARRL World Map*. Mount it on the wall within sight of the operating position. It's a good idea to be able to see where countries are in relationship to one another. For example, if you wanted to work Japan, it helps to know that it's around the world from Europe, so that if you're hearing many loud European signals on a band, you're not as likely to hear Japanese stations coming from the same direction at the same time. Over time, you'll learn how to use this to your advantage. For example, experience will tell you what time and band it would be expected to hear the Philippines (DU). So, if a DXpedition to very rare Scarborough Reef (BS7H) was known to be on the air, it could be expected that it could be found at the same time and frequency band as the Philippines, since it's only about 150 miles from Subic Bay. To use this for any of the rare countries, find another more common one that's very close. The times and frequencies won't vary that much, and you'll be prepared.

The *ARRL DXCC List* contains not only the list itself, but also a list of the ITU call-sign allocation series. Between the two lists, a DXer should be able to identify any call sign heard on the air. In fact, the DXer should begin early on memorization of the entire DXCC list, and the ITU prefix allocation table. This should be coupled with memorizing the short path beam headings as well. The DXer can then identify any call sign and be prepared to point a directional antenna.

Antennas for DXing are many, and varied. While most anything will work somewhere at some time, the secret to DXing is making sure that whatever antenna system is in use is installed properly. This means using the right feed line, having the connections properly prepared, and simple things such as making sure the coaxial cable and connectors are dry. If you are a new ham, or new to the HF bands, you might want to get a copy of *Your Ham Antenna Companion*, published by the ARRL. It will give you an overview of HF antennas and some suggestions for antennas you might build yourself. Nothing quite matches the thrill of working DX with a homebrewed antenna!

One of the first antennas a DXer might use is a dipole. Properly cut and installed, a dipole is a suitable antenna for DXing on any band. A minimum of 35 to 40 feet would be a good height for 10 to 30 meters. Of course, for most installations the higher, the better. The dipole may be mounted as an Inverted V, but would generally perform better for DX work with both ends at the same height. The dipole may also be mounted in a sloping position, with one end much higher than the other. This is often done for DXing on 10, 80 and 160 meters. There are other configurations for the dipole. These may be found in the *ON4UN's Low-Band DXing* by John Devoldere, published by ARRL. The configurations are as valid for the higher frequencies as the low bands. All things considered, a full size dipole will usually outperform a trapped single element antenna, especially if that single element antenna isn't properly installed.

Vertical antennas are also suitable for DXing. They can provide the lower takeoff angles that are often superior for DXing. However, installation is more critical for the vertical than the horizontal. Radials should be used with any vertical. If ground mounted, at least 40 should be used, and the more the merrier. If elevated, it is best to provide at least four full size radials for each desired band. This will provide increased performance over an installation without radials. Excellent references for installation of radial systems for verticals may be found in *The Antenna Compendium Volume 5* and *Vertical Antenna Classics*.

If the space is available, a small tribander is also a good antenna system. Properly mounted, it will provide superior performance over a unity gain antenna system.

LISTENING

Remember this if you remember nothing else about DXing: Listening carefully is the best thing one can possibly do to improve his DXing ability. It's sometimes hard to remember that, with the excitement of the chase, but over a period of time, the DXer will realize this is very true. It's the difference between having, for example, 10 years of experience, or having one year's experience 10 times.

As an example, suppose it is morning, during a period when the sunspots are low to moderate. For these condi-

Brian "Joe" Poole, enjoys CW QSOs from 9Q5MRC with his 100 watts and wire antennas. (*Photo courtesy of G3MRC*)

tions, the band of choice would be 20 meters. Begin on SSB, near the bottom of the band. You do have your headphones on, don't you? Remember now, we're just listening, so it doesn't matter where we are. Of course when transmitting, it will be necessary to be in the part of the band equivalent to your license class. Tune slowly up the band. Listen to the QSOs going on. If the signals are steady, and the accent sounds like your own countryman, pass it up now, and move slowly up the band. A signal is heard speaking in accented English. Stop here and listen. The signal may be strong and steady, or it may be strong but occasionally dipping in strength. This may well be a DX station. Stick around long enough for him to ID, or otherwise give a clue as to his location. See what style of operating he may be using. If he's in a QSO, mark the frequency and his call down in your notebook. If it is a European, then you know the band is open in that direction. But if he's in QSO, don't call him until he is definitely finished. To do otherwise would be rude, and usually only LIDs are rude. You can write the call and frequency on your notepad, and check back in a few minutes.

Keep moving up slowly. Be sure to pay attention to weak signals. Many times they are passed over because people either can't hear them through their speakers (but you're wearing headphones, remember?) or think they're too weak to be workable. More than once a rare DX station has appeared on the band, called a few CQs with no response, and gone away. With 100 watts and a simple antenna, it's to your advantage to try to find stations before anyone else is calling and the pileup starts. It's likely that the DX station may be running 100 watts to a dipole, sometimes even an indoor dipole. If you can hear him, he can most likely hear you, too, especially if no one else is calling.

The important thing while you're tuning is to listen to as many different DX stations as possible. Notice the sound, the accents, audio quality. With a little practice, you can tune the band quickly, and immediately identify the stations that are DX. Later, you will be able to do the same thing on CW.

OPERATING

Now it's time to begin calling some of the DX stations you've been hearing. You hear the DX station sign. You know his call. Now call him. If it's a station who has been working short QSOs, give his call one time only (he *knows* his call!), and yours twice, using standard phonetics. Then wait. If you're lucky, he'll come back to you! If he comes

Pirates and Policemen, LIDs and Jammers

The DXer does encounter a few problems in the pursuit of his avocation. Those nefarious denizens above are not there to make life easier for the DXer. However, all DXers will ultimately have in common the fact that they have overcome and persevered through the jungle created by the antics of these creatures of the ether. These cretinous individuals should not diminish the pleasure of the DXer, but allow him to puff out his chest with pride, as he overcomes the obstacles put into his path by such miscreants.

Pirates have been with us always. We can't be sure if Marconi heard one when he first fired up, but certainly the first signal received across the Atlantic by Godley was one. Probably the most famous pirate was named Slim. At a time when a volcanic island had popped up out of the sea near Iceland, Slim turned up from Cray Island as 8X8AA, claiming that the island had just popped up in the North Atlantic, and would qualify for a new country. He held forth for several days, then disappeared forever. Since then, many pirates over the years have been tagged as Slim. So, if an operation comes onto the air, and it's so improbable as to be suspect, it may be Slim, back for another run.

One thing the DXer must do when encountering a suspected pirate: *Work Him!* Yes, always work him, because sometimes the improbable is true. The rule is *Work 'em First, Worry Later (WFWL)*. Two things are accomplished by doing so: first, the DXer can practice his pileup technique, and second, if it is for real, it's in the log. If it's suspect, the QSL need not be sent until later. But if it's not in the log, it's hard to get a QSL. Some pirates and bootleggers do QSL, so it's not unusual to get cards rejected for awards credit. Everybody does, so it's best to continue working stations until one sticks.

Every pileup will have Policemen and Jammers. The Policemen may be well-meaning souls, but they really are in the same category as the Jammers. They just do it in a less sophisticated way. The DX is calling, working a few hundred an hour by split frequency. Meanwhile, on his transmit frequency, the LIDs, Policemen, and Jammers may be all heard at once, each pursuing his own route to DX infamy. The LID will start with "Who's the DX?", "Where's he listening?" or the inevitable, "Is the frequency in use?" All these questions could be answered by listening a little, but either the LID's time is too valuable, or his intelligence level is too low to think of that.

This is, of course, an excuse for the Policemen to jump in, with varying responses. Ten Policemen will, one after the other (never in unison) give them all the information anyone would care to know about the DX station, meanwhile totally obliterating the DX station and stopping the pileup in its tracks. This is immediately followed by 10 more who attack the LID, questioning his parentage, or alluding to his intelligence level. Finally come the last 10, who are telling the previous ones and the LID to just shut up. Ah yes, the musical chaos of the pileup!

Did we say music? That must be the cue for the Jammer to appear, as music is one form of jamming. The Laughing Box is also good, although not in much use in recent years. Finding someone calling in the pileup and recording it in the DVP and replaying it on the DX frequency is also a frequently used technique. Sometimes the call of their favorite DX net control is used this way. Most Jammers these days are not that sophisticated. Now it's mostly just tuning up or calling CQ on the DX frequency, or running a couple of minutes of unsquelched 2 meters onto the air. One of the easiest ways for the Jammer to enjoy himself is to just ask who the DX is, and get the Policemen started.

Meanwhile, the DXer can rejoice in the fact that, although all these obstacles are placed in the way, he's still working the DX! So let the others moan and groan, let them complain to the utmost. The fact remains, those who learn their skills well will succeed.

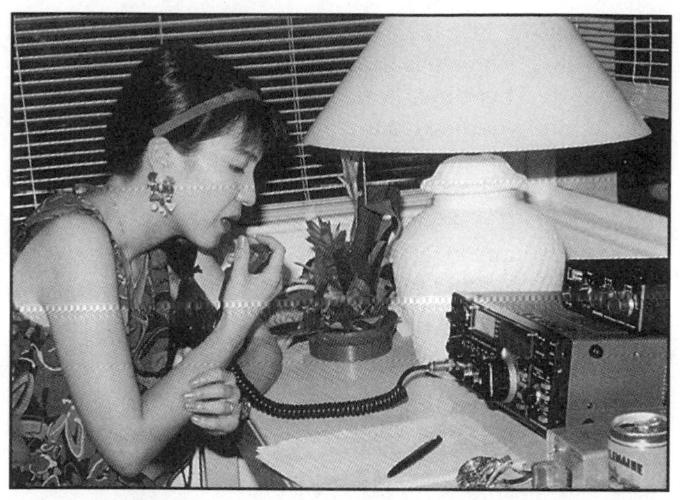

Sylvie, JP1LAB, operates 3D2LA from Fiji. Sylvie and OM Mike, JH1KRC, often take DX vacations in the Pacific. (*Photo courtesy of JH1KRC*)

back to someone else, wait patiently until he's finished. DXing is a game of patience. But turn off your VOX while waiting, Tripping your VOX while waiting on the frequency is the behavior of a LID.

You're lucky!! He's come back to you! You're now in QSO with the DX station. The first QSO with a DX station can be somewhat like dancing. He leads, you follow. As a rule, on your first transmission work by a formula. Give your name, state and signal report. A good form would be, "My name is Bill, in the state of Connecticut. Your report is 5 by 9, over." The DX station's English may be limited, he may be working many stations, or maybe is waiting on frequency for a schedule. In any case, what he does on his next transmission will determine whether he will say 73, or want to have an hour's discussion of the state of the fish in the river in your state. Just follow his lead and enjoy the QSO, no matter how long or short. Yes, you'll want to QSL. We'll talk about that later.

Many times you'll hear DX stations who are running stations at a fairly high rate of speed. They'll just answer with a call sign and signal report, then go on to the next station. When calling a running station, sign your call sign *one time*, using phonetics. If you're beaten in the pileup, wait until the next over, and call again. Don't ask the station for his QSL information in this type of pileup. If it's a European who has a QSL bureau in his country, he will most likely answer cards by the bureau. If he's somewhat rare, he will announce his QSL information on frequency, or it may be found from several sources off the air.

For the beginning DXer (or any DXer), contests are a very good way to pad your DX score. Contacts are often easier due to the number of pileups on the bands. There are a great many stations on the air, so it is often easy to work many stations from a given country. The more stations you work, the more QSLs you will receive, and the best way to QSL many of these contest operators is via the bureau. Either they are happy to QSL as the price they pay for your contest QSO, or they wouldn't QSL anyway, direct or otherwise. So go for the quantity.

THE INTERMEDIATE DXER

The pure pleasure of DXing may be best enjoyed by those who have worked a DXCC level of between 100 to 250. It is at this point that The DXer has become confident in what he is doing, has developed operating proficiency, and can still find plenty of exciting DX to chase. This Golden Time should best be enjoyed to the utmost, for once it passes, it cannot be reclaimed.

EQUIPMENT

While some of the requirements for equipment are the same, some areas may be upgraded for even better results.

- The transceiver can remain the 100-watt unit, but may be upgraded to one with a few more features. Cascaded filters are more desirable, and DSP, either outboard or inboard, has some advantages. Nothing like that DSP autonotch to remove those heterodynes—but be prepared to spend many hours learning to use it quickly and correctly!
- The headset is still essential, but now with an added boom microphone. It's nice to have both hands free for something else, like rotating the antennas while sending your call.
- A footswitch is a very handy thing to have, too. Not only are your hands left free, but there's no sneezing and tripping your VOX just as the 7O is coming back to you!
- A good keyer is nice, because the DXer has to be versatile. If the needed DX works only CW, the DXer must either work CW or maybe wait another five to ten years for him to come up again.
- An amplifier is a good purchase at this level. The rarer the DX, the deeper the pileups. With an amplifier, there's a reasonable chance of getting through. Without, better wait until the fourth or fifth day to begin calling.
- Desirable before, the second receiver or VFO is now essential. You will need to be capable of operating split frequencies most of the time.

By now, the DXer should have considered some sort of tower and beam antenna. This is a great investment, one of the best the DXer could ever make. A small tribander at 40 to 50 feet will provide a noticeable increase in results on both transmit and receive. Wires may still be used for the low bands, or the tower itself may be fed as a vertical.

LISTENING

Listening will always be the most important operating strategy the DXer can practice. The DXer has already learned what a DX station sounds like. The over-the-pole flutter is recognizable, as well as the QSB from a station far away. Weak signals stand out from the crowd for the DXer, who sees this as the sign of rare DX on the band. Now it's time to progress to listening for more than just the DX itself. It's time to listen to the pileup, and capture the dynamics in memory, put back for later use.

There are two kinds of pileups, transceive and split frequency. These are exactly what their names imply—transceive on the same frequency, and split by using one frequency for transmitting and another for listening. Both are easily manageable if proper technique is used. So, it's important to see how other operators apply that technique.

Tuning the band, a DX station is heard passing out rapid fire QSOs, signal reports only. First, as a rule, listen for a moment or two to see who it is. If needed, great, it's time to call. If not, stick around for some time to listen. The pileup builds, almost to the point of getting unmanageable. Notice who's getting through and who's not. The guys transmitting their last two letters on phone seem to be having a more difficult time. That's because many DX stations these days won't work anyone sending the last two as long as they can hear a full call sign. So, even if they have a good signal, the guys doing that are spinning their wheels to some degree. Full calls are always best. On CW this isn't much of a problem . . . yet. But you may hear a few in a CW pileup, too.

Notice that occasionally you will hear a station signing his call sign while extraneous information is being given to the DX station. This may be something like the second station signing his call on top of the signal report. This is a technique called *tailending*. Although it is a good technique, one needs to listen to a few good practitioners of the art, for it is an art, before trying it. Otherwise it can blow up in your face. Notice that often the calling station is on a slightly different frequency, and different speed on CW. Sometimes he's weaker. Some practitioners of the art actually turn off their amplifier before trying this. Doing so avoids being obnoxious. Notice also that if the station trying the technique tries it once or twice and it doesn't work, it isn't tried again. To continue could annoy the DX station, and he may be deliberately not taking tailenders when they call, but one or two calls later just to keep everyone from trying it at once. A pileup can quickly turn into bedlam if two or three tailenders are taken in a row.

Continue listening to the pileup, and see where callers are positioning themselves in the pileup. On CW, they may have their carrier up a hundred cycles or two from where the last caller was to make their call stand out. Sometimes slower speed is better if everyone else is fast. This also works somewhat on Phone. By varying the RIT, it's possible to make the voice on SSB take on a different pitch that may cut through the pileup.

Massimo, IK1GPG, and Betty, IK1QFM, at their beautiful station in Mondovi. (*Photo courtesy of IK1GPG*)

Tuning up the band, somewhere around the "usual DX frequencies," a signal is heard, again passing out rapid signal reports. This time, however, no one is heard coming back to him, yet the signal reports keep on coming. He is obviously working DXers on a split frequency. Now is when that second receiver pays for itself. Split frequency operation can be, and is, done most frequently with two VFOs, but can be even more effective when done with two receivers. On CW, start looking for the stations calling about one kHz up. Keep tuning up until the calling stations are found. The pileup may be as much as 5 to 10 kHz away. Then begin to look for the calling stations. Start tracking the stations that are working the DX station to see how far, and in which direction, he's moving between QSOs. See if he's taking tailenders (easier to do split than transceive). Do the same thing on SSB if it's a Phone pileup. *Listen carefully to what the DX station is saying. If you can't hear, wait until you can.* Surprisingly, when these pileups get really loud and raunchy, the DX station will often call out a specific spot frequency where no one is calling. Those who are listening will catch it. Those who don't hear will continue to call in vain. Look also for operators who operate at or below the edge of the announced calling frequency range. Sometimes the DX stations will announce something like "200 to 210." At that moment, he may have moved down to 200 to start tuning up again. Often a caller at 199.5 can get the QSO because of the difference in pitch, or because his sidebands make it difficult to hear on 200. This only works once. If he moves away from 200, then it's back to square one. Find and follow, then get ahead.

Have you heard the guy at the DX club who comes into the discussion on the latest big DXpedition? "Yep, worked him with one call," he says. You'll know that he may have spent half an hour or more using just the techniques above to set up that one call.

INFORMATION SYSTEMS

The DXer thrives on information. When is the next DXpedition? Who are the operators? Where are they going? All these questions can be answered from information sources available to the DXer, either by print, PacketCluster, or over the Internet. Frequently, in a pileup you might hear someone asking who the DX is or how to QSL. If that DXer had proper information sources, he would know without asking, and wouldn't be adding QRM to a pileup. Gathering information is as important to a DXer as any other facet of operating, because knowledge really is power.

The weekly DX newsletter, such as *QRZ DX* or other publications of that type, have been around for almost as long as there have been DXers. While some say their time is past, actually it isn't. They provide a hard copy record of events upcoming as well as events past, a record of QSL routes and corrections, lists of stations worked (very important if you're looking for specific resident DX stations), and provide this all in short factual tidbits. These publications are best used for upcoming events that have been announced well in advance. They also have the ability to provide chronicles of ongoing events, especially with resident operator situations.

PacketCluster is another important tool for the DXer. In fact, it has been said that many owe their Honor Roll plaques to the PacketCluster network. Add spot filtering software that alerts the operator if a needed one comes on the air, and you have a most formidable tool.

The PacketCluster network now stretches around the world. Through the use of Internet connections, it's possible to see what's being worked in Italy or Japan. This is actually more useful than one may think. First, it's possible to know that stations from certain areas are actually on the air. It's possible to identify these stations, and even write for schedules. There have been some who have found that spots from another area of the country actually can be used. It's sometimes possible to hear spotted stations where previously no

What to Look For

Your receiver is a critical element in being an effective DXer. The following description of receiver characteristics was written by former ARRL Lab Test Engineer Mike Gruber, W1DG. Mike performed the tests for each piece of equipment reviewed in *QST*. This material was taken from *QST Product Reviews: A Look Behind the Scenes*, in the October 1994 issue of *QST*, page 35. A full description of receiver performance tests is included in recent editions of *The ARRL Handbook For Radio Amateurs*, Chapter 26.

The ARRL Technical Information Service (TIS) offers a package titled *RIG*—a set of reprints from *QST*, answering the question: "Which rig should I buy?" *The ARRL Radio Buyer's Sourcebooks* are another good source of information on ham gear.

CW and SSB Sensitivity

One of the most common sensitivity measurements you'll find for CW and SSB receivers is Minimum Discernible Signal (MDS). It indicates the minimum discernible signal that can be detected with the receiver (although an experienced operator can often copy somewhat weaker signals). MDS is the input level to the receiver that produces an output signal equal to the internally generated receiver noise. Hence, MDS is sometimes referred to as the receiver's "noise floor."

You'll find MDS expressed in most spec sheets as μV or dBm. The lower the number in μV, or more negative the number in dBm, the more sensitive the receiver. (For example, a radio with an MDS of –139 dBm (0.022 μV) is more sensitive than one with an MDS of –132 dBm (0.055 μV).)

When making sensitivity comparisons between radios, keep in mind that more is not always better. The band noise (not the receiver noise) floor often sets the practical limit. Once you've reached that point, greater sensitivity simply amplifies band noise. Also, too much sensitivity may make the receiver more susceptible to overload and decrease dynamic range.

A typical modern HF transceiver has an MDS of between –135 and –140 dBm (or 0.0398 to 0.0224 μV). You can easily make a dB comparison between two radios if the MDS is expressed in dBm. Simply subtract one MDS from the other. If, for example, one has an MDS of –132 and the other –139 dBm, the latter radio has a sensitivity that is 7 dB better than the first—if both measurements are made at the same receive bandwidth.

Dynamic Range

Dynamic range is a measure of the receiver's ability to tolerate strong signals outside of its band-pass range. Essentially, it's the difference between the weakest signal a receiver can hear, and the strongest signal a receiver can accommodate without noticeable degradation in performance. Two types are considered in *QST* product reviews: blocking dynamic range (blocking DR) and third order intermodulation distortion dynamic range (IMD DR).

Blocking dynamic range describes a receiver's ability to maintain sensitivity (or not to become desensitized) while tuned to a desired signal due to a strong undesired signal on a different frequency. IMD dynamic range, on the other hand, is an indication of a receiver's ability to not generate false signals as a result of the two strong signals on different frequencies outside the receiver's passband. Both types of dynamic range are normally expressed in dB relative to the noise floor.

Meaningful dynamic range comparisons can only be made at equal frequency spacings. Also, as in the case of our hypothetical receiver, the IMD DR is usually 20 dB or more below the Blocking DR. This means false signals will usually appear well before sensitivity is significantly decreased. It's not surprising that IMD DR is often considered to be one of the more significant receiver specifications. It's generally a conservative evaluation for other effects that may or may not be specified.

Third Order Intercept Point

Another parameter used to quantify receiver performance in *QST* product reviews is the third-order intercept point (IP3). This is the extrapolated point at which the desired response and the third-order IMD response intersect. Greater IP3 indicates better receiver performance.

Second Order IMD Dynamic Range and Intercept Point

We've recently added a new test to the battery—second-order IMD distortion dynamic range. These products, like third-order products, also are generated within a receiver. Their relationship to the offending signals that cause them is f1 ± f2. In today's busy electromagnetic environment, they can become truly offensive under certain conditions. Consider the case of two strong shortwave stations, on two different bands, that sum to the frequency of the weak DX you're trying to copy. If their intermod product is strong enough, you may not be able to copy the station through the interference. As an example, a high seas coast telephone station on 4400 kHz might sum with a shortwave broadcast station on 9800 kHz to produce an intermod product on 14200 kHz.

IF and Image Rejection

A station at a receiver's IF frequency can be truly offensive if its signal is strong enough and the radio's IF rejection is insufficient. The station appears across a radio's entire receiving range! Be sure to consider IF rejection when considering a receiver and a nearby transmitter is at its IF frequency.

We're now testing both image and IF rejection in the Lab. We first measure the level that causes the unwanted signal to be at the MDS. Then, we reference this level to the MDS in order to express the measurement in dB.

DXpeditioners often travel to rare and remote islands, such as Mellish Reef, where the crew of VK9MM provided DXcitement in 1993. (*Photo courtesy of V73C*)

one would have even listened. And it has tipped operators to band openings not thought possible before. Real time, PacketCluster is probably the most superior method of intelligence gathering for ongoing operations currently in use. It is a truly amazing system, and the hybrid radio/Internet system is one of the best forms of useful ham ingenuity.

The Internet has some capabilities for information transfer as well. However, one really has to separate the wheat from the chaff with the various reflectors and newsgroups available. There is a lot of information, but a greater amount of misinformation, available. This is because everyone with a keyboard and a connection can put in their opinion on anything. It's best used for specific information posted to the various reflectors by those who are involved with DXpeditions, or by resident stations. It's been excellent for stations such as XV7SW or EL2WW, who have often provided information on what they're doing or what their times and conditions are on 160 meters. Specialty reflectors can often provide much in the way of information. See the Online Resources chapter of this book for more information on electronic newsletters and reflectors.

PROPAGATION

Here we can start by breaking one of the paradigms of DXing: Whatever the propagation is, *It Doesn't Matter!* That's right, it doesn't matter at all. Because if the DX station comes on the air on his chosen schedule, if the contest weekend falls at the worst propagation time of the year at the bottom of the sunspot cycle, the DXer still has to be there to work it! So rather than learning about corona holes and solar winds, it's better to concentrate on what you need to know about practical uses of propagation no matter what the solar conditions are. The basic difference between high sunspots and low ones is simply the choice of bands in use. The higher the spots, the higher the band.

So let's start a typical day on the bands in fall and winter. At sunrise, you have many choices. From the US, you can beam east on the high bands, or work west on the lower bands. In general, your antenna can follow the sun, starting southeast, then northeast, and by noon down to the south. In the early afternoon, look either east or southwest, moving up to northwest by late afternoon to early evening. From noon on, the southerly propagation should always be there, even into the late evening.

Another type of propagation is the *long path*, where propagation comes from an almost opposite direction. For example, in the morning especially in fall and spring, it's possible to work African stations by beaming west rather than east. The path seems to fall near the terminator (the area where darkness and light meet), and can sometime be more effective than the short path. Long path may even be used on 10 meters during times of high sunspot activity, with morning openings to Asia beaming southeast.

For the low bands, follow the darkness as you would follow the sun. Look toward the east to southeast at sunset, the south anytime after dusk, and the west from midnight on. Long path also works here, as Asia and Australia may be heard on the East Coast at sunset, and Europe on the West Coast in the mornings.

The 80,000+ QSOs made by the VKØIR team from Heard Island in the winter of 1997 (near the bottom of solar cycle 22) merely illustrates the principle that when the DX is out there, it will be worked, regardless of propagation. It's just a matter of band selection.

ADVANCED DXING

The Advanced DXer will be nearing the DXCC 300 level. By now, needed "New Ones" may be coming largely from DXpeditions, as only perhaps one or two will still be available from resident operators. By now, the DXer may have run into an unusual phenomenon, wherein a Needed One that seems quite common to other DXing friends becomes almost impossible to find. The DXer may even have found one or two others on his way up, countries where there are resident operators that seem to elude him. Sometimes his friends report working two or three in a contest, or sometimes he happens to wander onto the frequency of one who suddenly goes QRT before he can even give a call. One day the spell will be broken, however, and suddenly the DXer will be deluged with contacts from the former Elusive One, as they become QRM.

The Advanced DXer will find he spends less time transmitting than listening. He will find himself sometimes jumping into large pileups just to maintain those all-important pileup skills. Now is the time to develop the analytical skills to climb to the top, or develop new interests in news bands and modes.

EQUIPMENT

The Advanced DXer now can depend more on the skills he's developed rather than on the equipment. Still, equipment is important. It won't be so different from the Intermediate DXer's rig, and in fact there are some top DXers who are reluctant to change from a tried and true old rig because they don't want to miss an opportunity. However, the top of the line transceivers now are offering dual receive, which is almost a license to steal in a split frequency pileup. A receiver with a strong front end, able to stand nearby strong signals without creating internal distortion products is highly desirable, as he will be in pileups with everyone. After all, what he needs is most likely needed by everyone in the entire world, and the pileups will usually be deepest at this level.

If not using a stereo headset already, the DXer should add one with a boom mike. This makes it easier to have the DX in one ear, the pileup in the other while using two receivers.

While it's possible to work DX without an amplifier at all, or one with only 500-1000 watts, a full power amplifier is desirable. While some may say it represents only a decibel, remember that a decibel is sometimes defined as the smallest difference one can hear. Thus, in a competitive situation, it can make a difference. However, there are some who could have twice the legal power and still wouldn't get through. Still, it's nice to have strength and skill.

Antennas may range from the ubiquitous tribander at 50 feet to 200 pounds of aluminum at 200 feet. Usually the bigger the better. A height preference would run to at least a minimum of one wavelength at 20 meters or higher, and at least a half wavelength or a full size vertical for the lower bands. Local ordinance or deed restrictions may preclude this. DX can still be worked, but it does become a little harder. The Truly Serious will most likely have a multi-element monobander on 20 meters, and perhaps a two-element 40-meter beam. There are more than a few DXers who have multiple towers and monoband beams. From a city lot, the DXer will have to have good skills to beat these setups with a good operator on even a semi-regular basis, but he should know they're out there, and those guys will win more than their share of the pileups. It's best to be one of those guys, but not totally necessary.

LISTENING

Listening will always be an important part of the DXer's strategy at any level. In fact, the more experience the DXer gets, the more his listening skills will truly develop. Experience also brings a different perception of what he hears. For example, weak, watery signals may bring the thought that he should try another band. Or, hearing an expected station from an unexpected direction may signal an unusual opening, and tuning the band may find more stations from that part of the world, but maybe an even more rare country. The rule that most people would rather work loud stations rather than weak ones always holds true, and sometimes tuning away from the big pileup may bring another station from the same country, but weaker, and with no pileup. For example, in 1992 the YA5MM operation from Afghanistan was very successful. Big signals at seemingly all times of the day or night, good operators, and most especially large pileups. Meanwhile, OK1IAI/YA held forth each evening near 14.036 MHz, with a weak but workable signal. Seldom was there any pileup at all. The listeners were finding and working this station, while many others were waiting on a DXpedition with a loud signal.

Dudley Kaye-Eddie, Z22JE, relaxing at his shack in Harare, Zimbabwe. (*Photo courtesy of Z22JE*)

Listening at this level now also includes surveillance activities. The DXer has learned the players by now, so listening on the bands should include some conversations. You'll hear such as "Our friend from the East is planning to go south for the summer." Does this mean a major Japanese DXpedition to Bouvet? Perhaps, if the DXer knows who he's listening to, and who the speaker may have close contacts with. More on intelligence gathering later.

Listening also becomes important while in the dynamics of the pileup. Often the DX station will say or do something that will give you a clue as to what will happen next. For example, in 1991, XU1JA would suddenly disappear from 15-meter CW with no warning if the pileup got too large. A quick search of 10 or 20 meters right after that happened would net a QSO before the pileup got there and he returned to his original band. Many times the pileup is told that the operator is going QRT. Sometimes that's true. Sometimes it also means he's running to a different band or frequency for a schedule. The DXer hasn't anything to lose, so going to look can't hurt. Trying the higher end of the band may turn him up. To use this technique, the DXer *must* be able to identify his quarry by the sound of his signal, as call signs will be used as little as possible. Also, the DXer must avoid calling the DX station until the schedule is over, or risk the wrath of never appearing in the log.

The DX station quite often gives the instructions on his next band move over the air, yet the pileup never hears. The DXer who listens will. For his entire DXing career, the DXer needs to be aware of everything that's going on around him on the air. He must hear everything, and process the information at high speed to make his decision about how to do the most important thing, which is to get into the DX station's log with a good contact.

OPERATING

By now, the DXer has experienced tailending. He's worked many successful contacts using split frequency techniques. Yet he will continually find new ways to pick his way through the pileups, for as one DXer's successful techniques are propagated to others, then that technique, for a time, becomes less usable. So, a new technique, or dusting off an old one, may be necessary with any given pileup. The DXer will become adaptable, and will do so almost instinctively.

One of the most important techniques is to make short calls and listen after each one. For example, the DXer signs his call once, either Phone or CW, and listens for 1.5 to 2 seconds before signing his call again. He should not sign his call more than three times for each DX station's over. He remembers that if he's not the QSO, he's just QRM. The DXer will get beat in a pileup, and needs to remember that and not get frustrated about it. Sometimes it's just impossible to beat better propagation or better antennas. It is better to let the pileup proceed at as fast a pace as possible. Strangely, sometimes the DXer can actually improve his chances of getting through by not transmitting. Letting the pileup proceed at a faster pace will get more people through. Less QRM means a faster pace.

The DXer can also take advantage of a knowledge of propagation. It's pointless to waste kW hours of electricity while getting totally frustrated when there's no possible way to break a pileup. Since it's reasonable to expect propagation to flow from east to west as it follows the sun, then it's also reasonable to expect to take advantage of that knowledge. A 1981 pileup on 5H3KS on 20-meter SSB illustrates the point. He was somewhat audible, with an East Coast pileup making it through with some difficulty in the early afternoon. It was evident from the call sign prefixes that only the East Coast was making it through. Slowly over time, he became louder, and stations from the Midwest began joining in the calling. Still, only East Coast stations were getting through. Finally, he reached over S9. Mostly East Coast stations were making it through still. However, as his signal peaked and began to fall off, stations from the 0 and 5th call districts were the dominant stations getting through the pileup. This demonstrates two concepts: first, that patience is rewarded, and second, that an easterly station is often best worked *after* his signal peaks.

ADVANCED INTELLIGENCE GATHERING

As the DXer nears the 300 zone, the coveted "New One" seems to be harder and harder to find. In some cases, a DXpedition is required. In others, a little research into the operating habits of the station is required. In both cases, sometimes more is needed than just an occasional foray to the bands in the hopes that they may be there (although that sometimes happens too!). So, a little intelligence gathering is necessary.

The DX weeklies, such as *QRZ DX*, begin to pay off in the search for the operations of resident DX operators. The DXer may get a bulletin board, and begin keeping a chart of the activities of these operators by using the QSN (I heard on __kHz) reports. For example, it's noted that a VKØ in Macquarie is beginning to show up on a certain band, on or near a certain frequency, and usually around the same time of day. Further charting may observe that this occurs on the same day, but every other week. Thus, there's a high probability that if the DXer is there at the right time on the next likely day, he will get a shot.

The bulletin board may also be used to post notes about DXpeditions that will occur on certain days in the future. Sometimes the announcements are made weeks ahead of time, and not repeated. So, it's a good idea for the DXer to keep his own notes. That way, he checks his bulletin board daily for anything that may come up that day. Even information gleaned from over the air conversations or other not readily available sources may be tracked this way.

Gene Neill, ZA5B, operates from his shack in Shkodra, Albania. Gene, whose US call is WA7NPP, previously operated as TL8NG from the Central African Republic. (*Photo courtesy of WA1ECA*)

Earlier, it was mentioned that listening to conversations could pay off. Sometimes these are just rumors, but sometimes there's real information passed over the air about setting up DXpeditions. When this is done, it's often in seemingly meaningless conversation. Make no mistake, however, those in conversation know what they're talking about. With a little practice and thought, the DXer will too.

Many of these type conversations have now moved to e-mail. Private e-mail has greatly facilitated the logistical support of international DXpeditions. Private e-mail has also helped those DXers with friends who want him to know what's going on. The DXer will have acquired plenty of those on his way up. While not yet totally replacing the telephone, e-mail has made big gains for intelligence gathering.

Finally, keep track of national holidays and customs in the area of interest. Know when the DXer from other lands might be taking his cup of coffee (or whatever) into the shack to enjoy a few minutes of radio before work. Read a newspaper with a good international section to keep track of world events that may affect DXing.

BASIC DXPEDITIONING

By now the DXer may be thinking of the pileups, of what it may be like to be on the other side. He has seen the best and worst of DXpeditions, and would like to attempt something of his own. Great! The world will be waiting for him, because somebody needs anyplace he might go, and as for the rest, well, who can resist the siren's call?

To start planning a DXpedition, there must first be an objective in mind. For the first time DXpeditioner, it would be best to start with something with a minimum of hassle. The DXer wouldn't want to look up someplace like Heard Island and try to go there by himself to get started! Far better to begin by picking a destination where licenses are obtainable, where a hotel room can be arranged, and where airlines fly. A location with propagation to population centers is best, as part of the mission is to make some contacts in quantity as practice.

After site selection (and that includes finding a place friendly to antennas and transmitters), inquiries need to be made about the license. Unless the location selected is under FCC jurisdiction, the DXer will need to obtain a license from the governing body of the DX location. In some cases, this may take months, especially if he is not on site. Check with the Regulatory Information Branch at ARRL HQ for reciprocal licensing information. Most forms are on file there, along with a recap of necessary procedures to follow. *ARRLWeb* (**http://www.arrl.org/field/regulations/**) has operating permit information by country and other information.

The DXer should have a passport. While not strictly necessary in some countries he may visit, it's always a good idea to have it handy. Also, a photocopy should be carried on his person at all times. In the event the passport is lost, or the DXer is stopped by someone in authority, it may be necessary to produce satisfactory papers instantly.

The next step would be to determine what to take. Transceiver selection in many cases boils down to what is owned. However, the DXer will find that the transceiver needed for the average DXpedition need not be quite as full featured as what is owned. Dual receive is not a requirement, but a built-in keyer is. RIT or receiver offset is necessary to work split. Light weight and small size is important, as it is best to carry the transceiver on board. As checked baggage it may never reach the destination.

Antennas will be determined by what can be carried. The DXer would never go wrong by selecting an all band vertical to start. Wire dipoles are also good, especially if one or both ends can be placed on high supports. Carry two or three hundred foot lengths of small coax with connectors installed, along with several barrel connectors. Some spare connectors may also be handy. Take nylon or dacron rope of some type to use for antenna installation.

It is very important to check the line voltage available at the chosen location. More countries use 220 or 208 Vac than the usual 120 Vac found in the US. The DXer must have a power supply available for whatever voltage is found.

By packing the carry-on baggage correctly, the DXer will be assured his DXpedition will come off even if his baggage is lost. The transceiver, power supply, headset with boom mike, and wire antenna and coax should all go in the carry-on. A lap-top computer, if available, should go too. Everything else, including clothes should go in the checked baggage. If the checked baggage doesn't make it, at least the station is there. Log sheets, etc., may be obtained locally (most parts of the world have blank sheets of paper and writing instruments), serviceable clothes will be obtainable, and most hotels will have laundry services.

Upon arriving, it may be necessary for the DXer to make some arrangements with customs for admitting the radio equipment. This may in some cases require a large deposit, although sometimes the gear can be talked through. The DXer should try to find out as much about customs procedures in advance as possible.

When possible, the DXer should try to find an accommodating high rise hotel overlooking the ocean. The antennas should go as near to the edge of the roof as possible, to the DXer's top floor room. Of course it's possible that those kinds of accommodations will be impossible to obtain. In that case, the DXer will have to work with obtaining the best compromise he can.

Once the station is set up and on the air, it's time for the DXer to enjoy the pileups he came to create. After a few minutes of operation, the pileup will build. It is here the DXer should make a transition in operating style that will let him make the maximum number of contacts in the minimum amount of time. On SSB, try to operate transceive if the pileup isn't too deep. If it begins to get large and continues for a long time, it may be time to listen up. On CW, as soon as there are multiple stations calling, it's time to start listening about two up. Working split allows the pileup to hear the DX.

The DXer should be very consistent in working the contacts. The call of the station worked should be given, and a report, as in "K5FUV, five-nine." Every contact should be ended uniformly as in "Thanks, XZ1A." A minimum of verbiage should be used, only enough to make the contact and get on to the next one. On CW, it's similar, as in "K5FUV 5NN" and "TU XZ1A." QSL information may be given about every 10 minutes.

The DXer will find that practice helps. Rates will increase, and so will his enjoyment. DXpeditioning is a fever that may even be worse than the regular DX Bug.

ADVANCED DXPEDITIONING

At this point the DXer is advised to seek company, for DXpeditioning at this level requires more work in all areas, both in logistics and in political savvy. These are the kinds of places where one doesn't just waltz in and request a license. In some cases, two or three trips may be required before even getting a decent hearing from the guy who can say yes. The DXer who wants to play at this level must be prepared to spend some bucks.

First, letter writing is required. If there is someone the DXer knows who might know someone with some influence in government in some country, it may be possible that they could help facilitate contacts. In many cases, the hardest job

Edwin Hartz, K8VIR, operates from the main hut during the 1997 ZL9DX DXpedition to Auckland Island. (*Photo courtesy of K8VIR*)

is getting to the right person. Letters should be written. In many cases, the DXer will be rejected. Perhaps after an exchange of letters, it may be possible to go and meet with officials. Sometimes this can take months or years. The techniques or responses for doing this are not well defined, and nothing ever works the same way twice. Sometimes a check with the DXCC Desk may give some insight as to what's required. In most cases, getting the permission is the hardest part. Sometimes, it never happens.

The logistics of such a trip are more difficult as well. Since this is likely a trip to a highly needed country, it is necessary to take enough equipment to put a loud signal on the bands, possibly on two or three bands at the same time. This means amplifiers and beam antennas. More operators allow for more equipment. Some things may be purchased locally. A 6-meter-long steel tube provided the mast for a beam in Myanmar. From the top of a hotel, that's usually enough. The DXer should heed the Scout's motto: "Be Prepared!"

Being prepared also means checking into State Department travel advisories for the area, as well as checking with the Centers for Disease Control in Atlanta for any health concerns in the region. The DXer should be sure he has any required inoculations before going. Some are for his health; some are required by the country visited. Also, the DXer may want to take a suit. One can never tell when he may need to go meet a high government official!

In 1977, DXers focused their attention on tiny Kingman Reef, where KP6BD, operated by K4SMX, WB9KTA, N9MM and K6NA made 11,000 QSOs in three days. (*Photo courtesy of K4SMX*)

Arriving is sometimes exciting. After all, the DXer may be bringing several boxes of otherwise contraband equipment into the country! Many countries don't look at people with small radio transmitters in quite the same way that they do in the US. If at all possible, the DXer should be met at the port of entry by someone who can help get him through.

Operating from a country such as this will be different. The DXer should determine from a Great Circle map centered on the location the areas of the world that will be most difficult to work. These will normally be those where signals must pass through the auroral zones. An operating strategy should be planned that will allow for working only those areas when the band is open. Other areas may be worked at a later time. This technique is called *targeting*. For a more extensive discussion of targeting, see *DXpeditioning Basics* by Wayne Mills. It's available from INDEXA, PO Box 607, Rock Hill, SC 29731, for $5 postpaid.

With several operators, scheduling should be arranged so that the prime station is on the air at all times, and if available, a second station should be manned as well. The prime objective of a DXpedition such as this should be to maximize the number of different call signs in the log, since it may be a long time before anyone goes back.

The DXer should plan to operate split frequency almost as soon as he hits the band. By announcing likely frequencies, he will find people waiting for him to arrive, and may need no more than one CQ for as long as he remains on the frequency. However, while operating split, efforts should be made to listen on only 10 to 20 kHz at most. It is tempting to keep going up looking for a clear receive frequency, but it isn't really necessary. Just move back to the bottom and start back up. Remember, by the time the DXer gets to the operating part of the project, most of the real work has already been done. The operating is the fun part.

ULTIMATE DXPEDITIONING

There are many obstacles in the path of the Ultimate DXpedition. First, permission must be obtained. This is sometimes easy, sometimes difficult. Often, the Permittor wants to be sure the DXpeditioner is capable of pulling this off without being killed or stranded, a very real possibility. Once that is assured, there's always the transportation problem, Then, equipment, food, survival gear, and other necessities. Oh yes, there's radio stuff too. The costs for the Ultimate Class DXpedition can run into hundreds of thousands of dollars. So, funding is needed too. The Ultimate DXpedition is not a task to be taken lightly, and certainly not without some experience.

OPERATING PERMISSION AND DOCUMENTATION

Operating permission is most certainly needed for these destinations. In many cases it consists only of landing permission. For example, some French and American possessions have limited access, but no need for a special license. These destinations will require special landing permission, and it must be in writing from the proper agency. This may be a difficult task, but it is not insurmountable, as proven by the fact that it has been done before. But it may take many letters and a few visits before the goal is accomplished. Nothing difficult is ever easy.

The copies of operating and landing permissions will document that the DXer had permission to be there for the purpose of Amateur Radio. There is a need, however, to further document presence at the location. Documents may consist of transportation receipts, ship's logs or certifications signed by the captain, and pictures of the operators at known locations. Postcards mailed from the location, or, if that's impossible, from the nearest port to the DXCC Desk also help document the operation.

ADVANCED TRANSPORTATION

The DXer will find that getting to the Ultimate DX location requires a bit more ingenuity than a trip to Aruba. These locations are mostly exotic because they're hard to reach! They will require special charter flights or boat transportation. The DXer should use care in choosing the provider. It's best to find someone who has used transportation in the area before, and ask them about a provider and the quality of service provided. In reading the details of past DXpeditions, the DXer may find that the same boat was used by several different DXpeditions. Charter flights are usually very difficult to arrange. Distances involved and lack of landing usually preclude the use of an airplane at all, although DXpeditions to both Willis Island and Annobon have on occasion used airplanes. But in reality, the DXer should plan a long boat trip.

LIFE SUPPORT

Whether in the Antarctic or a desert island in the Pacific, life support for the DXpedition team is very important. The shelters chosen should provide shelter from whatever elements may be found at the DXpedition site. For polar regions, this means shelters that can protect against cold without being blown apart by high winds. Some form of heating may be required, and exposure suits for the operators may also be essential.

In hot, sunny areas, it's necessary to protect against wind, rain, and sun. It should be stressed that long term exposure to the sun is harmful, and so the operators should take precautions to prevent permanent damage. Prevailing winds may bring salt spray to damage equipment, and provisions should be made to block any spray that may appear.

Food is always a consideration. Meals should be planned to make sure that there are adequate provisions not only for the DXpedition period itself, but also for the trip both ways, and a several day safety factor should be allowed, in the event the group is trapped on the island due to weather conditions.

The Ultimate DXpedition will take place days from civilization. Medical aid will not be available for the team members should they need it. Exposure to hazardous plants, fish, and animals is possible. Therefore it is a good idea to choose a doctor as part of the operating crew when possible. He will know what medications to take for the area and potential health threats involved.

EQUIPMENT

The Ultimate DXpeditioner should do as much as he possibly can to ascertain the conditions he may find at the operating site, and plan for any possibility. This is especially true in the area of antennas, where selections should be made based upon what may be carried and landed for their support.

Antennas at this level should consist of beams, as many as can be found for the number of stations, along with a spare. Masts will have to be taken to get them into the air. Often this can consist of heavy duty TV masts with multiple guy wires. A TV mast will allow up to a 20-meter beam to be placed 30 to 40 feet high when properly installed. Taking an aluminum step ladder is necessary, as the masts need to be pushed up vertically. A rotator is not necessary, as the antenna may be rotated by a rope tied to the director end of the boom, and tied off in the desired direction. A DXpedition isn't required to rotate an antenna as often, since they *are* the DX, and don't have to look for it. Remember that all peripheral equipment, such as coax, connectors, and soldering equipment will have to be carried.

Amplifiers will be needed. The Ultimate Expedition won't be back here for five to ten years, so needs must be satisfied. The only way to do that satisfactorily is to be *loud*. The amplifier chosen should provide the necessary output, and be flexible about running on the varying voltages sometimes found with a generator. Sophisticated control systems sometimes won't take this type of service. It is recommended that an amplifier taken on a DXpedition be tested with its generator. Any other electrical devices to be used with that generator should also be operating during the test. Simpler is often better when it comes to DXpedition choice.

Since the DXpedition will use generators, it is best to select those with adequate capacity for the job at hand. Figure maximum current requirements for the equipment to be used with a particular generator, and add a safety factor. Generators should use diesel fuel, as the transportation operator will feel better about carrying this than more volatile gasoline. Lack of spark-plug noise is also nice. You don't want to import QRN!

PACKING

The Ultimate DXpeditioner may decide that the best way to get the equipment to the embarkation point is by shipping. The equipment should be well packed so that it won't be destroyed by shipping, and sent to an agent at the embarkation point. It should be sent well in advance so that it will be in place before the DXpeditioner gets there. It is possible to take smaller loads as checked baggage, but usually only if one can get the gear in hand between airplane transfers. Checking through to final destination is sometimes risky. It's hard for an airline to bring lost baggage out to Heard Island if it should go astray.

Clothes should be taken with a view of the needs at the DXpedition site. For the Pacific, bathing suits and shorts may be the order of the day. For cold regions, one may change clothes every several days rather than daily. Pack accordingly.

If any equipment is to be carried onto an airplane, it should be items such as lap top computers, keyers, headsets, etc, that would be needed by the individual operator. Don't forget the camera, and books are nice for airplanes and boats.

QSLING

For DXers who are pursuing awards, QSL cards are almost as important as the contact itself. The QSL has been a part of Amateur Radio and DXing for as long as radio itself has existed. It is the QSL that actually proves the contact. As many have found, pirates and bootleggers enjoy having a good time, and the DXer would never know if the station was legitimate without the attempt to get a QSL. While electronic QSLs may be on the horizon, the majority of DXers currently show strong support for the old-fashioned paper QSL, so it seems likely that QSL cards will be with us for a while longer.

OUTGOING

Yes, the DXer wants to receive that all-important piece of cardboard. However, in order to receive, one must first send. Strangely enough, most Rare DX stations are not eagerly waiting for the DXer's QSL card. They already have a few, all bands, all modes. So, it is up to the DXer to present his card in such a way that it will get returned. This is very important, as otherwise the card might as well be going into a Black Hole.

First, the card itself should be designed for maximum ease of the DX station or his manager. After processing 20 to 30,000 QSLs, it's easy to understand how a manager could get a little upset with having to check around the card, or get a case of carpal tunnel syndrome from turning a card over to check out both sides. QSL card design, while not making the chore totally painless, can at least prevent the manager or DX station from learning the DXer's call sign in a negative context! So, the QSL card should have all the information on one side only. It should be easy to read, and in a logical order.

There is certain information required on a QSL card that is to be submitted for DXCC credit. Confirmation for two-way communication must include the call signs of both stations, the DXCC country, mode, date, time and the frequency band used. Desirable information includes the county and grid square. If the DXer lives on an island, an Islands on the Air (IOTA) identifier number is also desirable. This way, if the DX station happens to be pursuing awards of his own, he will be able to use the DXer's QSL card for his own purposes.

The card should have the information printed plainly on the card. After seeing a few of the optical illusion cards, or ones with overly embellished lettering, it's easy to understand how confusing this can be. Plain lettering on a plain card is much better. If a picture card is desirable, it may be better to have a plain card printed on one side, and the picture card on the other. Print your call on both sides of the card. That way the card can be displayed by the DX station, but processed rapidly as well.

There are three ways to get a card to its destination: By bureau, by QSL service, and direct mail. Each has its advantages, although there are some differences in the speed. There's also some differences in how a DX station will handle them. The method the DXer uses depends upon his personal requirements.

The ARRL Outgoing QSL Service (see the end of this chapter) provides economical service to the countries that have incoming bureaus. At $6 per pound, this is indeed a bargain for those who don't mind waiting for a QSL. While slow, with turnaround time sometimes exceeding a year or more, still this may be considered an efficient and cost effective method for QSLing. This is especially true when compared with pouring money down black holes with direct QSLs to some parts of the world. The disadvantages are that in some countries served by bureaus, only cards to members are delivered. This method is highly recommended when QSLing North and South America, Japan, Europe and the CIS countries in particular.

To use the bureau, the cards should be sorted in country order. Cards to managers should have the call of the manager on both the front and back side of the card. Turn the card over and put the call sign of the station you want it to go to on the back in block letters. This makes it easier for the sorters at either end.

Next up the scale is a QSL service, such as the WF5E QSL Service. For only a small amount per card, the QSL service will send the cards by the bureau, direct, or to a manager. Cards are returned via the DXer's own incoming bureau unless special arrangements are made. The DXer will notice the cards coming back this way, as they are usually marked with the service's stamp. Turnaround time is improved as cards are sent to bureaus in smaller quantities, and cards are sent to managers with return postage. In some cases, it isn't even necessary to know the manager, as the QSL service keeps track of managers, and will get the card to the right place. The success rate over time with this method is very good.

Direct QSLs are used by many. The caveat here is that QSLing direct isn't cheap, so be prepared to shell out some money if this method is used. Presentation is worth a lot too, so one must be prepared to go the whole way for any chance

William Wu, BV2VA, running BV9P from Pratas Island in 1995. Operators from the CTARL operated for 10 days. *(Photo courtesy of BV2VA)*

of a return. This method is generally best not used for resident amateurs if a bureau is available. The bureau is often more convenient for resident amateurs.

Direct QSL returns start with a good address. Sometimes the station will give his complete address over the air. If not, then *The Radio Amateur Callbook*, the Buckmaster CD ROM (available from the ARRL), or *QSL Routes* may have the information. *QRZ DX* often has newer addresses for stations who have recently been active. It is best to get the complete name of the operator, as it is always best that call signs *not* be placed on the envelope.

The envelopes themselves are important. Other countries do not use the same envelope sizes that are commonly found in the United States. This means that using a standard US #10 envelope will create the undesirable situation of sending something through the mail that's crying "Steal me! Steal me!" It is far better to obtain the proper size envelope for the job at hand rather than trying to make do. The best size is a $4^3/_4 \times 6^1/_2''$ outer Air Mail envelope, and a $4^1/_2 \times 6^1/_4''$ inner Air Mail envelope for returns. If possible, the addresses should be typed directly on the envelope, or printed labels. It is important to remember that *no* call signs should go on the outside of any envelope.

Inside the flap of the return envelope the DXer should put his call and date, time, and mode of the QSO. In the event the card gets separated from the return envelope, it is still possible for the manager to look up the QSO and send the card. Also, either cards or a list of all QSOs made with the station, if a DXpedition, should be sent so that the manager won't have to guess about whether to provide cards for all QSOs he finds in the log. Chances are he *won't* if he's not asked, as he would then have to determine if the DXer worked him or if what's in the log is the result of a miscopied call sign.

The old saying goes "The final courtesy of a QSO is a QSL." That may have been so years ago when postcards were a penny, but that time is long past. Perhaps the saying should be changed to "It is discourteous to send a QSL card *without* return postage provisions." QSL cards aren't cheap these days, with even the cheapest around $3^1/_2$ cents each. As this is written, inside the United States it costs 33 cents for mailing in an envelope or 20 cents without. Multiply this by a thousand, and you'll find that QSLing could easily cost $225 to $355! Make it air mail, and the price jumps even further. Then consider that United States postal rates are among the cheapest in the world, which could put the tab for a thousand QSLs without return postage well over $1,000! That's an awful lot of money for something the DX station likely doesn't need. So it's easy to see why there is poor or no response to cards received without return postage. This even applies to Stateside cards these days. A word to the wise: if you really want that exotic QSL card, be sure that return postage is provided in whatever form is necessary.

The best way to provide return postage these days is the International Reply Coupon (IRC). They may be obtained from a Post Office, and are redeemable for the lowest air mail rate in any UPU country. IRCs can often be obtained from a QSL manager in the DXer's own country. IRCs purchased this way trade somewhere between the value of new

DXers may be found anywhere, including the Great Wall of China. DXers Kan Mizoguchi, JA1BK, Petri Laine, OH2KNB, and Martti Laine, OH2BH, are ready for operation from BT1X. (*Photo courtesy of JA1BK*)

IRCs and the redemption value. At the present, in the US this would mean somewhere between 60 cents and $1.05. Usually 75 cents will get an IRC.

A caveat is that in some countries, the lowest Air Mail rate won't provide enough postage to return an envelope and QSL card. Sometimes it takes two. For example, until recently, Germany's lowest Air Mail rate provided for less weight than the average QSL card and envelope. So it became necessary to go to the next highest rate, which was 3 Marks. One IRC exchanged for the lowest amount of postage, 2 Marks. So it took two IRCs to provide enough postage. At the same time, it took about $1.25 US to exchange for 2 Marks. $1 US was not enough for return postage.

Another way to provide return postage is through the use of mint stamps from the DX station or QSL manager's country with sufficient value to provide return postage. This allows the manager to fill out the card and put it in the envelope with no trip to convert money or IRCs, purchase postage, and affix it to the envelope. It also removes the temptation to keep the IRCs or money and think about returning the card later. If this method is used, it's best to put the postage on the return envelope so that it may only be used for returning the DXer's cards.

The least effective, but most often used method of providing return postage is the ubiquitous Green Stamp, the United States one dollar bill. There are several problems associated with using Green Stamps, not the least of which is mail pilferage. Even mail passing through the US is not immune to it, and in some countries if the envelope is identified as going to an Amateur Radio operator, it is almost certain to disappear. One card sent as registered mail to the Middle East was returned as undeliverable. It was still sealed, but the $5 bill enclosed had been removed! This is often the fate of the Green Stamp. In addition to exchange problems at the country of destination, there are some countries around the world where hard currency is so controlled it is just flat illegal to have it. Two countries that quickly come to mind where this is so are South Africa and India. There are many others.

In other countries, Green Stamps trade much like IRCs; that is, they are not exchanged at all, but traded for the Amateur's own QSLing needs. They are often sent with visitors from the US back to the States to buy US postage and forward the cards on from there. One Green Stamp no longer will buy a sufficient amount of postage in many countries. So two are required if the DXer expects to get a return. However, they are convenient. For most DXers, a reach into the pocket is all it takes. So, the Green Stamp is used mostly for convenience. The price paid may well be loss of the card.

With the QSL card, return postage in whatever form, and envelopes, the DXer is ready to prepare the QSL for mailing. First be sure the QSL is completed properly, with all QSOs made with the station listed on the card. The card should then be inserted into the envelope, which should have the DXer's return address typed or printed on it. Be sure that the country name is the bottom line of the return address. If a Green Stamp or IRC is used, insert that into the return envelope. The idea is to make as flat a package as possible. If the card is going to an area of high humidity, putting a piece of waxed paper under the return envelope flap can prevent it from sticking together before it is used. All this should then be placed into the outer envelope and sealed. Remember, no call signs or references to Amateur Radio should be on the outer envelope. With some luck, the DXer will see his desired rare QSL coming back to him in a few months.

One word about QSL turnaround: Many DXers have much too high an expectation about how fast they should receive QSLs. It depends upon the operation and whether a special effort is put into quick turnaround, but six months is very reasonable for QSL return. Whereas the Mega-DXpedition may have a small army of helpers to spring into action to complete the QSLing, most smaller DXpeditions depend upon one individual. Looking at an operation that made 10,000 QSOs, and allowing 3 minutes per card for opening, finding and checking the log entry, writing or attaching a label to the card, inserting the card and sealing the envelope, and obtaining and affixing stamps, figure on the manager handling 100 cards per day. That's five hours per day for 100 days without a break! If the manager also had a full time job and family, the DXer can see where he might want a little time off from this routine. The DXer should allow a minimum of at least six months before even thinking about a second QSL, even if his friends are receiving theirs. His card may have been near the bottom of the pile. Also, the DXer shouldn't send a request via the bureau at the same time as sending the direct card. QSL managers note all cards received and sent in the log. When they find a DXer using this practice, they make a note of the call sign. The next time, that DXer's card often goes out with the last batch mailed, or through the bureau.

INCOMING

Incoming cards are easy: the DXer just needs to have a mailing address! However, he also needs to be prepared to receive cards by the incoming bureau. Each call district in the US has its own Incoming QSL Bureau. The DXer should send cards or envelopes with postage to his own bureau, according to the rules established by the bureau in his district. DXers should also note that the bureau depends upon their call sign, not upon where they live. *With the Vanity Call system allowing choices of calls anywhere within the 48 states, some DXers living in Arkansas will now find that they should be receiving cards from the W9 Bureau in Illinois.* The latest addresses of the various bureaus can be found at the back of this chapter, and they are printed periodically in *QST*.

Coming back from the mailbox with a new batch of QSL cards is one of the joys of DXing. Following these instructions should assure the DXer that his chances of making many happy trips are increased.

PROPAGATION

While this was stated before, it is worth saying again: Whatever the propagation is, *It Doesn't Matter!* The DXpedition will still start on time, the contest will arrive on its appointed weekend, and the DXer will be there to work it, whether there are 50 spots on the sun or none. Yet it is good to know how each band works in order to take advantage of what is offered. Herein is offered a band-by-band breakdown of what to expect.

160 METERS

Often called the "Top Band," 160 offers one of the great DX challenges. Openings are sometimes very short in duration, and usually provide shorter distance openings than other bands. During high sunspot levels, the band is open, often with greater intensity. But the sun's activity causes an increase in noise level, and absorption of signals increases as the D layer,

DXpedition operating contrasts with operations from home, but Karl, PS7KM, enjoys the 1997 action from ZYØSK. ***(Photo courtesy of PS7KM)***

which forms at sunrise and begins to dissipate at sunset, is more intense in years of high solar activity. This effect is somewhat lessened during times of low sunspot activity, but for the most part, 160 meters is still a nighttime band.

Propagation on 160 begins in the late afternoon to early evening, in the direction of the approaching darkness. As stations in the east approach their sunrise, there will likely be a short but noticeable increase in their signal strength. This is the opportunity to work the longest distances, and the greatest effect (*gray line*) occurs when it is sunrise at one end of the path, and sunset at the other. Propagation continues to flow east to west, as the DXer follows the sun. As the sun rises, the DXer can expect to work stations to his west until they finally sink into the noise as the D layer begins to absorb the signal again. See the Antenna Orientation chapter of this book for information on software that generates gray line path plots.

Also worth more than a little mention are the aurora belt zones that occur around the magnetic poles. Charged particles discharged by the sun are attracted by the magnetic poles, and create a dense layer that attenuates any signals passing through it. Of course, the longer the wave, the greater the attenuation. Thus, low band signals are affected more than others by the aurora belts. This will be apparent on northerly paths, as stations in the Yukon and Alaska often report one way propagation where they can hear stations to the south, yet are not able to communicate with them.

Due to shorter periods of sunlight and less thunderstorm activity, 160 meters provides better propagation in the fall and winter. However, one should not overlook possible openings to the Southern Hemisphere in the summer, as the seasons are reversed, and those in the Southern Hemisphere are experiencing their best propagation of the year.

80 METERS

Many of the same phenomena that affect 160 meters also affect 80 meters. The aurora belts, D-layer absorption, and thunderstorm activities also have limiting effects on this band. It usually begins to open in the early afternoon. The gray line is especially useful at the beginning of the opening, but will require a good antenna system and power to be able to take full advantage of it. The band opens to the east, and the gray line stations will often come in from the southeast. The opening will continue toward the east until after sunrise at the eastern end of the path, then to the south, finally ending as sunrise approaches with westerly openings.

Eighty meters is seasonal, with better conditions in the wintertime, but with DX still available in the summer. In times of low sunspots, 80 (or 75, as the phone part of the band is commonly called) becomes the prime nighttime phone band, as 20 closes and 40 turns into a morass of broadcast stations. While less used in times of high sunspots, it's still open, although usually later in the evening.

XF4CI took place from this small shack on Socorro Island in the Revillagigedo group. This is the second operating location. (*Photo courtesy of XE1CI*)

40 METERS

If 20 meters is the King of DX bands, then 40 meters surely must be the Queen. While affected by many of the same conditions as 80, the D-layer absorption problem is reduced because of the shorter wavelength, and it becomes possible to work intercontinental DX during the late afternoon. In fact, at the sunspot minimum, it's possible to hear DX all day long, and some of it may be worked if the signals are not absorbed too badly.

In the late afternoon, the band is likely to provide good long path openings to the southeast. Openings are again basically toward the east until well after eastern sunrise. Afterward, the opening will swing south, then toward the west as sunrise approaches. An opening to the southwest is very common in the morning, and stations that would normally be found with a northwest Great Circle path will be found there, with no opening at all to the northwest! An example of this is the USA to Hong Kong path, which will often be open at a heading of 210-225°, rather than the expected 330°.

30 METERS

If ever there was a DX band that provided 24 hour coverage at most stages of the sunspot cycle, 30 meters would be it. It shares many of the characteristics of both 20 and 40 meters, with long nighttime openings, while providing DX throughout most of the daytime as well. Propagation is hard to describe because of this, as 30 meters can almost be open to anywhere at anytime as the seasons change. The DXer should check this band frequently for some pleasant surprises.

20 METERS

The undisputed King of the DX bands is 20 meters. For long distance propagation at any time of the sunspot cycle, 20 will be the band of choice. During low sunspots, it's likely to be the *only* band open during the daytime. At sunspot maximums, 20 will be open to somewhere almost 24 hours each day. Almost every form of propagation can be found here.

During the winter, 20 will open to the east at sunrise. The opening will last until stations at the eastern end of the path are past sunset. In the early afternoon, the opening will extend to the south and to the southwest. Early evening should find the band opening to the west to northwest. In the spring and fall, long path propagation exists, with opening to the east coming from a southwesterly to western path just

DXing, DXpeditioning and . . . Contesting!

This sidebar was written by Wayne Mills, N7NG, at the combined request of the chapter author and this book's editor. Wayne has been a DXer and contester for over 40 years. He is currently at the top of the DXCC Honor Roll and holds WAZ and 5BDXCC awards as well. After participating in contests for many years from home, Wayne began his ventures to offshore locations in the late eighties. Since then he has been involved in numerous world-class contest operations including several world records in multi-operator classes and has a single operator world record as well. Since 1985, he has participated in over 15 major DXpeditions to every continent of the world, including ZA1A—the operation that launched the Amateur Radio service in Albania, and significant operations from Myanmar, Pratas Island, Scarborough Reef, and the Temotu Province of the Solomon Islands. Wayne was elected to the CQ DX Hall of Fame in May 1999.

This chapter offers excellent insight into how to become a seasoned DXer. Learning, however, is best accomplished by doing. Practice with the lessons of this chapter can be obtained by participating in contests. Contesting offers a great opportunity to practice DXing skills. Contesting can also point to ways to improve your station. DXing is an art, which requires considerable skill, skill that often is acquired only after years of practice, practice, and more practice. The more you work at it, the more proficient you will become. Contesting can help the DXer gain important operating skills and increased station effectiveness.

Many years of experience as a DXer and contester convinces me that this is true. My contesting experiences began shortly after I began DXing. Contests created many opportunities to practice chasing relatively non-rare DX in a short period of time. The skills acquired in the course of many contest efforts have made DXing easy for me. Emulating the contester who is successful leads to greater efficiency on your part. Listening is a particularly important ability. Paying attention to what is happening and deciding how to approach a pileup is particularly important. Is the DX working split? Is there a distinctive listening pattern? Where and when should I call next? In major pileups, it is easy to spot those who lack the techniques to be successful. I credit contesting with teaching me the timing, placement and rhythm to be a successful DXer.

From home, contesting offers many opportunities to snag a rare DX station in a major-league pileup. You may be competing with some of the largest stations in the country, so it's not necessarily easy. However, this is great practice for getting through to the very rare DX, and it's also a good gauge of your abilities and your station's capabilities. At the same time, pileups in a contest are not always as large and as unruly as those on the rarest of DX stations. Because there are so many DX stations in a contest, many DXers are off chasing one or the other, leaving this one for you.

Success in contests requires a high quality station. The stronger signal you can deliver to the DX station, the better your chance to bust the pileup. Big power

after sunrise, and lasting for several hours. In the evening, the path opens to the southeast. It is not unusual to find long path openings to areas of the world that are actually stronger than the short path opening.

Summertime often finds openings to the Far East in the mornings, while European openings continue through the early evening and on into the night. There is always a southern path, extending well into the night after other paths have closed. The band may seem to close, then open up again later, around midnight, on certain paths.

This single band is so productive that more than a few serious DXers restrict their antenna choice to one large monoband 20-meter antenna. All DX passes through 20 meters at some point; it's impossible to be a truly effective DXer without it.

17 METERS

Another of the newer bands, 17 meters hasn't been with us for a full sunspot cycle yet. It has many of the characteristics of 15 meters, but due to its lower frequency it's open more often, especially during times of low sunspots. Again, look for morning openings from the east, and evening openings to the west.

15 METERS

This is somewhat of a transition band. It has many of the characteristics of 20 meters at times, including possible long path openings, and in addition some of the characteristics of 10 meters. It often provides very long distance openings, and would be a good choice for working Africa in the afternoon during times of moderate to good solar activity. At times of low sunspots, it may not open at all on east-west paths, but is usually open to the south.

Typically 15 opens after sunrise, if it opens at all. At the bottom of the sunspot cycle, these openings are likely to be of very short duration, but during cycle peaks are likely to last from sunrise to long after sunset. Propagation begins to move noticeably from east to west, with absorption around local noon often making signal levels go down for a while, only to return later in the afternoon. The band will begin to isn't always necessary, and good antennas are often better than power, as antenna gain improves receiving as well as transmitting. A knowledge of propagation often substitutes for big power. Calling when propagation favors your area will likely result in success. All of these factors, which are crucial to successful contesting, will make you a better DXer.

Want to learn DXpeditioning? Just what exactly do you need to know to be a DXpeditioner? One of the best places to start learning is in a DX contest. Contesting offers the opportunity to operate on the other end of the pileup. As a DXpeditioner, you need to learn how to dig callsigns out of big pileups while maintaining control and minimizing the impact on the band. You need to know how to work specific regions when everyone from everywhere is wanting to call you. Stamina, you'll need *lots* of stamina! Most successful DXpeditioners are also well-known contesters.

Elementary contesting is often learned at a local club's ARRL Field Day site. Field Day includes many of the elements of contesting and DXpeditioning. For many hams, the ARRL Field Day is their introduction to contesting. The ARRL Sweepstakes Contest is a great next step since good results can be had with a smaller station and antenna system. For the DXer, the ultimate experience is the DX contest. In addition to providing DXing experience, DX contests can add many counters to your DX total. Success in these contests may require bigger signals and a good deal of skill, but good results can still be had with a modest station.

When you feel comfortable participating in contests from home, you may be ready to graduate to participating in contests outside the 48 states. At some point, many DXers wonder what it would be like to operate at the other end of a pileup. With travel to distant points on the globe easier than ever before, the opportunity to operate at "the other end" is a reality for many DXers. If you would like this type of activity, why not take your wife on a vacation to the Caribbean and spend a weekend working a DX contest. You will probably find yourself inundated with callers in a way you have never have heard. Read some of the literature beforehand, though. Such an effort will give you experience in logistics, licensing, and travel as well as DXpeditioning operating. Setting up a station in a remote hotel accommodation can be an experience. This can be a low-pressure affair designed to give you a real DXpedition experience. Have you learned the basics? When the action starts, you will soon find out!

DXpeditioning requires different types of ability. Rather than putting your signal in the right place at the right time, you must identify signals in that huge mess we call a pileup. You must not only pick out a call, but you must do it quickly, with a predictable pattern and rhythm. DXpeditioning is mostly about operating. To some extent, the name of the game is *rate*. How many stations can you work in an hour? For rare locations, it may also be about how to work as many different stations as possible. At the operating position of a DXpedition, you will usually find a DXer who is also a contester. Contesters have the operating skills required to be successful DXpeditioners. For DXpeditioning, you must also set up a station in some far-off place, often with few resources. Everything is done "Field-Day" style. Logistics and planning are very important. Most of all, however, you must be familiar with working a pileup. This is DXpeditioning.

If you are getting the idea that contesting can help in developing your DXing and DXpeditioning skills, you are right. There are two steps in contesting that can help develop these skills. The first is simply to become familiar with contest-style operating by participating in contests from home. Many of the skills required can be learned by observing how it is done. This experience will greatly enhance your DXing skills. The second step, which is the easy and fun way to learn DXpeditioning, is to actually participate in contests from a location "off shore." Contesting offers a condensation of DXing opportunities into a number of short but highly concentrated operating periods throughout the year. For intense experience, practice in breaking pileups and developing DXing skills, try contesting!

open to the west as the sun moves across the Pacific.

At times of high solar activity, 15 meters will provide openings to Asia as late as local midnight. However, at times of low solar activity, the band isn't likely to open to Asia at all, unless the DXer happens to live on the West Coast. When the bands are open, no band provides more access to exotic DX than 15 meters.

12 METERS

As you might expect, 12 meters closely resembles 10 meters. However, because of the lower frequency, it is open more often than 10, but slightly less than 15. If 15 is open, it is worthwhile to try 12. Another full sunspot cycle will allow its characteristics to be explored more fully.

10 METERS

At this time in 1999, newer hams are only now hearing 10 meters in all its glory. With a strong opening during a contest, it may well be impossible to find a clear frequency from 28.300 MHz all the way to 29.000 MHz! Band openings here are incredible, although worldwide band openings are somewhat more rare. Yet it is possible to cover the world, and with low power, when conditions are right.

Unfortunately, when sunspots are low, 10 meters stays in hibernation for years, with only an occasional flash of its greatness. That's why it is necessary to know all its propagation well. Of all the HF bands, 10 meters has more propagation modes than any other. Each has its place, depending upon the time of year, time of the sunspot cycle, or time of day.

During low sunspots, 10 meters often opens in the *sporadic E* mode. E-layer clouds form rather quickly, and often break up just as fast. While sometimes intense enough to even provide propagation on 20 meters, it is most noticeable on 10 and 6 meters. Openings can be from 1500 to 3000 miles, depending upon whether the cloud is formed well enough to permit one or two hops. Sporadic E often occurs during the summer, and sometimes in mid-winter. Openings can hold in as long as 24 hours. When this occurs, it's likely to be open all night long. Summertime openings can provide openings to Europe or the Pacific, and can occur at either low or high sunspot numbers. The best seasons are May through early August, and December.

Scatter is another form of 10-meter propagation. Using high power and high antennas, stations from 400 to 800 miles away can be worked, albeit with weak signals. Occasionally the ping of a meteor can be heard, as signal strengths jump suddenly, then fall. Scatter occurs during high sunspot or low sunspot periods.

Another form of scatter involves beaming away from the desired direction. Often called *side scatter*, this propagation mode can be used by beaming southeast in the morning, and southwest in the afternoon. At the early stages of a sunspot cycle, this is a main propagation path to Europe, open when the direct path never does. It will sometimes seem that all signals come from this one direction. When in doubt about which way to point the beam, this is the mode to try.

Back scatter is still another of the scatter modes that can be useful. Sometimes 10 meters is an extreme long distance band. At times like that, it's often difficult to work close-in stations. Back scatter can be used for these situations. To use this mode, *both* stations should point their antennas in the *same* direction. The stations then receive each other by bouncing signals back to each other. The band needs to be somewhat quiet for this to occur. As an example, suppose a station in Texas wants to work a station in Mexico. The Mexican station is working the East Coast or Europe. The Texan's best chance is to turn his antenna in the direction of the stations being worked by the Mexican station.

During high sunspots, the preferred propagation mode is the F layer. This produces the best long distance propagation of all. Due to the nature of 10 meters and the long skip involved, there is less interference from stations on the same frequency. This can be deceptive, as the DXer is often beaten in pileups by stations he can't hear at all! Still, this is preferable to hearing the pileup hammer his ears.

North-south trans-equatorial propagation is available at almost any time in the sunspot cycle. It is possible to work Africa, New Zealand, Australia, or South America at almost any time of the sunspot cycle. Paths south of 90° or 270° are open weakly throughout most of the sunspot cycle.

With its very unique characteristics, 10 meters gets attention from amateurs of all interests. When 10 meters is open again, most of the DX will be here most of the time.

6 METERS

A DX band? Yes! It can provide some of the most fun DX of any band. Again, many modes of propagation are available, but sporadic E is probably the most popular. Occurring in the spring and early summer, and again in December, sporadic E can provide DX contacts up to 4000 miles, and sometimes more.

This band is the lowest practical frequency for EME (moonbounce) operations. Stations utilizing four or more high gain antennas and high power have been able to communicate internationally even though the band is closed for other propagation modes. At present over 200 DXCC awards have been issued for 6 meters. No longer a curiosity, 6 meters is a legitimate DX band.

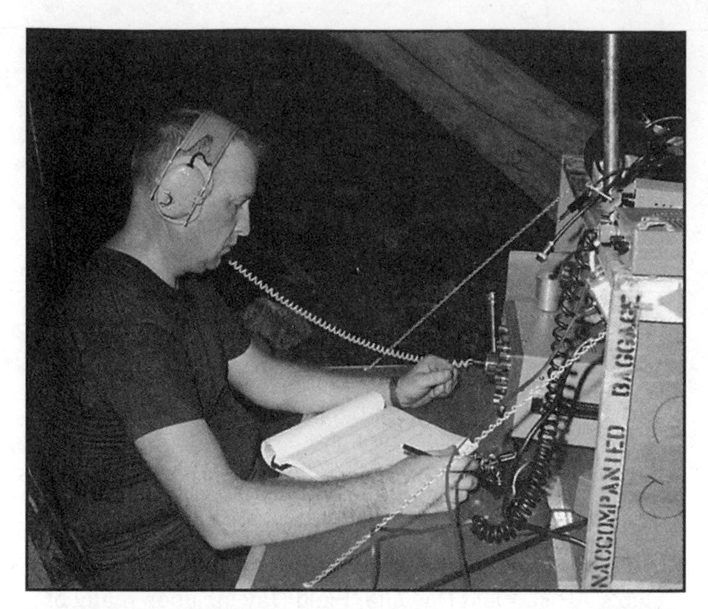

Dave Bowker, WØRJU (now K1FK), operates the night shift from WØRJU/KP1 on Navassa Island. He's using a unique shipping crate which turns into an operating desk. (*Photo courtesy of NØTG*)

2 METERS

Two meters is also a DX band. Through the use of meteor scatter, many Europeans have earned credits for their Mixed DXCC on 2 meters before upgrading their license and finishing up on the HF bands. However, most DXers who have earned DXCC on 2 meters did so through the use of EME. Over 150 DXCC entities have been worked on 2 meters.

PICK A BAND, FROM 160 TO 2

This is not a scientific guide to propagation on the DX bands. It is only to alert the DXer to the possibilities of the DX bands and how to use them. The DXer should note the time of day, solar activity, and the direction of the DX, and make a band choice that may give him an opportunity to find the DX. No attempt has been made to fully explain any phenomenon, only to point out that it does exist, and the DXer should adjust his plans to the possibility. For example, he may have to calculate the sun's position at the other end of the desired path in order to make a band choice. The important thing to remember is that the DXpedition doesn't usually sit at home waiting for the sun. So propagation doesn't matter; how to use it does.

RESOURCES FOR THE DXER

The following are some of the resources available to DXer over the Internet:

The Arkansas DX Association home page: **http://www.qsl.net/ad5xa/**. This page, maintained by Wayne Beck, K5MB, has links to many DX resources online. The DXer may also make Telnet connections to many PacketClusters worldwide. This may be the only URL the serious DXer will ever need.

ARRL Home Page: **http://www.arrl.org/**. Source of information about ARRL award programs, including DXCC

The beauty of a summer day in the South Orkney Islands greets the operators of LU6Z in 1996. (*Photo courtesy of GACW*)

Rules and forms. Check the members-only page for DXCC monthly listings.

WF5E DX QSL Service: **http://www.qsl.net/wf5e/**. The story of the WF5E QSL Service, and how to use it.

The DX Notebook: **http://www.dxer.org/**. Links to much DX related information. If it wasn't on the ADXA page, it'll likely be here.

The Top Band Reflector: **topband@contesting.com**. This reflector deals with 160 meter information, including DX, antennas, propagation, and conditions. Moderator W4ZV does well in keeping threads under control.

QSLing resources:

QSL Routes
Theuberger Verlag GmbH
PO Box 73
10122 Berlin
Germany
20 IRC or US $20 for 1997 edition
Air mail 30 IRC or US $30.

One of the best resources for the serious QSLer. This book contains routes for many of the very old QSL cards. An ideal source for tracing operators from long ago.

DX News Service:

Bernie McClenny, W3UR
Editor of *The Daily DX* and *How's DX?*
3025 Hobbs Road
Glenwood, MD 21738
Fax 301-854-5105
http://www.dailydx.com

QRZ DX
Box DX
Leicester, NC 28748-0249
Phone/Fax 828-683-0709
E-mail: **qrzdx@dxpub.com**
http://www.dxpub.com

Envelopes and Mint Stamps:

James E. Mackey, PO Box 270569, West Hartford, CT 06127-0569

William Plum, 12 Glenn Road, Flemington, NJ 08822, Fax 908-782-2612

Both of these sources provide mint stamps and European Air Mail inner and outer envelopes.

The ARRL Outgoing QSL Service

Note: The ARRL QSL Service cannot be used to exchange QSL cards within the 48 contiguous states.

One of the greatest bargains of ARRL membership is being able to use the Outgoing QSL Service to conveniently send your DX QSL cards overseas to foreign QSL Bureaus. Your ticket for using this service is proof of ARRL Membership and following the fee schedule below. For those of you who are not quite so DX active, you can send 10 cards or less for just $1.00. You can't even get a deal like that at your local warehouse supermarket! And the potential savings over the substantial cost of individual QSLing is equal to many times the price of your annual dues. Your cards are sorted promptly by the Outgoing Service staff, and cards are on their way overseas usually within a week of arrival at HQ. Approximately two million cards are handled by the Service each year!

QSL cards are shipped to QSL Bureaus throughout the world, which are typically maintained by the national Amateur Radio Society of each country. While no cards are sent to individuals or individual QSL managers, keep in mind that what you might lose in speed is more than made up in the convenience and savings of not having to address and mail each QSL card separately. (In the case of DXpeditions and/or active DX stations that use US QSL managers, a better approach is to QSL directly to the QSL manager. The various DX newsletters, the GOLIST QSL manager directory, and other publications, are good sources of up-to-date QSL manager information.)

The gang's all here on Sable Island, CYØXX, in 1996. (*Photo courtesy of WA4DAN*)

As postage costs become increasingly prohibitive, don't go broke before you're even halfway towards making DXCC. There's a better and cheaper way—"QSL VIA BURO" through the ARRL Outgoing QSL Service!

HOW TO USE THE ARRL OUTGOING QSL SERVICE

1) Presort your DX QSLs alphabetically by parent call-sign prefix (AP, C6, CE, DL, ES, EZ, F, G, JA, LY, PY, UN, YL, 5N, 9Y and so on). NOTE: Some countries have a parent prefix and use additional prefixes, i.e., G (parent prefix) = M, 2E, 2I, 2M, 2W . . . When sorting countries that have multiple prefixes, keep that country's prefixes grouped together in your alphabetical stack.

Addresses are not required. DO NOT separate the country prefixes by use of paper clips, rubber bands, slips of paper or envelopes.

2) Enclose proof of current ARRL Membership. This can be in the form of a photocopy of the white address label from your current copy of *QST*. You can also write on a slip of paper the information from the label, and use that as proof of Membership. A copy of your current Membership card is also acceptable.

3) Members (including foreign, QSL Managers, or managers for DXpeditions) should enclose payment of $4.00 for the first 1/2 pound of cards or portion thereof—approximately 75 cards weigh 1/2 pound. $8.00 for one pound, the fee rate increases at the rate of $4.00 for each additional 1/2 pound (i.e. a package containing 1 1/2 pounds of cards should include the fee of $12.00 and so on). A package of only **ten (10)** cards or fewer sent in a single shipment costs only $1.00. Please pay by check (or money order) and write your callsign on the check. Send "green stamps" (cash) at your own risk. **DO NOT** send postage stamps or IRCs. (**DXCC** credit **CANNOT** be used towards the QSL Service fee.)

4) Include only the cards, proof of Membership, and fee in the package. Wrap the package securely and address it to the ARRL Outgoing QSL Service, 225 Main Street, Newington CT 06111.

5) Family members may also use the service by enclosing their QSLs with those of the primary member. Include the appropriate fee with each individual's cards and indicate "family membership" on the primary member's proof of Membership.

6) Blind members who do not receive *QST* need only include the appropriate fee along with a note indicating the cards are from a blind member.

7) ARRL affiliated-club stations may use the service when submitting club QSLs by indicating the club name. Club secretaries should check affiliation papers to ensure that affiliation is current. In addition to sending club station QSLs through this service, affiliated clubs may also "pool" their members' individual QSL cards to effect an even greater savings. Each club member using this service must also be a League member. Cards should be sorted "en masse" by prefix, and proof of Membership enclosed for each ARRL member.

RECOMMENDED QSL CARD DIMENSIONS

The efficient operation of the worldwide system of QSL Bureau requires that cards be easy to handle and sort. Cards of unusual dimensions, either much larger or much smaller than normal, slow the work of the Bureaus, most of which is done by unpaid volunteers. A review of the cards received by the ARRL Outgoing QSL Service indicates that most fall in the following range: Height = 2 3/4 to 4 1/4 in. (70 to 110 mm), Width = 4 3/4 to 6 1/4 in. (120 to 160 mm). Cards in this range can be easily sorted, stacked and packaged. Cards outside this range create problems; in particular, the larger cards often cannot be handled without folding or otherwise damaging them. In the interest of efficient operation of the worldwide QSL Bureau system, it is recommended that cards entering the system be limited to the range of dimensions given. [Note: IARU Region 2 has suggested the following dimensions as optimum: Height 3 1/2 in. (90 mm), Width 5 1/2 in. (140 mm).]

COUNTRIES NOT SERVED BY THE OUTGOING QSL SERVICE

Approximately 260 DXCC countries are served by the ARRL Outgoing QSL Service, as detailed in the *ARRL DXCC List*. This includes nearly every active country. As noted previously, cards are forwarded from the ARRL Outgoing Service to a counterpart Bureau in each of these countries. In some cases, there is no Incoming Bureau in a particular country and cards therefore cannot be forwarded. However, QSL cards can be forwarded to a QSL manager, e.g., 3C1MB via (EA7KF). The ARRL Outgoing Service cannot forward cards to the following countries:

Prefix	Country	Prefix	Country
A5	Bhutan	TY	Benin
A6	United Arab Emirates	V6 (KC6)	Micronesia
D2	Angola	VP2E	Anguilla
J5	Guinea-Bissau	VP2M	Montserrat
KHØ	Mariana Is.	XU	Kampuchea
KH1	Baker and Howland Is.	XW	Laos
KH4	Midway I.	XZ (1Z)	Myanmar (Burma)
KH5	Palmyra and Jarvis Is.	YA	Afghanistan
KH7K	Kure I.	ZD9	Tristan da Cunha
KH8	American Samoa	ZK1	North and South Cook Islands
KH9	Wake I.	3CØ	Pagalu I.
KP1	Navassa I.	3C	Equatorial Guinea
KP5	Desecheo I.	3W, XV	Vietnam
P5	North Korea	3X	Guinea
S7	Seychelles	5A	Libya
SU	Egypt	5R	Madagascar
T2	Tuvalu	5T	Mauritania
T3	Kiribati	5U	Niger
T5	Somalia	7O, 4W	Yemen
T8	Palau	7P	Lesotho
TJ	Cameroon	7Q	Malawi
TL	Central African Republic	8Q	Maldives
TN	Congo	9N	Nepal
TT	Chad	9U	Burundi
		9X	Rwanda

Countries that currently restrict the forwarding of QSL cards to anyone other than members of that country's national radio society include the following:

France	Germany	Japan	Monaco
Morocco	Poland	Portugal	

ADDITIONAL INFORMATION:

We no longer hold cards for countries with no Incoming Bureau. Only cards indicating a QSL manager for a station in these particular countries will be forwarded.

When sending cards to Foreign QSL Managers, make sure to sort these cards using the Manager's callsign, rather than the station's callsign.

SWL cards can be forwarded through the QSL Service.

The Outgoing QSL Service **CANNOT** forward stamps, IRCs or "green stamps" (cash) to the foreign QSL bureaus.

Please direct any questions or comments to the ARRL Outgoing QSL Service, 225 Main Street, Newington CT 06111-1494. Inquiries via email may be sent to **buro@arrl.org**.

THE ARRL INCOMING QSL BUREAU SYSTEM

PURPOSE

Within the U.S. and Canada, the ARRL DX QSL Bureau System is made up of numerous call area bureaus that act as central clearing houses for QSLs arriving from foreign countries. These "incoming" bureaus are staffed by volunteers. The service is free and ARRL membership is not required.

HOW IT WORKS

Most countries have "outgoing" QSL bureaus that operate in much the same manner as the ARRL Outgoing QSL Service. The member sends his cards to his outgoing bureau where they are packaged and shipped to the appropriate countries.

A majority of the DX QSLs are shipped directly to the individual incoming bureaus where volunteers sort the incoming QSLs by the first letter of the call sign suffix. One individual may be assigned the responsibility of handling one or more letters of the alphabet. Operating costs are funded from ARRL membership dues.

CLAIMING YOUR QSLS

Send a $5 \times 7^1/_2$ or 6×9 inch self-addressed, stamped envelope (SASE) to the bureau serving your callsign district. Neatly print your callsign in the upper left corner of the envelope. Place your mailing address on the front of the envelope. A suggested way to send envelopes is to affix a first class stamp and clip extra postage to the envelope. Then, if you receive more than 1 oz. of cards, they can be sent in the single package.

Some incoming bureaus sell envelopes or postage credits in addition to the normal SASE handling. They provide the proper envelope and postage upon the prepayment of a certain fee. The exact arrangements can be obtained by sending your inquiry with a SASE to your area bureau. A list of bureaus appears below.

HELPFUL HINTS

Good cooperation between the DXer and the bureau is important to ensure a smooth flow of cards. Remember that the people who work in the area bureaus are volunteers. They are providing you with a valuable service. With that thought in mind, please pay close attention to the following DOs and DON'Ts.

DOs

- DO keep self-addressed $5 \times 7^1/_2$ or 6×9 inch envelopes on file at your bureau, with your call in the upper left corner, and affix at least one unit of first-class postage.
- DO send the bureau enough postage to cover SASEs on file and enough to take care of possible postage rate increases.
- DO respond quickly to any bureau request for SASEs, stamps or money. Unclaimed card backlogs are the bureau's biggest problem.
- DO notify the bureau of your new call as you upgrade. Please send SASEs with new call, in addition to SASEs with old call.
- DO include a SASE with any information request to the bureau.
- DO notify the bureau in writing if you don't want your cards.
- DO inform the bureau of changes in address.

DON'Ts

- DON'T send domestic US to US cards to your call-area bureau.
- DON'T expect DX cards to arrive for several months after the QSO. Overseas delivery is very slow. Many cards coming from overseas bureaus are over a year old.
- DON'T send your outgoing DX cards to your call-area bureau.
- DON'T send SASEs to your "portable" bureau. For example, NUØX/1 sends SASEs to the WØ bureau, not the W1 bureau.
- DON'T send SASEs or money credits to the ARRL Outgoing QSL Service.
- DON'T send SASEs larger than 6 × 9 inches. SASEs larger than 6 × 9 inches require additional postage surcharges.

ARRL INCOMING DX QSL BUREAU ADDRESSES

First Call Area: All calls**

W1 QSL Bureau
P.O. Box 7388
Milford, MA 01757-7388

Second Call Area: All calls**

ARRL 2nd District QSL Bureau
NJDXA, P.O. Box 599
Morris Plains, NJ 07950

Third Call Area: All calls

Pennsylvania DX Association
P.O. Box 100
York Haven, PA 17370-0100

Fourth Call Area: All single-letter prefixes (K4, N4, W4)

Mecklenburg Amateur Radio Club
P.O. Box DX
Charlotte, NC 28220

Fourth Call Area: All two-letter prefixes (AA4, KB4, NC4, WD4, etc.)

Sterling Park Amateur Radio Club
Call Box 599
Sterling, VA 20167

Fifth Call Area: All calls*-

W5 Incoming QSL Bureau
Magnolia DX Assn.
P.O. Box 999
Wiggins, MS 39577-0999

Sixth Call Area: All calls* +-

ARRL Sixth (6th) District DX QSL Bureau
P.O. Box 900069
San Diego, CA 92190-0069

Seventh Call Area: All calls*

Willamette Valley DX Club, Inc.
P.O. Box 555
Portland, OR 97207

Eighth Call Area: All calls

8th Area QSL Bureau
P.O. Box 182165
Columbus, OH 43218-2165

Ninth Call Area: All calls*

Northern Illinois DX Assn.
W9 Incoming QSL Bureau
P.O. Box 273
Glenview, IL 60025-0273

Zero Call Area: All calls*

WØ QSL Bureau
P.O. Box 4798
Overland Park, KS 66204

Puerto Rico: All calls*

Puerto Rico QSL Bureau
P.O. Box 9021061
San Juan, PR 00902-1061

U.S. Virgin Islands: All calls

Virgin Islands ARC
GPO Box 11360
Charlotte, Amalie
Virgin Islands 00801

Hawaiian Islands: All calls*

Wayne Jones, NH6K
P.O. Box 860778
Wahiawa, HI 96786

Alaska: All calls*

Alaska QSL Bureau
P.O. Box 520343
Big Lake, AK 99652

Guam:

Guam QSL Bureau
Marianas A.R.C.
P.O. Box 445
Agana, Guam 96932

SWL:

Mike Witkowski, WDX9JFT
4206 Nebel St.
Stevens Point, WI 54481

QSL Cards for Canada may be sent to:

RAC National Incoming QSL Bureau
Loyalist City Amateur Radio Club
P. O. Box 51
Saint John, NB E2L 3X1
Canada

QSL cards for Canada may also be sent to the individual bureaus:

VE1, VEØ*
Brit Fader Memorial QSL Bureau
P. O. Box 8895
Halifax, NS B3K 5M5

VE2
J. Dube, VE2QK
875 St. Severe St.
Trois-Rivieres, QC G9A 4G4

VE3
The Ontario Trilliums
PO Box 157
Downsview, ON M3M 3A3

VE4
Adam Romanchuk, VE4SN
26 Morrison St.
Winnipeg, MB R2V 3B4

VE5*-
Bjarne Madsen, VE5FX
Box 2860
Tisdale, SK S0E 1TO

VE6*-
VE6 Incoming QSL Bureau
Box 57205
2020 Sherwood Drive
Sherwood Park, AB T8A 5L7

VE7*-
Dennis Livesey, VE7DK
8309 112th St.
Delta, BC V4C 4W7

VE8*-
Rolf Ziemann, VE8RZ
2 Taylor Road
Yellowknife, NWT X1A 2K9

VE9, VY2-
VE9, VY2 QSL Bureau
Box 12-255
1633 Mountain Rd
Moncton, NB E1G 1A5

VO1, VO2
Rick Burke, VO1SA
Box 23099
St. John's, NF A1B 4J9

VY1
Hugh Henderson, VY1HH
P.O. Box 33062
Whitehorse, YT Y1A 5Y5

***These bureaus sell envelopes or postage credits. Send an SASE to the bureau for further information.**

+These bureaus can only accept specific sized envelopes. Send an SASE to the bureau for further information.

Many incoming QSL bureaus now have email addresses. You can find these listed on *ARRLWeb* at http://www.arrl.org/qsl/qslin.html.

♣These bureaus will not accept SASE's. Send money credits only.

Chapter 10

Contesting

The Appeal of Competition

Bill Kennamer, K5NX and Dan Henderson, N1ND
ARRL Headquarters

He hoped it would to be out there. Once again he slowly turned the VFO dial up the ten-meter band. "Not that signal," he thought. "Already worked that one," and then on the next signal... He had turned on the radio during this ARRL November Sweepstakes just to find the five states he needed. Texas was already in the log on 15 meters. Alaska and Hawaii on 40 meters had been easier than expected. Vermont on 75 meters last night left him on the edge of success.

Now all that was needed to finish his five-band Worked All States was Arkansas on 10.

After returning from church Sunday morning he checked to see if the band was open. As he tuned past 28.430 he heard a big signal and caught the end of a CQ: "AB5SE QRZ?" "A five-land call. That could be it!" he thought as his heart started to raced a bit in anticipation. Keying the microphone, he calmly spoke his call—"NUØX"—and held his breath.

The call came back. "NUØX, number 1675 Bravo AB5SE Check 93 Arkansas." Bingo, he thought, I've done it! "AB5SE QSL, my number 36 Alpha NUØX Check 66 North Dakota. 73"

"Thanks for the sweep. 73, QRZ from AB5SE."

In that simple exchange lasting no more than 20 seconds, both NUØX—a casual contester—and AB5SE—one of the stalwarts of contesting—had done something very special. NUØX had used the contest to reach his goal—the coveted Five-Band WAS from the ARRL. AB5SE had worked every multiplier available for the ARRL November Sweepstakes Contest—a "clean sweep" as it is known. By adding that contact to his log, he greatly improved his chance of finishing high, and possibly winning his Section, in this annual event.

Contests attract operators from all across the amateur spectrum. During a contest weekend you might run into a station working VHF/UHF bands trying to add to his Grid Square total for a VUCC award. Or you could find an old-timer who likes to get on the air and enjoy working good code on the ARRL's Straight Key Night. You could work a non-contester who tuned up just to enjoy making some QSOs to add to her WAS totals. Or you could even make the acquaintance of a rare DX station from Africa looking to win his category.

Some aspect of contesting appeals to almost all segments of the amateur population. There are hard-core operators who spend hundreds of hours and thousands of dollars designing and putting together a competitive station for the ARRL 160 Meter Contest. Or it could be a group of guys from the local radio club who go out and set up a portable station and generator on a mountain to operate during the ARRL Field Day. Almost all hams find something exciting in the thrill of competition.

Contesting in some form has been around from the early days of Amateur Radio. Field Day has its roots as an emergency preparedness exercise in the 1930s. The goal? Test your individual operating capability under less-than-perfect conditions. When you look at the exchanges sent during the annual November Sweepstakes, there is no doubting their link to the traffic handled by the National Traffic System. The exchange contains all of the parts of a preamble to a formal NTS message.

The International DX contest will find its heritage in the early days of radio when amateurs experimented and vied for the honor of being the first to complete transatlantic QSOs in different frequency ranges. Specialty contests such as the EME Contest or the RTTY Round-Up spring from attempts to encourage experimentation with new modes or concepts within the hobby.

Contests are held throughout the year. Looking to work some DX to build up your totals? The major DX contests are the ARRL International DX Contest, with the CW weekend in February and Phone in March, and the CQ World Wide DX Contest, which holds its Phone weekend in October and CW weekend in November. If you are chasing states for your Five-Band WAS award, you should be checking out the ARRL November Sweepstakes. The CW weekend is always the first weekend and Phone the third weekend in November.

WHY CONTEST?

Why does it always seem there is a contest on the air? Almost every IARU national society sponsors at least one event during the year. These are generally intended to achieve the same goals as ARRL contests for those national societies. They promote activity among the licensed amateurs in those countries and encourage the development of operating skills. The contest calendar included in this chapter is a good place to find the major contests which are held globally. You can also find listings of upcoming contests monthly in *QST* and linked from the ARRL Contest Branch homepage.

What inspires normally mild-mannered hams to develop a "warrior" mentality during contest weekends? There are no cash prizes or awards that lead to great fame and reward. What causes even casual operators to hunker down during certain times of the year? What motivates otherwise normal operators to plan a year in advance to build and erect new antennas, to add station equipment, and clear time to participate in these on-the-air events?

Thousands of hams in the US and around the globe spend some time during the year contesting. Why go to the effort? The answers to that are as varied as their operators. There is no single reason, but most responses will fall into one of several categories.

Honing Operator Skills

Contesting is a great way to develop and hone operating skills. Contests require accuracy, on both the sending and receiving side. Many contesters have a background in traffic handling. With modern computer techniques to verify contacts made in a contest, accuracy and thoroughness are requirements to be a top-flight contester. Where do you find these operators in emergencies? They can be found using their experience and stations to assist with emergency traffic and communications. Many of the top contesters hold ARRL Field Organization appointments, placing their talents at the disposal of communities in crisis.

If you contest on the Top Band for long you will run across Jeff Briggs, K1ZM. Hours of preparation and planning, as well as years of experience, have made Jeff a dominant force in the annual ARRL 160 Meter Contest. He operates from either his QTH in New York or the one on Cape Cod, Massachusetts.

Demonstrating Use of Amateur Frequencies

Contesting, especially on the VHF, UHF and microwave bands, is valuable as a means of demonstrating our use of frequencies and the need to preserve allocation on those parts of the radio spectrum. In today's age of frequency auctions and increasing commercial demand for spectrum, activity on the higher amateur bands provides a good argument for protecting and preserving our allocation. By encouraging the use of microwave bands during contest periods, and providing an increased, concentrated period of activity during contest weekends, contesting increases both our use and knowledge of this valuable radio spectrum.

Advancing Amateur Technological Development

Contesting has been the inspiration for some of the technological advances during recent years. Many hams today use packet as a regular part of their daily station operation. From spotting rare DX to putting out the word about band openings on 6 and 2 meters, packet radio has become an important part of amateur activity. But how many hams know that the PacketCluster was created to allow contesters to spot contest multipliers? From antenna designs and station enhancements to integration of computer technology in the shack, you can see the work of members of the contest community in many areas of technological improvements in the hobby.

It's Fun!

Yet, despite the technological advancements and operating skill development, maybe the best reason amateurs contest is the simplest: it's FUN! Whether they are running stations at a rate of 120 per hour during the ARRL International DX Contest or experimenting with earth-moon-earth communications during the annual ARRL EME Competition, thousands of hams annually use contesting as their outlet to enjoy the hobby. The invested hours of study to earn your license and the dollars you spend in building and improving your shack and equipment come back to you with the rewards of enjoying a successful contest.

TYPE OF CONTESTS

Whom do I work?

Contests generally can be categorized in three different ways. First a contest will categorize *WHOM* you may work. Some contests only allow point credit for working stations in the same country or region. These are referred to as *domestic contests*. For an ARRL domestic contest, stations may work for QSO credit other stations in the US or Canada (known in contest lingo as W/VE.) The ARRL November Sweepstakes and the various ARRL VHF QSO parties are examples of domestic contests.

Other contests allow you to work only stations outside

Listening for the weak signals of fellow EME competitors requires intense concentration. Here Alan, BA1DU (l) and Chen, BA1HAM operate their club station, BY1QH, in an ARRL EME Contest.

of your own DXCC entity. The ARRL International DX Contest is an example of what is referred to as a *DX contest*. In this event, you receive QSO point credit for working non-W/VE stations. Finally, there are some contests, such as the ARRL 10-Meter Contest, which allow you to work any station anywhere for credit.

Where Do I Work Stations?

After *whom* you may work in a contest, you then focus on *WHERE* you may work these stations. Some contests focus on high-frequency (HF) contacts while others will focus on VHF, UHF or microwave bands. No ARRL contests are held on 30, 17 and 12 meters. The only ARRL activity featuring both HF and VHF/UHF in the same event is the annual ARRL Field Day, held in June.

How Do I Work Them?

The third part of the equation will be *HOW* you work stations; that is what modes of operation are valid. Some contests, such as the ARRL November Sweepstakes and ARRL International DX Contest have separate weekends for CW and phone operation. For the ARRL Ten Meter Contest and Field Day, you work both CW and phone during the same weekend. ARRL VHF/UHF contests allow both CW and phone QSOs during the same event. Be sure to check the rules for each specific contest to verify the "who, where and how" for each contest.

One more contest type needs to be discussed. These are referred to as "specialty" contests and would include such contests as the ARRL RTTY Round-Up and the ARRL International EME Contest. The RTTY Round-Up is the annual ARRL event to promote non-CW digital operation, not just Baudot RTTY. The EME contest challenges participants to use moonbounce to complete their QSOs.

Field Day is a *non-contest* in a contest-like environment-and dear to the hearts of any ham who has ever participated. It incorporates all modes, including satellite communications, and all bands. More US licensed amateurs participate in Field Day annually than any other on-the-air operating event. It allows contacts with any station worldwide, on any mode and on all bands (except 30, 17 and 12 meters). Even diehard non-contesters seem to enjoy participating in this all-inclusive operating event.

One other specialty contest which you will frequently see are the various state QSO parties. Set up and run by groups in the various states, the goal is to encourage activity among the hams of a particular state with hams from across the globe. Some of these state QSO parties offer interesting prizes. Be a winner in the California state QSO party and you may find yourself with a bottle of vintage California wine for your efforts. Top scores in another state QSO party may receive jams, apples, or even salmon (as well as some beautiful certificates to hang on the wall).

Contests are held during all seasons of the year. ARRL-sponsored HF events are primarily held between the months of November and March, while VHF/UHF contests are scattered throughout the seasons.

Frequently you will hear the comment, "There always seems to be a contest on the air." When one considers that most IARU national Societies sponsor at least one event annually, this is probably true. The ARRL sponsors two all-band CW and two all-band Phone contests annually: the November Sweepstakes and the International DX Contest. The ARRL also sponsors three HF specialty contests: the RTTY Round-Up, the 160 Meter Contest and the 10 Meter Contest. The only ARRL-sponsored all-band, all-mode, single weekend event on the HF bands is Field Day.

THE BEGINNING CONTESTER

Like to try a contest or two? It's easy, and you will be welcomed by everyone, especially the serious participants. They are always looking for additional contacts, and most will guide you through your first contest QSOs. Don't be bashful; you'll do fine.

Reading the Rules

With so many contests, what should any ham—serious contester or casual operator—know before *lighting up the ether* and making scores of contacts during a contest?

To begin, the best thing to do is find the current issue of *QST*, and turn to the *Contest Corral* column. Look for the next contest date and notice which contest might be running on that weekend. There's usually one or two every weekend. *Contest Corral* has a synopsis of the contest rules. While it might be nice to be familiar with them later, for the beginner's purpose it's enough to chose a contest which serves his interest. For example, someone interested in DXing would chose a contest that features working DX, while a VHFer would choose a VHF contest, and someone chasing WAS or the All Counties award might choose one of the many state QSO parties. Pick one that might be interesting to you, and read the rules.

Most of the time, it is easy to determine which modes are allowable during an event. You would obviously use phone for the ARRL International DX Phone contest. But in the ARRL RTTY Round-Up, you can operate using any digital mode, from Baudot and packet to AMTOR and PSK31. In the ARRL 10 Meter contest, you can make both

CW and phone contacts, while the ARRL 160 Meter contest is a CW-only event. Being familiar with the rules will make your operating more enjoyable and help you avoid embarrassing situations on the air.

Knowing the rules also will help you with planning your operating times. When the event starts, and how long you are allowed to operate, are important components. Does the contest require *off-times* during the event? Is it a 12, 24 or 48-hour contest? Not knowing the rules beforehand has forced many operators to delete multipliers and QSOs that they worked late in a contest because they had exceeded the maximum permissible operating time under the rules.

Learning the Exchange

After you choose the contest, learn the exchange for the contest. This is the most important thing to know for every contest you may enter, no matter how much experience you might gain over the years. If you memorize the exchange, it becomes so automatic that you'll find you can do other things while sending your exchange, without even thinking about it. That can become valuable in the future. After you've learned the exchange, you're ready to go on the air.

As you begin to tune around the band, you should be hearing a few stations calling "CQ Contest" or even simply "CQ Test." It's time to dive in. Call them by sending your call, the full call, one time and one time only. Why the full call? Because sending only a partial call cuts the QSO rate of the CQing station, and is considered a breach of contest etiquette. And only one time? Because if you're heard, you'll get through with only one call. If not, you'll just be contributing to the interference, with the end effect of slowing you down in getting the contact. But wait, he's heard you, and responds with a report. Write down his report to you, while returning to him with your report. For example, if you were at W1INF, the whole sequence would go something like this:

Lamar, W9LT, tunes through the band as he employs the "search and pounce" technique for working stations during a recent ARRL 10 Meter Contest.

CQ Contest, X-Ray Zulu One Alpha
Whiskey One India November Foxtrot
W1INF, 59 49
59 08
Thanks. QRZ, X-Ray Zulu One Alpha.

In this exchange, you've just exchanged ITU zones with XZ1A during the IARU HF Championship. You're on the way to becoming an experienced contester.

Remember to log the QSO, using time and date in UTC. Then continue to tune up or down the band, chasing more contest contacts as they become available. Go ahead and get your feet wet. There's still a lot to learn, but the basic fun of contesting begins right now.

THE CONTEST CLUB

If you've tried a contest or two and liked it, you might want to consider rubbing shoulders with like-minded individuals. One of the most important phenomena available to today's amateur is the development of the Contest Club. While such clubs have been around for many years, the formation of new clubs has been on the upswing. What better way to improve your operating and contesting skills, to learn the tricks of the trade, or to gain knowledge and experience in contesting than by joining together with other amateurs with the same interests? In every region of the country you will find solid contest clubs, all working toward a similar goal: being the best at what is called in some parts of the world *radiosport*.

Just as different hams have different interests, you will find a wide variety of contest clubs, each with a special focus. Some clubs concentrate their efforts of VHF, UHF and microwave events. These clubs often form portable multi-op expeditions, or seek to have Rover stations active during the contest. They can provide technical assistance in the utilization of microwave bands during the contest, and can assist you in getting on the air for the contest.

Other clubs will concentrate on HF events, and possibly will be a joint club with DXing as an additional focus of the group. Some of these clubs specialize in DX contests, while others specialize in domestic contests. More than a few of these clubs also sponsor some specialty contests of their own, such as state QSO parties. Programs often show new and simple ideas for improving the station's capabilities.

What challenges these clubs to compete? The ARRL sponsors affiliated club competition in six major operating events: the January VHF Sweepstakes, the International DX Contest, the September VHF QSO Party, the November Sweepstakes, the 160 Meter Contest and the 10 Meter Contest. Each year dozens of clubs—in Local, Medium or Unlimited categories—enlist hundreds of their members to win one for the club. The criteria for the ARRL Affiliated Club Competition can be found each year in the *General Rules For All ARRL Contests*, usually published in the November issue of *QST*. These clubs also compete against one another in other major contests, such as the CQ World Wide DX Contest.

At meetings of the contest club, you'll have the opportunity to meet like-minded individuals at all experience levels. Many clubs have some sort of program for helping new contesters, and programs for helping staff club multi-operator stations. Since multi-operator stations in the same club are often on different levels competitively, it's often easy to get onto a multi-op crew to learn contesting from experienced individuals. Camaraderie in a contest club makes the contester feel that he's not alone, and sharing of contest experiences at meetings (official and otherwise) helps bring the contester along to new heights.

Many clubs also run some sort of intra-club competition to help foster higher club scores. This allows contesters within a club to compete for awards that might not be available on a national level, and provides motivation for sticking with it. The value of a contest club to the individual contester cannot be denied, and it is rare that anyone is found in the Top Ten of any contest who is not a member of a recognized contest club.

THE CASUAL CONTESTER

At some point, you will probably want to get a little deeper into contesting. This stage doesn't require the full-blown effort of the serious contester, but does require a little more effort than the beginner. At this stage, you will probably want to turn in a log and see your call in the results, or add your score to the scores of others in club competition. That's great, and you'll always be welcome in any contest, regardless of how much or how little time you can put in.

It doesn't take a kilowatt of power to be successful and have fun in a contest. Here Cam, HP1AC, runs QRP in an ARRL International DX CW Contest. Cam managed a second place world single-op QRP finish with this station in 1998.

You'll also want to look at equipment and antennas for contesting. Sticking with it for several hours to turn in a score is different from getting your feet wet for a few contacts. There are several things to do to make sure the experience isn't frustrating.

First, realize that there will always be QRM during a contest. Sometimes it will be quite severe, other times only minor. It is how you handle the QRM that can make a difference in whether you can develop your contesting prowess to its fullest. Be prepared for it, and don't complain, except for obvious cases of frequency theft.

Know the Limitations and Prepare For the Contest

One more key element to know is simple: know your limitations. It is possible to enjoy a contest, even with limited station resources. If you live in a mountain valley of western North Carolina, the odds of making a large score in the June VHF QSO party are slim, no matter how many elements your 2-meter beam has or how much power you are running. In this case, you know the limitations you face, and you start planning a strategy that will allow you to meet some reasonable goals. Maybe the solution here is to develop some type of portable station, and participate in the contest as a QRP portable station from the top of the mountain rather than from the valley floor. Knowing the limitations of your station can allow you to develop strategies that allow you to develop your skills and talents.

Minimal Contest Equipment

While at this level equipment doesn't have quite the same priority as it would to the contester who's striving for a high placing or a win, having equipment to help your contest effort is still important.

A good receiver is needed for even the casual contester.

Many hams find great pleasure in participating in contests as part of a well-planned multi-operator station. Here (clockwise from back left) N3BB, K5NA, WD5N, N5ZC and AB5EB relax after working the 1998 IARU International HF World Championship and finishing as the second place W/VE multi-operator team at N3BB's station.

Computers in Contests

I can still remember the dark ages of contesting. No computers. An automatic pencil with a fresh supply of lead was the high-tech logging device. The routine was simple; call the station, write the call in the log, send the report, write the call in the dupe sheet, hit the CQ button on the keyer (or keep talking) and scratch off another multiplier. Then spend a week after the contest rechecking the dupes to avoid the dreaded three additional QSO penalty. Scores bandied about just after the contest were only a rough approximation, and the real ones weren't known for days.

Progress came when the home computer arrived, and after-the-contest duping programs were developed. As part of the N5AU contest crew, we used a program for the Apple II called *SoDups*, developed for us by Bruce Hubanks, WD5FLK. The procedure was simple: I would read the calls to Randy, K5ZD, and he would type them into the computer. This easily saved us at least a week in log preparations for a large multi-multi operation, but it still didn't give us the real time scores and multiplier checks we wanted.

Around 1988, Ken Wolff, K1EA, began developing a contest program called *CT*. This program was designed for real-time logging on an IBM PC, and it was apparent almost from the start that users of this program had a significant advantage in the contest over those who weren't using it. The program at that time was developed only for DX contests, so Dave Pruett, K8CC, developed a program having the look and feel of *CT* for domestic contests. Dave called his program *NA*. Eventually both programs were developed for more contests, and had some different functions, so even though the screens were similar, each program had some features the other one did not. Larry Tyree, N6TR, then developed a program called *TR Log* that had a different look and feel, and used the logical sequence of events in a contest to determine keystrokes. By then, contest logging had become a reality, with most serious competitors using computers to their best advantage.

Your Advantage

So why should you use a computer, and what can it do for you? The easiest answer is that it will improve your score and relieve you of after-the-contest paperwork. Given two options like that, it's hard to imagine why anyone would want to operate without one.

To begin using a computer, you should buy one of the popular contest logging programs. These programs are developed by very active (and usually high scoring) contesters who know how to use a computer to improve their scores. A general logging program is just not efficient enough to even consider. Use one of those for your general logging, and buy a real contest program for contesting. Install the program on your hard drive, and be ready for increased enjoyment and scores.

The contest program will step you through the set up routine for a particular contest. Just follow the instructions. It's important to do this correctly since several things that can happen during the contest are based upon the data input at set up. Be especially sure to complete the name, address, and most importantly the information about the location from which you are participating. These are important for the exchange information in the contest, as well as for the contest sponsor to know where to send your contest award. You will have to set up COM ports and parallel ports for the program.

On CW, one of the COM ports, or a parallel port, can be used to key the radio. In some programs, a keyer paddle can be connected through the same port, and the computer will function as an electronic keyer for those who prefer sending non-repetitive data with a paddle. The COM ports may also control frequencies on the radio, a rotator, a voice keyer, a Packet TNC or antenna relays, all from the keyboard input. This is a significant advantage, especially in split-frequency phone operations. The ports can even control two radios with one computer, or be networked together to share log information in a multi-multi operation.

In operation, a call sign is entered in the call sign field. A key can be pressed for a duplicate contact check, a multiplier check, or even a partial call check. After making the contact, pressing the key sends the exchange and logs the contact, then moves on to the next line. A window will usually show previous contacts with a station on other bands, allowing the multiplier to

As with anything, the better it is, the better your results will be, all else being equal. High dynamic range is essential. Headphones are also needed, as you just can't catch all of the weak calls in the QRM without them. Others near your operating position will appreciate their use too.

The transmitter should have a good, properly adjusted speech processor, and the microphone should have a restricted frequency range. The emphasis is not on high fidelity, but on intelligibility. The ability to punch through pile-ups is important. Remember that overdriving your transceiver or amplifier is not only bad manners but also hurts your intelligibility on the air. It is important to have a crisp, clear signal. It is not important that the meters move excessively if such movement contributes to a lousy signal.

A memory keyer, or a computer with a keying interface, is a must on CW. You will want to use the keyer memory to send as much information as possible. A lot of repetitious information will be sent, so using the keyer leaves you time for housekeeping chores, such as keeping up a band map, multiplier sheet, or other things that will contribute to raising your score.

For antennas, you should have a good one available. That doesn't necessarily mean a beam, but whatever you can afford will work. Just make sure that it's safely and properly installed. Full sized antennas are preferred. A dipole isn't a bad antenna, and will actually beat a trap vertical most of the time. However, it is necessary that any soldered connections be done properly, and that all connectors are installed correctly. If you can't install it high, then try sloping it from as high as possible on one end and lower on the other end. It will probably work better than you'd expect, and don't be afraid to call anyone you can hear with it. They'll probably hear you better than you can hear them.

Tribanders, verticals, and other antennas are desirable too.

be moved to needed bands. In most programs, the contact is entered into a *band map*, which lists the stations on the band by frequency (including incoming Packet spots, if connected), allowing those stations to be skipped on subsequent S & P passes, or the band map may be loaded with calls, and they may be worked later. Some programs allow this to happen with a simple key stroke or mouse click.

On CW, all exchanges are sent by the computer, allowing the contester to spend more time with things other than sending. For SSB, a voice keyer may be set up to save the contester's voice. On RTTY, some programs generate the transmitted tones and decode the received tones. This allows operation to be almost totally automatic.

Having trouble copying a call sign? The *check partial* allows you to be prompted with possible calls after two letters of any call sign are entered. Don't use these cues to guess. The idea is much like finding a picture hidden in a picture—once you know the pattern, it's easier to find. This feature can often help when communications are marginal.

Want to know how you're doing? The *rate meter* will tell you what your rate looks like over whatever time period you specify. The score summary can always be displayed. A multiplier table will show missing multipliers. All of these things will improve your score over what might be possible without it.

After the contest, the logging program will prepare your log for you. While you do need to review it for typos, the preparation itself is automatically taken care of by the program. No need to second guess; the programmers know what is needed by the sponsors and how to prepare the log files as requested by the sponsors. So, if you send the proper file, there's no need to worry about what the log form looks like. Just use the file created by the program.

After the contest, *data mining* is useful for future contest operations. The logging program facilitates this with information about QSO rates per hour (including what bands are being used), points per hour, when each multiplier was worked for the first time, continent and country breakdowns for DX contests, and all manner of important information for after-contest review. Carefully studying these results files should help devise strategies for improved future contest scores.

Contest logging programs are sometimes available as freeware (in an obsolete, unsupported version) so that a contester may try one and make a decision on which new version to buy. Most are DOS programs (and don't like to be run under Windows, although *Writelog* was developed for Windows from the start, and *CTWin* is now in beta test.

To find out more, check out *CT* at **http://www.k1ea.com**. On the order page you can download CT6.26 as freeware (unsupported) for demonstration purposes, or order the latest version from XX Towers, 814 Hurricane Hill Road, Mason, NH 03048, tel 603-878-4600, fax 603-878-1102, or e-mail kc1xx@rcn.com. Note hardware requirements. The dominant version runs under DOS, and there is a Windows version available.

Information about *NA* software may be found at **http://www.contesting.com/datom**. It may be ordered from the Radio Bookstore, PO Box 209, Rindge NH 03461, tel 800-457-7373, fax 603-899-6826. This is a DOS program.

TR Log information is found at **http://www.qth.com/tr/**. A free trial version is available for download. It may be purchased from the website or call Geo Distributing at 830-868-2510. This is also a DOS program.

Writelog for Windows has information posted on **http://www.contesting.com/writelog/**. This is a Windows only program that can use the soundcard for RTTY or as a voice keyer. Ordering information is on the website, or you may contact Ron Stailey, K5DJ at k5dj@contesting.com or 504 Dove Haven Dr, Round Rock TX 78664-5926, tel 512-255-5000.

Super-Duper by EI5DI is another DOS based program for contest logging. It is a single-op program, with many edit features, and separate modules for many different contests. It covers the DX side of many contests very well, and some versions are available for download from the website as freeware. Information and online ordering may be done at **http://www.ei5di.com**.—*Bill Kennamer, K5NX*

Without specific recommendations, one thing will always hold true: proper installation is the key. Ultimately contesting is the true test of any station's potential. It doesn't do any good to have the best, highest antenna in the world if it's fed with a piece of junk coaxial cable and unsoldered connectors.

As a casual contester, you may want to consider purchasing contest software. This is a good investment, as the contest software makes it easy to calculate scores and turn in logs. Back in the pencil and paper days, contesters could often spend as much as a week removing duplicate contacts from the log. Now, the computer marks them for you, making it possible to turn in the log within minutes of the end of the contest, if you wish. It's still a good idea to examine the log for typographical errors or other obvious signs of brain fade during a long contest. Remember, it's usually best to start with one of the mainstream programs. The more esoteric ones should be tried only after you have an experience baseline to work from. Rest assured—you'll be a lot happier with a computer and software than without them, even if you're not a typist.

Maximizing Score for Minimum Hours

Operating for the casual contester usually consists of utilizing the peak hours of the contest for the maximum benefit of score. Or, it could be that you utilize the hours available within other commitments for the weekend. Either way, you should operate to get the maximum enjoyment out of the contest consistent with maximum score. To do that, you should consider what a contest really is.

Any contest should be considered as something similar to a scavenger hunt. The object is to collect something (QSOs in this case) from a list, with the most complete list winning. Each contest has a different objective (list) with different strategies in mind. For example, in a DX contest

Single Operator Assisted—Using Packet Radio Power for Higher Scores

Within the past 10 years, many contests have added a Single Operator Assisted (SOA) or Unlimited (SU) category. This category usually has the same rules at the Single Operator category with one major exception. The entrant may use spotting nets such as Packet Radio DX clusters. This has opened up a whole new world for many contesters, allowing access to information about band conditions and stations and multipliers available during the contest like never before.

This new category allows operators who enjoy monitoring their Packet network for new countries to participate in contests without missing anything. Other operators like working cooperatively with their friends while operating contests. Still others feel that the use of spotting nets can help them score more points for their club by increasing their multiplier totals. Being part of the network can keep some operators motivated to push on to bigger scores. Being "connected" in many ways makes SOA closer to a multi-op category than single op.

The only special equipment necessary to participate in the SOA category is a link to a spotting network. Originally, the links were by VHF radio. Later, contest software was developed to integrate VHF Packet networks. Today, it is possible for those in remote areas to connect by the Internet. Many VHF cluster systems are connected to the Internet. Most logging programs will accommodate Packet connections, and will interact with the radio, computer and logging in such a way that getting to a Packet spot is only a mouse click or keystroke away.

If you decide to try to make the best scores you can, some station design features can help you and make SOA much more fun. The first thing you should have is a transceiver that can be controlled by your contest logging program. This allows you to move quickly to new multipliers and back to your original frequency, where you can resume your tuning or calling CQ. (Finding stations by tuning around is usually called S & P—search and pounce—by contesters, and finding them by calling CQ is called *running*.)

You may choose to use an amplifier, since most SOA categories do not have a separate low power division. An auto tuning amplifier and automatic antenna switch allows you to quickly jump from one band to another, or you may turn off your amplifier when making a quick contact on another band.

As in the Single Op category, the use of a second radio can improve your performance as a SOA. If your station is capable of running at good rates on some bands, you can use the second station to study Packet spots. When the opportunity is right, make quick contacts between answers to your CQ. Since many SOA stations are not the biggest on the bands, being able to stay on a frequency and catch some extra multipliers without leaving for more than a few seconds at a time can help get the most out of your station. Having a second radio also decreases the importance of auto-tuning amplifiers and rapid band change antenna switching. Of course the more automated everything is, the more flexible your station can be.

Historically the top Single Op entrants usually outscore the top SOA. There are several reasons for this. Traditionally the top operators with stations capable of winning at the highest levels enter the Single Op category. After competing in the SOA category for a few years, and watching the performance of other participants, I coined the term "Single Op Distracted."

There are many elements involved in building a big score, and access to spotting information is only one. Being distracted by all the Packet spots is the biggest nemesis of the serious SOA entrant. If you are actually trying to make the best score you can, reverting to being a DXer can ruin your effort. In order to chase Packet spots, it is usually necessary to give up some of the other activities that can build a big score. Even if you work every spot with one call, you will lose running time chasing the spots.

Most operators will find it difficult to hold a run (CQ) frequency and enter into a lot of large Packet pileups. It is very easy to become distracted by all of the information available on Packet, and end up forgetting to do all the other things that Single Ops do to make big scores. At some point, you must be able to ignore beautiful, hard-to-work multipliers and concentrate on making lots of easy-to-make contacts.

Knowing your objective, regardless of your entry category, is important in establishing a game plan. If you want to work some new countries, or make a clean sweep, monitoring the spotting nets can improve your chances. If you want to make more points for your club, you will need a different strategy. The most effective game plan for the SOA participant, would be to do everything that you would do as a Single Op, and

Charles Fulp, K3WW, operates in the Single Op Assisted category from his well organized ham shack. You can count on him to finish in the top 10.

judiciously pick up extra multipliers without detracting from your most efficient practices.

Calling CQ (running) is still the fastest way to build a large QSO total for most operators. When we can't run effectively, searching and pouncing can produce good QSO totals. With Packet spotting, S & P can be even more effective. As you tune the band, you can skip over duplicates quickly, and depending on your abilities, may get some help in identifying some of the stations you come to. It pays to tune in an orderly manner, even if you do run off to a different portion of the band or even to another band to grab a new multiplier. On the other hand, you may set a goal of working the most multipliers possible, and not be concerned about optimizing your score.

On some bands you will find stations working split. An unfortunate complication of Packet spotting is that many stations now announce their listening frequency less often. For the Packet user, this information is often available, making it faster to work these stations. Another situation involves stations that do not identify very often. There is a strong temptation to assume that the spot is correct and work these stations without ever hearing the call. This is a risky proposition, as a fair percentage of spots are wrong; and even when they are correct, sometimes a new station is on the frequency when you arrive.

The cluster information will let you know which bands are most active and should allow you to choose the best band at any time. Nothing replaces knowledge of what is going on around you. Lists of contest DXpeditions are usually available from Packet or the Internet.

It is important to be patient, especially early in the contest when there are large numbers of spots of stations that will be active and much easier to work later in the contest. Unless you especially like the challenge of entering pileups with many of the biggest stations in your area, you may find it easier to wait for the smoke to clear before you go after a new multiplier.

If a station in an area where you have long periods of propagation is spotted early, and the pileup is large, don't waste time unless you have an incredible signal. Even if you do, it may be more productive to wait until later to pick up many of these stations.

If you want to maximize your enjoyment of the entire contest experience, give the Single Op Assisted category a try.—*Charles Fulp, K3WW*

Charles Fulp, K3WW, has been entering DX contests since 1962. In 1965 he joined the Frankford Radio club and participated mostly in multi-op efforts through the early 1980s. He has concentrated on the Single Op Assisted category since its inception in the fall of 1989, and has finished in the top 10 of the SOA Class in all 40 ARRL DX and CQ-WW DX events reported since then. This includes 35 top 5 , 23 top 2 and 6 first place finishes (W/VE in ARRL DX, World in CQWW). Charles maintains a Packetcluster node, and only on rare occasions turns off the monitor and enters as a single op. He frequently enters as a 1-man Multi Op in events that do not have an assisted category.

you will make the maximum score by working as many stations as possible, remembering to fill out your scavenger hunt list (multipliers). So, this means that a possible strategy would be to work as many stations in as many countries as possible, concentrating your efforts in those hours that would produce that result. From the US, the most active, available multipliers are in Europe. From the Eastern US, the most QSOs will also be found in Europe. Thus, if you live in the Eastern US, you would concentrate your operating hours on Europe. However, if you live in the Western US, you are likely to work the most QSOs in Asia, but there will be a lesser number of multipliers there. So, you would have to try to work European multipliers, but plan on making maximum QSOs by working Asia. This requires an entirely different operating strategy.

The casual contester may make many of his contest QSOs by Searching and Pouncing (often called S & P). This technique is just what it says, searching for a potential QSO, and Pouncing upon it. In many cases, just following the pileups and calling into them is what it takes to make a QSO. Over time, however, you will learn some discretion in which pileup to attack. Time Calling in Pileups (TCIP) is important. The less time spent per QSO, the better. A big pileup is harder to crack than a small one. By the same token, a G or DL station may be just as rare as a P5, if you haven't already worked one on a particular band that weekend, and probably won't have as big a pileup. You'll often find that checking back on a pileup later in the contest will produce a quick QSO with the same station, as his pileup diminishes. Remember, the object is to work as many countries and QSOs on that particular band as possible, not to work the rarest ones to the exclusion of making more QSOs and multipliers.

Checking back on a pileup is a specialty in itself. One of the best things to do is making a band map. The band map is simply a list of frequencies with a list of who is calling CQ on that frequency (which some contest software does for you). Beside the frequency, write the call sign of the station with the pileup, and his exchange. Then go on to the next one. After you work the station, draw a line through his call. That way you'll know which pileup it is if you tune across it later. If your radio has memories, you can also stuff the pileups in memory and come back later. If you have dual receive, you can even continue exploring the band while receiving the pileup. When the time is right to call, just push the split button and drop your call, and continue tuning with the main receiver looking for another one.

How you go through the bands can make a difference. The natural thing to do is to start at the bottom of the band and go up. However, you may want to get out of sync with other stations doing the same thing. There's nothing more demoralizing than following the same station up the band and working all your QSOs after he does. So try starting at the top of the band and working down. Listen carefully as you tune the bands. Sometimes you'll find a rare one, weak but workable, between the loud stations. Call every station that would count for points as you go through the bands. If you remain tuned by them long enough to find out who they are, you may as well call, since it will take some amount of time to find someone new.

When you use S & P, at some point you will run out of stations to work. There are two obvious choices to make. One is to change bands. The other is to call CQ. Which you do may depend upon what point in the contest it is. If it's the first day, and other bands are open, you might want to continue your S & P activity on another band, especially if you're running low power or have to fight for a place on the band. Remember that balancing your QSOs and multipliers probably means more for a casual effort than it does for a major effort. If your plan calls for making 200 QSOs, then perhaps it's better to make them with 50 multipliers each on four bands than with ninety on one band. If you can call CQ with some degree of success, that's by far the fastest way to make QSOs. If you can't call CQ with a reasonable rate of callers, it may be best to continue to S & P on another band. The key is doing whatever it takes to continue to make QSOs during the time allotted for operating.

Increased Productivity

As an intermediate contester, you'll want to use a computer for contest logging. The reasons are explained in the sidebar Computers in Contests.

VHF/UHF and Microwave contesters go to great lengths to put competitive stations on the air during the various contests in those frequency ranges. Here KB6WKT supervises the deployment of a 432 MHz and 1.2 GHz antenna array at the W6TOI operating site in the 1999 ARRL June VHF QSO Party.

Another major increase in scoring can be realized by entering the Single Operator Assisted (SOA) category. (In some events it's called the Single Op Unlimited category.) The rules differ from the Single Operator category in one way—the entrant may use Packet Radio spotting assistance.

Most operators who enter contests have no expectation of a winning score. They're looking for fun. Perhaps they want to increase their club's aggregate score. The assistance they receive from the PacketCluster increases their score. They can also give assistance to their fellow club members by the spots that they put out.

You'll need a computer and a contest logging program to take maximum advantage of the SOA category. You'll also want to interface your radio to the computer. As an intermediate contester, there are two software features that you'll want to learn to use. These are the announce window and the band map.

The announce window shows needed multipliers or QSOs on the current band or all bands. You'll probably want to look only at needed multipliers for the band on which you're operating. Use this window to find new multipliers, but don't spend much time calling in huge pileups unless you have high confidence of making the QSO.

The band map works best when your radio is interfaced to your computer. As you work and spot stations, they fill in the band map (a table of call signs and frequencies). Spots from the PacketCluster also fill in the band map. As you change your radio's frequency, the band map scrolls. Previously worked stations, needed QSOs and needed multipliers are all clearly marked as such. This allows you to more efficiently find needed stations, and to spot the stations that do not appear on the band map by looking for holes in-between spots.

Don't become dependent on Packet Radio assistance. Use it to aid you and your own abilities. Remember it's an assistant, not a replacement. Make sure that you copy each call sign that you work. Spots are helpful, but they may be–and frequently are–wrong.

THE ADVANCED CONTESTER

For some, contesting has become quite a serious hobby. They have progressed beyond making a few contacts for awards chasing, or to pass the time for a few hours over the weekend. For them, often many hours are spent just in preparation for a contest. This group often makes contests their only Amateur Radio activity. While it is not necessary to join that group to enjoy contesting, you can take advantage of some of their knowledge to improve your own contesting experience.

The advanced contester takes advantage of knowledge of his equipment, propagation, techniques, and even sleep cycles to maximize his score. It is important to concentrate on certain areas to do well, and even if you do not always have time to do a complete contest, operating in a serious manner will maximize your score for the time you put into the contest. It always gives something to talk about during lunch with the boys after the contest. Remember, knowledge is power.

You don't have to have the fanciest equipment and modern logging computers to enjoy contesting. Here C6A/W7DRA operates during the 1999 ARRL International DX CW Contest using a circa 1973 Heathkit HW-16 transceiver with matching HG-10 VFO. Notice the bottom of his antenna mast on the corner of the deck which supported a ½ wave long wire antenna.

EQUIPMENT

For optimum contest operations, it is necessary to optimize the radio station as much as possible within your budget. A contest station can be as cheap or expensive as you want to make it, but selecting the best you can afford, being sure that the quality will enhance your contest operations. It's often said that you can't work 'em if you can't hear 'em. So you'll need a good receiver or transceiver. Many of the top of the line transceivers of the past, such as the TS-930 or FT-107, are available on the used market at reasonable prices. Of course, any of the competition grade transceivers currently available will do the job. Just be sure it has adequate dynamic range and second and third order intercept points (see the sidebar in the DXing chapter). Good filters are necessary too, both phone and CW, regardless of any extras, such as DSP.

How you adjust and use your receiver can be critical. The ability to hear through QRM makes the difference between the good contester and the great one. The better you can hear, the better the results. Sometimes you should use the receiver attenuator to reduce overload. Today's receivers have more than enough gain for most conditions. Sometimes the receiver's AGC can be overpowered by strong signals. Try adjusting the RF gain control as this often will help you hear weak signals through strong signals. Offsetting the variable bandwidth or shift controls will help too. Experiment with the settings in the face of QRM, but remember the original settings so that you can return to them.

A modern high-end transceiver with dual receive may often be the ultimate secret weapon in any pileup. Further, it allows you to work two pileups simultaneously or to search for multipliers while running stations. If you don't have such a transceiver, it's possible to use an outboard receiver with a two-port power divider coupled through the external antenna port of the transceiver. Two receivers are effective for finding new stations to work on the same band or on different bands, or for working split frequency pileups on 40 and 75 meter phone.

Choice of filters is nice, especially on CW. While there are some who prefer using the wider filter option, such as 500 Hertz, on CW, others prefer the narrower (typically 250 Hertz) option, if available. There are two schools of thought on this; one thinks that it is better to hear stations calling if they are not exactly on your frequency, while the other school feels that in today's crowded band conditions it's better to use the narrow filters—that station that sounds off frequency to you may not be calling you anyway. The narrow filters also allow for more detailed S & P operations, with a possibility to work as many as four CW stations per kHz. DSP filters are an added extra and seem most attractive as automatic notch filters for SSB, especially on 40 meters. They can also be used to tailor audio response in any mode.

Any receiver needs a way to get the audio to your ears, and of course headphones are the only choice for the serious contester. Anything that sounds good to you is fine. Some prefer headphones with a restricted audio range, while others prefer hi-fi type phones with a good low frequency response for hearing low pitched CW signals. The main thing to look for is that the phones are wired for stereo if you have a dual channel receiver, and that they are comfortable to wear for 48 hours. While some prefer the lightweight foam padded headsets, most prefer the full muff headsets that help reduce sounds from outside the radio. Sometimes it's best to have a change for those periods when you feel like your head is in a vise.

A headset/boom microphone combination is best for SSB operation. Fatigue is always a factor in any contest, and not having to keep your head in a certain position the whole time during a contest can provide great relief. It's also essential to have a way to do T/R switching without having to hold something in your hand all weekend. Some folks prefer the freedom of movement that VOX control permits. Others rely on a footswitch so that if they should sneeze just when the weak P5 answers their call, they won't have to ask him for a repeat as he's drowned out by the pileup.

Amplifiers should be chosen for high duty cycle usage. Cooling is important, so don't expect the fan to be quiet. A good amplifier should be able to loaf along at full legal output power, because it's likely to have to run it for 48 hours at a time. Today's amplifiers usually seem to have a total plate dissipation at least equal to maximum legal power, thus most of them can easily handle the power requirements.

It's important that the exciter and the amplifier be operated within the manufacturer's recommendations, as a clean signal is very important. Not only must you be heard clearly without distortion, but you should also have a clean signal, one that does not annoy your neighbors on the band, thus subjecting you to attacks which ultimately cost QSOs. Watch the grid current. Don't overdrive your amplifier.

High Band Antennas

Antennas for serious contesting are usually bigger, higher, and more efficient than for any other use in Amateur

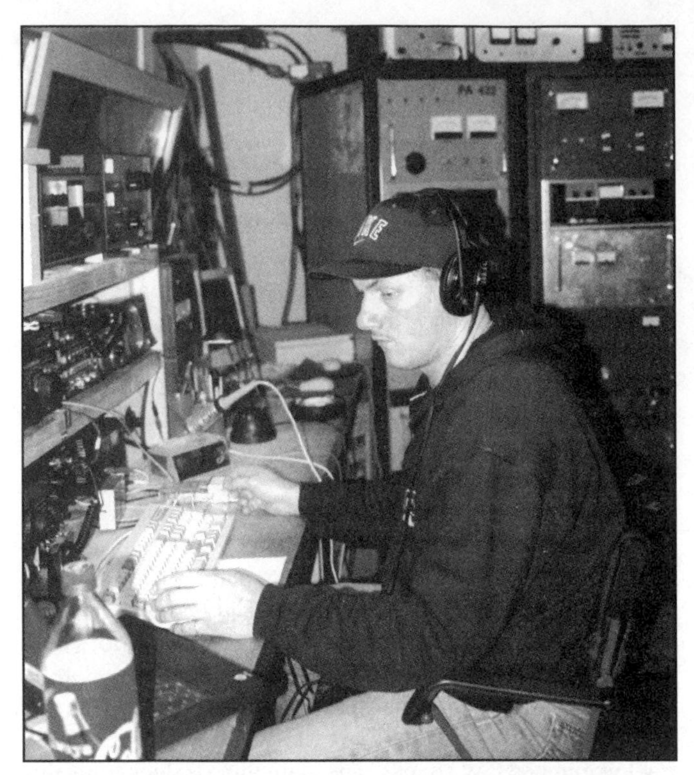

Uwe Dowidat, DL8UD, is a model of determination as he participates in the 1999 ARRL International DX CW Contest. Notice that even with the computer for controlling the equipment and logging that he still has need for a keyer while making contacts.

Radio. Design should be based upon the need for a complete system for the intended purpose, with the needs of the single operator varying from the needs of the multi-op station. For example, the single band or multi-multi operator will need antennas capable of opening and closing the bands while providing good coverage during the peak band openings for each band operated. The single operator all-band station will need antennas that provide good coverage during the peak band openings, but need not necessarily provide coverage for marginal band openings. The single operator will find it more profitable to be working lots of QSOs rather than chasing marginal openings.

The first place to start in planning a contest antenna system is with *The ARRL Antenna Book*. The propagation chapter and the companion disk shows the angles of radiation for maximum probability for most areas of the world. By using that information, an effective antenna strategy can be developed. For example, the path from Cincinnati, Ohio, to Europe on 20 meters always requires an arrival angle between 1° and 28°. However, 90% of the time, the angle is from 3° to 12°. The peak signal angle is 8°, which occurs 26% of the time. So the antenna system for 20 meters would be designed to provide angles which approach 8° as closely as possible. This study can be carried out for each band of interest, and antennas optimized to provide patterns within the range of angles specified in the tables for your area.

Choosing a site for your antenna is also important. If you haven't bought that dream QTH yet, site considerations could come into play. Even if you're working with an existing location, some properties are large enough that where the antenna is placed could make a difference. For choosing an antenna site, refer to the chapter on effects of the earth in *The ARRL Antenna Book*. There you will find information about site selection and the use of the *YT* program included in the book. Through use of this program, a USGS 7.5 minute topographical map of your QTH, and the files of arrival angles for the various bands and regions, you can determine the most useful angles for propagation for your area. You can then apply those studies to generate ideas for your own antenna farm. If you can't choose a site, at least you can check the potential from your present site.

As an example, suppose you live in Cincinnati, Ohio, where a 20-meter antenna should provide a take-off angle of 8° for peak propagation to Europe on 20 meters. Using a profile for your terrain (and each QTH is different), you might find that your terrain is actually almost 2 dB better than flat terrain for an equivalent antenna at 100 ft. Many of you may not be so lucky, and your terrain may have an apparent disadvantage. By working with either the site or the height of the antenna, you can help bring the angle to as favorable an angle as is possible for the site. You may also find that raising the antenna may actually hurt your situation. In this example, changing the antenna from 100 to 150 feet might bring the optimum angle performance at 8° down by about 1 dB, but actually the overall pattern is hurt even more, with higher angles being introduced, and some of the angles between 8° and 15° being suppressed. On the other hand, a multi-multi station would want at least one antenna this high, or maybe even higher, to use for marginal band opening and closing conditions, with other choices for other conditions.

Comparisons of antenna heights should of course be made based upon the type of terrain. But what differences can one expect by changing antenna heights given the same terrain? Given a flat ground site, the difference between 100 feet and 70 feet can be about 2 dB. But the difference between 100 feet and 50 feet on flat ground can be almost 5 dB, and even over 2 dB between 50 and 70 feet. That is not the case on some other terrain, however, as it is possible to show that steeply sloping terrain in front of the antenna can provide such an advantage that a 50 foot high antenna will perform almost the same as a 100 foot high antenna, but both will be outperformed by an antenna at 70 feet, which may be exactly the perfect height for that terrain. The difference in such a case might even be as much as 2 dB, certainly worth the extra effort to get there, however, the 50 foot high antenna would be reasonable, while the 100 foot high antenna in that situation would be an unnecessary ego trip, and not at all worth the added complexity and expense.

Type of antenna also makes a difference in choice of heights as well. A dipole over flat ground at 70 feet would have a gain of approximately 7.7 dBi at 14°. A three-element beam at the same height would have a gain of 14.2 dBi at 14°. However, at the desired angle of 8°, the dipole's apparent gain would have dropped to 5.8 dBi gain, while the beam's gain would be 12.1 dBi. The beam has a substantial gain difference at the same height as the dipole, and would be equal to it at a lower height. To have the advantage at 8°, the beam could be as low as 40 feet and still be marginally better than a dipole at 70 feet. By the same token, it shows

Baron Thomas, K6VWL, operates QRP from a modest station during the annual ARRL Field Day held every June.

that a high dipole may be as good as a low beam. Again, using modeling programs and studying the angles can make a difference in the antenna.

Stacked antennas are now almost standard equipment for the competitive contest station. Not only do they allow choices of height and direction, they also broaden the footprint of the transmitted signal, and provide anywhere from 1 to 3 dB extra gain. For contest stations, loud is good, so it's worth studying the possible gains available from stacked antennas of varying numbers and heights when considering station improvements. Many have successfully used stacked tribanders, so the limitations of using only monobanders no longer exists. Using stacked tribanders on a city lot will often increase the flexibility and competitiveness of the smaller contest station.

Obviously some compromises must be made in antenna selection and antenna heights, but these choices should be made from an informed position, and not without some understanding of the basic principles involved. Possibly the biggest improvement in contest scores can be obtained from an efficient, competitive antenna system.

Low-Band Antennas

The low bands (160, 80 and 40 meters) have many of the same problems as the high bands and require just as much study. However, the physical size of low-band antennas makes for a large difference in what one can do about it. Being on a city lot is different from being in the country, yet it is still possible to get out a decent signal from a city lot if you pay attention to detail.

For 40 meters, the easiest solution these days seems to be the two-element shortened beam, often called the shorty 40. Offered by several companies, these antennas work well at 70 feet or higher. Phased verticals can also be used, and a simple vertical can be made of push-up TV mast with a base insulator.

You'll find that 80 and 160 meters are similar—it's just the size that's different. While three-element 80-meter beams are nice, the odds of your running into competition with one in a pileup are slim. You can be competitive with well planned, but simple, arrays. Again, take into consideration the desired wave angles, then do what it takes to get there. In most cases, that's likely to mean vertical antennas, but not always. You would do well to buy a copy of *ON4UN's Low Band DXing* written by John Devoldere and published by ARRL. This book has everything you need to know about low-band transmit and receive antennas. The winning contester has a copy and studies it, and you should too.

If you are looking for a fully equipped contest station, K3KYR's shack shows a wide range of versatility of equipment. From radios to power supplies to amplifiers, the basic needs are readily at hand.

ERGONOMICS

Ergonomics is the study and practice of human engineering in station design. Simply put, if you're not comfortable, you won't stay in the chair. If you don't stay in the chair, you won't be competitive. One of the main formulas of contesting is that scores are directly proportional to time spent in the chair, and inversely proportional to time spent sleeping.

Speaking of chairs, this is one of the main areas where you can make or break your weekend. Choose an office chair designed for someone who works at a computer keyboard all day. It should be firm but comfortable, adjustable for height and back support, and sturdy enough to last. You'll find that big, comfortable and expensive executive model unsuited to your needs. Chairs have been the subject of discussion on the CQ-Contest Reflector, and it might be worth digging through some archives on the Internet at **http://www.contesting.com/_cq-contest/**.

The desk or table will usually be 29 inches high. That's the standard for office furniture in the US, and an ideal height for a working surface. However, the standard height for a typing surface is 26 inches. That's the height you would want for a computer keyboard. It's important that the keyboard be lower to prevent repetitive motion injuries, and to prevent fatigue from using a keyboard at the wrong height. The desk's surface should be large enough to hold all of the operating equipment, usually with multiple tiers for holding the various items in their most convenient locations.

Equipment should be arranged so that everything used frequently during the contest is immediately at hand or in front of the operator. Usually, this would mean the keyboard, radio and monitor should be directly in front, with all peripheral equipment surrounding the operator in accordance with how it will be used.

Rotator control boxes and antenna switches should be at hand, although consideration should be given to having these items controlled by the radio or computer. Rotator boxes in particular should be types that don't require a hand to be on them throughout rotation. Boxes with presets are much preferred. It's tiring to hang onto a rotator control box throughout a contest, and you can concentrate more on making QSOs without having to do that. Some antennas' switches, such as high/low or stacks, are probably best controlled by hand, but the band switching of antennas is best done by some sort of controller slaved to the radio.

Unless your amplifier has automatic band switching, you should mark the dials so that you can quickly set the controls when you make band changes. Amplifiers should be placed in your line of sight so that you can check for proper operation and where you can make occasional adjustments.

The computer is a major factor in today's contest operation. As such, the creation of the operating position should revolve around the placement of the radio and the computer monitor and keyboard. The CPU should be placed off the operating desk, with the keyboard being only a short hand movement away from the radio, and the monitor within the sight path of the operator. The operator should not have to raise his head to see the monitor, and should be in a comfortable viewing position when his hands are on the keyboard. You should have a wristpad or forearm supports for the keyboard. The keyer paddle should be located next to the keyboard, so that your hand can move quickly to it.

A clock should be placed at the operating position, even though the computer clock will actually be the one used. The clock provides a reference for checking the computer time throughout the contest.

Phone operators should definitely use a voice keyer of some sort. Best might be a sound card program for the computer that allows the use of the computer function keys so that hands do not need to leave the keyboard. Many operators prefer the style of keyboard that has the function keys down the side rather than on top. Over the course of a contest, this can make a difference in the comfort factor.

Wattmeters should be placed in visual range. You'll want to check on the functions of the transmitter and antennas during the course of the contest.

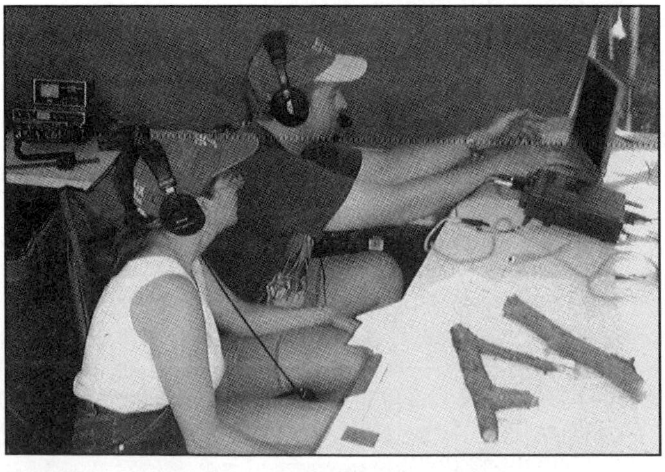

Teamwork is a must for contests like Field Day. Here Rob, KZ5RW, and Melissa, KG4CRK, put their 20 meter RTTY station on the air during ARRL Field Day 1999.

CONTEST STRATEGY

Setting Goals

Without some kind of plan, it is difficult to be successful in any contest. While there are some experienced contesters who appear never to have any plan, that's not really true. It's just that their experience is such that they can estimate where they will be, what kind of multiplier they will have and what kind of QSO totals will be required before the contest starts.

Category Selection

To plan your contest, first select your category. Base the selection on your equipment limitations, operating time available, or desired non-contest needs, such as single band DXCC. Even if the category is one where you do not think you will win, you should still have some sort of plan and goals set before the contest. Find the results of last year's contest, and see what it took to win your section, division,

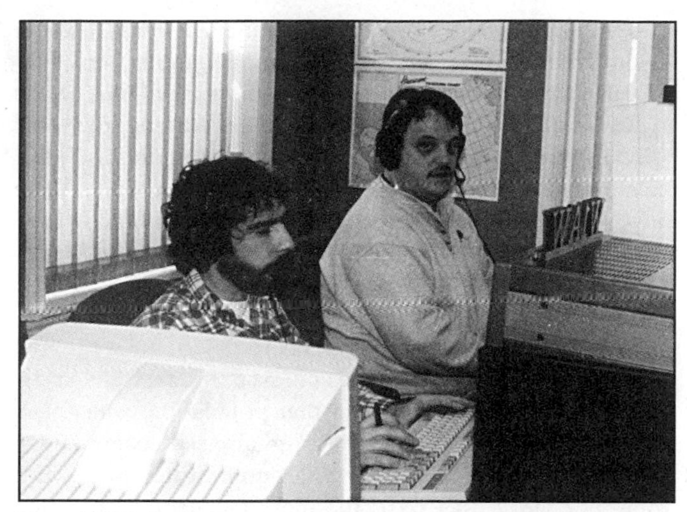

Even ARRL headquarters staff can catch the contesting bug. Here *QST* Assistant Technical Editor Joe Bottiglieri, AA1GW, and ARRL Contest Manager Dan Henderson, N1ND, put W1AW on the air during the 1999 ARRL International DX Phone Contest.

or call area. Then see what the leaders in your chosen category did in last year's running of the contest. Evaluate this year's potential by the expected propagation. For example, in years of rising sunspots, expect 10 meters to be better than the year before. In declining sunspots, expect 10 meters to be worse. The other bands should see an increase or decrease in activity, based upon what 10 meters will do. Most single operator participants will naturally migrate to the highest band open.

Checking Results From Past Contests

After evaluating the potential of the bands, check the breakdowns of your category. Check the number of multipliers made on each band and the QSO totals of each band. It's best to pick a station near to you for this evaluation. Another way to do this is to go to **http://www.contesting.com** and look for the 3830 archives for a year ago, or a comparable contest from 6 months ago, or both, as you may find a trend in propagation. On 3830, you may be more likely to find a station in your area with similar equipment. See where the contacts fell, and make an estimate based upon these numbers for QSOs and multipliers for each band.

Operating the Contest

Operating the station is the most important factor to be successful in contesting. A good operator can do well with a marginal station. A bad operator couldn't make a great score with the best station. Much of the operating of any station comes through some experience and intuition, but there are some simple rules to follow that will improve the performance of any operator.

Pre-Contest Preparation

Of primary importance is learning enough of the contest rules the to be able to perform properly during the contest. This includes knowing whom to work during the contest and how the scoring system works. All of these will play a part in strategic planning for the contest. The exchange should be memorized to the point that you could send the exchange while concentrating on some other contest chore, such as logging, checking multipliers, or equipment conditions. You should know your grid square, CQ zone, ITU zone, state, county, precedence, section, or anything else required for any contest you might enter. A seasoned contester can give the exchange for any contest that he has operated more than once.

Be sure that you have made the necessary plans to prevent interruptions to your contest effort. This means being in front of the station at the start of the contest period, and staying there as much as possible during the contest. Food preparation should be such that there is sufficient food available, but you should not plan on having a sit-down dinner during the contest. Sleep time will be minimal, but the sleeping area should be close by, and comfortable. Be sure to have a good alarm clock, and be able to motivate yourself to get out of bed when it goes off.

Getting a Fast Start

Making a plan based upon expected results is the easiest way to develop a strategy for your contest. Going to the band expected to be open at the start of the contest is the best thing to do. Getting some success by working stations in the first few minutes of the contest is important to your overall success. While you can't win a contest in the first few minutes, it can have an effect on your attitude for the whole weekend. Don't sit on a frequency waiting to be someone's first QSO unless you're already talking to him before the contest begins. There are probably hundreds lurking on frequency to do the same thing. Start making the band map at the beginning, make a few quick contacts, and settle in. After the first thirty minutes of the contest, things should begin to get a little more reasonable. Try to find your best band making contacts in the first few minutes. This will possibly improve your score more than anything you will do later.

Economy in speech and action serves the contester well during the contest. When calling CQ, call very short ones. A long CQ is a guarantee that anyone tuning around will go right past. Something in the format of "CQ contest, Whisky One India November Foxtrot" with a short interval for listening is efficient and quick, and makes for a better rate. Answering a call should be similarly efficient. For the November SS, a typical answer to a calling station would be something like "W1AW, number 1068 B W1INF CK 45 CT." Notice that this exchange is in the format of the logsheet for the contest. (Some operators don't bother to send the CK on CW.)

During the contest you should never say anything twice. Repeat only if asked, and only what is asked. If you are asked for your prefix, give only your prefix. If asked for your suffix or any letter of your call sign or report, give only what is asked. The receiving station already has most of what he's asking for, or will confirm a full call sign after getting a partial. To give him anything extraneous may cause confusion under marginal receiving conditions.

There are several ways to improve your rate over the course of the contest. It is extremely important to be able to get the call signs sent to you the first time, every time. Repeats not only take your time, they also mean that other stations waiting to work you may move on. One of the best things to learn is to be able to receive a call sign phonetically and give it back to the other station without phonetics. This only works, however, when you're absolutely, positively sure that you have the correct call. Remember that contest sponsors remove points for incorrect call signs, and that points are important. This will speed up your rate considerably over the course of the contest.

The good contest operator develops a strategic awareness of what is happening around him during the contest and changes his strategy as necessary. For example, while you may have a band plan made before the contest, it could all go out the window on any given weekend if the propagation, other participants and the like don't cooperate. If so, then do whatever works. Remember that the object is to make QSOs, so do whatever it takes to do that. Many operators have gone into a contest with some sort of basic game plan, and stubbornly stayed with it even if it wasn't working. During the contest you must always watch your progress. Take what the bands give you. If 15 meters was your band of choice and nothing is happening for you there, go to 10 or 20, or possibly even lower. If 10 meters works best, don't try to force 15. Be constantly aware of where the propagation is, and how your rate is progressing. Either a marked lessening of the rate or a lack of new stations to work is a sign that you need to be somewhere else for a while.

If calling CQ doesn't work, then go to S & P. If the band is open and everyone answers on your first call, then try CQing. If a multiplier calls in, remember to have the presence of mind to move him or make a schedule for a different band. If you overhear another station making a schedule or moving a juicy multiplier to another band, try to follow the move, or show up on the schedule, and work that multiplier too. Always be listening and thinking during the contest.

Chase or Race?

Experience over time will tell you when to break way and chase multipliers on a different beam heading. For example, when the sunspots are lower, there's often a propagation lull around local noon. You can look to the south or possibly even the southwest at that time for new contacts. Being around for marginal openings can often have unexpected benefits, but you should use that strategy only for very short time periods.

The decision as to when to break away to chase multipliers or to stay with making QSOs largely depends on both your rate and the point value of a new multiplier. Ellen White, W1YL, in a 1955 *QST* article, stated a formula that still works well today. This formula can help determine whether it's more worthwhile to work a multiplier or a new QSO. The formula is: Value of Multiplier (in equivalent QSOs) = Number of Contacts Worked ÷ Number of Multipliers worked + 1.

As an example, in a DX contest, 20 QSOs ÷ 10 Mults + 1= 3 QSOs equivalent to 1 additional multiplier. In another example, 1200 QSOs ÷ 350 Mults + 1 = 4.4 QSOs equivalent to 1 additional multiplier. As you can see, the further into the contest you go, the greater value of a multiplier. Further calculations may be made that will take your current ten-minute rate into account to determine how much time can be spent in pursuit of one multiplier versus continuing to work QSOs. Many contest logging programs will actually do this calculation for you in a real-time manner. Of course, if you have all the multipliers available, such as in SS, then no consideration should be given to additional multipliers.

Remember the Population Centers

One strategy that never fails is to determine where the populations centers are for the contest of interest, and maximize your efforts in reaching that population center. DX contests are of course different from domestic contests. On the East Coast of the US, for a DX contest that means Europe. For the West Coast, the most reachable population centers are in Asia and Oceania, but some time must be devoted to working Europe. The Mid-South often has the best of both worlds, but with shorter openings to either. For domestic contests, the object is to work the population centers of the eastern seaboard while getting multipliers from the west. This even applies to eastern stations, who often devote more time to lower bands. Set your strategy around population centers.

By the same token, pointing your antennas away from a population center for extended periods of time will not increase your contest scores. Yet valuable multipliers often come from these directions. The best solution to that problem is determining when there is a high likelihood to work a station in that direction, devoting a few minutes to the pursuit of multipliers in that direction, and returning to the primary directions. As an example, often 10 meters is open for northern hemisphere stations in a southerly direction around local noon. The higher bands sometimes open in a more southerly direction, and stations can sometimes be worked on that path just prior to the opening in the usual (direct short-path) direction.

MISCELLANEOUS CONTEST TIPS

QRM and contests go hand in hand. That's part of what contesting is about, being able to copy and make yourself heard through QRM. So don't worry about trying to find an absolutely clear frequency; that's just wishful thinking. The thing to remember is to not get into frequency fights. It's not worth it, and costs you QSOs. Either get the other station to move on, or move yourself. If two CW stations move 500 Hz, sometimes that's enough for peaceful coexistence.

If you're not getting answers to your calls, or are being asked for repeats a lot, think about why this might be. Are you sending too fast on CW? Are you transmitting clear, undistorted audio on SSB? Either can prevent you from making QSOs, so it's counterproductive to have poor audio or to send too fast. On CW, it's sometimes more productive to go slower late in the contest, when you've already worked most of the active contesters. Remember, you cannot win many contests by working only the hard core, experienced contesters. You must work casual operators who turn on their radios, make a few contacts, and go off to enjoy other pursuits during the weekend. Remember that when you're

Bill Weed, K5BJW, loves to contest and chase DX from his home in Las Cruces, New Mexico. Bill named his cats (l-r) "CQ" and "DX." His children say they're glad their father was not into radio when *they* were being named!

tempted to take your keyer up to 40 WPM, or say your call sign as rapidly as possible. Temper your voice or keying speed to what you're try to accomplish at that minute. If you have a large pileup, go fast to get through them. If only one or two call per minute, be slower and friendlier.

Sending your call sign clearly and completely is sometimes a misunderstood part of contesting. Standard phonetics should always be used. You should, however, know three sets of phonetics, because sometimes in marginal conditions it pays to be able to change them for more audio penetration or to prevent confusion. Your call sign should always be sent completely. If only one particular letter is requested send only that letter. Never, ever send only two letters of your call sign. You want the station at the other end to be able to work you fast so you can move on, and it can't be done if two exchanges are required to pass the call sign. Some stations won't work a station calling with a partial call sign unless there is absolutely no one else calling.

ARRL Contest E-Mail Addresses

10 GHz and Up Contest	**10GHZ@arrl.org**
ARRL 10 Meter Contest	**10Meter@arrl.org**
ARRL 160 Meter Contest	**160Meter@arrl.org**
August UHF Contest	**AugustUHF@arrl.org**
ARRL International DX CW Contest	**DXCW@arrl.org**
ARRL International DX Phone Contest	**DXPhone@arrl.org**
International EME Contest	**EMEContest@arrl.org**
Field Day	**Fieldday@arrl.org**
IARU HF World Championships	**IARUHF@arrl.org**
January VHF Sweepstakes	**JanuaryVHF@arrl.org**
June VHF QSO Party	**JuneVHF@arrl.org**
RTTY Round-UP	**RTTYRU@arrl.org**
September VHF QSO Party	**SeptemberVHF@arrl.org**
November Sweepstakes CW	**SSCW@arrl.org**
November Sweepstakes Phone	**SSPhone@arrl.org**
Straight Key Night	**Straightkey@arrl.org**

Results Analysis

After the contest, results may be analyzed to see what might have been done differently. Comparing notes with friends is a good idea, as you will often find they did something totally different from you, and it may come in handy to have the benefit of their experience in the future. (See sidebar, "Log Preparation.")

Using the Rate Sheet

If you use a computer to log, you'll find that most programs have several post-contest features for use in log analysis. One of the best is the simple rate chart. This chart shows your hourly rate for each hour of the contest, and the bands used during that hour. Log files often show the frequency used for each contact. This can tell you if you were CQing or using S & P during the time of higher rates. It could be either one, depending upon station size. Other breakdowns will show the general area where the contacts came from, who the first contact in a particular multiplier was, or how many contacts came from each individual multiplier.

A points-per-hour sheet may also be found as part of most analysis packages. Using this data, one can draw a graph to see how his score progresses hourly, or which hours seem to be most productive. You will also notice that the score rises at a faster rate the farther into the contest you go. The more contacts, the more total point value to multiply, and the score starts to increase geometrically. This is further graphic proof that it pays to stay with it.

Log Preparation

The contest doesn't end when the final QSO is logged. Post-contest work at times can be as challenging as the contest itself. As with anything in contests, it's a job that requires attention to detail. A good contest effort can easily be spoiled by sloppy post-contest work.

Contest sponsors have criteria for submitting entries for their events. You should know what forms and electronic files the contest sponsor requires for officially entering their event. Deadlines are also important. The contest sponsor has one to meet, and he knows the latest date he must receive logs in order to meet it. Each contest has a deadline for submission of entries. For all ARRL contests, the deadline is 30 days after the end of the event. The deadline for other contests may be found in the rules announcement in *QST*'s Contest Corral or on *ARRL Web* (**http://www.arrl.org/contests/**). Because of the amount of work involved in processing and verifying/checking entries, contest sponsors are cracking down on meeting the contest deadlines. Make certain that your entry is either e-mailed or postmarked by the deadline date.

As part of your preparations, you should have acquired copies of the official contest entry forms, usually known as summary sheets. If you log by hand, these summary sheets are a necessary part of your entry. Without them, the contest sponsor may not be able to determine how you intend to enter the contest. Don't rely on photocopies of forms you have from the same contest several years earlier. Contest rules change, and those changes will be reflected on the summary sheets.

Make certain you *legibly* complete the form, including all information. If you don't complete the required forms, your entry is going to be difficult to process by the contest sponsor. If you don't include the little things, such as a correct entry category or power level, the sponsor may have no alternative but to code you into their databases at default levels. For example, your failure to include the fact that you were QRP in the contest may mean your entry is coded as high power—and could cost you a chance to win your category, division or section.

Just because your computer logged an event, don't assume that you don't need to look at the output data file you are submitting. In setting up the preliminary information before the contest you may have made a simple typographical error which ends up causing you to be entered as something other than the category in which you competed. The logging software will catch dupes and make certain you pick up multipliers: it can't catch data input errors. Finding typographical errors is the responsibility of the contest operator, not the contest sponsor. Remember, the log you submit is the log that will be checked.

When preparing an electronic file for submission, remember to open the log file itself to make certain that the file is in required file format and is readable. Glitches in copying files can easily corrupt the file and make the data it contains unusable. Make sure the files are named correctly, according to the contest rules. In ARRL contests, the filename should always be your call sign. It should never be the name of the contest you're submitting. The file extension, such as .log, .sum, .prn, or .cbr will tell the contest sponsor what the file format is. Use .sum for a summary sheet file and .log for a log file. It may be helpful to you to name your 1999 ARRL Ten Meter Log file 10Meter99.log, but that causes a problem for the contest sponsor, especially when 500 other entrants choose the same filename. Renaming one file to be submitted correctly may not be a big chore for you; renaming hundreds of files takes a lot of time and can lead to errors for the contest sponsor. By naming the file with your call sign.sum, the chances of another file overwriting it when they are saved is negligible.

Starting with the 2000 ARRL November Sweepstakes, all electronic files for ARRL (and many other) contests must be in the *Cabrillo format*. This format specifically lays out what data is contained in specific

Determining Sleep Times

Analysis of the hourly rate also shows when it is most productive to sleep if you can't stay awake for the entire contest. For example, if your rate went to 5 per hour at 0800 UTC, and only one of those was a multiplier, perhaps an hour's rest would have helped you more the next day. There are always tradeoffs in any contest, but analysis of results can show how those tradeoffs may affect your overall score.

When determining times of low activity, and whether it might be more profitable to sleep for the higher rates ahead, it's good to think about how long to sleep as well. Two *NCJ* articles on the subject suggest sleep periods should be 90 minutes in length.[1] Use the rate sheets to determine when during the contest period these sleep periods should fall. If you can get a rate sheet from a multi-multi on the www.contesting.com web page, look for their time of lowest activity. That may be a good time for your sleep period.

[1]"Sleep—A Contest Prescription," T. Scott Johnson, KC1JI, *NCJ*, Nov/Dec 1988 and "A Sleep Strategy for DX Contesting," Randall A. Thompson, K5ZD, *NCJ*, Sept/Oct 1994. Both recommend 90 minutes, or multiples thereof, as the most effective sleep periods.

Analysis of Band Breakdowns From Past Contests

Breakdowns from last year's contest can be used to prepare for this year's contest. Use those breakdowns to set goals for this year's contest. It's often helpful to compare the rate from the previous year to find out where you were and what you were doing, especially if you were wrong last year. By looking at the band breakdowns and looking at your own goals, you can determine the probability of meeting them, and make an educated guess as to what to do if you need to correct something in your operation. You should strive to stay ahead of last year's score, even to the point of computing how much you want to be ahead in this year's data fields within the file. This will greatly speed the process of log conversions, the greatest time consumer in the checking of electronic files. This allows the contest sponsors to have more time available to accurately determine the scores of the entrants. File specifications for *Cabrillo* will be found in the "General Rules for All ARRL Contests" in *QST* and online at: **http://www.arrl.org/contests**.

If you use a computer to log your contest entry, ARRL rules (and many others) require that a copy of the log file in required ARRL file format be submitted. Sending in a paper log of the electronic file is no longer an acceptable substitute. This allows all electronic entries to be processed through the log checking software. Even hand-written logs which are possible high scores and award winners will be converted to electronic files as necessary and processed.

After you have verified the data files and the information in your log, where do you submit them? For ARRL contests, the easiest and most reliable means of submission is to send an e-mail with the required files sent as attachments to the e-mail, not as the text of the message. Each ARRL contest has a specific e-mail address, and only entries for that contest should be sent to it (see sidebar). General information and queries should be sent to the Contest Manager's e-mail address. Other contests will have submission requirements along with their published rules.

If you don't have e-mail capability, you still need to submit those electronic files. If you submit via the postal service, you should send in a 3.5-inch floppy disk that is clearly marked with your call sign and the contest name. Include on that floppy disk the properly named electronic files. However, if you can do it, e-mail submission is greatly preferred.

If you use one of the commercially available logging programs, you should only submit those files which are needed by the contest sponsor—the file with the summary sheet information and the log file of your contacts. The support files which those logging programs produce—rate sheets, band by band breakdowns, etc—help you in the post-contest analysis of your effort and to plan your next contest strategy. They should not be submitted to the contest sponsor.

All of this talk about electronic files does not mean hand logged entries are a thing of the past or are unacceptable. Even though the number of paper logs is declining, they will continue to be accepted by the sponsors of most contests.

The mailing address for ARRL contest entries is: ARRL Contest Branch, 225 Main St, Newington, CT 06111. Please write the name of the contest on the outside of the envelope. You may also use that postal address to request copies of the latest contest summary sheets, log sheets and rules. Please include an SASE with sufficient postage to cover your request (about one unit per 4 sheets of information.)

Finally, here are a couple of "DO's" and a "DON'T" for submission of entries. DO submit entries only after you're sure that all required information is contained on the summary sheets and that the log or log files are ready to be examined by the log checkers. DO submit your entries for each contest, even if you are only a casual operator. The more logs received, the better able the sponsor is to gauge participation and to more fairly determine the winners.

While it is tempting, DON'T submit entries for more than one contest in the same envelope or e-mail. Contest sponsors are looking for a single entry in each envelope or e-mail. If you include more than one in a submission, it can be overlooked or stapled together as a single entry and can be misfiled. When the error is finally discovered, it could be well past the submission or publication deadline for that contest.

Submitting the contest entry is really quite painless, and is appreciated by the contest sponsors, no matter how big or small the log may be. *—Dan Henderson, N1ND*

contest. Watching the rate sheet and comparing will keep you on track for an increased performance. Improvement is always the name of the game in contesting, and results analysis will help get you there.

Multi-Operator Contesting

Nowhere in Amateur Radio will you find the camaraderic that exists in operating a multi-op station. Shared experiences are always more enjoyable, and what better way to share the contest experience than to do so with a group of like-minded friends. Multi-ops provide an ideal training ground for operators new to contesting, allowing the new guy to mix with operators having years of experience. Often contest careers are prolonged by multi-ops, allowing those who have neither the time nor the endurance, but still have the desire and skills, to participate in the contest.

There are generally two types of multi-ops: the multi-op single-transmitter and the multi-op multi-transmitter.

Donn Baker, WA2VOI, uses this neat antennamobile to enter the Rover category in VHF contests.

The ARRL multi-op two-transmitter class is actually an offshoot of the multi-single class designed for those stations who are capable of putting two signals on the air at the same time. Each contest's multi-single rules are different, and should be studied carefully before operating the contest.

Multi-Single

The multi-single class was originally conceived as a way to keep a single station on the air for the entire contest by allowing more than one operator. It has evolved over time into a very competitive category. In many cases more than one station is used during the course of the contest.

Strategy plays an important part in the planning and operation of the multi-single station. The strategy revolves around the permissible activities of the station during the contest and how a second station and operator can be used for maximum effectiveness.

The multi-single stations are generally limited to a certain number of band changes per hour, or by a time limit on a particular band before changing bands. The multi-single participant accepts this as part of the rules, and therefore uses those band changes strategically. For example, before a band change is made for purposes of working a multiplier, the operator manning the multiplier station would make a band map of stations to work before changing bands. He might even record those stations in the memory of his transceiver. At the band change, he becomes the main transmitter for a few minutes (usually a minimum of ten minutes). He works any multipliers in his band map, any other QSOs possible, even calling CQ for a while, and then returns to the hot band and the main station. This process repeats again, possibly on another band.

While the limitations on the multi-single stations (and multi-two) may seem somewhat restrictive, in fact they are there to create a class of competition for multi-op stations that are not set up with individual stations for each band. This has worked well over the years. Easing of the restrictions would definitely up the ante in the equipment required to play this game. Now, it can be done successfully with two or three independent stations at one QTH.

Multi-Multi

The multi-multi station does exactly what the name implies, many operators, many transmitters, with only one per band transmitting at any time. The successful stations often have at least two-station capability on each band. One chases multipliers on the band while the other one calls CQ throughout the contest. Only one transmitted signal per band at any time is allowed, but that's easy to work out. It's no wonder that these stations make very high scores in the contests where they are permitted.

These stations provide excellent training grounds for new operators, as staffing a multi-multi is one of the more difficult tasks in contesting. You can frequently get a chair in one if you make the acquaintance of one of the operators. You may be calling "CQ Canada" on 160 meters during the daytime from Texas, but it's worth it to gain the experience at a multi-op station. You'll always value the opportunity to watch and hear the operators go about their business at a top multi-op.

The multi-multi is also useful to the beginning operator for learning what is possible on each band. The successful multi-multi is required to work all bands at all times. By doing so, one can learn about those midnight over the Pole openings on 15 or 20, how early 40 meters opens to Asia, how 80 meters opens before sunset, and stays open an hour or two after sunrise, and what might be available at those times. It's a great way to gain useful experience in every phase of contest operating, probably the only way to really learn under fire with experienced teachers of the contest art.

TWO-RADIO CONTESTING

The use of two radios in a single-operator station has been around for many years, but has grown significantly in recent years. The wider use coupled with new innovations in logging software has made it almost necessary for the winning single-op contester to use two radios.

This subject could almost be a book in itself. Design of a station capable of two-radio contesting is similar to the design of a multi-multi station. One transceiver might listen on one band while the second is transmitting on another. The use of band pass filters and harmonic stubs is necessary for successful high-power operation. However, if adequate separation is available between antennas, a two-radio station can be as simple as adding another transceiver and a dipole or vertical antenna. There is an advantage to using the second station. The degree of the advantage depends on the operator and how he uses the station.

The purpose of the second radio is to allow an operator who might be calling CQ on one band to be listening for activity or multipliers on another band. Once found, a multiplier may be called and worked on the second station during a listening period for the first station. Over the course of the contest, a neophyte to two radio operations may add up to 100 QSOs to his totals. An accomplished operator can add as many as 400 QSOs to his totals. Needless to say, this is a tremendous advantage in close competitions.

There are several ways to listen to the second radio. Some listen to one radio in one ear, the other in the other ear on a stereo headset. Others like to mix the audio into both ears, but with one radio running a higher volume. A third and less confusing method is to shut off the monitor on the main radio, and listen to the second radio only during transmit periods for the first. This can be done with a relay to switch the audio. Designs for two-radio switching have appeared in the *National Contest Journal* in recent years, and an advanced design appears in *The 2000 ARRL Handbook.*

PROPAGATION

When it's all said and done, the fact is that the contest starts at 0000 UTC on the given date, and regardless of the propagation, you still have to compete. Knowing what's likely to happen will give you an edge, however.

General Rules

In general, your antenna on 10 and 15 meters should follow the sun. It should be east to southeast at the morning band opening, moving south by local noon, then west, southwest and northwest until after sunset. Openings to the east may last longer, so use discretion. Twenty meters will be open to the west at sunrise, but may be more productive to the east. The lower the sunspot activity, the more important 20 will be.

You'll find that 160, 80 and 40 are generally nighttime bands. All three may be open at sunset, generally to the east, but sometimes to the southeast. The southeast opening may yield some Asian or Africans in a DX contest. In the morning, the bands are open to the west, although a southwesterly path will again yield Asian DX. Watch European sunrise. The bands will be open in that direction, with 160 and 80 closing before 40, which is likely to stay open much longer.

Twenty meters will often share characteristics with both the low and high bands. It is more likely to be like the high bands in periods of low sun activity, and like the low bands during high sunspot activity.

Someday, you will experience a solar flare, especially during times of high solar activity. If it happens, it will almost feel as though someone turned the bands off with a switch. If you find that condition, try going to a higher band and possibly beaming south. Sometimes nothing works until the band comes back, but more often than not you may find that there is less absorption on the higher bands, and there may be some propagation to somewhere.

Develop your band plans from your own experience. It's best to listen around all of the bands at different times in the weeks leading up to a contest. See who's coming in on which band, and how the band sounds. PacketCluster is invaluable for doing this, since you can watch spots for a week or two before and see what area and what bands seem to be working to what area of the world at a given time.

VHF CONTESTS

VHF contests are by nature a different animal from an HF contest. However, the same people who do well in VHF contests have often had past experience with HF contesting, and may have been winners there as well.

Many VHF contesters set up portable stations for the contest period. This is especially true of the multi-op stations, where the locations on mountain tops are necessary to extend line of sight communications on those microwave bands that are so essential in VHF contests. There are enough well-equipped home stations around to make it interesting for all participants.

While you can get your feet wet on 2-meter FM, the beginning participant should be on as many bands as possible. Now it's possible to be on as many as three bands with one of the new transceivers that cover the 50, 144, and 432-MHz bands. Other bands may require the use of transverters. Transverters are available for most of the microwave bands.

Operating in a VHF contest is a little more casual and relaxed. Stations tend to come in waves, as the casual entrants come up to operate for an hour or two and go away. In VHF contests, you will be asked to change bands often. You will learn to do some of the asking yourself. That's the best way to gain maximum points. Frequently, if you can work someone on 2 meters, you should be able to work them up through most of the microwave bands too. Obviously that's not the case with an E skip or F2 opening on 6 meters. On that band, you should take advantage of the skip while it's there.

Strategy in a VHF contest is different than in an HF contest. Part of this is because of the point structure of the VHF contest. The higher band you use, the higher the point value. A contact on 432 MHz is worth twice as much as on 144 MHz, and a contact on 2304 MHz is higher still. So, whenever possible get a contact on the higher bands.

The June contest has totally different propagation than the September or January contest, and the bands should be attacked accordingly. In June, sticking with 6 meters to expand your multipliers is a good idea. In January, when 6 meters is often dead, it is usually better to change bands frequently. You should balance your efforts between working new multipliers and making more QSO points in a way that maximizes your score.

REMEMBER, CONTESTING IS FUN!

Of course there are many good reasons for contest participation, all of them perfectly valid. But the main reason it is so popular (and when did you ever hear as much activity on the bands as during a contest?) is that it's fun.

Over the years, you'll find that you work a lot of the same people each year, and it's fun to say hello by giving them a contact. You'll develop new topics of conversation as you run into them on the air between contests.

Meeting other contesters in person adds another dimension to contesting. It provides the opportunity to see the faces of your on-air friends. Contesting builds a special camaraderie between contesters. Try it and see!

Contest Calendar

Month	*Contest Name*	*Scope*	*Exchange*	*For More Information*	*Usually Held*
Jan	ARRL RTTY Round-Up	Int	W/VE: RST and state/province; DX: RST and serial #	Dec *QST*	First full weekend but never Jan 1
Jan	CQ Worldwide 160 Meter Contest (CW)	Int	W/VE: RST and state/province; DX: RST and country	Dec *CQ*, Jan Contest Corral	First full weekend
Jan	North American QSO Party (CW)	Dom	Name and location (state/province/country)	*NCJ* Nov/Dec; Jan CC	Second weekend
Jan	North American QSO Party (PH)	Dom	Name and location (state/province/country)	*NCJ* Nov/Dec; Jan CC	Third weekend
Jan	ARRL VHF Sweepstakes	Dom	Grid Square locator	Dec *QST*	Weekend before the Superbowl
Feb	North American Sprint (CW)	Dom	Callsigns, consecutive serial #, name and location	*NCJ* Jan/Feb; Feb CC	Second Sunday
Feb	North American Sprint (PH)	Dom	Callsigns, consecutive serial #, name and location	*NCJ* Jan/Feb; Feb *QST*	First Sunday
Feb	CQ Worldwide 160-Meter (PH)	Int	W/VE: RST and state/province; DX: RST and country	Dec *CQ*, Jan CC	Last full weekend
Feb	ARRL International DX Contest (CW)	Int	W/VE: RST and state/province; DX: RST and power	Dec *QST*	Third full weekend
Mar	ARRL International DX Contest (PH)	Int	W/VE: RST and state/province; DX: RST and power	Dec *QST*	First full weekend
Mar	CQ WPX Contest (PH)	Int	RST and consecutive serial #	Jan *CQ*, Feb CC	Last full weekend
May	Russian CQ-M Contest (PH and CW)	Int	RST and consecutive serial #	Apr CC	Second full weekend
May	CQ WPX Contest (CW)	Int	RST and consecutive serial #	Jan *CQ*; Feb CC	Last full weekend
Jun	All Asian DX Contest (CW)	Int	Signal report and age	June CC	Third full weekend
Jun	ARRL June VHF QSO Party	Dom	Grid Square locator	May *QST*	Second full weekend
Jun	ARRL Field Day	Dom	Transmitter classification and ARRL Section	May *QST*	Fourth full weekend
Jul	IARU HF World Championships (PH and CW)	Int	RST and ITU zone	Apr *QST*	Second full weekend
Jul	CQ Worldwide VHF Contest	Int	Call Sign and Grid square	*CQ* July, July CC	Second full weekend
Aug	ARRL 10 GHz and Up Cumulative Contest	Dom	Grid Square Locator	June *QST*	Third Full Weekends Aug & Sep
Aug	ARRL UHF Contest	Dom	Grid square locator	July *QST*	First full weekend
Aug	Worked All Europe (CW)	Int	RST and consecutive serial number	May CC	Second full weekend
Jul/Aug	North American QSO Party (RTTY}	Dom	Name and location (state/province/country)	*NCJ* Jan/Feb; Aug CC	As announced
Aug	North American QSO Party (CW)	Dom	Name and location (state/province/country)	*NCJ* Nov/Dec; Aug CC	First weekend
Aug	North American QSO Party (PH)	Dom	Name and location (state/province/country)	*NCJ* Nov/Dec; Aug CC	Third weekend
Sep	North American Sprint (CW)	Dom	Name and location (state/province/country)	*NCJ* Jan/Feb, Sep CC	First Sunday
Sep	North American Sprint (PH)	Dom	Name and location (state/province/country)	*NCJ* Jan/Feb, Sep CC	Second Sunday
Sep	CQ/RJ Worldwide DX Contest (RTTY)	Int	RST and CQ Zone	*CQ* Sep	Last full weekend
Sep	ARRL September VHF QSO Party	Dom	Grid square locator	Aug *QST*	Second full weekend
Sep	Worked All Europe (PH)	Int	RST and consecutive serial #	May CC	Second full weekend
Sep	All Asian DX Contest (PH)	Int	Signal report and age	Sep CC	First full weekend
Oct	CQ Worldwide DX Contest (PH)	Int	RST and CQ Zone	Sep *CQ*, Oct CC	Last full weekend
Nov	ARRL November Sweepstakes (CW)	Dom	Serial #, category, Call, year licensed, section	Oct *QST*	First full weekend
Nov	ARRL November Sweepstakes (PH)	Dom	Serial #, category, Call, year licensed, section	Oct *QST*	Third full weekend
Nov	CQ Worldwide DX Contest (CW)	Int	RST and CQ Zone	Sep *CQ*, Oct CC	Last full weekend
Dec	ARRL 160 Meter Contest (CW)	Int	W/VE: RST and ARRL Section, DX: RST	Nov *QST*	First full weekend
Dec	ARRL 10 Meter Contest (CW and PH)	Int	W/VE: RST and state/province; DX: RST and serial #	Nov *QST*	Second full weekend
Various	State QSO Parties	Dom	As announced	monthly in CC	
Varies	ARRL International EME Contest	Int	Callsign and signal report	Sep *QST*	2 full weekends in fall as announced

Scope indicates Domestic or International.

Chapter 11

Operating Awards

Achievement Recognition

Steve Ford, WB8IMY
ARRL Headquarters

Awards hunting is a significant part of the life-support system of Amateur Radio operating. It's a major motivating force of so many of the contacts that occur on the bands day after day. It takes skillful operating to qualify, and the reward of having a beautiful certificate or plaque on your ham-shack wall commemorating your achievement is very gratifying. (On the other hand, you don't necessarily have to seek them actively; just pull out your shoebox of QSLs on a cold, winter afternoon, and see what gems you already have on hand.) Aside from expanding your Amateur Radio-related knowledge, it's also a fascinating way to learn about the geography, history or political structure of another country, or perhaps even your own. This chapter provides information on awards sponsored by ARRL plus some other awards that may be of interest to you.

There are some basic considerations to keep in mind when applying for awards. Always carefully read the rules, so that your application complies fully. Use the standard award application if possible, and make sure your application is neat and legible, and indicates clearly what you are applying for. Official rules and application materials are available directly from the organization sponsoring the particular award; always include an SASE (self-addressed, stamped envelope) or, in the case of international awards, a self-addressed envelope with IRCs (International Reply Coupons, available from your local Post Office) when making such requests. Sufficient return postage should also be included when directing awards-related correspondence to Awards Managers, many (if not most) of whom are volunteers. So above all, be patient!

As to QSL cards, if they are required to be included with your application, send them the safest possible way and always include sufficient return postage for their return the same way. It is vital that you check your cards carefully before mailing them; make sure each card contains your call and other substantiating information (band, mode, and so on). Above all, don't send cards that are altered or have mark-overs, even if such modifications are made by the amateur filling out the card. Altered cards, even if such alterations are made in "good faith," are not acceptable on this no-fault basis. If you are unsure about a particular card, don't submit it. Secure a replacement.

None of the above is meant to diminish your enthusiasm for awards hunting. Just the opposite, since this chapter has been painstakingly put together to make awards hunting even more enjoyable. These are just helpful hints to make things even more fun for all concerned. Chasing awards is a robust facet of hamming that makes each and every QSO a key element in your present or future Amateur Radio success.

ARRL AWARDS

To make Amateur Radio QSOing more enjoyable and to add challenge, the League sponsors awards for operating achievement, some of which are the most popular awards in ham radio. Except for the RCC and Code Proficiency awards, applicants in the US and possessions, Canada and Puerto Rico must be League members to apply.

Rag Chewers' Club (RCC)

New hams often go for RCC as their first operating award. This award is designed to encourage friendly contacts of more substance than the hello-goodbye type QSO. RCC has just one requirement: "Chew the rag" over the air for at least one solid half hour. If you want to obtain the RCC certificate, report the QSO to ARRL HQ, with a fee of $5, and you'll soon be issued the distinctive award. If you want to nominate someone else for membership, send the nomination to him or her, for forwarding to ARRL HQ. This way, no one gets an unwanted certificate, and confirmed rag chewers can still nominate those they think are qualified. RCC is available to all amateur licensees.

Friendship Award

The purpose of this award is to encourage friendly conversational contacts between radio amateurs (hams) and thereby discover new friends through personal communication with others. The ARRL Friendship Award is available

to any ARRL member who submits log extracts that show two-way communications with 26 stations whose call signs end with each of the 26 letters of the alphabet. (For example: W4RA, KØORB, W3ABC . . . K1ZZ.) QSL cards are not required. Applications sent to ARRL HQ must include a fee of $5. (ARRL affiliated clubs may issue Friendship, Rag Chewers' Club, or Old-Timers' Club awards.)

Logs must indicate the contact date, call sign, name, location and another fact about the person contacted. (For example, age, other interests and occupation, etc.) Contacts must be made since November 1993 or after and may be made on any frequency and mode. (Contacts made through repeaters or satellites are both permitted and welcomed.) Contacts may be made from any number of locations. Special endorsements (band, mode, etc) are not available for this award.

Complete the application and submit it to any participating ARRL-affiliated club or to ARRL HQ. The application form is included in the References chapter or is available at **http://www.arrl.org/awards/friend/**. The clubs issue these at no cost beyond postage to applicants.

Code Proficiency (CP)

You don't have to be a ham to earn this one. But you do have to copy one of the W1AW qualifying runs. (The current W1AW operating schedule is printed in *QST*, listed on *ARRLWeb* (**http://www.arrl.org**) or available from ARRL HQ for an SASE.) Twice a month, five minutes worth of text is transmitted at the following speeds: 10-15-20-25-30-35 WPM. For a real challenge, W1AW transmits 40 WPM twice yearly. To qualify at any speed, just copy one minute solid. Your copy can be written, printed or typed. Underline the minute you believe you copied perfectly and send this text to ARRL HQ along with your name, call (if licensed) and complete mailing address. Your copy is checked directly against the official W1AW transmission copy, and you'll be advised promptly if you've passed or failed. If the news is good, you'll soon receive either your initial certificate or an appropriate endorsement sticker. Please include an SASE with your submission: 9 × 12 inches with two units of First-Class postage for a certificate or a business-size envelope for an endorsement.

Worked All States (WAS)

The WAS (Worked All States) award is available to all amateurs worldwide who submit proof of having contacted each of the 50 states of the United States of America. The WAS program includes the numbered awards and endorsements listed below.

Two-way communication must be established on amateur bands with each state. Specialty awards and endorsements must be two-way (2×) on that band and/or mode. There is no minimum signal report required. Any or all amateur bands may be used for general WAS. The District of Columbia may be counted for Maryland.

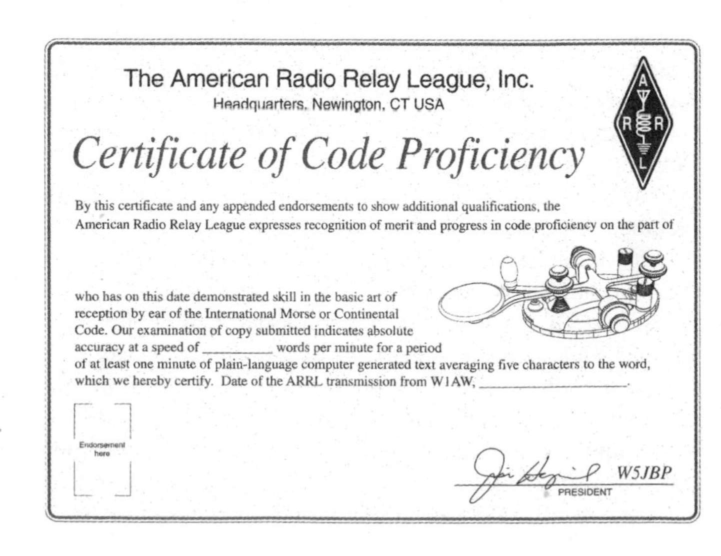

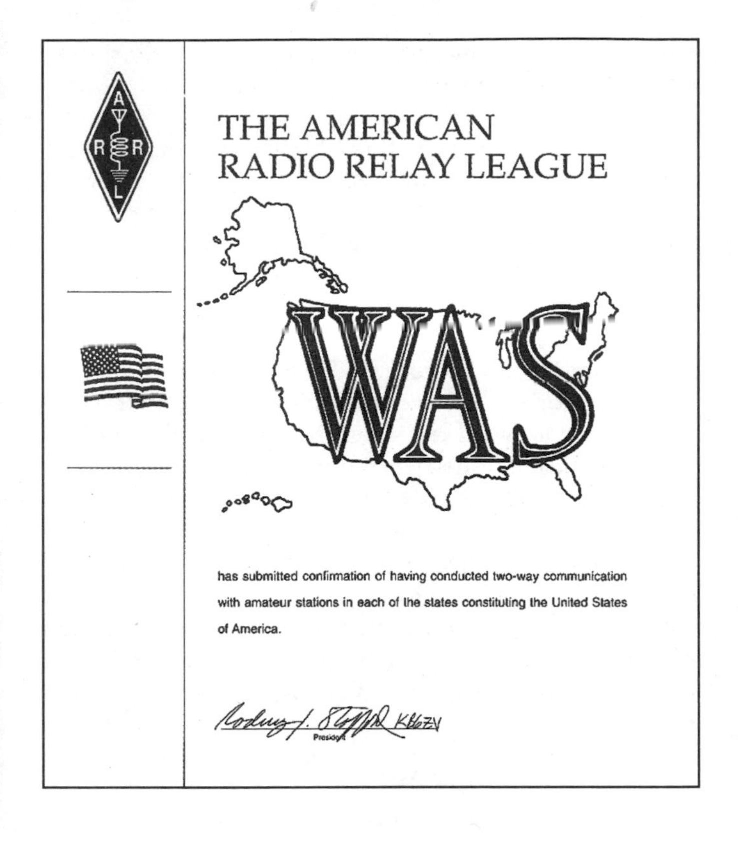

Contacts must all be made from the same location or from locations no two of which are more than 50 miles apart, which is affirmed by signature of the applicant on the application. Club-station applicants, please include clearly the club name and call sign of the club station (or trustee).

Contacts may be made over any period of years. Contacts must be confirmed in writing, preferably in the form of QSL cards. Written confirmations must be submitted (no photocopies). Confirmations must show your call and indicate that two-way communication was established. Applicants for specialty awards or endorsements must submit confirmations that clearly confirm two-way contact on the specialty mode/band. Contacts made with Alaska must be dated January 3, 1959 or later, and with Hawaii dated August 21, 1959 or after.

Specialty awards (numbered separately) are available for OSCAR satellites, SSTV, RTTY, PSK31, 432 MHz, 222 MHz, 144 MHz, 50 MHz and 160 meters. Endorsements for the basic mixed mode/band award and any of the specialty awards are available for SSB, CW, EME, Novice, QRP, packet and any single band except 30 meters. The Novice endorsement is available for the applicant who has worked all states as a Novice licensee. QRP is defined as 5-watts output as used by the applicant only, and is affirmed by signature of the applicant on the application.

Contacts made through "repeater" devices or any other power relay method cannot be used for WAS confirmation. A separate WAS is available for OSCAR contacts. All stations contacted must be "land stations." Contact with ships (anchored or otherwise) and aircraft cannot be counted.

Applicants must be ARRL members to participate in the WAS program. DX stations are exempt from this requirement.

HQ reserves the right to "spot call" for inspection of cards (at ARRL expense) of applications verified by an HF Awards Manager. The purpose of this is not to question the integrity of any individual, but rather to ensure the overall integrity of the program. More difficult-to-be-attained specialty awards (222 MHz WAS, for example) are more likely to be so called. Failure of the applicant to respond to such a spot check will result in nonissuance of the WAS certificate.

Disqualification: False statements on the WAS application or submission of forged or altered cards may result in disqualification. ARRL does not attempt to determine who has altered a submitted card; therefore do not submit any marked-over cards. The decision of the ARRL Awards Committee in such cases is final.

Application Procedure (please follow carefully): Confirmations (QSLs) and application form (MSD-217) may be submitted to an approved ARRL Special Service Club HF Awards Manager. ARRL Special Service Clubs appoint HF Awards Managers whose names/addresses are on file at HQ. If you do not know of an HF Awards Manager in your local area, call a club officer to see if one has been appointed, contact HQ or visit *ARRLWeb* **http://www.arrl.org/was/wasfield.html**. If you can have your application verified locally, you need not submit your cards to HQ. Otherwise, send your application, cards, and required fees to HQ, as indicated on the application form (reproduced in the References chapter).

Be sure that when cards are presented for verification (either locally or to HQ) they are sorted alphabetically by state, as listed on the back of application form MSD-217.

All QSL cards sent to HQ must be accompanied by sufficient postage for their safe return, and the required fee of $5 for each WAS certificate (which includes any endorsements with the same 50 QSL cards), or $3 per endorsement application.

Five-Band WAS (5BWAS)

This award is designed to foster more uniform activity throughout the bands, encourage the development of better antennas and generally offer a challenge to both newcomers and veterans. The basic WAS rules apply, including cards being checked in the field by Awards Managers; in addition, 5BWAS carries a start date of January 1, 1970. Unlike WAS, 5BWAS is a one-time-only award; no band/mode endorsements are available. Contacts made on 10/18/24 MHz are not valid for 5BWAS. The $10 application fee includes the certificate and lapel pin. In addition, a 5BWAS plaque is available at an additional charge.

Pins

WAS and 5BWAS pins may be purchased through the ARRL Publication Sales Department.

Worked All Continents (WAC)

In recognition of international two-way Amateur Radio communication, the International Amateur Radio Union (IARU) issues Worked All Continents certificates to Amateur Radio Stations of the world. Qualification for the WAC award is based on examination by the International Secretariat, or a Member-Society of the IARU, that the applicant has received QSL cards from other amateur stations

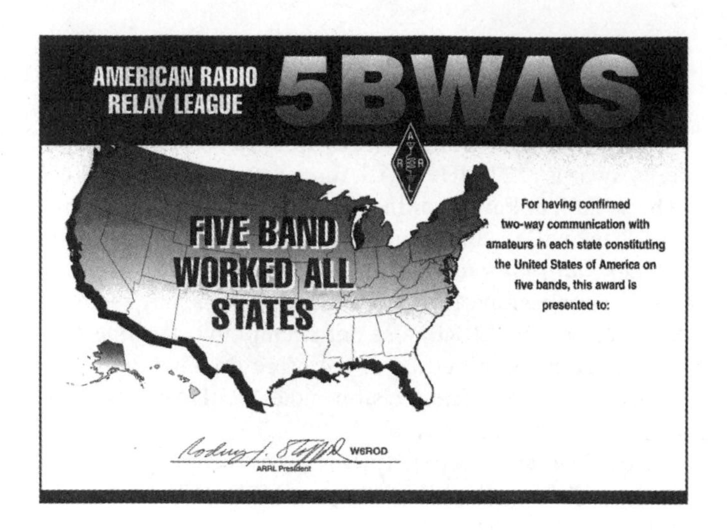

in each of the six continental areas of the world (see the ARRL DXCC List in the References chapter for a complete listing of continents). All contacts must be made from the same country or separate territory within the same continental area of the world. All QSL cards (no photocopies) must show the mode and/or band for any endorsement applied for.

WAC Certificates

The following WAC certificates are available:

- Basic Certificate (mixed mode)
- CW Certificate
- Phone Certificate
- SSTV Certificate
- RTTY Certificate
- FAX Certificate
- Satellite Certificate
- 5-Band Certificate

WAC Endorsements

The following WAC endorsements are available:

- 6-Band endorsement
- QRP endorsement
- 1.8-MHz endorsement
- 3.5-MHz endorsement
- 50-MHz endorsement
- 144-MHz endorsement
- 430-MHz endorsement
- Any higher-band endorsement

Contacts made on 10/18/24 MHz or via satellites are not allowed for the 5-band certificate and 6-band sticker. All contacts for the QRP endorsement must be made on or after January 1, 1985, while running a maximum power of 5-watts output.

For amateurs in the United States or countries without IARU representation, applications should be sent to the IARU International Secretariat, PO Box AAA, Newington, CT 06111, USA. After verification, the cards will be returned, and the award sent soon afterward. There is a $5 fee for US applicants. Sufficient return postage for the cards, in the form of a self-addressed, stamped envelope or funds is required. US amateurs must have current ARRL membership. [A sample application form appears in the References chapter.] All other applicants must be members of their national Amateur Radio Society affiliated with IARU and must apply through the Society only. Note: The DXCC List in the References chapter includes a continent designation for each DXCC country.

A-1 Operator Club (A-1 OP)

Membership in this elite group attests to superior competence and performance in the many facets of Amateur Radio operation: CW, Phone, procedures, copying ability, judgment and courtesy. You must be recommended for the certification independently by two amateurs who already are A-1 Ops. This honor is unsolicited; it is earned through the continuous observance of the very highest operating standards.

Old-Timers' Club (OTC)

In recognition of amateurs who have held an amateur license 20-or-more years (lapses permitted), a suitable award is available—OTC. If you qualify as an "old-timer," you'll find the necessary paperwork pretty easy. Drop a note to HQ (with a fee of $5) with the date of your first amateur license and your call then and now. HQ will verify the information, and if you're eligible, you'll soon receive your OTC certificate by return mail.

The American Radio Relay League, Inc.

A-1 Operator Club

This certifies that:

is a member of the ARRL A-1 Operator Club and is authorized to nominate other deserving qualified radio amateurs for membership.

Membership in the A-1 Operator Club represents adherence to several principles of good operating: careful keying, good voice operating practice, correct procedure, copying ability, judgement and courtesy.

PRESIDENT

VHF/UHF Century Club Award

The VHF/UHF Century Club (VUCC) is awarded for contacts on 50 MHz and above with Maidenhead 2 degree by 1 degree grid locators. Grid locators are designated by a combination of two letters and two numbers. More information on grid locators can be found in the VHF/UHF Operating chapter. *The ARRL World Grid Locator Atlas* ($5) and the *ARRL Grid Locator for North America* ($1) are available from the ARRL Publication Sales Department.

The VUCC certificate and endorsements are available to amateurs worldwide. ARRL membership is required for US hams, possessions and Puerto Rico.

The minimum number of grid locators needed to qualify for a certificate is as follows:

- for 50 MHz, 144 MHz and Satellite—100 Credits;
- for 222 MHz and 432 MHz—50 Credits;
- for 902 MHz and 1296 MHz—25 Credits;
- for 2.3 GHz—10 Credits;
- for 3.4 GHz, 5.7 GHz, 10 GHz, 24 GHz, 47 GHz, 75 GHz, 119 GHz, 142 GHz, 241 GHz and
- Laser (300 GHz)-5 Credits.

VUCC rules appear in the References chapter. Complete information on the VUCC Program can be found on *ARRLWeb* at: **http://www.arrl.org/awards/vucc/**. This includes all the forms, rules and copy of the January 1983 *QST* article.

DX Century Club (DXCC)

DXCC is the premier operating award in Amateur Radio. The DXCC certificate is available to League members in the US and possessions, and Puerto Rico, and all amateurs in the rest of the world. There are several DXCC awards available and fall roughly into four categories:

Mixed bands and modes: **Mixed**
Mode specific: **Phone, CW, RTTY**
Band specific: **160 meter, 80 meter, 40 meter, 20 meter, 10 meter, 6 meter, 2 meter**
Via Satellite: **Satellite**

The complete DXCC rules appear in the References chapter.

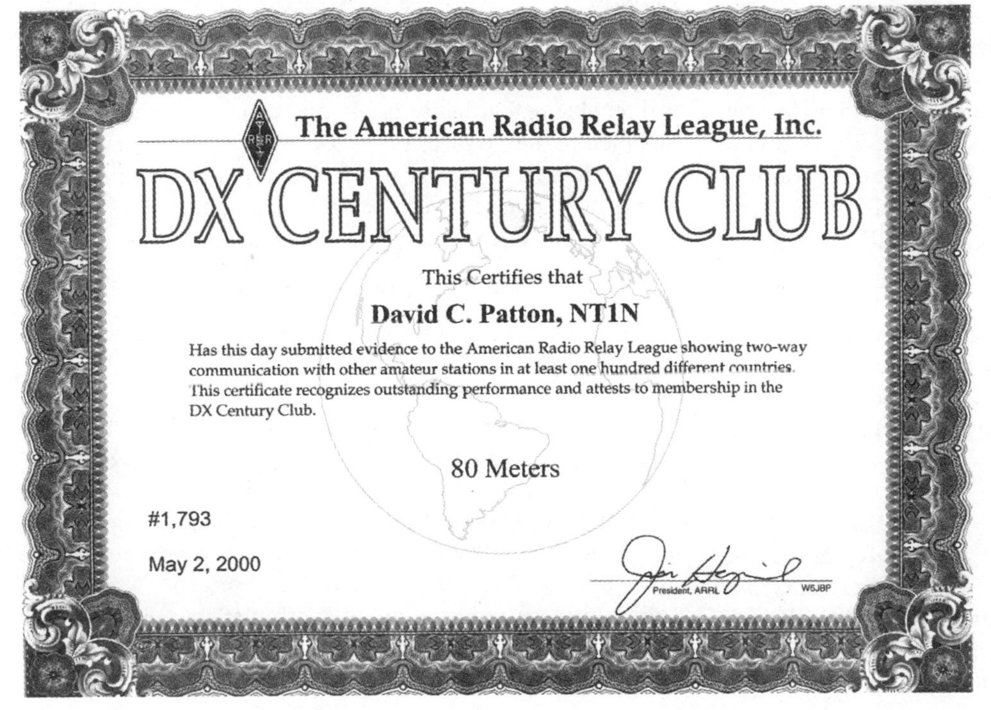

5BDXCC

The Five-Band DXCC Award has been established to encourage more uniform DX activity throughout the amateur bands, encourage the development of more versatile antenna systems and equipment, provide a challenge for DXers, and enhance amateur-band occupancy.

The 5BDXCC certificate is issued after the applicant submits QSLs representing two-way contact with 100 different DXCC countries on each of the 80, 40, 20, 15 and 10-meter Amateur Radio bands. 5BDXCC is endorsable for additional bands: 160 meters,

17 meters, 12 meters, 6 meters, 2 meters. In addition to the 5BDXCC certificate, a 5BDXCC plaque is available at an extra charge.

The DXCC Challenge Award is given for working and confirming at least 1000 DXCC Entities on any Amateur bands, 1.8 through 54 MHz (except 30 meters), for which a separate DXCC is available. The Challenge award is in the form of a plaque, which can be endorsed in increments of 500. Entities for each band are totaled to give the Challenge standing. Deleted entities do not count for this award. All contacts must be made after November 15, 1945. Effective January 1, 2000, credits on the 160, 80, 40, 20, 10 and 6 meter bands may be counted for the Challenge Award. Beginning July 1, 2000, credits for contacts on 15 meters may be included. The DXCC Challenge plaque will be available beginning January 1, 2001.

The DeSoto Cup is presented to the DXCC Challenge leader as of the 30th of September each year. The DeSoto Cup is named for Clinton B. DeSoto, whose definitive article in QST for October, 1935 forms the basis of the DXCC award. Only one cup will be awarded to any single individual.

DXCC List Criteria

The ARRL DXCC List is the result of progressive changes in DXing since 1945. Each Entity on the DXCC List contains some definable political or geographical distinctiveness. While the general policy for qualifying Entities for the DXCC List has remained the same, there has been considerable change in the specific details of criteria which are used to test Entities for their qualifications. See the DXCC rules in the References chapter.

Plaques

- Those who qualify for either 5BWAS and/or 5BDXCC are eligible for a handsome individually engraved 9 × 12-inch walnut plaque. Further information, including required fee, is included in the 5BWAS or 5BDXCC application materials.
- ARRL International DX Contest Awards Program. Beautiful plaques also can be won for specific achievements in the ARRL DX Contest. Details appear in *QST*.

Islands on the Air—IOTA

The IOTA Program was created by Geoff Watts, a leading British shortwave listener, in the mid-1960s. When it was taken over by the RSGB in 1985 it had already become, for some, a favorite award. Its popularity grows each year and it is highly regarded among amateurs world-wide.

The IOTA Program consists of 18 separate awards. They may be claimed by any licensed radio amateur eligible under the General Rules, who can produce evidence of having made two-way communication, since 15 November 1945, with the required number of amateur radio stations located on the islands both worldwide and regional. Part of the fun of IOTA is that it is an evolving program with new

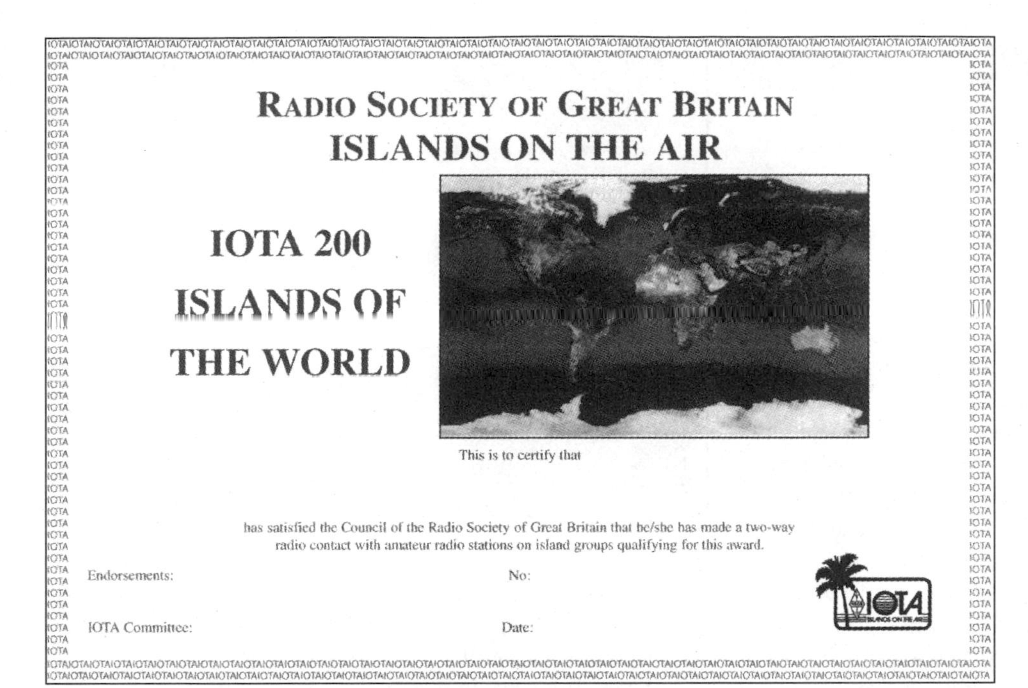

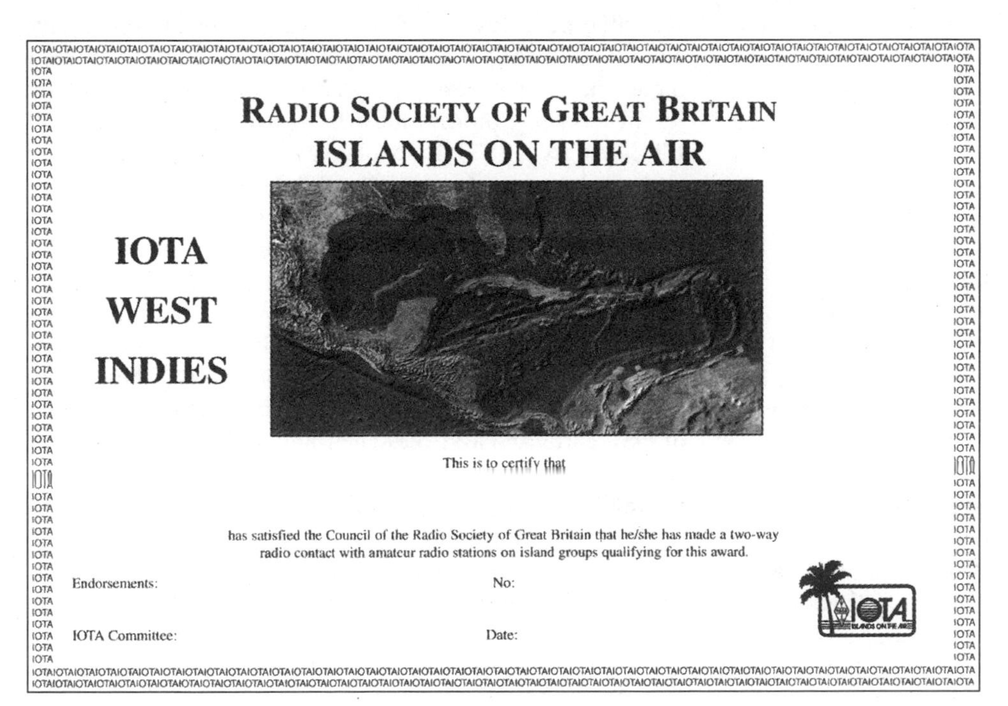

islands being activated for the first time.

The basic award is for working stations located on 100 islands/groups.There are higher achievement awards for working 200, 300, 400, 500, 600 and 700 islands/groups. In addition there are seven continental awards (including Antarctica) and three regional awards—Arctic Islands, British Isles and West Indies—for contacting a specified number of islands/groups listed in each area. The IOTA World-wide diploma is available for working 50% of the numbered groups in each of the seven continents. A Plaque of Excellence is available for confirmed contacts with at least 750 islands/groups. Shields are available for every 25 further islands/groups.

The rules require that QSL cards be submitted to nominated IOTA checkpoints for checking. These checkpoints are listed in the *RSGB IOTA Directory and Yearbook.*

A feature of the IOTA program is the annual Honor Roll which appears in the July edition of *RadCom* and on their Web page at: **http://www.rsgb.org.uk/operate/iota/iota.htm**.

RSGB IOTA Directory

The official source of IOTA information is the IOTA Directory. This lists thousands of islands, grouped by continent and indexed by prefix, details the award rules, and provides application forms and a wealth of information and advice for the island enthusiast. The colorful new IOTA certificates are also shown.

The latest *IOTA Directory* is an essential purchase for those interested in island-chasing.

The RSGB will accept orders by most credit cards, and you can order online by following the links on their Web page. The mail address is: RSGB IOTA Programme, PO Box 9, Potters Bar, Herts, EN6 3RH, United Kingdom. Advance orders can be taken.

RULES FOR *CQ* MAGAZINE AWARDS

Worked all Zones (WAZ)

The CQ WAZ Award will be issued to any licensed amateur station presenting proof of contact with the 40 zones of the world. This proof shall consist of proper QSL cards, which may be checked by any of the authorized CQ checkpoints or sent directly to the WAZ Award Manager. Many of the major DX clubs in the United States and Canada and most national Amateur Radio societies abroad are authorized CQ check points. If in doubt, consult the WAZ Award Manager or the *CQ* Magazine DX editor. Any legal type of emission may be used, providing communication was established after November 15, 1945.

The official CQ WAZ Zone Map and the printed zone list will be used to determine the zone in which a station is located. [The DXCC List, which includes CQ Zones, appears in the References chapter of this book.] Confirmation must be accompanied by a list of claimed zones, using CQ form 1479, showing the call letters of the station contacted within each zone. The list should also clearly show the applicant's name, call letters and complete mailing address.

The applicant should indicate the type of award for which he or she is applying, such as All SSB, All CW, Mixed, All RTTY.

All contacts must be made with licensed, land-based, amateur stations operating in authorized amateur bands, 160-10 meters. All contacts submitted by the applicant must be made from within the same country. It is recommended that each QSL clearly show the station's zone number. When the applicant submits cards for multiple call signs, evidence should be provided to show that he or she also held those call letters. Any altered or forged confirmations will result in permanent disqualification of the applicant.

A processing fee ($4 for subscribers—a recent *CQ* mailing label must be included; $10 for nonsubscribers) and a self-addressed envelope (with sufficient postage or IRCs to return the QSL cards by the class of mail desired and indicated) must accompany each application. IRCs equal in redemption value to the processing fee are acceptable. Checks can be made out to the WAZ Award Manager.

In addition to the conventional certificate for all bands and modes, specially endorsed and numbered certificates are available for phone (including AM), SSB and CW operation. The phone certificate requires that all contacts be two-way phone, the SSB certificate requires that all contacts be two-way SSB and the CW certificate requires that all contacts be two-way CW.

If at the time of the original application, a note is made pertaining to the possibility of a subsequent application for an endorsement or special certificate, only the missing confirmations required for that endorsement need be submitted with the later application, providing a copy of the original authorization signed by the WAZ Manager is enclosed.

Decisions of the CQ DX Awards Advisory Committee on any matter pertaining to the administration of this award will be final.

All applications should be sent to the WAZ Award Manager after the QSL cards have been checked by an authorized CQ checkpoint. Zone maps, printed rules and application forms are available from the WAZ Award Manager. Send a business-size (4 × 9-inch), self-addressed envelope with two units of First-Class postage, or a self-addressed envelope and 3 IRCs.

Single Band WAZ

Effective January 1, 1973, special WAZ Awards will be issued to licensed amateur stations presenting proof of contact with the 40 zones of the world on 80, 40, 20, 15 and 10 meters. Contacts for a Single Band WAZ award must have been made after 0000 hours UTC January 1, 1973. Single-band certificates will be awarded for both two-way phone, including SSB and two-way CW.

5 Band WAZ

Effective January 1, 1979, the CQ DX Department, in cooperation with the CQ DX Awards Advisory Committee, announced a most challenging DX award—5 Band WAZ. Applicants who succeed in presenting proof of contact with the 40 zones of the world on the five HF bands—80, 40, 20, 15 and 10 meters (for a total of 200)—will receive a special certificate in recognition of this achievement.

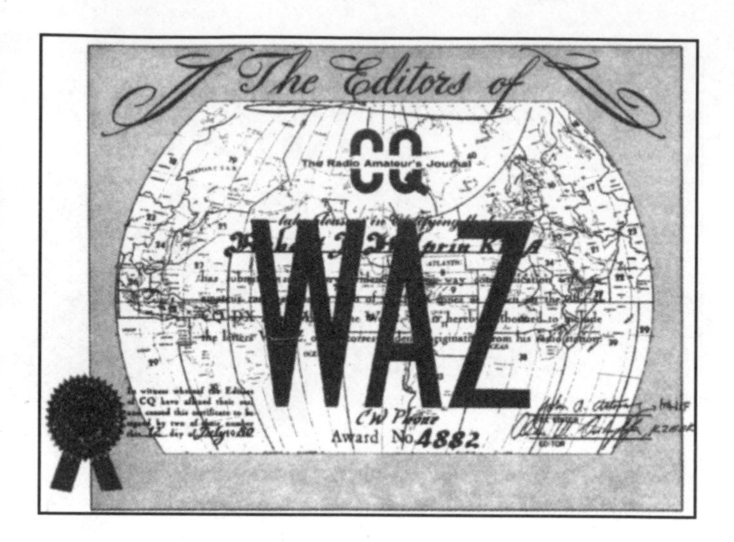

These rules are in effect as of July 1, 1979, and supersede all other rules. Five Band WAZ will be offered for any combination of CW, SSB, phone or RTTY contacts, mixed mode only. Separate awards will not be offered for the different modes. Contacts must have been made after 0000 UTC January 1, 1979. Proof of contact shall consist of proper QSL cards checked only by the WAZ Award Manager. The first plateau will be a total of 150 zones on a combination of the five bands. Applicants should use a separate sheet for each frequency band, using CQ Form 1479.

A regular WAZ or Single Band WAZ is a prerequisite for a 5 Band WAZ certificate. All applications should show the applicant's WAZ number. After the 150 zone certificate is earned, each 10 additional zones requires the submission of QSL cards and a $1 fee. The final objective is 200 zones for a complete 5 Band WAZ. The applicant has a choice of paying a fee for a plaque and/or applying for an endorsement sticker commemorating this achievement.

All applications should be sent to the WAZ Award Manager. The 5 Band Award is governed by the same rules as the regular WAZ Award and uses the same zone boundaries.

WARC Bands WAZ

Effective January 1, 1991, single band WAZ Awards were issued to amateurs presenting proof of contact with the 40 zones of the world on any *one* of the WARC bands: 30, 17 or 12 meters. (Each band constitutes a separate award and may be applied for separately.) This award is available for Mixed Mode, SSB, RTTY or CW. Contacts for each WARC WAZ Award must have been made after each station involved in the contact had permission from its licensing authority to operate on the band and mode.

RTTY WAZ

Special WAZ Awards are issued to Amateur Radio stations presenting proof of contact with the 40 world zones using RTTY. For the mixed band award, QSL cards must show a date of November 15, 1945 or later. The RTTY WAZ is also available with a single band endorsement for 80, 40, 20, 15 or 10 meters. QSL cards submitted for single band endorsements must show a date of January 1, 1973 or later.

WNZ

WNZ stands for "Worked Novice Zones" and is available *only* to holders of a US Novice or Technician license. Proof of contact with at least 25 of the 40 CQ zones as described by the WAZ rules is required. All contacts must be made using the 80, 40, 15 and 10-meter Novice bands. In addition, all contacts must be made while holding a Novice or Technician license, although the application may be submitted at a later date. Contacts must be made prior to receiving authorization to operate with higher class privileges. The WNZ is available as a Mixed Mode, SSB or CW award. It may also be endorsed for a single band. The WNZ award may be used to fulfill part of the application requirement for the WAZ Award when all 40 zones are confirmed.

The basic award can be obtained by submitting QSL cards for 25 zones. The processing fee is $5 for all applicants. All QSL cards must show a date of January 1, 1952 or later. Use CQ Form 1479 to apply for this award.

160 Meter WAZ

The WAZ Award for 160 meters requires that the applicant submit directly to the WAZ Manager QSL cards from at least 30 zones. All QSL cards must be dated January 1, 1975 or later and a $5 fee must accompany all applications. The 160 WAZ is a mixed mode award only. The basic 160 WAZ Award may be secured by submitting QSLs from 30 zones. Stickers for 35, 36, 37, 38, 39 and 40 zones can be obtained from the WAZ Manager upon submission of the QSL cards and $2 for each sticker.

Satellite WAZ

The Satellite WAZ Award is issued to Amateur Radio stations submitting proof of contact with all 40 CQ zones through any Amateur Radio satellite. The award is available for mixed mode only. QSL cards must show a date of January 1, 1989 or later.

The CQ DX Awards Program

The CQ CW DX Award and CQ SSB DX Award are issued to any amateur station submitting proof of contact with 100 or more countries on CW or SSB. The CQ DX RTTY Award is issued to any amateur station submitting proof of contact with 100 countries on RTTY. Applications should be submitted on the official CQ DX Award application (Form 1067B). All QSOs must be 2× SSB, 2× CW or 2× RTTY. Cross-mode or one-way QSOs are not valid for the CQ DX Awards. QSLs must be listed in alphabetical order by prefix, and all QSOs must be dated after November 15, 1945. QSL cards must be verified by one of the authorized checkpoints for the CQ DX Awards or must be included with the application. If the cards are sent directly to the CQ DX Awards Manager, Ted Melinosky, K1BV, 65 Glebe Road, Spofford, NH 03462-4411, postage for return by First Class mail must be included. If certified or registered mail return is desired, sufficient postage should be included. Country endorsements for 150, 200, 250, 275, 300, 310 and 320 countries will be issued.

Any altered or forged confirmations will result in permanent disqualification of the applicant. Fair play and good sportsmanship in operating are required for all amateurs working toward CQ DX Awards. Continued use of poor ethics will result in disqualification of the applicant. A fee of $4 for subscribers (subscribers must include a recent *CQ* mailing label with their application) or $10 for nonsubscribers, or the equivalent in IRCs, is required for each award to defray the cost of the certificate and handling. An SASE or 1 IRC is required for each endorsement.

The ARRL DXCC List (see the References chapter) constitutes the basis for the CQ DX Award country status. Deleted countries will not be valid for the CQ DX Awards. Once a country has lost its status as a current country, it will automatically be deleted from our records. All contacts must be with licensed land-based amateur stations working in authorized amateur bands. Contacts with ships and aircraft cannot be counted. Decisions of the CQ DX Advisory Committee on any matter pertaining to the administration of these awards shall be final.

To promote multiband usage and special operating skills, special endorsements are available for a fee of $1 each:

- A 28-MHz endorsement for 100 or more countries confirmed on the 28-MHz band.
- A 3.5/7-MHz endorsement for 100 or more countries confirmed using any combination of the 3.5 and 7-MHz bands.
- A 1.8-MHz endorsement for 50 or more countries confirmed on the 1.8-MHz band.
- A QRP endorsement for 50 or more countries confirmed using 5-watts input or less.
- A Mobile endorsement for 50 or more countries confirmed while operating mobile. The call-area requirement is waived for this endorsement.
- An SSTV endorsement for 50 or more countries confirmed using two-way slow scan TV.
- An OSCAR endorsement for 50 countries confirmed via amateur satellite.

(After the basic award is issued, only a listing of confirmed QSOs is required for these seven special endorsements. However, specific QSLs may be requested by Award Manager N4UF.)

The CQ DX Honor Roll will list all stations with a total of 275 countries or more. Separate Honor Rolls will be maintained for SSB and CW. To remain on the Honor Roll, a station's country total must be updated annually.

USA-CA Rules and Program

The United States of America Counties Award, also sponsored by *CQ*, is issued for confirmed two-way radio contacts with specified numbers of US counties under rules and conditions below. [Note: A complete list of US counties appears in the References chapter.]

The USA-CA is issued in seven (7) different classes, each a separate achievement as endorsed on the basic certificate by use of special seals for higher class. Also, special

endorsements will be made for all one band or mode operations subject to the rules.

Class	*Counties Required*	*States Required*
USA-500	500	Any
USA-1000	1000	25
USA-1500	1500	45
USA-2000	2000	50
USA-2500	2500	50
USA-3000	3000	50

USA 3076-CA for ALL counties and Special Honors Plaque [$40]

USA-CA is available to all licensed amateurs everywhere in the world and is issued to them as individuals for all county contacts made, regardless of calls held, operating QTHs or dates. All contacts must be confirmed by QSL, and such QSLs must be in one's possession for identification by certification officials. Any QSL card found to be altered in any way disqualifies the applicant. QSOs via repeaters, satellites, moonbounce and phone patches are not valid for USA-CA. So-called "team" contacts, wherein one person acknowledges a signal report and another returns a signal report, while both amateur call signs are logged, are not valid for USA-CA. Acceptable contact can be made with only one station at a time.

The National Zip Code & Directory of Post Offices will be helpful in some cases in determining identity of counties of contacts as ascertained by name of nearest municipality. Publication No. 65, Stock no. 039-000-00264-7, is available at your local Post Office or from the Superintendent of Documents, US Government Printing Office, Washington, DC 20402, but will be shipped only to US or Canada.

Unless otherwise indicated on QSL cards, the QTH printed on cards will determine country identity. For mobile and portable operations, the postmark shall identify the county unless information stated on QSL cards makes other positive identity. In the case of cities, parks or reservations not within counties proper, applicants may claim any one of adjoining counties for credit (once).

The USA-CA program will be administered by a *CQ* staff member acting as USA-CA Custodian, and all applications and related correspondence should be sent directly to the custodian at his or her QTH. Decisions of the Custodian in administering these rules and their interpretation, including future amendments, are final.

The scope of USA-CA makes it mandatory that special Record Books be used for application. For this purpose, *CQ* has provided a 64-page $4^1/_4 \times 11$-inch Record Book which contains application and certification forms and which provides record-log space meeting the conditions of any class award and/or endorsement requested. A completed USA-CA Record Book *constitutes medium of basic application* and becomes the property of *CQ* for record purposes. On subsequent applications for either higher classes or for special endorsements, the applicant may use additional Record Books to list required data or may make up his own alphabetical list conforming to requirements. Record Books are to be obtained directly from *CQ*, 25 Newbridge Rd, Hicksville, NY 11801, for $2.50 each. It is recommended that two be obtained, one for application use and one for personal file copy.

Make Record Book entries necessary for county identity and enter other log data necessary to satisfy any special endorsements (band-mode) requested. Have the certification form provided signed by two licensed amateurs (General class or higher) or an official of a national-level radio organization or affiliated club verifying the QSL cards for all contacts as listed have been seen. The USA-CA custodian reserves the right to request any specific cards to satisfy any doubt whatever. In such cases, the applicant should send sufficient postage for return of cards by registered mail. Send the original completed Record Book (not a copy) and certification forms and handling fee. Fee for nonsubscribers to *CQ* is $10 US or IRCs; for subscribers, the fee is $4 or 12 IRCs. (Subscribers, please include recent *CQ* mailing label.) Send applications to USA-CA Custodian, Norm Van Raay, WA3RTY, Star Route 40, Pleasant Mount, PA 18453.

For later applications for higher-class seals, send Record Book or self-prepared list per rules and $1.25 or 6 IRCs handling charge. For application for later special endorsements (band/mode) where certificates must be returned for endorsement, send certificate and $1.50 or 8 IRCs for handling charges. Note: At the time any USA-CA award certificate is being processed, there are no charges other than the basic fee, regardless of number of endorsements or seals; likewise, one may skip lower classes of USA-CA and get higher classes without losing any lower awards credits or paying any fee for them. Also note: IRCs are not accepted from US stations.

[The Mobile Emergency and County Hunters Net meets on 14,336 kHz SSB every day and on 3866 kHz evenings during the winter. The CW County Hunters Net meets on 14,066.5 MHz daily.]

Information on Other Awards

One of the handiest reference books for the awards chaser is *The K1BV DX Awards Directory*. The latest edition comprises 250 loose-leaf, $8^1/_2 \times 11$ inch, 3-hole-punched pages. You'll find information for 3030 different awards from 123 countries.

The K1BV DX Awards Directory also contains a special chapter with hints and suggestions for both the beginning and advanced collector of awards. The price in the USA and Canada is $24 US and includes shipping by Priority Mail. For further information, contact: Ted Melinosky, K1BV, 65 Glebe Road, Spofford, NH, USA 03462-4411. You can also visit Ted's Web page at **http://www.dxawards.com/**.

Reminiscent of the *original* FCC Amateur Radio Extra Class License Certificate (no longer available), this beautiful certificate allows the Extra Class Amateur to display evidence of his achievement. The Extra Class License certificate indicates the name and callsign of the operator as well as the date he or she achieved this top grade. Send your name and address and a copy of your Amateur Extra Class license indicating the year you received your Extra Class license, along with $7.50 ($10.00 for non-League members) to the Awards Branch at ARRL HQ.

Chapter 12

Image Communications

ATV
Art Towslee, WA8RMC
180 Fairdale Ave
Westerville, OH 43081

SSTV/Fax
Dennis Bodson, W4PWF
233 N Columbus St
Arlington, VA 22203

SSTV/Fax
Steven L. Karty, N5SK
8709 Southern Pines Ct
Vienna, VA 22182

Amateur radio operators or hams communicate via radio. That's what our hobby is about. A natural extension to radio communication is television or image communication. Sound interesting? That's what this chapter is all about.

Thousands of ham radio operators in the USA have ventured beyond *talking* and regularly transmit and receive pictures on the ham bands using modes such as slow scan (SSTV), facsimile (FAX) and other experimental digital forms. In the first part of this chapter, we will concentrate on what is called Fast Scan Amateur Television (FSATV) or simply ATV for short. Did you know that at least 5000 hams find enjoyment in ATV?

If you're skeptical, you may ask "What practical use does ATV have, and why would I ever want to do it?" Why communicate via ham radio when you can simply call a person on the phone? It doesn't *have* to be practical, but it is indeed practical—and fun to boot! Are you interested in public safety and emergency communications? You can assist with security by transmitting video of local events such as parades, marathons, airport disaster drills, etc. Maybe you'd like to share videos with fellow ATVers. Perhaps you would just like to see how many video contacts you can make.

In general, ATV is like the television that broadcasters have been transmitting for over 50 years. However, much simpler equipment is used. At first you may think that ATV is overly complex and expensive, but it is quite the contrary on both counts. Yes, it is rather complex, but it is easy and

W8SJV (right) and WA8RMC (left), atop a building with cameras and a 915-MHz FM ATV transmitter sending crowd-observation video to police emergency operation headquarters.

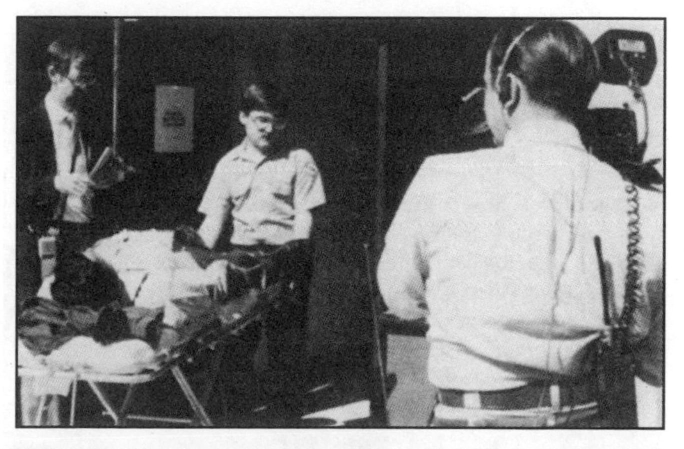

During the National Disaster Medical System drill, ATV was used to send pictures of simulated injury victims arriving at a Virginia hospital to drill coordinators at a USAF base command post in Maryland. (*WA9GVK photo*)

not too expensive mainly because of commercially available home-entertainment equipment.

If you choose to pursue FAX or SSTV, you'll find that communication via great distances is possible, but surplus or consumer equipment is generally not available. However, for ATV, you may already own 2/3 of the main components in a beginner's station, namely, the receiver and camera. Your standard cable-ready TV set will work as a receiver without modification of any kind. The required camera is the same camcorder that you use to record your family and vacation memories.

For SSTV, the camcorder video output must be converted to a slow scan rate via a special converter. For ATV the video output can drive the transmitter directly.

How about the ATV transmitter? Well, *that* can be the hardest part because usually there are no ATV transmitters in use for other purposes. More about that later.

Admittedly, there is one down side. ATV requires a large bandwidth signal to reproduce the picture with sufficient detail and in color. For that reason, among other factors, the communication distance is usually no greater than an average TV broadcast signal in your area. It is important for you to understand this now so you don't create an ATV station with hopes of regularly communicating thousands of miles. But even with the normally limited range of around 50 miles or less, there's a lot of fun and satisfaction to be had. When propagation on the band becomes enhanced and that 100-500 mile DX rolls in, the excitement becomes intense.

The most popular ATV band, by far, is 70 cm (420-450 MHz) where 439.25 MHz (cable channel 60) sees the most activity. To check for local activity, you can monitor this frequency with a cable-ready TV set connected to an outside UHF antenna. However, it's best if you can locate someone in your area who will transmit an ATV picture to you first. The 910-920 MHz, 1250-1280 MHz or 2410-2450 MHz portions of these bands are also in use in many areas, but reception is slightly more involved. A detailed description will follow later.

Hams don't use the same high power levels nor have the same antenna heights that the broadcasters use. Therefore, we must use creativity to help make up some of the difference. While the broadcasters use hundreds of thousands of watts of power and antennas a thousand feet or so above ground, a typical ATV station use less than 50 W with antenna heights of less than 50 feet.

Broadcasters use omnidirectional antennas with less than 10-dB gain; hams can build directional antennas with more than 16-dB gain—at each end. This, combined with the fact that they are usually willing to be satisfied with less than a snow-free signal most of the time, makes two-way communication feasible. In fact, given the overall circumstances, it is surprising how good the communication link can be. If it were not for the challenge to achieve a good two-way communication link, it wouldn't be nearly as much fun. Without doubt this is the challenge that motivates most ATVers.

ATV HISTORY

ATV generally appeals to hams more interested in the technical portion of the hobby. Therefore, many have little concern with contests but lean toward building, modifying and working on equipment. Initially, if you didn't build it, you weren't successful in ATV because there was very little affordable and readily available commercial equipment. Today that's changed dramatically so more and more non-technical people are entering this fascinating hobby and learning to be quite technical as they progress.

ATV in the '50s saw very few people experimenting

Richard Logan, WB3EPX, is an active ATV operator. His station is designed with space limitations in mind.

Doug McKinney, KC3RL, checks out his portable 8-W ATV station powered by a motorcycle battery.

with amateur television. Those who did tended to be TV broadcast engineers who had the knowledge and equipment to get started.

In the '60s there was very little commercial equipment available. Most of the experienced ATVers were busy converting the transmitter portion of taxicab UHF radios trying to squeeze those extra few milliwatts out of a 2C39 or 5894 tube. They got, if they were lucky, a few watts of RF. Feedlines were lossy, so if a watt was delivered to the antenna, that was good.

On the receiving side, the situation was even more difficult. The only way to use commercially available stuff was to modify (retune) a UHF tube-type tuner so that it would cover 439 MHz. Bell Telephone had a special low-noise tube known as the 416B that could be used for a receiving preamp if one of these could be found. If located, you then had to find a machinist willing to fabricate a socket and housing for the circuitry.

Transmitters and receivers were difficult enough but the real stumbling block was the camera. Black and white vidicon cameras were available but generally not on the surplus market. The cheapest way to generate video was with a photo multiplier tube using a TV raster as the scanning device (another story for another day). This method *only* produced still pictures.

The '70s introduced those tiny solid-state devices called transistors and a new era of experimentation began. Many types of low-noise devices became available, but not necessarily affordable for all. Numerous ATV articles appeared in ham magazines for both transmitting and receiving equipment. Black and white vidicon cameras were now becoming more available.

The '80s and early '90s were transitional, as more and more modifiable commercial units became available. When this happened, more hams got involved because it no longer required an engineer to make things work.

Today, individual transistors of earlier times are reduced dramatically in size, combined into entire circuits on a single integrated circuit (IC) and packaged into complete radios. Even more dramatic is the cost reduction. A complete radio circuit is cheaper now than a single transistor was in the '70s. As a result, low-cost, high-efficiency circuitry for ATV without modification is available to everyone. Those expensive color camcorders of the '80s and early '90s are now available at hamfests for $25 or less, putting the cost of this hobby well below many others!

Twenty years ago, no one even dreamed this could happen. So what's next? Have we approached the limit? No, we're still on a nearly vertical learning curve with many more exciting things yet to come. Total digital processes are now emerging, so get involved and join the excitement of the adventures of amateur television.

HOW TO GET STARTED

What comes first? Well, that depends. Many factors influence where you go from here so no one can tell you exactly how to go about it. Instead, the best way is to outline the information resources available so you can choose the best avenue. After all, the best help is to know where to find information, and you'll learn a lot along the way. Here are some places to start for details.

1. **The Internet**. This is a wonderful resource of activity, because many ATV clubs post activity details on their Internet homepage. Area activity, as well as frequencies used, provides details of where and when to look for signals. Some of the most active web sites are:

Columbus, Ohio ATCO group:
http://psycho.psy.ohio-state.edu/atco
Baltimore, Maryland BRATS Group:
http://www.bratsatv.org
Houston, Texas HATS Group:
http://www.hats.stevens.com
Detroit, Michigan DATS Group:
http://www.icircuits.com/dats
Southern California ATN Group:
http://www.qsl.net/atn/atn.htm

2. **Hamfests**. Check *QST* magazine or visit *ARRLWeb* (**http://www.arrl.org/hamfests.html**) for a list of hamfests or ham conventions in your area. Most of the larger hamfests post signs about ATV activity and have ATV forums.

3. ***ARRL Repeater Directory***. Check this directory for ATV repeater listings.

4. **Local Ham Store**. Ask about ATV activity in the area. Most ATVers are well known in these places so a list of individual hams involved in ATV can be compiled. Contact these individuals. Most active ATVers are willing to help newcomers and frequently invite potential ATVers to their hamshack to see firsthand what it is all about.

5. **Mail Order ATV Dealers**. Many are willing to share information about where their equipment is sold. If you are interested in a given manufacture's equipment, ask about it and who owns that item in your area. Most will help you. A few of the larger original equipment ATV dealers that have been in business for a number of years are:

PC Electronics—Transmitters, receivers, preamps (**http://www.hamtv.com**)
Downeast Microwave—Transmitters, converters, preamps (**http://downeastmicrowave.com**)
Directive Systems—Antennas (**http://www.directivesystems.com**)
M^2—Antennas (**http://www.m2inc.com**)
Wyman Research—Transmitters, receivers (**http://www.svs.net/wyman**)
DCI—Filters (**http://www.dci.com**)

6. **Ham Magazines**. *QST* regularly publishes ATV related articles and carries ads for ATV dealers. *ATVQ* magazine is dedicated solely to amateur television. Check **http://www.hampubs.com/atvq.htm** for details. *The ARRL Handbook for Radio Amateurs* ATV section also contains good reference information.

That should keep you busy for awhile. But sooner or later, a transition from planning to implementation must be made. So, a recommendation is in order at this point. Start by *receiving* an ATV signal. After all, if you own a cable ready TV set and an outside UHF antenna with rotation capabilities, you're ready for action. Tune in and enjoy. The rest will follow in the order you prefer.

LICENSING, LIMITS, RESTRICTIONS . . . DON'T THINK TOO BIG!

ATV can be used by any ham with a Technician class license or higher on any ham band 420 MHz and above. Some regional limitations exist for frequency and power levels in parts of the United States, so check the *ARRL Repeater Directory* or local sources before operating. Activity concentrates on different frequencies in different parts of the USA. However, most ATV activity can be found in the 420-440 MHz portion of the 70-cm band. ATV repeaters usually transmit on 421, 426 or 427 MHz and receive on 434 or 439 MHz (cable channel 59 or 60). A few ATV repeaters transmit on 439 MHz and receive on 421, 426 or 427 MHz. If you prefer simplex operation, 439.25 MHz is the most popular transmit and receive frequency.

The only formal FCC imposed frequency restriction is that no operation is allowed outside the allocated ham bands. That includes sideband energy, so it is important to know that television requires a large bandwidth. If not adequately suppressed with appropriate filters, it may extend beyond the band limits. Insufficient space is available here to define and detail the specifics of the various modulation types, so be sure you know the characteristics of your signal before selecting an operating frequency. For example, many ATV repeaters transmit on 421.25 MHz. The lower band limit here is 420.0 MHz so special filtering of the sound/color subcarrier is required to prevent the signal from extending below the 420-MHz limit.

Although the full legal limit of 1500 W is allowed on any of these bands (check for some local exceptions), very few operators exceed about 50 W. One quickly learns that the complexity, technical expertise, mechanical ability and cost all go up exponentially past this point so don't get carried away. There's plenty of time to dream later!

Other than the FCC imposed restrictions, the remaining allocations are mostly by *gentleman's agreement* or locally agreed to band plans. If the local frequency coordinators have assigned a repeater frequency within a given ATV bandpass to someone else, it should be honored. Cooperation is necessary. As the operating frequency is increased to the higher ham bands, i.e. 902-928, 1240-1300, 2390-2450, interference becomes less of a problem because of directivity, activity and losses. If it becomes too crowded in your area, serious consideration should be given to one or more of these higher bands. One last note on the interference issue—ATV is a wide band signal with energy dispersion that decreases rapidly as it departs from center carrier. The ATV receiver must tune this entire bandwidth so it will see all signals within its passband. Therefore, a *foreign* narrow-band signal is much more likely to interfere with an ATV signal than the other way around, so be understanding. We must all share.

SIGNAL DETAILS . . . A BROADCAST STANDARD COMPARISON

At this point, we need a brief description of the type of signals we are dealing with for ATV. The description is broken into two parts, as there are two basic transmission modes in common use.

The first mode is called *vestigial sideband* and is a form of amplitude modulation. This is most common and is used by all broadcast TV stations. In this system, an all white screen will generate a minimum RF signal. An all black screen is a maximum RF signal. Therefore, if you are transmitting a signal and put your hand over the lens, the average power level will be greater than if you display a bright scene.

The amplitude-modulated signal has upper and lower sidebands plus the carrier frequency in the middle. The sidebands result from the modulation process, which produces the sum and difference of the carrier and video frequencies. Only one of these sidebands is needed for picture reconstruction because each is identical.

In broadcasting, the lower sideband is *partially* eliminated to conserve spectrum space. Because part of it remains, it is called vestigial sideband. However, if the lower sideband wasn't partially suppressed, the TV receiver would ignore it anyway. Since the sideband suppression process tends to be complex and expensive, ATV transmitters rarely do it since the receiver doesn't care.

These (AM) signals can be detected with relatively low transmitted power levels (compared to FM modulation, which is discussed later). However, it does take a very strong signal to get a snow-free picture. If you receive a signal that is just discernible (sync bars are visible but the content is not recognizable) and the power level at the transmitter end is 0.1 W (100 mW), the transmitted signal must be boosted by 30 dB or to roughly 100 W to obtain a snow-free signal. Likewise, the received signal can be improved by using better coax, a better antenna or a more sensitive receiver. However, the transmitted signal is relatively easy to create (if you don't use vestigial sideband like the broadcasters) and it's easy to receive on the 70-cm (420-450 MHz) band mainly because it allows the use of a standard TV receiver for reception without modification of any kind.

The second mode is FM (frequency modulation), and it is employed with almost all satellite systems. If you watch satellite TV (either 4-GHz C band or the newer 11/12-GHz Ku band systems) the signal is first downconverted and then FM detected at a 70-MHz intermediate frequency. The resulting video is either fed directly to an ATV video input or used to modulate a signal on TV channel 2 or 3 for reception on your TV set. Because of this, a number of commercially available converters can be used for ATV.

As discussed earlier, a weak AM signal can be detected easily. It is not so with FM. As the FM signal strength is increased, nothing is seen until the limiter circuit in the receiver sees a significant signal. Once that limiter threshold is crossed, from the point of signal recognition to snow-free is, in many cases, only 6-8 dB. As in the earlier example, instead of 100 mW producing the barely recognizable signal, 25 W would be needed if 100 W produced a snow-free picture. In practice, the comparison may or may not be quite as dramatic, but hopefully you get the point.

There are many variables involved, so true comparisons are difficult. On the up side, an FM signal tends to produce a higher resolution picture. However, it's mainly limited by the monitor you use. It is also less subject to fading than its AM counterpart. Also, since noise is fundamentally AM, the FM detection process results in lower noise. Haven't you noticed static crashes on your AM radio but not on FM?

There's no free lunch. Each system has its advantages and disadvantages. You may want to experiment and determine for yourself, if both are used in your area or if you have a friend or two who is willing to work with you.

Because of increased available bandwidths, FM systems dominate the 2.4 GHz and higher bands while 900 and 1200-MHz bands are mixed. By contrast, 420 MHz is the only place where AM enjoys exclusive use.

The differences are shown in the experience of the ATCO (Amateur Television in Central Ohio) ATV group. They originally used 920 MHz with AM modulation for a repeater link. That link is located in a metropolitan area plagued with intermodulation problems from nearby TV broadcast and FM commercial radio transmitters. The link suffered from fading, herringbone bars in the picture and frequent noise bursts. They changed to FM, and the improvement was miraculous—even though the power levels remained about the same! However, if power levels were reduced, there would have been a point where AM could have been recognized and FM wouldn't. The rule is if you have adequate power, are willing to spend more money for increased complexity and want a better quality picture . . . use FM.

When you watch an ATVer's picture, you want to tell the sender how well it is being received. You could say, "Your picture is 20% snow." But that terminology is vague and wordy. The only exceptions are "I can't see it at all," or "You're perfectly snow-free," which we all understand. It's analogous to the digital 1s or 0s indicating on or off, but it's the shades of gray that become a bit more arbitrary. To solve this the *P system* was developed for AM signal reception. It goes like this: P stands for picture level and is divided into six levels from P0 to P5.

A signal received as P0 is recognizable as to its existence only. No detail is discernible and usually only sync bars can be seen in the snow. Since the minimum recognizable signal change is about 3 dB, 6 dB steps are easily recognized and they represent a convenient increment. The numbers continue in 6 dB steps from P0 to P5, which is a snow-free signal and 30 dB greater than P0.

Beyond that, we tend to say "P5 plus," or "broadcast quality." Everyone likes compliments and ATVers are no exception, so if you like what you see, tell the sender about it. However, try not to overdo it.

P-unit reporting is universal across the USA and in other countries as well. This system is accurate only for AM because of the near-linear levels. P-unit reporting of FM signals can be used as long as it's understood that it will not be 6 dB per P unit because of the non-linear nature of the receiver detection system. For a visual representation of what the AM signal for each P-unit level looks like, see **Fig 12-1**.

BUILDING THE STATION

Let's start putting something together. If you have convinced the rest of the family to relinquish the main TV set, let's hope you have been able to see an ATV picture. In any case, now you need equipment for the ham shack . . . The receiver is a good start.

Receiving Equipment

There are many receiver possibilities and combinations available, mainly depending upon which band you want to receive. Let's start with 70 cm (420-450 MHz) because it's the easiest to receive and, universally, most popular.

70 cm

This band is home to much more than ATV operation, but as a rule the segment from 420 to 440 is used for ATV while 440-450 is used for narrowband FM simplex and repeater activity. Of this 420-440 segment, 439.25 MHz (cable channel 60) is the most popular ATV frequency for simplex and repeater inputs, while 434.25 MHz (cable channel 59) is used for simplex and repeater inputs to a lesser extent. Most repeater outputs are 421.25 (cable channel 57), 426.25 (receivable on cable channel 58) and 427.25 (cable channel 58). Other frequencies are in use, but they're generally avoided to preserve most bandplans.

ATV signals in the 70-cm band usually are horizontally polarized. This came about from two main factors: existing broadcast TV antenna systems are horizontal and many early ATVers were weak signal DXers first. They used horizontal antennas exclusively. In addition, horizontally polarized 439.25 MHz signals are better neighbors to the vertically polarized FM voice repeaters in the 440-450 MHz segment because of cross polarization isolation. If you also plan 70-cm narrowband FM activity, consider a cross-polarized antenna, separate antennas or a polarization rotor so vertical and horizontal signals can be accommodated.

One type of 70-cm ATV receiver is the cable-ready TV itself. Almost all TVs manufactured since 1990 are cable ready, so if you own one you have a good start! ATV activity on cable channels 57, 58, 59 and 60 is no coincidence; it is the result of the cable ready sets. Most tuners in these sets are quite good and will suffice to start but usually not as good as a separate receiver made specifically for ATV reception.

A preamp for any receiver is a good idea, but if you decide to purchase a separate receiver, it may already contain an acceptable preamp. Most preamps today are constructed with a GaAsFET transistor (rather than bipolar) which is a good idea for intermodulation rejection characteristics and almost a must for wideband operation. Simple GaAsFET preamps are not expensive, and on this band it's easy to achieve low-noise operation. Expensive microwave devices won't buy anything extra. A good preamp can be had for $25 (kit—assembled by you in your box) or $100 (assembled by manufacturer in a box with connectors).

When buying a preamp, make sure it has a tuned input (filter on the incoming line). This will help deal with the interference and overloading that is apt to occur in metropolitan areas because of narrowband FM activity nearby. Consider the purchase of an *interdigital* type of filter if interference is experienced. This filter has the wide passband characteristics needed for ATV while maintaining steep passband skirts for maximum rejection of unwanted out-of-band signals. The cost is about $100 for a 4-pole unit.

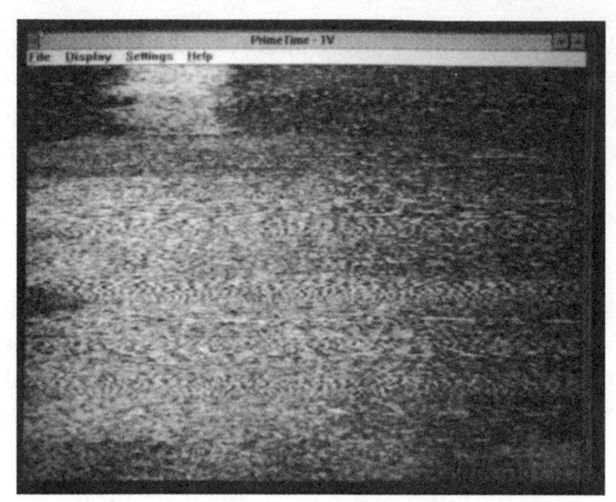

P0—Picture is barely recognizable. Only sync bars can be seen.

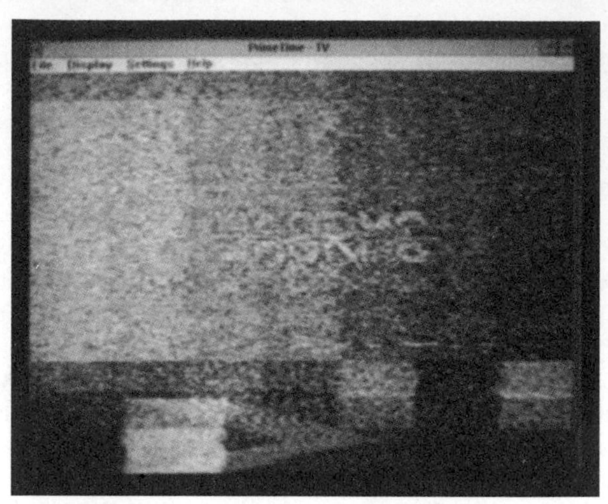

P1—6 dB > P0. Picture is recognizable, but extremely snowy.

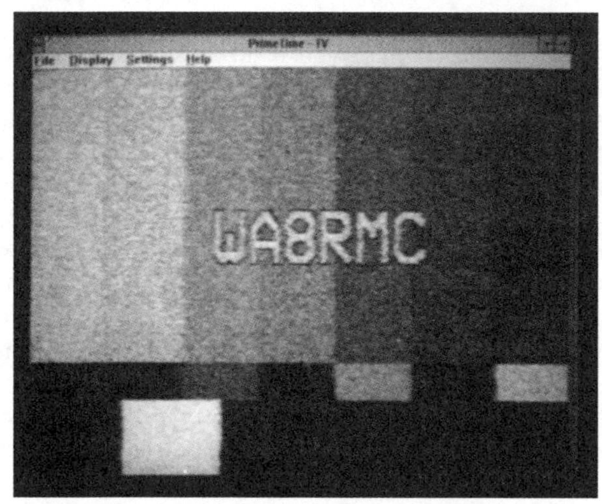

P2—12 dB > P0. Picture is easily recognizable, but lacks detail and still is quite snowy.

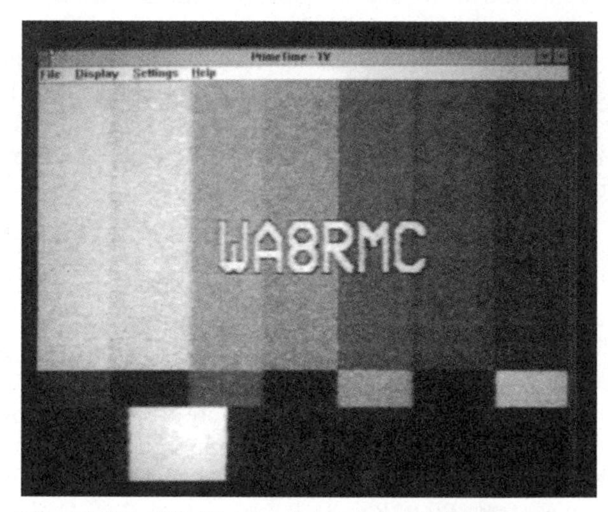

P3—18 dB > P0. Picture detail is much better, but snow is still visible

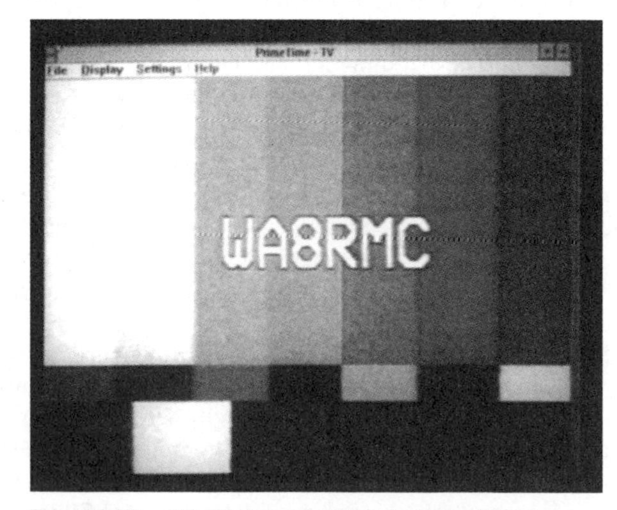

P4—24 dB > P0. Picture detail is better with very little snow.

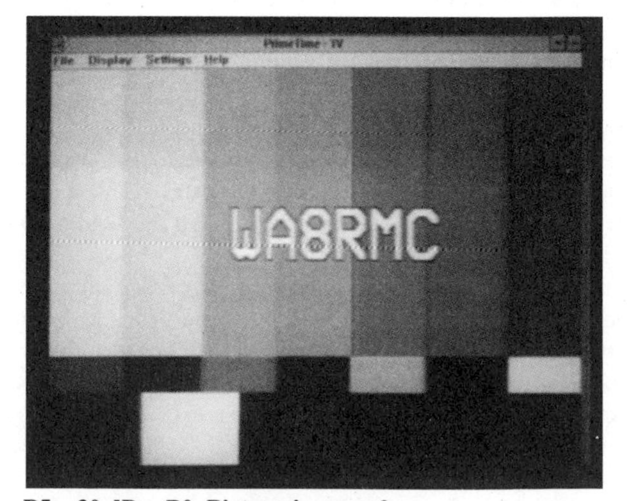

P5—30 dB > P0. Picture is snow free.

Fig 12-1—P-Level Reporting System

Fig 12-2—A complete 20-W 70-cm receiver transmitter combination ideal for the beginner, if an all-in-one package is desired. Its 20-W output may eliminate the need for a final amplifier. ***(Photo courtesy of PC Electronics, Inc.)***

You may ask, "If I buy a good preamp, can I mount it at the antenna and use cheap coax to go to the receiver?" Well, it *might* be a good idea, but let's continue with the receiver discussion first. More on that later.

If you purchase a separate receiver or use your existing TV, think about an additional factor. Sooner or later you are going to want to transmit. Therefore, some means is necessary to disconnect the receiver while transmitting and vice versa. This is usually accomplished with a coaxial relay. If a self-contained receiver/transmitter package is purchased, this feature is usually built in so that issue is simplified and may be more desirable even though the cost is higher. See a typical receiver/transmitter combined package in **Fig 12-2**. If you like to build things and are handy with tools, don't forget about the antenna relay. When the transmitter and receiver are purchased separately, this little item is often forgotten.

33 cm

This band (902-928 MHz) offers no easy opportunity to utilize unmodified commercial equipment. For that reason, building expertise is generally required. Surplus 850-MHz cellular equipment is available that could be modified for use on 902-928, but all of this equipment is for narrow-band FM so it may not work satisfactorily. RadioShack stores used to market what is called Rabbit radio units that operate with low power on 910 MHz. Some people have used these units with limited success but only for short ranges. However, it's a very low-cost way to experiment. There are a few mail order ATV dealers offering FM receiver/transmitter units also.

23 cm

This band (1240-1300 MHz) offers the best opportunity for those who want to get away from the sometimes-crowded 70-cm band. ATV operation here is primarily FM but there are some old timers still using AM. Check for activity in your area. If AM is chosen, a suitable downconverter that takes 1250 MHz and converts it to a suitable VHF TV channel is the best choice. Several are available. The overall activity on this band largely depends upon your location but in general, narrowband SSB and CW are found around 1290-1300 MHz. Simplex and repeater FM ATV inputs can be found around 1270-1280 MHz. FM ATV repeater outputs are generally located below this at around 1250-1270 MHz. Localities with UHF TV station channels 39 to 44 sometimes have second harmonic energy strong enough to cause interference in this band, so ATV activity is usually away from these signals.

The ideal receiver for FM ATV reception is a surplus LNB (low-noise block converter), the type used for a satellite receiver. These units sometimes have acceptable sensitivity, tune the entire band and output video directly for connection to a TV monitor or modulated channel 3 or 4 for connection to a standard TV set. The best part is the cost. They are rarely priced above $30 and if you're lucky, hamfest $5 bargains could happen! Here are some things to look for when searching for these units at your favorite hamfest or garage sale.

- The bad news. Some don't work which may be why they're there in the first place. Now, here's the good news. Most do, and the ones that are bad usually have defective satellite rotator/polarizer circuitry not needed anyway. If you can, power it before purchase. If it lights, it's probably ok for ATV use.
- Make sure the unit that you choose has a switch on either the front or rear panel for *inverted video* operation. In its intended use, the LNB amplifier on the satellite dish converts the 3.8-4.2 GHz incoming signal to a 950-1450 MHz output frequency inverting the signal. It is then fed into the LNB receiver. When it is used for 1250-MHz ATV, the signal from the antenna goes directly to the receiver and no signal inversion takes place so the receiver must be in the inverted video mode if used for ATV. Some receivers don't have this switch and if not, it's useless for ATV use unless an inverting amplifier is added to the video output.
- There are two types of LNB receivers: those that tune 950-1450 MHz (which is what you *do* want) and those that tune 450-950 MHz (which you *don't* want–unless you use it for the 33 cm band). It's nearly impossible to tell which is which if it isn't marked on the rear of the unit at the input connector. Fortunately, very few 450-950 MHz units were in service, so it's very unlikely that you will get one by mistake. (Nevertheless, WA8RMC found one of these disguised as an *ideal* unit.)
- Select a unit with manual switch or pushbutton tuning if possible. Nothing fancy is needed, so don't get one with a lot of bells and whistles. These features will only be bothersome.
- Look at the unit closely. There are many look-alikes with only a 70-MHz input intended to operate with LNA satellite dish voltage-tunable converters. These won't work. (WA8RMC also made that mistake once.)
- Most LNB receivers are equipped to provide 12-18 V dc to power the LNB converter on the dish. They usually provide this dc voltage on the center conductor of the incoming coax, so check for this voltage at the input connector. If found, it must either be removed or disabled, or a dc isolator inserted to prevent a short when the antenna or preamp is connected.
- These receivers are intended for use with an LNB satellite dish low-noise amplifier/converter, which has high signal amplitude output. Therefore, the input sensitivity will

probably be low. Search for an IF gain pot and turn it up if it has one. If not, an input preamp may be needed.

Preamps–Most of the previous discussion about 70-cm preamps applies here also. However, many have untuned front ends and these require an input filter of some type for acceptable operation. Some people have reported that when an untuned preamp was first tried, the results were poorer than with no preamp at all. This usually results from front-end signal overload caused by out-of-band signals. A simple cavity filter will usually fix the problem. Again, the preamps are not expensive and run about the same price as the 70-cm ones.

2400 MHz

This band has grown very rapidly in popularity ever since Radio Shack introduced the Wavecom/Wavecom Jr. short-range video extender units. Other manufacturers have similar ones. They are purchased as a receiver/transmitter pair and operate on 1 of 4 pushbutton selected frequencies of 2411, 2434, 2453 and 2473 MHz. Since two of the channels are in the ham band, they're ideal for 2.4 GHz experiments and short range ATV communication. These units are also modifiable in a number of ways to increase power in the transmitters and provide more available frequencies. Quite a few ATV dealers stock these units as well as modification kits. If you like to experiment, here is a wonderful opportunity to have some fun at minimum cost. The only bad part is the limited range. It is difficult to increase the transmitter power beyond 50 mW or so because of available devices and high cost, so this usually becomes a *buddy project* with normal ranges being a mile or so (and then only if a good line-of-sight path is available).

However, because of the receiver/transmitter compactness, Wavecoms have become the favorite for remote control projects such as radio-controlled airplanes, radio-controlled cars and balloons. Here the low power is an advantage because a very good signal can be achieved with a small and lightweight package. The line-of-sight restriction is not normally a problem because it's a good idea to be able to see these devices while operating or bad things could happen.

Above 2400 MHz

Very little ATV activity is found on the bands above 2400 MHz. They are generally reserved for the more experienced and dedicated ATVer. Most activity is centered around 10 GHz because of the availability of Gunnplexer radar units. You may have seen some of these at hamfests sold as brake-light testers because they can generate signals that activate automotive radar detectors. In any case, these bands are generally reserved for experienced hams willing to experiment.

Transmitting Equipment

Let's investigate transmitter possibilities. As before, let's start with 70 cm.

70 cm

A number of ATV manufacturers make transmitters both in kit and preassembled form. Your choice depends upon your level of expertise. For this band, the *minimum* usable power level is approximately 100 mW. This coupled to a reasonably large antenna with 50 feet or less of low loss feedline will yield a communication range of about 5 miles over flat terrain. However, many entry-level transmitters output about 2 to 5 W average. This is a good starting point because a reasonable picture can be transmitted about 20 miles. In addition, if it is decided to add more power later, most high power amplifiers (50 to 100 W) require about 2-5 W of drive so it's a final amplifier today and an exciter for the high power final amplifier tomorrow. If you settle on a 10-15 W unit now and later want to add a power amp, you'll probably have to attenuate the signal to lower it enough to be compatible with the maximum allowable power level into the amplifier.

Solid state final amplifiers that output up to 100 W usually require about 2-5 W of drive and operate from a 12-VDC power supply. For power output levels beyond that, tubes are normally used. If this range interests you, bear in mind that you'll probably have to build it yourself or buy a home-brew unit from someone else. Amplifiers with one or two 4CX250 tubes are common and can output up to 500 W (for 2 tubes); but remember, beyond about 75 W, the cost goes up exponentially with improved signal strength. Keep in mind that doubling the power level will only yield 1/2 P-unit better picture at the other end. Most hams start with 5-10 W transmitters and then get solid state power amps later to boost the signal to 50-60 W (but seldom beyond that).

33 cm

There's not too much to say about this band for the beginner as far as transmitting equipment is concerned. A few manufacturers provide FM transmitters with outputs in the 1-5 W range and additional amplifiers with outputs in the 15 W range but not much beyond that. It's possible to convert surplus cellular transmitting equipment for use on this band, but most are designed for narrowband service and converting them to accept a 5-MHz-wide video signal may be quite a chore. As mentioned earlier, Rabbit transmitters are available, but the output power of a couple of milliwatts puts them in the toy category.

23 cm

Now here's a neat band with lots of opportunity for adventure. Unfortunately, again there are few, if any complete packaged solutions; only individual modules are available. A few manufacturers have put together units for security purposes that can output up to 100 mW. They are great for ham use because they're either within the amateur band as is or they can be switched there. This is an excellent starting point because with the availability of higher gain antennas, a transmission distance up to 20 miles is possible. Inexpensive FM power amps that require 100 mW or so of drive and output 10-15 W are relatively easy to obtain.

Higher power on this band is not very common and again you must use tubes, with the 2C39 variety the most common. If you go this route, most likely you're going to have to roll your own.

2400 MHz

At the present, this band presents a real transmitter challenge if power levels in the 1 W range and higher are desired. For the beginner, stick with the Wavecom or equal units and be happy. The unmodified Wavecoms output about 0.5 mW but a simple internal attenuator modification will bring the output up to approximately 3 mW. Add an internal MMIC amplifier to easily boost it to about 50-60 mW. At that level, it serves as a very good transmitter for short-range communications of up to 1 mile. The required modifications are easily found in a number of ATV publications as well as on the Internet and are not difficult, even for the beginner.

Above 2400 MHz

Although a number of articles are written about transmitters in this range, little exists for ATV applications except for 10 GHz. These bands are normally reserved for the serious, experienced individuals and definitely not beginner material. They always are buddy projects because of the short ranges involved.

Fig 12-3—W8SJV's antenna arrays. Sandwiched between his HF beam (below) and his 2-meter Yagi (above) is a 21-element 15-dBd Yagi on a 14-foot boom made by M^2. John uses this for his ATV work, and has reported excellent results.

Antennas

What a potentially huge subject! This is an area where a lot of us like to experiment and rightfully so, because there are more articles on this subject than any other in the Amateur Radio (including television) field. However, it is also the least understood. You can find antenna designs of almost every imaginable size and shape with emphasis on many variables for a variety of reasons.

Size is probably the most important factor. For example, many of us live in apartments or in antenna-restricted locations so a small one is anticipated. Then again some lucky people live in rural areas and the spouse doesn't mind looking outside at a tower with multiple antennas on it, so a much larger array may be your choice.

Whatever the size, try to stay with one designed for 50-ohm transmission lines, as they are most popular and will make the transmission line choice much easier. Let's summarize with the things to look for when selecting an antenna specifically for ATV.

Concentrate on 70-cm antennas first because that's where most start. Most ATV signals in this band are horizontally polarized. Antennas with gains below 10 dBd (referenced to a dipole) should not be considered unless repeater-only operation in a metropolitan area is anticipated. These Yagi antennas are compact with boom lengths of about 6 feet or less. Antennas with gains of 10 to 15 dBd are considered intermediate and are most common (8-14 foot booms). Gains greater than 15 dBd are found only on very large antennas with boom lengths in the 15-20 foot category. One last point: Beware of those manufacturers that claim gains that are hard to believe. This is where many stretch the truth. For instance, it is common to state the gain without stating the reference or state it compared to an isotropic radiator, which is about 2 dB less than a dipole. Therefore, 10 dBi is about the same as 8 dBd but it looks much better if dBi values are listed on the data sheet. Sometimes it is just stated as 10 dB so you can guess whether they mean dBi or dBd.

Without a doubt, the Yagi or its derivatives is the antenna of choice for most ATV applications on 70 cm, even if size and space is no factor. It's easy to build if you like that sort of thing but is also the most available commercially. Just try to keep these things in mind:

a) Yagis can be high-Q antennas. That is, they're sensitive to construction variables, surrounding objects and weather conditions for consistent optimum performance. In general, Yagis don't do well in the rain. The SWR goes up and performance goes down.

b) Because of the inherent narrow bandwidth properties of most Yagis, they probably won't be acceptable for optimum ATV use on both 421 MHz and 439 MHz. If a compromise must be made here, optimize it for 439 MHz (your transmit and simplex receive frequency) and take whatever you get at 421 MHz (repeater transmit frequency).

c) The bandwidth limitation at a given frequency *may* degrade ATV performance but usually is not a consideration. Be more concerned about the achievable communication distance than signal quality. If the signal at the receive end is 50% snow (about P3) it won't matter if the resolution is good or fair. In fact, a narrower bandwidth antenna generally has a higher gain for a given size.

d) The physical size is relatively small for a given gain. However, if a really large Yagi is used to get as much gain as possible, difficulty may be found keeping it intact because of wind and ice-loading problems. We all learn this fact the hard way and you probably will be no exception.

e) Polarization arrangements are easily accommodated. If you decide to mount a polarization rotator in the boom, this will work OK unless you decide on a monster antenna. If you decide on two antennas, it's easy to mount two on the same mast.

The next most popular 70-cm ATV antenna is the collinear or broadside array. This antenna is not usually found commercially so home construction is in order. *The ARRL Antenna Book*, as well as other publications, contains good designs.

There are rewards for rolling your own so this avenue should not be ignored. Here are some of the pros and cons:

a) Broadside arrays are low-Q antennas. As a result, construction errors, nearby objects and tolerances generally are not critical. In fact, it's a good idea to stagger the element lengths slightly in construction to help broaden the bandwidth with no significant gain sacrifice. Neat huh?

b) Rain has little impact on its performance.

c) It is more difficult to erect because of vertical construction. It looks similar to bedsprings standing on end fastened to the mast. It takes a lot of vertical mast area, which many folks don't have, but it is very tolerant of nearby antennas.

d) This antenna is very broadband and usually works well for the *entire* 70-cm band. In fact, it is an acceptable antenna for UHF commercial television up to about channel 20 so it can be used to check for band openings by watching the lower UHF TV station signal strength.

e) If all antennas with equal claimed gain are compared, this one will perform better because of its increased capture area.

A number of other antennas stand out for particular individual situations or for the desire to just try something new but are normally not used for ATV purposes. For instance, omnidirectional antennas are commonly employed for ATV repeaters, but shouldn't be used by an individual because of low gain. Remember that you don't get something for nothing. For a given desirable feature, something else must be sacrificed. The loop Yagi and helical antennas are also used but primarily on the higher bands.

The 900 and 1200-MHz bands have more antenna choices because the ones that are too large on 70 cm work just fine here. The Yagi is the first choice, but it usually shows up in a slightly different configuration. It has become attractive to fold each straight element around in a circle to form loops (which actually occupy a little less area for a given antenna) creating what we commonly call the Loop Yagi. It normally has a 1 to 2-dB gain advantage for the same number of elements, so it seems we actually *do* get something for nothing. WA8RMC has used them on both the 900 and 1200 bands with good success.

The helix antenna is worth noting at this point because of its circular polarization properties. Because it's circularly polarized, a 3-dB cross-polarization loss will occur when transmitting to or receiving from a vertical or horizontally polarized antenna. However, if a vertically polarized antenna was receiving a horizontally polarized signal (or vice versa), a cross polarization loss of approximately 25 dB would result. So, if a single antenna must be selected for both polarizations and it can't be rotated for polarity, the 3-dB loss for each could be the better choice.

The collinear or broadside array also can be used on this band, and in many cases is easier to construct because of the small size. The supports for the elements can be fashioned from 1/4-inch dowel rods so if you like to construct model airplanes and cars, this is for you. The bad news is that there are very few published articles on this; the good news is they can be scaled down from existing lower frequency designs.

Antennas for 2400 MHz are similar to 900 and 1200 MHz ones except they are smaller. Loop Yagis dominate this band and are now small enough to be placed inconspicuously almost anywhere. Parabolic dishes are also sometimes used here but the size on this band tends to be rather large. However, as the frequency goes up, the antenna gain must also go up to maintain the same signal level. With all other factors equal, if the frequency is tripled, the antenna gain at each end must be increased by about 5 dB to maintain the same signal strength.

Feedlines

All too often the feedline is not considered, or if it is, not nearly enough emphasis is put on its importance. Since the coax passes energy from the antenna to the receiver as well as from the transmitter, any losses here will be noticed on both transmitting and receiving. Therefore, buy the best coax you can afford, as it will improve both. Coax losses are determined by many factors but it's important to understand a few of the major contributors. *The ARRL Antenna Book* treats the theory and detailed applications very well but here's what you should know about selecting for the UHF bands.

Belden and others make a coax cable (type 9913) that looks similar to RG-8 on the outside which is approximately 0.4-inch in diameter but much stiffer than RG-8. Times Microwave makes similar coax identified as LMR400 Ultraflex cable which is slightly better than 9913. The real difference is on the inside where the dielectric takes on a spiral form to reduce losses. It's pretty good stuff and is the minimum recommended. As the frequency goes up the losses get worse so in general, reserve its use for the 70-cm band and then only for runs of 50 feet or less. Of course, if you have RG-8 and can't find anything better, it will work but expect significant losses. Refer to **Fig 12-4** for a comparison.

An improvement, but not by much, is surplus 75-ohm solid aluminum outer jacket coax. It is available in both 1/2-inch and 3/4-inch sizes, the latter being better, but it has drawbacks. First, the fittings are a little hard to find. Second, the impedance probably doesn't match a selected antenna so it will introduce a mismatch if not compensated. Fortunately, most 50-ohm antennas can be reasonably matched to 75-ohm line. If both coax and fittings are *free*, (ATVers love that word) because it was discarded by the local cable company, go ahead and use them. Even without the matching sections, it will probably be as good as 9913 or LMR cable. Otherwise, buy something better.

Next we move up to the Heliax category. This cable, made by Andrew Corp, has a ribbed solid copper outer conductor and vinyl jacket and is readily available on the surplus market in both 1/2-inch and 7/8-inch sizes. It's used commercially for industrial communications including police, fire and cellular applications. Since new cell sites are springing up at a rapid rate, old installations as well as new cable end runs are available at bargain prices. New 1/2-inch Heliax can cost over $3/foot new from the supplier but it's generally available at hamfests for $1/foot or less, frequently with connectors attached.

The connectors are not easy to work with but they are manageable so be careful and patient. Connectors generally

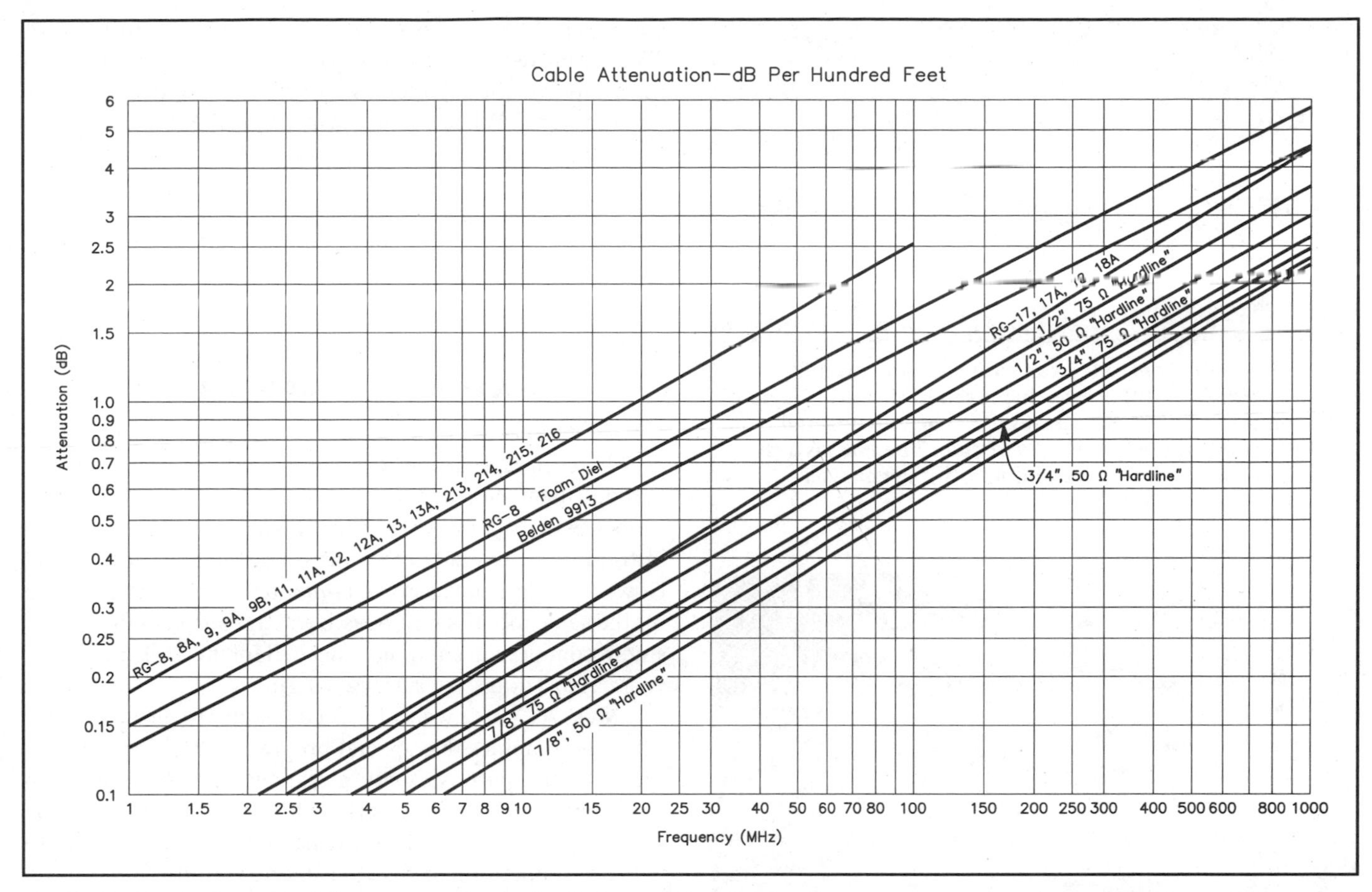

Fig 12-4—Nominal matched-line attenuation in decibels per 100 feet of various common transmission lines that might be used for ATV. Total attenuation is directly proportional to length.

cost about $6 each when purchased separately as cut offs. If you can afford the 7/8-inch size, the losses become greatly less, especially on 1200 MHz and up, so the investment is well worth it. Study the chart in Fig 12-4 for a good comparison.

Oh yes, here's a short message about connectors. Try, if at all possible, to use type N connectors and not the PL-259 type. PL-259 connectors are okay for HF, but they have objectionable losses at 70 cm and above. Also, the N connectors have gaskets to make them waterproof. Do yourself a favor and get the best from the start. They don't cost that much more.

ATV PROJECTS

Here's where it gets fun. After the station is built and a number of contacts have been made, the imagination starts to roam. "What else can I get involved in?" After all it's great talking with and seeing the ATVers but sometimes that's not enough. Many other activities could use ATV involvement; but they sometimes just don't know it! That helps form groups and then clubs dedicated to helping other activities. After all, it's a hobby so it's supposed to be fun. Get involved and make it just that.

Balloon launches are now becoming popular. Hams fill up a 6-foot, or larger, weather balloon with helium, attach an ATV camera with transmitter, put in some avionics and direction locating equipment and launch it. These balloons can reach heights beyond 100,000 feet (19 miles) and usually send back spectacular pictures of Earth's curvature and near outer space. Also, at that altitude the reception distance extends beyond 500 miles, so it provides a show for many viewers. It's exciting, but it's also expensive.

If ballooning is not your fancy, try radio controlled airplanes with a camera and ATV transmitter aboard. The transmitters available in the Wavecom units mentioned earlier operating in the 2.4 GHz band are small enough to fit quite nicely inside a typical model plane and consume little power so small batteries can be used. The antenna is also small so this item is seldom a problem. I've seen the monitor video from a model airplane and it's just like being in a real one. The range with this configuration can exceed one mile!

Don't like planes but cars are your bag? Easier yet! Radio controlled cars don't have as much of a weight restriction, so more hardware can be incorporated. A car constructed by N8QPJ from the Detroit Amateur Television Society (DATS) uses a camera connected to an ATV transmitter for the *eyes* and *ears* (the Wavecom units transmit stereo sound also). It has a 2-meter hand-held receiver for the *voice* so he can literally drive the car around the block and talk to people as it goes. See **Fig 12-5**. Many were surprised and astounded that this unit could actually communi-

Fig 12-5—Radio controlled car with camera and ATV transmit equipment that was displayed at the 1999 Dayton Hamvention by N8UDK and N8QPJ. The antenna and 2.4-GHz transmitter is partially visible. The inset is a close-up view of the camera and pan/tilt mechanism. ***(Photos courtesy of DATS.)***

Fig 12-6—Aerial view of more than 500,000 people waiting the start of the 1999 Columbus, Ohio July 4th fireworks. ATCO members provided video coverage of the event for the police.

cate with them with no one else around.

Want a club activity? It's fun and constructive to supply surveillance video of public-service events such as marathons, airport disaster drills and 4th of July celebrations. To provide police with video at these events is a valuable service to the community. It helps to draw new hams into this segment of the hobby and makes the general public aware of the existence of this hobby. Oh by the way, it's lots of fun for the participants too. **Fig 12-6** is a view of the crowd at the 4th of July fireworks show in Columbus, Ohio. ATCO ATVers helped the local police by supplying surveillance video of the crowd.

Are you interested in being a weather spotter? You know, they are the people who chase tornadoes to give the local weather service details about the incoming storm. Well, chasing severe weather is not what many consider fun, but ATVers frequently provide video to assist the spotters. Here is an activity that a local ATV amateur repeater group can get involved with. More on ATV repeaters later but this too can be a group project. The ATV group in central Ohio (ATCO) provides this service by retransmitting local TV station radar on the ATV repeater upon command to assist the weather spotters with up to date video of an incoming storm. When not in service, any ATVer can bring up the radar signal on command to check on weather conditions.

Here's another one for you. N8UDK attached a camera and transmitter to a miniature blimp (about 6 feet long) and flew it inside a shopping mall. He used a monitor to show mall-goers what it looks like from the ceiling. It's quite an attention getter.

ATV AND COMPUTERS

Computers are now part of our daily lives. They're everywhere. In fact, this article would not have been practical without them. So, why not use them for ATV? For many years we've used cards with the call sign or graphics on them in front of a camera lens for the identification mechanism. Now, an unlimited variety of computer graphics is available for the same job. It's now possible at a fraction of the cost of studio commercial equipment needed for this task only a few years ago. Things are surely changing.

Most ATVers own and use a computer to generate identification graphics. No special program is required specifically for the purpose, only a standard graphics program commonly used for presentation purposes. That way, a slide show of sorts can be shown. It has been my experience to witness competition in this area among various ATVers who compete for the best and most unusual graphic presentation.

The problem, then, is not how to do the graphics razzle dazzle but how to convert the VGA monitor output to NTSC (USA TV broadcast standard) video for ATV transmission. A number of suppliers have come to the rescue with VGA to video converter units. They're not specifically to assist the ATVers but to supply those who want a remote monitor typically needed in the classroom. As these became affordable (purchase price under $100), we started using them for ATV purposes also. So, here again is another inexpensive way to enhance ATV video quality.

ATV REPEATERS

Here is a subject of great interest and becoming very popular among the ATVers. The design, construction and cost of these devices are beyond the scope of the beginner, but communicating through them is simple and can be quite exciting. A number of groups, clubs and individuals around the country have decided to construct and operate a repeater to receive ATV signals and retransmit them at higher power to an omnidirectional antenna mounted in a high location. (This is the same way FM repeaters operate with your hand-held radio. The difference is that video and audio signals are repeated instead of just the audio.) A list of ATV repeaters is available on many Internet homepages or in *The ARRL Repeater Directory*. You might want to check if any exist in your area before starting station construction. The first ATV signal you see most likely will come to you through a repeater.

It also is a great tool to use for fine tuning your transmitter because you can transmit on one frequency and, at the same time, watch your signal on the repeater output frequency. Some repeaters have outputs on different bands specifically for this purpose because it's often very difficult to transmit and receive on the same band without interference problems. In Columbus, Ohio for instance, the ATCO repeater has an ATV input on 439.25 MHz and ATV output on 427.25 MHz with auxiliary ATV outputs on 1250 MHz FM and 2433 MHz FM. A signal sent to it on 439 can easily be watched on 1250 or 2400 with only a very small separate antenna pointed toward the repeater during the 439 signal transmission. Neat huh?

Once you become involved with the people using the repeater, you'll find that help with equipment selection, troubleshooting and all around technical information will be plentiful. These individuals are willing to assist because quite often they had to learn the hard way. There was no help available for them. It's gratifying to be able to help others who are just starting. So make yourself known to the others—the rewards are great.

Once an ATV repeater is located, look for a talk frequency because most repeaters have a 2 meter or 440-MHz FM frequency to serve as a discussion channel and act as a function control for the ATV repeater. The frequency varies from locality to locality but the most popular national frequency is 144.34 MHz simplex. Most areas monitor more than one frequency, so check for others also. For instance, in the Columbus, Ohio area, 147.45 MHz is the primary local talk frequency, which many monitor most of the time. When the band opens up, however, 144.34 MHz becomes primary for out-of-town signals.

Summary

These are some of the main concerns encountered by hams just entering or considering this interesting portion of Amateur Radio. There is no single best recommendation for all. With the information presented here and that gathered from other sources, including the Internet, you will be well on the road toward a successful ATV station. It's a lot of fun and you'll amaze your friends with your capability.

Special thanks to the many folks who helped with needed information and assistance in the preparation of this material.

SLOW-SCAN TELEVISION

Nestled among the CW, phone and RTTY operators in the Amateur Radio bands is a sizable following of hams who regularly exchange still pictures in a matter of seconds virtually anywhere on Earth. They are using a system called slow-scan television (SSTV), which was originally designed in the early 1960s by an amateur—Copthorne Macdonald, VE1BFL. Over the years, the amateur community has been continually refining and improving the quality of SSTV. Amateur success with SSTV for almost four decades has led to its application by the military and commercial users as a reliable long-range, narrow-bandwidth transmission system. The worldwide appeal of SSTV is evident by the many DX stations that are now equipped for this type of picture transmission. Several amateurs have even worked more than 100 DXCC entities on SSTV!

Fig 12-7—SSTV picture as seen on a standard TV set using a digital scan converter.

Just as the name implies, SSTV is the transmission of a picture by slowly transmitting the picture elements, while a television monitor at the receiving end reproduces it in step. The most basic SSTV signal for black and white transmission consists of a variable frequency audio tone from 1500 Hz for black to 2300 Hz for white, with 1200 Hz used for synchronization pulses. Unlike fast-scan television, which uses 30 frames per second, a single SSTV frame takes at least eight seconds to fill the screen. Additionally, the vertical resolution of SSTV is 480 lines or less compared with 525 lines for fast scan. These disadvantages are offset by the fact that SSTV requires less than $^{1}/_{2000}$ of a fast-scan TV's bandwidth. Thus, the FCC permits it in any amateur phone band.

The basic SSTV format represents a tradeoff among bandwidth, picture rate and resolution. To achieve practical HF long-distance communications, the SSTV spectrum was designed to fit into a standard 3-kHz voice bandwidth through a reduction in picture resolution and frame rate. Thus, SSTV resolution is lower than FSTV and is displayed in the form of still pictures. A sample SSTV picture is shown in **Fig 12-7**.

Extraordinary new developments are emerging from the amalgamation of consumer color cameras, digital techniques and personal computers. The greatest advancements currently being made in the realm of color SSTV are increased resolution and improved noise immunity. These efforts have spawned over 40 SSTV modes that are not interoperable. For that reason interface and software developers are designing products with a multiple mode capability. **Table 12-1** lists many of the formats. Scottie and Martin are the two most popular modes.

LICENSE REQUIREMENTS AND OPERATING FREQUENCIES

Since SSTV operation is restricted to the phone bands, license requirements are identical to those for voice operations. While most activity occurs on HF using SSB, it is also well suited to FM at VHF or UHF frequencies. In the US, slow-scan TV using double-sideband AM or FM on the HF bands is not permitted.

The commonly accepted SSTV calling frequencies are 3.845 MHz (Advanced Class), 7.171 MHz (Advanced Class), 14.230 MHz, 14.233 MHz (General Class), 21.340 MHz (General Class), 28.680 MHz (General Class) and 145.5 MHz (All Classes). Traditionally, 20 meters has been the most popular band for SSTV operations, especially on Saturday and Sunday mornings. A weekly international SSTV net (International Visual Communications Association, IVCA), is held each Saturday at 1500 UTC on one of the active 14-MHz frequencies. Many years ago, when SSTV was first authorized, the FCC recommended that SSTVers not spread out across the band even though it was legal to do so. A *gentlemen's agreement* has remained to this day that SSTVers operate as close as possible to the above calling frequencies to maintain problem-free operation.

IDENTIFYING

On SSTV, the legal identification must be made by voice or CW. Sending "This is N5SK" on the screen is not sufficient. Most stations intersperse the picture with comments anyway, so voice ID is not much of a problem. Otherwise SSTV operating procedures are quite similar to those used on SSB.

Table 12-1 SSTV Formats

Mode	*Designator*	*Color Type*	*Scan Time (sec)*	*Scan Lines*	*Notes*
AVT	24	RGB	24	120	D
	90	RGB	90	240	D
	94	RGB	94	200	D
	188	RGB	188	400	D
	125	BW	125	400	D
Martin	M1	RGB	114	240	B
	M2	RGB	58	240	B
	M3	RGB	57	120	C
	M4	RGB	29	120	C
	HQ1	YC	90	240	G
	HQ2	YC	112	240	G
Pasokon TV	P3	RGB	203	16+480	
	P5	RGB	305	16+480	
	P7	RGB	406	16+480	
Robot	8	BW	8	120	A,E
	12	BW	12	120	E
	24	BW	24	240	E
	36	BW	36	240	E
	12	YC	12	120	
	24	YC	24	120	
	36	YC	36	240	
	72	YC	72	240	
Scottie	S1	RGB	110	240	B
	S2	RGB	71	240	B
	S3	RGB	55	120	C
	S4	RGB	36	120	C
	DX	RGB	269	240	B
Wraase SC-1	24	RGB	24	120	C
	48	RGB	48	240	B
	96	RGB	96	240	B
Wraase SC-2	30	RGB	30	128	
	60	RGB	60	256	
	120	RGB	120	256	
	180	RGB	180	256	
Pro-Skan	J120	RGB	120	240	
WinPixPro	GVA 125	BW	125	480	
	GVA 125	RGB	125	240	
	GVA 250	RGB	250	480	
JV Fax	JV Fax Color	RGB	variable	variable	F
FAX 480	Taggart	B&W	138	480 (512 Pixels)	
	Truscan	B&W	128	480 (546 (Pixels)	H
	Colorfax	RGB	384	480 (546 Pixels)	I

Notes

RGB—Red, green and blue components sent separately.

YC—Sent as Luminance (Y) and Chrominance (R-Y and B-Y).

BW—Black and white.

A—Similar to original 8-second black & white standard.

B—Top 16 lines are gray scale. 240 usable lines.

C—Top 8 lines are gray scale. 120 usable lines.

D—AVT modes have a 5-second digital header and no horizontal sync.

E—Robot 1200C doesn't really have B&W mode but it can send red, green or blue memory separately. Traditionally, just the green component is sent for a rough approximation of a b&w image.

F—JV Fax Color mode allows the user to set the number of lines sent. The maximum horizontal resolution is slightly less than 640 pixels. This produces a slow but very high resolution picture. SVGA graphics are required.

G—Available only on Martin 4.6 chipset in Robot 1200C.

H—Vester version of FAX 480 (with Vertical-Interval-Signaling instead of start signal and phasing lines).

I—Trucolor version of Vester Truscan.

EQUIPMENT

All you need to get started is an amateur radio transceiver, a computer or a scan converter, and a video monitor. Like RTTY, SSTV is a 100%-duty-cycle transmission. Most sideband rigs will have to run considerably below their voice power ratings to avoid ruining the final amplifier or power supply.

Early SSTV monitors used long-persistence CRTs much like classical radar displays. In a darkened environment, the image remained visible for a few seconds while the frame was completed. This type of reception is no longer used, and has been replaced with digital scan converters which convert SSTV to FSTV so that conventional television monitors can be used to display the images.

An example of a commercial scan converter is the Robot 1200C. Scan converters are being replaced with innovative software programs based on the work of Ben Vester, K3BC. These computer programs include *Vester Truscan*, *Pasokon TV Lite*, *ProScan*, *JVFAX* and *HamComm*; they all use a simple *clipper* hardware interface, which can easily be built with less than $15 worth of RadioShack parts, because the computer program does all of the processing work previously done by more expensive hardware. See the July 1998 issue of *QST* for "FAX 480 and SSTV Interfaces and Software" beginning on page 32. IBM PC compatible computers with VGA video display monitors can also be used with their existing SoundBlaster compatible sound boards for the interfaces, if other software programs (including *WinSkan* and *WinPix Pro*) are used.

The video source can be a black-and-white or color camera, video recorder or audio tape recorder, or a PC. In addition, page scanners as well as the direct connection of a computer video camera plugged into the USB port of a PC are also excellent sources of pictures. Since the signal emanating from the transmit side of the scan converter is audio, it can be recorded on a cassette tape recorder. The frequency response is not critical, but some consideration should be given for the wow and flutter specifications to minimize picture skew. The combinations of computer, camera and tapes (both audio and video) are endless. Before computers were so prevalent, hams would often tape record some pictures for later on-the-air transmissions. Now that computers are everywhere, the transmitted and received pictures are usually stored as computer files on PC disks.

The results of a QRP contact on SSTV between Ralph Taggart, WB8DQT, in Michigan and Robert McSpadden, WB5UZR, in Texas.

Setting Up Your Studio

Standard photographic lighting arrangements work well for TV. Outdoors, of course, the light level is pretty much predetermined. Care must be taken, however, to prevent direct sunlight from shining into the lens, since TV camera pickup tubes can be damaged. Adjust the brightness and contrast controls on the camera so that the darkest screen area comes out black while the brightest areas reach maximum brightness without *blooming* on the monitor screen. This assumes that the monitor brightness and contrast controls are set correctly. Many of the newer TV cameras are fully automatic and do not need any adjustment, while most of the computer cameras are set up by their computer software driver programs.

It may be useful to run a test audiotape consisting of a test pattern with various gray shades through your system to test the operation of your equipment. Such a tape may be obtained by an on-the-air recording from another amateur. Many computer programs also provide the test signals to adjust your equipment properly.

Operating

Prior to sending a picture, it is important to announce the specific SSTV mode that will be used. Many operators then follow with descriptive verbal comments about the picture.

An SSTV signal must be tuned in properly so the picture will come out with the proper brightness and synchronization. If the signal is not *in sync*, the picture will appear wildly skewed. The easiest way to tune SSTV is to wait for the transmitting operator to say something on voice and then fine tune for proper pitch. With experience, you may find you are able to zero in on an SSTV signal by listening to the sync pulses and by watching for proper synchronization on the screen. Many SSTV computer programs display built-in tuning aids.

If you want to record slow-scan pictures off the air, there are several ways of doing it. One is to tape record the audio signal for later playback. Another is to take a photographic picture of the image right from the SSTV screen. An instant print camera equipped with a close-up lens enables you to see the results shortly after the picture is taken. If you want to do this without darkening the room lights, you'll have to fabricate a light-tight hood to fit between the camera and monitor screen. If you use a scan converter, it is also possible to feed the converter's fast-scan output to a videotape recorder for later viewing. In SSTV systems using computers, the most popular storage medium is the computer's hard disk or floppy disk.

Picture Subjects

The selection of things to show is endless, but you will find that high contrast black-and-white or high-saturation color pictures, that are not too cluttered, work best. A live close in shot of you in front of the operating position is probably the most desirable picture. Don't forget a couple of frames with your call, name and QTH. A close up of your QSL card is ideal, and how about showing off your prized QSLs? Illustrating a technical conversation with a simple schematic diagram can often make clear a point that would be very difficult to get across on voice alone. There is a wealth of good SSTV material in newspapers. Home-brew cartoons are also very popular subjects.

Some SSTVers find it convenient to mount the camera in front of an easel onto which various prepared cards can be inserted. These cards can be photographs, drawings or lettered signs. A kitchen-type note board or menu board with press on removable letters is handy for making headings. Today it is more convenient to use the keyboard on a computer or other specialized captioning devices to produce these letters.

It is fun to make up *programs* of related pictures. Some operators have dozens of programs on floppy disks that can be selected and put on the air in seconds. People might enjoy seeing a pictorial tour of your shack and QTH. Do you have another hobby? Maybe that person you are working would like to learn about your model train collection. A close up of your *1862 Cannonball Express* could be a lot more interesting than a few routine comments about the weather.

Spectacular pictures of Saturn and Jupiter taken by the *Voyager 2* spacecraft and received at the Jet Propulsion Laboratory in California have been relayed via SSTV instantaneously to amateurs around the world. Color SSTV using the ROBOT 1200C scan-converter format came into its own when Space Shuttle *Challenger* carried aloft SSTV equipment on the STS-51F mission and exchanged pictures with Amateur Radio clubs throughout the world (**Fig 12-8**).

Of course, there are more down-to-earth applications of SSTV. Slow-scan television has good potential for rendering service to the public by means of third-party traffic. For example, the scientists working on the Antarctic ice pack

Fig 12-8—SSTV picture using the ROBOT 1200C format transmitted from the space shuttle during the STS-51F mission.

An SSTV view of the St Lawrence Seaway as transmitted from the Russian *Mir* space station. *(Image provided by Farrell Winder, W8ZCF)*

often don't see their families for months—except by amateur SSTV. This was one of the demonstrations that was used by slow-scan pioneers in 1968 to convince the FCC that slow-scan television should be permanently authorized in the ham bands. Anyone can write or call almost anywhere in the world these days, but grandma and grandpa would certainly enjoy seeing pictures of the kids from 3000 miles away by SSTV.

In whatever you show, use your imagination. But remember never to pass up the opportunity to transmit your own image occasionally to let everyone on the frequency know with whom they are communicating!

FACSIMILE

Facsimile (fax) is a method for transmitting very high-resolution still pictures using voice-bandwidth radio circuits. The narrow bandwidth of the fax signal, equivalent to SSTV, provides the potential for worldwide communications on the HF bands. Fax is the oldest of the image-transmitting technologies and has been the primary method of transmitting newspaper photos and weather charts. Facsimile is also used to transmit high-resolution cloud images from both polar-orbit and geostationary satellites. Many of these images are retransmitted using fax on the HF bands.

The resolution of typical fax images greatly exceeds what can be obtained using SSTV or even conventional television (typical images will be made up of 800 to 1600 scanning lines). This high resolution is achieved by slowing down the rate at which the lines are transmitted, resulting in image transmission times in the 4 to 10-minute range. Prior to the advent of digital technology, the only practical way to display such images was to print each line directly to paper as it arrived. The mechanical systems for accomplishing this are known as *facsimile recorders* and are based on either photographic media (a modulated light source exposing film or paper) or various types of direct printing technologies including electrostatic and electrolytic papers.

Modern desktop computers have eliminated bulky fax recorders from most amateur installations. Now the incoming image can be stored in computer memory and viewed on a standard TV monitor or a high-resolution computer graphics display. The use of a color display system makes it entirely practical to transmit color fax images when band conditions permit.

The same computer-based system that handles fax images is often capable of SSTV operation as well, blurring what was once a clear distinction between the two modes. The advent of the personal computer has provided amateurs with a wide range of options within a single imaging installation. SSTV images of low or moderate resolution can be transmitted when crowded band conditions favor short frame transmission times. When band conditions are stable and interference levels are low, the ability to transmit very high-resolution fax images is just a few keystrokes away!

Hardware and Software

In the past few years, electromechanical fax equipment has been replaced by personal computer hardware and software. This replacement allows reception and transmission of various line per minute rates and indices of cooperation by simply pressing a key or clicking a mouse. Many fax programs are available as either commercial software or shareware. Usually, the shareware packages (and often trial versions of the commercial packages) are available by downloading from the Internet.

It is a good idea to check and software programs that you download from the Internet for viruses before—and after—unzipping them. Also before downloading a program from any Internet source, check other sources for newer versions of that same program. It is not uncommon to have older versions posted in one place and newer versions in another.

The *FAX480* computer software program by Ben Vester can also be used with fax as well as SSTV. For more information on this program and others including Web site addresses, see the July 1998 *QST* article "FAX 480 and SSTV Interfaces and Software" starting on page 32. The URL for downloading Vester's software is: **ftp://oak.oakland.edu/pub/hamradio/arrl/bbs/programs**. The file vester_n.zip is freeware. A simple interface for use with this software is shown in **Fig 12-9**. If you use this interface with other soft-

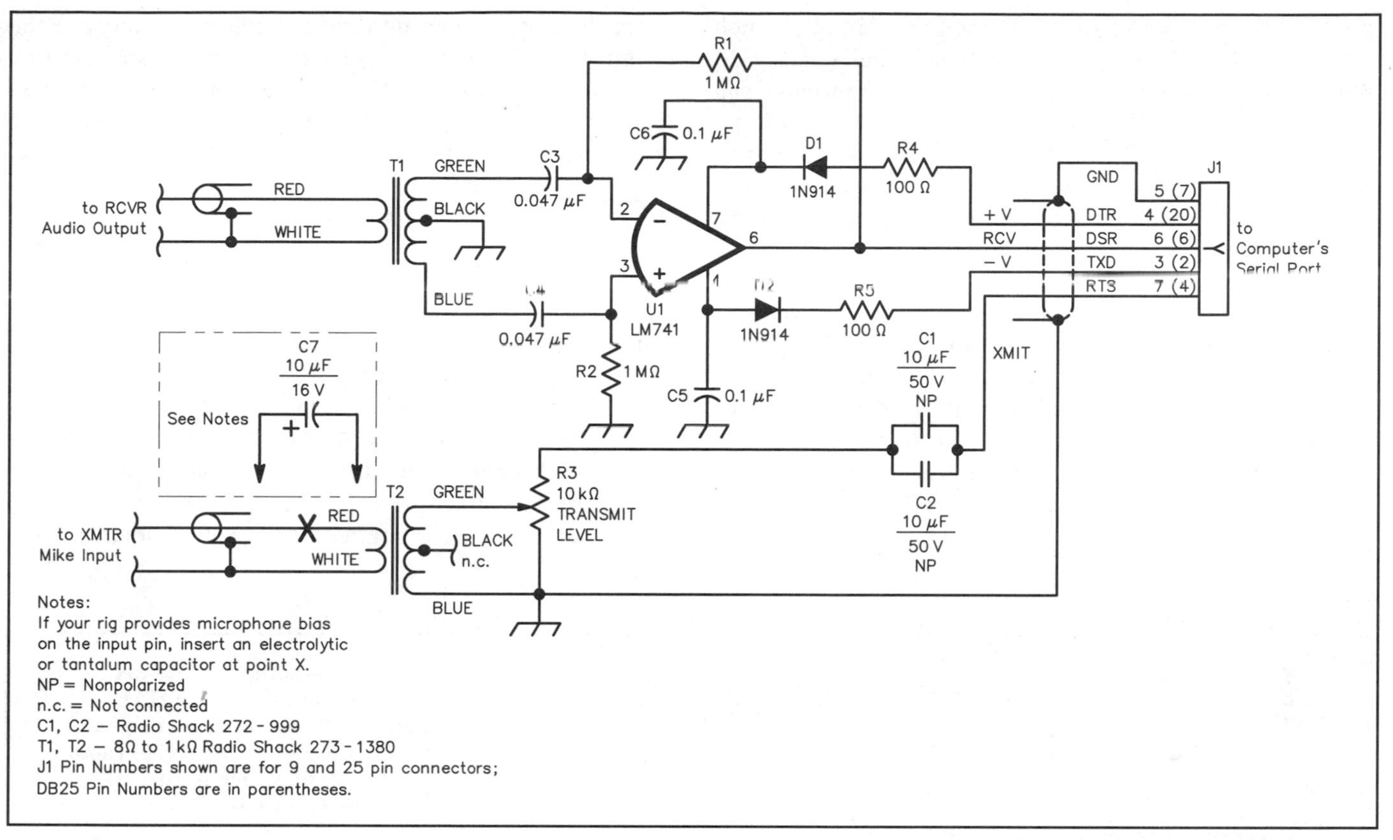

Fig 12-9—Schematic of the Vester simple clipper interface (see text). Unless otherwise specified, resistors are 1/4 W, 5% tolerance carbon-composition or film units. The transformers are used to protect the PC from stray current that might damage motherboard signal or ground traces.

C1, C2—10 µF, 50 V nonpolarized (Radio Shack 272-999)
J1—DB9 or DB25 connector
R3—10 kΩ potentiometer
T1, T2—8Ω to 1 kΩ audio transformer (Radio Shack 273-1380)

ware, you may need to swap the TXD and RTS lines.

Weatherman is a DOS-based program, using a *Sound-Blaster* (or compatible) card as the interface. The program is shareware, receive only, and a single shielded wire from your receiver audio output to the computer audio input is the only connection needed.

WXSat was written to operate under *Windows 3.X*. While specifically set up to decode and store weather satellite APT pictures, it can also be used for HF fax reception.

Both *Weatherman* and *WXSat* are samples of what you can find during a search on the Internet. Often, programs are offered and then either withdrawn or improved over the versions previously distributed. To get the latest and greatest you have to periodically search and see what comes up. If you use an online service such as CompuServe or AOL, they are another source of fax software. Check their ham forums or sections for listings.

Many commercial multi-mode controllers either contain software to receive and transmit fax, or are compatible with PC-hosted software. Available controller suppliers include Hal Communications, **http://www.halcomm.com**, MFJ, **http://www.mfjenterprises.com**, Kantronics, **http://www.kantronics.com**, PacComm, **http://www.paccomm.com** and Timewave Technology (formerly AEA), **http://www.timewave.com**. Check the advertising pages of *QST* for the latest units available.

One well known fax page on the Internet, complete with downloadable software, is **http://www.hffax.de/**. Posted and maintained by Marius Rensen, it contains listings of commercial fax transmissions for you to test your software or just SWL for interest.

To test your fax receive setup, you can usually find fax transmitted weekdays on 3357 (0000 to 1200 UTC), and 10865 kHz (1200 to 2400) from NAM in Cutler, ME; 8682 and 12730 kHz from NMC in Point Reyes, CA. You may have to set your receiver approximately 1.9 kHz higher or lower than the listed frequencies, depending on the station and your selection of USB or LSB.

New Version of HF Fax

A brand new way of sending conventional fax over HF radio is just getting started. It uses a standard Group 3 (G3) fax machine with a HAL CLOVER 2000 (DSP-4100) HF radio data modem, which in turn connects to an HF transceiver. HAL Communications Corporation (**http://www.halcomm.com**) has developed this system which enables a standard G3 fax machine to send fax images without phone lines or satellite terminals. HAL Communications accomplishes this with just two small ancillary devices, which connect between a standard fax machine and an ordinary HF radio transceiver (see **Fig 12-10**). Any G3 fax machine can be connected to the HAL FAX-4100 controller with just a standard RJ-11 modular connector. The FAX-4100 controller connects directly to the HAL CLOVER-2000 (DSP-4100) radio data modem, which in turn connects to the HF transceiver. This entire setup is duplicated at the opposite end of the link.

A *call* is initiated from the fax machine keypad just as if the fax machine were connected to a phone line. The FAX-4100 controller includes a built-in 9600-baud G3 modem, which emulates the telephone system. The controller at the initiating end answers the ring from the originating fax machine, establishes the HF radio link (based on the *phone number*), and handshakes with the controller at the other end to start the receiving fax machine. Fax image data then passes from the fax machine into the controller's memory at the originating end. The controller also establishes a data link between the CLOVER-2000 modems at both ends, then passes the fax data through them and the controller at the receiving end, and finally into the receiving G3 fax machine.

The entire link set up and maintenance procedure is transparent to the fax operator, who need not know nor care that an HF radio system is part of the fax link. It all works just like a standard fax telephone transmission.

To the user, sending a fax over HF radio is a simple three-step process:

1. Lay the page(s) on the fax machine.
2. Enter the ID number of the distant station.
3. Push GO on the fax machine.

Housekeeping control functions and indications are also automated, feeding messages back to the fax machine whenever possible (link failed, other station not available,

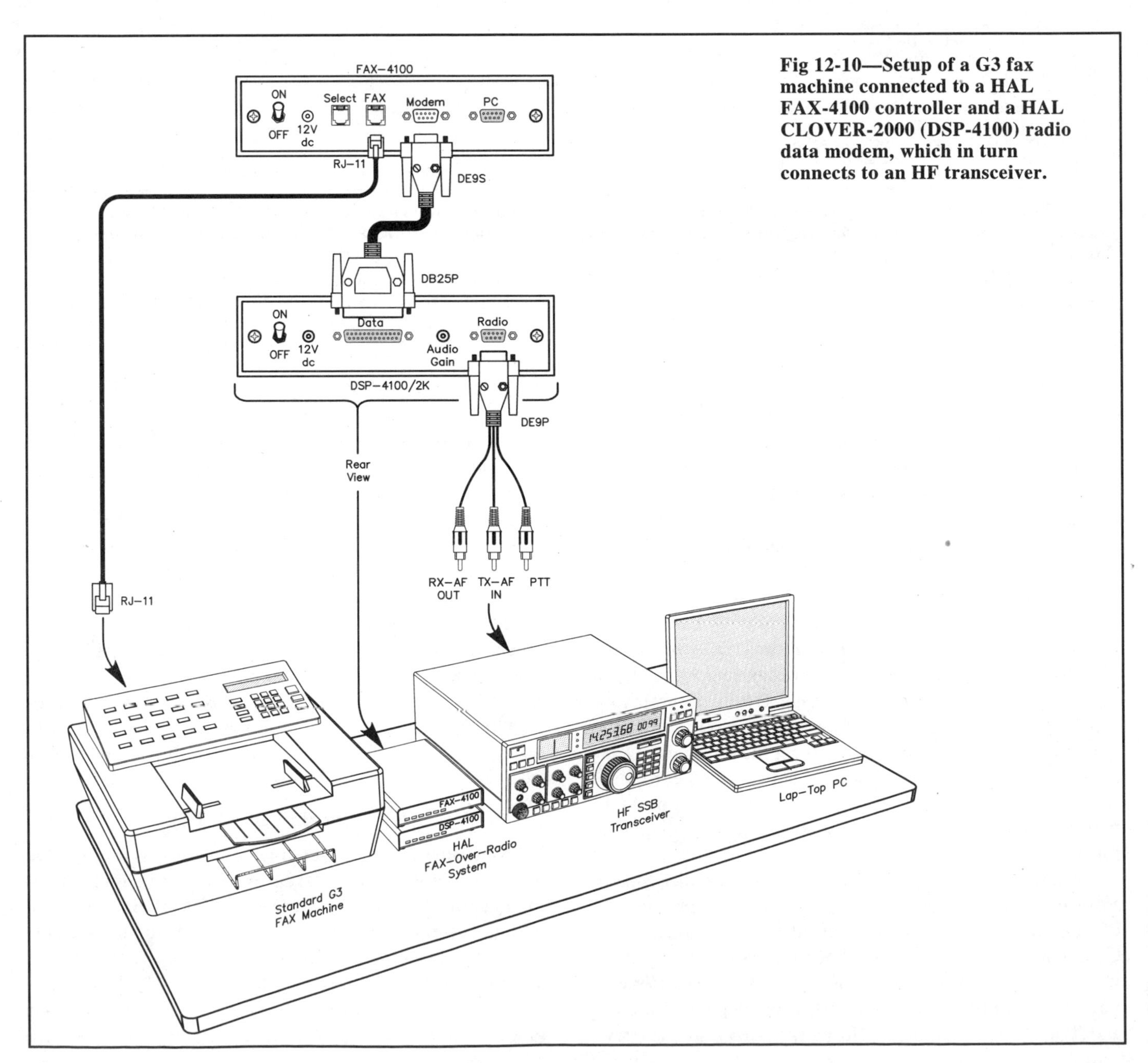

Fig 12-10—Setup of a G3 fax machine connected to a HAL FAX-4100 controller and a HAL CLOVER-2000 (DSP-4100) radio data modem, which in turn connects to an HF transceiver.

etc.). A full page can be sent in 2 to 6 minutes, depending upon ionospheric conditions and density of the page to be transmitted.

FORMATS AND STANDARDS

All facsimile transmissions using voice transmitters are accomplished by feeding a video-modulated tone into the transmitter's audio input. On frequencies below 30 MHz this tone or subcarrier is frequency modulated between 1500 (black) and 2300 Hz (white). This permits the use of audio limiters at the receiving end, making reception relatively immune to fading. Accurate tuning of the SSB receiver is critical to obtaining faithful image reproduction. Many fax adapters incorporate circuits to indicate when the signal is properly tuned.

On the VHF and microwave frequencies used by weather satellites, the subcarrier is amplitude modulated (maximum amplitude = white, minimum amplitude = black). This is practical because the transmissions are made using FM modulation of the carrier, avoiding the effect of Doppler frequency shifts caused by the rapid movement of the spacecraft in orbit.

Unlike conventional television and SSTV, fax transmissions do not employ line synchronization pulses. Proper synchronization of both the transmitter and receiver in a mechanical system is maintained by operating the drums using precision synchronous motors locked to a crystal oscillator frequency standard. Computer-based systems derive their timing from crystal-locked counter chains.

One of the primary standards in facsimile is the rate at which lines are transmitted, expressed as the number of lines per minute (LPM). Successful reception of signals requires that the same line rate be used at both ends of the circuit. Commonly encountered speeds and the associated services include 90 LPM (wirephotos), 120 LPM (HF weather charts and satellite images and polar orbit satellite transmissions), 180 LPM (wirephotos) and 240 LPM (polar orbit and geostationary WEFAX satellites). Some old wirephoto equipment can still be found operating at 60 LPM, but this is too slow for practical amateur work where transmissions need to be kept shorter than 10 minutes. Some 360 and 480-LPM systems are also in use. At those speeds the video bandwidth becomes excessive for use on amateur HF bands, although they would be usable on VHF and UHF links. In general, 120 to 240 LPM systems are probably optimum for amateur use, providing the best trade-off between resolution and image transmission time.

Selection of a suitable surplus fax recorder or conversion of such a system requires dealing with the transmission rate (LPM), the scanning density (lines per inch) relative to the size of the image and the modulation format. These subjects are discussed at length, complete with conversion data, in the Modulation Sources chapter of any recent edition of *The ARRL Handbook*.

COLOR FAX

Amateur fax systems based on the use of high-resolution computer displays are capable of displaying color images using techniques similar to those in use for color SSTV. In such cases, the original color image must be scanned as separate red, green and blue images and then stored in memory for transmission.

Transmission is accomplished in one of two ways. Each red, green and blue version of the complete image can be transmitted in its entirety in sequence. This is known as *frame sequential* transmission. Assuming that the images are formatted correctly at the receiving end, a reproduction of the original color image will be displayed on the color graphics monitor. Alternatively, the red, green and blue versions of each line of the original image can be transmitted in sequence (*line sequential* transmission) such that the image will appear, line by line, in full color as it is received.

At present, two factors limit the extensive use of color on amateur facsimile. First, high-resolution fax images use a great deal of memory. (A color image requires three times the memory capacity of a grayscale image of the same resolution.) The second factor is time. A grayscale fax image that requires 2 minutes for transmission would take three times as long in color. HF band conditions and interference levels are not suited for such extended transmissions. Experimentation on VHF and UHF is quite practical, however, and the use of digital image compression techniques can be expected to make color more practical in the years to come.

WEATHER-SATELLITE FAX RECEPTION

At present, the area of greatest amateur fax activity involves the reception of images transmitted by weather satellites. All of these spacecraft use AM video subcarrier modulation with line rates of 120 and 240 LPM. There are two major categories of weather satellites: those in near-polar orbits at relatively low altitudes (600-800 miles) and geostationary spacecraft located 22,000 miles above the equator.

Most polar-orbiting spacecraft are capable of transmitting both visible light and infrared (IR) images. The US NOAA spacecraft transmit simultaneous visible and IR views during daylight passes with two channels of IR data available at night. Transmissions from these spacecraft occur in the 137-138 MHz range (137.50 and 137.62 MHz in the case of the NOAA satellites) using frequency modulation of the carrier.

Reception requires the use of a receiver with an IF bandwidth of 30-40 kHz and receivers designed for this service, such as the Vanguard Labs WEPIX 2000, are among the most popular options. Modern low-noise RF preamplifiers make it possible to use the omnidirectional turnstile of helix antennas without the need to track the spacecraft. Optimum passes, 10-15 minutes in duration, occur twice a day and an orbital prediction program can be a great help in determining when to listen for a specific spacecraft. AMSAT (the Radio Amateur Satellite Corporation) provides such programs on disk for a number of different personal computers. Check the AMSAT Web page: **http://www.amsat.org**.

Image data are transmitted continuously from polar orbit spacecraft. If the entire pass is recorded, you can expect images to cover an area about 1500 miles north to south and 800 miles east to west in the case of an optimum pass. Not many years ago, home-built or surplus mechanical fax recorders were the primary display systems, but most sta-

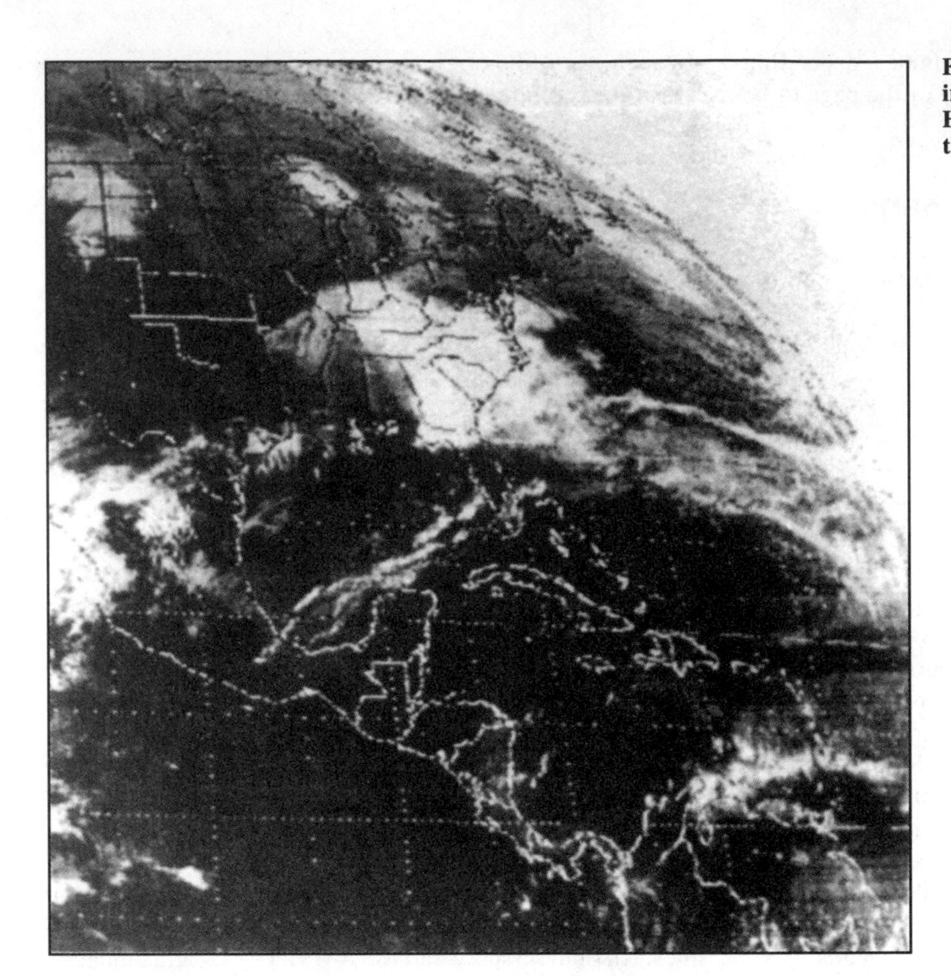

Fig 12-11—A northeast quadrant infrared image obtained from the GOES satellite. High, cold clouds are visible in white in the infrared format.

tions today are using dedicated scan converters or systems using computers with high-resolution graphics capabilities.

Geostationary spacecraft obtain images of the entire globe and transit them to earth in a series of quadrants. In addition, many of these spacecraft also relay weather charts and mosaics prepared from polar-orbit spacecraft data. The US GOES satellite system ideally consists of two primary spacecraft, one over South America and the other over the eastern Pacific, often with a third spacecraft devoted entirely to image relay functions. (Premature failure of spacecraft may reduce the number of available satellites.) A consortium of European nations operates the METEOSAT spacecraft stationed over Africa while the Japanese operate their GMS satellite over the western Pacific.

All of the geostationary satellites have FM transmission formats similar to those of the polar-orbiters, but transmissions are made at 1691 MHz. Reception typically involves the use of a small dish antenna (4-foot models will provide a reasonable gain margin) operating into a low-noise RF preamplifier/downconverter designed to convert the signal to the 137 - 138 MHz range. This permits the VHF receiver to serve as an IF for the geostationary system as well as provide reception of polar-orbit transmissions.

The *Weather Satellite Handbook*, published by the ARRL, provides comprehensive coverage of the various aspects of amateur weather satellite activity, including chapters on antennas, receivers, digital display systems and tracking programs.

In addition to the *ARRL Handbook* and *QST*, *Amateur Television Quarterly* (*ATVQ*) regularly contains information on image communications: You can reach ATVQ at: 5931 Alma Dr, Rockford, IL 61108.

Chapter 13

Amateur Satellites
Signals from Space

Steve Ford, WB8IMY
ARRL Headquarters

Even amateurs who've held their licenses for decades are astonished when they discover the size and sophistication of the Amateur Radio space "fleet." This isn't a new development. Hams were present at the dawn of the Space Age, creating the first amateur satellite in 1961, and we've been active on the "final frontier" ever since. When this chapter was written, nearly a dozen Amateur Radio satellites were in orbit around the Earth!

Satellite-active hams compose a relatively small segment of our hobby, primarily because of an unfortunate fiction that has been circulating for many years—the myth that operating through amateur satellites is difficult and expensive.

Like any other facet of Amateur Radio, satellite hamming is as expensive as you allow it to become. If you want to equip your home with a satellite communication station that would make a NASA engineer blush, it will be expensive. If you want to simply communicate with a few low-Earth-orbiting birds using less-than-state-of-the-art gear, a satellite station is no more expensive than a typical HF or VHF setup.

What about difficulty? Prior to 1982 hams calculated satellite orbits using an arcane manual method that many people found unfathomable. In truth, the manual method taught you a great deal about orbital mechanics, but it was viewed by some as being too difficult. Today computers do all of the calculations for you and display the results in easy to understand formats (more about this latter). Satellite equipment also has become much easier for the average ham to use.

SATELLITES: ORBITING RELAY STATIONS

Most amateurs are familiar with repeater stations that retransmit signals to provide coverage over wide areas. Repeaters achieve this by listening for signals on one frequency and immediately retransmitting whatever they hear on another frequency. Thanks to repeaters, small, low-power radios can communicate over thousands of square miles.

This is essentially the function of an amateur satellite as well. Of course, while a repeater antenna may be as much as a few thousand meters above the surrounding terrain, the satellite is hundreds or thousands of kilometers above the surface of the Earth. The area of the Earth that the satellite's signals can reach is therefore much larger than the coverage area of even the best Earth-bound repeaters. It is this characteristic of satellites that makes them attractive for communication. Most amateur satellites act either as analog repeaters, retransmitting CW and voice signals exactly as they are received, or as packet store-and-forward systems that receive whole messages from ground stations for later relay.

Linear Transponders

Most analog satellites are equipped with *linear transponders*. These are devices that retransmit across a band of frequencies, usually 50 to 100 kHz wide, known as the *passband*. Since the linear transponder retransmits the entire band, many signals may be retransmitted simultaneously. For example, if four SSB signals (each separated by 20 kHz) were transmitted to the satellite, the satellite would retransmit all four signals—still separated by 20 kHz each. Just like a terrestrial repeater, the retransmission takes place on a frequency that is different from the one on which the signals originally were received.

In the case of amateur satellites, the difference between the transmit and receive frequencies is similar to what you might encounter on a crossband terrestrial repeater. In other words, retransmission occurs on a different *band* from the original signal. For example, a transmission received by the satellite on 2 meters might be retransmitted on 10 meters. This crossband operation allows the use of simple filters in the satellite to keep its transmitter from interfering with its receiver.

Don't Be a Transponder Hog

A linear transponder retransmits a faithful reproduction of the signals received in the uplink passband. This means that the loudest signal received will be the loudest signal retransmitted. If a received signal is so strong that retransmitting it would cause an overload, automatic gain control (AGC) within the transponder reduces the transponder's amplification. However, this reduces all of the signals passing through it!

Strong signals don't receive any benefit from being loud since the downlink is "maxed out" anyway. Instead, all of the other users are disrupted since their signals are reduced. Avoiding this condition is just common sense—and good operating practice. It's easy, too: just note the signal level of the satellite's beacon transmission and adjust your transmit power to a level just sufficient to make your downlink appear at the same level as the beacon. Any additional power is too much!

Some linear transponders invert the uplink signal. In other words, if you transmit to the satellite at the *bottom* of the uplink passband, your signal will appear at the *top* of the downlink passband. In addition, if you transmit in upper sideband (USB), your downlink signal will be in lower sideband (LSB). Transceivers designed for amateur satellites usually include features that cope with this confusing flip-flop.

Linear transponders can repeat any type of signal, but those used by amateur satellites are primarily designed for SSB and CW. The reason has to do with the problem of generating power in space. Amateur satellites rely on batteries that are recharged by solar cells. "Space rated" solar arrays and batteries are expensive. They are also heavy and tend to take up a substantial amount of space. Thanks to meager funding, hams don't have the luxury of launching satellites with large power systems such as those used by commercial birds. We have to do the best we can within a much more limited "power budget."

So what does this have to do with SSB or any other mode?

Think of *duty cycle*—the amount of time that a transmitter operates at full output. With SSB and CW the duty cycle is quite low. A linear transponder can retransmit many SSB and CW signals while still operating within the power generating limitations of an amateur satellite. It hardly breaks a sweat.

Now consider FM. An FM transmitter operates at a 100% duty cycle, which means it is generating its *full output* with every transmission. Imagine how much power a linear transponder would need to retransmit, say, a dozen FM signals—all demanding 100% output!

Having said all that, there *are* a few FM repeater satellites, and we'll discuss them in this chapter. However, these are very low-power satellites (typically less than 1 W output) and they do not use linear transponders. They retransmit only one signal at a time, not many signals simultaneously.

FINDING A SATELLITE

Before you can communicate through a satellite, you have to know when it is available. As I said at the beginning of this chapter, computers make this task very easy. First, however, we need to understand a little bit about how ham satellites behave in orbit.

Amateur satellites, unlike many commercial and military spacecraft, do not travel in geostationary orbits . Satellites in geostationary orbits cruise above the Earth's equator at an altitude of about 22,000 miles. From this vantage point the satellites can "see" almost half of our planet. Their speed in orbit matches the rotational speed of the Earth itself, so the satellites appear to be "parked" at fixed positions in the sky. They are available to send and receive signals 24 hours a day over an enormous area.

Of course, amateur satellites *could* be placed in geostationary orbits. The problem isn't one of physics; it's money

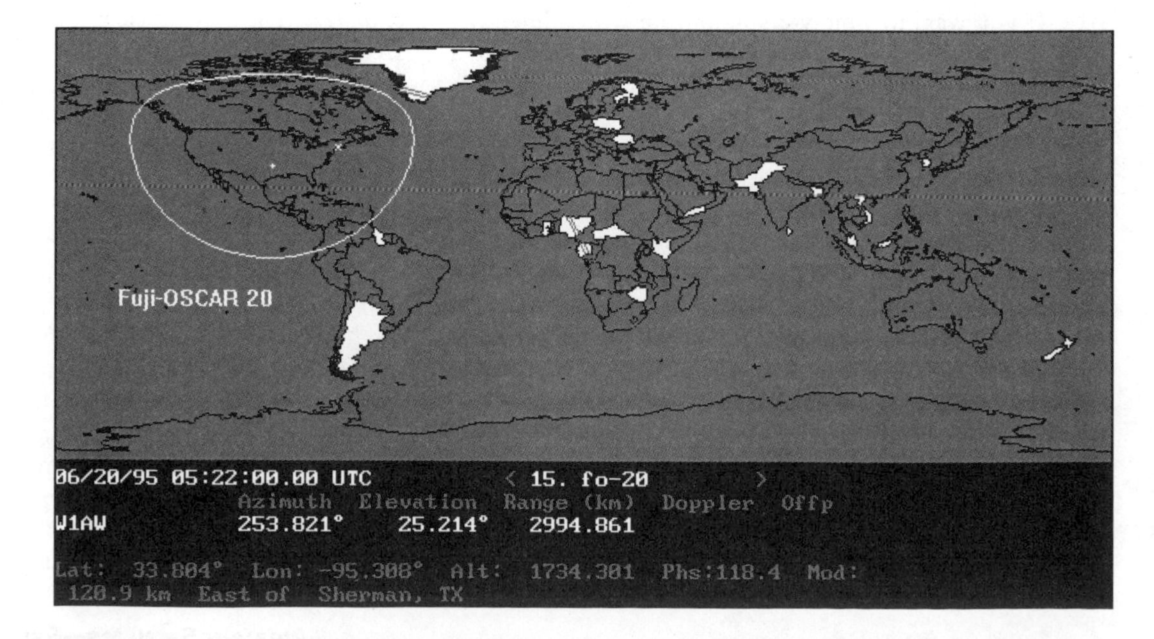

This is one of two tracking screens provided by the *Instantrack* satellite-tracking software (available from AMSAT). We're looking at the position of Fuji-OSCAR 20 over North America. Most of the stations within the white circle (the satellite's *footprint*) should be able to communicate with each other.

and politics. Placing a satellite in geostationary orbit and keeping it there costs a great deal of money—more than any one amateur satellite organization can afford. An amateur satellite group could ask similar groups in other areas of the world to contribute funds to a geostationary satellite project, but why should they? Would you contribute large sums of money to a satellite that may never "see" your part of the world? Unless you are blessed with phenomenal generosity, it would seem unlikely!

Instead, all amateur satellites are either low-Earth orbiters (LEOs), or they travel in high, elongated orbits. Either way, they are not in fixed positions in the sky. Their positions relative to your station change constantly as the satellites zip around the Earth. This means that you need to predict when satellites will appear in your area, as well as what paths they'll take as they move across your local sky.

A bare-bones satellite-tracking program will provide a schedule for any satellite you choose. A very simple schedule might look something like this:

Date	*Time*	*Azimuth*	*Elevation*
10 OCT 01	1200	149°	4°
10 OCT 01	1201	147°	8°
10 OCT 01	1202	144°	13°
10 OCT 01	1203	139°	20°

The date column is obvious: 10 October 2001. The time is usually expressed in UTC. This particular satellite will appear above your horizon beginning at 1200 UTC. The bird will "rise" at an azimuth of 149°, or approximately southeast of your station. The elevation refers to the satellite's position above your horizon in degrees—the higher the better. A zero-degree elevation is right on the horizon; 90° is directly overhead.

By looking at this schedule you can see that the satellite will appear in your southeastern sky at 1200 UTC and will rise quickly to an elevation of 20° by 1203. The satellite's path will curve further to the east as it rises. Notice how the azimuth shifts from 149° at 1200 UTC to 139° at 1203.

The more sophisticated the software, the more information it usually provides in the schedule table. The software may also display the satellite's position graphically as a moving object superimposed on a map of the world. Some of the displays used by satellite prediction software are visually stunning!

Satellite prediction software is widely available on the Web. Some of the simpler programs are freeware. My recommendation is to browse the AMSAT-NA site at **www.amsat.org**. They have the largest collection of satellite software for just about any computer you can imagine. AMSAT software isn't free, but the cost is reasonable and the funds support amateur satellite programs.

Whichever software you choose, you must provide two key pieces of information before you can use the programs:

(1) **Your position**. The software must have your latitude and longitude before it can crank out predictions for your station. The good news is that your position information doesn't need to be extremely accurate. Just find out the latitude and longitude and plug it into the program. You can obtain the longitude and latitude of your town by calling your public library or nearest airport. It is also available at the Geographic Nameserver at **http://www.mit.edu:8001/geo/**on the web.

(2) **Orbital elements**. This is the information that describes the orbits of the satellites. You can find orbital elements (often referred to as *Keplerian elements*) at the AMSAT Web site, and through many other sources on the Internet. You need to update the elements every few months. Many satellite programs will automatically read in the elements if they are provided as ASCII text files. The less sophisticated programs will require you to enter them by hand. I highly recommend the automatic-update software. It's too easy to make a mistake with manual entries.

GETTING STARTED WITH OSCARS 27 AND 14

Would you like to operate through an FM repeater with a coverage area that spans an entire continent? Then check out AMRAD-OSCAR 27 and UoSAT-OSCAR 14, the 2-m to 70-cm FM repeater satellites. From their near-polar orbits at altitudes of approximately 500 miles, these satellites can hear stations within a radius of 2000 miles in all

Down with the Doppler

The relative motion between you and the satellite causes *Doppler shifting* of signals. As the satellite moves toward you, the frequency of the downlink signals will increase by a small amount as the velocity of the satellite adds to the velocity of the transmitted signal. As the satellite passes overhead and starts to move away from you, there will be a rapid drop in frequency of a few kilohertz, much the same way as the tone of a car horn or a train whistle drops as the vehicle moves past the observer.

The Doppler effect is different for stations located at different distances from the satellite because the relative velocity of the satellite with respect to the observer is dependent on the observer's distance from the satellite. The result is that signals passing through the satellite transponder shift slowly around the calculated downlink frequency. Your job is to tune your uplink transmitter—*not your receiver*—to compensate for Doppler shifting and keep your frequency relatively stable on the downlink. That's why it is helpful to hear your own signal coming through the satellite. If you and the station you're talking to both compensate correctly, your conversation will stay at one frequency on the downlink throughout the pass. If you don't compensate, you will drift through the downlink passband as you attempt to "follow" each other's signals. This is highly annoying to others using the satellite because your drifting signals may drift into their conversations.

Doppler shift through a transponder becomes the sum of the Doppler shifts of both the uplink and downlink signals. In the case of an inverting type transponder (as in OSCAR 40), a Doppler-shifted increase in the uplink frequency causes a corresponding decrease in downlink frequency, so the resultant Doppler shift is the *difference* of the Doppler shifts, rather than the *sum*. The shifts tend to cancel.

directions. OSCAR 27's FM repeater has one channel with an uplink frequency of 145.850 MHz and a downlink frequency of 436.795 MHz, both plus/minus the Doppler effect (sidebar "Down with the Doppler"). OSCAR 14 listens on 145.975 MHz and repeats on 435.070 MHz.

You can operate through these satellites with a basic dual-band FM transceiver. Assuming that the transceiver is reasonably sensitive, you can use an omnidirectional antenna such as a dual-band collinear ground plane or something similar. Some amateurs even have managed to work it with H-Ts, but to reach AO-27 or UO-14 with an H-T you'll need to use a multi-element directional antenna. Of course, this means that you'll have to aim your antenna at the satellites as they cross overhead.

Start by booting up your satellite tracking software. Check for a pass during daylight hours where the satellite rises at least 45° above your horizon. As with all satellites, the higher the elevation, the better. If you plan to operate outdoors or away from home, either print the schedule or jot down the times on a piece of scrap paper that you carry with you.

When the satellite comes into range, you'll be receiving its signal about 10 kHz higher than the designated downlink frequency, thanks to Doppler shifting. So, begin listening there. At about the midpoint of the pass you'll need to shift your receiver down 10 kHz or more; and as the satellite is heading away, you may wind up stepping down to as much as 10 kHz below the downlink frequency. Some operators program these frequencies into memory channels so that they can compensate for Doppler shift at the push of a button.

Table 13-1
Active Amateur Satellites: Frequencies and Modes

Satellite	***Uplink (MHz)***	***Downlink (MHz)***
SSB/CW		
AMSAT-OSCAR 10	435.030—435.180	145.825—145.975
Fuji-OSCAR 20	145.900—146.000	435.800—435.900
Fuji-OSCAR 29 (available biweekly)	145.900—146.000	435.800—435.900
RS-12	21.260—21.300	29.460—29.500
	145.960—146.00	
RS-15	145.858—145.898	29.354—29.394
Packet—1200 bit/s		
(FM FSK uplink, PSK downlink except as noted)		
AMSAT-OSCAR 16	145.90, .92, .94, .96	437.0513
Packet—9600 bit/s		
(FM FSK uplink and downlink.)		
UoSAT-OSCAR 22	145.900, .975	435.120
KITSAT-OSCAR 25	145.98	436.50
TMSAT-OSCAR 31	145.925	436.925
UoSAT-OSCAR 36	145.960	437.400
FM Voice Repeaters		
AMRAD-OSCAR 27	145.850	436.790
UoSAT-OSCAR 14	145.975	435.070

Technicians make final adjustments to OSCAR 27 before the satellite is shipped to the launch site.

AO-27 and UO-14 behave just like terrestrial FM repeaters. Only one person at a time can talk. If two or more people transmit simultaneously, the result is garbled audio or a squealing sound on the output. The trick is to take turns and keep the conversation short. Even the best passes will only give you about 15 minutes to use the satellite. If you strike up a conversation, don't forget that there are others waiting to use the bird.

The FM repeater satellites are a great way to get started. They are easy to hear and easy to use. Once you get your feet wet, however, you'll probably wish you could access a satellite that wasn't so crowded, where you could chat for as long as the bird was in range.

MOVING UP TO THE FUJIS AND RADIO SPUTNIKS

The RS—*Radio Sputnik*—satellites were built and launched by the former Soviet Union. There have been a number of RS satellites in orbit. At the time of this writing, only RS-12 and RS-15 were operating.

RS-12 is by far the more popular of the two active Radio Sputniks. It is actually a transponder module riding piggyback, so to speak, on a much larger navigational bird. RS-12 carries a *Mode K* transponder, which means that it receives signals on the 15-meter band and retransmits on the 10-meter band. RS-12 also is equipped with a *Mode A* transponder, which receives on 2 meters and retransmits on 10 meters. (The use of "mode" terminology is one of the most confusing aspects of amateur satellite operation to the newcomer. Don't let it scare you away. The sidebar "Satellite Modes Demystified" should clear up any confusion.) When this edition went to press, RS-12 was operating.

When using the RS-12 Mode-K transponder, you don't need to know precisely where the satellite is positioned in the sky. After all, you aren't likely to be using narrow beamwidth antennas. (Often heard on RS-12: "Antenna here is a dipole.") Mainly, you want to know when the satellite will be in view. Of course, 15- and 10-meter signals are subject to ionospheric bending, so it pays to listen for the satellite before and after the predicted visibility period.

Once you've determined when the satellite is due to rise above the horizon at your location, listen for the satellite's CW telemetry beacon (Table 13-1). This signal

From East Harbor State Park on the south shore of Lake Erie, Mike, WB8ERJ, takes the relaxed approach to working OSCAR 27, one of the FM repeater satellites. He is using a dual-band hand-held transceiver with a multielement hand-held Yagi antenna.

is transmitted constantly by the satellite and carries information about the state of the satellite's systems, such as its battery voltage, solar-panel currents, temperatures and so on. As soon as you can hear the beacon, start tuning across the downlink passband.

On an active day you should pick up several signals. They will sound like normal amateur SSB and CW conversations. Nothing unusual about them at all—except that the signals will be slowly drifting downward in frequency! That's the effect of Doppler shift. It's not too serious on the 10-meter downlink, but it can be a challenge when the downlink is at 70 cm because the shift is proportional to the transmitted frequency.

Now tune your transmitter's frequency to the satellite's uplink passband. RS-12 does not use an inverting transponder. If you transmit at the low end of the uplink passband, you can expect to hear your signal at the low end of the downlink passband. Generally speaking, CW operators occupy the lower half of the transponder passband while SSB enthusiasts use the upper half.

RS-15 operates in the same fashion but listens only on 2 meters and retransmits on 10 meters. See Table 13-1. Unfortunately, RS-15 has been suffering from a damaged power system. As a result, its signal is often very weak.

Fuji OSCARs 20 and 29 are also linear transponder birds that function much like RS-12 and RS-15. The main difference is that they listen on 2 meters and retransmit on 70 cm. (They also use inverting transponders.) Relatively few amateurs own receivers that can listen for 70-cm CW and SSB, so these satellites are not very active. During weekend passes, however, you should be able to hear two to four conversations taking place.

STATION REQUIREMENTS FOR THE RS AND FUJI SATELLITES

To work RS-12 in Mode K you'll need at minimum a multiband HF SSB or CW transceiver. You *do not* need an amplifier; 100 W is more than enough power for the uplink. In fact, even 100 W may be too much in many instances. The rule of thumb is that your signal on the downlink never should be stronger than the satellite's own telemetry beacon.

A 15-meter wire dipole is adequate for sending and receiving with RS-12. (Yes, you'll be listening on 10 meters, but the 15-meter dipole should function well as a 10-meter receiving antenna.) The ideal situation is to have *separate* 15- and 10-meter radios and antennas so that you can listen on 10 meters while you are transmitting on 15 meters. The ability to hear your own signal is a tremendous asset for satellite work. It allows you to listen to the Doppler shifting of your own signal, giving you an opportunity to tweak your transmit frequency to compensate.

To work RS-12 and RS-15 in Mode A, you'll need a

Satellite Modes Demystified

Since there are a number of amateur satellites operating, and since they operate on a variety of bands, some way of distinguishing the various combinations of bands became necessary. The first amateur transponder in space was the one carried by OSCAR 6. It received uplink signals at 2 meters and retransmitted them on the downlink at 10 meters. When OSCAR 7 was launched, it included a transponder with a 70-cm uplink and a 2-meter downlink. To distinguish between the two, the 2-meter-to-10-meter transponder operation was called "Mode A" and the 70-cm-to-2-meter operation "Mode B."

The launch of OSCAR 8 complicated matters by using a transponder with a 2-meter uplink and a 70-cm downlink, just the reverse of the Mode-B transponder of OSCAR 7. Japanese amateurs built this new transponder, and in their honor the new setup was dubbed "Mode J."

OSCAR 10 presented yet another possibility with its 23-cm-to-70-cm transponder. The 23-cm band is in that portion of the spectrum called "L band" by microwave engineers, so this uplink/downlink setup acquired the name "Mode L." A similar situation occurred when OSCAR 13 deployed a 70-cm-to-13-cm transponder. Since the 13-cm band is in the S-band part of the microwave spectrum, this mode is called—you guessed it—"Mode S."

The digital satellites, the first of which was the Japanese OSCAR 12, threw a monkey wrench into this (relatively) simple naming scheme. For the first time, amateurs had to contend with transponders that were different in kind rather than just frequency. FO-12 (and its successor, FO-20) carried both an analog transponder and a digital packet-radio transponder. Both transponders received uplink signals on 2 meters and transmitted on 70 cm, conforming to the Mode J band selection. To indicate the band selection while distinguishing between the analog and digital transponders, the analog operation is called Mode JA and the digital operation Mode JD.

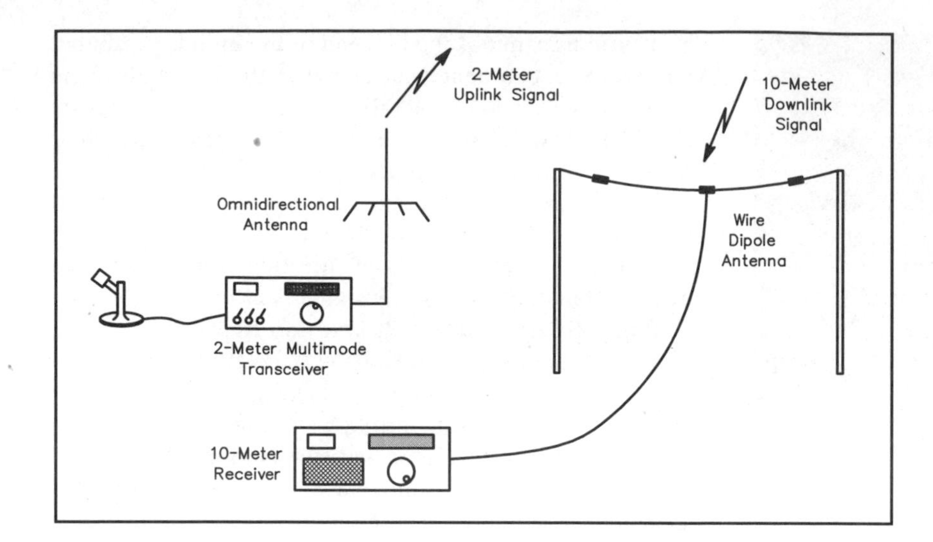

To work RS-15, or RS-12 using its 2-meter uplink, you'll need a 2-meter multimode transceiver, a 10-meter receiver and antennas for 2 meters and 10 meters. (Omnidirectional antennas will do nicely.)

2-meter multimode transceiver that can operate in CW or SSB. Remember that in Mode A RS-12 and RS-15 are listening for signals on 2 meters and retransmitting on 10 meters. This means that you'll still need a 10-meter SSB receiver. Choose your radios carefully. A number of modern HF transceivers also include 2 meters and even 70 cm. The problem, however, is that these radios usually do not allow *crossband splits* between VHF and HF. That is, they won't allow you to transmit on 2 meters and receive on 10 meters. At the very least they won't allow you to do this simultaneously. Check the features and specifications carefully.

Omnidirectional antennas for 2 meters are sufficient for transmitting to RS-12 and RS-15. A beam on 2 meters would be even better, but then you incur the cost of an antenna rotator that can move the antenna up and down as well as side to side—the so-called *azimuth/elevation rotator*.

For the Fuji OSCARs the ability to transmit and receive simultaneously is a must. The Doppler effect is pronounced on the 70-cm downlink. You need to listen to your own signal continuously, making small adjustments to your 2-meter uplink so that your voice or CW note does not slide downward in frequency. To achieve this you will need separate 2-meter and 70-cm transceivers (such as a couple of used rigs), or a transceiver that is specifically designed for satellite use. Kenwood, ICOM and Yaesu have such radios in their product lines. These wondrous rigs make satellite operating a breeze, although their price tags may give you a bit of sticker shock (about $1600+). They feature full crossband duplex, meaning that you can transmit on 2 meters at the same time you are listening on 70 cm. They even have the ability to work with inverting transponders automatically. That is, as you move your receive frequency down, the transmit VFO will automatically move up!

Although beam antennas and azimuth/elevation rotators are not strictly necessary to work the Fujis (I've done it myself with omnidirectional antennas on both bands), they vastly improve the quality of your signal. If you decide to go the omnidirectional route, you'll need to add a 70-cm receive preamp at the antenna to boost the downlink signal.

TAKING THE HIGH ROAD WITH OSCAR 10

The limitations of LEO satellites, especially their brief periods of availability, are overcome by a class of satellites called "Phase 3." The name comes from the various phases in the development of amateur satellites. The earliest ones, during Phase 1, contained beacon and telemetry transmitters, but not transponders. These early satellites were all in circular, low-Earth orbits—as were the Phase 2 satellites, which carried communication transponders.

Phase 3 satellites are not in low-Earth orbits. Rather, their orbits describe an ellipse. These satellites swing within a few hundred kilometers of the Earth's surface at one end of the ellipse (the *perigee*) and streak out to 30,000 km or so at the other end (the *apogee*). The physics of an orbiting

This simple eggbeater antenna is adequate for communicating with the low-Earth orbiting analog satellites. You can install it outdoors or in your attic.

body dictates that the satellite spends much more of its time near apogee than perigee. Therefore, the Phase 3 satellites spend most of their time at very high altitudes. From a typical point in the Northern Hemisphere, a particular Phase 3 satellite is available for more than 10 hours per day. This is a remarkable improvement over the LEO satellites! And because the Phase 3 satellite is so much higher, it is also visible from a greater fraction of the Earth's surface. The result is a vast improvement in the communications capability of the satellite.

There is a downside, however. The greater distance to the Phase 3 satellite means that more transmitted power is needed to access it, and a weaker signal is received from the satellite at the ground station. (This problem is alleviated somewhat by the use of gain antennas on the satellite.) The signal levels are such that operation using the first generation of Phase 3 satellites is feasible only with ground-station antennas that exhibit significant gain (10 dBi or more). Such directional antennas must be pointed directly at the satellite. The satellite's position in the sky changes over time, however, so the antenna's position must change as well. In fact, the satellite changes both its bearing from the ground station (azimuth) and its height above the horizon (elevation). Most ground stations that access Phase 3 satellites use antenna systems that can be rotated in both the azimuth and elevation planes.

At the time of this writing, the only Phase 3 satellite in orbit is OSCAR 10. It is only intermittently available; its control computer suffered accumulated radiation damage that rendered the satellite uncontrollable. It occasionally operates when it gets sufficient sunlight.

When OSCAR 10 is working, however, it is a hot DX satellite! At apogee OSCAR 10 can see half of the globe. This means that you can enjoy transatlantic and transpacific conversations for hours at a time.

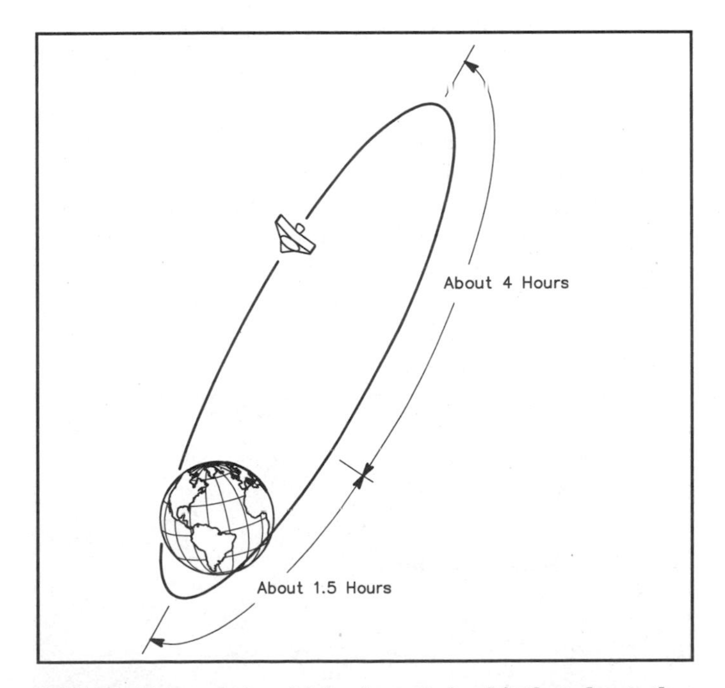

OSCAR 10 travels in a high, elongated orbit that shoots the satellite far into space. From our perspective here on Earth, OSCAR 10 appears almost to hang in the sky when it is at the highest point of its orbit. OSCAR 40 travels in a similar orbit.

Such astonishing capability comes at a price in terms of station hardware. Not only will you need the Yagi antennas and the rotator we've already discussed, you will also need a VHF/UHF dual-band "satellite ready" SSB transceiver, a 150-W amplifier and a receive preamplifier. If you buy brand-new equipment, the cost of a station for OSCAR 10 could approach $3000. Some careful shopping at flea-markets and on the Web can bring the cost down to about $1500.

OSCAR 40—THE SUPERSAT

During November 2000 the AMSAT-OSCAR 40 satellite, an ambitious spacecraft 10 years in the making, rocketed into orbit. Champagne corks were popping as hams throughout the world heard its 2-meter beacon blasting through their receivers. Then, weeks later, AO-40 went ominously silent during an engine burn.

The command team scrambled for answers, trying all possible uplink frequencies and software commands to bring the satellite back to life. Their collective prayers were answered at Christmas when OSCAR 40 finally responded and began transmitting on its 2.4-GHz downlink. According to what could be deduced from the telemetry, something went very wrong with the main engine. The command team is still studying the incident, but it appears as though a "combustion event" may have damaged the 400-Newton motor and possibly some of antennas. For example, the 2-meter beacon hasn't been heard since the event, even though the transmitter seems to respond to "on" commands.

The good news—and there is really quite a bit of good news—is that the remaining satellite systems appear to be functioning normally. And although OSCAR 40 did not reach the inclination the team desired, its current orbit is stable and very useful. At the highest altitude of its orbit (the *apogee*), AO-40 seems to "hover" in the sky for hours at a time. We may have indeed lost our *downlinks* (satellite-to-ground transmissions) on 2 meters and 70 cm, but OSCAR 40's microwave downlink capability is intact, as are the 2-meter and 70-cm *uplinks* (ground-to-satellite). This means that OSCAR 40 still has the potential to be a fantastic DX satellite. The only difference is that you will need to set up equipment to receive on the microwave bands, which isn't nearly as difficult as you might think.

Communications Possibilities

Despite the damage, AO-40 still has a wide variety of operating frequencies and modes. (See **Table 13-2**.) To keep everything as unambiguous as possible, a new naming convention has been created. An alpha designator refers to each frequency band. As this article goes to press, AO-40 has downlink capabilities on two sections of 13 cm (2400 MHz: S1 and 2401 MHz: S2), 3 cm (10450 MHz: X), 1.5 cm (24050 MHz: K) and a 360-THz IR laser. Uplinks are on 70 cm (U), two uplinks on 23 cm (1269 MHz: L1 and 1268 MHz: L2), two uplinks on 13 cm (2400 MHz: S1 and 2446 MHz: S2) and 6 cm (5668 MHz; C).

Where Do I Start?

My suggestion is that you start by assembling the components you'll need for a Mode U/S station—transmitting on 430 MHz and receiving on 2.4 GHz. This is likely to become the most popular configuration.

Antennas

There are a number of 2.4 GHz receive antennas available to you. Check out the Down East Microwave Web site (**www.downeastmicrowave.com**) and take a look at some of their 2.4 GHz loop Yagis. Other excellent choices include the Andrews 26T-2400 grid-style dish antenna and AirLink parabolic dishes. Both assemble in about 15 minutes—no tuning or other adjustments required. PC Electronics sells the Andrews dish. You'll find them online at **www.hamtv.com**.

Table 13-2
Transponder Frequency Band Plan for AMSAT-OSCAR 40

Note: Frequencies shown are for transponders that were known to be functional when this book went to press. All signals are digital, SSB or CW. FM is not permitted on AO-40.

Uplink Frequencies

Band	*Digital*	*Analog*
70 cm	435.300 – 435.550 MHz	435.550 – 435.800 MHz
23 cm(1)	1269.000 – 1269.250 MHz	1269.250 – 1269.500 MHz
23 cm(2)	1268.075 – 1268.325 MHz	1268.325 – 1268.575 MHz
13 cm(1)	2400.100 – 2400.350 MHz	2400.350 – 2400.600 MHz
13 cm(2)	2446.200 – 2446.450 MHz	2446.450 – 2446.700 MHz
6 cm	5668.300 – 5668.550 MHz	5668.550 – 5668.800 MHz

Downlink Frequencies

Band	*Digital*	*Analog*
13 cm(1)	2400.650 – 2400.950 MHz	2400.225 – 2400.475 MHz
13 cm(2)	2401.650 – 2401.950 MHz	2401.225 – 2401.475 MHz
1.5 cm	24048.450 – 24048.750 MHz	24048.025 – 24048.275 MHz

Telemetry Beacons

Band	*General (GB)*	*Middle (MB)*	*Engineering (EB)*
13 cm(1)	2400.200 MHz	2400.350 MHz	2400.600 MHz
13 cm(2)	2401.200 MHz	2401.350 MHz	2401.600 MHz
1.5 cm	24048.000 MHz	24048.150 MHz	24048.400 MHz

The impressive I8CVS satellite station in Naples, Italy is usable on all uplinks and downlinks from 430 MHz through 24 GHz.

For transmitting on 430 MHz, you don't need a monstrous antenna; it depends on how much transmitter output you can generate. The less transmitter output you have available, the bigger the antenna you'll need. For example, if you run 100 W output on 430 MHz, you probably will be able to get away with a small 6- or 8-element Yagi for your uplink antenna. Don't worry too much about antenna polarization at this point. Yes, it would be nice to have a circularly polarized antenna for your uplink, but AO-40 is so sensitive, you aren't likely to notice the 3-dB loss you'll incur by using just horizontal or vertical polarization.

What about an antenna rotator? An *azimuth/elevation* rotator that can track OSCAR 40 across the sky sure is handy, but a new model off the shelf can be expensive (about $500). If you can't afford this investment now, don't worry about it. Use your tracking software and follow AO-40 through a typical pass. You'll notice that it rises to a certain point in your local sky, then doesn't move very much. Aim your antenna at the point and you should be able to enjoy quite a bit of "talk time" before you need to adjust them again. Magazine ads aside, beam antennas *aren't* laser beams. Signal strength falls off gradually with imperfect pointing, so signals will simply get weaker as the satellite's elevation above the horizon increases. A directional antenna may provide satisfactory performance even if the satellite is significantly off axis. In practical operation, beam antennas cover a lot of sky, in azimuth and elevation.

Receiving and Transmitting

To operate Mode U/S remember that you need to transmit to the satellite on 70 cm and receive on 2.4 GHz.

The uplink is straightforward. You can use a 70-cm multimode rig as your uplink transmitter (possibly paired with a "brick" amplifier to generate a little more "oomph").

To receive on 2.4 GHz you will need to convert the signals from microwave to something lower in frequency—such as 2 meters or even 10 meters. For this you'll need a

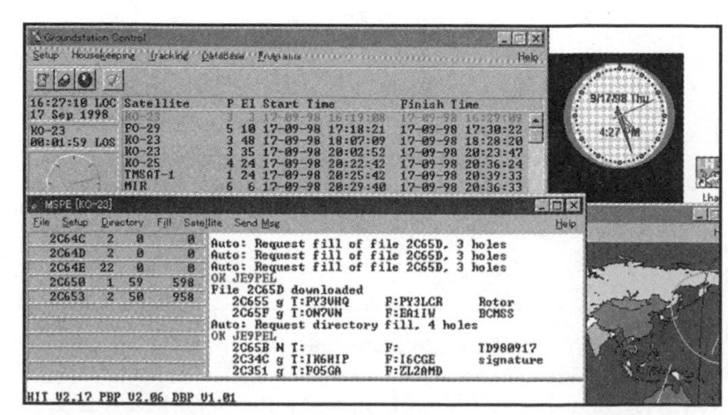

***WISP* software in action as it communicates with the KITSAT-OSCAR 23 packet satellite.**

receive converter (sometimes referred to as a *downconverter*) such as the Down East Microwave 2400RX or the SSB Electronic UEK-3000. You install the unit right at your antenna so that the signal is converted right away before making the trip through the coaxial cable to your station (more about this in a moment). Most 2.4-GHz receive converters convert to 2 meters. So, you'll need a 2-meter all-mode receiver at your station. A number of HF transceivers now include 2 meters and these will work just fine in this application. Another alternative is an all-mode-scanning receiver. If you only have HF receive capability, contact the suppliers for converters that will convert 2.4-GHz signals to 10 meters.

As long as we're talking about hardware, another popular approach to setting up the foundation of a microwave satellite station is to simply buy a multiband (2 meters/70 cm or even HF through 70 cm) transceiver that includes satellite features. These rigs take a lot of the guesswork out of the whole operation. The ICOM IC-910H transceiver generates all the output you'll need to uplink on 70 cm and includes the 2-meter receiver for use with a downconverter. The same is true of the Kenwood TS-2000 and the Yaesu FT-847. If you prefer to buy used gear, check out the ICOM IC-821H, Yaesu FT-736 or Kenwood TS-790.

Digital Operation

AO-40 has a variety of digital experiments in dedicated subbands. There are two hard-wired 9600-baud modems and 16 "agile" (programmable) modems attached to the RUDAK computer. Downlinks up to 138 kbits/s are planned. The satellite also has several cameras for the SCOPE and YACE (Yet Another Camera Experiment) experiments, whose digitized pictures will be broadcast. The YACE camera already has taken pictures of the second stage AR-507 (launch rocket) separation.

THE PACSATS

If you enjoy packet operating, you'll love the PACSATs! The PACSATs work like temporary mail boxes in space. You upload a message or a file to a PACSAT and it is stored for a time (days or weeks) until someone else—possibly on the other side of the world—downloads it. Many PACSATs are also equipped with on-board digital cameras. They snap fascinating images of the Earth which are stored as files that you can download and view.

Which PACSAT is Best?

You can divide the PACSATs into two types: The 1200- and 9600-baud satellites. OSCAR 16 is presently the only operational 1200-baud PACSAT. You transmit packets to it on 2-meter FM and receive on 435-MHz SSB. The remaining PACSATs operate at 9600 baud. You send packets to them on 2-meter FM and receive on 435-MHz FM.

So which PACSATs are best for beginners? There's no easy answer for that question. You can use any 2-meter FM transceiver to send data to a 1200-baud PACSAT, but getting your hands on a 435-MHz SSB receiver (or transceiver) could put a substantial dent in your bank account. In addition, you need a special PSK (*phase-shift keying*) terminal

The Red Sea as seen by camera aboard TMSAT-OSCAR 31.

node controller (TNC). These little boxes are not common and could set you back about $250.

So the 9600-baud PACSATs are best for the newbie, right? Not so fast. It's true that you don't need a special packet TNC. Any of the affordable 9600-baud TNCs will do the job. The catch is that not any FM transceiver is usable for 9600-baud packet. Both the 2-meter *and* 440-MHz FM radios must be capable of handling 9600-baud signals; and not all 440-MHz FM rigs can receive down to 435 MHz.

Broadcasting Data

Despite the huge amounts of data that can be captured during a pass, there is considerable competition among ground stations about exactly *which* data the satellite should receive or send! There are typically two or three dozen stations within a satellite's roving footprint, all making their various requests. If you think this sounds like a recipe for chaos, you're right.

The PACSATs produce order out of anarchy by creating two *queues* (waiting lines)—one for uploading and another for downloading. The upload queue can accommodate two stations and the download queue can take as many as 20. Once the satellite admits a ground station into the queue for downloading, the station moves forward in the line until it reaches the front, whereupon the satellite services the request for several seconds.

For example, let's say that OSCAR 16 just accepted me, WB8IMY, into the download queue. I want to grab a particular file from the bird, but I have to wait my turn. OSCAR 16 lets me know where I stand by sending an "announcement" that I see on my monitor. It might look like this:

WB8ISZ AA3YL KD34GLS WB8IMY

WB8ISZ is at the head of the line. The satellite will send him a chunk of data, then move him to the rear.

AA3YL KD34GLS WB8IMY WB8ISZ

Now there are only two stations ahead of me. When I reach the beginning of the line, I'll get my share of "attention" from the satellite.

Unless the file you want is small, you won't get it all in one shot. If the satellite disappears over the horizon before you receive the complete file, there's no need to worry. Your PACSAT software "remembers" which parts of the file you still need from the bird. When it appears again, your software can request that these "holes" be filled.

And while all of this is going on, *you're receiving data that other stations have requested!* That's right. Not only do you get the file you wanted, you also receive a large portion of the data that other hams have requested. You may receive a number of messages and files without transmitting a single watt of RF. All you have to do is listen. That's why they call it "broadcast" protocol.

Station Software

You must run specialized software on your station PC if you're going to enjoy any success with the PACSATs. If your computer uses DOS only, you need a software package known as *PB/PG*. *PB* is the software you'll use most of the time to grab data from the satellite. *PG* is only used when you need to upload.

If you're running Microsoft *Windows* on your PC, you'll want to use *WISP*. *WISP* is a *Windows* version of *PB/PG* that includes such features as satellite tracking, antenna rotator control and more. Both software packages are available from AMSAT at the address mentioned elsewhere in this chapter. While you're contacting AMSAT, pick up a copy of the *Digital Satellite Guide*. It gets deep into the details of PACSAT operation far beyond the scope of what we can discuss here.

Chapter 14

Radio Monitoring

Tune In to the Action

Curt Phillips, W4CP
114 Castle Circle
Smithfield, NC 27577
w4cp@arrl.net

On the ham bands, from the microwave spectrum to 160 meters there is a myriad of interesting things to do. Although nearly all of these activities require receivers and listening, for most of us the focal point of having an amateur radio license is the ability to talk back and participate in the communications.

If you examine the radio-frequency spectrum, you'll see that the ham bands occupy just a tiny portion of it. On the frequencies between the ham bands there are also a myriad of interesting things to do, but unless you are covered by some other license besides a ham license, your participation will be limited solely to listening.

So why would anyone who knows about ham radio be interested in frequencies where they would be limited to just listening? As an Amateur-Radio operator, you can transmit RF, not just simply receive it. Wouldn't this be a step backward?

The short answer is no. The diverse communications that you have access to on the frequencies between the ham bands provide a wealth of possibilities that supplement the choices provided by Amateur Radio.

So where and what are these diverse communications?

The spectrum under consideration is so vast that to begin to answer the *where* question requires an explanation of the natural lines of demarcation within the radio spectrum. Although many people (hams and non-hams alike) enjoy listening to the full range of anything a radio can be tuned to receive, in general the radio monitoring hobby is broken into two major pursuits: shortwave listening and radio scanning (as compared to scanning images into computers).

There is no official definition of the shortwave bands, but they are most often considered to comprise the frequencies between 3 and 30 MHz. For hobbyist purposes, the monitoring of frequencies between 100 kHz and 30 MHz can be considered shortwave listening, notwithstanding that the AM broadcast band is in the medium wave spectrum and frequencies below that are technically considered long wave.

The ham bands within the shortwave bands are also referred to as the high frequency or HF bands, as opposed to the very high frequency (VHF) and ultra high frequency (UHF) bands that are above 30 MHz.

And it is on the VHF and UHF frequencies above 30 MHz where the scanning part of radio monitoring takes place.

Many hams experienced the delights of monitoring either or both of these frequency ranges prior to obtaining their license. But whether you are renewing your acquaintance with frequencies forgotten or sampling these frequencies for the first time, there is an amazing number of intriguing signals that can be heard on the shortwave and scanner bands. Indeed, since no license is required to listen, everyone in the family can enjoy the diverse offerings of these bands. And it is also an excellent way to introduce non-ham friends to hobby radio.

SHORTWAVE LISTENING

In the old days, many new recruits entered the hobby through direct exposure to the shortwave bands. Since in the days before satellites shortwave radio provided the only way to directly receive news from overseas, many people owned and listened to shortwave radios who were not hobbyists as we would consider a shortwave listener today.

Then before they took their exams, they spent hours listening to shortwave as they honed their CW skills. The code practice sessions provided by the ARRL flagship station W1AW and the other CW activity on the shortwave bands were the prime source of CW practice fodder. Bear in mind that this was before the days of cassette tape machines and personal computers.

After they were on the air, their shortwave experience was limited. For a period of time from about the mid-1950s until the late 1970s, ham transceivers tuned the Amateur Radio portions of the shortwave spectrum only. Why? The technology of radio equipment during that period was such

Table 14-1
Shortwave Frequency Guide
All frequencies shown in kHz

Frequencies	Service	Comments
535 —1705	AM Broadcast	Standard North America AM
1705 —1800	Fixed Service	Land/Mobile/Marine
1800 —2000	Amateur 160 Meters	
2000 —2107	Maritime Mobile	
2107 —2170	Fixed Service	Land/Mobile/Marine
2170 —2194	Land Mobile Service	
2194 —2300	Fixed Service	
2300 —2495	Shortwave Broadcast	120 Meters
2495 —2505	Time Standard	
2505 —2850	Fixed Service	Land/Mobile/Marine
2850 —3155	Aeronautical Mobile	Transoceanic Flights
3155 —3200	Fixed Service	
3200 —3400	Shortwave Broadcast	90 Meters
3400 —3500	Aeronautical Mobile	Transoceanic Flights
3500 —4000	Amateur 80/75 Meters	
3900 —4000	Shortwave Broadcast	75 Meters, Not in Region 2
4000 —4000	Time Standard	
4000 —4063	Fixed Service	
4063 —4438	Maritime Mobile	Ship/Shore
4438 —4650	Fixed Service	
4650 —4750	Aeronautical Mobile	Transoceanic Flights
4750 —5060	Shortwave Broadcast	60 Meters
5005 —5450	Fixed Service	
5450 —5730	Aeronautical Mobile	Transoceanic Flights
5730 —5950	Fixed Service	
5950 —6200	Shortwave Broadcast	49 Meters
6200 —6525	Maritime Mobile	Ship/Shore
6525 —6765	Aeronautical Mobile	Transoceanic Flights
6765 —7000	Fixed Service	
7000 —7300	Amateur 40 Meters	
7100 —7300	Shortwave Broadcast	41 Meters, Not in Region 2
7300 —8195	Fixed Service	
8195 —8815	Maritime Mobile	Ship/Shore
8815 —9040	Aeronautical Mobile	Transoceanic Flights
9040 —9500	Fixed Service	
9500 —9900	Shortwave Broadcast	31 Meters
9775 —9995	Fixed Service	
10005—10100	Aeronautical Mobile	Transoceanic Flights
10100—10150	Amateur 30 Meters	CW/Data Only
10100—11175	Fixed Service	
11175—11400	Aeronautical Mobile	Transoceanic Flights
11400—11650	Fixed Service	
11650—12050	Shortwave Broadcast	25 Meters
12050—12330	Fixed Service	
12330—13200	Maritime Mobile	Ship/Shore
13200—13360	Aeronautical Mobile	Transoceanic Flights
13360—13600	Fixed Service	
13600—13800	Shortwave Broadcast	New WARC Allocation
13800—14000	Fixed Service	
14000—14350	Amateur 20 Meters	
14350—14995	Fixed Service	
15010—15100	Aeronautical Mobile	Transoceanic Flights
15100—15600	Shortwave Broadcast	19 Meters
15600—16460	Fixed Service	
16460—17360	Maritime Mobile	Ship/Shore
17360—17550	Fixed Service	
17550—17900	Shortwave Broadcast	16 Meters
17900—18030	Aeronautical Mobile	Transoceanic Flights
18030—18780	Fixed Service	
18068—18168	Amateur 17 Meters	
18780—18900	Maritime Mobile	Ship/Shore
18900—19680	Fixed Service	
19680—19800	Maritime Mobile	Ship/Shore
19800—21000	Fixed Service	
21000—21450	Amateur 15 Meters	
21450—21850	Shortwave Broadcast	13 Meters
21850—22000	Aeronautical Mobile	
22000—22720	Maritime Mobile	Ship/Shore
22720—23200	Fixed Service	
23200—23350	Aeronautical Mobile	
23350—24990	Fixed Service	
24890—24990	Amateur 12 Meters	Shared with Fixed Service
25010—25330	Petroleum Industry	
25330—25600	Government Frequency	
25600—26100	Shortwave Broadcast	11 Meters
26100—26480	Land Mobile Service	
26480—26950	Government	
26950—26960	International Fixed Service	
26960—27410	Citizen's Band	Channels start at 26965 kHz
27410—27540	Land Mobile Service	
27540—28000	Government	
28000—29700	Amateur 10 Meters	
29700—29800	Forestry Service	
29800—29890	Fixed Service	
29890—29910	Government	
29910—30000	Fixed Service	

that cost-effective optimum performance was only realized in rigs that covered a small portion of the radio spectrum. So hams that may have started with a general coverage receiver quickly downgraded to a ham-band only receiver.

In the past few years shortwave listening (also known as world band listening) has experienced a resurgence of interest. One reason for this is the increased availability of good performing, moderately priced shortwave receivers and, among hams, the fact that most ham radio HF transceivers include full receive coverage of the frequencies from 0.5-30 MHz. Also contributing to this is the widening realization that, in an information society, the unique information available on the shortwave bands cannot be ignored.

Unique Perspective on the News

Perhaps the most popular reason for listening to the shortwave spectrum is to monitor unfolding world events and news.

For the past several years, the United States has been sending troops on peace keeping and humanitarian missions around the world. The United States Army and Marines have served on extended missions in Bosnia, Haiti and Somalia. Air Force and naval assets are shuttled back and forth from the area around Iraq, as the threat there ebbs and flows. Boatloads of Cuban expatriates flood into Florida and are relocated to such places as Guantanamo Bay.

The shortwave listener has a ringside seat to events such as these. The military frequencies are full of traffic as troops and equipment are moved. The Voice of America broadcasts the American view of the world situation. Opposing viewpoints are heard from such voices as Radio Iraq International and Radio Habana Cuba. The shortwave stations of other nations weigh in with their analyses.

Despite all the high tech, satellite equipment available, HF is still used by the government. Transmissions on the

ITU

Radio waves cannot be kept within a country's borders and radio is a means of international communication, so it was evident very early in the development of radio that international treaties and agreements must be established to prevent chaos on the radio waves.

The organization that evolved into the radio coordinating council for the world is the International Telecommunication Union. They allocate frequency ranges for various uses, and the assignment of specific operating frequencies is left to the governments of the individual nations.

As a part of their authority, the ITU sets the frequency ranges where the various services operate, but many countries are not bashful about moving their national broadcasting services outside the specified bands. That's why you need to scan outside the listed frequency ranges. You'll find many fascinating signals lurking there.

The various frequency ranges are traditionally referred to using wavelength in meters of one of the component frequencies. For example, 49 meters is the wavelength of 6122 kHz. However, the 49-meter band extends from 5950-6200 kHz.

"Mystic Star" frequencies from VIP aircraft can be heard as they shuttle about on their diplomatic missions.

Even for events occurring within the United States, shortwave radio can provide information substantially before it is released by the traditional media. When President Ronald Reagan was shot, listeners monitoring the radio traffic between the White House and Vice President George Bush heard reports on the President's condition long before it was broadcast on the major networks.

Listeners to the Israeli broadcast station Kol Yisroel heard the warning for the gas attack from Iraq as the first missiles were detected in flight—and the warning was broadcast to the Israeli public on all medium and shortwave transmitters. Many parts of the US news media were unaware of the attack until an hour or two later.

In our turbulent world there seem to always be hot spots where shortwave radio provides a direct access to the information. The broadcast news of the major networks tells you what they think is important, filtered through their perspective and edited to fit into a brief newscast. The shortwave bands bring you news direct from the source, and in as much detail as they care to provide.

During the Russian coup in August 1991 and the armed uprising of old-line Soviets in the legislature during September 1993, those monitoring Radio Moscow got quite a different perspective on the internal situation in Russia than those confined to the popular media. In both cases, it was evident that the pro-democratic forces maintained control of Russia's international broadcasting apparatus, portending a favorable outcome for them while the American broadcasters were very negative on their chances.

After the end of the Cold War, leaders in several Eastern Bloc countries told of how the Voice of America, Radio Free Europe, Radio Liberty and other Western shortwave broadcasts had helped them overcome the news blackouts and censorship that had been common in their countries.

The ability to listen directly to broadcasts from countries around the world can encourage students to have greater interest in history, geography and foreign languages. Therefore, shortwave listening is an excellent way to both encourage the younger generation in their educational pursuits and get them interested in radio as a hobby.

But it's not just wars and military actions where the shortwave bands offer special news access. In December 1986, shortwave listeners were treated to the exciting communications supporting the flight of the Voyager, the first non-stop, non-refueled airplane flight around the world.

In the week before Quebec voted on the issue of secession from Canada, the media in the United States was dominated by predictions that it would pass while those listening to Radio Canada International heard the more accurate prediction that it would fail. And those following the reports on the BBC after the automobile crash of English Princess Diana in Paris heard the pronouncement of her death almost one-half hour before that news was reported on American television.

Radio Drama

One art form that might have died without the support of the international shortwave broadcasters is the radio drama. On these stations, radio dramas are not only surviving but thriving.

The BBC's weekly offerings, which include *Play of the Week* are excellent, and have been consistently chosen as some of the ten best shows on shortwave by a major shortwave publication.

The religious show *Unshackled* (broadcast on HCJB and several other stations) is the longest-running shortwave radio drama, and sounds about the same today as it did when first heard almost 30 years ago. *Unshackled* retains an old-time aura (swelling organ music, a terse announcer) about it, so listening to it is to be time-warped back to the heyday of radio dramas in the 1930s. *Unshackled* is produced by Pacific Garden Mission, located in Chicago.

Music

On shortwave you can hear African rhythms, the sitar of India and other exotic music direct from the source. The national shortwave broadcasts present a range of regional music that is not available via any other media, along with commentaries and information on its composition and history. Often, unique music from a country is used as a station's signature just prior to its identification, or broadcast for a few minutes just before the start of scheduled programming.

Language Lessons

With the prevalence of the English language in the world, almost every international broadcaster offers English language broadcasts. But if you are fluent in another tongue, the shortwave bands will give you ample opportunity to exercise your skill. For the would-be fluent, many stations offer introductory courses on their native language.

Economics

Many countries around the world maintain standard time and frequency stations. For over a year, I have attempted to compile a list of those operating on HF. However, with the worldwide trends of decreased government budgets and political changes, the list seemed to undergo revisions almost weekly. Major changes were either carried out or promised for 1997 and 1998. Areas of the world that have undergone major political changes, such as the CIS (previously USSR) also have made major changes in their time and frequency stations, with more forecast for the near future.

Many non-US stations broadcast on multiples of 5 MHz, or close to these frequencies, and are covered up by the US stations. Therefore, this chapter has concentrated on the major US and Canadian stations, which seem to stay fairly stable in their broadcast schedule and immediate future plans.

To identify one of these non-US time and frequency stations, a general-purpose frequency guide such as *Ferrell's Confidential Frequency List*, published by Listening In, P.O. Box 123, Park Ridge, NJ 07656, may help. —*N1II*

DXing

The activities we have discussed so far focus on the content of the shortwave signals. But there is a sector of the shortwave listening hobby that concentrates on the technical challenge of receiving hard to hear, distant (DX) stations. Shortwave DXers pride themselves on obtaining official verification of their reception (QSL cards) and competing for various awards.

THE IONOSPHERE

Someone accustomed to the line-of-sight nature of the VHF/UHF bands (and who may have forgotten the HF theory they studied for their license exam!) might wonder how the shortwave signals make it around the world. To answer this question we need to briefly examine the nature of the ionosphere.

The ionosphere is a layer of the Earth's atmosphere approximately 80 to 300 miles above Earth's surface where free ions and electrons can cause some radio waves to be reflected or refracted. In brief, at the VHF region and above (above 30 MHz) the radio waves will most often pass through the ionosphere and out into space. Within the shortwave spectrum, radio waves are often reflected or refracted (bent) back to the Earth's surface. As you drop below the shortwave frequencies, the ionosphere tends to absorb and attenuate the signal so as to essentially eliminate any return signal. It is the *bouncing* of radio signals off the ionosphere (sometimes via several bounces) that allows shortwave signals to travel (propagate) around the world.

The ability of radio waves to be refracted over long distances by the ionosphere is affected by the time of day, time of year and the number of sunspots. Speaking very generally, higher frequencies (closer to 30 MHz) work best during the daylight hours and during the summer. During late evenings and in the winter, the lower frequencies (close to 3 MHz and even below) begin to peak in efficiency. This phenomenon can be observed on the medium wave AM broadcast band (0.5 MHz-1.7 MHz). During the day you hear mainly local stations. In the evening, distant stations can be clearly heard. During the summer, however, evening reception of distant AM broadcast stations, even at night, is often difficult and complicated by high noise levels created by thunderstorms and other weather phenomena.

These daylight/darkness propagation patterns are amplified by the sunspot cycle. During the peak of the 11-year sunspot cycle, the high frequencies are at their best for long distance communication. At sunspot minimums, the capabilities of the lower frequencies become more prominent. As the 1900s yielded to the dawn of a new century, sunspots were peaking and the entire shortwave spectrum was alive with activity.

WWV: Propagation and Time Information

If you are interested in detailed propagation information and forecasts, or simply want to know the exact time, station WWV is the station for you. Located in Fort Collins, Colorado, it is operated by the Time and Frequency Division of the National Institute of Standards and Technology (formerly the National Bureau of Standards). Their 24-hour-a-day broadcast provides a precise and accurate time signal around the clock and technical information on propagation conditions.

Additional services include standard time intervals, standard frequencies, geophysical alerts, marine storm warnings, Global Positioning System (GPS) information, UT1 time corrections and BCD Time Codes.

Another station, WWVH is located in Hawaii. Both stations broadcast on 2.5, 5, 10, 15 and 20 MHz.

Voice announcements are made from WWV and WWVH once each minute. Since both stations can be heard in some locations, a man's voice is used on WWV, and a woman's voice is used on WWVH to reduce confusion. The WWVH announcement occurs first, at about 15 seconds before the minute. The WWV announcement follows at about 7.5 seconds before the minute. Though the announcements occur at different times, the tone markers are transmitted at the exact same time from both stations. However, they may not be received at exactly the same instant due to differences in the propagation delays from the two station sites. The announced time is Coordinated Universal Time (UTC).

The international agreement that established UTC in 1972 also specified that occasional adjustments of exactly 1 second will be made to UTC so that UTC should never differ from a particular astronomical time scale, UT1, by more than 0.9 second. This was done as a convenience for some time-broadcast users, such as boaters, using celestial navigation, who need to know time that is based on the rotation of the Earth. These occasional 1 second adjustments are known as *leap seconds*. When deemed necessary by the international Earth Rotation Service in Paris, France, the leap seconds are inserted into UTC, usually at the end of

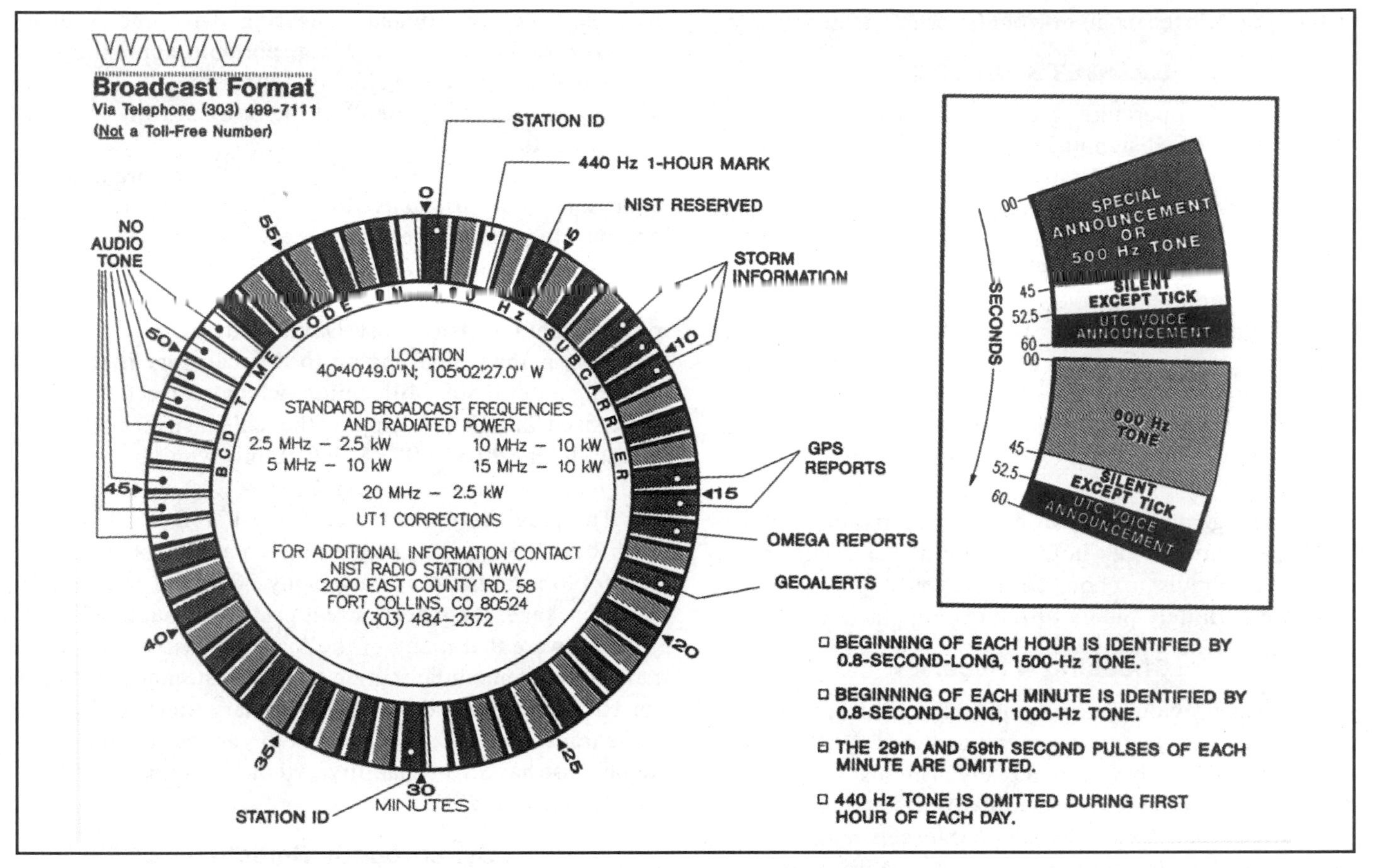

WWV broadcasts continuously. They have a Web site at http://www.boulder.nist.gov that includes a virtual reality tour of WWV and WWVH. ***(Chart courtesy of the National Institute of Standards and Technology)***

June or at the end of December, making that month 1 second longer than usual. Typically, a leap second has been inserted at intervals of 1 to 2 years.

The broadcasts from WWV and WWVH include a time code. The time code signal is 100 Hz away from the main carrier and is called a *subcarrier*. The code pulses are sent out once per second. With a good signal from a fairly high-quality receiver, you can hear the time code as a low rumble in the audio. HF receivers that receive and decode this signal can automatically display the time of day.

WWV/WWVH also broadcasts propagation information at 18 minutes past each hour. Unfortunately, the information they provide isn't quite as simple as "Propagation will be good today" or even "Propagation will be fair under 15 MHz today." Instead, they give you indices. With a little effort, you can turn the indices into a meaningful prediction of radio conditions.

Understanding the indices does take a bit of study. Both the *ARRL Handbook* and the *ARRL Antenna Book* offer some details on how to interpret the propagation data provided by WWV.

Since WWV and other time and frequency stations broadcast with substantial power around the clock on a variety of frequencies, just using them as a beacon can provide you with information on propagation. A quick listen to how strong the signal from Boulder, Colorado, is on the East Coast on 5, 10 and 15 MHz gives an immediate, if imprecise, reading on how the ionosphere is doing. Another good source for time signals and as a propagation beacon is Canada's CHU. Located in Ontario, it transmits on 3.330, 7.335 and 14.670 MHz.

Without doing any calculations or checking sunspot cycles, there is one propagation characteristic on which you can depend; propagation tends to get substantially worse just before the top of the hour. As you listen to an unidentified station and anxiously await the station identification at the top of the hour, a law of the universe known as Murphy's Law comes into play. Murphy's Law is well known to hams and engineers. A corollary to that law predicts that the more rare the station is, the worse propagation will become as the station ID nears. For predicting propagation at other times, however, the WWV information can be very helpful.

Propagation also can be humbling at other times. When propagation is at its worst, Radio Fredonia can put a megawatt into a 20-dB gain curtain array antenna and you can crank up your $20,000 ex-CIA receiver connected to the best receiving antenna and you'll still hear nothing! When the signal's not there, it's just not there. All the radio horsepower in the world can't change that. When this occurs, it's time to either switch to another frequency (which may help) or switch to another hobby for a while, like fishing or commiserating with the locals on your favorite 2-meter repeater.

GADGETS GALORE

Ham radio operators tend to be inveterate gadgeteers, and the shortwave listening hobby brings a whole new set of interesting radios for you to consider.

The past decade has seen the introduction of ever smaller and more conveniently sized shortwave radios. And the new generation of handy-talkie sized UHF/VHF scanners includes the shortwave spectrum (see sidebar, The DC-to-Daylight Radios Are Here!). In addition to their gadget appeal, the new generation of portable radios have sufficient sensitivity and selectivity (and the sideband reception capability) to allow you to monitor the activities on the HF ham bands when you're away from home (and your transmitter).

The range and variety of portable shortwave receivers available today offer a whole new world of electronic goodies for the hobbyist to choose from. If you've got the money, there are definitely plenty of interesting places to put it.

Choosing A Receiver

As was previously mentioned, many current HF amateur transceivers feature a general coverage receiver (usually 0.5-30 MHz). If you own one you're just a spin of the dial away from the joys of shortwave listening. If you're planning to purchase an amateur transceiver in the near future, make general coverage receive capability a requirement. You won't regret it.

However, if you're just interested in browsing the shortwave bands, one of the new portables on the market may be for you.

Widely considered the best of the low cost portables are the mid-sized (approximately 7.5" × 4.75" × 1.5", 1.5 pounds) Sony ICF-SW7600G with synchronous detection (note the G) and the Grundig Yacht Boy 400. Both of these are available at a street price of under $200 each.

More money buys more features and performance, such as the DXers' delight, Sony ICF-2010, renowned for its weak signal capability and pioneering synchronous detection. This radio is a relatively large portable, approximately 12" × 6.5" × 2.5" and weighing approximately 4 pounds. The street price (early 2000) of the Sony 2010 is approximately $350.

The SW-8, made by Drake, is designed to be used on a tabletop, with comfortable-sized controls. *(Photo courtesy of The R.L. Drake Co.)*

The Drake SW-8A and the Lowe HF-150 are in a category known as *porta-tops*. Both of these offer performance approaching that of a desk-top receiver in a portable (albeit a bit bulky) package. These two receivers are in the $700 price range. The Lowe HF-150 is now famous as the choice of late night television host, David Letterman.

Also perhaps appropriate to this category is the new Grundig Satellit 800 Millennium, which replaced the Sony 2010 sized Satellit 700. Tipping the scales at over 14 pounds and stretching 20.5" × 9" × 8" it certainly won't be confused with its more petite predecessor.

The previously mentioned Sony ICF-SW100 (4.5" × 0.7" × 2.9"; 8 ounces) offers performance reputed by its owners to rival the renowned Sony 2010 in a remarkably small package. It carries a street price of about $350.

Be aware that many of the least expensive shortwave radios can't demodulate single sideband, the mode of choice for HF ham radio and for utility/military transmissions. If these transmissions are of interest to you, be sure the radio you choose has SSB capability. All of the radios listed here can demodulate SSB.

Synchronous Detection

So what is this much-hyped synchronous detection? It's a radio technology that can stabilize signals with fluctuating signal strengths, and provide a clearly readable signal in the presence of severe adjacent channel interference, with a minimum degradation in audio quality. Since it is only used for AM reception, even a well informed ham who doesn't listen to shortwave broadcasting probably won't know about it (perhaps that's an oxymoronic phrase... "a well informed ham who doesn't listen to shortwave broadcasting.") Even hams who once transmitted AM may not know about it, since in the AM days, synchronous detection didn't exist.

Amplitude modulation has three constituent parts: two identical sidebands and a carrier. If the shortwave receiver can't get an adequate reception of both sidebands and the carrier, the signal quality can quickly deteriorate and become unreadable. The synchronous detection circuit automatically locks onto the transmitted carrier, mutes it and substitutes a carrier generated internally in the radio. Since this carrier is strong and stable, the effects of fluctuations on the received signal are reduced. Also as a part of this process it is now possible to receive a clear signal using only one of the sidebands. Therefore, the user can tune to the sideband opposite to any station whose close-by transmission is splattering over on the station you want to hear (adjacent channel interference), which improves readability in a crowded band. Also, as synchronous detection substitutes its own carrier, any signal whose carrier is not precisely on the same frequency is substantially attenuated, without the loss in fidelity that a narrow filter would entail.

In actual use, the results can be dramatic. Signals that

are unintelligible because of adjacent channel interference can become easily and clearly readable with synchronous detection. Weak stations can be made much more readable when their fading carrier is replaced internally. There can be a difference of opinion about any technology, but few hobbyists who have heard a radio with a well designed synchronous detection system deny its substantial benefit for the reception of AM signals.

From "Death Before Digital" to "Digital Forever"

Many long time hams once believed that real hams don't use digital readouts. However, with one exposure to a good digital readout, they quickly changed their minds. Particularly for someone new to shortwave listening, a digital readout is a tremendous aid. The ability to accurately determine the frequencies of unidentified signals makes looking them up in frequency guides a breeze. Also, digital readout assists in *repeatability*—the ability to be sure that you can return to precisely the same frequency you were listening to last week.

With the vagaries of propagation, the signal might not be strong enough to listen to, but at least you will be sure you are on the correct frequency. With the old, less precise analog frequency readouts, when you couldn't hear the signal you were seeking, it was difficult to determine whether it was bad propagation or you were just off frequency.

Getting to the frequency you want is often easier with digital receivers. The calculator or telephone-type keypad on many of them allows you to directly punch in the frequency you want and instantly be on frequency. This makes jumping from frequency to frequency to check propagation or different programs an easy matter.

But it can get easier still. Typical among the modern digital shortwave receivers are memories, where frequencies and modes can be stored for quick access. Thus your favorite BBC frequency and other often-used frequencies can be accessed by pressing a couple of buttons. This makes manual scanning of multiple frequencies (like the low-traffic utility frequencies) a breeze.

Another advantage of radios with digital readouts is their use of frequency synthesizers, which most often all but eliminate drift. Drift is the tendency of a radio to change the frequency as its internal components warm up. While a digital readout does not ensure that the radio uses frequency synthesis, the two are found together so commonly as to be virtually synonymous.

Oldies But Goodies?

Some well meaning hams may advise you to get an older shortwave receiver at a hamfest to get your start in shortwave listening. They may speak with nostalgia about spinning the dials on a Hammarlund, Hallicrafters, National, etc back in their early days in radio.

Although it sounds good at first, there are a couple of problems with this advice. Those older receivers don't have digital readouts or synchronous detection. They also don't have the stability of frequency synthesis; it was a common practice in the old days to let receivers warm up for an hour or so before doing any serious listening, to get the drift stabilized.

Another situation negatively impacting the attractiveness of older receivers is that in recent years they have become the targets of collectors. What this means is that the good ones aren't cheap and the cheap ones aren't good, and weren't very good even when they were new. Given the appetite of collectors, increasingly even the bad ones aren't cheap.

Certainly, these old receivers have a special appeal and that special glow that comes from a hernia-inducing big box full of tubes. They can also provide much of the heat your shack requires in the winter. A late 1950s Hammarlund HQ-180 and a 90-pound military surplus Collins/EAC R-390A may have a place of honor in many shacks. But while these older receivers have excellent sensitivity and selectivity, you can usually achieve better performance and ease of use (not to mention lower cost) with modern receivers.

If there is any exception to this rule, it is in medium wave and long wave spectrum. Many of the newer shortwave receivers tend to suffer from performance drop-off at the AM broadcast band and below. The best of the oldies are great performers on these frequencies. If you intend to be especially active in medium or longwave monitoring, or if you must have that *real* radio look, a well selected oldie can be a goodie. Just don't expect these radios to be inexpensive, and the cost of replacement tubes might make you gulp—twice!

Antennas

Put as plainly as possible, if you're using an HF transceiver, an 80-meter or 40-meter dipole will work well for shortwave listening. If you use a portable shortwave receiver, adequate reception of the international broadcasters is often possible using the attached whip antenna.

If you want maximum performance, put up a wire outside as high as possible and as long as possible. Practically, a wire between 25 feet and 75 feet long and 20 feet (perhaps between two trees) will work well. If your portable receiver doesn't have an antenna jack, just use an alligator clip to connect the external antenna to the whip. Connecting an outside antenna to the whip will sometimes overload the receiver, so you may have to experiment to determine the optimum combination for the signals you want to hear.

While you may achieve acceptable results with just the whip attached to a portable radio, don't underestimate the value of a good antenna system. Many people like to spend most of their money on the radio. It's certainly the element of the station that will draw *oohs and aahs* from the uninitiated. However, the more experienced operator knows that, while the antennas may not be as photogenic as radios, it is easier to get high quality results with a good radio and a good antenna than with the best radio and a lousy antenna!

Living in a place where it is difficult to erect an outside antenna shouldn't prevent you from enjoying shortwave listening. An invisible outside antenna (constructed using a wire so thin as to be nearly invisible) can bring useful results. Invisible antennas are covered in many ham radio and

DC to Daylight

The *DC-to-Daylight Receiver* Is Here! Almost.

"If the broadcast receiver were a very special one that could continue to tune higher in frequency (there are technical reasons why this is impractical without switching), you would find many different groups, or bands, of frequencies, used by many different services." This quote from the ARRL's *How to Become a Radio Amateur* from the distant past has been used to begin a section called "The Super-Special Wide-Range Receiver Is Here" in past editions of the *Operating Manual.*

These editions have discussed the evolution of receiver technology up to the point of continuous coverage receivers spanning from 3 to 30 MHz, and indeed this technology is now so prevalent that virtually all ham transceivers incorporate a full-coverage shortwave receiver within them.

However the ultimate in receiver coverage was always called DC-to-daylight. The receivers that are out today may not hit that impossible goal, but they do cover a frequency range so wide as to boggle the mind. As amazing as it may seem, several models come in small hand-held size packages, including a few as small as a pack of cigarettes!

At the low frequency end, these wide-range receivers bottom out at 30 to 500 kHz, which is about typical for a general coverage shortwave receiver. But at the high frequency end, these receivers soar past the 30 MHz shortwave boundary deep into the VHF/UHF region. All of them go well past 1 GHz (1000 MHz) and the most advanced of them exceed 2 GHz.

Actually, these receivers have sprung from an explosion in UHF/VHF scanner technology. Monitors of the multitudinous services in this spectrum have sought portable radios that had an expansive number of memories and covered all the popular voice modes. For years scanners have encroached upon shortwave territory, going as low as 25 MHz. By the mid-1990s, scanners began to expand into full shortwave spectrum coverage.

One of the first entries into this category was the AOR AR-1000, covering 500 kHz to 1300 MHz in 5 kHz steps with 1000 memories. It is small in size (6.7" × 2.6" × 1.4") and provides AM, narrow-band FM (NBFM) and wide-band FM (WBFM). The AR-1500 was essentially the same radio, but with a beat frequency oscillator (BFO) for reception of single side-band (SSB) and continuous wave (CW) signals.

Soon thereafter, ICOM introduced the R1, an even smaller radio (4" × 1.9" × 1.4"), which covered 100 kHz to 1300 MHz in steps as small as 0.5 kHz. However, it only demodulated AM, FM and WBFM and had 100 memories.

As early entries in this super high-tech arena, each of these radios had some deficiencies like poor sensitivity over part of the specified range, intermodulation problems and complicated programming. Although each of these has their ardent defenders, their well publicized liabilities limited their appeal.

It wasn't long afterward before American readers of the British publication *Shortwave Magazine* noticed some ads for a small (6.1" × 2.6" × 1.5") radio from a little-known manufacturer, the Yupiteru MVT-7100. This radio covered 100 kHz to 1650 MHz in 50 Hz steps. Because it used true carrier injection to provide SSB reception, the full range of modes (AM/FM/WBFM/LSB/USB) were programmable in its 1000 memories. A conversation about the radio commenced on one of the national on-line services, and soon several enthusiasts had ordered 7100s from an accommodating English dealer.

In the wake of the rave reviews that followed on the on-line services, more of the Yupiteru 7100s were ordered from England and a phenomenon had started. Finally here was a radio that was reasonably simple to use (considering all it would do) and that provided good sensitivity throughout its frequency range. A review on the shortwave show *Spectrum* actually said it had as good sensitivity on shortwave as the Sony 2010, but that may be a bit of an overstatement. However, it does provide shortwave reception approximately commensurate with the respected

Hand held, this ICOM IC-R10 covers 500 kHz to 1300 MHz, with AM, FM, WBFM, USB, SSB and CW receive capabilities.

shortwave radio antenna texts.

Reading and experimenting will help you achieve the optimum performance for your location and equipment. Two books that contain many ideas for invisible—or at least inconspicuous—antennas are: *Your Ham Antenna Companion*, by Paul Danzer, N1II and *Stealth Amateur Radio*, by Kirk Kleinschmidt, NTØZ. Both are published by the ARRL.

The International Shortwave Broadcasters

The international broadcasters are the best known and easiest to find stations on the shortwave bands. They are often sponsored by governments, who are willing to spend substantial sums to spread their viewpoint around the world. Their booming signals, not uncommonly in the half-megawatt range, provide music, news (sometimes tainted with propaganda), educational and other entertainment programming. Not only are their signals usually strong, their transmissions tend to be confined to identifiable frequency ranges.

Another reason that the international broadcasters may be easy to hear is that the transmitter may not be located where you think it is. Though you may be listening to Radio Netherlands, Deutsche Welle or Radio Taiwan, the transmitter might be located off the north coast of South America (the Netherlands Antilles is a popular location) or even from within North America, via a rented transmitter site! Those listening for program content won't care where the transmitter is located, but DXers often specifically seek those transmissions that originate far from their locations.

Interval Signals

If you listen to an international shortwave broadcaster at the beginning of their transmission or at the top of the hour, you may hear a repeating series of tones. These tones

Sangean ATS-803 (which, like it, doesn't have synchronous detection) in a much smaller package. And it's also a full-feature VHF/UHF scanner!

The Yupi's position alone at the top of the heap was relatively short-lived. About a year after the advent of the 7100, AOR introduced the AR-8000, which covers 100 kHz to 1.9 GHz in 50 Hz steps, with 1000 channels and true single sideband, as with the MVT-7100 and in almost exactly the same size package. However, it also features a four line alphanumeric display and computer interface capability.

More recent entrants into this race are the Yupiteru MVT-9000 and the ICOM IC-R10. The MVT-9000 covers 530 kHz to 2039 MHz and has twin VFOs. The R-10 is a little smaller than the MVT-9000 or AR-8000, and covers from 500 kHz to 1300 MHz. The R-10 also features a computer interface, while the 9000 does not. Both of these rigs have 1000 memories and an alphanumeric display with a real-time bandscope.

But now the hobbyist wanting this frequency coverage doesn't have to lug those large handie-talkie sized radios around. The size standard for a shirt pocket sized radios has always been the (politically incorrect) pack of cigarettes, and this size barrier has now been met by the latest generation of offerings from the ever inventive radio companies.

ICOM's R-2 covers 500 kHz- 1.3 GHz (less the evil cellular phone frequencies) in a 2.3" × 3.4" × 1.1" package, with 400 memories and computer interface capability. Yaesu yanked out coverage of frequencies from 16-48 MHz and can't get above the big 1 GHz barrier in their slightly larger VX-5, but added transmit capability on 6 meters, 2 meters and 440 MHz. With all this the VX-5 only had room for 220 memories but does provide for a computer interface. Sadly, neither of these units offers SSB reception. Back to the drawing board, guys.

As covered in the section on Scanning and alluded to above, the FCC requires that new radios with coverage in the 800-MHz range have the cellular phone frequencies exorcised from them. Yupiteru has not chosen to comply with this yet, so only the ICOM, Yaesu and AOR units are readily available in the United States, although at this writing some Yupiteru radios are being shipped to the U.S. from other countries.

Increasingly, table-top radios are being introduced with wide-ranging frequency coverage. Two of the standards in this arena are the ICOM IC-R8500 and the AOR AR-5000. With the hand-held radios, although their shortwave capability was acceptable, it was well known that their performance had been optimized for the VHF/UHF frequencies. But these two table-top radios defy description. From all reports, they are excellent performers throughout their frequency coverage. As with their smaller brothers, cellular frequency coverage is deleted.

Recently a totally new category in super-wide range receivers has emerged; the receiver-in-a-computer. The pioneering WinRadio WR-1000i and WR-1550i and ICOM's PCR-100 and PCR-1000 lead an expanding field of competitors in this category. In addition to having all of the functions of top-level desktop receivers and such extras as frequency spectrum displays, this class of radios can be enhanced with an ever-increasing number of third-party software offerings. With a full powered desktop computer running the front-end of the radio, the variety of scanning and signal analysis functions possible are virtually limitless.

So hobbyists now have numerous choices in the DC-to-daylight category. The hand-helds sell in the $500 to $600 range, and the ICOM and AOR desktop units are both about $1900.

Admittedly, receivers with a DC-to-daylight type frequency range have been available to the government and military for years, but in desk crushing sizes and weights and at a cost approaching that of a Rolls-Royce. Even those ostensibly marketed to hobbyists carried price tags in the $5000 range.

For those without such a deep bank account, the advent of these affordable, full-frequency receivers gives a ham the capability to monitor everything from 2 meter repeaters, local police and fire activity, the BBC and 80 meter sideband nets from one radio small enough to easily hang from a belt. The radio hobbyist never had it so good!

are known as the *interval signals*. They're designed to help you identify the station when they are transmitting in a language foreign to you, or when the signal is weak. The familiar tones that the National Broadcasting Company (NBC) still uses are a domestic example of an interval signal. In addition to interval signals, there is another type of musical identification: the musical signature.

Interval signals are short and often played repeatedly (sometimes to the point of irritation) for 2-5 minutes just before the top of each hour. The instrumentation of an interval signal is simple, usually just a piano, celesta, carillon or electronic keyboard. The musical signature is typically more fully orchestrated and is played just once per period. Most often, interval signals are played hourly and musical signatures only at sign-on and sign-off, but there are numerous exceptions.

The tones or music are often related to the culture and history of the originating nation. The Voice of America uses "Yankee Doodle Dandy" and Radio Australia uses "Waltzing Matilda," while the BBC uses a very regal sounding tune called "Lilliburlero." Many of the European nations use classical music for their musical signatures. Not surprisingly, the religious stations often use hymns.

Hearing and identifying interval signals is another subset of the shortwave listening hobby. The level of interest is indicated by the fact that one pirate radio station uses the bogus call WLIS—We Love Interval Signals.

World Time

If your operating has been confined to the VHF bands, you may not have much experience with what is known as World Time. Since shortwave signals reach around the world, the listing of the time of the transmissions can be a problem. Whose time do you use?

It "only" covers 100 kHz to 2000 MHz, so there is not much you will miss with this ICOM IC-R8500. Scanning and 1000 memories included, of course!

To eliminate this confusion, the time at the Royal Observatory in England at the zero degree meridian (formerly at Greenwich) is used. It was called Greenwich Mean Time (GMT) for years, but now it is known as Coordinated Universal Time (UTC), World Time, Universal Time or Zulu (from the phonetic for the letter Z, which the military uses to indicate UTC).

UTC was established by international agreement in 1972, and is governed by the International Bureau of Weights and Measures (BIPM) in Paris, France. It differs from local time by a specific number of hours. The number of hours depends on the number of time zones between your location and the location of the zero meridian. For time zones in the continental United States, UTC is 5 hours ahead of EST, 6 hours ahead of CST, 7 hours ahead of MST and 8 hours ahead of PST. That is, at 5 AM EST, the UTC time is 10 AM. When local time changes from daylight saving to standard time, or vice versa, UTC does not change. However, the difference between UTC and local time does change—by 1 hour.

UTC is expressed using the 24-hour clock system, sometimes called *military time*. The hours are numbered beginning with 00 hours at midnight through 12 hours at noon to 23 hours and 59 minutes just before the next midnight. So 7 PM is 1900 hours.

Another attribute that you must remember when using UTC is that (for the United States at least), since UTC reaches midnight before the American time zones, the date also changes earlier. Thus, since 7:00 PM EST (1900 EST) is midnight UTC, at 7:01 PM (1901) EST on December 31st, it is 0001 UTC on January 1st. Many people keep 24 hour clocks set to UTC at their listening post, which makes logging the correct time a breeze. But even experienced hobbyists sometimes forget to advance the date after 0000 UTC.

THE MAJOR PLAYERS

Before we delve into some of the major international broadcasters, a word about names is in order. Since we are dealing with foreign countries and foreign languages, these stations can be called by either the native name or the English variant of it. There are no definite rules, but some standards have arisen. The Voice of Germany is almost always referred to as Deutsche Welle (and as with the classical composer Wagner, the "w" is pronounced as a "v"). The national stations of Cuba and Holland are referred to almost equally by their native and English names, Radio Habana Cuba/Radio Havana Cuba and Radio Nederland/Radio Netherlands, respectively. The French are very sensitive about maintaining the influence and purity of their language; therefore, their national broadcaster is always called Radio France Internationale and never Radio Francaise or some other variant. (A note to our French readers... although true, that was a joke.) HCJB's nickname The Voice of the Andes is sometimes given in Spanish as, *la Voz de los Andes*. So if you see or hear a name unfamiliar to you, be sure it's not just a native-language rendition of a station whose name you already know.

As the post cold war period evolves, shortwave stations may change their names (see Radio Moscow, below) and may also change the focus of their broadcasts. Almost all of the major shortwave broadcasters are faced with fighting cutbacks in funding, and the reduction in services that these would entail. Nonetheless, even if the most severe cutbacks proposed actually occur, there will still remain a tremendous amount of interesting programming on the shortwave frequencies.

BBC

If you can't hear the BBC on your shortwave receiver, the radio isn't working. The British Broadcasting Corporation's worldwide network of shortwave transmitters was established to serve the British Empire back in the days when the sun never set on it (it was known then as the Empire Service), and they continue to put out powerhouse signals all across the shortwave spectrum. Even with the budget cuts resulting from political battles within the House of Parliament, the BBC remains the standard by which all shortwave broadcasting operations are measured.

Long before the days of cable TV and CNN, the 24-hour-a-day news reports of the BBC from around the globe provided listeners with virtually constant access to late-breaking information of important events. Even in today's news-rich environment, those without cable TV or wanting an alternative viewpoint depend on the World Service (its name today) of the BBC. The program content does, however, often represent the British government point of view, particularly that of the conservative Foreign Office.

Reliance on the BBC extends to more than news. If you want to know the proper pronunciation of a word, listen for it on the BBC; reportedly, they have several people whose full-time job it is to determine the correct English (as opposed to American) pronunciation of any new word that comes up on their broadcasts.

Voice of Russia (Radio Moscow)

If you ever listened to Radio Moscow prior to 1987, you should listen to a few broadcasts from Radio Moscow today, even if you aren't interested in shortwave listening. Borrow a portable shortwave receiver or go to a friend's shack if you don't have shortwave broadcast receive capability. The change in the content of their newscasts and programming will flabbergast you. If fact, they've even changed their name. They are now called the Voice of Russia.

In the old days, the only times they weren't slamming

the United States and the western world in general was when they were broadcasting fascinating reports on The Wheat Harvest in the Ukraine, or some other such thrilling subject. They also were reported to be very tough on announcers whose tongues were loose. One announcer who slipped and called the Soviet action in Afghanistan "the invasion" was never heard from again. (Evidently, the official line was that they were in Afghanistan for humanitarian reasons, using Soviet battle tanks to plow the fields and the heavily armed Hind helicopters for crop spraying.)

But as changes took place in Russia from 1987 to 1991, so did Radio Moscow change. As previously mentioned, monitoring Radio Moscow during the August 1991 coup and the insurrection of the legislature in September 1993 provided more accurate information than virtually any of the outside news sources.

Now the Voice of Russia sounds very much like many of the western broadcasters. Most shocking to those who listened years ago, is hearing the frank discussions of problems within Russia. Not to make light of their problems (every country has its share), but to go from such a controlled purveyor of the party line to a relatively independent broadcaster has been an amazing transformation.

A downside of the changes is that Russia, and the Commonwealth of Independent States (CIS) that replaced the Soviet Union, are reducing the amount of programming on the Voice of Russia. Some of the transmitter sites are being used by the newly independent states for their independent broadcasts. Other transmitter sites have been rented to anyone who will pay their price. At present, the Voice of Russia is still a major presence on the shortwave bands, but substantially less so than in its heyday. What the future holds is very uncertain, but whatever happens, monitoring the evolution of the Voice of Russia is a fascinating experience.

Voice of America

The Voice of America (VOA) was established immediately after the end of World War II. After spending the war years listening to the German propaganda broadcasts orchestrated by Joseph Goebbels, the US government was especially sensitive to broadcast content. As a result, the Voice of America was forbidden to create broadcasts for consumption within the United States. The idea was to remove the temptation for the government to propagandize the general population.

The VOA (Voice of America) site near Greenville, NC remains a tourist attraction. Wouldn't it be nice to rent these antennas for a contest? *(Photo courtesy of VOA)*

However, in recent years this view has moderated and now VOA provides an excellent amount of programming information, even providing a site on the Internet. Though the VOA definitely broadcasts from the American point of view, it is generally respected as an authoritative and respected source of information. VOA broadcasts were a critical lifeline to the outside world for those behind the Iron Curtain during the Cold War.

The end of the Cold War also has been difficult for the VOA. The programming priority has shifted from Eastern Europe to emerging democracies of Africa and Latin America, but there is less urgency (and less money) supporting these broadcasts. The large VOA transmitter site in Bethany, Ohio, was decommissioned in late 1994, and a transmitter and receiver site near Greenville, North Carolina were decommissioned in the late 1990s. The remaining site near Greenville is the VOA's sole remaining voice on the East Coast.

The Voice of America facility in Greenville supports worldwide broadcast of VOA programming material. *(Photo courtesy of VOA)*

Should you be in the North Carolina area and have the time to drive to Greenville, a trip to the remaining transmitter site will surprise and astound you. If you call them (919-752-7115) they can set up a tour, but just driving by one of the sites will allow you to see an antenna field that no radio enthusiast will forget. Cut into the North Carolina pines is an open field approximately 3 miles long and 3 miles wide. The huge transmitter building sits about 1/2 mile off the narrow secondary road, and off in the distance are groups of four 400 foot towers set in a semi-circle with about a 1-mile radius spanning over 180 degrees of arc. Each of these sets of 4 towers holds a massive curtain array antenna. To change the direction of their signal beam, they don't rotate an antenna; they just switch to an antenna pointing in the needed direction. Visitors never fail to be impressed.

The Voice of America headquarters in Washington, D.C. also has an excellent tour, albeit one that is less impressive from a radio hardware point of view.

Radio Netherlands

The Happy Station, home of the famous Happy Station program for well over half a century, is one of the most popular shortwave broadcasters in the world. Radio Netherlands International dates back to 1927 when Philips Radio Laboratories began broadcasting to the Dutch West Indies. In 1994, Radio Netherlands celebrated the 75th anniversary of Dutch radio, contending (in a friendly way) that the Dutch radio pioneer Steringa Idzerda invented radio. . . or radio broadcasting. . . or at least regularly scheduled radio broadcasts.

Even though not many encyclopedias recall Idzerda's work back in 1918, Radio Netherlands had a fun celebration of his work and the radio equipment and techniques of the early days. Their goal was to prove that 75 years on, there's a lot of life left in international radio. That's a sentiment sure to increase their popularity.

HCJB, Quito, Ecuador

Broadcasting from the equator at an elevation of over 10,000 feet, the evangelical Christian "Voice of the Andes" dates back to the early 1930s. Compared to the sometimes strident type of Christian broadcasting common within the United States, the style of HCJB is relatively understated. HCJB also has many programs popular with those who do not necessarily agree with the beliefs of their sponsors, the World Radio Missionary Fellowship of Colorado Springs, Colorado.

To combat antenna arcing problems caused by high power broadcasting at such a high altitude, engineers at HCJB invented the cubical quad antenna in the late 1930s, an antenna that has become very popular with ham radio operators. HCJB sends a booming signal into the United States, and is typically an early catch for a new shortwave listener

Radio Habana Cuba

The Cold War lives on at RHC, so don't tune in here to hear any praise of the western world in general and the United States in particular! But for all their enmity, they seem obsessed with the USA. You will seldom listen to RHC for more than a few minutes without hearing about the USA, of course in the most unflattering terms. Because of its proximity to the United States, it generally puts in a solid signal. Over the past few years, there have been times when the signal suffered from distortions and weaknesses that have been attributed to power brownouts and lack of spare parts. Russia no longer pours money and resources into the island nation, which has led to fuel and resource crises. The former Soviet Union is not infrequently criticized on the air for abandoning the Communist philosophy (and with it, their substantial support for Cuba).

This will be an interesting station to monitor to observe the changes that are sure to come in Cuba. And given its location, the implications for its northern neighbor (us!) give listeners in the United States substantial motivation to follow the trends.

Deutsche Welle

The Voice of Germany was for many years the voice of *West* Germany. In the many years since World War II it has emerged from the shadows of the propaganda broadcasts of Radio Berlin. It is a consistent source of news and analysis of events on the continent. Though its world-wide resources aren't as vast as those of the BBC, it is an emerging leader in international reporting and provides the political perspective of this country in the post-Cold War world.

Radio France Internationale

Anchored by its amazing ALLISS rotating monster antenna and recently increased transmitter output, RFI is on its way to establishing itself as a power in international broadcasting. (ALLISS refers to two French transmitting locations—Allouis and Issoudun—which were the first to be equipped with this rotatable HF curtain antenna system.)

RFI retains strong African coverage from its colonial days and its clever programming treads a fine line between spontaneity and professionalism. For French language students, they offer an excellent source of practice direct from a most reliable source.

Shortwave Shows for the Hobbyist

To keep up with the fast changing world of shortwave radio, the serious listener requires access to information resources that can quickly disseminate updated data. One excellent resource for this is Internet (see the sidebar "Online Computer Resources for SWLs").

However, the shortwave broadcasters themselves offer shows on a regular basis (usually weekly) that help the SWL stay on top of what's happening. The only equipment the listener needs to access this information is their shortwave receiver.

The breadth of topics varies with each show, but among those covered are the latest happenings in the shortwave world, tips on improving the proficiency and enjoyment of shortwave listening, human interest stories relating to SWLing and other related communications and electronics information. Information on satellite broadcasting and reception has been added to many of these shows.

Online Computer Resources for SWLs

The books, magazines and shortwave shows listed in the main body of this chapter provide the shortwave listener with valuable information, but shortwave's ties to fast breaking events and the vast quantities of information make the type of communication provided by interconnected computers the ideal method of information exchange.

How else can you exchange detailed, exhaustive information with other hobbyists around the world, 24 hours a day, in a format that allows you to quickly turn it into convenient printed hard copy? Where else can you be in instant communication with the programmers at major international broadcasters overseas? All this and more can be found among the electronic information resources available today via computers and modems. Although well-equipped radio hobbyists almost always have a microcomputer at their disposal, only now are many of them becoming aware of how much additional usefulness connecting it to a telephone line can add.

Long before all the hype about the information superhighway has been networks of computer information resources that can substantially increase the enjoyment of hobby radio. The integration of telecommunications capability and microcomputers has provided an important new way to transmit and exchange information.

Although independent computer bulletin board systems (BBSs) and commercial services continue to provide a valuable source of radio information, the explosion of activity on the Internet makes it the primary source of information today. A chapter on using the Internet is provided elsewhere in this book.

Internet allows for access to information in several ways. The most common methods of information dissemination are newsgroups, electronic mail, the World Wide Web and files.

Newsgroups

No matter how knowledgeable you are about your special interests in radio or how long you have practiced your hobby, on occasion you will be confronted with a seemingly unsolvable question or need information on some topic unfamiliar to you. Electronic information services can provide you with a forum to get input on your questions from hobbyists from all over the country. A corollary to the Internet called USENET provides public message areas called "newsgroups."

On BBSs and the commercial services they may be called conference areas or SIGs (Special Interest Groups). The user chooses the newsgroup with the most appropriate title and types in (posts) the question, usually addressed to ALL. Anyone accessing the newsgroup (nationwide or even internationally on USENET) after that will see the message. Within days or hours, the original question will usually receive responses; the responses may produce more questions and responses and will often yield an in-depth exploration of the topic by a number of knowledgeable (and some not so knowledgeable) responders.

Of course, the responders are self-selecting, so there are no guarantees as to their accuracy and expertise. But this same caveat applies to many seminars, books, magazine articles and the like. In an electronics analogy, this is known as the signal-to-noise ratio, or noise factor. The number of unknowledgeable or belligerent responders (and there are a few belligerent ones) cause the noise factor, and the higher their number compared to knowledgeable and reasonable respondents, the higher the noise factor.

Over time reading the messages in a conference area, the discerning user will be able to determine which participants' information is trustworthy. Newsgroups of interest to shortwave listeners include:

rec.radio.shortwave
rec.radio.scanner
rec.radio.swap
alt.radio.pirate
alt.radio.scanner
rec.radio.amateur.antenna
rec.radio.amateur.boatanchors
rec.radio.amateur.digital
rec.radio.amateur.equipment
rec.radio.amateur.misc

Electronic Mail

The newsgroups are generally public posting areas (where everyone can read them), but if you have a private message you can send it somewhat confidentially to one person or several through e-mail (Electronic mail). The *somewhat* modifier is used because, in addition to the intended recipient(s), the administrators of the service, some people at Internet servers (and anyone they designate) can see the message.

Files

People being introduced to computer communications are usually surprised by the number of free programs available. Most of this software can be downloaded (transferred to your computer via the phone line) from the Internet for free. See, for example, **http://www.shareware.com/**. The range of this *shareware* is as broad as the range of regularly distributed commercial software and the quality and level of sophistication is often as good or better. Spreadsheet, communications, database and word processing programs are all available, as well as a seemingly endless supply of games.

The quality of this software varies widely, but you can't complain about the cost of acquiring it. Although the number of programs dealing with radio and electronics is not as large as those on some other topics, they are growing in number. Some of them can be very useful and some are worth exactly what you paid for them, but they all provide some insight in the interface of computers and radios.

Given the volume of shortwave information, sometimes it is too voluminous for the conference areas and often in these cases it is compiled and placed in a file area. In many cases the author of the software gives the rights of usage freely or only requests donations from satisfied users. Other authors consider shareware an alternative method of distributing and selling commercial software (allowing you to try the software instead of advertising), and require a modest payment if you use the program regularly. Of course, given the circumstances, you're on the honor system.

The World Wide Web

There are many Web pages on shortwave and other radio topics, and they come and go almost daily. To get their addresses, use a search engine (see the Online Resources chapter).

Getting Started in Telecomputing

The basic tools needed for telecomputing are a computer with a serial interface, communications (or terminal) software, a modem and a telephone line. A typical voice-type telephone line is all that is required from the phone company. Once you get on-line, you'll never again lack for reading material on your favorite hobby. In fact, you'll soon have to begin dealing with an excess of information, "information overload." That's a nice problem to have. For additional information on the Internet, see the Online Resources chapter of this book.

The Drake R8B world-band communications receiver features selectable sideband, synchronous AM detection, five built-in filters (6.0, 4.0, 2.3, 1.8, and 0.5 kHz), 1000 programmable memories and covers 100 kHz to 30,000 kHz (also 35 - 55 MHz and 108 - 174 MHz with optional VHF converter). Serious listeners may want to check out the other features of this high-end receiver. ***(Photo courtesy of The R.L. Drake Co.)***

In addition, some of the mailbag shows include SWL reports that have been mailed to the station. They also give listening tips, frequency changes and other information of interest to shortwave listeners.

Here's a selected list of the most popular shows:

MediaScan, on Radio Sweden (formerly called *Sweden Calling DXers*) is the world's oldest radio program about international broadcasting. Radio Sweden has presented this round-up of radio news, features, and interviews on Tuesdays since 1948. It is currently broadcast on the first and third Tuesdays of the month.

Media Network on Radio Netherlands is a staple of shortwave enthusiasts. If it's on the air in this solar system, producer Jonathan Marks and the Media Network team are listening! Each Thursday (repeated Fridays) this award-winning survey of communication developments draws from its network of more than 157 regular contributors around the globe, including Jim Cutler, Mike Bird, Arthur Cushen, Diana Janssen, Victor Goonetilleke and their secret weapon, Andy Sennitt. Andy Sennitt just happens to be editor of the venerable (and Amsterdam-based) *World Radio-TV Handbook*, the creation of which keeps him plugged into high-quality information sources around the world. *Media Network* also covers the latest in the world of technology.

DXers Unlimited is a program broadcast by Radio Habana Cuba. Host Arnie Coro, CO2KK, devotes a considerable amount of his time on topics of interest to those new to shortwave radio. *DXers Unlimited* is totally nonpolitical, and probably the only program on RHC about which that can be said. *DXers Unlimited* is broadcast twice a week, on Sundays and Wednesdays.

DX Party Line from HCJB (Saturdays) is for dial-twiddlers and those who like to listen to faraway countries just for the thrill of hearing a distant signal. Ken MacHarg hosts this popular show, which also has regular features for shortwave newcomers.

Ham Radio Today is also offered by HCJB. As its name suggests, *Ham Radio Today* focuses on issues of interest to hams, although in acknowledging the wide interests of hams, it does stray in to non-ham but related subjects including the history of radio. It is hosted by John Beck every Wednesday.

Spectrum is a weekly show covering the spectrum of electronics, ham radio, shortwave radio and related subjects. It is broadcast on independent shortwave station WWCR (Nashville, Tennessee) and a network of medium wave (AM broadcast) stations early Sunday mornings (Saturday evenings, in United States time zones).

World of Radio can be heard on various shortwave stations throughout the week, including WWCR and Radio of Peace International (RFPI in Costa Rica). Produced and hosted by Glenn Hauser, this show focuses rather narrowly but in depth on radio monitoring in shortwave and mediumwave. The program is a mix of updates on changes in programming and frequencies of the international broadcasters, with an occasional special program on a particular aspect of the shortwave hobby.

DXing with Cumbre is a relative newcomer to the scene, but it has already attracted a loyal audience. Host Marie Lamb provides the latest shortwave radio information for hard-core DXers. It is broadcast weekends on independent shortwave stations WHRI, WRMI and a few medium-wave outlets.

LISTENING AND THE LAW

The first rule of radio monitoring has always been that no license is required to listen. In general, there are virtually no rules regulating listening, so there was little to know, from a professional and legal standpoint.

The guiding rule of shortwave listening has always been the Communications Act of 1934 (Section 705), whose rule was simple: you could listen to anything you wanted to, but unless the transmission was by a broadcaster, a hobbyist (amateur, Citizens Band, etc.) or a ship in distress you were not to divulge the contents of the transmission nor use it for personal gain (don't get excited... the average hobbyist can listen for years and not hear anything that can be readily used for personal gain). In fact, the spirit of the law was observed such that no one was prosecuted for telling their listening friends about an interesting signal they heard last night, as long as there was no negative effect to the parties of the transmission.

In 1986, due to pressure from lobbyists from the cellular telephone industry, Congress passed a law called the Electronic Communications Privacy Act of 1986 (ECPA). This law made it illegal to listen to cellular phone calls and several other types of communication that normally take place in the VHF/UHF spectrum and not generally of interest to SWLs. However, it also made it illegal to listen to remote broadcast transmissions and studio to transmitter links, communications that sometimes take place in the 26-30 MHz range.

There have been numerous reports of the reception of studio to transmitter communications, often hundreds of miles from the studio site. In one publicized example during the peak in the sunspot cycle, a Florida radio station was receiving (and soliciting!) reception reports from across the United States from its 100-watt shortwave studio to transmitter link. Interesting and innocent though this occurrence may be, it is illegal now and prudence would indicate that the legally minded hobbyist avoid it—or at least avoid a confession that such illegal listening took place (which is what a reception report would be). To date there are no known instances of the casual reception of such stations being prosecuted, but just because such a silly law has been ignored so far does not ensure that some techno-noramus (technological ignoramus) lawyer won't seek to make his reputation on such a case in the future. Don't you be the guinea pig for such legal action.

UTILITIES AND MILITARY

Despite how the name sounds, shortwave utilities monitoring is not listening in to crews from your local electric and gas company. In the broad sense of the term, utility monitoring is listening to any of the vast array of non-Amateur Radio transmissions not intended for public consumption. The respected *Klingenfuss Guide to Utility Stations* calculates that 77% of the shortwave bands are dedicated to the utility services.

Within the realm of utility monitoring falls maritime communications, aeronautical communications, commercial fixed stations and military transmissions on the air, land and sea.

This activity requires a little more sophisticated level of equipment than shortwave broadcast monitoring. Because these transmissions are only intended for a select audience, the power level is lower than most broadcasters. Thus reception requires a better antenna, a more sensitive receiver—or both. In addition, most utility's voice transmissions are in upper sideband (USB), a mode not included in many low-end shortwave receivers. Radioteletype (RTTY) and the digital modes require outboard decoding equipment (usually including a computer) and a receiver with good selectivity and stability.

Table 14.2
Mystic Star Frequencies

Radio frequencies used by high-ranking military and government officials when traveling on official business.

All frequencies shown in kHz

3032	6817	9023	11413	13585	20053
3046	6803	9026	11441	13710	20154
3076	6918	9043	11460	13823	20313
3071	6927	9120	11466	13960	22723
3116	6993	9158	11484	14715	23265
3144	7316	9180	11488	14902	25578
4721	7690	9270	11498	14913	26471
4731	7735	9320	11545	15015	
4742	7765	9958	11596	15036	
4760	7813	9991	11615	15048	
5688	7858	10112	11627	15091	
5700	7997	10427	12324	15687	
5710	8040	10530	12317	16080	
5760	8060	10583	13201	16117	
5800	8162	10881	13204	16320	
5820	8170	11035	13214	16407	
6683	8967	11055	13215	17385	
6715	8992	11118	13241	17480	
6738	8993	11176	13247	17972	
6756	9007	11180	13412	17993	
6757	9014	11210	13440	18027	
6760	9017	11226	13455	18218	
6790	9018	11249	13457	19047	
6812	9020	11407	13485	20016	

Military monitoring is a subset of utilities monitoring. It encompasses the full spectrum of the utilities world: upper sideband, RTTY and digital modes as well as unique and secret transmission modes.

To those willing to go that extra step to acquire the equipment necessary for utilities monitoring, the rewards are many. The Mystic Star frequencies can yield very interesting listening as members of the Executive Branch and VIPs travel about the world. Although the planes that serve as Air Force One are equipped with sophisticated satellite communications gear, the shortwave equipment is still onboard and occasionally used for routine traffic or to achieve a certain diplomatic effect. Occasional, unencoded phone patches can be heard from a variety of aircraft, including Air Force One.

Monitoring the Land and the Sea

Despite all the high-tech equipment available to the U.S. Air Force, the Global High Frequency System remains a backbone of communications. Long ago, during the height of the Cold War, when many first heard the transmission, "Skyking, Skyking, this is McDill, do not answer...", they were ready to head for the bomb shelter. They knew these transmissions concerned the (now defunct) Strategic Air Command and the American global nuclear forces, but did not realize this type of transmission occurred frequently.

Even with the major reorganization of the military air commands that took place in June 1992, the Skyking... transmissions can still be heard. However, frequency lists compiled prior to June 1992 cannot be considered accurate. Those frequencies listed in the sidebar (K4ZAD's Shortwave Voice Utility Sampler) have been compiled since the reorganization. Though many of these transmissions are cryptic, those transmissions that are in the clear can be fascinating.

The civilian air communications also provide interesting listening, and they are a little easier to understand. When flying over land, civilian aircraft use VHF frequencies, but on transoceanic and transpolar flights the shortwave frequencies come into use. Since English is the language of international aviation, translation is seldom a problem for the American listener.

The frequencies listed as *Air Traffic Control* in the Utility Sampler are used to transmit in-flight information such as weather queries, fuel consumption, updated estimated times of arrival and other related data. These frequencies are not in constant use; if you want to hear some activity on these bands choose a busy time (a time that you have reason to believe that air traffic will be heavy on the route you have chosen) and *park* on the associated frequency or do a little scanning.

For instance, on the frequency list, 5598, 5616, 11279, 13306 and 17946 (all kHz) are listed for the North Atlantic. Remember, these transmissions are on upper sideband. Prime listening times (when the bulk of the jets are in the air) is generally 7 PM-12 AM EST (0000-0500 UTC) and 7 AM-12 PM EST (1200-1700 UTC).

During this period, just sit on one of these frequencies (below 10 MHz evenings, above 10 MHz mornings), or use

Typical of newer ham transceivers, this Kenwood TS-570D includes a 3-kHz to 30-MHz receiver. It is a full-featured ham transceiver, and you can make use of these features when listening on the shortwave bands.

the memory function of your receiver to occasionally switch between them. Since there is no continuous chatter on these types of frequencies it may take a while, but sooner or later you will come across some transmissions. Obviously, patience is required for this aspect of the hobby.

Some of the easiest catches in civilian shortwave aviation transmissions are the VOLMET or aviation weather transmissions. Listed as *Aero WX* (for Aero Weather) in the accompanying Utility Sampler, they also list the times of broadcasts. For example, New York Radio transmits on its assigned frequencies (3485 kHz, 6604 kHz, 10051 kHz, etc.) from the top of each hour to 20 minutes past each hour and from 30 minutes past each hour to 50 minutes past each hour.

As with aircraft, both VHF and shortwave frequencies are used for maritime communications. While most aircraft transmissions use *back-and-forth* simplex-type transmission typical on Amateur Radio, quite a bit of the maritime communications take place in full duplex, where two frequencies are used simultaneously, and only one side of the communication can be heard on any given frequency.

However, there are some simplex frequencies and the Utility Sampler contains some of both. The U.S. Coast Guard Information frequency of 2670 kHz is a good one to check when conditions are favorable for a frequency that low.

RTTY and Digital Modes

Many of the signals heard on the utility bands are radioteletype and other digital modes. If you are active on VHF packet radio, you may already have some of the equipment you need to monitor these signals.

Many popular packet TNCs such as the AEA PK-232MBX, PK-900 and the Kantronics KAM Plus can decode a variety of digital signal types, including radioteletype. These, along with your computer, will allow you to monitor transmission from news agencies, diplomatic messages, marine traffic, meteorological data and other information that is best transferred by means suitable for hard copy.

Although monitoring radio teletype and the digital modes can be complicated, the ham bands are the best place to start. This book has detailed information on monitoring RTTY and packet radio in the HF and VHF Digital Communications chapters.

Numbers Stations On Shortwave

If you tune around in the utility shortwave frequencies, sooner or later you will encounter strange stations sending groups of numbers or letters in Spanish, English, Russian or German. These stations are known as *numbers stations.*

Ever since these stations were first noticed in the early 1960s, people have speculated as to their purpose. Does the transmission contain encoded weather forecasts or shipping information? The possibilities are endless. Some people even thought the numbers transmissions were part of a secret project communicating with UFOs!

Over time a consensus developed that these transmissions were messages from various intelligence agencies to their agents in the field. As radio direction finding techniques located transmission sites in both (the former) East and West Germany, Nicaragua and Cuba, as well as on some military bases in the United States, this theory seemed largely confirmed. More recently, former intelligence agents have related how these transmissions were used to convey instructions to them. However, it continues to puzzle listeners that the level of numbers station activity has not appreciably decreased with the end of the Cold War, and that the stations on both sides of the former German border continue to operate.

Numbers stations are found as low as 2 MHz to as high as 26 MHz, using (CW) and voice transmissions in both AM and SSB. Though they can be heard at any time, activity tends to peak from about 0000 UTC to 0800 UTC.

PIRATES AND CLANDESTINE BROADCASTERS

Two unique types of shortwave stations are pirate and clandestine broadcasters. Though both generally operate illegally, they are quite different in their purposes.

Clandestine broadcasters are specifically political in nature. When a political group is out of power in a country, they will sometimes set up a shortwave station to broadcast to those sympathetic to their cause. These stations may be located within the target country, often illegally, or may be located in a nearby sympathetic country. In this case they may be operating legally, even though the host country may not officially acknowledge their existence (to avoid official diplomatic hostility with the target country).

Central America has traditionally been a hotbed of clandestine broadcasts, with stations appearing and disappearing as the fortunes of their sponsoring groups change within the target countries. Africa and the Middle East also have had a sizable number of clandestine stations over the years, but since North America is far from their target audiences, they can be very hard to hear.

These stations generally broadcast in the native language of the target country, so for Central American stations Spanish is the norm. However, some clandestine stations want to obtain money and support from within the United States. Such stations may use English, even when English is not widely spoken within the target country.

Pirate broadcasters are sometimes politically oriented, but typically they are people broadcasting for the joy of

K4ZAD'S Shortwave Voice Utility Sampler

(All frequencies are in kHz)

kHz	Service
2182	Marine Emergency Calling Channel
2598	Canadian CG Marine Information Broadcasts
2670	USCG Marine Information Broadcasts
3413	Aero Weather—Shannon Ireland
3485	Aero WX—New York, NY and Gander, Newfoundland
4065	Inland River Towboats—WCM—Cincinnati
4125	Marine Ship Calling
4149	Marine Simplex Utility Channel 4D
4372	US Navy
4381	Great Lakes Ore Boats—WLC—Rogers City, MI (Ships on 4089)
4582	Civil Air Patrol—Emergency Channel
4722	RAF Aero Weather—Continuous
4725	US Air Force—Global High Frequency System
4742	RAF—Architect
5015	US Army Corps of Engineers—Net at 8:00 AM ET, M-F
5211	Federal Emergency Management Agency —Primary Night Channel
5505	Aero Weather—Shannon Ireland—Continuous
5598	Air Traffic Control—North Atlantic—NY, Gander, Shanwick
5616	Air Traffic Control—North Atlantic—Gander, Shanwick
5680	Search & Rescue Channel—Worldwide
5692	US Coast Guard—Chopper Ops.
5696	US Coast Guard—Air Ops.
5841	US Anti-Drug Agents
6215	Marine Ship Calling Ch./Utility Ch. (Ch. 606) —Shore on 6516
6230	Marine Simplex Utility Channel 6C
6510	River Towboats—WCM—Cincinnati (Skeds at 14:30-15:00 ET)
6577	Air Traffic Control—Caribbean—NY
6604	Aero WX—New York, NY and Gander, Newfoundland
6628	Air Traffic Control—North Atlantic—NY, Santa Maria
6676	Aero WX—Sydney, Singapore, Bangkok, Bombay
6679	Aero WX—Honolulu, Tokyo, Auckland, Hong Kong
6697	US Navy
6720	US Navy
6738	US Air Force—Global High Frequency System
6753	Canadian Military WX—Edmonton, Trenton, St. Johns
6812	USAF—A prime frequency for Air Force One
7527	US Anti-Drug Agents
7535	US Navy
7635	CAP—Nationwide Frequency (Command Net Weekdays at 1600 UTC)
8125	FAA—Eastern Net (Wednesdays at 10:45 AM ET)
8176	Sydney, Australia. Marine Radio —VIS—(early mornings)
8213	River Towboats—WCM—Cincinnati (Skeds at 13:00-14:00 ET)
8255	Marine Ship Calling (Ch. 821) —Shore Stations on 8779
8297	Marine Simplex Utility Channel 8B
8794	Great Lakes Ore Boats—WLC— Rogers City, MI—Ships on 8270
8825	Air Traffic Control—North Atlantic—NY, Gander, Shanwick
8828	Aero WX—Honolulu, Tokyo, Auckland, Hong Kong
8846	Air Traffic Control—Caribbean—NY
8864	Air Traffic Control—North Atlantic—Gander, Shanwick
8867	Air Traffic Control—Pacific—Honolulu, Auckland, Sydney, Nandi
8903	Air Traffic Control—Pacific & Africa
8906	Air Traffic Control—North Atlantic—NY, Santa Maria
8912	US Anti-Drug Agents
8957	Aero Weather—Shannon Ireland—Continuous
8967	US Air Force—Global High Frequency System
8980	US Coast Guard—Chopper Ops.
8984	US Coast Guard—Air Ops.
8993	US Air Force—Global High Frequency System
9023	Canadian Military & USAF NORAD
9032	RAF—Architect
10051	Aero WX—New York, NY and Gander, Newfoundland
10493	Federal Emergency Management Agency— Primary Day Channel
10780	USAF—NASA Support—Cape Radio— Primary Day Channel
11176	US Air Force—Global High Frequency System
11191	US Navy—Air Operations—Hershey at Key West, FL
11195	US Coast Guard—Air Ops.
11198	US Coast Guard—Chopper Ops.
11200	RAF Aero Weather—Continuous
11205	US Navy
11233	Canadian Military
11234	RAF—Architect
11255	US Navy
11267	US Navy
11279	Air Traffic Control—North Atlantic—NY, Gander, Shanwick
11282	Air Traffic Control—Pacific—San Francisco, Honolulu
11309	Air Traffic Control—North Atlantic—NY, Santa Maria
11384	Air Traffic Control—Pacific—Honolulu, Tokyo, Hong Kong
11387	Aero WX—Sydney, Singapore, Bangkok, Bombay
11396	Air Traffic Control—Caribbean—NY
11494	US Anti-Drug Agents
12290	Marine Ship Calling Ch. (Ch. 1221)—Shore Stations on 13137
12359	Marine Simplex Utility Channel 12C
13201	US Air Force—Global High Frequency System
13257	Canadian Military
13261	Air Traffic Control—Pacific—Honolulu, Auckland, Sydney, Nandi
13264	Aero Weather—Shannon Ireland—Continuous
13270	Aero WX—New York, NY and Gander, Newfoundland
13282	Aero WX—Honolulu, Tokyo, Auckland, Hong Kong
13288	Air Traffic Control—Pacific—San Francisco, Honolulu
13297	Air Traffic Control—Caribbean—NY
13300	Air Traffic Control—Pacific—Honolulu, Tokyo, Hong Kong
13306	Air Traffic Control—North Atlantic—NY, Gander, Shanwick
13312	Anti-Drug Agents and FAA Flight Tests and Commercial Flight Tests
13330	Air—Long Distance Operational Control —NY, Houston
13354	Air Traffic Control—Pacific—San Francisco, Honolulu
13457	FAA—Western Net (Wednesdays at 10:30 AM MT)
15015	US Air Force—Global High Frequency System
15867	US Anti-Drug Agents
16420	Marine Ship Calling Channel (Ch. 1621) —Shore Stations on 17302
16534	Marine Simplex Utility Channel 16C
17904	Air Traffic Control—Pacific—Honolulu, Tokyo, Hong Kong
17946	Air Traffic Control—North Atlantic—NY, Gander, Shanwick
17975	US Air Force—Global High Frequency System
18009	US Navy
22060	Marine Ship Calling Channel (Ch 2221) —Shore Stations on 22756
22171	Marine Simplex Utility Channel 22E
23287	US Navy

Frequencies courtesy of Tom McKee, K4ZAD, from his book, *The Other Shortwave* (Key Research, PO Box 846G, Cary, NC 27512) broadcasting. Their enthusiasm notwithstanding, the quality of pirate broadcasts ranges from that rivaling the professionals to sounding like a drunk turned loose in a radio studio—and indeed, they sometimes are inebriated!

The most common programming format of pirate stations is satire and comedy (sometimes unintentional). When political views are expressed, they tend to be extreme (at either end of the spectrum). Pirate stations are most active evenings on weekends and holidays. They tend to keep erratic schedules to help them avoid discovery by the FCC. To prevent FCC monitors from getting an accurate fix on their location, their transmissions tend to be no longer than 30 minutes to 1 hour in length. The most common frequencies for pirate operations have been 7415 kHz and 7485 kHz, but recently legitimate broadcast stations have begun transmissions on these frequencies causing the pirates to move.

Remember, pirates are generally using old Amateur Radio transmitters from the 1950s running 100-200 watts into temporary antennas. The major international broadcasters may run from 100,000 to 500,000 watts into sophisticated antenna arrays. When the two clash on a frequency, there is little doubt whose signal will triumph.

Now, most pirate stations are operating between 6925 and 6970 kHz (6955 is the favorite) so that's the best place to start tuning. Other fruitful places to tune are around 13900 and 15050 kHz on weekend afternoons and 1620 kHz evenings. Most signals are AM, but occasionally upper sideband broadcasts will be heard. Most weekends there will be between 2 and 20 North American pirates to be found (the most on holiday weekends), but since they don't have powerful signals, a good antenna and receiver are needed to hear them. Since pirates are already operating illegally, they aren't bound by FCC obscenity rules, so be forewarned!

While pirates would contend that they operate in the spirit of good fun, the FCC does *not* take a casual attitude towards pirate broadcasters. Fines in excess of $10,000 are not uncommon and the revocation of any FCC licenses held (including Amateur Radio licenses) is virtually automatic. Whatever the appeal of running your own shortwave broadcast station may be, it's not worth your Amateur Radio career.

KEEPING TRACK—SHORTWAVE LOGS

One of the traditional rules of scientific experimentation has been, "If you don't record it, it didn't happen." The tradition of keeping a list of the stations heard, the date and time, the frequency, conditions and transmission content in a *logbook* has been continued since the days when radio transmission and reception *were* scientific experiments.

For many years, the keeping of a logbook by licensed ham radio operators was the law. The Federal Communications Commission used to mandate that licensed ham radio operators keep a meticulous log of their transmissions and communications, but those regulations have long since been relaxed.

Despite the less restrictive regulations, the logbook still has many useful purposes in both ham radio and shortwave listening. The log is a permanent record of the reception achievements of the operator and the station. It is a reference resource for active frequencies, propagation and the comparison of different radios and antennas.

QSLs

So you've heard a rare station. How do you prove it? One way is to send a report of the reception to the station, including the date, time, frequency and information on the reception quality and programming content. This is essentially the same information you record in a logbook.

If they find your information to be accurate, often they will send you a confirmation. Sometimes this will be in the form of a letter (especially with a utility type station), but the international broadcasters traditionally send a picture postcard in verification. These cards have become known as *QSL cards* (from the ham radio Q signal meaning, acknowledging receipt). Collecting QSL cards is another subset of shortwave listening, along with collecting the sometimes exotic stamps that come with them.

The QSL card is a courtesy of the transmitting station, and to increase the chances of receiving one and to defray the mailing costs of the broadcaster, many SWLs include an IRC (International Reply Coupon) to help defray the postage cost. An IRC can be exchanged for first class postage in any country and is available from most larger post offices.

The SINPO Code

The SINPO code is a way of quantifying reception conditions in a five-digit code, especially for use in reception reports to broadcasters. The ham radio equivalent is the RST system for Readability, Strength and Tone. The SINPO components cover Signal strength, Interference (from other stations), Noise (from atmospheric conditions), Propagation disturbance (or Fading due to propagation, which sometimes leads to it being called the *SINFO code*), and Overall merit. The code is as follows:

In recent years, many broadcasters have tried to steer listeners away from the SINPO code and toward the simpler SIO code. SIO deletes the extremes (1 and 5) and the noise and propagation categories, which were confusing to too many people to be useful. In sending reports to stations other than large international broadcasters, who are likely to understand the codes, it is better to simply describe reception conditions in words.

SHORTWAVE VS SATELLITES

Some people have predicted the end of shortwave radio as satellite technology becomes more prevalent. Certainly many countries have cut back on their shortwave operations, both to conserve funds and because they've switched their programming to satellites. Military and diplomatic operations have switched much of their communications to satellites. So is shortwave listening on its deathbed?

SINPO Code

Value	*(S)ignal*	*(I)nterference*	*(N)oise*	*(P)ropagation*	*(O)verall*
5	excellent	none	none	none	excellent
4	good	light	slight	slight	good
3	fair	moderate	moderate	moderate	fair
2	poor	severe	severe	severe	poor
1	barely audible	extreme	extreme	extreme	unusable

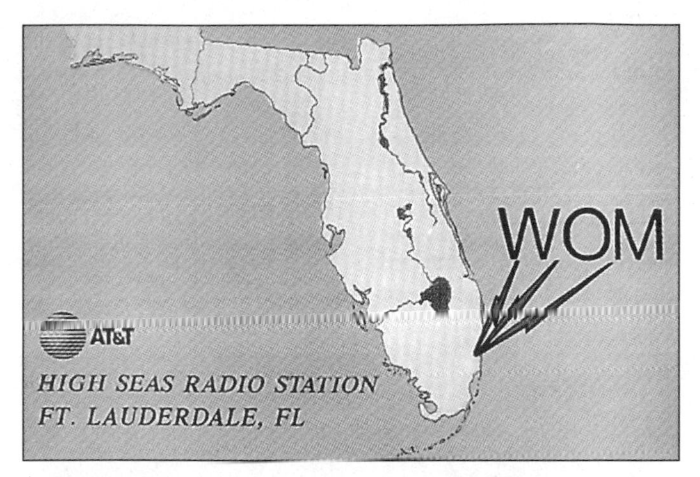

WOM handles high-seas radiotelephone traffic from a computerized control station in Florida. Log one of their many transmissions or those of their sister stations (KMI in California and WOO in New Jersey) and they will QSL.

No! One major difference between shortwave transmissions and satellite transmissions is infrastructure. For satellite operations to work, a lot of complex equipment has to perform in concert. Satellite dishes have to be aimed precisely at both ends of the signal path. To make matters worse, the satellites have to function in one of the harshest environments imaginable. As the old saying goes, "The more complicated they make the plumbing, the easier it is to clog up the pipes." That's why even the military, with their desire for secure communications, can still be heard frequently on the shortwave bands.

The need for infrastructure also requires approval from those who control the infrastructure. When Operation Desert Storm commenced, the Iraqi government immediately took control of the satellite uplinks. From that moment onward, they had exclusive control of all the information leaving their country.

In the past few years, both Iraq and Saudi Arabia have outlawed the private ownership of satellite dishes, to better control the information being received by their citizens.

Shortwave communication, on the other hand, is not so easy to control. Relatively simple equipment is all you need: a transmitter, a receiver and the appropriate antennas (and a little help from the ionosphere). While it would be hard for someone in a controlled country to hide a satellite dish, a portable shortwave receiver and antenna can be very easily concealed.

So for all the convenience and quality that satellite operations provide, there will always be a need for shortwave operations. Your shortwave receivers won't lack for signals for a long time.

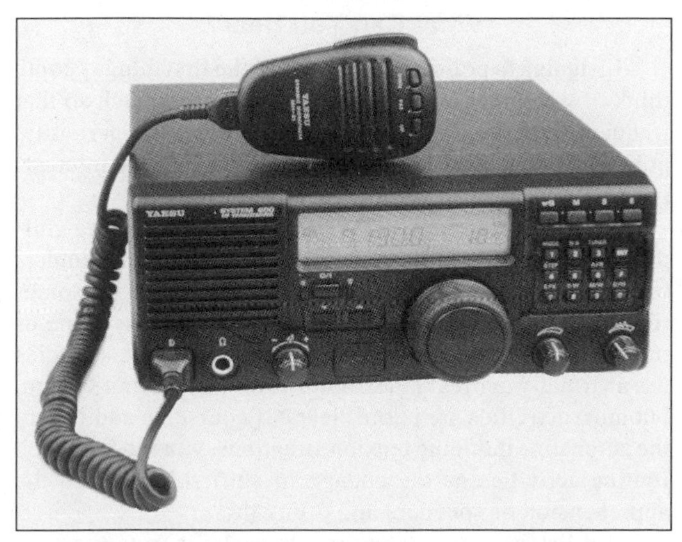

It looks deceptively simple, but there is a lot of capability packed into this Yaesu FT-600 HF transceiver, including 50 kHz to 30 MHz receive coverage and a computer interface.

HOW LOW CAN YOU GO?

Finally, an old mode—dating back to the days before HF was technically feasible—has received increased attention. The frequencies, below the broadcast band, provide many people with interesting listening. Although many countries have abandoned the ship-to-shore communication, previously found in the 400-500 kHz region, beacons, arctic and sub-arctic area and European broadcasting, weather and some data services still use these bands. Contact the Longwave Club of America, 45 Wildflower Road, Levittown, PA 19057 for information on this specialized SWL activity.

SCANNING THE VHF/UHF FREQUENCIES

The hobby of scanning, the monitoring of the VHF/UHF public service frequencies with multi-channel scanning receivers, has exploded from the minor sub-set of shortwave listening that it was two decades ago into the primary activity of many hobbyists. Many Amateur Radio operators have been enticed to sample these by the increasingly common ability of 2 meter and 440-MHz transceivers to receive frequencies substantially above and below the confines of these two ham bands.

THE LAW

It is unfortunate, but appropriate given recent events, that an overview of the legal aspects of scanning should be discussed prior to delving into other aspects of the hobby.

For years scanning was relatively unregulated, governed primarily by the same law as shortwave listening, namely the Communications Act of 1934. It was illegal for the listener to divulge or profit by monitored communications not intended for the general public.

But it also is now illegal to listen to cellular phone transmissions. It is not illegal to own radios that cover the cellular frequencies, but you're not supposed to listen to them. This law, the Electronic Communications Privacy Act of 1986 (ECPA '86), also makes it illegal to "intercept (listen to) ...all forms of common carrier communication, except cordless phones and tone-only paging communications, and of any non-common carrier or private radio communications when they are encrypted [or] scrambled..." This law was primarily passed based on lobbying pressure from the cellular phone manufacturers and service providers, to give their users a (false) sense that their unscrambled conversations were private and secure. In 1994, a law was passed that

Where are the knobs, buttons, lights and dials? There aren't any. The Ten-Tec RX-320 is a PC-based shortwave (100 kHz to 30 MHz) receiver in a black box. You'll need a PC running *Windows* software and a COM port to unleash the power of this nifty receiver, which was reviewed in the March 1999 issue of *QST*.

made it illegal to listen to cordless phone activity, but not baby monitor transmissions that share many of the same frequencies.

As of April 1994, it is illegal to import or manufacture scanners for resale that can receive or can be easily modified to receive the cellular phone frequencies. What constitutes easily modified has been the topic of much discussion, but in general it has to be harder than just clipping a diode (as was required on the old RadioShack PRO-2004/5/6 series). It is still not illegal to own or resell a used scanner that can cover these frequencies, possibly because many old television sets can receive some of the cellular phone frequencies with their UHF tuner. Since the ban, some people have successfully ordered uncensored scanners from sources in Canada or England, but this is a chancy proposition; such a scanner might possibly be confiscated or returned by Customs agents at the border.

The monitoring of transmissions from state, federal and local law enforcement agencies, public safety services, aviation, business, the military and most other sources that are readily accessible to the public is not illegal. It is up to you to determine why our law makers consider a police transmission on the 800 MHz band readily accessible to the public, and the cellular phone transmission a few megahertz up the band not readily accessible.

The practical effect of this law has been minimal, since the pertinent law enforcement agencies have made it clear that they have more important things to do than to become the radio Gestapo. Generally it would be difficult to prove that a person had listened to a prohibited transmission unless they confessed, or recorded their eavesdropping session but this is the law and as with all laws, you ignore it at your own risk.

Of course, this is exactly what happened in early 1997. A couple in Florida cruising in their automobile monitored a cell phone transmission of a conference call of Republican politicos and happened to have a tape recorder available that they used to record the activity. They then forwarded the tape to their Democratic congressman, who gave it to the press and embarrassed the Republicans.

It was a Democratically controlled House of Representatives that passed the laws of 1986 and 1994, so when the embarrassed Republicans now in charge called for hearings on cellular phone eavesdropping, the hobby had few (no?) friends among the attending members of Congress. Although no further laws on this have been passed yet, the FCC has become increasingly strict in their interpretation of existing law. Now they contend that the 1994 legislation allows them to delve even more intricately into the electronics used inside and outside of scanners. To this end, a well known scanner author and experimenter was arrested for selling a *data slicer* circuit consisting of an op-amp, four diodes, three capacitors and three resistors. This simple circuit can be used to (among other things) decode signals sent to pagers. So anyone interested in scanning should pay close attention to legal developments.

Another area of scanning law that the prospective hobbyist should investigate is the legality of scanning while mobile. Using a scanner in an automobile is subject to restrictions in many locales (including being prohibited in Florida). If you are interested in scanning on the move, you definitely should check your local laws.

In some places where mobile scanners are prohibited, licensed Amateur Radio operators are exempted from the law. And by mandate and exemption of the FCC, any ham radio transceiver can be legally used in mobile operation by an Amateur Radio operator, even if it also receives frequencies outside the ham bands. This has caused some scanner enthusiasts to obtain their ham license, just so they can enjoy their scanner hobby with fewer restrictions. Unfortunately, some overzealous local police have seized ham VHF/UHF transceivers because they were capable of scanning. Recovery by the hams involved proved difficult and tedious—even in one well publicized case where the scan capability was used to help find an individual who persistently jammed the local police radio system!

What Can You Hear?

Listening to police calls are one of the first things people think of when they hear the word scanner, so much so that frequently they are called police scanners. And certainly, police and law enforcement operations are a very interesting part of scanner listening.

You can listen to car chases, the hunt for missing children, hostage negotiations, the breathless voice of an undercover cop chasing a suspect on foot and shoot-outs, sometimes hearing the gunfire in the background. For some of their most secretive activities, law enforcement agencies use a virtually unbreakable digital voice scrambling system, but most activities are in the clear. Of course, in addition to the adrenaline inducing tension situations, you can hear such routine activities as the change of shift roll-call and the apprehension of speeders and drunk drivers.

But listening to law enforcement is far from the only use for scanners. Today the fire department is not just fighting fires, although listening to those operations can be as

Print Resources for the Shortwave Listener

With the vast frequencies in the shortwave spectrum, the ever-changing international implications of shortwave radio and the constant changes in technology, many information sources can be helpful. The previously mentioned shortwave shows for the hobbyist can be helpful and on-line computer services are available for those on the much touted information highway (see "On-line Resources for SWLs"), but what if you're a beginner with no radio or modem? For those people and many others, the multitudinous print resources for the shortwave listener come to the rescue.

The two print mainstays of shortwave broadcast listening are the *WRTH* or *World Radio TV Handbook* (BPI Communications, 1515 Broadway, New York, NY 10036) and *Passport to World Band Radio* (available from the ARRL), both published yearly. *Passport* is considered easier to navigate for the beginner, while the *WRTH* is considered the bible, the definitive source for shortwave information.

Passport offers a listing of broadcasts by frequency, and a 24-hour tour of selected transmission to North America. *WRTH* provides a listing of transmissions by language and details of everything you could possibly want to know about each individual broadcaster. If you're new to shortwave, you should get a copy of each. Don't let your $300-$500 receiver go under-utilized for lack of the information in a couple of $20 books. If you are like many people you may find that they are both so useful in their own special ways that they each warrant a yearly purchase.

Introductory books on the subject include *Radio Monitoring: The How-to Guide*, by T.J. "Skip" Arey N2EI (available via Barnes and Noble) and *The Complete Shortwave Listener's Handbook* by Andrew Yoder et al (TAB Books, Division of McGraw-Hill, Inc., 860 Taylor Station Rd., Blacklick, OH 43004). Helms' book (first published in 1990) is an excellent survey of all aspects of shortwave listening, and is highly recommended.

Those of us with a copy of Len Buckwalter's *The Fun of Short-Wave Radio Listening* (Copyright 1965, now out of print) might question the attempt to promote *The SWL's Handbook* (the edition by Hank Bennett was first published in 1974) as the first complete text on shortwave listening. Nevertheless, its position as a founding informational resource on shortwave listening is undisputed. The editions during the transitional period between Hank Bennitt's authorship and Andrew Yoder's authorship (with patchwork editing from four different authors) could cause reader schizophrenia. However the latest edition has been totally revised by Andrew Yoder and speaks coherently with just his voice. Still, in any edition *The Complete Shortwave Listener's Handbook* is a valuable reference.

Monthly publications can help the SWL stay on top of the ever-changing shortwave world, and the two standard monthlies are *Popular Communications* (25 Newbridge Road, Hicksville, NY 11801) and *Monitoring Times* (P.O. Box 98, Brasstown, NC 28902). *PopComm* is a slickly produced shortwave newsmagazine that can be found on many newsstands, while *Monitoring Times'* use of newsprint allows for slightly more current information in each issue. Both magazines cover the full gamut of shortwave listening interests, from international broadcasting to military to utilities to digital modes.

For those interested in specific information on utilities and the digital modes, two frequency guides stand out. *Ferrell's Confidential Frequency List* (Listening In, P.O. Box 123, Park Ridge, NJ 07656) carries a full listing by frequency of the usage, mode, call, location and type signal from 1602 kHz to 25545 kHz. It also has a reverse listing sorted by call.

The *Klingenfuss Guide to Utility Stations* (Klingenfuss Publications, Hagenloher Str. 14, D-72070 Tuebingen, Germany) has its own frequency list, in addition to detailed listings of RTTY and FAX services and schedules and other miscellaneous but useful information.

For those interested in the content of the shows on the international broadcasting stations, *The Guide To Shortwave Programs* by Kannon Shanmugan (Grove Enterprises, P.O. Box 98, Brasstown, NC 28902) is an invaluable resource. It lists the frequencies and programs that are on the air at any time of the day or year. Though many have tried to lay claim to the title, this is perhaps the closest approximation of a *TV Guide* for the shortwave bands. It also has a special section on shortwave shows for the hobbyist.

Should you decide to delve deeply into propagation theory and the use of the indices, *The New Shortwave Propagation Handbook* by George Jacobs, W3ASK, Theodore J. Cohen, N4XX and Robert Rose, K6GKU (CQ Publishing, Inc., 25 Newbridge Road, Hicksville, NY 11801) is an excellent text on the subject.

thrilling as the law enforcement radio traffic. The fire department is also the first responder for many diverse types of incidents including hazardous waste spills, people trapped both high and low, and UFO incidents (yes, even those).

Listening to the fire department sometimes can give you information crucial to your health and well-being. Messages on the Internet told how scanner monitors in the San Francisco Bay area were able to monitor the progress, direction and danger areas of the widespread fires that swept through that area a few years ago, giving them vital information long before it was broadcast on the local news.

Fire departments are particularly interesting, since often they use a low-power repeater located on a fire truck. Thus, personnel equipped with low-power hand-held radios on a fire scene work though the fire-truck repeater, parked at the scene. The local repeater connects both the local personnel and those at the fire dispatch station. You may never hear one of these low-power fire repeaters unless you are located within a few blocks of a fire or fire investigation.

Listening to rescue-squad activities and the Lifeflight helicopters transporting the critically injured to the life-saving facilities of a hospital puts you right in the middle of real life-and-death situations. . . who needs the doctor shows on television?

Yet there is much more.

You can listen to the Blue Angels as they coordinate their intricate air ballet at an air show, or to race car drivers as they discuss strategies with their pit crews at the races. At major airports, the scanner's frequencies are full of radio traffic from planes arriving and departing.

During inclement weather, monitoring can give you information on road conditions from the Highway Patrol, radio traffic reporters (before it is broadcast) and state and city road crews. If the electricity is out for any reason, a battery powered scanner can help you track the utility company workers as they try to restore power.

Since so many businesses use radio in their activities, you can hear construction crews, security forces, taxi companies, and other business in your area. You can even intercept the important communications between the order-taker and the customer at fast food restaurants, and learn if the car ahead of you is ordering a medium or large order of fries!

Frequencies

The specific frequencies used by agencies vary with the locality, and there are books published with this information. One of the best and best known is *POLICE CALL* (now called *POLICE CALL PLUS*); it is available at RadioShack. Often a local RadioShack or other electronics stores will have a photocopied list of frequencies active in the area. If they have one, they will give you a copy of the list if you buy a scanner from them (and sometimes even if you don't). Try to find a scanner buff at a local Amateur Radio club meeting: this can be a good way to learn the hot frequencies. For those online, the Internet and the World Wide Web are a source for many frequency lists.

For general information on what signals are where, here's a sampling of the frequencies ranges with the most activity between the ham bands:

30-50 MHz—This is called the VHF low-band. Most users are gradually leaving this band because it requires a relatively large antenna and, as with Citizens Band (CB is at 26.9-27.4 MHz, just below it), long range propagation via the ionosphere can be a problem. Still, some highway patrol (including North Carolina's) and sheriff's offices continue operations in this frequency range on FM, so it's good that most scanners include it. Quite often, police officers don't realize that *skip* can bring in signals from other parts of the country, and it can be amusing to hear them argue about who *owns* the frequency.

Baby monitors (with which people essentially bug their own homes) and older cordless telephones also transmit within this band. Given their very low power, if you hear them they are almost certainly close by, but if it's a cordless phone you should immediately tune past it since listening to it would be against the law.

108-136 MHz—This is the aviation band and almost all operations use AM. Primarily used by civilian aircraft, this band is omitted from the many scanners that only have FM demodulators. This can be a very interesting band, even if you don't live near an airport. A commercial airliner at 30,000 feet has quite a range (line of sight). Several of the selectable mode AM/FM scanners expand the upper end of this tuning range to 144 MHz, to receive the FM military mobile operations often found between 136-144 MHz, and some restrict the lower end of this tuning range to 118 MHz, since there is very little voice activity from 108-118 MHz. If you have a choice, get a scanner that covers this range since many airports broadcast continuous weather (ATIS—automated terminal information system) on some of these frequencies.

150-174 MHz—This is called the VHF High band and is the most active and diverse of the scanner bands. Within this range FBI, Secret Service, police, fire, and rescue squad stations can be found, as well as less exciting business, taxi, paging, railroad and other general purpose radio services. In addition, the marine VHF frequencies fall within this range. Transmissions are primarily FM.

225-400 MHz—Military operations that use AM, primarily aviation, are found in this range, which is omitted on almost all but the most expensive and continuous coverage scanners.

406-420 MHz—Operations by the federal government can be found in this frequency range, including transmissions from Air Force One, the Bureau of Alcohol, Tobacco and Firearms and the FBI. This FM band is common on most recent vintage scanners.

450-470 MHz—Law enforcement agencies predominate on FM in this range, although some general purpose radio services can also be found here. There is some migration from this band to 800-MHz trunked systems (discussed later), but an enormous amount of activity remains.

470-512 MHz—This is called the T-band, because the primary allocation in this range is for UHF TV. Rarely used for general radio service, scanner coverage here is most often used to receive the FM audio for TV channels 14-20. Occasionally a wireless microphone's transmissions can be heard on these frequencies.

806-960 MHz—This is probably the fastest growing area of the scanning spectrum, but some of the less expensive scanners don't cover it. Coverage of this range is quite rare in scanners more than 10 years old.

Cellular phones occupy parts of this range, along the high end of the UHF-TV channels. As was previously mentioned, it is illegal to listen to cellular phone transmissions, which is why so many people are now so interested in it. Informed sources claim that most cellular phone conversations are boring, although occasionally details of a drug deal or an illicit romantic rendezvous might be overheard.

The Personal Communication Systems (PCS) that are currently being marketed in metropolitan areas also operate in this frequency range. However, their signals are digital in nature and not decipherable with equipment that is readily available.

The latest (and most expensive) cordless phones also operate in this range. Some of these units use a digital or spread-spectrum transmission method, which makes them almost impossible to monitor. Regardless of the mode, however, it is illegal to listen to them.

TRUNKED RADIO

Many municipalities are installing trunked radio systems

that operate in the 800-MHz frequency range. For reasons beyond the scope of this article, a trunked system uses radio spectrum more efficiently than conventional systems and can provide for better communication between participating agencies. In a trunked system, many agencies may share a group of frequencies, but through a sophisticated data transmission system (much like packet radio), each agency hears only the radio traffic considered appropriate for their mission.

Trunked systems can be difficult to monitor because frequencies are often switched. For example, in a trunked system, a given conversation between a police officer and a dispatcher may switch frequencies every time the mike button is released. If there is only one conversation taking place, a fast scanner may find the next active frequency rapidly enough for the conversation to be monitored essentially continuously, but if other conversations are underway the scanner may stop on a different conversation. If the listener quickly presses the *scan* button and can recognize the voices, the monitoring can be continued reasonably well.

The trunking system by GE/Ericsson transmits a series of tones after each transmission specifically to make it more difficult for scanners to follow the activity. Again, if the listener remains actively involved in operating the scanner, the conversation can be followed, but it isn't easy.

Recently both Uniden and RadioShack have begun offering scanners that can decode the data signals and follow the conversations of the trunked systems manufactured by both Motorola and Ericsson. Before making a purchase you should become informed about the specifics of the trunked system you want to monitor, but these new radios will solve the trunked radio problem for many hobbyists.

EQUIPMENT

Multi-band, programmable scanners are so prevalent and reasonably priced now that crystal-controlled scanners almost have to be given away at hamfests. Even if someone gave you a 16-channel crystal-controlled scanner, by the time you bought the crystals for it (if you bought them for all channels) you would have spent almost enough to purchase a good low-end programmable rig.

The one exception to this is the avid scanner listener who wants to constantly monitor just one frequency. For instance, the hobbyist may want to hear every transmission on the police or fire dispatch channel, and yet continue to listen to other activity. In this case, a crystal-controlled scanner with just one crystal can be used to cost-effectively monitor that critical frequency continuously, and programmable scanners can be used for all other listening.

As part of their Bearcat line, Uniden manufactures this BC245XLT. It is a 300-channel programmable hand-held scanner capable of monitoring trunked radio systems. It covers 12 bands including aircraft and 800 MHz. Another handy feature is the computer interface.

Programmable scanners are generally priced from $100 to $500, but well-heeled enthusiasts are generally opting for a rig in the burgeoning *DC-to-daylight* category (see sidebar). The favorites among the traditional desktop scanners (with no shortwave capability) are the RadioShack PRO-2052 and the Uniden Bearcat BC-9000XLT.

Within the moderate price range, more expensive units are characterized by broader frequency coverage and more channels. At the top of this range, common features include a priority channel and frequency search capability, which allows for general band scans to seek out new active frequencies. Most of the popular scanners have reasonable sensitivity and selectivity, so the choice of which to buy is reduced to frequency coverage, number of channels and the user convenience features. A number of amazingly high performance scanners are now available as hand-held units, and definitely merit your consideration if portability would be useful. A few to consider are the Bearcat Sportcat 200, and the RadioShack PRO-94, both of which can be found for under $200.

Some good sources for scanners, frequency lists and scanner accessories are RadioShack, Communications Electronics (PO Box 1045, Ann Arbor, MI 48101), Electronic Equipment Bank (137 Church St. NW, Vienna, VA 22180 and Scanner World (10 New Scotland Ave., Albany, NY 12208). Advertisements from these and other scanner dealers can be found in *Popular Communications* and *Monitoring Times*. The addresses for these two magazines can be found in the section on "Print Resources for the Shortwave Listener."

ANTENNAS

Some radio hobbyists show a pronounced tendency to spend lots of time and money on their equipment, and a relatively small amount of time and money on their antennas. Most scanners are sold with a whip antenna that attaches to the radio and many scanner listeners never get beyond using it. This type of antenna would be only marginally effective in the best of locations, but because they are attached to the radio they can't be placed to best effect. At least they usually are oriented vertically, which corresponds to the vertical polarization of most VHF transmissions.

An outdoor antenna can enable you to listen to transmissions that are totally impossible to receive on an indoor whip. An outdoor antenna at 30 feet may provide a range of 50 to 60 miles, compared to an indoor whip's range of 5 to 10 miles.

If you are active on the UHF/VHF ham bands, you may already know that selecting the proper coax is increasingly important as you go to the higher frequencies. Certain types of coax, such as the popular RG-58, literally soaks up (attenuates) signals in the UHF/VHF range, with signal loss getting worse the higher in frequency you go. Coax that is marginally useful on the 2-meter band can be terrible on 440 MHz and virtually worthless if you are monitoring in the 800-MHz range. If you are going to need more that 30 feet or so of coax, consult a ham knowledgeable in UHF operation or check the *ARRL Handbook* or *ARRL Antenna Book* for information on what types of coax are best for the UHF ranges.

Despite their usefulness, there are times when you just can't use an outdoor antenna. Apartment dwellers and residents of areas with restrictive covenants have to contend with this situation constantly, but even for those with outdoor access there are times when using an outdoor antenna isn't prudent.

Thunderstorms present a particular dilemma. Leaving the scanner connected to an outside antenna during a thunderstorm is an open invitation for lightning damage, but the police, fire and rescue frequencies are particularly active then. The solution is an effective indoor antenna. In this case, the antenna supplied with the radio can allow you to monitor the activity while providing some protection for your valuable radio.

The News While It Happens

Monitoring the VHF/UHF scanner frequencies can keep you in touch with breaking events, inform you about local happenings that might not make the news and open up a whole new world of radio activity. Today, no radio shack can be considered well equipped unless it includes at least one scanner.

THIS IS JUST THE BEGINNING

If the foregoing has sparked your interest in scanner and shortwave listening, then this is just the beginning for you. Amateur Radio—in its broadest sense, the enjoyment of all facets of radio—is a tremendously rewarding and educational hobby, and the scanner and shortwave listening portion of the hobby is a vital part of it.

Shortwave listening has the appeal to attract people of all types, from nationally known figures to the kid down the block. Many hams were once the kid down the block, and shortwave listening was their entrance to Amateur Radio—and remains a continuing and immensely satisfying part of it.

Experience all that the frequencies between the ham bands have to offer. Find a radio and give these frequencies a try. The world awaits you.

Chapter 15

Antenna Orientation

Which Direction to Aim

Chuck Hutchinson, K8CH
ARRL Headquarters

You may already know that true direction from one place to another is not what it appears to be on the old Mercator school map. On such a map, if you start "west" from Wichita, Kansas (the approximate center of the continental US) you wind up in the neighborhood of Beijing, China. Actually, as a minute's experiment with a strip of paper on a small globe will show you that, a signal starting due west from Wichita never hits China at all but rather passes near Perth, in Western Australia.

"The shortest distance between two points is a straight line" is true only on a flat surface. The determination of the shortest path between two points on the surface of a sphere is a bit more complicated. Imagine a plane that intersects two points on the surface and the center of the sphere. The intersection of the plane and the sphere describes a circle on the surface of the sphere that is defined as a great circle. The shortest distance between the points follows the path of the great circle. The direction or bearing from your location to another point on the Earth is the direction of a great circle as it passes through your location on its way to the other point.

If, therefore, you want to determine the direction of some distant point from your own location, the ordinary Mercator projection alone is utterly useless. True bearing, however, may be found in several ways: by using a special type of world map that does show true direction from a specific location to other parts of the world, by working directly from a globe or by using mathematics.

DETERMINING TRUE NORTH

Determining the direction of distant points is of little use to amateurs erecting a directive array unless they can put up the array itself in the desired direction. This, in turn, demands a knowledge of the direction of true north (as against magnetic north), since all our directions from a globe or map are worked in terms of true north.

A number of ways may be available to amateurs for determining true north from their location. Frequently, the streets of a city or town are laid out, quite accurately, in north-south and east-west directions. A visit to the office of your city or county engineer will enable you to determine whether or not this is the case for the street in front of or parallel to your own lot. Or from such a visit it is often possible to locate some landmark, such as a factory chimney or church spire, that lies true north with respect to your house. If you cannot get true north by such means, three other methods are available: compass, pole star and sun.

BY COMPASS

Get as large a compass as you can; it is difficult, though not impossible, to get satisfactory results with the "pocket" type. In any event, the compass must have not more than 2 degrees per division.

It must be remembered that the compass points to magnetic north, not true north. The amount by which magnetic north differs from true north in a particular location is known as variation. Your city engineer's office or the flight office at a nearby airport can tell you the magnetic variation for your locality. The information is also available from US Geological Survey topographic maps for your locality. These may be available in the orienteering equipment section of your local sporting goods store, or may be on file at your local library.

If you have access to the World Wide Web, you can determine the magnetic variation for your area. Point your browser to **http://www.airnav.com/airports/** and you'll have access to data, including magnetic variation, for a nearby airport or heliport. Another web page that will give you your magnetic variation (they call it "declination" at this site) is at **http://www.ngdc.noaa.gov/cgi-bin/seg/gmag/fldsnth1.pl**. You'll need to know your latitude, longitude and elevation. The same site will offer to let you download a copy of *pgeomag3.exe*, a self-extracting compressed file. Run it, and you will be able to get your variation off-line.

When correcting your "compass north," do so *opposite* to the direction of the variation. For instance, if the variation for your locality is 12° west (meaning that the compass points 12° west of north), then true north is found by counting 12° east of north as shown on the compass.

When taking the bearing, make sure that the compass is located well away from ironwork, fencing, pipes, etc. Place the instrument on a wooden tripod or support of some sort, at a convenient height as near eye level as possible. Make yourself a sighting stick from a flat stick about 2 feet long with a nail driven upright in each end (for use as "sights") and then, after the needle of the compass has settled down, carefully lay this stick across the face of the compass—with the necessary allowance for variation—to line it up on true north. *Be sure you apply the variation correctly.*

This same sighting-stick and compass rig can also be used in laying out directions for supporting poles for antennas in other directions—provided, of course, that the compass dial is graduated in degrees.

BY THE POLE STAR

Many amateurs in the Northern Hemisphere use the pole star, Polaris, in determining the direction of true north. An advantage is that the pole star is never more than 0.8° from true north, so that in practice no corrections are necessary. Disadvantages are that some people have difficulty identifying the pole star, and that because of its comparatively high angle above the horizon at high northerly latitudes, it is not always easy to "sight" on it accurately. Further, Polaris is not a very bright star. Once you've sighted it, you can use a string with a weight tied on the end and held high at arm's length to look along and to identify a landmark that is north of your position. Polaris is not visible in the Southern Hemisphere.

BY THE SUN

The sun can be used for determination of true north. The method is based on the fact that exactly at noon, local time (not Standard Time), the sun bears due south (in the northern latitudes), so at that time the shadow of a vertical pole or rod will bear north. The resulting shadow direction is true north.

Clock or Standard Time for local noon is halfway between calculated sunrise and sunset times for your location. For example, calculations show that the sun rises in Newington, Connecticut, at 1144 UTC on February 20. It sets at 2225 UTC. The day is 10 hours and 41 minutes long, and local noon is 5 hours and 20.5 minutes after sunrise. This is 17:04:30 UTC or four and a half minutes after noon EST.

Many local newspapers publish sunrise and sunset times. A number of popular Amateur Radio oriented software packages also include this information. On the Internet you can point your Web browser to: **http://www.almanac.com/rise/rise.html**.

Here's another way to use the shadow of a vertical pole or rod. Mark the end of the shadow at some convenient time around mid-morning. When the shadow is the same length in the afternoon, mark that spot. The line between the marks runs east/west.

An Alternative Method for Antenna Orientation

It is not necessary to use true north for orienting your antenna. Any convenient landmark at a known bearing can be used for this purpose. The method is explained in the following example.

W1AW has four towers with rotating antennas—three at 60 feet, and one at 120 feet. A neighbor's chimney serves as a south reference for the 120-foot tower. (It's a bit close, but it works.) There are no north/south reference landmarks for the three 60-foot towers.

There is a landmark over a mile away—a small structure atop a tall building on a hill. From a US Coast and Geodetic Survey Map, W1AW staff determined the bearing to be 118 degrees true. Corrected compass sightings verified the bearing.

Today it's a simple matter to set a rotator indicator at 118 degrees and align the antenna to point at the reference structure. Because the reference point is over a mile away, the same bearing works for all four towers. Any errors are small enough to be insignificant—especially for W1AW's HF antennas.

AZIMUTHAL MAPS

While the Mercator projection does not show true directions, it is possible to make up a map that will show true bearings for all parts of the world from any single point. Three such maps are reproduced in this chapter. **Fig 15-1** shows directions from Washington, DC, **Fig 15-2** gives directions from San Francisco and **Fig 15-3** gives directions from the approximate center of the United States—Wichita, Kansas.

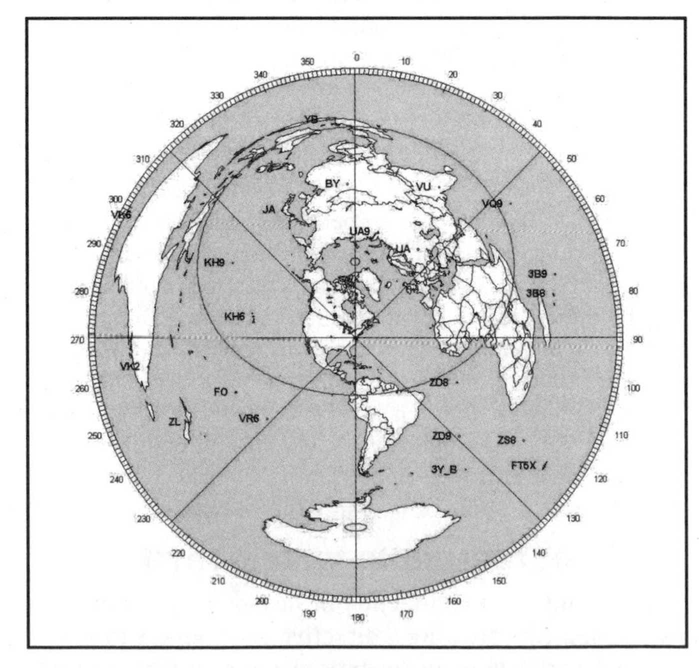

Fig 15-1—Azimuthal map centered on Washington, DC. (Map generated using *GCMWin* version 2.3.)

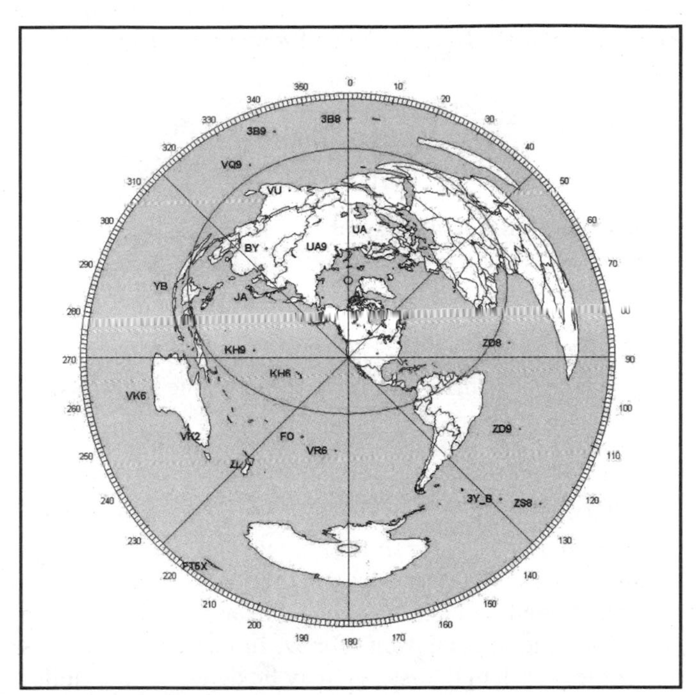

Fig 15-2—Azimuthal map centered on San Francisco, California.

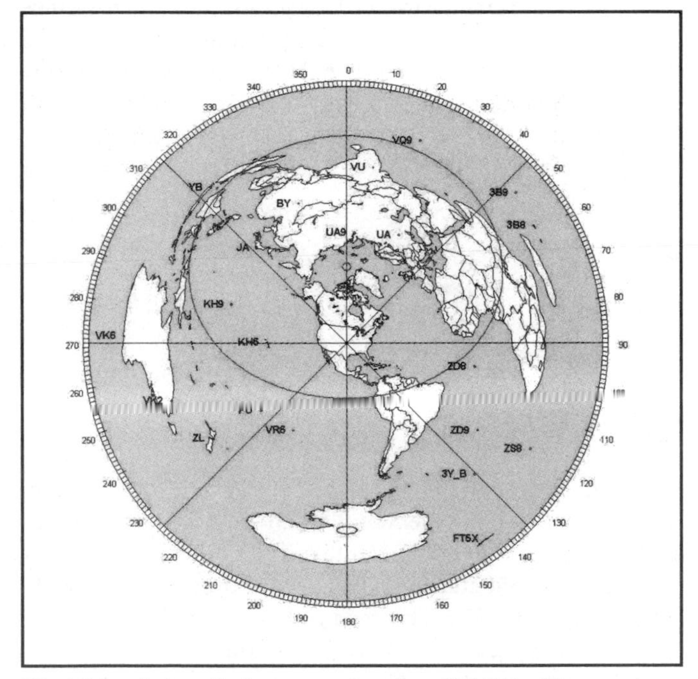

Fig 15-3—Azimuthal map centered on Wichita, Kansas.

For anyone living in the immediate vicinity (within 150 miles) of any of these three reference points, the directions as taken from maps will have a high degree of accuracy. However, one or another of the three maps will suffice for any location in the United States for all except the most accurate work; simply choose the map whose reference point is nearest you. Greatest errors will arise when your location is to one side or the other of a line between the reference point and the destination point; if your location is near or on the resulting line, there will be little or no error.

The current edition of the ARRL Amateur Radio Map of

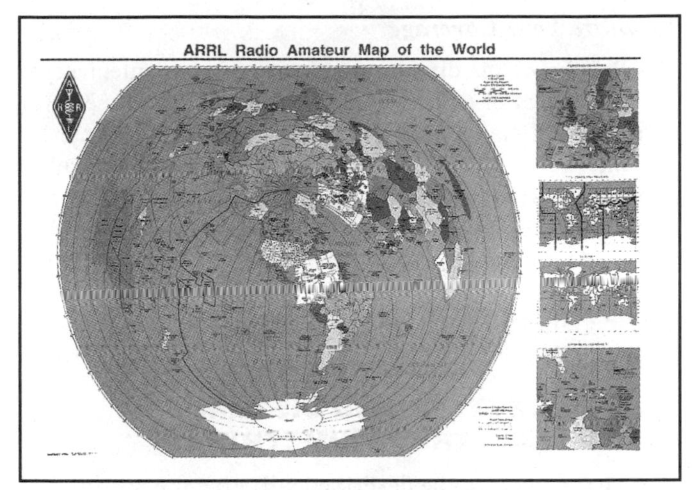

Fig 15-4—The ARRL Amateur Radio Map of the World is based on an azimuthal projection. See text.

the World contains a wealth of information especially useful to amateurs. It's based on an azimuthal projection made by Mapquest.com to ARRL specifications and gives great-circle bearings from the geographical center of the continental United States. The map shows all DXCC entities, principal cities of the world, local time zones, and amateur prefixes throughout the world. Inset maps show CQ Zones, ITU Zones and ITU Regions, as well as a detailed look at Europe and the Caribbean. The map is large enough to be easily readable from the operating position, 27×39 inches, and is printed in full color and laminated for durability. See **Fig 15-4**. The map is available for sale on *ARRL Web*, **http://www.arrl.org/catalog/**.

Bill Johnston, K5ZI, offers computer-calculated and computer-drawn great-circle maps; an extensive selection of these fine maps for various areas of the world appears in the References chapter. An 11×14-inch map can be custom made for your location. Write to PO Box 640, Organ NM 88052-0640.

You can find a number of excellent programs available for download over the Web. Use your favorite search engine to find them. See the Online Resources chapter for additional help.

One of my favorite programs is a freeware offering from Roger Hedin, SM3GSJ. Roger's program, *GCMWin* makes great circle maps. You can set the center point to anyplace on earth (except the poles) with latitude/longitude or by the Maidenhead locator system. You can view the whole earth or as many km (or miles) from center point as you wish. You can have meridians or fields (or fields and squares) and optionally add field letters and square figures. You can add points, prefix or text from a database. You can also have line data in a database so you can add your own borders, call zones or whatever. The program prints with color—if you have a color printer. Click with your mouse on the map, to see the bearing and distance from the center to that point.

Version 2.3 for Windows is available for download at Roger's home page: **http://hem.passagen.se/sm3gsj/**. Some of the maps in this book were prepared using *GCMWin*.

Visualize Your Coverage

By tracing the directional pattern of your antenna system on a sheet of tissue paper, then placing the paper over an azimuthal map with the origin of the pattern at your location, the "coverage" of the antenna will be readily evident. This is a particularly useful technique when a multi-lobed antenna, such as any of the long single-wire systems, is to be laid out so the main lobes cover as many desirable directions as possible. Often a set of such patterns will be of considerable assistance in determining what length antenna to put up, as well as the direction in which it should run.

DIRECTION AND DISTANCE BY TRIGONOMETRY

The methods to be described will give the bearing and distance as accurately as one cares to compute them. All that is required is a table of latitude and longitude information, such as **Table 15-1** (at the end of this chapter), and a calculator or computer with trigonometric functions. The latitude and longitude for any other location can be taken from a map of the area in question.

Fig 15-5 will help you to visualize the nature of the situation. That sketch represents the path between points situated relatively such as Wichita, Kansas, USA (at point A), and Perth, Western Australia (at point B). In using these equations, northerly latitudes are taken as positive, and southerly latitudes are taken as negative. Also, westerly longitudes are taken as positive, and easterly longitudes are taken as negative. *In all calculations*, the appropriate signs are to be retained. *All additions and subtractions throughout the procedure are to be made algebraically*. Thus, if a negative-value number is subtracted from a positive-value number, the resultant will be positive, and it will be the sum of the two absolute values, and so on.

THE CALCULATIONS

The two equations we'll be using for these calculations are:

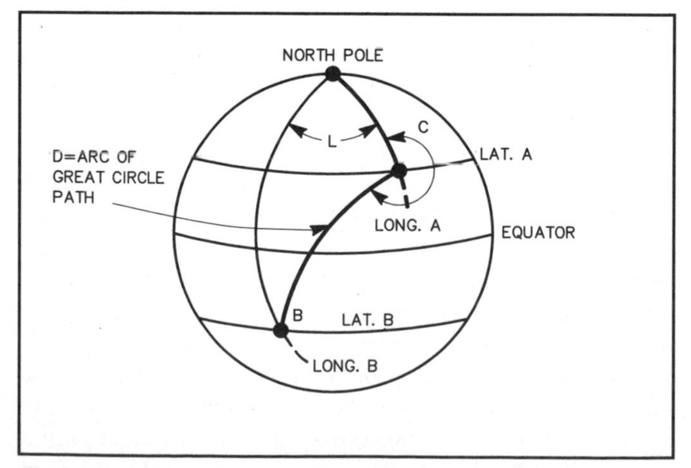

Fig 15-5—The various terms used in the equations for determining bearing and distance. North latitudes and west longitudes are taken as positive, while south latitudes and east longitudes are taken as negative.

$$\cos D = \sin A \sin B + \cos A \cos B \cos L \quad \text{(Eq 1)}$$

$$\cos C = \frac{\sin B - \sin A \cos D}{\cos A \sin D} \quad \text{(Eq 2)}$$

where

A = *your* latitude in degrees
B = latitude of the other location in degrees
L = *your* longitude minus that of the other location (algebraic difference)
D = distance along path in degrees of arc
C = true bearing from north if the value for sin L is positive.
If sin L is negative, true bearing is 360 – C.

The term *cos* is an abbreviation for cosine, and the term *sin* is an abbreviation for sine. A knowledge of the meanings of these terms isn't necessary for their use here.

The actual calculating procedure uses, first, Eq 1 to determine the angular value for D, in degrees. From this value the path-length distance may be determined in miles or kilometers. Next, Eq 2 is used to determine the bearing angle.

Using the Wichita-to-Perth example mentioned earlier, refer to Fig 15-5 to see how the equations are used. From Table 15-1, it can be seen that the location of Wichita is 37.7° north latitude, 97.3° west longitude. Similarly, Perth is located at 31.9° south latitude, 115.8° east longitude. Your location is in Wichita. Values for use in the equations are as follows:

A = lat A = +37.7°
B = lat B = –31.9°
L = long A – long B
= 97.3° – (–115.8°) = 213.1°

Solving Eq 1, $\cos D = \sin 37.7° \times \sin(-31.9°) + \cos 37.7° \times \cos(-31.9°) \times \cos 213.1°$. D = 152.4°. Each degree along the path equals 60.0359 nautical miles. Therefore, 152.4° of arc is equivalent to 60.0359 × 152.4 = 9,147 nautical miles. To convert to statute miles, multiply degrees by 69.0826. If the distance is desired in kilometers, multiply degrees by 111.1775. This means that the Wichita-to-Perth distance is 10,525 miles or 16,939 kilometers.

Solving Eq 2

$$\cos C = \frac{\sin(-31.9°) - \sin 37.7° \cos 152.4°}{\cos 37.7° \sin 152.4°}$$

C = 87.9°. Because the sin of L (213.1°) is negative, however, the correct value for C is 360° – 87.9° = 272.1°. Thus, the true bearing from Wichita to Perth is 272.1° and the distance is 10,525 statute miles. If the bearing from Perth were desired, it would be necessary only to work through Eq 2, interchanging latitude values for A and B. Because of the way L is defined (now –213.1°), sin L will be positive, and it will not be necessary to subtract from 360° to get the true bearing at Perth, which is 68.6°.

These equations give information for the great-circle bearing and distance for the shortest path. For long-path work, the bearing will be 180° away from the answers obtained.

The equations described above may be used for any two points on the Earth's surface—both locations in the Northern Hemisphere, both locations in the Southern Hemisphere, either or both on the equator, and so on. The equations themselves are exact, not being based on any approximations. However, there are some cases where practical limitations exist in the accuracy of the results obtained from Eq 2, in relation to the number of significant figures used during calculations. (Round-off errors in calculators and computers during computations will effectively reduce the number of significant figures in the resulting answers.) These cases are where both locations are near or at exact opposite points on the Earth (antipodes), where the locations are close together, or where your location is at or near one of the poles. (At the poles, all directions are either south or north, anyway.) More specifically, these situations exist when lat A is near +90°, or where D is near 0° or 180°.

Other Computer Resources

GeoClock was one of the first shareware programs widely used by hams. As this is written, it is available both as shareware and in an enhanced version from GeoClock, 2218 N Tuckahoe St, Arlington, VA 22205. Telephone 703-241-2661, e-mail joe@geoclock.com. The current shareware versions of the program and detailed information are available on their web page at **http://www.geoclock.com**. With registration you get the latest Windows and DOS versions of the programs (includes screen saver) and 44 maps.

The basic program generates maps with the night (dark) and twilight areas shown—in other words a gray-line plot. The sunlight and twilight settings are adjustable, with an explanation included in the program on-line help. If you want the gray-line plot for a selected date and time, simply change the clock setting and the new map is generated. Various display types are accommodated by changes made with a word processor to a configuration file. The ham add-in generates distance, short-path bearing and long-path bearing to any prefix you enter. In **Fig 15-6** G3RED was entered (left center of the screen). The resulting bearing line running from Connecticut to England corresponds to a short-path bearing of 52° with a distance of 3417 miles. Long-path bearing and direction, contest and certificate zones and sunrise/sunset times are also displayed on the screen.

In this screen picture, the gray-line runs through Alaska, Western Australia and around through Eastern Europe. If you are a ham in the continental US, you can see in one glance you are too late today for any gray-line propagation!

A wide variety of software is available for producing antenna bearings, distances and gray-line predictions. From time to time *QST* announces new offerings and prints reviews of them. The ARRL book *Personal Computers in the Ham Shack* summarizes a number of PC-based ham software packages. Frequent searches on the Internet will probably turn up new programs every few months.

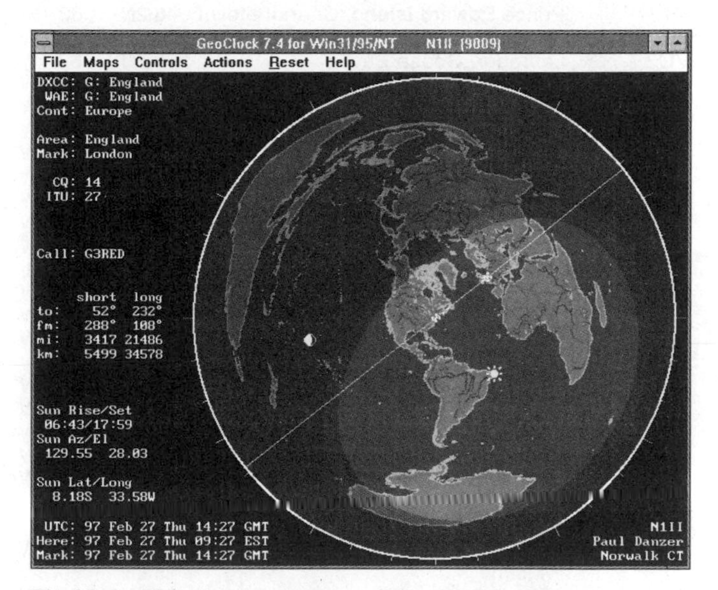

Fig 15-6—This screen capture of the *GeoClock* program gives the bearing from Connecticut to a G3. From the map center (your location), any spot on the globe may be selected.

Table 15-1
Latitude and Longitude of Various US/Canadian Cities and DX Locations with Bearing from East, Central and Western USA

Prefix	*State/Province/Country/City*	*Lat*	*Long*	*fm E USA*	*fm C USA*	*fm W USA*
VE1	New Brunswick, Saint John	45.3N	66.1W	58.0	62.6	64.8
	Nova Scotia, Halifax	44.6N	63.6W	63.8	64.5	65.4
VE2	Quebec, Montreal	45.5N	73.6W	38.4	59.7	65.7
	Quebec City	46.8N	71.2W	40.3	57.3	63.0
VE3	Ontario, London	43.0N	81.3W	342.3	63.2	71.9
	Ottawa	45.4N	75.7W	28.9	58.8	66.1
	Sudbury	46.5N	81.0W	353.9	50.4	63.9
	Toronto	43.7N	79.4W	6.9	62.0	70.1
VE4	Manitoba, Winnipeg	49.9N	97.1W	315.1	2.8	47.0
VE5	Saskatchewan, Regina	50.5N	104.6W	309.6	341.5	33.5
	Saskatoon	52.1N	106.7W	312.4	339.5	24.8
VE6	Alberta, Calgary	51.0N	114.1W	306.4	324.0	6.2
	Edmonton	53.5N	113.5W	312.1	330.5	6.3
VE7	British Columbia, Prince George	53.9N	122.8W	310.4	321.2	344.0
	Prince Rupert	54.3N	130.3W	310.4	317.2	330.9
	Vancouver	49.3N	123.1W	301.7	310.2	333.9
VE8	Northwest Territories, Yellowknife	62.5N	114.4W	329.0	343.0	1.9
	Resolute	74.7N	95.0W	353.2	1.3	9.3
VY1	Yukon, Whitehorse	60.7N	135.1W	320.6	326.6	336.6
VY2	Prince Edward Island, Charlottetown	46.2N	63.1W	57.7	61.0	62.7
VO1	Newfoundland, St. John's	47.6N	52.7W	59.7	58.8	58.4
VO2	Labrador, Goose Bay	53.3N	60.4W	38.5	47.0	51.2
W1	Connecticut, Hartford	41.8N	72.7W	69.6	70.9	72.4
	Maine, Bangor	44.8N	68.8W	56.2	63.3	66.2
	Portland	43.7N	70.3W	59.7	65.9	68.4
	Massachusetts, Boston	42.4N	71.1W	67.4	69.3	70.9
	New Hampshire, Concord	43.2N	71.5W	60.5	67.0	69.6
	Rhode Island, Providence	41.8N	71.4W	71.7	71.0	72.1
	Vermont, Montpelier	44.3N	72.6W	49.5	63.6	67.8
W2	New Jersey, Atlantic City	39.4N	74.4W	96.1	78.3	77.4
	New York, Albany	42.7N	73.8W	57.9	68.0	71.0
	Buffalo	42.9N	78.9W	15.6	65.3	71.7
	New York City	40.8N	74.0W	78.1	73.9	74.7
	Syracuse	43.1N	76.2W	41.3	65.9	70.8
W3	Delaware, Wilmington	39.7N	75.5W	93.5	77.4	77.2
	District of Columbia, Washington	38.9N	77.0W	114.4	80.3	79.3
	Maryland, Baltimore	39.3N	76.6W	103.9	78.9	78.4
	Pennsylvania, Harrisburg	40.3N	76.9W	81.8	75.4	76.6
	Philadelphia	39.9N	75.2W	90.0	76.7	76.7
	Pittsburgh	40.4N	80.0W	15.4	74.6	77.3
	Scranton	41.4N	75.7W	65.4	71.8	74.0
W4	Alabama, Montgomery	32.4N	86.3W	215.7	116.9	98.3
	Florida, Jacksonville	30.3N	81.7W	188.7	114.9	98.4
	Miami	25.8N	80.2W	180.9	123.9	104.5
	Pensacola	30.4N	87.2W	213.7	127.2	103.3
	Georgia, Atlanta	33.8N	84.4W	210.9	106.8	93.8
	Savannah	32.1N	81.1W	186.8	108.1	94.7
	Kentucky, Lexington	38.0N	84.5W	241.7	85.8	84.5
	Louisville	38.2N	85.8W	250.1	85.0	84.6
	North Carolina, Charlotte	35.2N	80.8W	187.9	96.2	88.5
	Raleigh	35.8N	78.6W	164.8	92.1	86.1
	Wilmington	34.2N	77.9W	163.2	97.1	88.6
	South Carolina, Columbia	34.0N	81.0W	188.0	101.1	91.0
	Tennessee, Knoxville	36.0N	83.9W	218.8	95.8	88.7
	Memphis	35.1N	90.1W	241.7	112.2	95.2
	Nashville	36.2N	86.8W	236.8	98.0	90.1
	Virginia, Norfolk	36.9N	76.3W	135.7	87.0	82.8
	Richmond	37.5N	77.4W	140.1	85.4	82.2
W5	Arkansas, Little Rock	34.7N	92.3W	245.4	124.0	98.2
	Louisiana, New Orleans	29.9N	90.1W	222.4	138.7	107.5
	Shreveport	32.5N	93.7W	240.0	146.2	105.7
	Mississippi, Jackson	32.3N	90.2W	230.0	129.5	102.2
	New Mexico, Albuquerque	35.1N	106.7W	265.4	250.1	120.7
	Oklahoma, Oklahoma City	35.5N	97.5W	257.5	170.5	101.3
	Texas, Abilene	32.5N	99.7W	250.8	194.7	114.7
	Amarillo	35.2N	101.8W	261.4	228.6	108.6
	Dallas	32.8N	96.8W	247.2	168.9	109.0
	El Paso	31.8N	106.5W	257.3	230.9	133.9
	San Antonio	29.4N	98.5W	240.7	183.0	121.1
W6	California, Los Angeles	34.1N	118.2W	271.3	262.7	197.3
	San Francisco	37.8N	122.4W	280.1	277.0	248.2

Prefix	State/Province/Country/City	Lat	Long	fm E USA	fm C USA	fm W USA
W7	Arizona, Flagstaff	35.2N	111.7W	269.3	259.9	143.3
	Phoenix	33.5N	112.1W	266.0	252.8	153.1
	Idaho, Boise	43.6N	116.2W	289.6	297.9	357.2
	Pocatello	42.9N	112.5W	287.5	298.5	41.0
	Montana, Billings	45.8N	108.5W	295.0	318.4	40.9
	Butte	46.0N	112.5W	295.1	311.3	21.9
	Great Falls	47.5N	111.3W	298.9	318.6	22.8
	Nevada, Las Vegas	36.2N	115.1W	273.5	267.7	169.0
	Reno	39.5N	119.8W	282.1	281.9	261.5
	Oregon, Portland	45.5N	122.7W	294.4	300.1	320.4
	Utah, Salt Lake City	40.8N	111.9W	282.3	289.0	74.3
	Washington, Seattle	47.6N	122.3W	298.4	306.3	331.2
	Spokane	47.7N	117.4W	298.6	310.7	353.0
	Wyoming, Cheyenne	41.1N	104.8W	281.4	302.7	79.0
	Sheridan	44.8N	107.0W	292.5	318.1	51.2
W8	Michigan, Detroit	42.3N	83.0W	316.5	64.7	73.8
	Grand Rapids	43.0N	85.7W	307.0	58.0	72.5
	Sault Ste. Marie	46.5N	84.4W	335.2	45.4	63.6
	Traverse City	44.8N	85.6W	321.1	49.8	67.8
	Ohio, Cincinnati	39.1N	84.5W	256.9	79.9	81.9
	Cleveland	41.5N	81.7W	319.8	69.3	75.4
	Columbus	40.0N	83.0W	271.0	75.6	79.2
	West Virginia, Charleston	38.4N	81.6W	218.4	83.1	82.3
W9	Illinois, Chicago	41.9N	87.6W	290.8	60.7	75.6
	Indiana, Indianapolis	39.8N	86.2W	269.6	75.2	80.8
	Wisconsin, Green Bay	44.5N	88.0W	309.9	45.9	68.5
	Milwaukee	43.0N	87.9W	299.5	53.7	72.6
WØ	Colorado, Denver	39.7N	105.0W	277.2	289.5	88.5
	Grand Junction	39.1N	108.6W	276.9	280.8	96.6
	Iowa, Des Moines	41.6N	93.6W	283.2	41.8	77.3
	Kansas, Pratt	37.7N	98.7W	267.0	242.1	94.2
	Wichita	37.7N	97.3W	265.8	117.9	93.0
	Minnesota, Duluth	46.8N	92.1W	311.8	24.4	60.6
	Minneapolis	45.0N	93.3W	301.4	25.2	65.8
	Missouri, Columbia	39.0N	92.3W	267.9	75.6	85.5
	Kansas City	39.1N	94.6W	270.1	66.5	86.2
	St. Louis	38.6N	90.2W	263.2	82.0	85.7
	Nebraska, North Platte	41.1N	100.8W	280.7	326.0	79.6
	Omaha	41.3N	95.9W	281.3	25.5	78.6
	North Dakota, Fargo	46.9N	96.8W	305.1	5.3	57.2
	South Dakota, Rapid City	44.1N	103.2W	291.0	328.9	62.5
1AØ	S.M.O.M	41.9N	12.4E	54.4	45.5	35.8
1S	Spratly Is.	8.8N	111.9E	344.6	322.8	306.5
3A	Monaco	43.7N	7.4E	54.9	46.4	37.6
3B6	Agalega	10.4S	56.6E	64.6	46.2	14.5
3B7	St. Brandon	16.3S	59.8E	67.5	48.1	10.0
3B8	Mauritius	20.3S	57.5E	74.2	57.1	17.7
3B9	Rodriguez Is.	19.7S	63.4E	67.9	46.6	1.7
3C	Equatorial Guinea, Bata	1.8N	9.8E	88.7	77.4	63.7
	Malabo	3.8N	8.8E	87.9	76.5	63.0
3CØ	Annobon Is.	1.5S	5.6E	94.0	82.7	69.6
3D2	Fiji Is., Suva	18.1S	178.4E	263.0	251.8	240.5
3D2	Conway Reef	21.4S	174.4E	262.5	251.5	240.8
3D2	Rotuma Is.	12.3S	177.7E	268.2	256.8	245.1
3DA	Swaziland, Mbabane	26.3S	31.1E	99.0	90.1	73.5
3V	Tunisia, Tunis	36.8N	10.2E	60.1	50.5	40.3
3W, XV	Vietnam, Ho Chi Minh City (Saigon)	10.8N	106.7E	351.5	329.5	312.4
	Hanoi	21.0N	105.8E	353.8	335.0	319.2
3X	Guinea, Conakry	9.5N	13.7W	98.1	85.9	74.8
3Y	Bouvet	54.5S	3.4E	139.1	135.1	131.0
3Y	Peter I Is.	68.8S	90.6W	184.0	177.2	170.5
4J, 4K	Azerbaijan, Baku	40.4N	49.9E	35.8	24.0	10.8
4L	Georgia, Tbilisi	41.7N	44.8E	38.0	26.9	14.3
4S	Sri Lanka, Colombo	7.0N	79.9E	26.2	2.9	339.0
4U	ITU Geneva	46.2N	6.2E	52.8	44.9	36.6
4U	United Nations Hq.	40.8N	74.0W	78.1	73.9	74.7
4X, 4Z	Israel, Jerusalem	31.8N	35.2E	50.4	38.7	24.9
5A	Libya, Tripoli	32.5N	12.5E	62.7	52.5	41.4
	Benghazi	32.1N	20.0E	59.0	48.4	36.4
5B	Cyprus, Nicosia	35.2N	33.4E	49.1	37.9	25.0
5H	Tanzania, Dar es Salaam	7.0S	39.5E	75.7	62.1	40.2
5N	Nigeria, Lagos	6.5N	3.4E	89.2	77.8	65.2
5R	Madagascar, Antananarivo	18.9S	47.5E	80.8	67.2	38.7
5T	Mauritania, Nouakchott	18.1N	16.0W	92.0	80.2	69.8
5U	Niger, Niamey	13.5N	2.0E	84.6	73.3	61.2
5V	Togo, Lome	5.8N	1.2E	91.2	79.8	67.3
5W	Western Samoa, Apia	13.5S	171.8W	260.7	249.4	236.6

Prefix	State/Province/Country/City	Lat	Long	fm E USA	fm C USA	fm W USA
5X	Uganda, Kampala	0.3N	32.5E	74.9	62.0	43.4
5Z	Kenya, Nairobi	1.3S	376.8E	86.6	75.2	60.3
6W	Senegal, Dakar	14.7N	17.5W	96.2	83.9	73.3
6Y	Jamaica, Kingston	18.0N	76.8W	171.9	131.3	111.5
7O	Yemen, Aden	12.8N	45.0E	56.5	41.9	22.6
	Sanaa	15.4N	44.2E	55.3	41.1	22.5
7P	Lesotho, Maseru	29.3S	27.5E	103.9	95.9	81.7
7Q	Malawi, Lilongwe	14.0S	33.8E	85.7	74.0	54.1
	Blantyre	15.8S	35.0E	86.5	74.9	54.5
7X	Algeria, Algiers	36.7N	3.0E	63.6	54.3	44.7
8P	Barbados, Bridgetown	13.1N	59.6W	140.6	115.7	102.0
8Q	Maldive Is.	4.4N	73.4E	35.2	12.6	346.7
8R	Guyana, Georgetown	6.8N	58.2W	143.7	120.6	106.5
9A	Croatia, Zagreb	45.8N	16.0E	49.3	40.7	31.4
9G	Ghana, Accra	5.5N	0.2W	92.3	80.9	68.6
9H	Malta	36.0N	14.4E	58.7	48.9	38.0
9J	Zambia, Lusaka	15.4S	28.3E	90.6	79.6	61.9
9K	Kuwait	29.5N	47.8E	43.7	30.5	14.9
9L	Sierra Leone, Freetown	8.5N	13.2W	98.6	86.4	75.2
9M2	West Malaysia, Kuala Lumpur	3.2N	101.6E	357.7	331.7	312.1
9M6,8	East Malaysia, Sabah, Sandakan(9M6)	5.8N	118.1E	335.7	314.4	299.3
	Sarawak, Kuching (9M8)	1.6N	110.3E	344.7	320.0	302.8
9N	Nepal, Kathmandu	27.7N	85.3E	13.9	356.8	340.1
9Q	Zaire, Kinshasa	4.3S	15.3E	89.9	78.7	64.0
	Kisangani	0.5N	25.2E	79.7	67.7	51.0
	Lubumbashi	11.7S	27.5E	87.9	76.6	59.0
9U	Burundi, Bujumbura	3.3S	29.3E	79.9	67.6	49.6
9V	Singapore	1.3N	103.8E	354.3	327.8	308.6
9X	Rwanda, Kigali	2.0S	30.1E	78.3	65.9	47.7
9Y	Trinidad & Tobago, Port of Spain	10.5N	61.3W	145.5	120.2	105.7
A2	Botswana, Gaborone	24.8S	25.9E	100.5	91.4	76.2
A3	Tonga, Nukualofa	21.1S	175.2W	256.6	245.7	234.0
A4	Oman, Masqat	23.6N	58.6E	39.0	23.6	5.5
A5	Bhutan, Thimpu	27.3N	89.4E	10.2	352.8	336.3
A6	United Arab Emirates, Abu Dhabi	24.5N	54.2E	41.9	27.3	9.8
A7	Qatar, Ad-Dawhah	25.3N	51.5E	43.5	29.3	12.3
A9	Bahrein, Al-Manamah	26.2N	50.6E	43.7	29.7	13.0
AP	Pakistan, Karachi	24.9N	67.1E	31.2	15.0	356.9
	Islamabad	33.7N	73.2E	22.6	7.7	352.1
BS7H	Scarborough Reef	15.1N	117.8E	339.5	320.6	306.0
BV	Taiwan, Taipei	25.1N	121.5E	339.0	323.0	309.9
BV9P	Pratas Is.	20.7N	116.7E	342.4	324.7	310.5
BY	Peoples Rep. of China, Beijing	40.0N	116.4E	347.4	334.2	322.6
	Harbin	45.8N	126.7E	341.7	330.6	320.6
	Shanghai	31.2N	121.5E	341.0	326.3	313.8
	Fuzhou	26.1N	119.3E	341.4	325.3	312.0
	Xian	34.3N	108.9E	352.4	337.4	324.2
	Chongqing	29.8N	106.5E	354.0	337.7	323.5
	Chengdu	30.7N	104.1E	356.3	340.1	325.8
	Lhasa	29.7N	91.2E	8.1	351.4	335.6
	Urumqi	43.8N	87.6E	9.0	355.9	343.2
	Kashi	39.5N	76.0E	18.5	4.7	350.6
C2	Nauru	0.5S	166.9E	284.9	272.8	261.2
C3	Andorra	42.5N	1.5E	58.5	50.1	41.6
C5	The Gambia, Banjul	13.5N	16.7W	96.7	84.4	73.7
C6	Bahamas, Nassau	25.1N	77.4W	170.9	120.5	103.0
C9	Mozambique, Maputo	26.0S	32.6E	97.8	88.8	71.4
	Mozambique	15.1S	40.7E	82.0	69.3	45.9
CE	Chile, Santiago	33.5S	70.8W	172.0	156.9	143.5
CEØY	Easter Island	27.1S	109.4W	207.3	191.1	173.6
CEØZ	Juan Fernandez	33.6S	78.8W	179.0	163.4	149.4
CEØX	San Felix	26.3S	80.1W	180.1	162.5	146.9
CM,CO	Cuba, Havana	23.1N	82.4W	187.6	133.7	110.7
CN	Morocco, Casablanca	33.6N	7.5W	71.7	62.1	53.1
CP	Bolivia, La Paz	16.5S	68.4W	166.8	147.3	131.8
CT	Portugal, Lisbon	38.7N	9.2W	66.9	58.3	50.1
CT3	Madeira Islands, Funchal	32.6N	16.9W	77.4	67.5	59.1
CU	Azores, Ponta Delgada	37.7N	25.7W	75.0	66.1	59.2
CX	Uruguay, Montevideo	34.9S	56.2W	160.2	146.5	134.8
CYØ	Sable Is.	43.8N	60.0W	69.1	66.3	65.7
CY9	St. Paul Is.	47.2N	60.1W	56.8	59.3	60.6
D2	Angola, Luanda	8.8S	13.2E	94.7	83.8	69.6
D4	Cape Verde, Praia	14.9N	23.5W	100.3	87.3	76.9
D6	Comoros, Moroni	11.8S	43.7E	76.7	62.8	38.1
DA-DL	Fed. Rep. of Germany, Bonn	50.7N	7.0E	47.9	40.7	33.1
	Berlin	52.5N	13.4E	43.9	36.6	28.8
DU	Philippines, Manila	14.6N	121.0E	335.9	317.4	303.2

Prefix	State/Province/Country/City	Lat	Long	fm E USA	fm C USA	fm W USA
E3	Eritrea, Asmara	15.3N	38.9E	59.3	45.8	28.2
E4	Palestinian Authority, Gaza City	31.5N	34.5E	51.0	39.4	25.6
EA	Spain, Madrid	40.4N	3.7W	62.8	54.3	45.9
EA6	Balearic Is., Palma	39.5N	2.6E	61.0	52.2	43.1
EA8	Canary Is., Las Palmas	28.4N	14.3W	80.7	70.2	60.9
EA9	Ceuta & Melilla, Ceuta	35.9N	5.3W	68.2	59.0	50.1
	Melilla	35.3N	3.0W	67.8	58.5	49.2
EI	Ireland, Dublin	53.3N	6.3W	48.6	42.9	37.1
EK	Armenia, Yerevan	40.3N	44.5E	39.1	27.8	14.9
EL	Liberia, Monrovia	6.3N	10.8W	90.7	86.7	75.3
EP	Iran, Tehran	35.8N	51.8E	37.2	24.6	10.1
ER	Moldova, Kishinev	47.0N	28.8E	42.6	33.4	23.2
ES	Estonia, Tallinn	59.4N	24.8E	33.5	26.8	19.4
ET	Ethiopia, Addis Ababa	9.0N	38.7E	63.9	50.1	31.3
EU,EV,EW	Belarus, Minsk	53.9N	27.6E	37.4	29.5	20.7
EX	Kyrgyzstan, Bishkek	42.9N	74.6E	18.4	5.5	352.2
EY	Tajikistan, Samarkand	39.7N	66.8E	25.0	11.9	357.8
	Dushanbe	39.1N	68.8E	23.8	10.4	356.2
EZ	Turkmenistan, Ashkhabad	38.0N	58.4E	31.6	18.7	4.5
F	France, Paris	48.8N	2.3E	51.5	44.2	36.7
FG	Guadeloupe	16.0N	61.7W	141.2	114.5	100.8
FJ, FS	St. Martin	18.1N	63.1W	141.4	113.4	99.8
FH	Mayotte	13.0S	45.3E	76.6	62.5	36.6
FK	New Caledonia, Noumea	22.3S	166.5E	266.4	255.1	245.1
FM	Martinique	14.6N	61.0W	141.4	115.5	101.7
FO	Clipperton Is.	10.3N	109.2W	229.1	202.9	166.7
FO	Fr. Polynesia, Tahiti	17.6S	159.5W	249.9	238.3	224.1
	Rurutu, (Austral Is.)	22.5S	151.3E	276.2	263.4	254.0
	Hiva Oa, (Marquesas Is.)	9.9S	139.0W	241.5	227.5	208.3
FP	St. Pierre & Miquelon, St. Pierre	46.7N	56.0W	61.0	60.5	60.4
FR/G	Glorioso	11.5S	47.3E	73.6	58.7	32.0
FR/J,E	Juan de Nova	17.0S	42.8E	82.3	69.5	44.7
	Europa	22.3S	40.4E	89.3	78.3	55.5
FR	Reunion	21.1S	55.6E	76.8	60.9	23.2
FR/T	Tromelin	15.9S	54.4E	72.0	55.3	21.9
FT5W	Crozet	46.0S	52.0E	116.0	119.5	128.4
FT5X	Kerguelen	49.3S	69.2E	123.5	144.9	199.9
FT5Y	Antarctica, Dumont D'Urville	66.6S	140.0E	206.7	209.6	211.0
FT5Z	Amsterdam & St. Paul Is., Amsterdam	37.7S	77.6E	89.7	86.4	277.9
FW	Wallis & Futuna Is., Wallis	13.3S	176.3W	263.7	252.4	240.1
FY	Fr. Guiana, Cayenne	4.9N	52.3W	137.3	116.9	103.7
G	England, London	51.5N	0.1W	49.2	42.6	35.9
GD	Isle of Man	54.3N	4.5W	46.9	41.3	35.6
GI	Northern Ireland, Belfast	54.6N	5.9W	46.8	41.5	35.9
GJ	Jersey	49.3N	2.2W	52.3	45.5	38.5
GM	Scotland, Glasgow	55.8N	4.3W	45.0	39.8	34.2
	Aberdeen	57.2N	2.1W	42.9	37.7	32.2
GU	Guernsey	49.5N	2.7W	52.2	45.5	38.6
GW	Wales, Cardiff	51.5N	3.2W	50.0	43.7	37.3
H4	Solomon Islands, Honiara	9.4S	160.0E	282.7	269.9	259.0
H4Ø	Temotu Province	10.7S	165.8E	277.4	265.3	254.2
HA	Hungary, Budapest	47.5N	19.1E	46.5	38.0	28.7
HB	Switzerland, Bern	47.0N	7.5E	51.5	43.7	35.3
HBØ	Liechtenstein	47.2N	9.6E	50.6	42.6	34.1
HC	Ecuador, Quito	0.2S	78.0W	176.9	149.5	129.6
HC8	Galapagos Is.	0.5S	90.5W	195.9	168.1	143.7
HH	Haiti, Port-Au-Prince	18.5N	72.3W	160.6	123.9	106.8
HI	Dominican Republic, Santo Domingo	18.5N	70.0W	155.3	120.8	104.8
HK	Colombia, Bogota	4.6N	74.1W	169.9	141.0	122.0
HKØ	Malpelo Is.	4.0N	81.1W	181.9	151.4	129.6
HKØ	San Adreas	12.5N	81.7W	183.6	146.0	122.7
HL	Korea, Seoul	37.5N	127.0E	338.6	325.8	314.6
HP	Panama, Panama	9.0N	79.5W	179.0	145.3	123.6
HR	Honduras, Tegucigalpa	14.1N	87.2W	195.7	155.3	127.6
HS	Thailand, Bangkok	13.8N	100.5E	359.4	337.8	319.8
HV	Vatican City	41.9N	12.5E	54.4	45.4	35.7
HZ, 7Z	Saudi Arabia, Dharan	26.3N	50.0E	44.1	30.2	13.6
	Mecca	21.5N	39.8E	54.4	41.2	24.8
	Riyadh	24.6N	46.7E	47.5	33.8	17.1
I	Italy, Rome	41.9N	12.5E	54.4	45.4	35.7
	Trieste	45.7N	13.8E	50.3	41.9	32.7
	Sicily	37.5N	14.0E	57.6	48.0	37.4
IS	Sardinia, Cagliari	39.2N	9.1E	58.4	49.2	39.4
J2	Djibouti, Djibouti	11.6N	43.2E	58.7	44.3	25.1
J3	Grenada	12.0N	61.8W	145.1	119.1	104.7
J5	Guinea-Bissau, Bissau	11.9N	15.6W	97.3	85.1	74.2
J6	St. Lucia	13.9N	61.0W	142.1	116.2	102.3

Prefix	State/Province/Country/City	Lat	Long	fm E USA	fm C USA	fm W USA
J7	Dominica	15.5N	61.3W	141.0	114.7	101.0
J8	St. Vincent	13.3N	61.3W	143.1	117.2	103.1
JA-JS	Japan, Tokyo	35.7N	139.8E	328.5	316.6	306.1
	Nagasaki	32.8N	129.9E	334.6	321.1	309.6
	Sapporo	43.1N	141.4E	331.1	320.7	311.3
JD1	Minami Torishima	24.3N	154.0E	311.8	299.9	289.1
JD1	Ogasawara, Kazan Is.	27.5N	141.0E	323.4	310.3	299.1
JT	Mongolia, Ulan Bator	47.9N	106.9E	355.4	343.6	332.7
JW	Svalbard, Spitsbergen	78.8N	16.0E	14.2	12.2	9.8
JX	Jan Mayen	71.0N	8.3W	25.1	23.6	21.5
JY	Jordan, Amman	32.0N	35.9E	49.8	38.1	24.2
KC4	Antarctica, Byrd Station	80.0S	120.0W	187.6	184.2	180.8
	McMurdo Sound	77.7S	166.7E	195.7	195.6	194.9
	Palmer Station	64.8S	64.0W	173.0	165.6	158.7
KC6	Belau, Yap	9.5N	138.2E	315.7	299.7	287.5
	Koror	7.3N	134.5E	317.9	300.9	288.4
KG4	Guantanamo Bay	19.9N	75.2W	167.0	126.1	107.8
KHØ	Mariana Is., Saipan	15.2N	145.8E	312.5	298.4	286.9
KH1	Baker, Howland Is.	0.5N	176.0W	274.2	262.9	250.0
KH2	Guam, Agana	13.5N	144.8E	312.2	297.9	286.2
KH3	Johnston Is.	17.0N	168.5W	282.3	272.2	258.8
KH4	Midway Is.	28.2N	177.4W	296.5	287.5	276.7
KH5	Palmyra Is.	5.9N	162.1W	269.5	258.2	243.0
KH5K	Kingman Reef	7.5N	162.8W	271.2	260.1	245.0
KH6	Hawaii, Hilo	19.7N	155.1W	276.4	266.4	250.4
	Honolulu	21.3N	157.9W	279.5	269.9	254.9
KH7	Kure Is.	28.4N	178.4W	297.2	288.2	277.5
KH8	American Samoa, Pago Pago	14.3S	170.8W	259.4	248.2	235.3
KH9	Wake Is.	19.3N	166.6E	299.7	288.5	277.5
KL7	Alaska, Adak	51.8N	176.6W	316.5	311.9	307.1
	Anchorage	61.2N	150.0W	321.3	323.2	327.0
	Fairbanks	64.8N	147.9W	326.2	329.1	334.0
	Juneau	58.3N	134.4W	316.8	322.7	333.4
	Nome	64.5N	165.4W	327.3	326.9	327.4
KP1	Navassa Is.	18.4N	75.0W	167.3	127.9	109.3
KP2	Virgin Islands, Charlotte Amalie	18.3N	64.9W	144.6	115.0	100.9
KP4	Puerto Rico, San Juan	18.5N	66.2W	147.0	116.2	101.7
KP5	Desecheo Is.	18.3N	67.5W	150.0	118.0	103.0
LA-LJ	Norway, Oslo	60.0N	10.7E	36.8	31.2	25.1
LU	Argentina, Buenos Aires	34.6S	58.4W	161.9	147.9	136.0
LX	Luxembourg	49.6N	6.2E	49.3	42.0	34.3
LY	Lithuania, Vilna	54.5N	25.5E	37.7	30.0	21.5
LZ	Bulgaria, Sofia	42.7N	23.3E	48.7	39.2	28.6
OA	Peru, Lima	12.1S	77.1W	176.4	154.4	136.6
OD	Lebanon, Beirut	33.9N	35.5E	48.8	37.3	23.8
OE	Austria, Vienna	48.2N	16.3E	47.0	38.8	29.8
OH	Finland, Helsinki	60.2N	25.0E	32.7	26.1	18.9
OHØ	Aland Is.	60.2N	20.0E	34.2	27.9	21.1
OJØ	Market Reef	60.3N	19.0E	34.4	28.2	21.5
OK,OL	Czech Rep., Prague	50.1N	14.4E	45.9	38.1	29.7
OM	Slovak, Rep., Bratislava	48.0N	17.0E	46.9	38.6	29.6
ON	Belgium, Brussels	50.9N	4.4E	48.5	41.6	34.2
OX	Greenland, Godthaab	64.2N	51.7W	25.0	31.0	34.6
	Thule	76.6N	68.8W	4.3	10.0	14.8
OY	Faroe Islands, Torshavn	62.0N	6.8W	37.3	33.8	29.8
OZ	Denmark, Copenhagen	55.7N	12.6E	40.9	34.3	27.1
P2	Papua New Guinea, Madang	5.2S	145.6E	298.0	282.7	271.4
	Port Moresby	9.4S	147.1E	293.0	278.1	267.1
P4	Aruba, Oranjestad	12.6N	70.1W	159.6	128.2	111.0
P5	North Korea, Pyongyang	39.0N	125.8E	340.1	327.4	316.4
PA-PI	Netherlands, Amsterdam	52.4N	4.9E	46.7	40.0	32.9
PJ2,4,9	Netherlands Antilles, Willemstad	12.1N	68.9W	157.6	127.2	110.5
PJ5-8	St. Maarten and Saba, St. Maarten	17.7N	63.2W	142.1	114.0	100.3
PY	Brazil, Brasilia	15.8S	47.9W	145.1	128.8	116.2
	Rio De Janeiro	23.0S	43.2W	144.5	130.2	118.3
	Natal	6.0S	35.2W	127.3	112.3	100.5
	Manaus	3.1S	60.2W	152.3	130.8	116.0
	Porto Alegre	30.1S	51.2W	154.4	140.3	128.4
PYØ	Fernando De Noronha	3.9S	32.4W	123.3	108.7	97.1
PYØ	St. Peter & St. Paul Rocks	1.0N	29.4W	117.1	102.8	91.4
PYØ	Trindade & Martin Vaz Is., Trindade	20.5S	29.3W	131.9	119.1	107.9
PZ	Suriname, Paramaribo	5.8N	55.2W	140.3	118.7	105.1
R1FJ_	Franz Josef Land	80.0N	53.0E	8.7	5.5	2.2
R1MV_	Malyj Vysotskij Is.	60.6N	28.6E	31.2	24.5	17.1
S0	Western Sahara, Smara	26.4N	11.4W	81.1	70.4	60.7
S2	Bangladesh, Dacca	23.7N	90.4E	9.8	351.3	334.0
S5	Slovenia, Ljubljana	46.0N	14.5E	49.8	41.3	32.1

Prefix	*State/Province/Country/City*	*Lat*	*Long*	*fm E USA*	*fm C USA*	*fm W USA*
S7	Seychelles, Victoria	4.6S	55.5E	60.5	42.5	14.4
S9	Sao Tome	0.3N	6.7E	91.9	80.6	67.3
SM	Sweden, Stockholm	59.3N	18.1E	35.6	29.3	22.5
SP	Poland, Krakow	50.0N	20.0E	43.9	35.7	26.8
	Warsaw	52.2N	21.0E	41.5	33.6	25.1
ST	Sudan, Khartoum	15.6N	32.5E	63.5	50.8	34.4
STØ	Southern Sudan, Juba	5.0N	31.6E	71.9	59.1	41.3
SU	Egypt, Cairo	30.0N	31.4E	54.0	42.4	28.7
SV	Greece, Athens	38.0N	23.7E	52.3	42.2	30.7
SV/A	Mount Athos	40.2N	24.3E	50.8	40.4	29.2
SV5	Dodecanese, Rhodes	36.4N	28.2E	51.1	40.5	28.3
SV9	Crete	35.4N	25.2E	53.5	43.0	30.9
T2	Tuvalu, Funafuti	8.7S	178.6E	270.6	259.1	247.1
T3Ø	West Kiribati, Bonriki	1.4N	173.2E	282.1	270.4	258.5
T31	Central Kiribati, Kanton	2.8S	171.7W	268.9	257.6	244.2
T32	East Kiribati, Christmas Is.	1.9N	157.4W	263.3	251.5	235.3
T33	Banaba Is.	0.5S	169.4E	283.2	271.2	259.6
T5	Somalia, Mogadishu	2.1N	45.4E	63.8	48.7	26.6
T7	San Marino	43.9N	12.3E	52.6	44.0	34.6
T9	Bosnia-Herzegovina, Sarajevo	43.9N	18.4E	50.0	41.0	31.0
TA	Turkey, Ankara	39.9N	32.9E	46.0	35.4	23.4
	Istanbul	41.2N	29.0E	47.1	37.0	25.6
TF	Iceland, Reykjavik	64.1N	22.0W	34.4	33.4	31.6
TG	Guatemala, Guatemala City	14.6N	90.5W	202.9	162.1	131.5
TI	Costa Rica, San Jose	9.9N	84.0W	187.8	152.3	127.8
TI9	Cocos Is.	5.6N	87.0W	192.2	160.1	135.1
TJ	Cameroon, Yaounde	3.9N	11.5E	86.0	74.6	60.8
TK	Corsica	42.0N	9.0E	55.8	47.1	37.7
TL	Central African Republic, Bangui	4.4N	18.6E	81.1	69.4	54.4
TN	Congo, Brazzaville	4.3S	15.3E	89.9	78.7	64.0
TR	Gabon, Libreville	0.4N	9.5E	90.0	78.7	65.1
TT	Chad, N'Djamena	12.1N	15.0E	77.5	66.0	52.2
TU	Ivory Coast, Abidjan	5.3N	4.0W	95.0	83.4	71.4
TY	Benin, Porto Novo	6.5N	2.6E	89.7	78.3	65.8
TZ	Mali, Bamako	12.7N	8.0W	91.6	79.9	68.7
UA	Russia, European, St Petersburg (UA1)	59.9N	30.3E	31.3	24.3	16.6
	Archangel (UA1)	64.6N	40.5E	24.0	17.4	10.3
	Murmansk (UA1)	69.0N	33.1E	22.3	17.1	11.4
	Moscow (UA3)	55.8N	37.6E	31.9	23.6	14.6
	Samara (UA4)	53.2N	50.1E	28.0	18.5	8.3
	Rostov (UA6)	47.5N	39.5E	37.1	27.2	16.3
UA2	Kaliningrad	55.0N	20.5E	39.1	31.8	23.8
UA9,Ø	Russia, Asiatic, Novosibirsk (UA9)	55.0N	82.9E	9.8	359.5	349.2
	Perm (UA9)	58.0N	56.3E	22.2	13.4	4.1
	Omsk (UA9)	55.0N	73.4E	15.0	4.9	354.6
	Norilsk (UAØ)	69.3N	88.1E	4.4	357.7	351.1
	Irkutsk (UAØ)	52.3N	104.3E	357.4	346.6	336.4
	Vladivostok (UAØ)	43.2N	131.9E	337.3	326.1	316.1
	Petropavlovsk (UAØ)	53.0N	158.7E	327.7	320.7	314.0
	Khabarovsk (UAØ)	48.5N	135.1E	337.6	327.6	318.6
	Krasnoyarsk (UAØ)	56.0N	92.8E	4.0	354.0	344.2
	Yakutsk (UAØ)	62.0N	129.7E	346.1	338.7	331.9
	Wrangel Island (UAØ)	71.0N	179.5W	337.1	335.8	335.2
	Kyzyl (UAØY)	51.7N	94.5E	3.4	352.3	341.6
UJ-UM	Uzbekistan, Bukhoro	39.8N	64.4E	26.6	13.7	359.7
	Tashkent	41.2N	69.3E	22.7	9.7	356.0
UN-UQ	Kazakhstan, Alma-Ata	43.3N	76.9E	16.6	3.8	350.6
UR-UZ,EM-EO	Ukraine, Kiev	50.4N	30.5E	39.1	30.4	20.7
V2	Antigua & Barbuda, St. Johns	17.1N	61.8W	140.2	113.3	99.8
V3	Belize, Belmopan	17.3N	88.8W	201.1	156.2	126.1
V4	St. Kitts & Nevis	17.3N	62.6W	141.4	113.9	100.2
V5	Namibia, Windhoek	22.6S	17.1E	103.6	94.2	80.6
V6	Micronesia, Ponape	6.9N	158.3E	296.8	283.9	272.6
V7	Marshall Islands, Kwajalein	9.1N	167.3E	291.9	280.1	268.5
V8	Brunei, Bandar Seri Begawan	4.9N	114.9E	339.5	317.1	301.3
VK	Australia, Canberra (VK1)	35.3S	149.1E	260.9	250.9	243.9
	Sydney (VK2)	33.9S	151.2E	261.8	251.6	244.2
	Melbourne (VK3)	37.8S	145.0E	258.8	249.6	243.4
	Brisbane (VK4)	27.5S	153.0E	269.1	257.5	248.8
	Adelaide (VK5)	34.9S	138.6E	267.1	255.8	249.3
	Perth (VK6)	31.9S	115.8E	297.5	272.2	264.2
	Hobart, Tasmania (VK7)	42.9S	147.3E	249.4	242.4	237.3
	Darwin (VK8)	12.5S	130.9E	306.6	287.0	275.1
VKØ	Heard Is.	53.0S	73.4E	134.6	161.1	203.1
VKØ	Macquarie Is.	54.7S	158.8E	228.8	225.0	221.3
VK9C	Cocos-Keeling Is.	12.2S	96.8E	6.7	329.0	304.7
VK9L	Lord Howe Is.	31.6S	159.1E	260.7	250.4	242.1

Prefix	State/Province/Country/City	Lat	Long	fm E USA	fm C USA	fm W USA
VK9M	Mellish Reef	17.6S	155.8E	278.1	265.3	255.2
VK9N	Norfolk Is.	29.0S	168.0E	258.8	248.5	239.1
VK9W	Willis Is.	16.3S	149.5E	284.3	270.3	260.1
VK9X	Christmas Is.	10.5S	105.7E	348.7	316.1	296.9
VP2E	Anguilla	18.3N	63.1W	141.2	113.1	99.6
VP2M	Montserrat	16.7N	62.2W	141.3	114.2	100.5
VP2V	British Virgin Is., Tortola	18.4N	64.6W	144.0	114.6	100.6
VP5	Turks & Caicos Islands, Grand Turk	21.4N	71.2W	155.5	118.1	102.5
VP8	Falkland Islands, Stanley	51.7S	57.9W	166.5	156.3	147.0
VP8	So. Georgia Is.	54.3S	36.8W	156.0	147.8	140.4
VP8	So. Orkney Is.	60.6S	45.5W	163.3	155.9	149.1
VP8	So. Sandwich Islands, Saunders Is.	57.8S	26.7W	153.4	146.8	140.8
VP8	So. Shetland Is., King George Is.	62.0S	58.3W	169.7	161.9	154.7
VP9	Bermuda	32.3N	64.7W	117.2	91.7	84.0
VQ9	Chagos, Diego Garcia	7.3S	72.4E	44.5	18.2	344.8
VR2,VS6	Hong Kong	22.3N	114.3E	345.2	327.6	313.3
VR6	Pitcairn Is.	25.1S	130.1W	224.9	210.8	193.9
VU	India, Bombay	19.0N	72.8E	28.7	10.3	350.3
	Calcutta	22.6N	88.4E	12.0	353.2	335.4
	New Delhi	28.6N	77.2E	21.0	4.6	347.7
	Bangalore	13.0N	77.6E	26.3	5.5	343.7
VU	Andaman Islands, Port Blair	11.7N	92.8E	8.9	346.3	326.3
VU	Laccadive Is.	10.0N	73.0E	32.7	11.8	348.5
XE	Mexico, Mexico City (XE1)	19.4N	99.1W	224.1	183.3	139.9
	Chihuahua (XE2)	28.7N	106.0W	250.1	218.0	140.9
	Merida (XE3)	21.0N	89.7W	206.4	154.8	122.5
XF4	Revilla Gigedo	19.0N	111.5W	241.4	215.5	168.2
XT	Burkina Faso, Ouagadougou	12.4N	1.6W	87.7	76.3	64.5
XU	Cambodia, Phnom Penh	11.7N	104.8E	354.0	332.1	314.6
XW	Laos, Viangchan	18.0N	102.6E	357.1	337.1	320.3
XX9	Macao	22.2N	113.6E	345.9	328.2	313.8
XZ	Myanmar, Yangon	16.8N	96.0E	4.6	343.8	325.7
YA	Afghanistan, Kandahar	31.0N	65.8E	29.5	14.7	358.4
	Kabul	34.4N	69.2E	25.4	11.0	355.6
YB-YD	Indonesia, Jakarta	6.2S	106.8E	348.0	318.5	299.8
	Medan	3.6N	98.7E	1.9	335.8	315.4
	Pontianak	0.0	109.3E	345.7	320.0	302.5
	Jayapura	2.6S	140.7E	304.8	288.4	276.6
YI	Iraq, Baghdad	33.0N	44.5E	43.7	31.3	16.8
YJ	Vanuatu, Port Vila	17.7S	168.3E	269.6	258.0	247.4
YK	Syria, Damascus	33.5N	36.3E	48.5	36.9	23.4
YL	Latvia, Riga	57.0N	24.1E	36.0	28.8	21.0
YN	Nicaragua, Managua	12.0N	86.0W	192.4	154.4	128.2
YO	Romania, Bucharest	44.4N	26.1E	46.0	36.6	26.0
YS	El Salvador, San Salvador	13.7N	89.2W	199.7	159.8	130.6
YU	Yugoslavia, Belgrade	44.9N	20.5E	48.2	39.2	29.2
YV	Venezuela, Caracas	10.5N	67.0W	155.1	126.5	110.3
YVØ	Aves Is.	15.7N	63.7W	145.1	117.0	102.7
Z2	Zimbabwe, Harare	17.8S	31.0E	91.0	80.2	61.7
Z3	Macedonia, (ex Yugoslav), Skopje	42.0N	21.4E	50.2	40.7	30.2
ZA	Albania, Tirane	41.3N	19.8E	51.6	42.1	31.6
ZB2	Gibraltar	36.1N	5.4W	68.1	58.9	50.0
ZC4	British Cyprus	34.6N	33.0E	49.7	38.5	25.5
ZD7	St. Helena	16.0S	5.9W	112.4	101.5	89.9
ZD8	Ascension Is.	8.0S	14.4W	112.3	100.2	88.7
ZD9	Tristan da Cunha	37.1S	12.3W	131.7	122.8	113.7
ZF	Cayman Is.	19.5N	81.2W	183.2	137.0	114.2
ZK1	No. Cook Is., Manihiki	10.4S	161.0W	256.3	244.5	229.9
ZK1	So. Cook Is., Rarotonga	21.2S	159.8W	247.3	235.9	222.3
ZK2	Niue	19.0S	168.9W	254.6	243.4	230.8
ZK3	Tokelaus, Atafu	8.4S	172.7W	265.3	253.9	240.9
ZL	New Zealand, Auckland (ZL1)	36.9S	174.8E	247.2	238.1	229.3
	Wellington (ZL2)	41.3S	174.8E	242.4	234.1	226.0
	Christchurch (ZL3)	43.5S	172.6E	240.7	233.0	225.5
	Dunedin (ZL4)	45.9S	170.5E	238.6	231.4	224.6
ZL5	Antarctica, Scott Base	77.9S	166.4E	195.4	195.4	194.7
ZL7	Chatham Is.	44.0S	176.5W	236.1	228.0	219.5
ZL8	Kermadec Is.	29.3S	177.9W	251.0	240.8	230.3
ZL9	Auckland & Campbell Is., Auckland	50.7S	166.5E	233.5	227.8	222.2
	Campbell Is.	52.5S	169.1E	230.5	225.1	219.6
ZP	Paraguay, Asuncion	25.3S	57.7W	158.4	142.5	129.4
ZS	South Africa, Cape Town (ZS1)	33.9S	18.4E	113.0	105.9	95.2
	Port Elizabeth (ZS2)	34.0S	25.7E	109.6	102.9	91.1
	Bloemfontein (ZS4)	29.2S	26.1E	104.6	96.6	82.7
	Durban (ZS5)	29.9S	30.9E	102.7	94.9	79.8
	Johannesburg (ZS6)	26.2S	28.1E	100.6	91.8	76.2
ZS8	Prince Edward & Marion Is., Marion Is.	46.8S	37.8E	120.3	119.7	118.4

Chapter 16

Online Resources

Enriching Your Radio Experience

Stan Horzepa, WA1LOU
One Glen Avenue
Wolcott, CT 06716-1442
E-mail wa1lou@arrl.net
http://www.tapr.org/~wa1lou

Unlike the rest of *The Operating Manual*, this chapter is about online communications, rather than on-the-air communications.

As its name implies, online communications uses lines and wires for communications. I know you're saying to yourself that the typical ham radio shack is a rat's nest of lines and wires already. "If that isn't online communications, then what is?"

Eventually all those lines and wires in the ham shack lead to an antenna. The communications being transferred over lines and wire is relayed by the antenna to the ether and at that point, becomes on-the-air communications.

With online communications, there is no antenna, no transfer to the ether. The lines and wires start at your computer and just keep on going! They exit your house and via telephone poles or underground conduits, the wires go down the street, cross town, cross state and maybe cross country until they interface with the worldwide network of lines and wires known as the Internet.

Besides bigger bombs and faster bomb delivery, the Cold War also was responsible for the birth of the Internet.

Back in the Sixties, when every radar blip crossing to this side of the DEW Line was considered a potential Russian invasion, the Department of Defense (DoD) decided that it had to protect its computer network from annihilation. The DoD wanted to build a "doomsday" network that would continue to function in the event of a nuclear war.

The Defense Advanced Projects Research Agency (DARPA, formerly ARPA) built the doomsday network and it was called ARPANET. When it was completed, the network crisscrossed the country interconnecting the DoD and the scientific and academic facilities doing DoD research.

Back in the Eighties, the National Science Foundation (NSF) used the ARPANET technology to build NSFNET, a network that interconnected scientific and academic facilities doing research for free.

Build it for free, and they will come!

They came and NSFNET grew tremendously. As the network grew, NSF stepped backed from running it as commercial firms took over building and maintaining the network. The commercialization of the network opened the door for everyone in the world to connect to the network, not just the scientific and academic researchers. That is how NSFNET became what is known today as "the Internet."

WHAT'S THE INTERNET?

The Internet is an aggregation of lines, wires, cables and communication devices that interconnect computers throughout the world to permit those computers and their users to interact with each other.

Internet interactions include:

- Sending and receiving mail electronically
- Obtaining and distributing information and computer software
- Real-time participation in round-table discussions

WHY THE INTERNET?

Internet interactions have been used for Amateur Radio applications even before the Internet was called "the Internet." After all, many of the computer network pioneers were ham radio operators. The computer networks they were pioneering provided a natural means of augmenting their ham radio operations.

In the late 1970s, the first home computers became available and hams, being the tinkerers we are, were among the first buyers, users and hackers of home computers. Build a gadget and hams will find a use for it. And that we did.

Today, computers have become an integral tool in many aspects of Amateur Radio. Like other tools we use in this hobby, computers have not replaced Amateur Radio; they have augmented Amateur Radio. And, like the computer, the Internet is another tool that augments our hobby. As the Internet expanded, its application to Amateur Radio also expanded and today, the Internet is an important implement in the toolbox of those hams choosing to use it.

To get the most out of their hobby, hams require a lot of information. Hams ask, "How do I modify my transceiver for 9600 packet? Where is the gray line? How do I QSL WA1LOU? What is my grid square? Where can I download the latest version of APRS? How can I make a schedule to work Clipperton on Christmas Eve?"

You can look it up if you have the right book or magazine, assuming that the information was published and you happen to have that publication handy. You can figure it out with your calculator if you have the correct formula. You can mail a blank disk and an SASE to the guy who wrote the software. You can write a letter to the Clipperton DX Club to set up a sked for December 24.

Or you can use the Internet!

The Internet is an information tool, probably the most powerful information tool in existence. And it is getting more powerful all the time.

It is powerful because it is so accessible. You can access the Internet from a computer sitting on your desk (right next to your radio) or from a computer sitting in your lap, which is remotely connected to the Internet via an RF link.

It is powerful because it accesses the world. The Internet literally puts the world at your fingertips. If the information you seek is stored on a computer anywhere in the world that is interfaced to the Internet, then that information is accessible to you. Similarly, if the information you seek is stored in the brain of a computer user anywhere in the world who has access to the Internet, then that information is accessible to you, too (assuming the user is willing to share that information).

It is powerful because you can access information that you cannot find anywhere else. The cost of publishing information on a Web page is relatively inexpensive when compared to publishing information in print. As a result, information whose cost-effectiveness precludes book or magazine publication may show up on a Web page instead. For example, not many publishers would be willing to devote a magazine article, much less a book on how to connect and configure brand X transceiver to brand Y TNC and brand Z GPS for APRS. However, you may be able to find a Web page or two devoted to that topic!

The Internet has Amateur Radio applications because there is a huge amount of Amateur Radio information available from those same computers (and computer users) that are interfaced to the Internet.

The purpose of this chapter is to describe how to use the Internet as a tool for Amateur Radio communications.

HOW TO INTERNET

Most hams use the Internet by connecting their computer to a modem, which is connected to a phone line. They subscribe to a service, which provides access to the Internet via this connection and often provides the software to do so.

Hardware Requirements

You can use any computer on the Internet, as long as it has a serial port or modem port. However, faster computers with oodles of read-only memory (RAM) provide better Internet performance than slower computers with minimal amounts of RAM. Let me explain why.

High-speed modems are inexpensive today (about 2 to 3 cents for each 10 bit/s). For less than $100 you can buy a 56k-bit/s modem, which is the data rate du jour. At that price, why buy a slower modem? And even if you decided to buy a slower modem to save a few dollars, you'd be hard pressed to find a slower one for sale unless you go the used modem route. (Perusing a handy computer catalog, I could not find a modem for sale that was slower than 56k.)

But if your computer is slow and RAM-impaired, it does not matter how fast your modem works. Slow computers can't handle the throughput of fast modems. A fast modem has to slow down while it waits for a slow computer to catch up. In order to take advantage of the high data rates provided by a fast modem, you need a fast computer, so don't scrimp on speed and memory if and when you are shopping for a computer.

Usually you can have your modem in two ways: outside the computer or inside the computer. Not all computers permit installation of a modem internally, but those that do, offer some advantages over external installation. An internal modem eliminates the need for a serial cable (for the computer to modem connection) and a power cable and power cube for powering the modem, not to mention eliminating the need for a power outlet for the power cube.

Online Services and Internet Service Providers

For most individuals, the two most common ways of accessing the Internet are by means of an online service or an Internet Service Provider (ISP).

Online services, such as America Online, CompuServe, Prodigy, and so on, are self-contained "Internets" that provide services in a user-friendly environment that is insulated from the real Internet. These services include e-mail, online chatting and shopping, software downloading, newsgroup and special interest groups, Web page hosting. . . services that are similar to the services that exist on the real Internet. Another service that an online service provides is access to the *real* Internet.

An ISP provides direct access to the real Internet. The main advantage of using an online service is that it allows a novice to get online without being familiar with the intricacies of the Internet. The disadvantages of an online service include:

- Difficulty accessing the service during peak hours because the service typically has a large subscriber base, as compared to most ISPs, which have a relatively small user base.
- Slow access to the real Internet through the service because Internet access is competing for computer time with all the other services that the online service provides. In contrast, an ISP's primary service is to provide access to the Internet, so there are no other significant services competing for computer time.
- Problems displaying Web pages on the real Internet are possible because the service typically requires you to use the service's proprietary software, which is not always compatible with all Web page bells and whistles. ISPs

allow their customers to use brand name browsers that are fully compatible with all the bells and whistles likely to be encountered on a Web page.

- Higher online service costs. In the past, most online services charged by the minute, whereas ISPs typically charge a flat fee for unlimited access. In order to be competitive with ISPs, many online services now provide unlimited access for a flat fee.

The main disadvantage of using an ISP is that it does not provide a safety net for novice users, ISP customers need to have some familiarity with the Internet. So, my recommendation is to cut your teeth on the Internet using an online service, then switch to an ISP as soon as possible after you feel comfortable using the Internet.

Try to find an ISP that provides local access in order to avoid long distance telephone toll charges. Check the yellow pages of your telephone book under "Internet Access Providers." Another place to find ISPs is the listings and advertisements appearing in the computer/Internet section that a lot of newspapers are publishing regularly.

Software Requirements

Whether you choose an online service or ISP, the software you need is usually provided by the service. An online service will likely provide you with an all-in-one program that allows you to use all of the services that the online service has available. An ISP will likely provide you with a brand name Web browser, like *Netscape Navigator* or *Microsoft Internet Explorer*, along with companion e-mail software, such as *Netscape Messenger* or *Microsoft Outlook*.

When you begin doing business with an ISP, there may or may not be a sign-up fee for setting up your account and providing you with the software you need to access the Internet from your computer. The quantity and quality of the ISP-provided software varies, but at a minimum, it should get you up and running on the Internet. If need be you can always upgrade the software later.

Your ISP should also provide you with the information you need for configuring the software (for example, IP address, host name, initial password, POP server, SMTP server) with directions on how to input the information into the software. Even on user-friendly computer systems like the *Mac OS* and *Windows*, this process is daunting to the first-time user. So, if you are having trouble, call the technical support department of your ISP and have them step you through the software configuration process.

You may not need all the software your ISP provides depending on how you plan to use the Internet. The following lists the type and function of software that your ISP may provide. Some or all these functions are combined in a Web browser supplied by your ISP.

Web Browser to locate and display Web pages. Most Web browsers include some or all the functions of the other software types in this list.

Mailer to send and receive e-mail via the Internet. Most Web browsers include a mailer function.

News Reader to read the messages published in the various special interest newsgroups that proliferate on the Internet. Pick a subject and there is likely to be a newsgroup for it. Most Web browsers include a news reader function.

FTP (file transfer protocol) software to transfer files to or from another computer on the Internet. Most Web browsers include this function.

Conferencing software allows you to participate in real-time conferences in which you exchange keyboarded messages or live audio and/or video. Some Web browsers incorporate conferencing features, but many people continue to use standalone conferencing software for these applications.

Telnet software to access the Internet with a command line interface. Using *UNIX* commands, you may perform some of the functions of the software types listed above. Telnet is recommended for the advanced Internet user. An important Internet capability in the past, Telnet is less so these days as fewer sites allow casual net occupants to really do very much on their computers via Telnet. As a result, few Web browsers include this function. Don't worry. You can probably live, and live very well, without Telnet capability!

You will also need software to log onto your ISP. With computers running *Windows*, your ISP should provide you with this "front-end" software. With Macintosh computers, the software is already included with the *Mac OS* and it is called *PPP* or *Remote Access*, depending on the version of *Mac OS* you are using.

USING THE INTERNET

Think of the Internet as another communications service, just like the telephone or the U.S. Postal Service. In order to use these services, you need not be concerned with the inner workings of the telephone network or postal system. In either case, communications commences by addressing the party you wish to communicate with (via a telephone number or street address). Similarly, communications commences on the Internet by addressing the party or Web site you wish to communicate with.

If you wish to communicate with an individual, you need to know the e-mail address of the individual. A typical e-mail address is composed of the following components:

- User name, for example, the user's actual name, a nickname, or any other identification that the user chooses to use that is acceptable by the entity providing the e-mail service. For example, there is often a limitation (eight characters) to the length of the user name.
- At-symbol (@), which divides the e-mail address in two.
- Name of the Internet site (known as the "domain name") where the individual has an e-mail account, followed by a decimal point/period (.), commonly known as a "dot." The domain name is typically an acronym of the entity that owns the computer handling the user's e-mail.
- A three-letter suffix that indicates the domain of the Internet site, that is, the type of entity that is operating the site; for example, **com** for a company, **edu** for an educational institution, **gov** for a government entity, **net** for a network or **org** for an organization.

For example, **quentin_pie@dot.com** may be Quentin Pie's e-mail address where he is employed (at the Dot Company), whereas **qt_pie@online.net** may be Quentin's e-mail address at home where he accesses the Internet via an ISP called Online.

If you wish to browse a Web page, you need to know the Uniform Resource Locator or URL, which is commonly pronounced "earl." Typically, the basic address of a Web page is composed of the following components:

- Initials http (for HyperText Transport Protocol) followed by a colon (**http:**). This indicates the type of protocol used for Internet communications. For browsing a Web site, you use **http:**, but for transferring files from a Web site, you would use **ftp://** (for File Transfer Protocol).
- Two forward slash bars (//) to indicate that the domain name follow.
- Domain Name: Prefix www (for World Wide Web) followed by a dot (**www.**) often is found at the beginning of the domain name, but this prefix is not mandatory.

The identification of the Web site, followed by a dot (**domain_name.**) The centerpiece of the domain name is unique and usually an acronym of the organization that owns the remote computer.

Suffix indicating the domain of Internet site followed by a forward slash (**com/**, **edu/**, **gov/**, **net/** or **org/**).

After the basic address, there may be additional components separated by forward slashes. These components indicate the path to specific Web pages at the Web site. For example, the URL **http://www.arrl.org/** is the address for the "home page" of *ARRLWeb*, whereas the URL **http://www.arrl.org/arrlletter/** is the address for the news page on *ARRLWeb*.

Electronic Mail

To send and receive electronic mail or "e-mail" to other users via the Internet, you use mailer software. If you access the Internet via an ISP, you can use the built-in or companion mailers of a Web browser such as *Netscape Navigator* or *Internet Explorer*. If you access the Internet via an online service, the service's proprietary software usually includes a mailer function.

If you plan to do little or no Web browsing, you can use a standalone mailer, such as *Eudora*, to send and receive mail via an ISP. Then, you can avoid the time it takes to load a Web browser just to do e-mail as well as conserve RAM and disk space occupied by a browser.

Sending e-mail is easy. Start by invoking the new mail message command. When a blank message template appears, enter the e-mail address of the intended recipient in the address field and the subject of the e-mail in the subject field, then enter the contents of the e-mail in the wide open field intended for the message contents. After you compose the message, you may send it immediately or save it for later, that is, until the next time you connect to the Internet.

Reading e-mail is easier than sending it. Invoke the receive or get mail command and if your computer is not connected to the Internet, the mailer initiates a connection. Once connected, the mailer checks the mail server for any new e-mail addressed to you. If new e-mail exists, it is transferred to your mailer and included in the list of e-mail that appears in the inbox of your mailer. Select a message to read from the inbox, then invoke the read mail command and the mailer displays the contents of the selected e-mail on your computer display. Typically, simply double-clicking on a message in the inbox will display its contents.

You can respond to received e-mail without entering an address or subject by using the reply function of the mailer. While the message you wish to respond to is displayed, invoke the reply command. This sets up a blank message template that is automatically addressed back to the sender of the original message. Enter your response and send it on its way.

When you use the reply function, most mailers automatically include all or a selected portion of the original e-mail contents in your reply in order that you may comment on something in the original e-mail. (This function is usually an option that you can configure in the mailer.) For clarity, the mailer differentiates the original e-mail contents from your response by placing a non-alphanumeric character, such as the greater-than symbol (>) at the beginning of each line of the original message.

Thus,

What — me worry?

in the original e-mail becomes

> What — me worry?

in your response.

Mailing Lists and Newsgroups

Mailing lists and newsgroups are ongoing discussion groups where ideas are exchanged about a particular topic of interest. There are thousands of these groups and as a result, if you can think of a topic, there is likely a mailing list or newsgroup devoted to that topic. In some cases, there may be more than one list or newsgroup devoted to a particular topic.

Mailing Lists

Mailing lists or *reflectors* are *private* ongoing discus-

ARRL E-Mail Address Alias Service

How would you like an e-mail address consisting of your call sign @arrl.net? If you are an ARRL member, it is possible to have such an address because the League now provides an e-mail forwarding service as a membership benefit.

This service is a forwarding (or "alias") service only. No messages will be stored on the ARRL servers. E-mail sent to you at arrl.net will be forwarded to the real e-mail address that you provide.

To sign up for this service, go to the ARRL Members Only Web page (**http://www.arrl.org/members/**). If you are accessing ARRL Members Only for the first time, you will need your ARRL membership number to log on (the number appears on your *QST* mailing label).

Not only will this new e-mail address identify you as a ham, but it likely will be easier to remember than your real e-mail address (for example, compare **wa1lou@arrl.net** to **stanzepa@ct2.nai.net**). Just think how simple it would be to send e-mail to another ham, if all hams used this service. If you know the other ham's call sign, you'd know his/her e-mail address as well.

sion groups that use e-mail to exchange messages concerning a particular topic of interest. The discussions are private because you must be a subscriber to the list in order to view its contents. In some cases, you must also be a subscriber to send e-mail to the list.

The power of the mailing list is that each message addressed to the list is disseminated to everyone who is a subscriber to the list. The message dissemination is performed automatically by the list server, which is software running on a computer. Whenever you want to send a message to the list, you address it to the list server, and the server resends it to all the list subscribers.

Usually, you do not have to be a subscriber to a mailing list to send a message to a list. One way to find out how to subscribe to a list is to send a message addressed to the list asking how to subscribe. In general, subscribing to a mailing list requires that you send e-mail addressed to the list server requesting a subscription. The exact subscription process varies with each mailing list and since you are dealing with a computer, you must follow the subscription process exactly; otherwise, the computer rejects your request.

After you e-mail your subscription request, the list server sends you a confirmation along with information on how to unsubscribe from the list, and how to send (post) messages to the list. Save this file for future reference.

Newsgroups

Newsgroups are *public* ongoing discussion groups concerning a particular topic of interest. The discussions are public because anyone on the Internet with news reader software can read the contents of the newsgroup as well as participate in discussions on the newsgroup.

There are thousands of newsgroups. You can use your news reader software to obtain a list of all the newsgroups or to search for newsgroups by name; for example, you can command the news reader to find all newsgroups whose names contain the word "radio."

Newsgroups fall into eight categories:

- **comp** for computer topics
- **news** for discussions about newsgroups
- **rec** for recreational topics
- **sci** for scientific topics
- **soc** for social science and cultural topics
- **talk** for controversial topics
- **misc** for everything else
- **alt** for newsgroups that are "alternative" to the newsgroups in the other seven categories

The name of each newsgroup begins with a category, followed by one or more dots that delineate a topic hierarchy. For example, the newsgroup **rec.radio** may be concerned with general radio topics, whereas, the newsgroup **rec.radio.amateur** may be concerned with Amateur Radio topics, and **rec.radio.amateur.antenna** may be concerned with Amateur Radio antenna topics.

After you find a newsgroup name that sounds like it may be of interest, you can use your news reader to obtain a list of the discussions or *articles* that are currently posted on the newsgroup. After you have a list of articles, you can view the contents of a particular article or article *thread*. A thread contains each newsgroup posting that is a reply or follow-up to a particular article. So, when you view a thread, you not only view the original article that started the thread, but you are able to read everyone's comments and replies to that article.

Your news reader allows you to add your two cents by posting your comments to an existing thread or by posting an article, thus starting a new thread.

You can also use your news reader to *subscribe* to one or more newsgroups. Subscribing to a newsgroup simplifies the process of obtaining lists of newsgroup articles. When you are subscribed to a newsgroup, your news reader automatically checks for new articles and threads posted to those newsgroups each time you start the news reader software.

Mailing Lists vs. Newsgroups

Mailing lists and newsgroups differ in that a mailing list, by virtue of its subscription feature, offers more privacy than a newsgroup. As a result, there is much less intentionally disruptive traffic (*spamming* and *flaming*) on a mailing list than on a newsgroup. The privacy of the mailing list deters transients, who intentionally spam and flame newsgroups, from doing the same on a list. Unless the spammer or flamer subscribes to a list, he cannot read the result of his spamming/flaming and thus has little incentive to spam and flame.

In defense of newsgroups, some newsgroups are *moderated*; that is, someone is responsible for reading and approving each article submitted to the newsgroup. As a result, a moderated newsgroup is less susceptible to spamming and flaming than a newsgroup that is not moderated. Another advantage to a moderated newsgroup is that the moderator can keep discussions on-topic. One of the prime causes of flaming is when an article is posted to a newsgroup that is off-topic. So, a newsgroup moderator is able to kill two birds with one stone.

Web Browsing

To view Web pages on the Internet, you use a Web browser. If your ISP provides you with any software, that software, at a minimum, likely includes a Web browser. The two most popular browsers are *Netscape Navigator* and *Internet Explorer*, which you can download from **http://www.netscape.com/** and **http://www.microsoft.com/** respectively.

If you are using an online service rather than an ISP, its proprietary software likely includes a Web browsing function. The purpose of the Web browser or Web browsing function is to connect your computer to a remote computer site and transfer the contents of a selected Web page stored at that site to your computer.

To connect your computer to a remote Web site and transfer the contents of a selected Web page stored at that site to your computer, you input the URL of the Web page into the browser. The URL is used to locate the remote Web site and the selected Web page stored at that site. Once the site and page are found, the contents of the selected page are transferred over the Internet to your computer.

Some Web pages appear almost instantly. Others seem

Computer on Fire

My computer caught fire the other night!

I have been writing for *QST* for over 20 years and during that time, I have received my share of less-than-complimentary correspondence. Typically, someone has a nit to pick and if I write contrary to the way they pick at that nit, they complain to me or go over my head and complain to my boss.

In the past, such complaints were few and far between, but lately, they have been arriving more frequently. Some may say that this indicates that I have become more offensive in my old age, but I think the cause of increased negative traffic is the result of our technology, which has made it too easy for human torching.

Someone reads something that gets him hot and bothered, so he jumps on his mailer, composes a flame and sends it on its merry way. Flames can be originated so quickly that the flamer has no time to think about what he/she is doing. Just write what is on your mind and push the send button. There is no time to think about whether the flame is rude. No time to think whether the flame is correct. No time for nothing!

When people had to put pen or typewriter keys to paper and compose a letter, there was more time to think about what one was doing. Even after you finished writing a complaint letter, you could still reconsider sending it because you still had to address the envelope and post the letter. You had a lot of time to reconsider your actions and decide whether you were doing the right thing.

Today, it is too easy to not do the right thing, to not think about what you are doing and the effect it will have. That's why *netiquette* is important.

Whenever I have the urge to flame someone, I may sit down at my computer keyboard and compose the flame, but I do not send it for 24 hours (more or less). During that 24-hour cooling-off period, I have time to reconsider what I wrote and if I really feel strongly about my actions and am cognizant of the results they may have, only then do I send it.

You know what? I seldom e-mail a flame. I estimate that I e-mail one flame for every twenty that I write. The 19 that I don't e-mail seem less important, less urgent, more frivolous and possibly rude after 24 hours. Did I write that? What a difference a day makes!

like they take an eternity to appear. The data rate that your modem uses to communicate with your ISP or online service is the most obvious factor affecting the display speed of a Web page. The faster the data rate, the faster the Web page appears. Note that just because you have a high-speed modem, there is no guarantee that it is operating at its top speed. Usually it is not. Instead, it adapts its data rate for the current telephone line conditions and the data rate of the modem at the ISP or online service.

The more graphical the Web page, the longer it takes to transfer its contents and display it on your computer. Web pages that contain only text appear very quickly. Web pages that contain graphics make you wait.

The configuration of your computer also affects the display speed of a Web page. The clock speed of your computer is a factor, but other factors also come into play. Some computers handle the video display functions more efficiently and quickly than other computers. A computer that is advertised as optimized for multimedia is likely to display a Web page faster than a plain vanilla computer. And so it goes.

Links

When Apple introduced its Macintosh computer in 1985, its graphical user interface (GUI) and hypertext-linked software received a lot of attention. The Macintosh GUI and hypertext links, as embodied in the Macintosh software called *Hypercard*, were very user-friendly when compared to the command line interface that was the run of the mill of the majority of computers up until then (this was long before *Windows* came along). Hypertext links were the epitome of user-friendliness.

Reread the previous paragraph pretending that what you are reading is displayed on a Macintosh computer and the term "GUI" has a hypertext link. Suppose you found that you were unsure as to what GUI meant and that you wanted more information about that term. You could use your mouse to click on GUI and the hypertext link would bring up a new window that provided more information, maybe even a graphic, explaining GUI.

This was magic. In 1989, a Swiss physicist, Tim Berners-Lee, wrote a paper proposing a protocol that used a graphical user interface and hypertext links to transfer, navigate and display the information stored in computers. His protocol became the HyperText Transfer Protocol (http) and within the following year, the term World Wide Web (www) was coined and the first Web browser software was written. It had a GUI and used hypertext links for navigation. Web pages began to appear on the Internet, and the rest is history.

Both text and graphics can have a hypertext link. Text links are easy to spot because they are typically underlined and appear in a different color than the surrounding text (you can select the color in your Web browser options). Graphic links are less obvious.

A sure-fire way of finding a link (text or graphic) is to move the mouse pointer around the Web page. When the mouse arrow changes to a hand with an extended index finger, a link is under that finger. Pressing the mouse button when the extended index finger is displayed activates the link and replaces the currently displayed Web page with a different Web page or a document.

In addition to links that open documents or new Web pages, you can use other types of links for navigation. For example, there are navigational links that permit you to navigate within the currently displayed Web page. These links work faster than the scroll function of your browser and are often contained in the Web page table of contents, if any, (to move you to a specific area of interest that is listed in the table of contents). A navigational link may be located at the bottom of a page (to get you back to the top).

By the way, most Web pages are larger than the default window provided by your Web browser, so open the window as much as possible (both vertically and horizontally) to permit the Web page to fill your monitor.

Another type of link you are likely to find on many Web

pages is an e-mail link. An e-mail link invites you to send e-mail to whoever or whatever is responsible for the Web page. You can use this link to comment or ask questions about the Web page. When you click on an e-mail link, your Web browser opens an e-mail message composition template already addressed to the party responsible for the Web page, waiting for you to enter your comments, questions, etc.

Saving URLs

Web browsers allow you to save the URL of any Web page you wish to revisit. In *Netscape Navigator*, the saved URLs are called *bookmarks*, while in *Internet Explorer*, they are called *favorites*.

To save the URL of a Web page that is currently displayed in your Web browser, use the Add Bookmark command in *Netscape Navigator* or use the Add to Favorites command in *Internet Explorer*. After you have added a bookmark or favorite, you can recall it from the Bookmark or Favorites menu.

Plug-Ins

Plug-ins are mini-applications that add functions to your Web browser like the ability to play music, view movies and multimedia. Typically, if you view a Web page that requires a plug-in in order to view the page or some component of the page, you are prompted if you would like to download the plug-in and install it (plug it into) your Web browser.

Many plug-ins are free. Some of the free plug-ins have enhanced versions that you must purchase. A free plug-in that I highly recommend obtaining is the Adobe *Acrobat* plug-in for displaying documents saved in Adobe's "portable document format" (pdf). This free plug-in may be downloaded from Adobe's Web site (**http://www.adobe.com/**).

The portable document format provides a universal means of viewing documents created by different applications on different computer platforms. For example, a file created in Microsoft *Word* on a Macintosh computer may be displayed on a *UNIX* workstation if the *Word* file was saved as a pdf file on the Mac and there is an Adobe *Acrobat* program installed on the *UNIX* workstation.

The universality of the portable document format makes it a natural format for the documents that you can access on a Web page. Sooner or later (most likely sooner), you are likely to encounter a pdf file, so be ready for it by downloading and installing the plug-in as soon as possible.

Searching

There is lots of stuff stored on the Internet. The problem is finding it! There are two ways: search engines and directories. Search engines prompt you to enter one, two or more words into a search field. In response, the engines return a list of the hypertext-linked Internet locations where the words you entered were found. Each engine has unique options for more intelligent searches, so check out the help information for each search engine before you initiate a search (and end up with a lot of useless information).

Directories present you with a hierarchy of topics. You begin a search by selecting a general topic and working down through a continuing narrower hierarchy of topics. At the end of the search, you find a list of the hypertext-linked Internet locations related to the topic.

Using search engines and directories is an art. Sometimes I have a good idea where the information is located, so I go to a particular Web site (such as **http://www.arrl.org/**) and use the search facility of that site to find the answer. Other times, I don't have a clue where to look and start my search using an Internet search engine or directory.

Just the other day, someone asked if a packet data rate of 9600 was permitted on 6 meters. I was sure it was, but could not remember the permissible maximum data rate on 6. So, I went to **http://www.arrl.org/**, and when the ARRLWeb page appeared, I clicked on its Search button, typed "data rate" in the Search field, and the *ARRLWeb* search engine came up with 53 matches. I clicked on the first match, titled "Part 97—Subpart D," and the search engine presented me with section 97.305 of the FCC rules, which enumerated the maximum data rates for all the Amateur Radio bands.

That was a piece of cake, but sometimes it is not that easy. Awhile back, I needed general information regarding packet radio meteor scatter on 6 meters. Since I am no expert on that subject and I know I did not have the answer in my computer or my archives, I decided to search the Internet.

My favorite Internet search engine is HotBot (**http://hotbot.lycos.com/**). I usually go there first when I am looking for a needle in a haystack. So, I took my 6-meter packet meteor scatter search there and the HotBot search engine came back with 98 matches, as illustrated in **Fig 16-1**. I also took my search to my favorite Internet directory, Yahoo (**http://www.yahoo.com/**). It came up with 222 matches when I typed "6-meter packet radio meteor scatter" in its search field, as illustrated in **Fig 16-2**.

Some of the HotBot and Yahoo matches were the same; that is, they listed links to the same Web pages. Most of the matches were not useful, but a handful of matches led me to

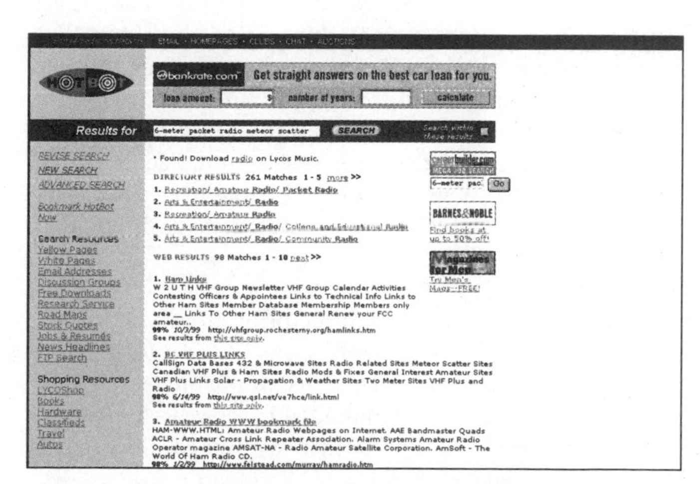

Fig 16-1—Wired Magazine's HotBot search engine found 98 Web pages on the topic of "6-meter packet radio meteor scatter."

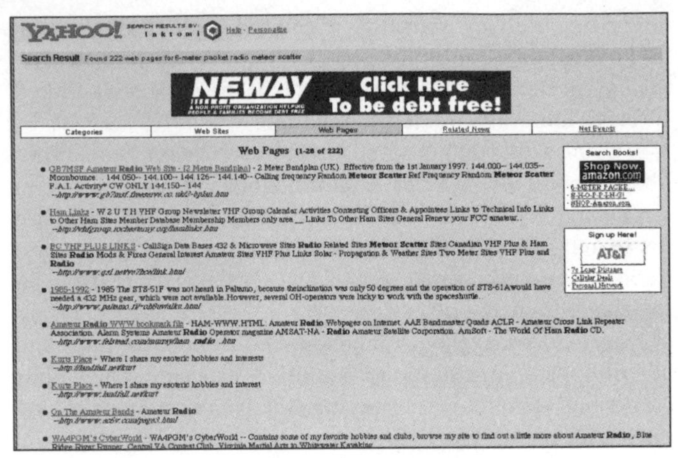

Fig 16-2—Yahoo's directory retrieved 222 entries from its database for "6-meter packet radio meteor scatter."

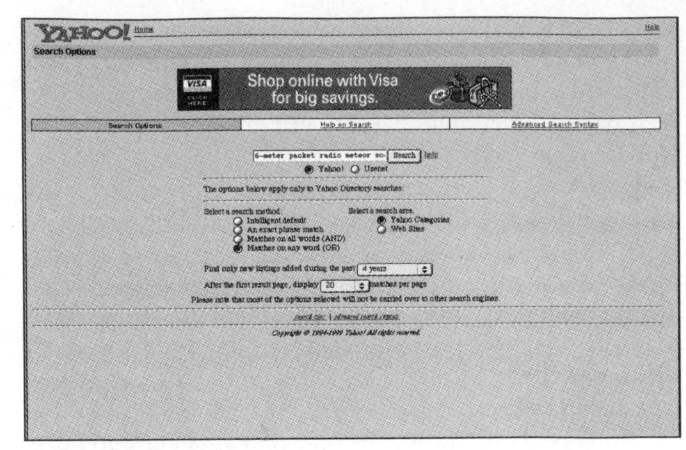

Fig 16-3—Some search engines and directories (like this one at Yahoo) have advanced search facilities that include the use of Boolean logic.

Web pages that contained pertinent information or links to other Web pages that were pertinent.

Search Engines vs Directories

As I said, the search facilities of *ARRLWeb* and HotBot use search engines. Search engines use *spider* software to capture information concerning the content of Web pages. This information is used to create an index of Web page contents. When you use a search engine, the engine searches the contents of its index for the information you seek and provides you with a list of links to the pertinent Web pages.

ARRLWeb and HotBot differ in that the *ARRLWeb* search engine indexes only the contents of all the Web pages on *ARRLWeb*, while the HotBot search engine indexes the contents of all the Web pages in the whole World Wide Web.

Another search engine that is similar to the *ARRLWeb* search engine is at QSL.NET (**http://www.qsl.net/**). QSL.NET is a site that provides free space for Amateur Radio Web pages, and its search engine indexes the thousands of pages at that site. Other search engines such as HotBot that index the whole World Wide Web are AltaVista (**http://www.altavista.com/**), Dogpile (**http://www.dogpile.com/**), Excite (**http://www.excite.com/**) and Lycos (**http://www.lycos.com/**).

World Wide Web "directories," such as Yahoo, are different than search engines. Instead of searching Web pages to create a searchable index, directories are created from information submitted by someone familiar with the Web page or pages. When you use a directory, you actually search the directory's database that is sorted according to the topic of the Web page. Some search engines, like AltaVista, Excite, HotBot and Lycos, also have directories.

Fine Tuning Your Search

Most search engines and directories allow you to fine-tune your search. For example, some sites allow you to search for the exact phrase that you enter in the search field; i.e., if that exact phrase is contained in the search facility's index or database for a particular Web page, then that is considered a match.

Some sites, including Yahoo, also allow you to use Boolean searches, as illustrated in **Fig 16-3**. AND Boolean searches attempt to match all the words you enter in the search field; i.e., if all the words appear somewhere in the index or database for a particular Web page, then that is considered a match. OR Boolean searches match any of the words you enter in the search field; i.e., if one word you enter is found in the index or database for a particular Web page, then that is considered a match.

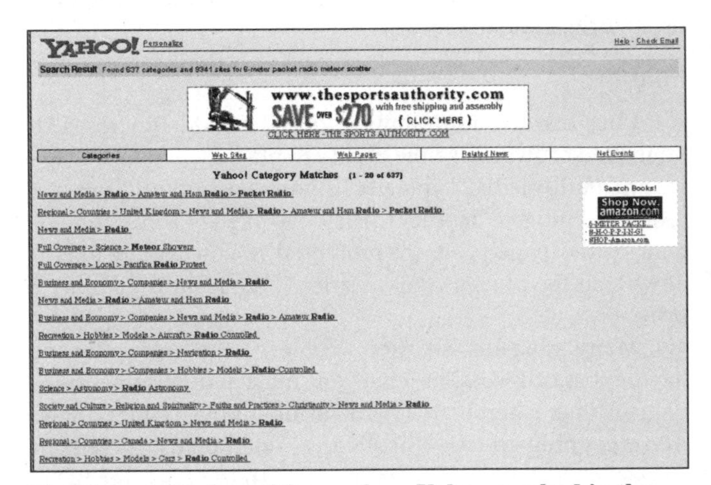

Fig 16-4—A Boolean OR search on Yahoo resulted in the retrieval of 637 database entries for "6-meter packet radio meteor scatter."

As you would expect, the exact phrase searches result in fewer matches than the Boolean searches, and AND Boolean searches result in fewer matches than OR Boolean searches. Searching for "6-meter packet radio meteor scatter" on Yahoo resulted in 0 exact phrase matches, 222 AND Boolean matches and 637 OR Boolean matches, as illustrated in **Fig 16-4**.

Using the exact phrase search, I found no matches, but what if the Yahoo database had an entry for "50-MHz packet radio meteor scatter"? Whereas the OR Boolean search would have matched the 50-MHz entry, the AND Boolean search would have missed it unless that entry also contained the phrase "6-meter" somewhere in the entry.

As this illustrates, you need a strategy when you search the

Web. For example, when I begin the hunt, I use the exact phrase search. If I get any matches, then I usually hit pay dirt. If not, I have a little more work to do. I can try exact phrase searching with a different but similar phrase or I can continue searching for the same phrase, but use the AND Boolean search instead, and if that fails, the OR Boolean search.

Like anything else, practice makes perfect, so, start searching!

File Transferring (FTPing)

Files are stored on computers that are connected to the Internet. These files may be computer programs or simply documents of interest. If the authors of these files desire to make them accessible to the public via the Internet, you may use the file transfer protocol (ftp) to copy those files to your computer.

You may use ftp software to perform the transfer or you may use ftp with a Web browser to perform the same task. In either case, you must know where the files are located; that is, you must know the computer directory path to the desired files.

To use ftp software to copy a file, you provide the software with the computer directory path to the files, then you command the software to connect the remote computer. If the software makes the connection, it will list all the files contained in the selected directory of the remote computer. Select the file you wish to transfer and command the software to copy or download it to your computer. When you are finished, close the connection.

To use a web browser to copy a file, you type **ftp://** in the Location or Address field above the page display window of the browser followed by the computer directory path to the files, then press **Enter**. If the browser makes the connection, it will list all the files contained in the selected directory of the remote computer. The file names will be hypertext links and clicking on the file name begins the process of copying or downloading the file to your computer.

For example, to download the latest version of *MacAPRS* from the ftp directories of the Tucson Amateur Packet Radio (TAPR) online computer, you first need the path to the files. In this example, the path is **ftp.tapr.org/aprssig/macstuff/MacAPRS**.

Using the *Windows* ftp software called *WS-FTP*, you click on the Connect button and type **ftp.tapr.org** in the Host Name/Address field of the General tab of the Session Properties window. Then, type **/aprssig/macstuff/MacAPRS** in the Initial Remote Host Directory field of the Startup tab of the Session Properties window. Click on the **OK** button to command *WS-FTP* to connect to the TAPR site. After the connection is made, *WS-FTP* displays the contents of the selected directory. Click on the file you wish to copy, the click on the arrow button that points to a directory on your computer to begin the file transfer. Click on the **Close** button to break the connection.

Using a Web browser, type **ftp://ftp.tapr.org/aprssig/macstuff/MacAPRS** in the Location or Address field above the page display window of the browser, then press **Enter** to command the browser to connect to the TAPR site. After the connection is made, the browser displays the contents of the selected directory. Click on the file you wish to copy and the browser will begin downloading the file as soon as you inform it where to store it on your computer.

Transferring files with a Web browser is easier than with ftp software when the computer directory path to the desired files is open to the public. However, in some cases the directory path is not open to the public and you must supply a user name and password in order to gain access to the files. In this case, you must use ftp software that allows you to type in the requisite user name and password.

Conferencing

Conference servers are computers connected to the Internet that allow you to participate in round-table discussions with other Internet users in real-time. Depending on the capabilities of the server, these discussions can be as simple as the exchange of text messages or can include live audio and/or video. The former only requires a computer keyboard and monitor, the latter requires a microphone and video camera interfaced to video conferencing software running on your computer.

With thousands of people using these servers, there is a tendency for the users to migrate into groups discussing topics of common interest. These groupings of common interest are known as *channels* or *chat rooms* and there are many devoted to various Amateur Radio topics.

IRC

Internet Relay Chat (IRC), with thousands of users exchanging real-time text messages, is the most popular conference server. To use IRC, you simply join an existing conference or start one of your own using a nickname to identify yourself. Hams typically use nicknames consisting of their call signs when participating in Amateur Radio-related conferences.

To find a conference that you may be interested in joining, you may use the search function of your conferencing software to look for conference names containing key words, like "ham" or "radio." The number sign (#) precedes the name of all conferences or channels. For example, **#hamradio** is the name of a channel that is always active. Normally, a channel disappears when there are no users logged into it, but some channels like **#hamradio** remain active by *robot* software when no human users are on the channel.

Telnetting

Before Web browsers, you used Telnet software to navigate the Internet. Using a command line interface, Telnet software is strictly text-based and, as a result, allows you to get around the Internet and perform tasks faster than the typical GUI intensive Web browser. However, this command line interface uses *UNIX* commands and thus turns off people who have no need to learn a computer language when they can use a relatively user-friendly Web browser instead.

Telnet still has some applications today. For example, you use Telnet software to access Internet-to-radio gateways discussed later in this chapter.

AMATEUR RADIO ON THE INTERNET

By now, you should have gotten the idea that Amateur Radio has a presence on the Internet. Considering the number of hams in the computer and communications fields, it is no surprise that Amateur Radio has had that presence for a long time. There are thousands of ham radio Web sites, mailing lists, newsgroups, software depositories, and links on the Internet (the Hot Bot search engine comes up with over 60,000 hits for "Amateur Radio," and over 48,000 hits for "ham radio"). This section describes what hams are doing on the Internet.

Mailing Lists and Newsgroups

There are a lot of Amateur Radio-related mailing lists and newsgroups. . .too many to list them all here. But, you can be sure there is a list or newsgroup that covers the niche of ham radio that interests you.

Rod Dinkins, AC6V, maintains a comprehensive list of Amateur Radio-related mailing lists and newsgroups on his Web page (**http://www.ac6v.com/**) and it is a good place to find a list or newsgroup for you.

There are Web sites that provide the resources for running mailing lists, in general. Among the hundreds of lists running on their computers, you will find numerous Amateur Radio-related lists. The University of California, San Diego (UCSD) site (**http://www-no.ucsd.edu/mail/listserv.html**) is non-commercial, whereas the Yahoo! Groups' site (**http://groups.yahoo.com/**) is commercial, so you can expect advertisements along with the messages you receive from such a list.

There are also Amateur Radio Web sites that provide mailing list resources: QSL.NET (**http://www.qsl.net**) and Tucson Amateur Packet Radio (TAPR at **http://www.tapr.org/tapr/html/sigf.html**). At QSL.NET, you will find ham radio lists devoted to a wide range of topics, whereas at TAPR, you will find lists devoted to digital ham radio topics, like packet, DSP, spread spectrum, APRS, etc.

You can also use your newsreader to search the Internet for newsgroups, whose names contain key words like "radio," "radio amateur," and "ham radio." A recent search resulted in the following list of newsgroups.

alt.ham-radio.amtor
alt.ham-radio.atv
alt.ham-radio.binaries
alt.ham-radio.digital-voice
alt.ham-radio.dxing
alt.ham-radio.eme
alt.ham-radio.exotic-modes
alt.ham-radio.fax
alt.ham-radio.flame
alt.ham-radio.fm
alt.ham-radio.hf
alt.ham-radio.marketplace
alt.ham-radio.mods
alt.ham-radio.morse
alt.ham-radio.nocode
alt.ham-radio.packet
alt.ham-radio.rtty
alt.ham-radio.slowscan
alt.ham-radio.spread
alt.ham-radio.ssb
alt.ham-radio.uwave
alt.ham-radio.vhf-uhf
alt.radio.amateur.club.clarc
alt.radio.digital
alt.radio.pirate
rec.antiques.radio+phono
rec.ham-radio
rec.ham-radio.packet
rcc.ham-radio.swap
rec.radio.amateur
rec.radio.amateur.antenna
rec.radio.amateur.boatanchors
rec.radio.amateur.digital
rec.radio.amateur.digital.misc
rec.radio.amateur.dx
rec.radio.amateur.equipment
rec.radio.amateur.homebrew
rec.radio.amateur.misc
rec.radio.amateur.packet
rec.radio.amateur.policy
rec.radio.amateur.space
rec.radio.amateur.swap
rec.radio.info
rec.radio.noncomm
rec.radio.scanner
rec.radio.shortwave
rec.radio.swap

Web Sites

The Internet is chock full of ham radio Web sites. And they are relatively easy to find.

In some cases, you can correctly guess the domain names of a Web site without much difficulty. For organizations such as the ARRL, TAPR and RSGB, **www.arrl.org**, **www.tapr.org** and **www.rsgb.org** would be pretty good guesses. For companies such as Alinco, Ten Tec and Yaesu, **www.alinco.com**, **www.tentec.com** and **www.yaesu.com** would also be right on the mark. Guess the domain name for the FCC? If you guessed **www.fcc.gov**, you're right!

Others are not as easy, but you can use a search engine to find the URLs that you can't reckon.

Organizations and Clubs

Most major Amateur Radio organizations and clubs have Web sites, including the ARRL, whose Web site is known as *ARRLWeb* (**http://www.arrl.org/**). As you would expect, *ARRLWeb* has a wealth of information that reflects the services the organization provides to the Amateur Radio community. Refer to the accompanying sidebar entitled "Tour of *ARRLWeb*," which describes the layout and contents of the site. To find other ham radio organization and club Web sites, click on the **Info & Services** link at *ARRLWeb*, then click on the **Off-Campus Links** link under the Reference Information category and you will be presented with a list of links to other ham radio organizations.

Other prominent organization Web sites include the Tucson Amateur Packet Radio (TAPR) Home Page at **http://www.tapr.org/**, which reflects the activities of the organization that is spearheading the Amateur Radio digital revolution. Another is The Radio Amateur Satellite Corporation (AMSAT) at **http://www.amsat.org/**, which is full of information concerning the launching and operation of ham radio birds in space.

Businesses

Many Amateur Radio businesses have sites on the Internet. At those sites, you can view new products, get information and specifications about those products, find product support and newsletters as well as software that you can download to use with those products. At some business Web sites, you can also order a product online using a secure system that assures that any information you send over the Internet, including credit card account numbers, arc safe from prying eyes. Typically, the information is encrypted before it is sent over the network.

If you cannot guess the domain name of an Amateur Radio business, you can use a search engine to find it or you can go to *ARRLWeb*, where there is a list of links to the Web sites of all ARRL advertisers. (Click on the *ARRLWeb*'s **Info & Services** link, then click on the **Links to ARRL Advertisers** link under the Reference Information category.)

Personal

Personal Web sites are plentiful, but are not as easy to find. A few hams have domain names containing their call signs, such as **www.ac6v.com**. The majority do not, however, so guessing a domain name with only a call sign at hand will not be too productive. However, using a search engine to locate a ham's personal Web site with only a call sign at hand would likely be more productive. Once you find one personal ham radio Web site, it may lead you to others because many people publish lists of their favorite links on their Web sites and the favorite links of a ham are likely to include links to other personal ham Web sites.

So, what will you find at an Amateur Radio operator's personal Web site? It depends on the ham's interests. If the ham chases DX, then you can expect to find exciting tales of the DX hunt at that ham's site. If the ham is an aficionado of boat anchors, then you can expect to find the trials and tribulations of resurrecting a boat anchor. You get the picture?

If a ham has non-ham interests, you might find out about those interests too at the Web site. For example, check out all the non-ham interests at **http://www.tapr.org/~wa1lou** and you will discover why my middle name is Jack, as in jack of all trades, master of. . . never mind!

CONFERENCING

Internet conferences are the most elusive aspect of the Internet—except for those like **#hamradio** that are kept alive by robots. The rest come and go unannounced. As a result, it is impossible to suggest where to go to participate in an interesting conference.

The best advice is to use your conference software to search for conferences with names that interest you (DX, packet, satellites, ARES, etc.), then join one of the conferences that your software finds. If your software can't find an interesting conference, then start your own!

Tour of *ARRLWeb*

ARRLWeb (**http://www.arrl.org/**, see **Fig A**) is the ARRL's presence on the World Wide Web. Its contents are a comprehensive source for Amateur Radio information on the Internet. If you are looking for something related to ham radio, this site is a good place to start.

ARRLWeb has seven links on its home page (at **http://www.arrl.org/**) that provide an outline for the information that can be found there.

News is where you can find *The ARRL Letter*, ARRL audio news, and ARRL/W1AW bulletins. If you are an ARRL member, you can also access *The ARRLWeb Extra* newsletter that provides late-breaking Amateur Radio news. News related to special interest topics (like license restructuring, ARRL board and committee reports, etc.) can also be found at this link.

Info & Services is the link for accessing all the information and services provided by the ARRL, including:

Member Services such as the Technical Information Service and Regulatory Information Branch, where you can find answers to your technical and FCC regulation questions; information concerning QSL bureaus, ARRL insurance, publications and ham radio awards can also be found here.

Licensing Information contains a cornucopia of data related to obtaining and maintaining your Amateur Radio license.

On the Air is a source for DX, propagation and Keplerian (satellite) bulletins, the W1AW schedule, VHF, UHF, microwave and EME standings, and information about special events, contests, and awards.

ARRL Field Activities contains everything you need to know about the ARRL Field Organization and affiliated clubs, public service, and the schedule of upcoming hamfests and conventions.

ARRL Educational Activities provides information regarding the educational aspects of our hobby such as the Space Amateur Radio Experiment (SAREX), scouting activities, Youth Skeds, school teacher and volunteer instructor support programs, and ham radio recruitment and instruction.

The ARRL as Advocate contains information about ARRL government and public relations.

Reference Information includes FCC forms, assistance and Part 97 (the Amateur Radio Service rules). Also there are links to call sign "look-up" Web sites, a grid square locator, Web sites of other ham radio organizations and ARRL advertisers, reciprocal licensing/operating information, *QST* Ham Ads, ARRL logo and photo clip art, plus diverse "info & services" that do not fit in the other categories.

Products is the link to the online ARRL publications catalog.

Join ARRL is where you will find everything you need to become an ARRL member.

Site Index provides a comprehensive alphabetical list of links to the highlights of *ARRLWeb*.

Search provides facilities for automatically finding information contained on *ARRLWeb*.

Members Only is the link for ARRL members where they find news and feature articles not published anywhere else, product reviews before they appear in *QST*, as well as an archive of reviews published in past *QST*s, and a searchable index of all issues of *QST* and *QEX*.

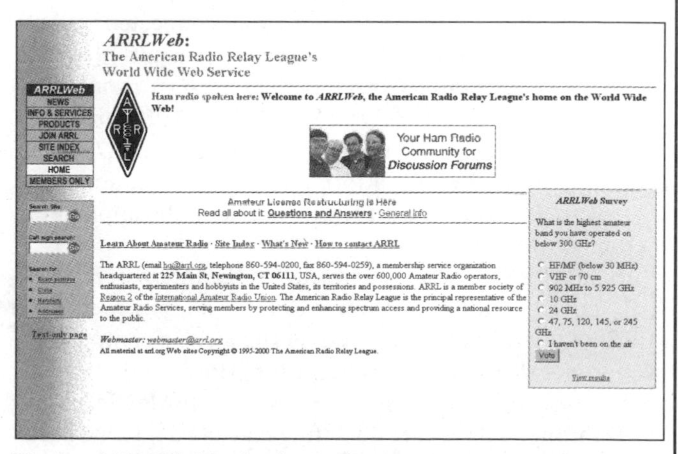

Fig A—*ARRLWeb* is the best place to start your quest for everything that's happening regarding Amateur Radio.

Gateways

Gateways are unique Internet applications in which two communications media are mated; that is, the online communication medium of the Internet and the RF communication medium of Amateur Radio. A gateway receives a packet off the air, looks at the address on the packet, then decides how to route the received packet over the Internet, ultimately to its intended recipient. As a result, gateways allow you to receive and transmit live ham radio activity by means of interfaces between the Internet and the amateur packet radio TCP/IP network (AMPRNET).

No two gateways are exactly alike. So, before using a gateway, familiarize yourself with its unique operating procedures. Do not assume that the procedures that worked at one gateway work at another. To find the gateway nearest you, use the QSL.NET search engine (**http://www.qsl.net/search/search.html**). It came up with 167 hits when I searched for "gateway."

WinLink

WinLink is a *Windows* application that permits PACTOR and PACTOR II stations to use the Internet for the transfer of non-commercial e-mail that complies with the existing Third Party Traffic agreements between the U.S. and other nations. *WinLink* is basically a bulletin board system (BBS) that provides for HF-to-HF and HF-to-VHF text message transfer as well as the HF/VHF-to-Internet e-mail transfer.

There are different versions of *WinLink*. For example, Steve Waterman, K4CJX, has authored *WinLink 2000*, that provides the automatic transfer of messages between mobile Amateur Radio operators and the Internet e-mail system.

Using the system is as simple as originating e-mail and the Internet connection is invisible to the user, as it should be. A web site for support of *WinLink 2000* is at **http://winlink.org/**.

APRS on the Internet

APRS is the acronym for Automatic Position Reporting System. APRS software uses the unconnected packet radio mode to graphically display the position of moving and stationary objects on maps displayed on a computer monitor. Unconnected packets are used instead of connected packets to permit all stations to receive each transmitted APRS packet on a one-to-all basis rather than the one-to-one basis afforded by connected packets. Versions of APRS are available as this is written for the *DOS*, *Mac OS*, *Palm OS*, *Windows* and *UNIX/LINUX* platforms. It is an understatement to say that APRS is the most popular packet radio application to come down the pike so far.

If an APRS station is connected to the Internet, it can be easily configured to act as an "IGate," and thus feed local APRS activity that it receives over the air to a computer on the Internet that acts as a conduit for APRS feeds from all over the world. This conduit is known as an APRServe and it sifts through all the feeds deleting bad and duplicate packets, then allows other stations to access that data in order to view APRS activity worldwide.

For example, refer to **Fig 16-5**, which is a screen save of APRS activity in the Northeast on a lazy Sunday afternoon as received on the air at WA1LOU. Now, refer to **Fig 16-6**, which is a screen save of that same lazy Sunday afternoon after the APRS software running at WA1LOU was connected via the Internet to APRServe. Notice how many more stations appear on the map, especially to the Southwest well beyond WA1LOU's normal reception area. These additional stations appear courtesy of the APRServe conduit.

Viewing worldwide APRS activity is not limited to those stations running APRS software with connections to the Internet. Anybody can view APRServe data by pointing their Java-capable Web browser to **http://www.aprs.net/**. **Fig 16-7** is another screen save of that same lazy Sunday afternoon except that this time the view is from the APRServe Web page.

In addition to viewing worldwide APRS activity, IGates permit you to originate one-line e-mail messages to legitimate Internet addresses. After addressing the message and

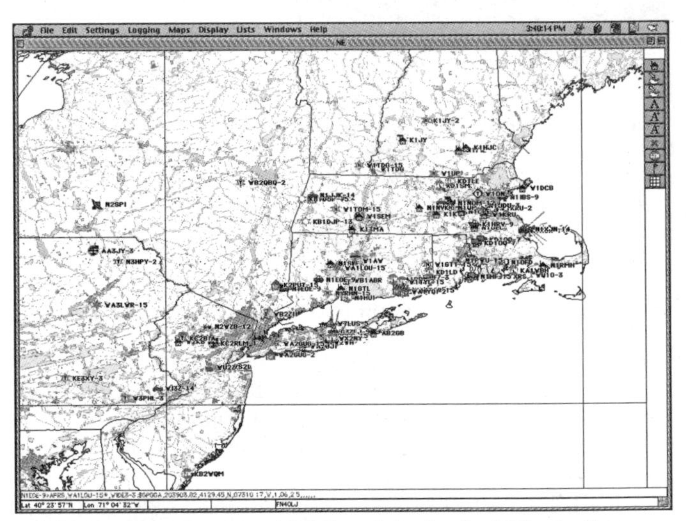

Fig 16-5—Here is what APRS activity looked like in the Northeast as received on the air by WA1LOU.

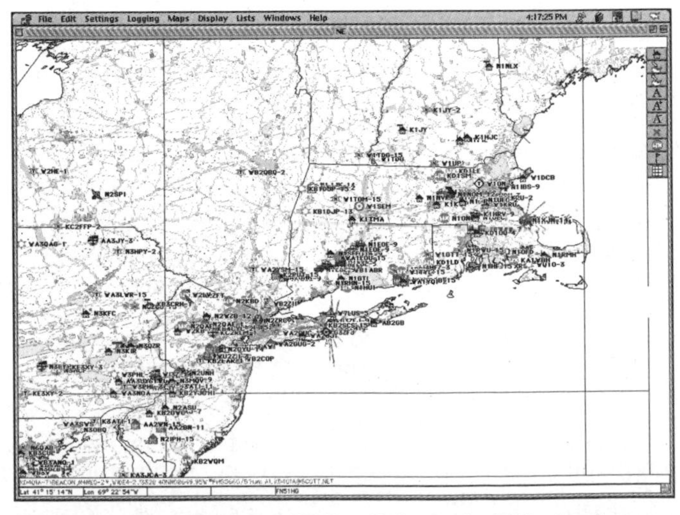

Fig 16-6—Here is what APRS activity looked like in the Northeast after WA1LOU connected MacAPRS to the APRServe network.

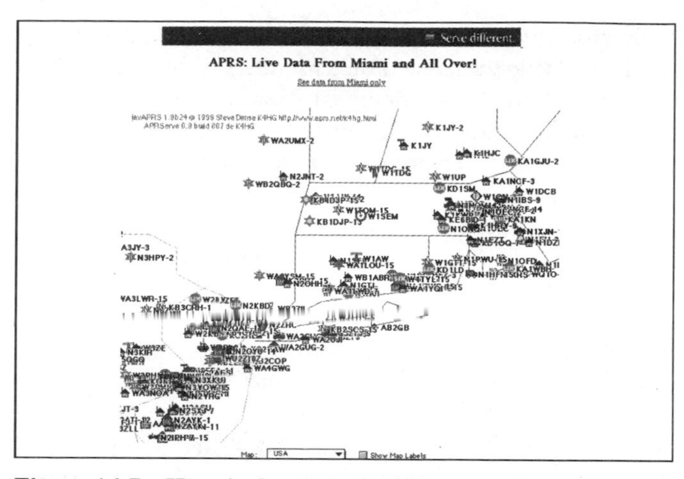

Figure 16-7—Here is the view of APRS activity in the North-east as viewed from the APRServe Internet site at http://www.aprs.net/usa1.html.

transmitting it on the air, it is relayed to the Internet by any IGate that receives it (and is properly configured to relay it). After the relay, the IGate sends an acknowledgment over the air to the originating station.

To send a message for Internet relay, simply address it to EMAIL and insert the Internet address as the first item in the message. For example, to send e-mail to k1efi@arrl.net, you address the message to EMAIL and the contents of the message would look something like this:

```
k1efi@arrl.net Have you worked the
Jabru DX-pedition yet?
```

After transmitting the message, you will receive an acknowledgment like "E-mail message delivered OK" from any IGate that handles it.

Call Sign Directories

One of the most useful Amateur Radio tools is the ability to lookup a call sign and find the name and address of the person holding the ham radio license. The ARRLWeb site (**http://www.arrl.org/**) provides this service. Call sign directories also exist at **http://www.qrz.com** and **http://callsign.ualt.edu/callsign.shtml**.

Some call sign servers only provide information regarding U.S. hams. Other servers also provide information regarding non-U.S. hams. Some servers also allow you to do searches to find hams that fall into specific categories. For example, you can perform a search for all the hams in a certain locality (search by state, city, or ZIP Code). You can also perform a search for all the hams with the same name (for example, search for hams with the last name of "Maxim" and you will get a list of six licensed hams).

CREATING YOUR OWN WEB SITE

Creating a Web page is not difficult. I've surfed a lot of Web pages and never found one built by a rocket scientist, so you don't have to be a rocket scientist to create a Web page. You do need three things, though: a site, software and a scheme.

Home Site

You need a place on the Web where to put your Web page. If you use an Internet Service Provider (ISP) or online service, it likely provides space for its customers' Web pages. Depending on your ISP or online service, this space may be included as part of your monthly fee or may be an option that costs extra.

Many ISPs and online services provide X-amount of Web page space to each customer for free, then charge extra for any space a customer needs beyond that X-amount. Check your ISP or online service for what they offer.

Other sources for sites are businesses on the Web that provide free space for your Web page. In exchange for this free space, the business inserts advertisements on your page. Like other Web space providers, X-amount of space is free and space beyond that X-amount costs extra.

Web-Spinning Tools

Hyper-Text Markup Language (HTML) is the language used to program Web pages. Web browsing software interprets HTML documents and the results of that interpretation are the pages you see in your browser. HTML documents are nothing more than files containing plain text, so if you know HTML, you can create an HTML document with nothing more than a text editor or a word processor. However, if you know HTML, you don't need me and have likely browsed to another page in this chapter or book by now. (This section is intended for folks who don't know HTML.)

If you don't know HTML, you can create an HTML document with Web page design software. Such software allows you to use tools similar to the ones you find in any graphic design software to create and manipulate the text and graphics on your Web page. As you create your page, the software interprets your design and outputs an HTML document representing the design.

I have used *Adobe PageMill* and *Microsoft FrontPage* for Web page design. They are moderately priced, similarly featured, and suitable for the novice Web page designer. Other Web page design programs are available at higher and lower prices with more and fewer features. These may be suitable for the novice, but I have not used them, so I can't recommend them.

You may already own software that outputs HTML and not even know it. Today, many word processors, graphic design, and page layout programs allow you to save your work in HTML. For example, recent versions of *Microsoft Word*, *Deneba Canvas* and *Adobe PageMaker* can save to HTML. Check the software on your computer and you may be surprised to find something that saves to HTML, too.

The most economical option for the novice Web page designer is to obtain a free evaluation copy of a browser that includes Web page design functions. *Netscape Communicator* Version 4 fits the bill, but I prefer its older cousin, *Netscape Navigator Gold* version 3, which is still available at **ftp://archive:oldies@archive.netscape.com/archive/index.html**.

I used *Gold* to create my first Web page and I still use it today to build the basic structure of a Web page. It is easy to use and provides the basic features I need in a Web page design program. After I build the basic structure of a Web page, I use *PageMill* and/or *Canvas* to add features that are unavailable in *Gold*.

In addition to Web page design software, you need ftp (file transfer protocol) software to transfer your completed Web page to the Internet site where it is to be. Just about any communications program that supports ftp is adequate. And there are freeware and shareware ftp programs, so you need not expend a lot to obtain one. Recommended are *WS-FTP LE* for *Windows* and *Fetch* for *Mac OS*, which are available from **http://tucows.cows.net**.

To save a thousand words by using a picture, you need software to create and manipulate the picture. A variety of graphic design/editing programs are available to do this and you probably already own one or two. The main consideration of such software is that it can save its output in a Web-compatible format like GIF and JPEG.

Site Scheming

Andy Warhol said that during our lifetime, each one of us experiences 15 minutes of fame. These days, the Internet seems to be the place for experiencing that 15 minutes as tens of thousands of people have personal Web pages, whose basic purpose is to publicize who they are, what they are doing, and how they are thinking.

Similarly, your Web page should publicize who, what, and how from the Amateur Radio perspective. You are likely to create an interesting Web page about subjects that interest you the most, so the focus of your page should be the ham radio niche that you itch. Whether your interest is traffic handling, DX chasing, or boat-anchor tinkering, your interest and involvement in it should be the subject of your page.

Building

Now, let's begin building a page.

Lou, my virtual friend, is an expert on modified trimmer capacitors and he wants to share his expertise by means of a Web page. So, I created a Web page for Lou called "Lou's Modified Trimmer Capacitor Page."

After starting *Netscape Navigator Gold*, I start building Lou's page by opening a blank document (File>New Document>Blank). Upon opening a blank page, *Gold* automatically enters the edit mode ready for my handiwork.

Before building, I save the page (File>Save) using the file name "index," and *Gold* appends the proper extension (.htm for *Windows*, .html for *Mac OS*). Most Web sites require that the file name of the first page in a set of pages be index.htm/html. Check to find out what your site requires.

While on the subject of file names, here are three rules you should adhere to:

- Use lowercase for all the file names that compose your Web page.
- Keep the file names short: eight characters maximum plus the extension.
- Save all the files that compose your page in the same directory location.

Next, I title the page "Lou's Modified Trimmer Capacitor Page" (Properties>Document>General>Title:). The file name and page title are not synonymous. The title appears at the top of your Web browser window. If you don't specify a title, the directory path for the page appears at the top of the window instead.

While in the Properties window, I select a color for the background of the page by selecting the Appearance tab, clicking on the Background button, and selecting pale yellow. After I click on the OK button, my selections are applied to the window.

At the cursor prompt, I enter the seven lines of text that are contained on the page: "Lou's Modified Trimmer Capacitor Page" through "E-mail Lou." Then, I format the text.

I select line 1, then select Heading 1 in the Paragraph Style window, +4 in the Font Size window, and hunter green after clicking the Font Color button.

I select line 2 and click on the Bold button.

I select line 3, then select Heading 3 in the Paragraph Style window, and hunter green after clicking the Font Color button.

Next, I select all the text (Edit>Select All) and click on the Align Right button.

Lines 4 through 6 are intended to be hypertext links to other Web pages. To add links, I select line 4, then click on the Make Link button. Then I enter the URL of the linked Web page (**http://www.etc.,etc.**) in the Link To box and click on the OK button. I repeat this for lines 5 and 6.

Line 7 ("E-mail Lou") is intended to be an e-mail hypertext link. Instead of jumping to another Web page, an e-mail link allows you to send e-mail to someone involved with the Web page. Typically, when you select an e-mail link, your e-mail software is activated with a preaddressed blank message form ready for the entry of your comments.

To create the e-mail link, I select line 7, then click on the Make Link button. Then I enter the HTML code *mailto:* in the Link To box followed by Lou's e-mail address. Note that there is no space between mail to: and the e-mail address. I click OK to apply the link.

While line 7 is selected, I format it to differentiate it from the other links; I select +4 in Font Size menu window, then click on the Bold button.

Now, I add a graphic to the page (a GIF of the modified trimmer capacitor icon). I place the cursor at the end of line 1, press Enter, then click on the Insert Image button. Next, I click on the Browse button next to the Image File Name box to locate the GIF file. When I locate "tc.gif," I click on the OK button.

To offset the hypertext links from the other text on the page, I insert line breaks by placing the cursor at the beginning of lines 3 and 7, then pressing Shift-Enter.

Finally, I save my work (File>Save) and switch to the browser mode (File>Browse Document) to view my handiwork, as illustrated in **Fig 16-8**. It is a good idea to save and switch to the browser mode often to make sure that your page is turning out as you planned. Be sure to click on the Reload button after you switch to the browser mode in order

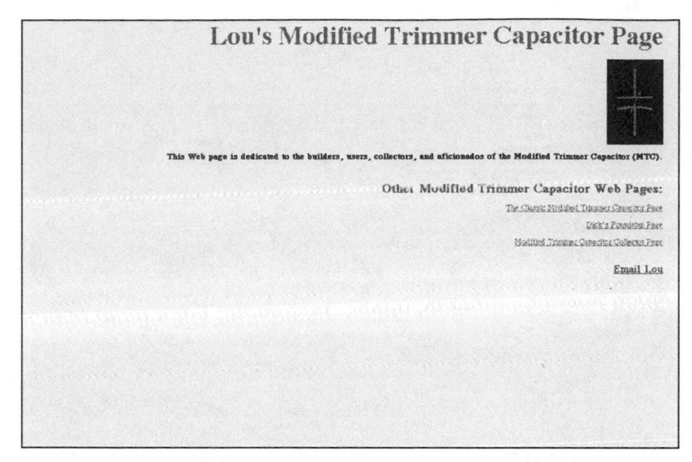

Figure 16-8—"Lou's Modified Trimmer Capacitor Page" is a simple Web page that anyone can build using the building tools provided by *Netscape Navigator*.

to update your page with any changes you made.

Adding a Table of Contents

I will build upon the page, so I open that page in the editor mode (Open > Open File in Editor...). This page is the first page of four pages, so I will add a table of contents, which will inform viewers what's on the other pages with links to get them to those pages.

I place the cursor at end of second line (the line ending with "...Modified Trimmer Capacitor (MTC).") and insert a horizontal line (Insert > Horizontal Line) to set the table of contents off from the top of the page. Then, I press the Enter key and type Contents:.

The table of contents will be contained in an "invisible" table; that is, there will be no visible horizontal or vertical lines separating the rows and columns of the table. However, the horizontal and vertical lines of the table will be visible initially to permit me to enter text (it is very difficult to enter text when the lines are invisible). I insert a visible table (Insert > Table > Table) with four rows (Number of Rows: 4) and I align it with the right margin (Table Alignment: Right) like everything else on the page.

Next, I select each cell of the table and enter the table of contents text ("Builders," "This page describes... ," etc.). After entering the text, it is obvious that the last row of the table needs space between each column to make the text more readable. I place the cursor at the beginning of the second column in the fourth row (before "Here are links...") and insert three spaces [Insert > Nonbreaking Space (three times)].

To adhere to the right margin alignment of the page, I realign the text contained in the table. To do so, I place the cursor in each row, then select Properties > Table. If the Row tab is not selected, select it, then click on the Right button for Text Alignment, Horizontal.

Now, I make the table invisible by placing the cursor anywhere in the table, then selecting Properties > Table. Next, I select the Table tab and enter 0 (pixels) in the "Border line width" box. Voila, the borders of the table disappear!

Just as I set the table off from the top of the page, I want to set it off from the bottom of the page, too. I place the cursor at the beginning of the line following the table (before "Other Modified Trimmer Capacitor Web Pages:" and insert a line (Insert > Horizontal Line).

Adding Links

The words Builders, Users, and Collectors in the table are intended to be links to the other pages. To configure them as links, I select each, select Insert > Link, and enter the file name for each page in the "Link to" box (for example, Builders may be linked to the file named page2.html).

"Links" in the last row of the table is intended to jump to the links at the bottom of the same page rather than to another page. This type of link is useful when a page is long. It permits viewers to jump to a particular point of interest instead of making them scroll through the whole page to get to that same spot.

Before I create this link, I must take a preliminary step: I must insert an anchor, an invisible spot on the page that can be the target of a link. I select "Other" in the line following the table, then insert an anchor [Insert > Target (Named Anchor)] which I name "Links" (in the "Enter a name for this target" box).

Now, I can set up the link. I return to the table and select the word "Links," then select Insert > Link, and choose "Links" in the "Select a named target in current document" box. After choosing "Links," the "Link to" box contains "#Links."

Finally, I finish up by formatting the newly entered text. First, I select the word Contents and make it green, bolder, and bigger (select Properties > Text, then select the Character tab, click the Choose Color button in the Color panel, and select hunter green. Click OK, select the Paragraph tab, and click on the Right button in the Align panel. Then select Properties > Paragraph > Heading 2). I also make the four links (Builders, Users, etc.) in the table bolder by selecting each, then selecting Properties > Character > Bold. **Fig 16-9** illustrates the completed Web page.

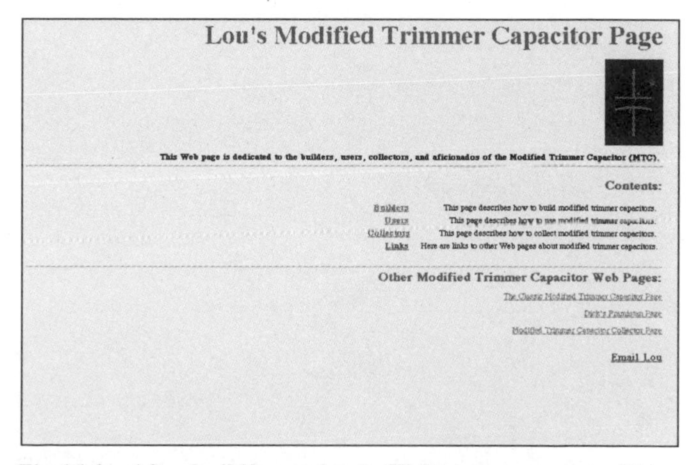

Fig 16-9—After building a simple Web page, you can add features as was done with "Lou's Modified Trimmer Capacitor Page."

Web Site Hosts

Drilling down the directory at Yahoo (**http://www.yahoo.com**), I found 114 free Web space providers. They come in a variety of shapes and sizes, so you should be able to find one that suits your needs. The URLs for popular free Web space providers are **http://geocities.yahoo.com/** and **http://www.tripod.com/**.

A very cool site that allows Amateur Radio operators to have Web pages with unlimited space for free is **http://www.qsl.net/**. QSL.NET is "dedicated to the sole purpose of furthering the abilities and interest of the Amateur Radio community. If you are a licensed Amateur Radio operator you are invited to reserve your free space on this server now. Sign up and you will receive free e-mail, with forwarding to your existing service, along with free server space to move your homepage to this server, mirror your existing one, or lose all those excuses and finally start one." (By the way, QSL.NET is funded by Alan Waller, K3TKJ, and by contributions from its users.)

The site seems too good to be true, but it really exists! There are hundreds of ham radio Web pages already there. And I know that if I was starting a ham radio Web page today, QSL.NET is the place I would build it.

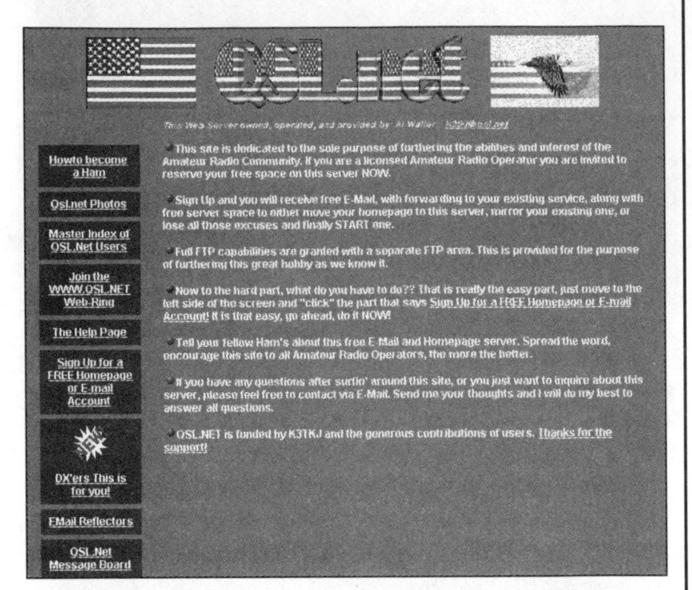

QSL.NET is dedicated to the ham radio community and provides free space for creating your own Amateur Radio Web page without advertising!

Your Turn

This concludes the assembly of Lou's Web page. What I did is only the tip of the cyberg. Surf the Internet, see what you like, and use those pages as a springboard for your own page.

Beyond the basics I have described, you can get help from the manuals and on-line documentation contained in the Web page design software you use. You can also get plenty of help on the Internet. The Home Page Improvement page (at **http://www.hpiweb.com/**) contains good information and great links that inform you how to do just about anything you want to do with your Web page.

THE FUTURE INTERNET

The Internet is evolving minute-to-minute. Today's Internet is different than yesterday's Internet, and tomorrow it will be different again. Change is good, but change also means that your favorite Web pages may disappear overnight!

To remain current with the ever-evolving Internet, you can read about Internet developments in the Digital Dimension column that appears in *QST*. Also, you can check *ARRLWeb*, which will alert you to new links and new information as soon as the ARRL staff finds them and posts them.

Have fun!

CHAPTER 17

References

AMERICAN RADIO RELAY LEAGUE DX CENTURY CLUB RULES

February 2002
Current Entities: Total 335

INTRODUCTION

The DXCC List is based upon the principle espoused by Clinton B. DeSoto, W1CBD, in his landmark 1935 *QST* article, "How to Count Countries Worked, A New DX Scoring System." DeSoto's article discussed problems DXers had in determining how to count the DX they had worked. He presented the solution that has worked successfully for succeeding generations of DXers.

In DeSoto's words, "The basic rule is simple and direct: *Each discrete geographical or political entity is considered to be a country.*" This rule has stood the test of time — from the original list published in 1937 to the *ARRL DXCC List* of today. For more than 60 years, the *DXCC List* has been the standard for DXers around the world.

DeSoto never intended that all DXCC "countries" would be countries in the traditional, or dictionary, meaning of the word. Rather, they are the distinct geographic and political Entities which DXers seek to contact. Individual achievement is measured by working and confirming the various Entities comprising the DXCC List. This is the essence of the DXCC program.

DXCC activity was interrupted by World War II. In 1947, the program started anew. Contacts are valid from November 15, 1945, the date US amateurs were authorized by the FCC to return to the air.

Over time, the criteria for the List has changed. However, Entities are not removed when changes are made. The List remains unchanged until an Entity no longer satisfies the criteria under which it was added. Thus, today's *DXCC List* does not fully conform with today's criteria. Changes are announced under DXCC Notes in QST.

SECTION I. BASIC RULES

1. The DX Century Club Award, with certificate and lapel pin is available to Amateur Radio operators throughout the world (see #15 below for the DXCC Award Fee Schedule). ARRL membership is required for DXCC applicants in the US, its possessions, and Puerto Rico. ARRL membership is not required for foreign applicants. All DXCCs are endorsable (see Rule 5). There are 17 separate DXCC awards available, plus the DXCC Honor Roll:

a) Mixed (general type): Contacts may be made using any mode since November 15, 1945.

b) Phone: Contacts must be made using radiotelephone since November 15, 1945. Confirmations for cross-mode contacts for this award must be dated September 30, 1981, or earlier.

c) CW: Contacts must be made using CW since January 1, 1975. Confirmations for cross-mode contacts for this award must be dated September 30, 1981, or earlier.

d) RTTY: Contacts must be made using radioteletype since November 15, 1945. (Baudot, ASCII, AMTOR and packet count as RTTY.) Confirmations for cross-mode contacts for this award must be dated September 30, 1981, or earlier.

e) 160 Meter: Contacts must be made on 160 meters since November 15, 1945.

f) 80 Meter: Contacts must be made on 80 meters since November 15, 1945.

g) 40 Meter: Contacts must be made on 40 meters since November 15, 1945.

h) 20 Meter: Contacts must be made on 20 meters since November 15, 1945.

i) 17 Meter: Contacts must be made on 17 meters since November 15, 1945.

j) 15 Meter: Contacts must be made on 15 meters since November 15, 1945.

k) 12 Meter: Contacts must be made on 12 meters since November 15, 1945.

l) 10 Meter: Contacts must be made on 10 meters since November 15, 1945.

m) 6 Meter: Contacts must be made on 6 meters since November 15, 1945.

n) 2 Meter: Contacts must be made on 2 meters since November 15, 1945.

o) Satellite: Contacts must be made using satellites since March 1, 1965. Confirmations must indicate satellite QSO.

p) Five-Band DXCC (5BDXCC): The 5BDXCC certificate is available for working and confirming 100 current DXCC entities (deleted entities don't count for this award) on each of the following five bands: 80, 40, 20, 15, and 10 Meters. Contacts are valid from November 15, 1945.

The 5BDXCC is endorsable for these additional bands: 160, 17, 12, 6, and 2 Meters. 5BDXCC qualifiers are eligible for an individually engraved plaque (at a charge of $35.00 US plus shipping).

q) The DXCC Challenge Award is given for working and confirming at least 1,000 DXCC band-Entities on any Amateur bands, 1.8 through 54 MHz (except 30 meters). This award is in the form of a plaque. Certificates are not available for this award. Plaques can be endorsed in increments of 500. Entities for each band are totaled to give the Challenge standing. Deleted entities do not count for this award. All contacts must be made after November 15, 1945. The DXCC Challenge plaque is available for $79.00 (plus postage). QSOs for the 160, 80, 40, 20, 17, 15, 12, 10 and 6 meter bands qualify for this award. Bands with less than 100 contacts are acceptable for credit for this award.

r) The DeSoto Cup is presented to the DXCC Challenge leader as of the 30th of September each year. The DeSoto Cup is named for Clinton B. DeSoto, whose definitive article in *QST* for October, 1935 forms the basis of the DXCC award. Only one cup will be awarded to any single individual.

s) Honor Roll: Attaining the DXCC Honor Roll represents the pinnacle of DX achievement:

i) Mixed—To qualify, you must have a total confirmed entity count that places you among the numerical top ten DXCC entities total on the current DXCC List (example: if there are 335 current DXCC entities, you must have at least 326 entities confirmed).

ii) Phone—same as Mixed.

iii) CW—same as Mixed.

iv) RTTY—same as Mixed.

To establish the number of DXCC entity credits needed to qualify for the Honor Roll, the maximum possible number of current entities available for credit is published monthly on the ARRL Members Only Web Page. First-time Honor Roll members are recognized monthly on the ARRL Members Only

Web Page. Complete Honor Roll standings are published annually in *QST*, usually in the July issue. See DXCC notes in *QST* for specific information on qualifying for this Honor Roll standings list. Once recognized on this list or in a subsequent monthly update of new members, you retain your Honor Roll standing until the next standings list is published. In addition, Honor Roll members who have been listed in the previous Honor Roll Listings or have gained Honor Roll status in a subsequent monthly listing are recognized in the DXCC Annual List. Honor Roll qualifiers receive an Honor Roll endorsement sticker for their DXCC certificate and are eligible for an Honor Roll lapel pin ($5) and an Honor Roll plaque ($35 plus shipping). Write the DXCC Desk for details or check out the Century Club Item Order Form at http://www.arrl.org/awards/dxcc/

v) #1 Honor Roll: To qualify for a Mixed, Phone, CW or RTTY Number One plaque, you must have worked every entity on the current DXCC List. Write the DXCC Desk for details. #1 Honor Roll qualifiers receive a #1 Honor Roll endorsement sticker for their DXCC certificate and are eligible for a #1 Honor Roll lapel pin ($5) and a #1 Honor Roll plaque ($50 plus shipping).

2. Written Proof: Except in cases where the rules of Section IV apply, written proof (e.g. QSL cards) of two-way communication (contacts) must be submitted directly to ARRL Headquarters for all DXCC credits claimed. Photocopies and electronically transmitted confirmations (including, but not limited to fax, telex and telegram) are not currently acceptable for DXCC purposes. Staff may accept electronic confirmations when procedures to do so are adopted.

The use of a ***current*** official DXCC application form or an approved facsimile (for example, ***exactly*** reproduced by a computer program) is required. Such forms must include provision for listing callsign, date, band, mode, and DXCC entity name. Complete application materials are available from ARRL Headquarters. Confirmations for a total of 100 or more different DXCC credits must be included with your first application.

Cards contained in the original received envelopes or in albums will be returned at applicant's expense without processing.

By action of the ARRL Board of Directors, 10-MHz confirmations may be credited to Mixed, CW and RTTY awards only.

3. The ARRL DXCC List is based on the DXCC List Criteria.

4. Confirmation data for two-way communications must include the call signs of both stations, the Entity name as shown in the DXCC List, mode, and date, time and band. Except as permitted in Rule 1, cross-mode contacts are not permitted for DXCC credits. Confirmations not containing all required information may be rejected.

5. Endorsement stickers for affixing to certificates or pins will be awarded as additional DXCC credits are granted. For the Mixed, Phone, CW, RTTY, 40, 20, 17, 15, 12 and 10-Meter DXCC, stickers are provided in exact multiples of 50, i.e. 150, 200, etc. between 100 and 250 DXCC credits, in multiples of 25 between 250 and 300, and in multiples of 5 above 300 DXCC credits.

For 160-Meter, 80-Meter, 6-Meter, 2-Meter and Satellite DXCC, the stickers are issued in exact multiples of 25 starting at 100 and multiples of 10 above 200, and in multiples of 5 between 250 and 300. Confirmations for DXCC credit may be submitted in any increment, but stickers and listings are provided only after a new level has been attained.

6. All contacts must be made with amateur stations working in the authorized amateur bands or with other stations licensed or authorized to work amateurs. Contacts made through "repeater" devices or any other power relay methods (other than satellites for Satellite DXCC) are invalid for DXCC credit.

7. Any Amateur Radio operation should take place only with the complete approval and understanding of appropriate administration officials. In countries where amateurs are licensed in the normal manner, credit may be claimed only for stations using regular government-assigned call signs or portable call signs where reciprocal agreements exist or the host government has so authorized portable operation. Without documentation supporting the operation of an amateur station, credit will not be allowed for contacts with such stations in any country that has temporarily or permanently closed down Amateur Radio operations by special government edict or policy where amateur licenses were formerly issued in the normal manner. In any case, credit will be given for contacts where adequate evidence of authorization by appropriate authorities exists, notwithstanding any such previous or subsequent edict or policy.

8. All stations contacted must be "land stations." Contacts with ships and boats, anchored or underway, and airborne aircraft, cannot be counted. For the purposes of this award, remote control operating points must also be land based. *Exception*: Permanently docked exhibition ships, such as the Queen Mary and other historic ships will be considered land based.

9. All stations must be contacted from the same DXCC Entity. The location of any station shall be defined as the location of the transmitter. For the purposes of this award, remote operating points must be located within the same DXCC Entity as the transmitter.

10. All contacts must be made using callsigns issued to the same station licensee. Contacts made by an operator other than the licensee must be made from a station owned and usually operated by the licensee and must be made in accordance with the regulations governing the license grant. Contacts may be made from other stations provided they are personally made by the licensee. The intent of this rule is to prohibit credit for contacts made for you by another operator from another location. You may combine confirmations from several callsigns held for credit to one DXCC award, as long as the provisions of Rule 9 are met. Contacts made from club stations using a club callsign may not be used for credit to an individual's DXCC.

11. All confirmations must be submitted exactly as received by the applicant. The submission of altered, forged, or otherwise invalid confirmations for DXCC credit may result in disqualification of the applicant and forfeiture of any right to DXCC membership. Determinations by the ARRL Awards Committee concerning submissions or disqualification shall be final. The ARRL Awards Committee shall also determine the future eligibility of any DXCC applicant who has ever been barred from DXCC.

12. Conduct:

Exemplary conduct is expected of all amateur radio operators participating in the DXCC program. Evidence of intentionally disruptive operating practices or inappropriate ethical conduct in *any* aspect of DXCC participation may lead to disqualification from all participation in the program by action of the ARRL Awards Committee.

Actions that may lead to disqualification include, but are not limited to:

a) The submission of forged or altered confirmations.

b) The presentation of forged or altered documents in support of an operation.

c) Participation in activities that create an unfavorable impression of amateur radio with government authorities. Such activities include malicious attempts to cause disruption or disaccreditation of an operation.

d) Blatant inequities in confirmation (QSL) procedures. Continued refusal to issue QSLs under certain circumstances may lead to disqualification. Complaints relating to monetary issues involved in QSLing will generally not be considered, however.

13. Each DXCC applicant, by applying, or submitting documentation, stipulates to:

a) Having observed all pertinent governmental regulations for Amateur Radio in the country or countries concerned.

b) Having observed all DXCC rules.

c) Being bound by the DXCC rules.

d) Being bound by the decisions of the ARRL Awards Committee.

14. All DXCC applications (for both new awards and endorsements) must include sufficient funds to cover the cost of returning all confirmations (QSL cards) via the method selected. Funds must be in US dollars using US currency, check or money order made payable to the ARRL, credit card number with expiration date. Address all correspondence and inquiries relating to the various DXCC awards and all applications to: ARRL Headquarters, DXCC Desk, 225 Main St., Newington, CT 06111, USA.

15. Fees.

Effective October 1, 1990, all amateurs applying for their very first DXCC Award will be charged a one-time registration fee of $10.00. This same fee applies to both ARRL members and foreign non-members, and both will receive one DXCC certificate and a DXCC pin. Applicants must provide funds for postage charges for QSL return.

a) A $10.00 shipping and handling fee will be charged for each additional DXCC certificate issued, whether new or replacement. A DXCC pin will be included with each certificate.

b) Endorsements and new applications may be presented in person at ARRL HQ, and at certain ARRL conventions. When presented in this manner, such applications shall be limited to 120 QSOs maximum, and a $5.00 handling charge will apply, in addition to other applicable fees.

c) Each ARRL member will be charged $10.00 for the first submission of the year, up to 120 QSOs and return postage.

d) A $0.15 fee will be charged for each QSO credited beyond the limits described in 15. b), 15. c), and 15. f).

e) Foreign non-ARRL members will be charged a $20.00 DXCC Award fee for their first yearly submission, in addition to return postage charges. Fees in 15. a), 15. b), and 15. d), and 15. f) also apply.

f) DXCC participants who wish to submit more than once per calendar year will be charged a DXCC fee for each additional submission made during the remainder of the calendar year. These fees are dependent upon membership status: ARRL Members: $20.00 (for the first 100 QSOs); Foreign non-members: $30.00 (for the first 100 QSOs). Additionally, return postage must be provided by applicant, and charges from 15. a), 15. b) and 15. d) also apply.

16. The ARRL DX Advisory Committee (DXAC) requests your comments and suggestions for improving DXCC. Address correspondence, including petitions for new listing consideration, to ARRL Headquarters, DXAC, 225 Main St., Newington, CT 06111, USA. The DXAC may be contacted by e-mail at dxac@arrl.org Note that this address is valid for the DXAC *only*. **Correspondence on routine DXCC matters should be addressed to the DXCC Desk, or by e-mail to dxcc@arrl.org**

SECTION II. DXCC LIST CRITERIA

Introduction:

The ARRL DXCC List is the result of progressive changes in DXing since 1945. Each Entity on the DXCC List contains some definable political or geographical distinctiveness. While the general policy for qualifying Entities for the DXCC List has remained the same, there has been gradual evolution in the specific details of criteria which are used to test Entities for their qualifications. The full DXCC List does not conform completely with current criteria, for some of the listings were recognized from pre-WWII or were accredited with earlier versions of the criteria. In order to maintain continuity with the past, as well as to maintain a robust DXCC List, all Entities on the List at the time the 1998 revision became effective were retained.

Definitions:

Certain terms occur frequently in the DXCC criteria and are listed here. Not all of the definitions given are used directly in the criteria, but are listed in anticipation of their future use.

Entity: A listing on the DXCC List; a counter for DXCC awards. Previously denoted a DXCC "Country."

Event: An historical occurrence, such as date of admission to UN, ITU, or IARU, that may be used in determining listing status.

Event Date: The date an Event occurs. This is the Start Date of all Event Entities.

Event Entity: An Entity created as the result of the occurrence of an Event.

Discovery Entity: An Entity "Discovered" after the listing is complete. This applies only to Geographic Entities, and may occur after a future rule change, or after an Event has changed its status.

Discovery Date: Date of the rule change or Event which prompts addition of the Entity. This is the Start Date for a Discovery Entity.

Original Listing: An Entity which was on the DXCC List at the time of inception.

Start Date: The date after which confirmed two-way contact credits may be counted for DXCC awards.

Add Date: The date when the Entity will be added to the List, and cards will be accepted. This date is for administrative purposes only, and will occur after the Start Date.

Island: A naturally formed area of land surrounded by water, the surface of which is above water at high tide. For the purposes of this award, it must consist of connected land, of which at least two surface points must be separated from each other by not less than 100 meters measured in a straight line from point to point. All of the connected land must be above the high tide mark, as demonstrated on a chart of sufficient scale. For the purposes of this award, any island, reef, or rocks of less than this size shall not be considered in the application of the water separation criteria described in Part 2 of the criteria.

Criteria:

Additions to the DXCC List may be made from time to time as world conditions dictate. Entities may also be removed from the List as a result of political or geographic change. Entities removed from the List may be returned to the List in the future, should they requalify under this criteria. However, an Entity requalified does so as a totally new Entity, not as a reinstated old one.

For inclusion in the DXCC List, conditions as set out below must be met. Listing is not contingent upon whether operation has occurred or will occur, but only upon the qualifications of the Entity.

There are five parts to the criteria, as follows:

1. Political Entities
2. Geographical Entities
3. Special Areas
4. Ineligible Areas
5. Removal Criteria

1. Political Entities:

Political Entities are those areas which are separated by reason of government or political division. They generally contain an indigenous population which is not predominantly composed of military or scientific personnel.

An Entity will be added to the DXCC List as a Political Entity if it meets any one of the following three criteria:

a) The entity is a UN Member State.

b) The entity has been assigned a callsign prefix bloc by the ITU[1]. A provisional prefix bloc assignment may be made by the Secretary General of ITU. Should such provisional assignments not be ratified later by the full ITU, the Entity will be removed from the DXCC List.

c) The Entity has a separate IARU Member Society.

New Entities satisfying any one or more of the three conditions above will be added to the DXCC List by administrative action as of their "Event Date."

Entities qualifying under this section will be referred to as the "Parent" when considering separation under the section "Geographical Separation." Only Entities in this group will be acceptable as a Parent for separation purposes.

2. Geographic Separation Entity:

A Geographic Separation Entity may result when a single Political Entity is physically separated into two or more parts. The part of such a Political Entity that contains the capital city is considered the Parent for tests under these criteria. One or more of the remaining parts resulting from the separation may then qualify for separate status as a DXCC Entity if they satisfy paragraph a) or b) of the Geographic Separation Criteria, as follows.

a) Land Areas:

A new Entity results when part of a DXCC Entity is separated from its Parent by 100 kilometers or more of land of another DXCC Entity. Inland waters may be included in the measurement. The test for separation into two areas requires that a line drawn along a great circle in any direction from any part of the proposed Entity must not touch the Parent before crossing 100 kilometers of the intervening DXCC Entity.

b) Island Areas (Separation by Water):

A new Entity results in the case of an island under any of the following conditions:

i) The island is separated from its Parent, and any other islands that make up the DXCC Entity that contains the Parent, by 350 kilometers or more. Measurement of islands in a group begins with measurement from the island containing the capital city. Only one Entity of this type may be attached to any Parent.

ii) The island is separated from its Parent by 350 kilometers or more, and from any other island attached to that Parent in the same or a different island group by 800 kilometers or more.

iii) The island is separated from its Parent by intervening land or islands that are part of another DXCC Entity, such that a line drawn along a great circle in any direction, from any part of the island, does not touch the Parent before touching the intervening DXCC Entity. There is no minimum separation distance for the first island Entity created under this rule. Additional island Entities may be created under this rule, provided that they are similarly separated from the Parent by a different DXCC Entity and separated from any other islands associated with the Parent by at least 800 km.

3. Special Areas:

The Special Areas listed here may not be divided into additional Entities under the DXCC Rules. None of these constitute a Parent Entity, and none creates a precedent for the addition of similar or additional Entities.

a) The International Telecommunications Union in Geneva (4U1ITU) shall, because of its significance to world telecommunications, be considered as a Special Entity. No additional UN locations will be considered under this ruling.

b) The Antarctic Treaty, signed on 1 December 1959 and entered into force on 23 June 1961, establishes the legal framework for the management of Antarctica. The treaty covers, as stated in Article 6, all land and ice shelves below 60 degrees South. This area is known as the Antarctic Treaty Zone. Article 4 establishes that parties to the treaty will not recognize, dispute, or establish territorial claims and that they will assert no new claims while the treaty is in force. Under Article 10, the treaty States will discourage activities by any country in Antarctica that are contrary to the terms of the treaty. In view of these Treaty provisions, no new Entities below 60 degrees South will be added to the DXCC List as long as the Treaty remains in force.

c) The Spratly Islands, due to the nature of conflicting claims, and without recognizing or refuting any claim, is recognized as a Special Entity. Operations from this area will be accepted with the necessary permissions issued by an occupying Entity. Operations without such permissions, such as with a self-assigned (e.g., 1S) callsign, will not be recognized for DXCC credit.

d) Control of Western Sahara (S0) is currently at issue between Morocco and the indigenous population. The UN has stationed a peacekeeping force there. Until the sovereignty issue is settled, only operations licensed by the RASD shall count for DXCC purposes.

e) Entities on the 1998 DXCC List that do not qualify under the current criteria remain as long as they retain the status under which they were originally added. A change in that status will result in a review in accordance with Rule 5 of this Section.

4. Ineligible Areas:

a) Areas having the following characteristics are not eligible for inclusion on the DXCC List, and are considered as part of the host Entity for DXCC purposes:

i) Any extraterritorial legal Entity of any nature including, but not limited to, embassies, consulates, monuments, offices of the United Nations agencies or related organizations, other inter-governmental organizations or diplomatic missions;

ii) Any area with limited sovereignty or ceremonial status, such as monuments, indigenous areas, reservations, and homelands.

iii) Any area classified as a Demilitarized Zone, Neutral Zone or Buffer Zone.

b) Any area which is unclaimed or not owned by any recognized government is not eligible for inclusion on the DXCC List and will not count for DXCC purposes.

[1] The exceptions to this rule are international organizations, such as the UN and ICAO. These Entities are classified under Special Areas, 3. a); and Ineligible Areas, 4. b).

5. Removal Criteria:

a) An Entity may be removed from the List if it no longer satisfies the criteria under which it was added. However, if the Entity continues to meet one or more currently existing rules, it will remain on the List.

b) An Entity may be removed from the List if it was added to the List:

i) Based on a factual error (Examples of factual errors include inaccurate measurements, or observations from incomplete, inaccurate or outdated charts or maps); and

ii) The error was made less than five years earlier than its proposed removal date.

c) A change in the DXCC Criteria shall not affect the status of any Entity on the DXCC List at the time of the change. In other words, criteria changes will not be applied retroactively to Entities on the List.

SECTION III. ACCREDITATION CRITERIA

1. Each nation of the world manages its telecommunications affairs differently. Therefore, a rigid, universal accreditation procedure cannot be applied in all situations. During more than 50 years of DXCC administration, basic standards have evolved in establishing the legitimacy of an operation.

It is the purpose of this section to establish guidelines that will assure that DXCC credit is given only for contacts with operations that are conducted with proper licensing and have established a legitimate physical presence within the Entity to be credited. Any operation that satisfies these conditions (in addition to the applicable elements of SECTION I., Rules 6, 7, 8, and 9) will be accredited. It is the intent of the DXCC administration to be guided by the actions of sovereign nations when considering the accreditation of amateur radio operation within their jurisdiction. DXCC will be reasonably flexible in reviewing licensing documentation. Conversely, findings by a host government indicating non-compliance with their amateur radio regulations may cause denial or revocation of accreditation.

2. The following points should be of particular interest to those seeking accreditation for a DX operation:

a) The vast majority of operations are accredited routinely without a requirement for the submission of authenticating documentation. However, all such documents should be retained by the operator in the unlikely event of a protest.

b) In countries where Amateur Radio operation has not been permitted or has been suspended or where some reluctance to authorize amateur stations has been noted, authenticating documents may be required before accrediting an operation.

c) Special permission may be required from a governmental agency or private party before entering certain DXCC Entities for the purpose of conducting amateur radio operations even though the Entity is part of a country with no amateur radio restrictions. Examples of such Entities are Desecheo I. (KP5); Palmyra I. (KH5); and Glorioso Islands (FR/G).

3. For those cases where supporting documentation is required, the following can be used as a guide to identify those documents necessary for accreditation.

a) Photocopy of license or operating authorization.

b) Photocopy of passport entry and exit stamps.

c) For islands, a landing permit and a signed statement of the transporting ship's, boat's, or aircraft's captain, showing all pertinent data, such as date, place of landing, etc.

d) For locations where special permission is known to be required to gain access, evidence of this permission must be presented.

e) It is expected that all DXpeditions will observe any environmental rules promulgated by the administration under whose authority the operation takes place. In the event that no such rules are actually promulgated, the DXpedition should leave the DXpedition site as they found it.

4. These accreditation requirements are intended to preserve the integrity of the DXCC program and to ensure that the program does not encourage amateurs to "bend the rules" in their enthusiasm, possibly jeopardizing the future development of Amateur Radio. Every effort will be made to apply these criteria uniformly and to make a determination consistent with these objectives.

Recent Changes to the DXCC List

Prefix	Entity	Start Date	Add Date
VP6	Ducie I.	Nov 16, 2001	June 1, 2002
4W	UNTAET (E. Timor)	Mar 1, 2000	Oct 1, 2000
FK/C	Chesterfield Is.	Mar 23, 2000	Oct 1, 2000
E4	Palestine	Feb 1, 1999	Oct 1, 1999
FO	Austral Is	Apr 1, 1998	Nov 1, 1998
FO	Marquesas Is	Apr 1, 1998	Nov 1, 1998
H4Ø	Temotu Province	Apr 1, 1998	Nov 1, 1998

Deleted from the DXCC List:

Prefix	Entity	Deleted Date
STØ	Southern Sudan	Apr 1, 1998

Note: New rules were adopted by the ARRL Board of Directors in their 1998 annual meeting. These went into effect on March 31, 1998.

SECTION IV. FIELD CHECKING OF QSL CARDS

QSL cards for new DXCC awards and endorsements may be checked by a DXCC Card Checker. This program applies to any DXCC award for an individual or station, except those specifically excluded.

1) Entities Eligible for Field Checking:

a) All cards dated ten years or less from the calendar year of the application may be checked, except for those awards specifically excluded from the program. 160 meter cards are currently excluded from this program. QSLs for years more than ten calendar years from the application date must be submitted directly to ARRL Headquarters. All deleted entities must be submitted to ARRL HQ.

b) The ARRL Awards Committee determines which entities are eligible for Field Checking.

2) DXCC Card Checkers:

a) Nominations for Card Checkers may be made by:

i) The Section Manager of the section in which the prospective checker resides.

ii) An ARRL affiliated DX specialty club with at least 25 members who are DXCC members, and which has DX as its primary interest. If there are any questions regarding the validity of a DX club, the issue shall be determined by the division director where the DX club is located. A person does not have to be a member of a nominating DX club.

b) Appointments are limited to one per section and one per DX club.

c) Qualifications:

i) Those nominated as a card checker must be of known integrity, and must be personally known to the person nominating them for appointment.

ii) Candidates must be ARRL members who hold a DXCC award endorsed for at least 150 entities.

iii) Candidates must complete an open book test about

DXCC rules concerning QSL cards and the Card Checker training guide.

iv) The applicant must be willing to serve at reasonable times and places, including at least one state or Division ARRL Convention each year.

v) The applicant must have e-mail and Internet capabilities, and must maintain current e-mail address with DXCC Desk.

d) Approval:

Applications for DXCC Card Checkers are approved by the Director of the ARRL Division in which they reside and are appointed by the Membership Services Manager.

e) Appointments are made for a two year period. Retention of appointees is determined by performance as determined by the DXCC Desk.

3) Card Checking Process:

a) Only eligible cards can be checked by DXCC Card Checkers. An application for a new award shall contain a minimum of 100 QSL confirmations from the list and shall not contain any QSLs that are not eligible for this program. Additional cards may not be sent to HQ with field checked applications. The application may contain any number of cards, subject to eligibility requirements and fees as determined by Section I, Basic, 15.

b) It is the applicant's responsibility to get cards to and from the DXCC Card Checker.

c) Checkers may, at their own discretion, handle members' cards by mail.

d) The ARRL is not responsible for cards handled by DXCC Card Checkers and will not honor any claims.

e) The QSL cards may be checked by one DXCC Card Checker.

f) The applicant and DXCC Card Checker must sign the application form. (See Section I no. 11 regarding altered, forged or otherwise invalid confirmations.)

g) The applicant shall provide a stamped no. 10 envelope (business size) addressed to ARRL HQ to the DXCC Card Checker. The applicant shall also provide the applicable fees (check or money order payable to ARRL-no cash, credit card number and expiration date is also acceptable).

h) The DXCC Card Checker will forward completed applications and appropriate fee(s) to ARRL HQ.

4) ARRL HQ involvement in the card checking process:

a) ARRL HQ staff will receive field-checked applications, enter application data into DXCC records and issue DXCC credits and awards as appropriate.

b) ARRL HQ staff will perform random audits of applications. Applicants or members may be requested to forward cards to HQ for checking before or after credit is issued.

c) The applicant and the DXCC Card Checker will be advised of any errors or discrepancies encountered by ARRL staff.

d) ARRL HQ staff provides instructions and guidelines to DXCC Card Checkers.

5) Applicants and DXCC members may send cards to ARRL Headquarters at any time for review or recheck if the individual feels that an incorrect determination has been made.

ARRL DXCC LIST

ALL ENTITIES ON THE CURRENT LIST ARE ELIGIBLE FOR FIELD CHECKING

NOTE: * INDICATES CURRENT LIST OF ENTITIES FOR WHICH QSLs MAY BE FORWARDED BY THE ARRL MEMBERSHIP OUTGOING QSL SERVICE.

NOTE: † INDICATES ENTITIES WITH WHICH U.S. AMATEURS MAY LEGALLY HANDLE THIRD-PARTY MESSAGE TRAFFIC.

Prefix	Entity	CONTINENT	ZONE		MIXED	PHONE	CW	RTTY	SAT	160	80	40	20	17	15	12	10	6
			ITU	CQ														
	Spratly Is.	AS	50	26														
1AØ[1]	Sov. Mil. Order of Malta	EU	28	15														
3A*	Monaco	EU	27	14														
3B6, 7*	Agalega & St. Brandon	AF	53	39														
3B8*	Mauritius	AF	53	39														
3B9*	Rodrigues I.	AF	53	39														
3C	Equatorial Guinea	AF	47	36														
3CØ	Annobon I.	AF	52	36														
3D2*	Fiji	OC	56	32														
3D2*	Conway Reef	OC	56	32														
3D2*	Rotuma I.	OC	56	32														
3DA†*	Swaziland	AF	57	38														
3V*	Tunisia	AF	37	33														
3W, XV	Vietnam	AS	49	26														
3X	Guinea	AF	46	35														
3Y*	Bouvet	AF	67	38														
3Y*	Peter I I.	AN	72	12														
4J, 4K*	Azerbaijan	AS	29	21														
4L*	Georgia	AS	29	21														
4P-4S*	Sri Lanka	AS	41	22														
4U_ITU†*	ITU HQ	EU	28	14														
4U_UN*	United Nations HQ	NA	08	05														
4W[44]	UNTAET (E. Timor)	OC	54	28														
4X, 4Z†*	Israel	AS	39	20														
5A	Libya	AF	38	34														
5B*	Cyprus	AS	39	20														
5H-5I*	Tanzania	AF	53	37														
5N-5O*	Nigeria	AF	46	35														
5R-5S	Madagascar	AF	53	39														
5T[2]	Mauritania	AF	46	35														
5U[3]	Niger	AF	46	35														
5V	Togo	AF	46	35														

Prefix	Entity	CONTINENT	ZONE		MIXED	PHONE	CW	RTTY	SAT	160	80	40	20	17	15	12	10	6
			ITU	CQ														
5W*	Samoa	OC	62	32														
5X*	Uganda	AF	48	37														
5Y-5Z*	Kenya	AF	48	37														
6V-6W[4]*	Senegal	AF	46	35														
6Y†*	Jamaica	NA	11	08														
7O[5]	Yemen	AS	39	21, 37														
7P	Lesotho	AF	57	38														
7Q	Malawi	AF	53	37														
7T-7Y*	Algeria	AF	37	33														
8P*	Barbados	NA	11	08														
8Q	Maldives	AS/AF	41	22														
8R†*	Guyana	SA	12	09														
9A[6]*	Croatia	EU	28	15														
9G[7]†*	Ghana	AF	46	35														
9H*	Malta	EU	28	15														
9I-9J*	Zambia	AF	53	36														
9K*	Kuwait	AS	39	21														
9L†*	Sierra Leone	AF	46	35														
9M2, 4[8]*	West Malaysia	AS	54	28														
9M6, 8[8]*	East Malaysia	OC	54	28														
9N	Nepal	AS	42	22														
9Q-9T*	Dem. Rep. of Congo	AF	52	36														
9U[9]	Burundi	AF	52	36														
9V[10]*	Singapore	AS	54	28														
9X[9]	Rwanda	AF	52	36														
9Y-9Z†*	Trinidad & Tobago	SA	11	09														
A2*	Botswana	AF	57	38														
A3	Tonga	OC	62	32														
A4*	Oman	AS	39	21														
A5	Bhutan	AS	41	22														
A6	United Arab Emirates	AS	39	21														
A7*	Qatar	AS	39	21														
A9*	Bahrain	AS	39	21														
AP-AS*	Pakistan	AS	41	21														
BS7[11]	Scarborough Reef	AS	50	27														
BV*	Taiwan	AS	44	24														
BV9P[12]	Pratas I.	AS	44	24														
BY,BT*	China	AS	(A)	23,24														
C2*	Nauru	OC	65	31														
C3*	Andorra	EU	27	14														

			ZONE															
Prefix	Entity	CONTINENT	ITU	CQ	MIXED	PHONE	CW	RTTY	SAT	160	80	40	20	17	15	12	10	6
C5†*	The Gambia	AF	46	35														
C6*	Bahamas	NA	11	08														
C8-9*	Mozambique	AF	53	37														
CA-CE†*	Chile	SA	14,16	12														
CEØ†*	Easter I.	SA	63	12														
CEØ†*	Juan Fernandez Is.	SA	14	12														
CEØ†*	San Felix & San Ambrosio	SA	14	12														
CE9/KC4▲*	Antarctica	AN	(B)	(C)														
CM, CO†*	Cuba	NA	11	08														
CN*	Morocco	AF	37	33														
CP†*	Bolivia	SA	12,14	10														
CT*	Portugal	EU	37	14														
CT3*	Madeira Is.	AF	36	33														
CU*	Azores	EU	36	14														
CV-CX†*	Uruguay	SA	14	13														
CYØ*	Sable I.	NA	09	05														
CY9*	St. Paul I.	NA	09	05														
D2-3	Angola	AF	52	36														
D4*	Cape Verde	AF	46	35														
D6†[13]*	Comoros	AF	53	39														
DA-DL[14]*	Fed. Rep. of Germany	EU	28	14														
DU-DZ†*	Philippines	OC	50	27														
E3[15]	Eritrea	AF	48	37														
E4[43]	Palestine	AS	39	20														
EA-EH*	Spain	EU	37	14														
EA6-EH6*	Balearic Is.	EU	37	14														
EA8-EH8*	Canary Is.	AF	36	33														
EA9-EH9*	Ceuta & Melilla	AF	37	33														
EI-EJ*	Ireland	EU	27	14														
EK*	Armenia	AS	29	21														
EL†*	Liberia	AF	46	35														
EP-EQ*	Iran	AS	40	21														
ER*	Moldova	EU	29	16														
ES*	Estonia	EU	29	15														
ET*	Ethiopia	AF	48	37														
EU, EV, EW*	Belarus	EU	29	16														
EX*	Kyrgyzstan	AS	30,31	17														
EY*	Tajikistan	AS	30	17														
EZ*	Turkmenistan	AS	30	17														
F*	France	EU	27	14														

			ZONE															
Prefix	Entity	CONTINENT	ITU	CQ	MIXED	PHONE	CW	RTTY	SAT	160	80	40	20	17	15	12	10	6
FG*	Guadeloupe	NA	11	08														
FJ, FS[1]*	Saint Martin	NA	11	08														
FH[13]*	Mayotte	AF	53	39														
FK*	New Caledonia	OC	56	32														
FK/C[45]	Chesterfield Is.	OC	56	30														
FM*	Martinique	NA	11	08														
FO[16]*	Austral Is.	OC	63	32														
FO*	Clipperton I.	NA	10	07														
FO*	French Polynesia	OC	63	32														
FO[16]*	Marquesas Is.	OC	63	31														
FP*	St. Pierre & Miquelon	NA	09	05														
FR/G[17]*	Glorioso Is.	AF	53	39														
FR/J, E[17]*	Juan de Nova, Europa	AF	53	39														
FR*	Reunion I.	AF	53	39														
FR/T*	Tromelin I.	AF	53	39														
FT5W*	Crozet I.	AF	68	39														
FT5X*	Kerguelen Is.	AF	68	39														
FT5Z*	Amsterdam & St. Paul Is.	AF	68	39														
FW*	Wallis & Futuna Is.	OC	62	32														
FY*	French Guiana	SA	12	09														
G, GX*#	England	EU	27	14														
GD, GT*	Isle of Man	EU	27	14														
GI, GN*	Northern Ireland	EU	27	14														
GJ, GH*	Jersey	EU	27	14														
GM, GS*	Scotland	EU	27	14														
GU, GP*	Guernsey	EU	27	14														
GW, GC*	Wales	EU	27	14														
H4*	Solomon Is.	OC	51	28														
H4Ø[18]*	Temotu Province	OC	51	32														
HA, HG*	Hungary	EU	28	15														
HB*	Switzerland	EU	28	14														
HBØ*	Liechtenstein	EU	28	14														
HC-HD†*	Ecuador	SA	12	10														
HC8-HD8†*	Galapagos Is.	SA	12	10														
HH†*	Haiti	NA	11	08														
HI†*	Dominican Republic	NA	11	08														
HJ-HK†*	Colombia	SA	12	09														
HKØ†*	Malpelo I.	SA	12	09														
HKØ†*	San Andres & Providencia	NA	11	07														
HL*	Republic of Korea	AS	44	25														

#Third-party traffic permitted with special-events stations in the United Kingdom having the prefix GB *only*, with the exception that GB3 stations are not included in this agreement.

Prefix	Entity	CONTINENT	ZONE		MIXED	PHONE	CW	RTTY	SAT	160	80	40	20	17	15	12	10	6
			ITU	CQ														
HO-HP†*	Panama	NA	11	07														
HQ-HR†*	Honduras	NA	11	07														
HS, E2*	Thailand	AS	49	26														
HV*	Vatican	EU	28	15														
HZ*	Saudi Arabia	AS	39	21														
I*	Italy	EU	28	15, 33														
ISØ, IMØ *	Sardinia	EU	28	15														
J2*	Djibouti	AF	48	37														
J3†*	Grenada	NA	11	08														
J5	Guinea-Bissau	AF	46	35														
J6†*	St. Lucia	NA	11	08														
J7†*	Dominica	NA	11	08														
J8†*	St. Vincent	NA	11	08														
JA-JS*	Japan	AS	45	25														
JD1[19]*	Minami Torishima	OC	90	27														
JD1[20]*	Ogasawara	AS	45	27														
JT-JV*	Mongolia	AS	32,33	23														
JW*	Svalbard	EU	18	40														
JX*	Jan Mayen	EU	18	40														
JY†*	Jordan	AS	39	20														
K,W, N, AA-AK†	United States of America	NA	6,7,8	3,4,5														
KG4†	Guantanamo Bay	NA	11	08														
KHØ†	Mariana Is.	OC	64	27														
KH1†	Baker & Howland Is.	OC	61	31														
KH2†*	Guam	OC	64	27														
KH3†*	Johnston I.	OC	61	31														
KH4†	Midway I.	OC	61	31														
KH5†	Palmyra & Jarvis Is.	OC	61,62	31														
KH5K†	Kingman Reef	OC	61	31														
KH6, 7†*	Hawaii	OC	61	31														
KH7K†	Kure I.	OC	61	31														
KH8†	American Samoa	OC	62	32														
KH9†	Wake I.	OC	65	31														
KL7†*	Alaska	NA	1, 2	1														
KP1†	Navassa I.	NA	11	08														
KP2†*	Virgin Is.	NA	11	08														
KP3, 4†*	Puerto Rico	NA	11	08														
KP5[22]†	Desecheo I.	NA	11	08														
LA-LN*	Norway	EU	18	14														
LO-LW†*	Argentina	SA	14,16	13														

Prefix	Entity	CONTINENT	ZONE		MIXED	PHONE	CW	RTTY	SAT	160	80	40	20	17	15	12	10	6
			ITU	CQ														
LX*	Luxembourg	EU	27	14														
LY*	Lithuania	EU	29	15														
LZ*	Bulgaria	EU	28	20														
OA-OC†*	Peru	SA	12	10														
OD*	Lebanon	AS	39	20														
OE	Austria	EU	28	15														
OF-OI*	Finland	EU	18	15														
OHØ*	Aland Is.	EU	18	15														
OJØ, OHØM*	Market Reef	EU	18	15														
OK-OL[23]*	Czech Rep.	EU	28	15														
OM[23]*	Slovak Rep.	EU	28	15														
ON-OT*	Belgium	EU	27	14														
OX*	Greenland	NA	5, 75	40														
OY*	Faroe Is.	EU	18	14														
OZ*	Denmark	EU	18	14														
P2[24]*	Papua New Guinea	OC	51	28														
P4[25]*	Aruba	SA	11	09														
P5[26]	Dem. People's Rep. Korea	AS	44	25														
PA-PI*	Netherlands	EU	27	14														
PJ2, 4, 9*	Bonaire,Curacao(Neth. Antilles)	SA	11	09														
PJ5-8*	St.Maarten,Saba,St.Eustatius	NA	11	08														
PP-PY†*	Brazil	SA	(D)	11														
PPØ-PYØF†*	Fernando de Noronha	SA	13	11														
PPØ-PYØS†*	St. Peter & St. Paul Rocks	SA	13	11														
PPØ-PYØT†*	Trindade & Martim Vaz Is.	SA	15	11														
PZ*	Suriname	SA	12	09														
R1FJ*	Franz Josef Land	EU	75	40														
R1MV	Malyj Vysotskij I	EU	29	16														
SØ[1,27]	Western Sahara	AF	46	33														
S2*	Bangladesh	AS	41	22														
S5[6]*	Slovenia	EU	28	15														
S7	Seychelles	AF	53	39														
S9*	Sao Tome & Principe	AF	47	36														
SA-SM*	Sweden	EU	18	14														
SN-SR*	Poland	EU	28	15														
ST*	Sudan	AF	47, 48	34														
SU	Egypt	AF	38	34														
SV-SZ*	Greece	EU	28	20														
SV/A*	Mount Athos	EU	28	20														
SV5*	Dodecanese	EU	28	20														
SV9*	Crete	EU	28	20														

Prefix	Entity	Continent	Zone		Mixed	Phone	CW	RTTY	SAT	160	80	40	20	17	15	12	10	6
			ITU	CQ														
T2[28]	Tuvalu	OC	65	31														
T3Ø	W. Kiribati (Gilbert Is.)	OC	65	31														
T31	C. Kiribati (Brit. Phoenix Is.)	OC	62	31														
T32	E. Kiribati (Line Is.)	OC	61,63	31														
T33	Banaba I. (Ocean I.)	OC	65	31														
T5	Somalia	AF	48	37														
T7*	San Marino	EU	28	15														
T8, KC6[21]	Palau	OC	64	27														
T9[29]*	Bosnia-Herzegovina	EU	28	15														
TA-TC†*	Turkey	EU/AS	39	20														
TF*	Iceland	EU	17	40														
TG, TD†*	Guatemala	NA	11	07														
TI, TE†*	Costa Rica	NA	11	07														
TI9†*	Cocos I.	NA	11	07														
TJ	Cameroon	AF	47	36														
TK*	Corsica	EU	28	15														
TL[30]	Central Africa	AF	47	36														
TN[31]	Congo (Republic of)	AF	52	36														
TR[32]*	Gabon	AF	52	36														
TT[33]	Chad	AF	47	36														
TU[34]*	Côte d'Ivoire	AF	46	35														
TY[35]	Benin	AF	46	35														
TZ[36]*	Mali	AF	46	35														
UA-UI1,3,4,6 RA-RZ*	European Russia	EU	(E)	16														
UA2*	Kaliningrad	EU	29	15														
UA-UI8, 9, Ø RA-RZ*	Asiatic Russia	AS	(F)	(G)														
UJ-UM*	Uzbekistan	AS	30	17														
UN-UQ*	Kazakhstan	AS	29-31	17														
UR-UZ, EM-EO*	Ukraine	EU	29	16														
V2†*	Antigua & Barbuda	NA	11	08														
V3†*	Belize	NA	11	07														
V4[37]†*	St. Kitts & Nevis	NA	11	08														
V5*	Namibia	AF	57	38														
V6[38]†	Micronesia	OC	65	27														
V7†*	Marshall Is.	OC	65	31														
V8*	Brunei	OC	54	28														
VE, VO, VY†*	Canada	NA	(H)	1-5														
VK†*	Australia	OC	(I)	29,30														
VKØ†*	Heard I.	AF	68	39														
VKØ†*	Macquarie I.	OC	60	30														
VK9C†*	Cocos-Keeling Is.	OC	54	29														

Prefix	Entity	CONTINENT	ZONE		MIXED	PHONE	CW	RTTY	SAT	160	80	40	20	17	15	12	10	6
			ITU	CQ														
VK9L†*	Lord Howe I.	OC	60	30														
VK9M†*	Mellish Reef	OC	56	30														
VK9N†*	Norfolk I.	OC	60	32														
VK9W†*	Willis I.	OC	55	30														
VK9X†*	Christmas I.	OC	54	29														
VP2E[37]	Anguilla	NA	11	08														
VP2M[37]	Montserrat	NA	11	08														
VP2V[37]*	British Virgin Is.	NA	11	08														
VP5*	Turks & Caicos Is.	NA	11	08														
VP6†*	Pitcairn I.	OC	63	32														
VP6[46]*	Ducie I.	OC	63	32														
VP8*	Falkland Is.	SA	16	13														
VP8, LU*	South Georgia I.	SA	73	13														
VP8, LU*	South Orkney Is.	SA	73	13														
VP8, LU*	South Sandwich Is.	SA	73	13														
VP8, LU, CE9, HFØ, 4K1*	South Shetland Is.	SA	73	13														
VP9*	Bermuda	NA	11	05														
VQ9*	Chagos Is.	AF	41	39														
VR*	Hong Kong	AS	44	24														
VU*	India	AS	41	22														
VU*	Andaman & Nicobar Is.	AS	49	26														
VU*	Lakshadweep Is.	AS	41	22														
XA-XI†*	Mexico	NA	10	06														
XA4-XI4†*	Revilla Gigedo	NA	10	06														
XT[39]*	Burkina Faso	AF	46	35														
XU	Cambodia	AS	49	26														
XW	Laos	AS	49	26														
XX9*	Macao	AS	44	24														
XY-XZ	Myanmar	AS	49	26														
YA	Afghanistan	AS	40	21														
YB-YH[40]*	Indonesia	OC	51,54	28														
YI*	Iraq	AS	39	21														
YJ*	Vanuatu	OC	56	32														
YK*	Syria	AS	39	20														
YL*	Latvia	EU	29	15														
YN†*	Nicaragua	NA	11	07														
YO-YR*	Romania	EU	28	20														
YS†*	El Salvador	NA	11	07														
YT-YU, YZ*	Yugoslavia	EU	28	15														
YV-YY†*	Venezuela	SA	12	09														

Prefix	Entity	CONTINENT	ZONE		MIXED	PHONE	CW	RTTY	SAT	160	80	40	20	17	15	12	10	6
			ITU	CQ														
YVØ†*	Aves I.	NA	11	08														
Z2*	Zimbabwe	AF	53	38														
Z3[41]*	Macedonia (former Yugoslav Rep.)	EU	28	15														
ZA*	Albania	EU	28	15														
ZB2*	Gibraltar	EU	37	14														
ZC4[42] *	UK Sov. Base Areas on Cyprus	AS	39	20														
ZD7*	St. Helena	AF	66	36														
ZD8*	Ascension I.	AF	66	36														
ZD9	Tristan da Cunha & Gough I.	AF	66	38														
ZF*	Cayman Is.	NA	11	08														
ZK1	N. Cook Is.	OC	62	32														
ZK1	S. Cook Is.	OC	62	32														
ZK2*	Niue	OC	62	32														
ZK3*	Tokelau Is.	OC	62	31														
ZL-ZM*	New Zealand	OC	60	32														
ZL7*	Chatham Is.	OC	60	32														
ZL8*	Kermadec Is.	OC	60	32														
ZL9*	Auckland & Campbell Is.	OC	60	32														
ZP†*	Paraguay	SA	14	11														
ZR-ZU*	South Africa	AF	57	38														
ZS8*	Prince Edward & Marion Is.	AF	57	38														

Notes

[1] Unofficial prefix.
[2] (5T) Only contacts made June 20, 1960, and after, count for this entity.
[3] (5U) Only contacts made August 3, 1960, and after, count for this entity.
[4] (6W) Only contacts made June 20, 1960, and after, count for this entity.
[5] (7O) Only contacts made May 22, 1990, and after, count for this entity.
[6] (9A, S5) Only contacts made June 26, 1991, and after, count for this entity.
[7] (9G) Only contacts made March 5, 1957, and after, count for this entity.
[8] (9M2, 4, 6, 8) Only contacts made September 16, 1963, and after, count for this entity.
[9] (9U, 9X) Only contacts made July 1, 1962, and after, count for this entity.
[10] (9V) Contacts made from September 16, 1963 to August 8, 1965, count for West Malaysia.
[11] (BS7) Only contacts made January 1, 1995, and after, count for this entity.
[12] (BV9P) Only contacts made January 1, 1994, and after, count for this entity.
[13] (D6, FH8) Only contacts made July 6, 1975, and after, count for this entity.
[14] (DA-DL) Only contacts made with DA-DL stations September 17, 1973, and after, and contacts made with Y2-Y9 stations October 3, 1990 and after, count for this entity.
[15] (E3) Only contacts made November 14, 1962, and before, or May 24, 1991, and after, count for this entity.
[16] (FO) Only contacts made after 23:59 UTC, March 31, 1998 count for this entity.
[17] (FR) Only contacts made June 25, 1960, and after, count for this entity.
[18] (H4Ø) Only contacts made after 23:59 UTC, March 31, 1998, count for this entity.
[19] (JD) Formerly Marcus Island.
[20] (JD) Formerly Bonin and Volcano Islands.
[21] (KC6) Includes Yap Islands December 31, 1980, and before.
[22] (KP5, KP4) Only contacts made March 1, 1979, and after, count for this entity.
[23] (OK-OL, OM) Only contacts made January 1, 1993, and after, count for this entity.
[24] (P2) Only contacts made September 16, 1975, and after, count for this entity.
[25] (P4) Only contacts made January 1, 1986, and after, count for this entity.
[26] (P5) Only contacts made May 14, 1995, and after count for this entity.
[27] (SØ) Contacts with Rio de Oro (Spanish Sahara), EA9, also count for this entity.
[28] (T2) Only contacts made January 1, 1976, and after, count for this entity.
[29] (T9) Only contacts made October 15, 1991 and after count for this entity.
[30] (TL) Only contacts made August 13, 1960, and after, count for this entity.
[31] (TN) Only contacts made August 15, 1960, and after, count for this entity.
[32] (TR) Only contacts made August 17, 1960, and after, count for this entity.
[33] (TT) Only contacts made August 11, 1960, and after count for this entity.
[34] (TU) Only contacts made August 7, 1960, and after, count for this entity.
[35] (TY) Only contacts made August 1, 1960, and after, count for this entity.
[36] (TZ) Only contacts made June 20, 1960, and after, count for this entity.
[37] (V4, VP2) For DXCC credit for contacts made May 31, 1958 and before, see page 97, June 1958 *QST*.
[38] (V6) Includes Yap Islands January 1, 1981, and after.
[39] (XT) Only contacts made August 5, 1960, and after, count for this entity.
[40] (YB) Only contacts made May 1, 1963, and after, count for this entity.
[41] (Z3) Only contacts made September 8, 1991, and after, count for this entity.
[42] (ZC4) Only contacts made August 16, 1960, and after, count for this entity.
[43] (E4) Only contacts made February 1, 1999 and after, count for this entity.
[44] (4W) Only contacts made March 1, 2000, and after, count for this entity.
[45] (FK/C) Only contacts made March 23, 2000, and after, count for this entity.
[46] (VP6) Only contacts made November 16, 2001, and after, count for this entity. This entity will be/was added to the List on June 1, 2002.

▲Also 3Y, 8J1, ATØ, DPØ, FT8Y, LU, OR4, R1AN, VKØ, VP8, ZL5, ZS1, ZXø, etc. QSL via entity under whose auspices the particular station is operating. The availability of a third-party traffic agreement and a QSL Bureau applies to the entity under whose auspices the particular station is operating.

Zone Notes can be found with Prefix Cross References.

Deleted Entities Total: 58

DELETED ENTITIES

Credit for any of these entities can be given if the date of contact in question agrees with the date(s) shown in the corresponding footnote. Entities deleted after 23:59 UTC March 31, 1998 will not be included in deleted list.

Prefix	Entity	CONTINENT	ZONE		MIXED	PHONE	CW	RTTY	SAT	160	80	40	20	17	15	12	10	6
			ITU	CQ														
[2]	Blenheim Reef	AF	41	39														
[3]	Geyser Reef	AF	53	39														
[4]	Abu Ail Is.	AS	39	21														
1M[1, 5]	Minerva Reef	OC	62	32														
4W[6]	Yemen Arab Rep.	AS	39	21														
7J1[7]	Okino Tori-shima	AS	45	27														
8Z4[8]	Saudi Arabia/Iraq Neut. Zone	AS	39	21														
8Z5, 9K3[9]	Kuwait/Saudi Arabia Neut. Zone	AS	39	21														
9S4[10]	Saar	EU	28	14														
9U5[11]	Ruanda-Urundi	AF	52	36														
AC3[1, 12]	Sikkim	AS	41	22														
AC4[1, 13]	Tibet	AS	41	23														
C9[14]	Manchuria	AS	33	24														
CN2[15]	Tangier	AF	37	33														
CR8[16]	Damao, Diu	AS	41	22														
CR8[16]	Goa	AS	41	22														
CR8, CR10[17]	Portuguese Timor	OC	54	28														
DA-DM[18]	Germany	EU	28	14														
DM, Y2-9[19]	German Dem. Rep.	EU	28	14														
EA9[20]	Ifni	AF	37	33														
FF[21]	French West Africa	AF	46	35														
FH, FB8[22]	Comoros	AF	53	39														
FI8[23]	French Indo-China	AS	49	26														
FN8[24]	French India	AS	41	22														
FQ8[25]	Fr. Equatorial Africa	AF	47,52	36														
HKØ[26]	Bajo Nuevo	NA	11	08														
HKØ,KP3,KS4[26]	Serrana Bank & Roncador Cay	NA	11	07														
I1[27]	Trieste	EU	28	15														
I5[28]	Italian Somaliland	AF	48	37														
JZØ[29]	Netherlands N. Guinea	OC	51	28														
KR6,8,JR6,KA6[30]	Okinawa (Ryukyu Is.)	AS	45	25														
KS4[31]	Swan Is.	NA	11	07														
KZ5[32]	Canal Zone	NA	11	07														
OK-OM[33]	Czechoslovakia	EU	28	15														
P2, VK9[34]	Papua Territory	OC	51	28														
P2, VK9[34]	Terr. New Guinea	OC	51	28														
PK1-3[35]	Java	OC	54	28														
PK4[35]	Sumatra	OC	54	28														
PK5[35]	Netherlands Borneo	OC	54	28														

Prefix	Entity	CONTINENT	ZONE ITU	ZONE CQ	MIXED	PHONE	CW	RTTY	SAT	160	80	40	20	17	15	12	10	6
PK6[35]	Celebe & Molucca Is.	OC	54	28														
STØ[36]	Southern Sudan	AF	47, 48	34														
UN1[37]	Karelo-Finnish Rep.	EU	19	16														
VO[38]	Newfoundland, Labrador	NA	09	02,05														
VQ1, 5H1[39]	Zanzibar	AF	53	37														
VQ6[40]	British Somaliland	AF	48	37														
VQ9[41]	Aldabra	AF	53	39														
VQ9[41]	Desroches	AF	53	39														
VQ9[41]	Farquhar	AF	53	39														
VS2, 9M2[42]	Malaya	AS	54	28														
VS4[42]	Sarawak	OC	54	28														
VS9A, P, S[43]	People's Dem. Rep. of Yemen	AS	39	21														
VS9H[44]	Kuria Muria I.	AS	39	21														
VS9K[45]	Kamaran Is.	AS	39	21														
ZC5[42]	British North Borneo	OC	54	28														
ZC6, 4X1[46]	Palestine	AS	39	20														
ZD4[47]	Gold Coast, Togoland	AF	46	35														
ZSØ, 1[48]	Penguin Is.	AF	57	38														
ZS9[49]	Walvis Bay	AF	57	38														

NOTES:

1 Unofficial prefix.

2 (Blenheim Reef) Only contacts made from May 4, 1967, to June 30, 1975, count for this entity. Contacts made July 1, 1975, and after, count as Chagos (VQ9).

3 (Geyser Reef) Only contacts made from May 4, 1967, to February 28, 1978, count for this entity.

4 (Abu Ail Is.) Only contacts made March 30, 1991, and before, count for this entity.

5 (1M) Only contacts made from July 15, 1972, and before, count for this entity. Contacts made July 16, 1972, and after, count as Tonga (A3).

6 (4W) Only contacts made May 21, 1990, and before, count for this entity.

7 (7J1) Only contacts made May 30, 1976, to November 30, 1980, count for this entity. Contacts made December 1, 1980, and after, count as Ogasawara (JD1).

8 (8Z4) Only contacts made December 25, 1981, and before, count for this entity.

9 (8Z5, 9K3) Only contacts made December 14, 1969, and before, count for this entity.

10 (9S4) Only contacts made March 31, 1957, and before, count for this entity.

11 (9U5) Only contacts made from July 1, 1960, to June 30, 1962, count for this entity. Contacts made July 1, 1962, and after, count as Burundi (9U) or Rwanda (9X).

12 (AC3) Only contacts made April 30, 1975, and before, count for this entity. Contacts made May 1, 1975, and after, count as India (VU).

13 (AC4) Only contacts made May 30, 1974, and before, count for this entity. Contacts made May 31, 1974 and after, count as China (BY).

14 (C9) Only contacts made September 15, 1963, and before, count for this entity. Contacts made September 16, 1963, and after, count as China (BY).

15 (CN2) Only contacts made June 30, 1960, and before, count for this entity. Contacts made July 1, 1960, and after, count as Morocco (CN).

16 (CR8) Only contacts made December 31, 1961, and before, count for this entity.

17 (CR8, CR10) Only contacts made September 14, 1976, and before, count for this entity.

18 (DA-DM) Only contacts made September 16, 1973, and before, count for this entity. Contacts made September 17, 1973, and after, count as either FRG (DA-DL) or GDR (Y2-Y9).

19 (DM, Y2-Y9) Only contacts made from September 17, 1973 and October 2, 1990 count for this entity. On October 3, 1990 the GDR became part of the FRG.

20 (EA9) Only contacts made May 13, 1969, and before, count for this entity.

21 (FF) Only contacts made August 6, 1960, and before, count for this entity.

22 (FH, FB8) Only contacts made July 5, 1975, and before, count for this entity. Contacts made July 6, 1975, and after, count as Comoros (D6) or Mayotte (FH).

23 (FI8) Only contacts made December 20, 1950, and before, count for this entity.

24 (FN8) Only contacts made October 31, 1954, and before count for this entity.

25 (FQ8) Only contacts made August 16, 1960, and before, will count for this entity.

26 (HKØ , KP3, KS4) Only contacts made September 16, 1981, and before, count for this entity. Contacts made September 17, 1981, and after, count as San Andres (HKØ).

27 (I1) Only contacts made March 31, 1957, and before, count for this entity. Contacts made April 1, 1957, and after, count as Italy (I).

28 (I5) Only contacts made June 30, 1960, and before, count for this entity.

29 (JZØ) Only contacts made April 30, 1963, and before, count for this entity.

30 (KR6, 8, JR6, KA6) Only contacts made May 14, 1972, and before, count for this entity. Contacts made May 15, 1972, and after, count as Japan (JA).

31 (KS4) Only contacts made August 31, 1972, and before, count for this entity. Contacts made September 1, 1972, and after, count as Honduras (HR).

32 (KZ5) Only contacts made September 30, 1979, and before, count for this entity.

33 (OK-OM) Only contacts made December 31, 1992, and before, count for this entity.

34 (P2, VK9) Only contacts made September 15, 1975, and before, count for this entity. Contacts made September 16, 1975, and after, count as Papua New Guinea (P2).

35 (PK1-6) Only contacts made April 30, 1963, and before, count for this entity. Contacts made May 1, 1963, and after, count as Indonesia.

36 (STØ) Only contacts made between May 7, 1972 and December 31, 1994, count for this entity.

37 (UN1) Only contacts made June 30, 1960, and before, count for this entity. Contacts made July 1, 1960, and after, count as European RSFSR (UA).

38 (VO) Only contacts made March 31, 1949, and before, count for this entity. Contacts made April 1, 1949, and after, count as Canada (VE).

39 (VQ1, 5H1) Only contacts made May 31, 1974, and before, count for this entity. Contacts made June 1, 1974, and after, count as Tanzania (5H).

40 (VQ6) Only contacts made June 30, 1960, and before, count for this entity.

41 (VQ9) Only contacts made June 28, 1976, and before, count for this entity. Contacts made June 29, 1976, and after, count as Seychelles (S7).

42 (VS2, VS4, ZC5, 9M2) Only contacts made September 15, 1963, and before, count for this entity. Contacts made September 16, 1963, and after, count as West Malaysia (9M2) or East Malaysia (9M6, 8).

43 (VS9A, P,S) Only contacts made May 21, 1990, and before, count for this entity.

44 (VS9H) Only contacts made November 29, 1967, and before, count for this entity.

45 (VS9K) Only contacts made on March 10, 1982, and before, count for this entity.

46 (ZC6, 4X1) Only contacts made June 30, 1968, and before, count for this entity. Contacts made July 1, 1968, and after, count as Israel (4X).

47 (ZD4) Only contacts made March 5, 1957, and before, count for this entity.

48 (ZSØ, 1) Only contacts made February 28, 1994, and before, count for this entity.

49 (ZS9) Only contacts made from September 1, 1977, to February 28, 1994, count for this entity.

PREFIX CROSS REFERENCES

A8 = EL
AC (before 1972) = A5
AH = KH
AL7 = KL7
AM-AO = EA
AT-AW = VU
AX = VK
AY-AZ = LU
CF-CK = VE
CL = CO
CQ-CS = CT
CR3 (before 1974) = J5
CR4 (before 1976) = D4
CR5 (before 1976) = S9
CR6 (before 1976) = D2
CR7 (before 1976) = C9
CR9 (before 1985) = XX9
CT2 (before 1986) = CU
CXØ = CE9/VP8
CY-CZ = VE
CYØ (before 1985) = CY9
D7 = HL
DM-DT (before 1980) = Y2-9
DS-DT = HL
E2 = HS
EAØ (before 1969) = 3C
EK, EM-EO, ER-ES, EU-EZ = U
ER (after 1992) = UO
EU (after 1991) = UC
FA-FF (after 1983) =F
FA (before 1963) = 7X
FB8 (before 1961) = 5R
FB8 (before 1985) = FT
FC (before 1985) = TK
FD8 (before 1961) = 5V
FE8 (before 1961) =TJ
FL (before 1978) = J2
FU8 (before 1982) = YJ
GB = G
GC (before 1977) = GJ or GU
H2 = 5B
H3 = HP
H5 (Bophutatswana) = ZS
H7 = YN
HE = HB
HM (before 1982) = HL
HT = YN
HU = YS
HW-HY = F
J4 = SV
KA1 = JD1
KA2AA-KA9ZZ = JA
KC6 (before 1990) = V6
KB6 (before 1979) = KH1
KC4 (Navassa) = KP1
KG6 (before 1979) = KH2
KG6I (before 1970) = JD1
KG6R, S, T (before 1979) = KHØ
KH7 (before 1996) = KH7K
KJ6 (before 1979) = KH3
KM6 (before 1979) = KH4
KP4 (Desecheo) = KP5
KP6 (before 1979) = KH5
KS6 (before 1979) = KH8
KV4 (before 1979) = KP2
KW6 (before 1979) = KH9
KX6 (before 1990) = V7
L2-9 = LU
M = G
M1 (before 1984) = T7
MP4B (before 1972) = A9
MP4M (before 1972) = A4
MP4Q (before 1972) = A7
MP4T, D (before 1972) = A6
NH = KH
NL7 = KL7
NP = KP
OQ (before 1961) = 9Q
P3 = 5B
P4 (before 1986) = PJ
PX (before 1970) = C3
RA, RN = UA
RB-RR = UB-UR
RS = U
RT = UB
RU-RX = U
S4 (Ciskei) = ZS
S8 (Transkei) = ZS
T4 = CO
T4 (Venda) = ZS
TH, TM, TO-TQ, TV-TX = F
UB (before 1994) = UZ
UC (before 1991) = EU
UD (before 1994) = 4J
UF (before 1994) = 4L
UG (before 1994) = EK
UH (before 1993) = EZ
UI (before 1994) = UJ
UJ (before 1993) = EY
UL (before 1994) = UN
UM (before 1993) = EX
UO (before 1994) = ER
UP (before 1991) = LY
UQ (before 1992) = YL
UR (before 1991) = ES
V9 (Venda) = ZS
VA-VG = VE
VH-VN = VK
VK9 (Nauru) = C2
VP1 (before 1982) = V3
VP2A (before 1982) = V2
VP2D (before 1979) = J7
VP2G (before 1975) = J3
VP2K (before 1984) = V4 or VP2E
VP2L (before 1980) = J6
VP2S (before 1980) = J8
VP3 (before 1967) = 8R
VP4 (before 1963) = 9Y
VP5 (Jamaica) = 6Y
VP6 (before 1967) = 8P
VP7 (before 1974) = C6
VQ2 (before 1965) = 9J
VQ3 (before 1962) = 5H
VQ4 (before 1964) = 5Z
VQ5 (before 1963) = 5X
VQ8 (before 1969) = 3B
VQ8 (Chagos) = VQ9
VQ9 (Seychelles) = S7
VR1 (before 1980) = T3
VR2 (before 1971) = 3D2
VR2 (after 1991) = VS6
VR3 (before 1980) = T32
VR4 (before 1979) = H4
VR5 (before 1971) = A3
VR6 (before 1998) = VP6
VR8 (before 1979) = T2
VS1 (before 1966) = 9V
VS5 (before 1985) = V8
VS6 (before 1997) = VR
VS7 (before 1949) = 4S
VS9M = 8Q
VS9O (before 1961) = A4
VX-VY = CYØ/VE
WH = KH
WL7 = KL7
WP = KP
XJ-XO = VE
XP = OX
XQ-XR = CE
XV = 3W
XX7 (before 1976) = C9
YU2 (before 1992) = 9A
YU3 (before 1992) = S5
YU4 (before 1992) = T9
YU5 (before 1992) = Z3
ZB1 (before 1965) = 9H
ZD1 (before 1962) = 9L
ZD2 (before 1961) = 5N
ZD3 (before 1966 = C5
ZD5 (before 1969) = 3DA
ZD6 (before 1965) = 7Q
ZE (before 1981) = Z2-9
ZK9 (1983) = ZK2
ZM6 (before 1963) = 5W
ZM7 (before 1984) = ZK3
ZS3 (before 1991) = V5
ZS7 (before 1969) = 3D6
ZS8 (before 1967) = 7P
ZS9 (before 1967) = A2
ZV-ZZ = PY
2D = GD
2E = G
2I = GI
2J = GJ
2M = GM
2U = GU
2W = GW
3B-3C (before 1968) = VE
3D6 (before 1988) = 3DA
3G = CE
3Z = SP
4A-4C = XE
4D-4I = DU
4J-4L = U
4J (after 1991) = EK
4J1F (before 1994) = R1MV
4K (before 1994) = UA
4K1 (before 1994) = CE9/KC4
4K2 (before 1994) = R1FJ
4K3 (before 1994) = UA
4K4 (before 1994) = UAØ
4L (after 1991) = UF
4M = YV
4N-4O = YU
4T = OA
4U1VIC = OE
4V = HH
5J-5K = HK
5L-5M = EL
6C = YK
6D-6J = XE
6K-6N=HL
6O = T5
6T-6U = ST
7A-7I = YB
7G (before 1967) = 3X
7J-7N = JA
7JI = JA1 or JD1
7S = SM
7Z = HZ
8A-8I = YB
8J-8N = JA
8O = A2
8S = SM
9A (before 1984) = T7
9B-9D = EP
9E-9F = ET

CONTINENT

AF = AFRICA
AN = ANTARCTICA
AS = ASIA
EU = EUROPE
NA = NORTH AMERICA
OC = OCEANIA
SA = SOUTH AMERICA

ZONE NOTES

(A) 33, 42, 43, 44
(B) 67, 69-74
(C) 12, 13, 29, 30, 32, 38, 39
(D) 12, 13, 15
(E) 19, 20, 29, 30
(F) 20-26, 30-35, 75
(G) 16, 17, 18, 19, 23
(H) 2, 3, 4, 9, 75
(I) 55, 58, 59

ALLOCATION OF INTERNATIONAL CALL SIGN SERIES

Call Sign Series	Allocated to
AAA-ALZ	United States of America
AMA-AOZ	Spain
APA-ASZ	Pakistan
ATA-AWZ	India
AXA-AXZ	Australia
AYA-AZZ	Argentina
A2A-A2Z	Botswana
A3A-A3Z	Tonga
A4A-A4Z	Oman
A5A-A5Z	Bhutan
A6A-A6Z	United Arab Emirates
A7A-A7Z	Qatar
A8A-A8Z	Liberia
A9A-A9Z	Bahrain
BAA-BZZ	China (People's Republic of)
CAA-CEZ	Chile
CFA-CKZ	Canada
CLA-CMZ	Cuba
CNA-CNZ	Morocco
COA-COZ	Cuba
CPA-CPZ	Bolivia
CQA-CUZ	Portugal
CVA-CXZ	Uruguay
CYA-CZZ	Canada
C2A-C2Z	Nauru
C3A-C3Z	Andorra
C4A-C4Z	Cyprus
C5A-C5Z	Gambia
C6A-C6Z	Bahamas
* C7A-C7Z	World Meteorological Organization
C8A-C9Z	Mozambique
DAA-DRZ	Germany
DSA-DTZ	Republic of Korea
DUA-DZZ	Philippines
D2A-D3Z	Angola
D4A-D4Z	Cape Verde
D5A-D5Z	Liberia
D6A-D6Z	Comoros
D7A-D9Z	Republic of Korea
EAA-EHZ	Spain
EIA-EJZ	Ireland
EKA-EKZ	Armenia
ELA-ELZ	Liberia
EMA-EOZ	Ukraine
EPA-EQZ	Iran
ERA-ERZ	Moldova
ESA-ESZ	Estonia
ETA-ETZ	Ethiopia
EUA-EWZ	Belarus
EXA-EXZ	Kyrgyzstan
EYA-EYZ	Tajikistan
EZA-EZZ	Turkmenistan
E2A-E2Z	Thailand
E3A-E3Z	Eritrea
**E4A-E4Z	Palestinian Authority
FAA-FZZ	France
GAA-GZZ	United Kingdom of Great Britain and Northern Ireland
HAA-HAZ	Hungary
HBA-HBZ	Switzerland
HCA-HDZ	Ecuador
HEA-HEZ	Switzerland
HFA-HFZ	Poland
HGA-HGZ	Hungary
HHA-HHZ	Haiti
HIA-HIZ	Dominican Republic
HJA-HKZ	Colombia
HLA-HLZ	Korea (Republic of)
HMA-HMZ	Democratic People's Republic of Korea
HNA-HNZ	Iraq
HOA-HPZ	Panama
HQA-HRZ	Honduras
HSA-HSZ	Thailand
HTA-HTZ	Nicaragua
HUA-HUZ	El Salvador
HVA-HVZ	Vatican City State
HWA-HYZ	France
HZA-HZZ	Saudi Arabia
H2A-H2Z	Cyprus
H3A-H3Z	Panama
H4A-H4Z	Solomon Islands
H6A-H7Z	Nicaragua
H8A-H9Z	Panama
IAA-IZZ	Italy
JAA-JSZ	Japan
JTA-JVZ	Mongolia
JWA-JXZ	Norway
JYA-JYZ	Jordan
JZA-JZZ	Indonesia
J2A-J2Z	Djibouti
J3A-J3Z	Grenada
J4A-J4Z	Greece
J5A-J5Z	Guinea-Bissau
J6A-J6Z	Saint Lucia
J7A-J7Z	Dominica
J8A-J8Z	St. Vincent and the Grenadines
KAA-KZZ	United States of America
LAA-LNZ	Norway
LOA-LWZ	Argentina
LXA-LXZ	Luxembourg
LYA-LYZ	Lithuania
LZA-LZZ	Bulgaria
L2A-L9Z	Argentina
MAA-MZZ	United Kingdom of Great Britain and Northern Ireland
NAA-NZZ	United States of America
OAA-OCZ	Peru
ODA-ODZ	Lebanon
OEA-OEZ	Austria
OFA-OJZ	Finland
OKA-OLZ	Czech Republic
OMA-OMZ	Slovak Republic
ONA-OTZ	Belgium
OUA-OZZ	Denmark
PAA-PIZ	Netherlands
PJA-PJZ	Netherlands Antilles
PKA-POZ	Indonesia
PPA-PYZ	Brazil
PZA-PZZ	Suriname
P2A-P2Z	Papua New Guinea
P3A-P3Z	Cyprus
P4A-P4Z	Aruba
P5A-P9Z	Democratic People's Republic of Korea
RAA-RZZ	Russian Federation
SAA-SMZ	Sweden
SNA-SRZ	Poland
SSA-SSM	Egypt
SSN-STZ	Sudan
SUA-SUZ	Egypt
SVA-SZZ	Greece
S2A-S3Z	Bangladesh
S5A-S5Z	Slovenia
S6A-S6Z	Singapore
S7A-S7Z	Seychelles
S8A-S8Z	South Africa
S9A-S9Z	Sao Tome and Principe
TAA-TCZ	Turkey
TDA-TDZ	Guatemala
TEA-TEZ	Costa Rica
TFA-TFZ	Iceland
TGA-TGZ	Guatemala
THA-THZ	France
TIA-TIZ	Costa Rica
TJA-TJZ	Cameroon
TKA-TKZ	France
TLA-TLZ	Central Africa
TMA-TMZ	France
TNA-TNZ	Congo (Republic of the)
TOA-TQZ	France
TRA-TRZ	Gabon
TSA-TSZ	Tunisia
TTA-TTZ	Chad
TUA-TUZ	Ivory Coast
TVA-TXZ	France
TYA-TYZ	Benin
TZA-TZZ	Mali

Call Sign Series	*Allocated to*
T2A-T2Z	Tuvalu
T3A-T3Z	Kiribati
T4A-T4Z	Cuba
T5A-T5Z	Somalia
T6A-T6Z	Afghanistan
T7A-T7Z	San Marino
T8A-T8Z	Palau
T9A-T9Z	Bosnia and Herzegovina
UAA-UIZ	Russian Federation
UJA-UMZ	Uzbekistan
UNA-UQZ	Kazakhstan
URA-UZZ	Ukraine
VAA-VGZ	Canada
VHA-VNZ	Australia
VOA-VOZ	Canada
VPA-VQZ	United Kingdom of Great Britain and Northern Ireland
VRA-VRZ	China (People's Republic of)-Hong Kong
VSA-VSZ	United Kingdom of Great Britain and Northern Ireland
VTA-VWZ	India
VXA-VYZ	Canada
VZA-VZZ	Australia
V2A-V2Z	Antigua and Barbuda
V3A-V3Z	Belize
V4A-V4Z	Saint Kitts and Nevis
V5A-V5Z	Namibia
V6A-V6Z	Micronesia
V7A-V7Z	Marshall Islands
V8A-V8Z	Brunei
WAA-WZZ	United States of America
XAA-XIZ	Mexico
XJA-XOZ	Canada
XPA-XPZ	Denmark
XQA-XRZ	Chile
XSA-XSZ	China
XTA-XTZ	Burkina Faso
XUA-XUZ	Cambodia
XVA-XVZ	Viet Nam
XWA-XWZ	Laos
XXA-XXZ	Portugal
XYA-XZZ	Myanmar
YAA-YAZ	Afghanistan
YBA-YHZ	Indonesia
YIA-YIZ	Iraq
YJA-YJZ	Vanuatu
YKA-YKZ	Syria
YLA-YLZ	Latvia
YMA-YMZ	Turkey
YNA-YNZ	Nicaragua
YOA-YRZ	Romania
YSA-YSZ	El Salvador
YTA-YUZ	Yugoslavia
YVA-YYZ	Venezuela
YZA-YZZ	Yugoslavia
Y2A-Y9Z	Germany
ZAA-ZAZ	Albania
ZBA-ZJZ	United Kingdom of Great Britain and Northern Ireland
ZKA-ZMZ	New Zealand
ZNA-ZOZ	United Kingdom of Great Britain and Northern Ireland
ZPA-ZPZ	Paraguay
ZQA-ZQZ	United Kingdom of Great Britain and Northern Ireland
ZRA-ZUZ	South Africa
ZVA-ZZZ	Brazil
Z2A-Z2Z	Zimbabwe
Z3A-Z3Z	Macedonia (Former Yugoslav Republic)
2AA-2ZZ	United Kingdom of Great Britain and Northern Ireland
3AA-3AZ	Monaco
3BA-3BZ	Mauritius
3CA-3CZ	Equatorial Guinea
3DA-3DM	Swaziland
3DN-3DZ	Fiji
3EA-3FZ	Panama
3GA-3GZ	Chile
3HA-3UZ	China
3VA-3VZ	Tunisia
3WA-3WZ	Viet Nam
3XA-3XZ	Guinea
3YA-3YZ	Norway
3ZA-3ZZ	Poland

Call Sign Series	*Allocated to*
4AA-4CZ	Mexico
4DA-4IZ	Philippines
4JA-4KZ	Azerbaijan
4LA-4LZ	Georgia
4MA-4MZ	Venezuela
4NA-4OZ	Yugoslavia
4PA-4SZ	Sri Lanka
4TA-4TZ	Peru
*4UA-4UZ	United Nations
4VA-4VZ	Haiti
*4WA-4WZ	United Nations
4XA-4XZ	Israel
*4YA-4YZ	International Civil Aviation Organization
4ZA-4ZZ	Israel
5AA-5AZ	Libya
5BA-5BZ	Cyprus
5CA-5GZ	Morocco
5HA-5IZ	Tanzania
5JA-5KZ	Colombia
5LA-5MZ	Liberia
5NA-5OZ	Nigeria
5PA-5QZ	Denmark
5RA-5SZ	Madagascar
5TA-5TZ	Mauritania
5UA-5UZ	Niger
5VA-5VZ	Togo
5WA-5WZ	Western Samoa
5XA-5XZ	Uganda
5YA-5ZZ	Kenya
6AA-6BZ	Egypt
6CA-6CZ	Syria
6DA-6JZ	Mexico
6KA-6NZ	Korea (Republic of)
6OA-6OZ	Somalia
6PA-6SZ	Pakistan
6TA-6UZ	Sudan
6VA-6WZ	Senegal
6XA-6XZ	Madagascar
6YA-6YZ	Jamaica
6ZA-6ZZ	Liberia
7AA-7IZ	Indonesia
7JA-7NZ	Japan
7OA-7OZ	Yemen
7PA-7PZ	Lesotho
7QA-7QZ	Malawi
7RA-7RZ	Algeria
7SA-7SZ	Sweden
7TA-7YZ	Algeria
7ZA-7ZZ	Saudi Arabia
8AA-8IZ	Indonesia
8JA-8NZ	Japan
8OA-8OZ	Botswana
8PA-8PZ	Barbados
8QA-8QZ	Maldives
8RA-8RZ	Guyana
8SA-8SZ	Sweden
8TA-8YZ	India
8ZA-8ZZ	Saudi Arabia
9AA-9AZ	Croatia
9BA-9DZ	Iran
9EA-9FZ	Ethiopia
9GA-9GZ	Ghana
9HA-9HZ	Malta
9IA-9JZ	Zambia
9KA-9KZ	Kuwait
9LA-9LZ	Sierra Leone
9MA-9MZ	Malaysia
9NA-9NZ	Nepal
9OA-9TZ	Democratic Republic of the Congo
9UA-9UZ	Burundi
9VA-9VZ	Singapore
9WA-9WZ	Malaysia
9XA-9XZ	Rwanda
9YZ-9ZZ	Trinidad and Tobago

* Series allocated to an international organization
**In response to Resolution 99 (Minneapolis, 1998) of the Plenipotentiary Conference

DXCC AWARD APPLICATION

(Required with Each New Submission and Endorsements)

Please print clearly Please complete all sections

I am applying for the following DXCC awards (check appropriate boxes)
REQUIRED: FILL OUT CHART BELOW

	MIX	PHO	CW	RTTY	SAT	160M	80M	40M	20M	17M	15M	12M	10M	6M	2M	5BDX
NewAward																
Endorse																
5B Endorse																

\# of QSL cards enclosed_______________
\# of QSOs_______________
5 Band Award Number _______________
5 Band Award Date _______________

Fees:
Initiation fee - $10 (First ever DXCC, includes one certificate and DXCC pin, up to 120 QSOs)

Application fee for endorsements and additional new awards (per calendar year)
- ARRL members 1st submission of the year - $10 (up to 120 QSOs)
- ARRL members, additional submissions - $20 each (up to 100 QSOs)
- Non-ARRL members (non-US only*) 1st submission of the year - $20 (up to 120 QSOs)
- Non-ARRL members (non-US only*), additional submissions - $30 each (up to 100 QSOs)

* Applicants in the US and possessions must hold current ARRL membership.

Additional Fees
- Certificate fee (new or replacement) - $10 (Includes one DXCC pin)
- A $0.15 fee applies for each QSO in excess of established limits
- All applications presented at ARRL HQ or conventions attended by DXCC staff - $5 surcharge (limit 120 QSOs)

Complete DXCC fees are shown in Rule 15 of the Basic DXCC Rules

- The use of a current DXCC application form is required
- Do not use this form for plaque or pin orders
- Return postage is required for the return of cards and all written requests
- DXCC accepts most credit cards. If you are not sure of the correct charges, you may use a credit card. This will allow us to charge the exact amount. You must clear previous balances (per your last credit slip) with this submission in order to avoid delays.
- **DXCC cannot bill you**

Call Sign _______________________

Ex Calls _______________________

Name _______________________
Last (Spanish, Apellido) First

Mailing Address _______________________

____

____
(City, State/Zip, Country)

↑ **This is where your cards, paperwork, & certificates will be shipped** ↑

____ Check here if this is a new address

Telephone #: _______________________

Email Address _______________________

Name on Certificate_______________________
(Print name exactly as you want it to appear on certificate)

Please provide the following information:

Award Fees: _______________

Estimated Postage: _______________

Total amount sent: _______________

Method of Payment:
___ U.S. Currency
___ Check or Money Order
___ Credit Card # _______________________

Credit Card Exp Date: _______________

Return My QSL Cards Via: *
___Registered Mail **(Recommended)**
___First Class Mail
___Certified Mail (US Only)
___Airmail
___United Parcel Service (US Only)
___Other (Please Specify) _______________

* **If left blank, we will ship via Registered Mail at your expense**

"I affirm that I have observed all DXCC rules as well as all governmental regulations established for Amateur Radio in my country. I understand that ARRL is not responsible for cards handled by DXCC Card Checkers and will not honor any claims. I agree to be bound by the decisions of the ARRL Awards Committee and that all decisions of the ARRL Awards Committee are final."

____

Signature (REQUIRED) Callsign Date ARRL Membership Expiration Date

Send application forms, QSL cards, fees, and return postage to: DXCC Desk, ARRL HQ, 225 Main Street, Newington, CT 06111, U.S.A.
For questions or clarifications, please write to the DXCC Desk at the above address, or via e-mail to **dxcc@arrl.org** To confirm the receipt of your application, go to this link: **www.arrl.org/awards/dxcc/appstatus.html**. The DXCC Desk can also be contacted as follows: Telephone: 860-594-0234, Fax: 860-594-0259 (24 hour direct line to ARRL HQ). For complete program information, please visit the DXCC web site at: **www.arrl.org/awards/dxcc**

For ARRL DXCC Card Checker Use Only

I affirm that I have personally inspected the confirmations and verify that this application is true and correct.

____

Signature Callsign Date

DXCC Card Checkers must forward the application and fees to HQ within 2 working days. **FIELD CHECKED APPLICATIONS MAY BE SUBMITTED ONLY BY CARD CHECKERS.**
MSD-505 (07/2001)

DXCC Record Sheet

____
Your Call

DIRECTIONS: (1) Sort cards and list below first by band, e.g. all the 80-meter cards together, then the 40-meter cards, etc. (2) Within each band, sort and list below by mode, e.g. all the 80-meter Phone cards, then the 80-meter CW, 40-meter Phone, etc. (3) Make one entry for each QSO credit. (4) Cards indicating multiple contacts should be placed at the end, listing each contact on a separate line below. (5) QSO Date = Day, Month, Year. (6) Bands = 160, 80, 40, 30, 20, 17, 15, 12, 10, 6, 2. (7) Modes = Phone, CW, RTTY, Sat. (8) Full entity name required (not prefix).

	CALL	QSO DATE (DD|MM|YY)	BAND	MODE	ENTITY
1		| |			
2		| |			
3		| |			
4		| |			
5		| |			
6		| |			
7		| |			
8		| |			
9		| |			
10		| |			
11		| |			
12		| |			
13		| |			
14		| |			
15		| |			
16		| |			
17		| |			
18		| |			
19		| |			
20		| |			
21		| |			
22		| |			
23		| |			
24		| |			
25		| |			

This side of form may be photocopied if more pages are needed.

Russian Oblasts

Updated January 1996

Call sign number and first letter of suffix	Russian	English	Designator
1A, 1B	Sankt-Peterburg[1]	St. Petersburg	SP
Also 1D, 1F, 1G, 1H, 1I, 1J, 1L, 1M			
1C,	Leningradskaya obl.[1]	Leningradskaya Oblast	LO
1N	Karel'skaya ASSR	Karel'skaya Autonomous Soviet Socialist Republic	KL
1O	Arkhangel'skaya obl.	Arkhangel'skaya Oblast	AR
1P	Nenetskiy AO	Nenetskiy Autonomous District (Okrug)	NO
1Q,1R 1S	Vologodskaya obl.	Vologodskaya Oblast	VO
1T,1U	Novgorodskaya obl.	Novgorodskaya Oblast	NV
1W, 1X	Pskovskaya obl.	Pskovskaya Oblast	PS
1Y, 1Z	Murmanskaya obl.	Murmanskaya Oblast	MU
2F	Kaliningradskaya obl.[2]	Kaliningradskaya Oblast	KA
3A, 3B	g. Moskva[3]	Moscow City	MA
Also 3C, 3F, 3H			
3D	Moskovskaya obl.[3]	Moskovskaya Oblast	MO
3E	Orlovskaya obl.	Orlovskaya Oblast	OR
3G	Lipetskaya obl.	Lipetskaya Oblast	LP
3I, 3J	Tverskaya obl.	Tverskaya Oblast	TV
3L	Smolenskaya obl.	Smolenskaya Oblast	SM
3M	Yaroslavskaya obl.	Yaroslavskaya Oblast	JA
3N, 3O	Kostromskaya obl.	Kostromskaya Oblast	KS
3P	Tul'skaya obl.	Tul'skaya Oblast	TL
3Q	Voronezhskaya obl.	Voronezhskaya Oblast	VH
3R	Tambovskaya obl.	Tambovskaya Oblast	TB
3S	Ryazanskaya obl.	Ryazanskaya Oblast	RA
3T	Nizhegorodskaya obl.	Nizhegorodskaya Oblast	NN
3U	Ivanovskay aobl.	Ivanovskaya Oblast	IV
3V	Vladimirskaya obl.	Vladimirskaya Oblast	VL
3W	Kurskaya obl.	Kurskaya Oblast	KU
3X	Kaluzhskaya obl.	Kaluzhskaya Oblast	KG
3Y	Bryanskaya obl.	Bryanskaya Oblast	BR
3Z	Belgorodskaya obl.	Belgorodskaya Oblast	BO
4A, 4B	Volgogradskaya obl.	Volgogradskaya Oblast	VG
4C, 4D	Saratovskaya obl.	Saratovskaya Oblast	SA
4F	Penzenskaya obl.	Penzenskaya Oblast	PE
4H. 4I	Samarskaya obl.	Samarskaya Oblast	SR
4L, 4M	Ul'yanovskaya obl.	Ul'yanovskaya Oblast	UL
4N, 4O	Kirovskaya obl.	Kirovskaya Oblast	KI
4P, 4Q	Respublika Tatarstan	Republic of Tatarstan	TA
Also 4R			
4S, 4T	Respublika Mariy-El	Republic of Mariy-El	MR
4U	Mordovskaya SSR	Mordovskaya Soviet Socialist Republic	MD
4W	Respublika Udmurtiya	Republic of Udmurtiya	UD
4Y, 4Z	Respublika Chuvashiya	Republic of Chuvashiya	CU
6A, 6B	Krasnodarskiy kray	Krasnodarskiy Kray	KR
Also 6C, 6D			
6E	Karachayevo-Cherkasskaya	Karachayevo-Cherkasskaya Respublika Republic	KC
6H, 6F	Stavropol'skiy kray	Stavropol'skiy Kray	ST
Also 6H			
6I	Respublika Kalmykiya	Republic of Kalmykiya	KM
6J	Severo-Osetinskaya SSR	North-Osetinskaya Soviet Socialist Republic	SO
6L, 6M	Rostovskaya obl.	Rostovskaya Oblast	RO
Also 6N, 6O			
6P	Chechenskaya Respublika	Chechenskaya Republic	CN
Also 6R			
6U, 6V	Astrakhanskaya obl.	Astrakhanskaya Oblast	AO
6W	Respublika Dagestan	Republic of Dagestan	DA
6X	Respublik Kabardino-Balkariya	Republic of Kabardino-Balkariya	KB
6Y	Respublika Adygeya	Republic of Adygeya	AD
8T	Ust'-Ordynskiy Buryatskiy AO	Ust'-Ordynskiy Buryatskiy Autonomous District (Okrug)	UO
8V	Aginskiy Buryatskiy AO	Aginskiy Buryatskiy Autonomous District (Okrug)	AB
9A, 9B	Chelyabinskaya obl.	Chelyanbinskaya Oblast	CB
9C, 9D	Sverdlovskaya obl.	Sverdlovskaya Oblast	SV
Also 9E			
9F PM	Permskaya obl.	Permskaya Oblast	
9G	Komi-Permyatskiy AO	Komi-Permyatskiy Autonomous District (Okrug)	KP
9H, 9I	Tomskaya obl.	Tomskaya Oblast	TO
9J	Khanty-Mansiyskiy AO	Khanty-Mansiyskiy Autonomous District (Okrug	HM
9K	Yamalo-Nenetskiy AO	Yamalo-Nenetskiy Autonomous District (Okrug)	JN
9L	Tyumenskaya obl.	Tyumenskaya Oblast	TN
9M, 9N	Omskaya obl.	Omskaya Oblast	OM
9O, 9P	Novosibirskaya obl.	Novosibirskaya Oblast	NS
9Q, 9R	Kurganskaya obl.	Kurganskaya Oblast	KN
9S, 9T	Orenburgskaya obl.	Orenburgskaya Oblast	OB
9U, 9V	Kemerovskaya obl.	Kemerovskaya Oblast	KE
9W	Respublika Bashkortostan	Republic of Bashkortostan	BA
9X	Respublika Komi	Republic of Komi	KO
9Y	Altayskiy kray	Altayskiy Kray	AL
9Z	Gorno-Altayskiy AO	Gorno-Altayskiy Autonomous District (Okrug)	GA
0A	Krasnoyarskiy kray	Krasnoyarskiy Kray	KK
0B	Taymyrskiy AO	Taymyrskiy Autonomous District (Okrug)	TM
0C	Khabarovskiy kray	Khabarovskiy Kray	HK
0D	Yevreyskaya AO	Jewish Autonomous Oblast[4]	EA
0E, 0F	Sakhalinskaya obl.	Sakhalinskaya Oblast	SL
Also 0G			
0H	Evenkiyskiy AO	Evenkiyskiy Autonomous District (Okrug)	EW
0I	Magadanskaya obl	Magadanskaya Oblast	MG
0J	Amurskaya obl.	Amurskaya Oblast	AM
0K	Chukotskiy AO	Chukotskiy Autonomous District (Okrug)	CK
0L, 0M	Primorskiy kray	Primorskiy Kray	PK
Also 0N			
0O, 0P	Respublika Buryatiya	Republic of Buryatiya	BU
0Q, 0R	Respublika Sakha	Republic of Sakha	YA
0S, 0T	Irkutskaya obl.	Irkutskaya Oblast	IR
0U, 0V	Chitinskaya obl.	Chitinskaya Oblast	CT
0W	Respublika Khakassiya	Republic of Khakassiya	HA
0X	Koryakskiy AO	Koryakskiy Autonomous District (Okrug)	KJ
0Y	Respublika Tuva	Republic of Tuva[5]	TU
0Z	Kamchatskaya obl.	Kamchatskaya Oblast	KT

[1]St. Petersburg is within, and separate from, Leningradskaya Oblast.

[2]Kaliningradskaya Oblast covers the northern part of the former East Prussia. Annexed by the USSR at the end of World War II and made part of the RSFSR, the area is separated from the rest of Russia by Lithuanian and Belarussian territory. The administrative center of the Oblast is Kaliningrad, formerly Konigsberg.

[3]Moscow city is within, and separate from, Moskovskaya Oblast.

[4]According to a U.S. reference dating back to 1961, the Jewish Autonomous Oblast (Yevreyskaya avtonomnaya oblast or YeAO) was established in 1934 in the Soviet Far East but attracted few Jews; its population was 165,000, including no more than 30,000 Jews.

[5]This is the former Tannu Tuva, which was part of the Chinese Empire until 1911, became a Russian Protectorate in 1914, and proclaimed its independence in 1921. It was annexed by the USSR in 1944 and given the status of an autonomous oblast within the RSFSR. In 1961, it was elevated to the status of an autonomous SSR, still within the RSFSR.

Morse Code Character Set[1]

A	didah	• –
B	dahdididit	– •••
C	dahdidahdit	– • – •
D	dahdidit	– ••
E	dit	•
F	dididahdit	•• – •
G	dahdahdit	– – •
H	didididit	••••
I	didit	••
J	didahdahdah	• – – –
K	dahdidah	– • –
L	didahdidit	• – ••
M	dahdah	– –
N	dahdit	– •
O	dahdahdah	– – –
P	didahdahdit	• – – •
Q	dahdahdidah	– – • –
R	didahdit	• – •
S	dididit	•••
T	dah	–
U	dididah	•• –
V	didididah	••• –
W	didahdah	• – –
X	dahdididah	– •• –
Y	dahdidahdah	– • – –
Z	dahdahdidit	– – ••
1	didahdahdahdah	• – – – –
2	dididahdahdah	•• – – –
3	didididahdah	••• – –
4	dididididah	•••• –
5	dididididit	•••••
6	dahdididit	– ••••
7	dahdahdididit	– – •••
8	dahdahdahdidit	– – – ••
9	dahdahdahdahdit	– – – – •
0	dahdahdahdahdah	– – – – –

Period [.]:	didahdidahdidah	• – • – • –	AAA
Comma [,]:	dahdahdididahdah	– – •• – –	MIM
Question mark or request for repetition [?]:	dididahdahdidit	•• – – ••	IMI
Error:	dididididididit	••••••••	HH
Hyphen or dash [–]:	dahdididididah	– •••• –	DU
Double dash [=]	dahdidididah	– ••• –	BT
Colon [:]:	dahdahdahdididit	– •••	OS
Semicolon [;]:	dahdidahdidahdit	– • – • – •	KR
Left parenthesis [(]:	dahdidahdahdit	– • – – •	KN
Right parenthesis [)]:	dahdidahdahddidah	– • – – • –	KK
Fraction bar [/]:	dahdididahdit	– •• – •	DN
Quotation marks ["]:	didahdididahdit	• – •• – •	AF
Dollar sign [$]:	didididahdididah	••• – •• –	SX
Apostrophe [']:	didahdahdahdahdit	• – – – – •	WG
Paragraph [¶]:	didahdidahdidit	• – • – ••	AL
Underline [_]:	dididahdahdidah	•• – – • –	IQ
Starting signal:	dahdidahdidah	– • – • –	KA
Wait:	didahdididit	• – •••	AS
End of message or cross [+]:	didahdidahdit	• – • – •	AR
Invitation to transmit [K]:	dahdidah	– • –	K
End of work:	didididahdidah	••• – • –	SK
Understood:	didididahdit	••• – •	SN

Notes:

1. Not all Morse characters shown are used in FCC code tests. License applicants are responsible for knowing, and may be tested on, the 26 letters, the numerals 0 to 9, the period, the comma, the question mark, AR, SK, BT and fraction bar [DN].

2. The following letters are used in certain European languages which use the Latin alphabet:

Ä, Ą	didahdidah	• – • –
Á, Å, À, Â	didahdahdidah	• – – • –
Ç, Ć	dahdidahdidit	– • – ••
É, Ȅ, Ę	dididahdidit	•• – ••
È	didahdididah	• – •• –
Ê	dahdididahdit	– •• – •
Ö, Ǒ, Ó	dahdahdahdit	– – – •
Ñ	dahdahdidahdah	– – • – –
Ű	dididahdah	•• – –
Ź	dahdahdidit	– – ••
Z	dahdahdididah	[illegible]
CH, Ş	dahdahdahdah	– – – –

3. Special Esperanto characters:

Ĉ	dahdidahdidit	– • – ••
Ŝ	didididahdit	••• – •
Ĵ	didahdahdahdit	• – – – •
Ĥ	dahdidahdahdit	– • – – •
Ŭ	dididahdah	•• – –
Ĝ	dahdahdidahdit	– – • – •

4. Signals used in other radio services:

Interrogatory	dididahdidah	•• – • –	INT
Emergency silence	didididahdah	•••• – –	HM
Executive follows	dididahdididah	•• – •• –	IX
Break–in signal	dahdahdahdahdah	– – – – –	TTTTT
Emergency signal	didididahdahdahdididit	••• – – – •••	SOS
Relay of distress	dahdididahdididahdidit	– •• – •• – ••	DDD

Morse Abbreviated Numbers

Numeral		*Long Number*		*Abbreviated Number*	*Equivalent Character*
1	didahdahdahdah	• – – – –	didah	• –	A
2	dididahdahdah	•• – – –	dididah	•• –	U
3	didididahdah	••• – –	didididah	••• –	V
4	dididididah	•••• –	dididididah	•••• –	4
5	dididididit	•••••	dididididit	••••• or •	5 or E
6	dahdididit	– ••••	dahdididit	– ••••	6
7	dahdahdididit	– – •••	dahdidit	– •••	B
8	dahdahdahdidit	– – – ••	dahdidit	– ••	D
9	dahdahdahdahdit	– – – – •	dahdit	– •	N
0	dahdahdahdahdah	– – – – –	dah	–	T

Note: These abbreviated numbers are not legal for use in call signs. They should be used only where there is agreement between operators and when no confusion will result.

Morse Code for Other Languages

Code	*Japanese*	*Korean*	*Arabic*	*Hebrew*	*Russian*	*Greek*
•	ヘ he	ㅏ a		ו vav	Е,Э E	Ε epsilon
—	ム mu	ㅓ ŏ	ت ta	תת tav	Т T	Τ tau
••	゛ nigori	ㅑ ya	ي ya	י yod	И I	Ι iota
•—	イ i	ㅗ o	ا alif	א aleph	А A	Α alpha
—•	タ ta	ㅛ yo	ن noon	נן nun	Н N	Ν nu
——	ヨ yo	ㅁ m	م meem	מם mem	М M	Μ mu
•••	ラ ra	ㅕ yŏ	س seen	שש shin	С S	Σ sigma
••—	ウ u		ط ta	ט tet	У U	ΟΥ omicron ypsilon
•—•	ナ na	ㅠ yu	ر ra	ר reish	Р R	Ρ rho
•——	ヤ ya	ㅂ p(b)	و waw	צץ tzadi	В V	Ω omega
—••	ホ ho		د dal	ד dalet	Д D	Δ delta
—•—	ワ wa	ㅇ -ng	ك kaf	כבך chaf	К K	Κ kappa
——•	リ ri	ㅅ s	غ ghain	ג gimmel	Г G	Γ gamma
———	レ re	ㅍ p'	خ kha	ה heh	О O	Ο omicron
••••	ヌ nu	ㅜ u	ح ha	ח chet	Х H	Η eta
•••—	ク ku	ㄹ r-(-l)	ض dad		Ж J	ΗΥ eta ypsilon
••—•	チ ti	ㄴ n	ف fa	פף feh	Ф F	Φ phi
••——	ノ no				Ю yu	ΑΥ alpha ypsilon
•—••	カ ka	ㄱ k(g)	ل lam	ל lamed	Л L	Λ lambda
•—•—	ロ ro		ع ain		Я ya	ΑΙ alpha iota
•——•	ツ tu	ㅈ ch(j)		פ peh	П P	Π pi
•———	ヲ wo	ㅎ h	ج jeem	ע ayen	Й Y	ΥΙ ypsilon iota
—•••	ハ ha	ㄷ t(d)	ب ba	בב bet	Б B	Β beta
—••—	マ ma	ㅋ k'	ص sad		Ъ,Ь mute	Ξ xi
—•—•	ニ ni	ㅊ ch'	ث tha	ס samech	Ц TS	Θ theta
—•——	ケ ke	ㅔ e	ظ za		Ы I	Υ ypsilon
——••	フ hu	ㅌ t'	ذ dhal	ז zain	З Z	Ζ zeta
——•—	ネ ne	ㅐ ae	ق qaf	ק kof	Щ SHCH	Ψ psi
———•	ソ so		ز zay		Ч CH	ΕΥ epsilon ypsilon
————	コ ko		ش sheen		Ш SH	Χ khi
••—••	ト to		ه he			
••—•—	ミ mi					
••——•	゜ han-nigori					
•—•••	オ o					
•—••—	ヰ (w)i					
•—•—•	ン n					
•—•——	テ te					
•——••	ヱ (w)e					
•——•—	ー hyphen					
•———•	セ se					
—•••—	メ me					
—••—•	モ mo					
—••——	ユ yu					
—•—••	キ ki					
—•—•—	サ sa					
—•——•	ル ru					
—•———	エ e					
——••—	ヒ hi					
——•—•	シ si					
——•——	ア a					
———•—	ス su					
•—•••—			لا lam-alif			

Spanish Phonetics

America	ah-MAIR-ika
Brasil	brah-SIL
Canada	cana-DAH
Dinamarca	dina-MAR-ka
Espana	es-PAHN-yah
Francia	FRAHN-seeah
Grenada	gre-NAH-dah
Holanda	oh-LONN-dah
Italia	i-TAL-eeah
Japon	hop-OWN
Kilowatio	kilo-WAT-eeoh
Lima	LIMA
Mejico	MEH-heeco
Norvega	nor-WAY-gah
Ontario	on-TAR-eeoh
Portugal	portu-GAL
Quito	KEY-toe
Roma	ROW-mah
Santiago	santee-AH-go
Toronto	tor-ON-toe
Uniforme	oonee-FORM-eh
Victoria	vic-TOR-eeah
Washington, Wisky	washingtone, wisky
Xilofono	see-LOW-phono
Yucatan	yuca-TAN
Zelandia	see-LAND-eeah

W		DOE-bleh-vay
0	cero	SEH-roe
1	uno	OO-no
2	dos	DOS
3	tres	TRAYCE
4	cuatro	KWAT-roe
5	cinco	SINK-oh
6	seis	SAYCE
7	siete	see-AY-teh
8	ocho	OCH-oh
9	nueve	new-AY-veh

—*John Mason Jr., EA4AXW*

A large selection of phonetic alphabets is at
http://www.cl.cam.ac.uk/users/bck1/menu.html
and
http://www.columbia.edu/~fuat/cuarc/phonetic. html

DX Operating Code

For W/VE Amateurs

Some amateur DXers have caused considerable confusion and interference in their efforts to work DX stations. The points below, if observed by all W/VE amateurs, will help make DX more enjoyable for all.

1) *Call* DX only after he calls CQ, QRZ? or signs $\overline{SK}$, or voice equivalents thereof. Make your calls short.

2) Do not call a DX station:

a) On the frequency of the station he is calling until you are sure the QSO is over ($\overline{SK}$).

b) Because you hear someone else calling him.

c) When he signs $\overline{KN}$, $\overline{AR}$ or $\overline{CL}$.

d) Exactly on his frequency.

e) After he calls a directional CQ, unless of course you are in the right direction or area.

3) Keep within frequency band limits. Some DX stations can get away with working outside, but you cannot.

4) Observe calling instructions given by DX stations. Example: 15U means "call 15 kHz *up* from my frequency." 15D means *down*, etc.

5) Give honest reports. Many DX stations *depend* on W/VE reports for adjustment of station and equipment.

6) Keep your signal clean. Key clicks, ripple, feedback or splatter gives you a bad reputation and may get you a citation from FCC.

7) *Listen* and call the station you want. Calling CQ DX is not the best assurance that the rare DX will reply.

8) When there are several W or VE stations waiting, avoid asking DX to "listen for a friend." Also avoid engaging him in a ragchew against his wishes.

For Overseas Amateurs

To all overseas amateur stations:

In their eagerness to work you, many W and VE amateurs resort to practices that cause confusion and QRM. Most of this is good-intentional but ill-advised; some of it is intentional and selfish. The key to the cessation of unethical DX operating practices is in your hands. We believe that your adoption of certain operating habits will increase your enjoyment of Amateur Radio and that of amateurs on this side who are eager to work you. We recommend your adoption of the following principles:

1) Do not answer calls on your own frequency.

2) Answer calls from W/VE stations only when their signals are of good quality.

3) Refuse to answer calls from other stations when you are already in contact with someone, and do not acknowledge calls from amateurs who indicate they wish to be "next."

4) Give *everybody* a break. When many W/VE amateurs are patiently and quietly waiting to work you, avoid complying with requests to "listen for a friend."

5) Tell listeners where to call you by indicating how many kilohertz up (U) or down (D) from your frequency you are listening.

6) Use the ARRL-recommended ending signals, especially $\overline{KN}$ to indicate to impatient listeners the status of the QSO. $\overline{KN}$ means "Go ahead (specific station); all others keep out."

7) Let it be known that you avoid working amateurs who are constant violators of these principles.

The Origin of "73"

The traditional expression "73" goes right back to the beginning of the landline telegraph days. It is found in some of the earliest editions of the numerical codes, each with a different definition, but each with the same idea in mind— it indicated that the end, or signature, was coming up. But there are no data to prove that any of these were used.

The first authentic use 73 is in the publication *The National Telegraphic Review and Operators' Guide*, first published in April 1857. At that time, 73 meant "My love to you"! Succeeding issues of this publication continued to use this definition of the term. Curiously enough, some of the other numerals used then had the same definition as they have now, but within a short time, the use of 73 began to change.

In the National Telegraph Convention, the numeral was changed from the Valentine-type sentiment to a vague sign of fraternalism. Here, 73 was a greeting, a friendly "word" between operators and it was so used on all wires.

In 1859, the Western Union Company set up the standard "92 Code." A list of numerals from one to 92 was compiled to indicate a series of prepared phrases for use by the operators on the wires. Here, in the 92 Code, 73 changes from a fraternal sign to a very flowery "accept my compliments," which was in keeping with the florid language of that era.

Over the years from 1859 to 1900, the many manuals of telegraphy show variations of this meaning. Dodge's *The Telegraph Instructor* shows it merely as "compliments." The *Twentieth Century Manual of Railway and Commercial Telegraphy* defines it two ways, one listing as "my compliments to you"; but in the glossary of abbreviations it is merely "compliments." Theodore A. Edison's *Telegraphy Self-Taught* shows a return to "accept my compliments." By 1908, however, a later edition of the Dodge Manual gives us today's definition of "best regards" with a backward look at the older meaning in another part of the work where it also lists it as "compliments."

"Best regards" has remained ever since as the "put-it-down-in-black-and-white" meaning of 73 but it has acquired overtones of much warmer meaning. Today, amateurs use it more in the manner that James Reid had intended that it be used—a "friendly word between operators."—*Louise Ramsey Moreau, W3WRE*

The RST System

READABILITY

1—Unreadable.
2—Barely readable, occasional words distinguishable.
3—Readable with considerable difficulty.
4—Readable with practically no difficulty.
5—Perfectly readable.

SIGNAL STRENGTH

1—Faint signals barely perceptible.
2—Very weak signals.
3—Weak signals.
4—Fair signals.
5—Fairly good signals.
6—Good signals.
7—Moderately strong signals.
8—Strong signals.
9—Extremely strong signals.

TONE

1—Sixty-cycle ac or less, very rough and broad.
2—Very rough ac, very harsh and broad.
3—Rough ac tone, rectified but not filtered.
4—Rough note, some trace of filtering.
5—Filtered rectified ac but strongly ripple-modulated.
6—Filtered tone, definite trace of ripple modulation.
7—Near pure tone, trace of ripple modulation.
8—Near perfect tone, slight trace of modulation.
9—Perfect tone, no trace of ripple or modulation of any kind.

The "tone" report refers only to the purity of the signal, and has no connection with its stability or freedom from clicks or chirps. If the signal has the characteristic steadiness of crystal control, add X to the report (e.g., RST 469X). If it has a chirp or "tail" (either or "make" or "break") add C (e.g., 469C). If it has clicks or noticeable other keying transients, add K (e.g., 469K). Of course a signal could have both chirps and clicks, in which case both C and K could be used (e.g., 469CK).

Q Signals

Given below are a number of Q signals whose meanings most often need to be expressed with brevity and clarity in amateur work. (Q abbreviations take the form of questions only when each is sent followed by a question mark.)

QRG Will you tell me my exact frequency (or that of ____)? Your exact frequency (or that of ____) is ____ kHz.
QRH Does my frequency vary? Your frequency varies.
QRI How is the tone of my transmission? The tone of your transmission is ____ (1. Good; 2. Variable; 3. Bad).
QRJ Are you receiving me badly? I cannot receive you. Your signals are too weak.
QRK What is the intelligibility of my signals (or those of ____)? The intelligibility of your signals (or those of ____) is ____ (1. Bad; 2. Poor; 3. Fair; 4. Good; 5. Excellent).
QRL Are you busy? I am busy (or I am busy with ____). Please do not interfere.
QRM Is my transmission being interfered with? Your transmission is being interfered with ____ (1. Nil; 2. Slightly; 3. Moderately; 4. Severely; 5. Extremely.)
QRN Are you troubled by static? I am troubled by static ____ (1-5 as under QRM).
QRO Shall I increase power? Increase power.
QRP Shall I decrease power? Decrease power.
QRQ Shall I send faster? Send faster (____ WPM).
QRS Shall I send more slowly? Send more slowly (____ WPM).
QRT Shall I stop sending? Stop sending.
QRU Have you anything for me? I have nothing for you.
QRV Are you ready? I am ready.
QRW Shall I inform ____ that you are calling on ____ kHz? Please inform ____ that I am calling on ____ kHz.
QRX When will you call me again? I will call you again at ____ hours (on ____ kHz).
QRY What is my turn? Your turn is numbered ____
QRZ Who is calling me? You are being called by ____ (on ____ kHz).
QSA What is the strength of my signals (or those of ____)? The strength of your signals (or those of ____) is ____ (1. Scarcely perceptible; 2. Weak; 3. Fairly good; 4. Good; 5. Very good).
QSB Are my signals fading? Your signals are fading.
QSD Is my keying defective? Your keying is defective.
QSG Shall I send ____ messages at a time? Send ____ messages at a time.
QSK Can you hear me between your signals and if so can I break in on your transmission? I can hear you between my signals; break in on my transmission.
QSL Can you acknowledge receipt? I am acknowledging receipt
QSM Shall I repeat the last message which I sent you, or some previous message? Repeat the last message which you sent me [or message(s) number(s) ____].
QSN Did you hear me (or ____) on ____ kHz? I did hear you (or ____) on ____ kHz.
QSO Can you communicate with ____ direct or by relay? I can communicate with ____ direct (or by relay through ____).
QSP Will you relay to ____? I will relay to ____
QST General call preceding a message addressed to all amateurs and ARRL members. This is in effect "CQ ARRL."
QSU Shall I send or reply on this frequency (or on ____ kHz)? Send a series of Vs on this frequency (or ____ kHz).
QSW Will you send on this frequency (or on ____ kHz)? I am going to send on this frequency (or on ____ kHz).
QSX Will you listen to ____ on ____ kHz? I am listening to ____ on ____ kHz.
QSY Shall I change to transmission on another frequency? Change to transmission on another frequency (or on ____ kHz).
QSZ Shall I send each word or group more than once? Send each word or group twice (or ____ times).
QTA Shall I cancel message number ____? Cancel message number ____
QTB Do you agree with my counting of words? I do not agree with your counting of words. I will repeat the first letter or digit of each word or group.
QTC How many messages have you to send? I have ____ messages for you (or for ____).
QTH What is your location? My location is ____
QTR What is the correct time? The time is ____

PHONETIC ALPHABET

A — Alfa (**AL** FAH)
B — Bravo (**BRAH** VOH)
C — Charlie (**CHAR** LEE or **SHAR** LEE)
D — Delta (**DELL** TAH)
E — Echo (**ECK** OH)
F — Foxtrot (**FOKS** TROT)
G — Golf (GOLF)
H — Hotel (HOH **TELL**)
I — India (**IN** DEE AH)
J — Juliet (**JEW** LEE ETT)
K — Kilo (**KEY** LOH)
L — Lima (**LEE** MAH)
M — Mike (MIKE)
N — November (NO **VEM** BER)
O — Oscar (**OSS** CAH)
P — Papa (PAH **PAH**)
Q — Quebec (KEH **BECK**)
R — Romeo (**ROW** ME OH)
S — Sierra (SEE **AIR** RAH)
T — Tango (**TANG** GO)
U — Uniform (**YOU** NEE FORM or **OO** NEE FORM)
V — Victor (**VIK** TAH)
W — Whiskey (**WISS** KEY)
X — X-Ray (**ECKS** RAY)
Y — Yankee (**YANG** KEY)
Z — Zulu (**ZOO** LOO)

ARRL NUMBERED RADIOGRAMS

The letters ARL are inserted in the preamble in the check and in the text before spelled out numbers, which represent texts from this list. Note that some ARL texts include insertion of numerals. *Example*: NR 1 R W1AW ARL 5 NEWINGTON CONN DEC 25 DONALD R SMITH AA 164 EAST SIXTH AVE AA NORTH RIVER CITY MO AA PHONE 733 3968 BT ARL FIFTY ARL SIXTY ONE BT DIANA AR. For additional information about traffic handling, see Chapter 7.

Group One—For Possible "Relief Emergency" Use

ONE	Everyone safe here. Please don't worry.
TWO	Coming home as soon as possible.
THREE	Am in _____ hospital. Receiving excellent care and recovering fine.
FOUR	Only slight property damage here. Do not be concerned about disaster reports.
FIVE	Am moving to new location. Send no further mail or communication. Will inform you of new address when relocated.
SIX	Will contact you as soon as possible.
SEVEN	Please reply by Amateur Radio through the amateur delivering this message. This is a free public service.
EIGHT	Need additional _______ mobile or portable equipment for immediate emergency use.
NINE	Additional _________ radio operators needed to assist with emergency at this location.
TEN	Please contact _______. Advise to standby and provide further emergency information, instructions or assistance.
ELEVEN	Establish Amateur Radio emergency communications with _____ on _____ MHz.
TWELVE	Anxious to hear from you. No word in some time. Please contact me as soon as possible.
THIRTEEN	Medical emergency situation exists here.
FOURTEEN	Situation here becoming critical. Losses and damage from _____ increasing.
FIFTEEN	Please advise your condition and what help is needed.
SIXTEEN	Property damage very severe in this area.
SEVENTEEN	REACT communications services also available. Establish REACT communications with _____ on channel _____.
EIGHTEEN	Please contact me as soon as possible at _____.
NINETEEN	Request health and welfare report on _______. (State name, address and telephone number.)
TWENTY	Temporarily stranded. Will need some assistance. Please contact me at _______.
TWENTY ONE	Search and Rescue assistance is needed by local authorities here. Advise availability.
TWENTY TWO	Need accurate information on the extent and type of conditions now existing at your location. Please furnish this information and reply without delay.
TWENTY THREE	Report at once the accessibility and best way to reach your location.
TWENTY FOUR	Evacuation of residents from this area urgently needed. Advise plans for help.
TWENTY FIVE	Furnish as soon as possible the weather conditions at your location.
TWENTY SIX	Help and care for evacuation of sick and injured from this location needed at once.

Emergency/priority messages originating from official sources must carry the signature of the originating official.

Group Two—Routine messages

FORTY SIX	Greetings on your birthday and best wishes for many more to come.
FIFTY	Greetings by Amateur Radio.
FIFTY ONE	Greetings by Amateur Radio. This message is sent as a free public service by ham radio operators here at _______. Am having a wonderful time.
FIFTY TWO	Really enjoyed being with you. Looking forward to getting together again.
FIFTY THREE	Received your ______. It's appreciated; many thanks.
FIFTY FOUR	Many thanks for your good wishes.
FIFTY FIVE	Good news is always welcome. Very delighted to hear about yours.
FIFTY SIX	Congratulations on your ______, a most worthy and deserved achievement.
FIFTY SEVEN	Wish we could be together.
FIFTY EIGHT	Have a wonderful time. Let us know when you return.
FIFTY NINE	Congratulations on the new arrival. Hope mother and child are well.
*SIXTY	Wishing you the best of everything on _______.
SIXTY ONE	Wishing you a very merry Christmas and a happy New Year.
*SIXTY TWO	Greetings and best wishes to you for a pleasant _______ holiday season.
SIXTY THREE	Victory or defeat, our best wishes are with you. Hope you win.
SIXTY FOUR	Arrived safely at ______.
SIXTY FIVE	Arriving _______ on _______. Please arrange to meet me there.
SIXTY SIX	DX QSLs are on hand for you at the ______ QSL Bureau. Send ______ self-addressed envelopes.
SIXTY SEVEN	Your message number ______ undeliverable because of ______. Please advise.
SIXTY EIGHT	Sorry to hear you are ill. Best wishes for a speedy recovery.
SIXTY NINE	Welcome to the ______. We are glad to have you with us and hope you will enjoy the fun and fellowship of the organization.

* Can be used for all holidays.

Note: ARL numbers should be spelled out at all times.

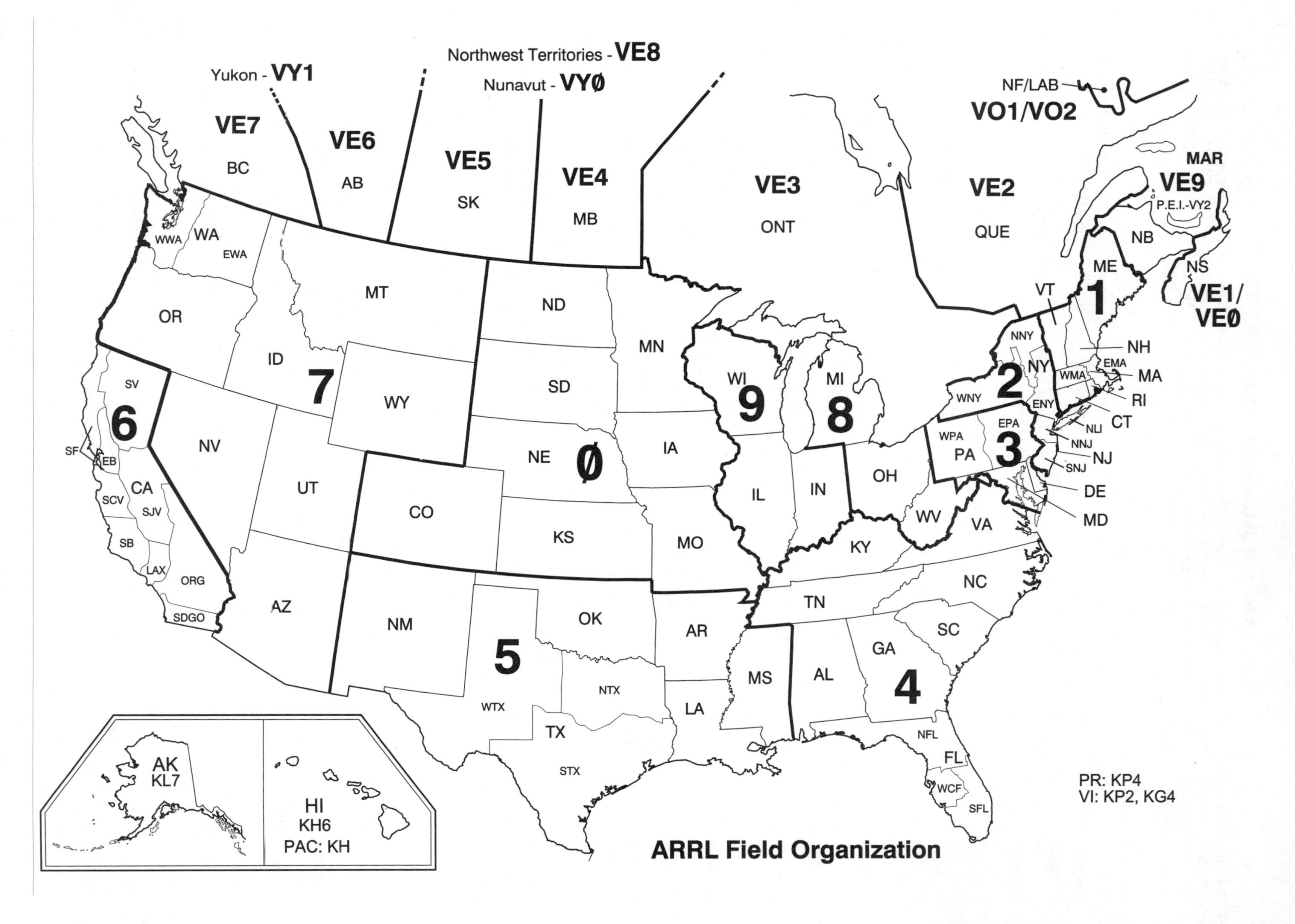
Yukon - VY1
Northwest Territories - VE8
Nunavut - VYØ
NF/LAB
VO1/VO2
VE7
BC
VE6
AB
VE5
SK
VE4
MB
VE3
ONT
VE2
QUE
MAR
VE9
P.E.I.-VY2
NB
NS
VE1/
VEØ
ME
1
VT
NH
EMA
MA
WMA
RI
CT
NNY
NY
2
WNY
ENY
NLI
NNJ
NJ
SNJ
EPA
WPA
PA
3
DE
MD
WWA
WA
EWA
OR
ID
7
MT
WY
SV
6
SF
EB
CA
SCV
SJV
SB
LAX
ORG
SDGO
NV
UT
AZ
ND
SD
NE
Ø
KS
CO
MN
IA
MO
WI
9
IL
MI
8
IN
OH
WV
VA
KY
TN
NC
SC
GA
4
AL
MS
NFL
FL
WCF
SFL
NM
5
OK
AR
LA
NTX
WTX
TX
STX
AK
KL7
HI
KH6
PAC: KH
PR: KP4
VI: KP2, KG4
ARRL Field Organization

CALIFORNIA COUNTIES/ARRL SECTIONS
HAM OUTLINE MAP No 27

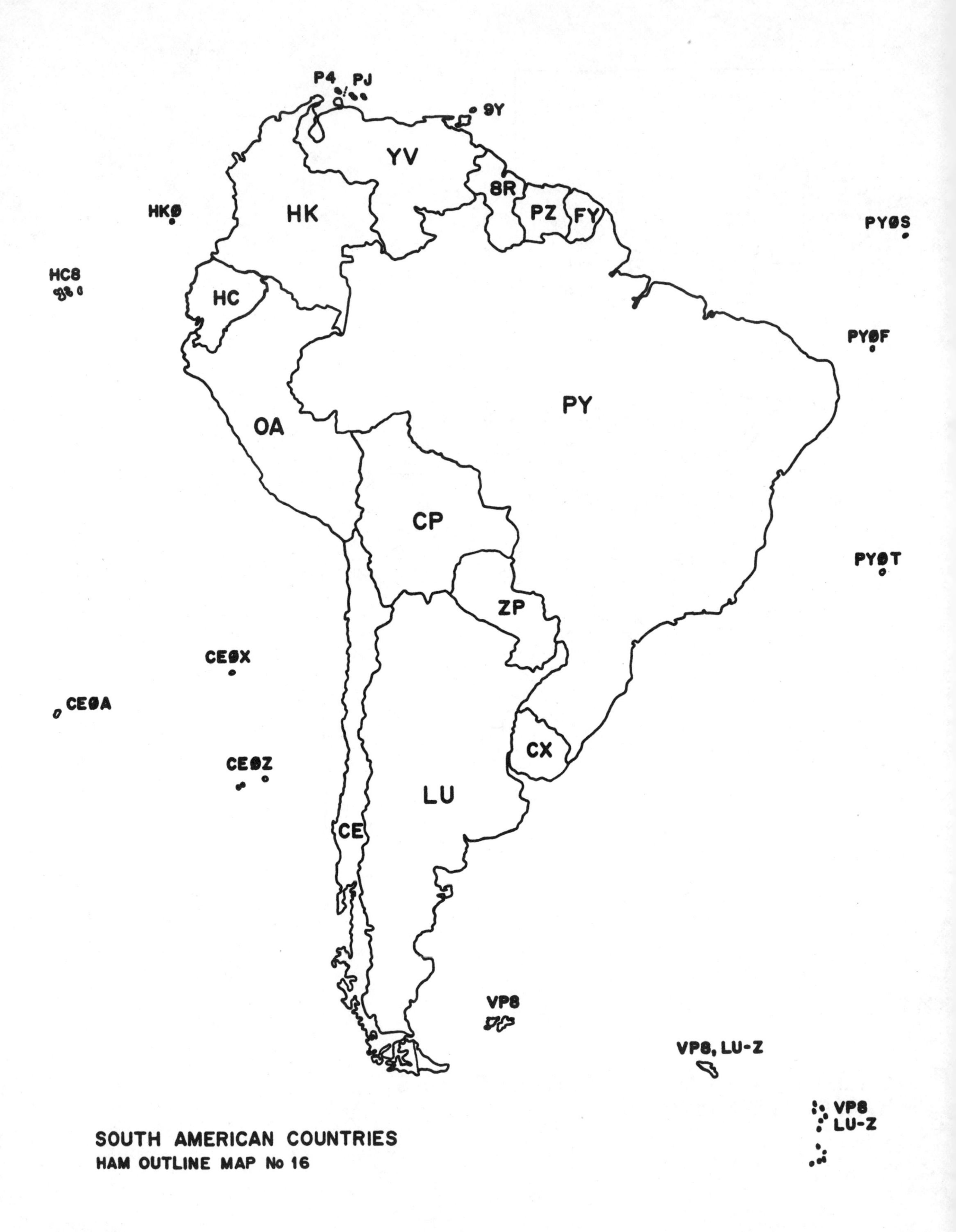
P4
PJ
9Y
YV
HK
HKØ
SR
PZ
FY
PYØS
HC8
HC
PYØF
PY
OA
CP
PYØT
ZP
CEØX
CEØA
CX
CEØZ
LU
CE
VP8
VP8, LU-Z
VP8
LU-Z
SOUTH AMERICAN COUNTRIES
HAM OUTLINE MAP No 16

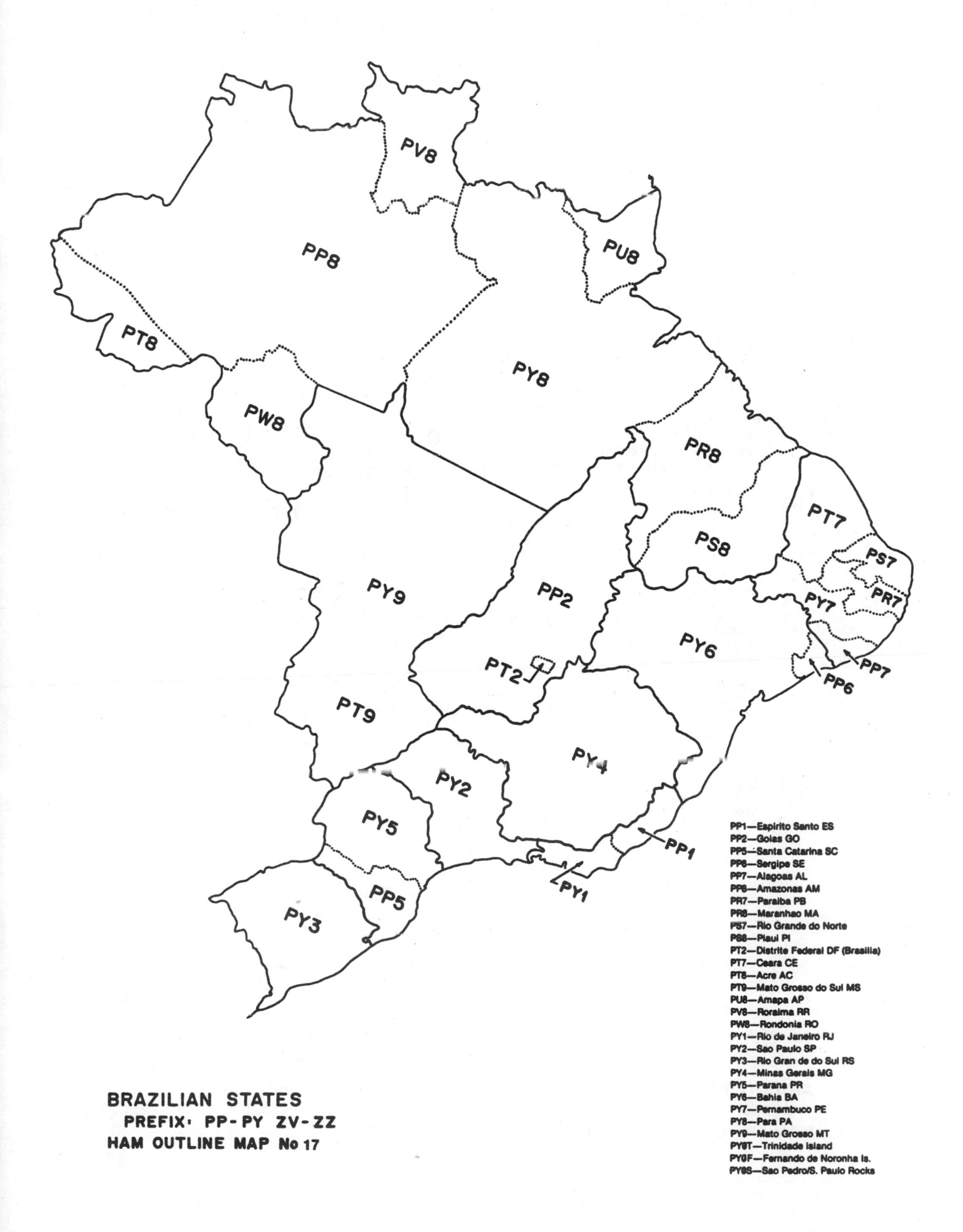
PV8
PU8
PP8
PT8
PY8
PW8
PR8
PT7
PS8
PS7
PP2
PY9
PR7
PY7
PY6
PT2
PP7
PP6
PT9
PY4
PY2
PY5
PP1
PY1
PP5
PY3
PP1—Espirito Santo ES
PP2—Goias GO
PP5—Santa Catarina SC
PP6—Sergipe SE
PP7—Alagoas AL
PP8—Amazonas AM
PR7—Paraiba PB
PR8—Maranhao MA
PS7—Rio Grande do Norte
PS8—Piaui PI
PT2—Distrite Federal DF (Brasilia)
PT7—Ceara CE
PT8—Acre AC
PT9—Mato Grosso do Sul MS
PU8—Amapa AP
PV8—Roraima RR
PW8—Rondonia RO
PY1—Rio de Janeiro RJ
PY2—Sao Paulo SP
PY3—Rio Gran de do Sul RS
PY4—Minas Gerais MG
PY5—Parana PR
PY6—Bahia BA
PY7—Pernambuco PE
PY8—Para PA
PY9—Mato Grosso MT
PY0T—Trinidade Island
PY0F—Fernando de Noronha Is.
PY0S—Sao Pedro/S. Paulo Rocks
BRAZILIAN STATES
PREFIX: PP-PY ZV-ZZ
HAM OUTLINE MAP No 17

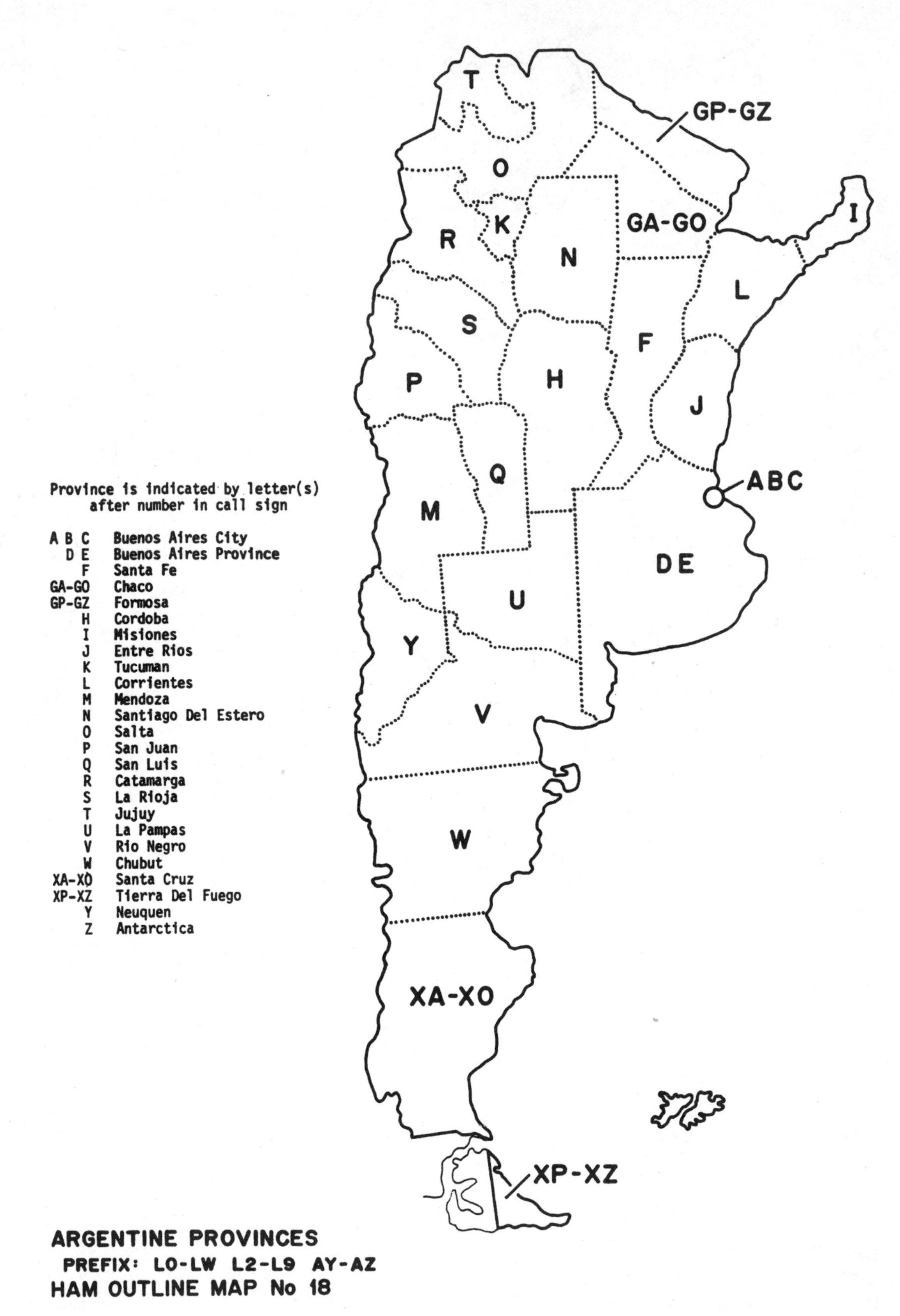

T
GP-GZ
O
I
K
GA-GO
R
N
L
S
F
P
H
J
ABC
Q
M
DE
U
Y
V
W
XA-XO
XP-XZ
Province is indicated by letter(s) after number in call sign
A B C Buenos Aires City
D E Buenos Aires Province
F Santa Fe
GA-GO Chaco
GP-GZ Formosa
H Cordoba
I Misiones
J Entre Rios
K Tucuman
L Corrientes
M Mendoza
N Santiago Del Estero
O Salta
P San Juan
Q San Luis
R Catamarga
S La Rioja
T Jujuy
U La Pampas
V Rio Negro
W Chubut
XA-XO Santa Cruz
XP-XZ Tierra Del Fuego
Y Neuquen
Z Antarctica
ARGENTINE PROVINCES
PREFIX: LO-LW L2-L9 AY-AZ
HAM OUTLINE MAP No 18

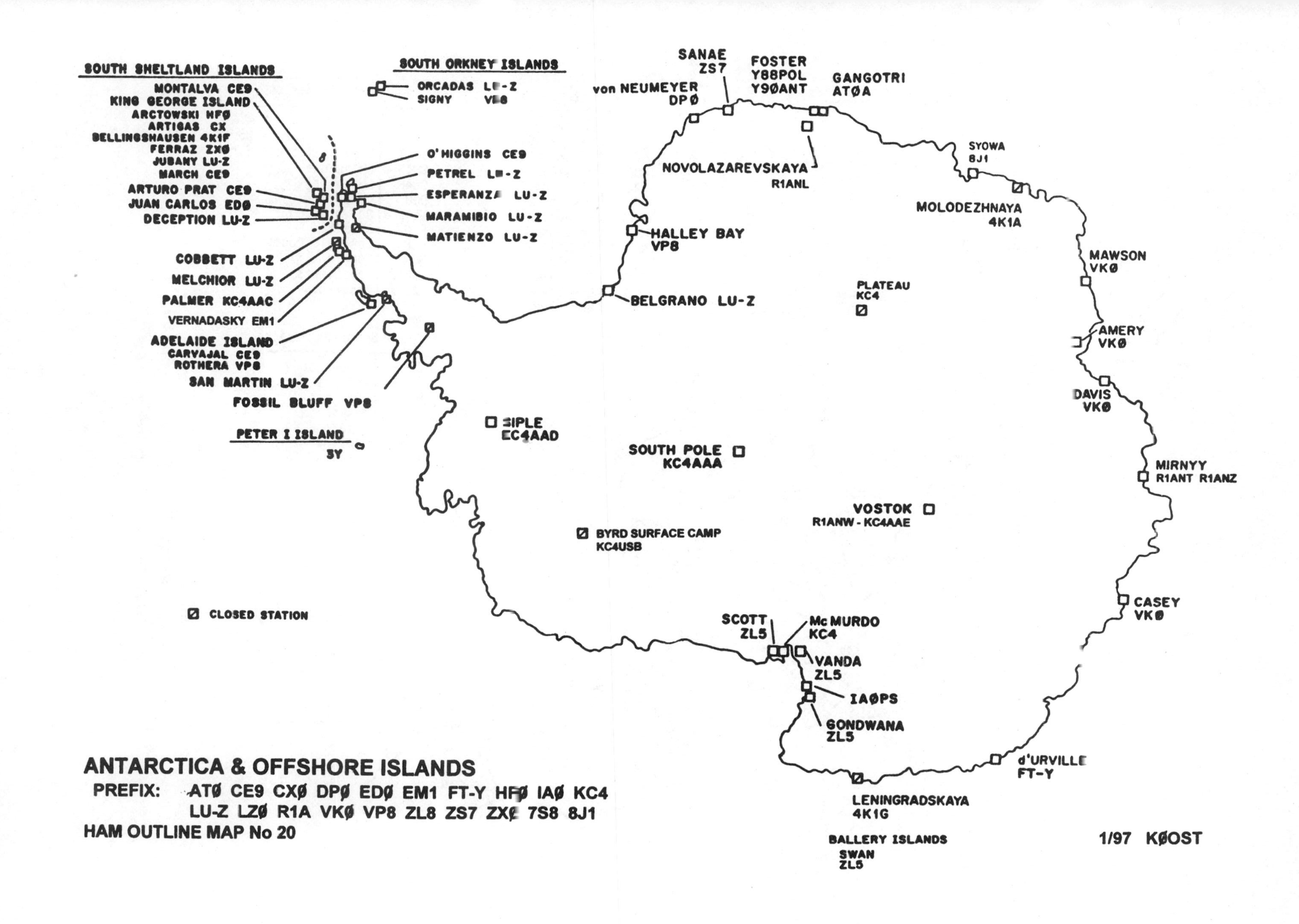
SOUTH SHELTLAND ISLANDS
MONTALVA CE9
KING GEORGE ISLAND
ARCTOWSKI HFØ
ARTIGAS CX
BELLINGSHAUSEN 4K1F
FERRAZ ZXØ
JUBANY LU-Z
MARCH CE9
ARTURO PRAT CE9
JUAN CARLOS EDØ
DECEPTION LU-Z
COBBETT LU-Z
MELCHIOR LU-Z
PALMER KC4AAC
VERNADASKY EM1
ADELAIDE ISLAND
CARVAJAL CE9
ROTHERA VP8
SAN MARTIN LU-Z
FOSSIL BLUFF VP8
PETER I ISLAND
3Y
SOUTH ORKNEY ISLANDS
ORCADAS
SIGNY
O'HIGGINS CE9
PETREL
ESPERANZA
MARAMIBIO LU-Z
MATIENZO LU-Z
von NEUMEYER
DPØ
SANAE
ZS7
FOSTER
Y88POL
Y9ØANT
GANGOTRI
ATØA
NOVOLAZAREVSKAYA
R1ANL
SYOWA
8J1
MOLODEZHNAYA
4K1A
HALLEY BAY
VP8
BELGRANO LU-Z
PLATEAU
KC4
MAWSON
VKØ
AMERY
VKØ
DAVIS
VKØ
SIPLE
EC4AAD
SOUTH POLE
KC4AAA
MIRNYY
R1ANT R1ANZ
VOSTOK
R1ANW - KC4AAE
BYRD SURFACE CAMP
KC4USB
CLOSED STATION
CASEY
VKØ
SCOTT
ZL5
Mc MURDO
KC4
VANDA
ZL5
IAØPS
GONDWANA
ZL5
d'URVILLE
FT-Y
LENINGRADSKAYA
4K1G
BALLERY ISLANDS
SWAN
ZL5
1/97 KØOST
ANTARCTICA & OFFSHORE ISLANDS
PREFIX: ATØ CE9 CXØ DPØ EDØ EM1 FT-Y IAØ KC4
LU-Z R1A VKØ VP8 ZL8 ZS7 7S8 8J1
HAM OUTLINE MAP No 20

MEXICAN STATES
PREFIX: XA-XI 4A-4C 6D-6J
HAM OUTLINE MAP No 4

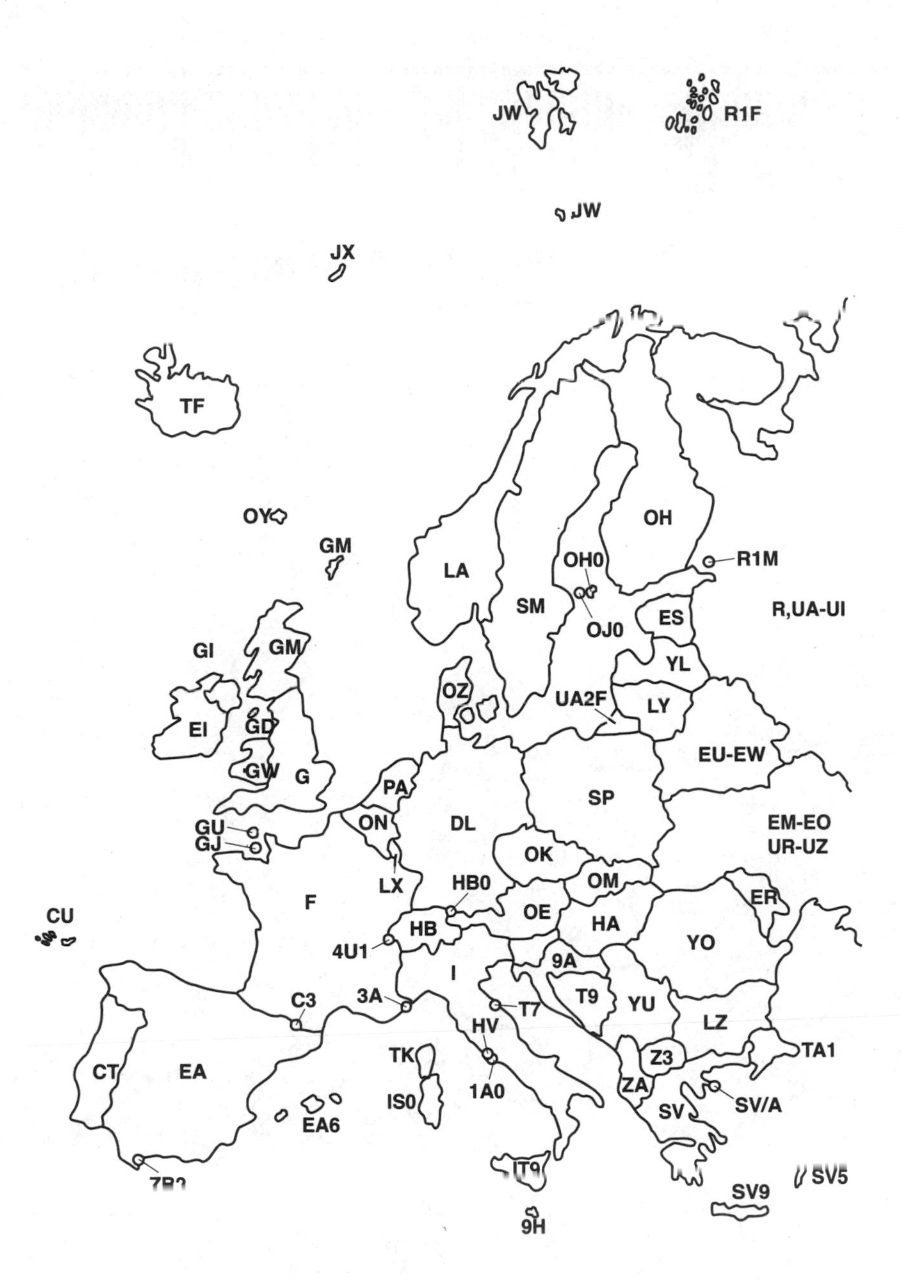

EUROPEAN COUNTRIES **12/95 KØOST**
HAM OUTLINE MAP No 6
FM-RTTY PUBLISHING • PO BOX 35742 • DALLAS, TEXAS 75235

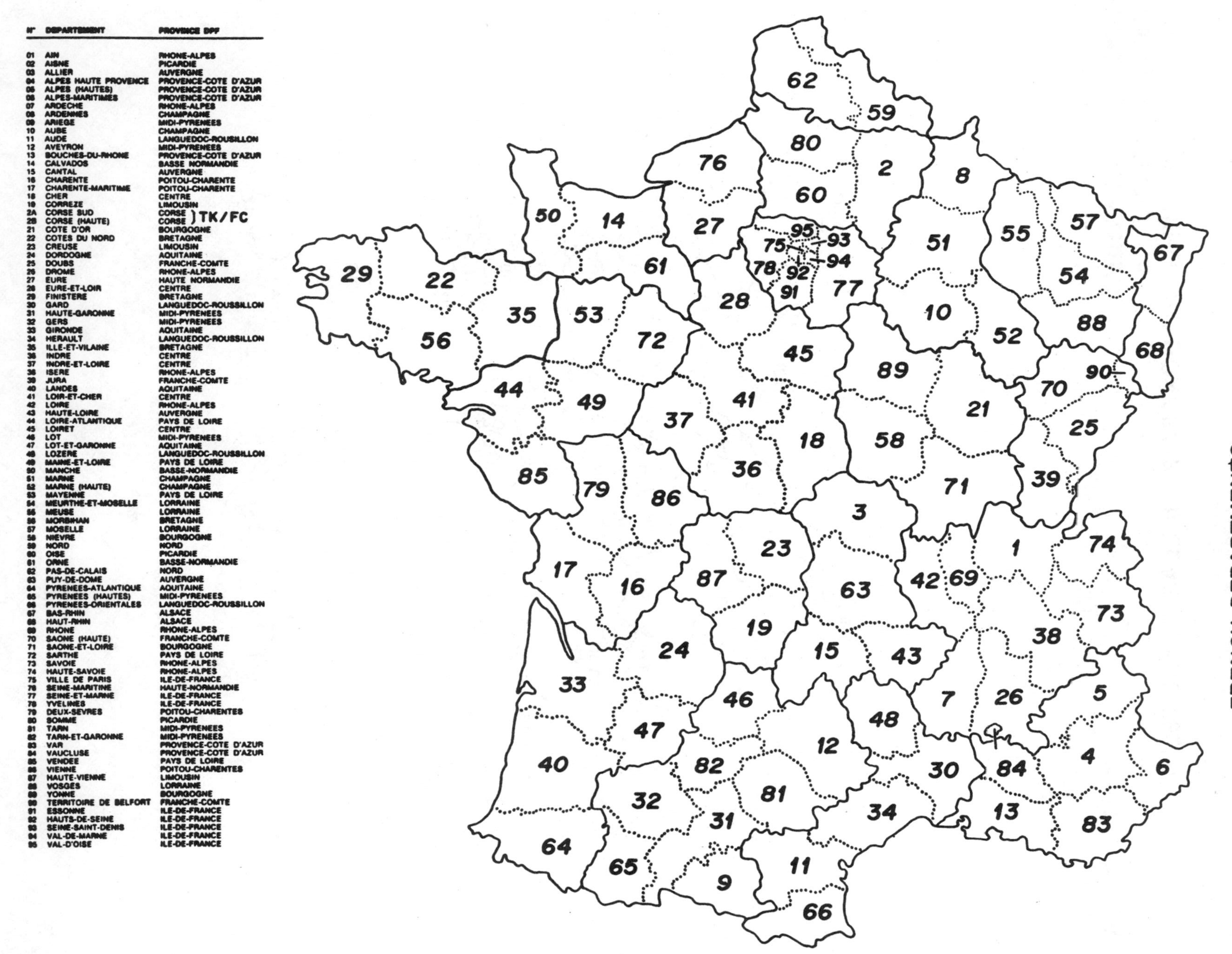

N°	DEPARTEMENT	PROVINCE DPF
01	AIN	RHONE-ALPES
02	AISNE	PICARDIE
03	ALLIER	AUVERGNE
04	ALPES HAUTE PROVENCE	PROVENCE-COTE D'AZUR
05	ALPES (HAUTES)	PROVENCE-COTE D'AZUR
06	ALPES-MARITIMES	PROVENCE-COTE D'AZUR
07	ARDECHE	RHONE-ALPES
08	ARDENNES	CHAMPAGNE
09	ARIEGE	MIDI-PYRENEES
10	AUBE	CHAMPAGNE
11	AUDE	LANGUEDOC-ROUSILLON
12	AVEYRON	MIDI-PYRENEES
13	BOUCHES-DU-RHONE	PROVENCE-COTE D'AZUR
14	CALVADOS	BASSE NORMANDIE
15	CANTAL	AUVERGNE
16	CHARENTE	POITOU-CHARENTE
17	CHARENTE-MARITIME	POITOU-CHARENTE
18	CHER	CENTRE
19	CORREZE	LIMOUSIN
2A	CORSE SUD	CORSE } TK/FC
2B	CORSE (HAUTE)	CORSE
21	COTE D'OR	BOURGOGNE
22	COTES DU NORD	BRETAGNE
23	CREUSE	LIMOUSIN
24	DORDOGNE	AQUITAINE
25	DOUBS	FRANCHE-COMTE
26	DROME	RHONE-ALPES
27	EURE	HAUTE NORMANDIE
28	EURE-ET-LOIR	CENTRE
29	FINISTERE	BRETAGNE
30	GARD	LANGUEDOC-ROUSSILLON
31	HAUTE-GARONNE	MIDI-PYRENEES
32	GERS	MIDI-PYRENEES
33	GIRONDE	AQUITAINE
34	HERAULT	LANGUEDOC-ROUSSILLON
35	ILLE-ET-VILAINE	BRETAGNE
36	INDRE	CENTRE
37	INDRE-ET-LOIRE	CENTRE
38	ISERE	RHONE-ALPES
39	JURA	FRANCHE-COMTE
40	LANDES	AQUITAINE
41	LOIR-ET-CHER	CENTRE
42	LOIRE	RHONE-ALPES
43	HAUTE-LOIRE	AUVERGNE
44	LOIRE-ATLANTIQUE	PAYS DE LOIRE
45	LOIRET	CENTRE
46	LOT	MIDI-PYRENEES
47	LOT-ET-GARONNE	AQUITAINE
48	LOZERE	LANGUEDOC-ROUSSILLON
49	MAINE-ET-LOIRE	PAYS DE LOIRE
50	MANCHE	BASSE-NORMANDIE
51	MARNE	CHAMPAGNE
52	MARNE (HAUTE)	CHAMPAGNE
53	MAYENNE	PAYS DE LOIRE
54	MEURTHE-ET-MOSELLE	LORRAINE
55	MEUSE	LORRAINE
56	MORBIHAN	BRETAGNE
57	MOSELLE	LORRAINE
58	NIEVRE	BOURGOGNE
59	NORD	NORD
60	OISE	PICARDIE
61	ORNE	BASSE-NORMANDIE
62	PAS-DE-CALAIS	NORD
63	PUY-DE-DOME	AUVERGNE
64	PYRENEES-ATLANTIQUE	AQUITAINE
65	PYRENEES (HAUTES)	MIDI-PYRENEES
66	PYRENEES-ORIENTALES	LANGUEDOC-ROUSSILLON
67	BAS-RHIN	ALSACE
68	HAUT-RHIN	ALSACE
69	RHONE	RHONE-ALPES
70	SAONE (HAUTE)	FRANCHE-COMTE
71	SAONE-ET-LOIRE	BOURGOGNE
72	SARTHE	PAYS DE LOIRE
73	SAVOIE	RHONE-ALPES
74	HAUTE-SAVOIE	RHONE-ALPES
75	VILLE DE PARIS	ILE-DE-FRANCE
76	SEINE-MARITINE	HAUTE-NORMANDIE
77	SEINE-ET-MARNE	ILE-DE-FRANCE
78	YVELINES	ILE-DE-FRANCE
79	DEUX-SEVRES	POITOU-CHARENTES
80	SOMME	PICARDIE
81	TARN	MIDI-PYRENEES
82	TARN-ET-GARONNE	MIDI-PYRENEES
83	VAR	PROVENCE-COTE D'AZUR
84	VAUCLUSE	PROVENCE-COTE D'AZUR
85	VENDEE	PAYS DE LOIRE
86	VIENNE	POITOU-CHARENTES
87	HAUTE-VIENNE	LIMOUSIN
88	VOSGES	LORRAINE
89	YONNE	BOURGOGNE
90	TERRITOIRE DE BELFORT	FRANCHE-COMTE
91	ESSONNE	ILE-DE-FRANCE
92	HAUTS-DE-SEINE	ILE-DE-FRANCE
93	SEINE-SAINT-DENIS	ILE-DE-FRANCE
94	VAL-DE-MARNE	ILE-DE-FRANCE
95	VAL-D'OISE	ILE-DE-FRANCE

FRENCH DEPARTMENTS
PREFIX: F HW-HY TH TK TM
HAM OUTLINE MAP No 7

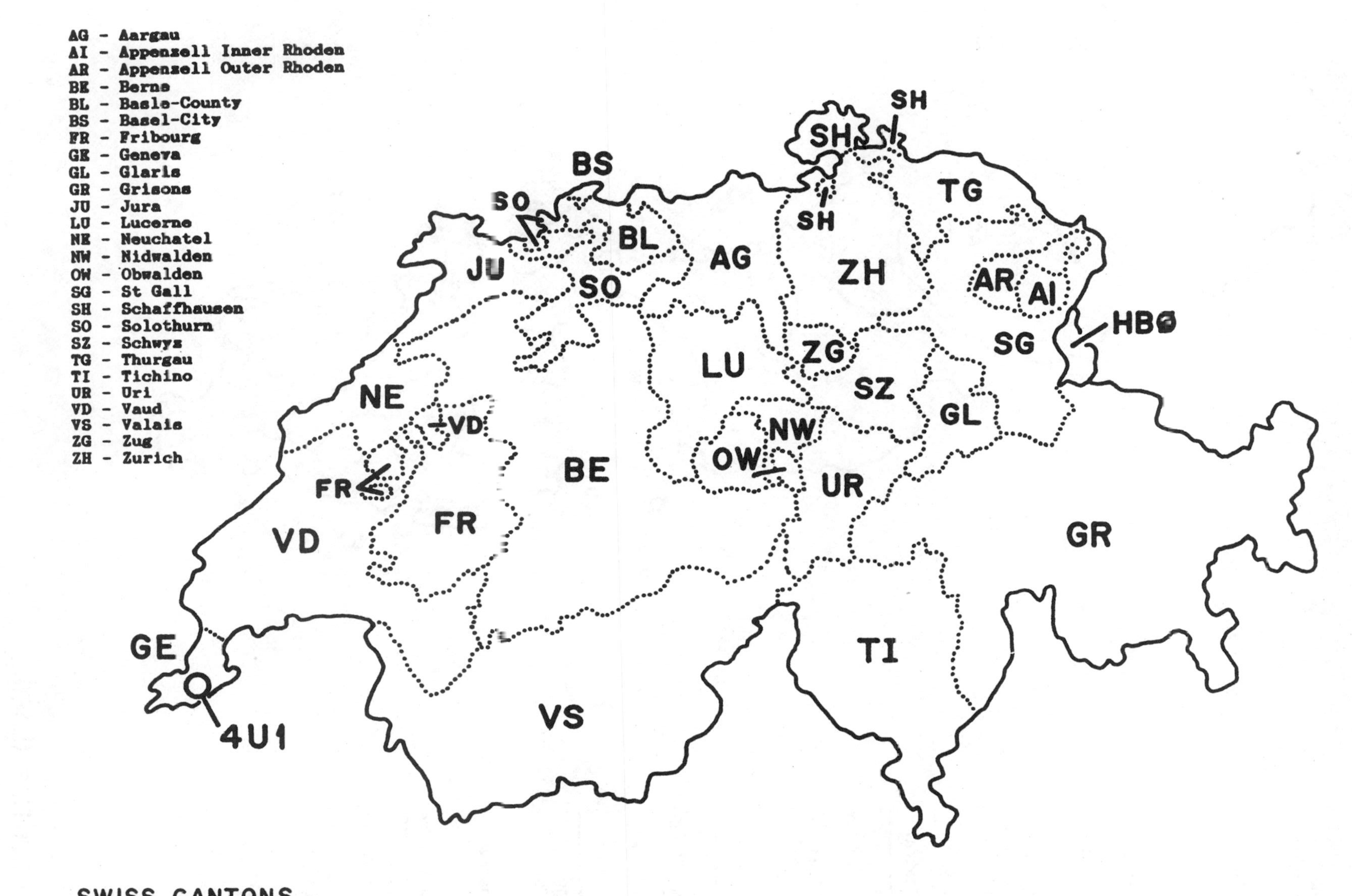
AG - Aargau
AI - Appenzell Inner Rhoden
AR - Appenzell Outer Rhoden
BE - Berne
BL - Basle-County
BS - Basel-City
FR - Fribourg
GE - Geneva
GL - Glaris
GR - Grisons
JU - Jura
LU - Lucerne
NE - Neuchatel
NW - Nidwalden
OW - Obwalden
SG - St Gall
SH - Schaffhausen
SO - Solothurn
SZ - Schwyz
TG - Thurgau
TI - Tichino
UR - Uri
VD - Vaud
VS - Valais
ZG - Zug
ZH - Zurich
SH
SH
SH
BS
SO
BL
AG
ZH
TG
AR
AI
HBØ
SG
JU
SO
ZG
LU
SZ
GL
NE
VD
NW
OW
BE
UR
FR
FR
GR
VD
GE
TI
4U1
VS
SWISS CANTONS
PREFIX: HB HE
HAM OUTLINE MAP No 44
KØOST

1 - Gotland

I	Gotland

2 - Northern

AC	Vasterbotten
BD	Norrbotten

3 - Northern

X	Gavleborg
Y	Vasternorrland
Z	Jamtland

4 - West Central

S	Varmland
T	Orebro
W	Kopparberg

5 - East Central

A	Stockholm City
B	Stockholm County
C	Uppsala
D	Sodermanland
E	Ostergotland
U	Vastmanland

6 - Southwestern

N	Halland
O	Goteborg och Bohus
P	Alvsborg
R	Skaraborg

7 - Southern

F	Jonkoping
G	Kronoberg
H	Kalmar
K	Blekinge
L	Kristianstad
M	Malmohus

Ø - Stockholm Area

A	Stockholm City
B	Stockholm County

SM8 Maritime Mobile
SJ9 Morokulien

SK Club
SL Military
SM Individual

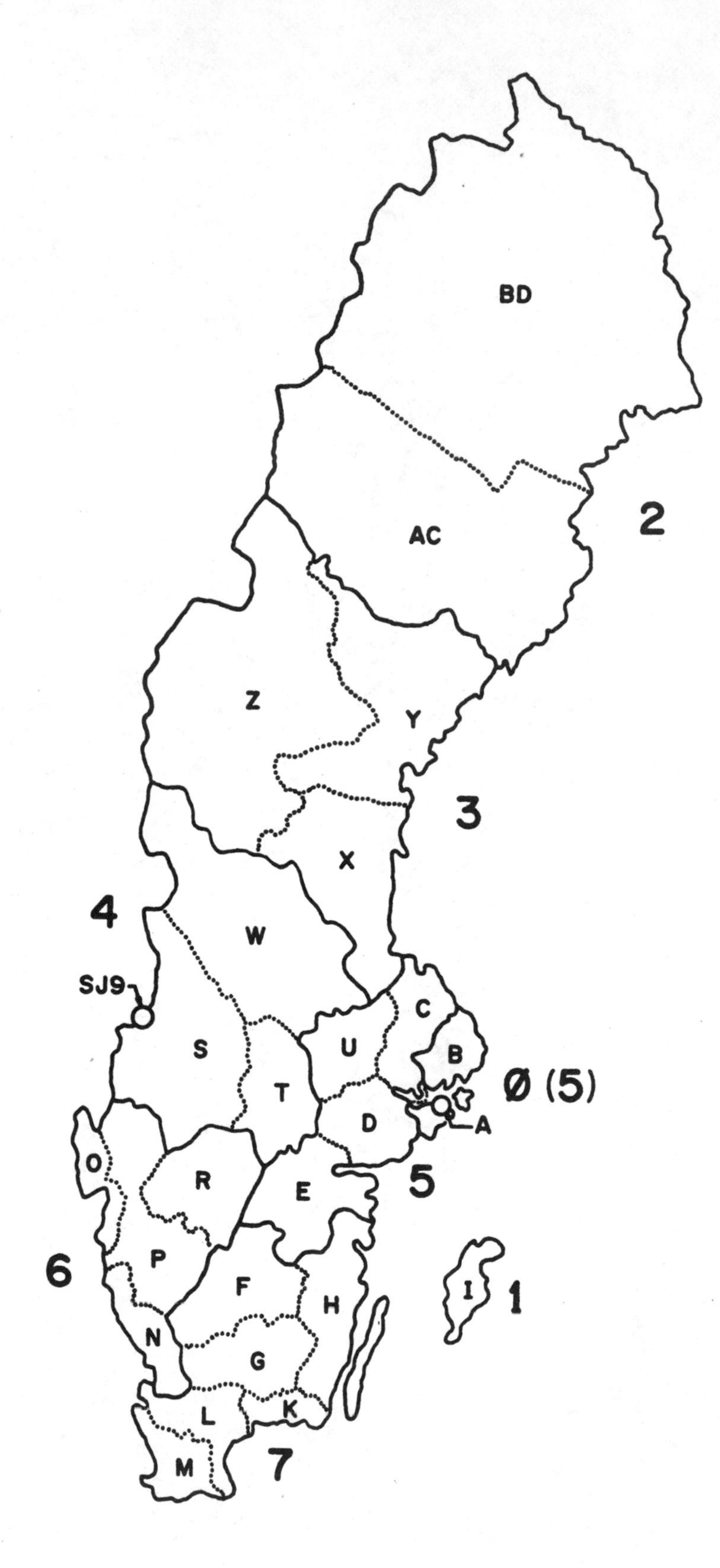

SWEDISH COUNTIES (LAENS)
PREFIX: SA-SM 7S 8S
HAM OUTLINE MAP No 8

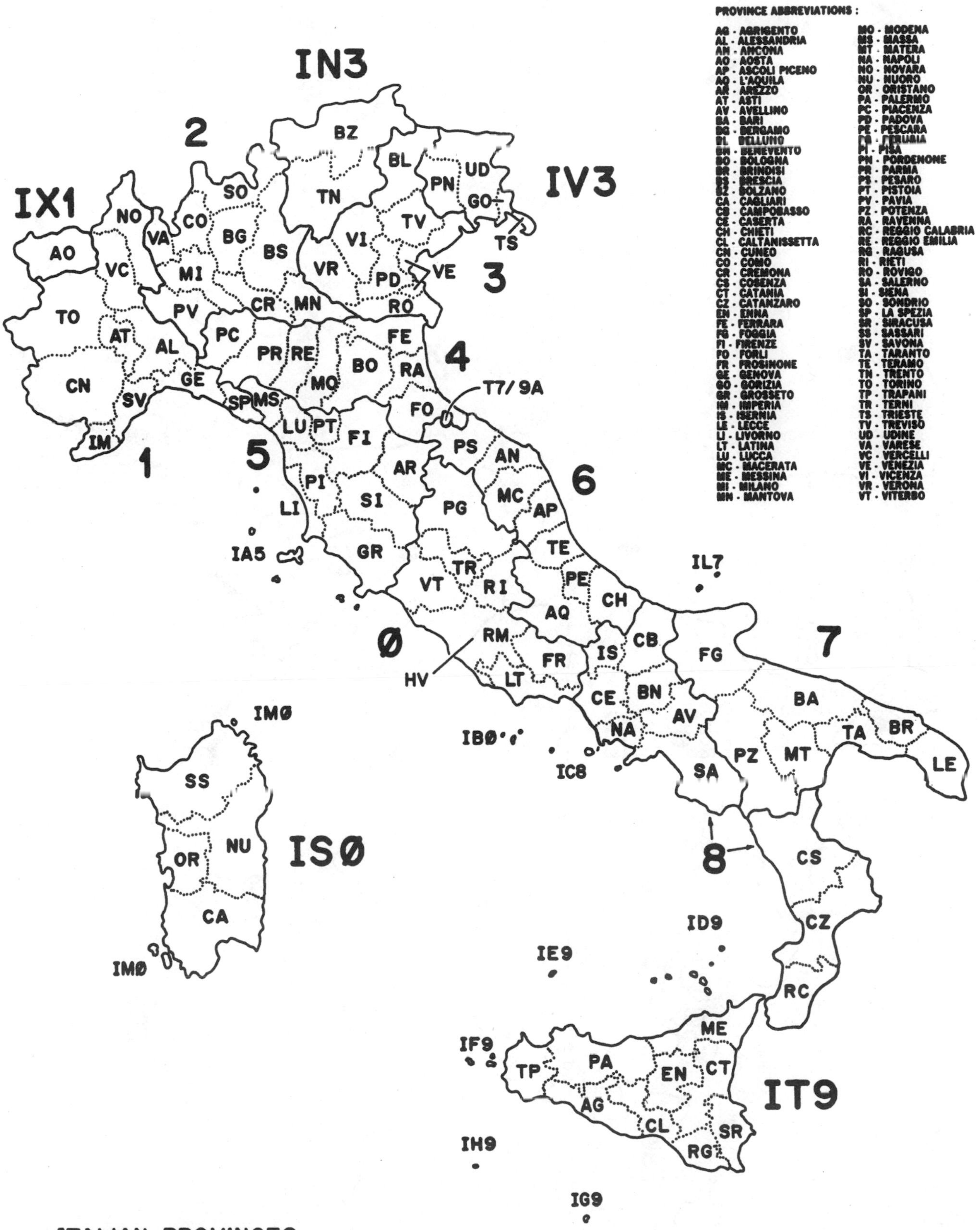

ITALIAN PROVINCES
PREFIX: I IA-IZ
HAM OUTLINE MAP No 9

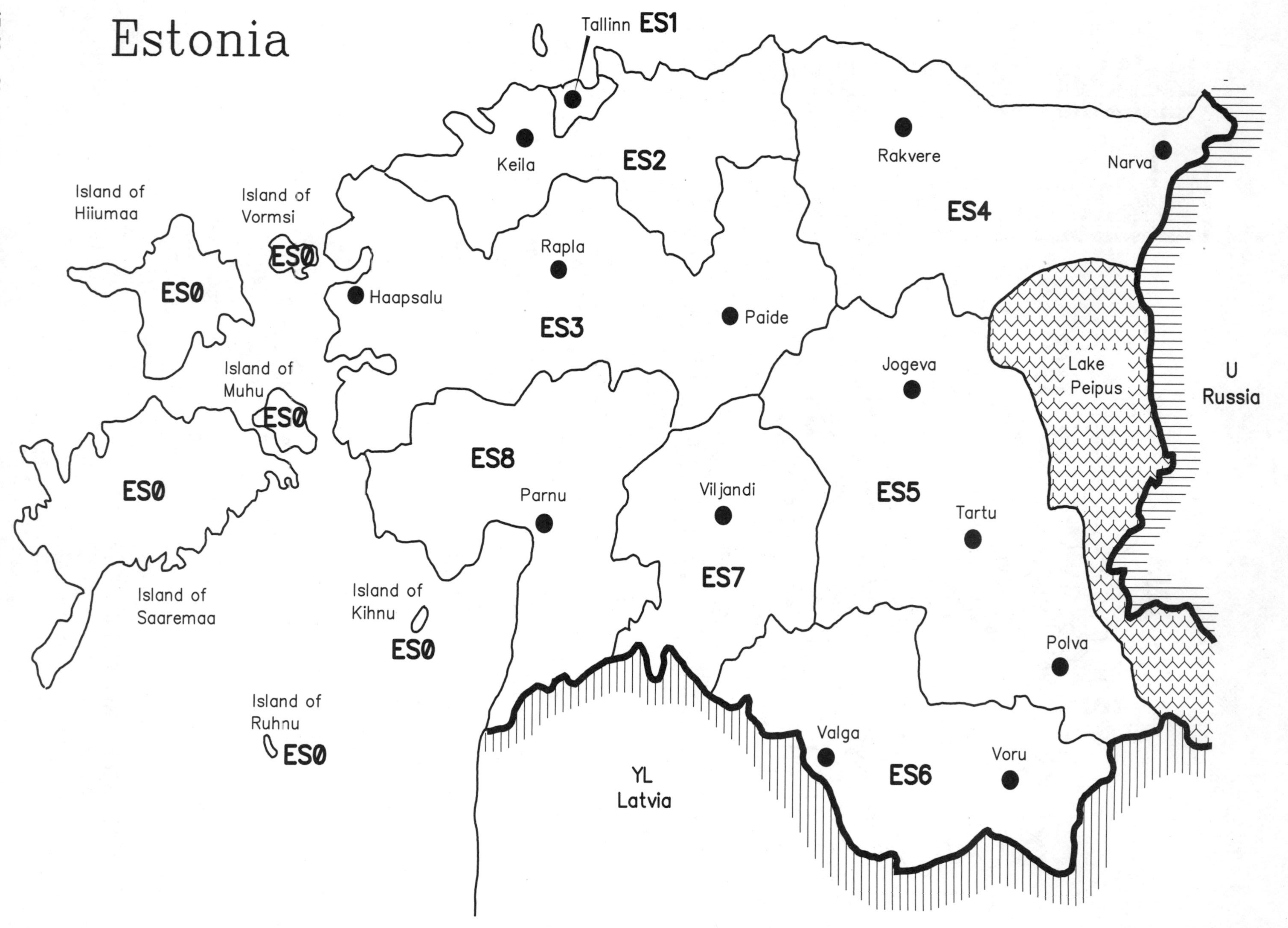

Estonia
Tallinn ES1
Keila
ES2
Rakvere
Narva
ES4
Island of
Hiiumaa
Island of
Vormsi
ESØ
ESØ
Rapla
Haapsalu
Paide
ES3
Island of
Muhu
ESØ
Jogeva
Lake
Peipus
U
Russia
ES8
ESØ
Parnu
Viljandi
ES5
Tartu
ES7
Island of
Saaremaa
Island of
Kihnu
ESØ
Polva
Island of
Ruhnu
ESØ
Valga
Voru
ES6
YL
Latvia

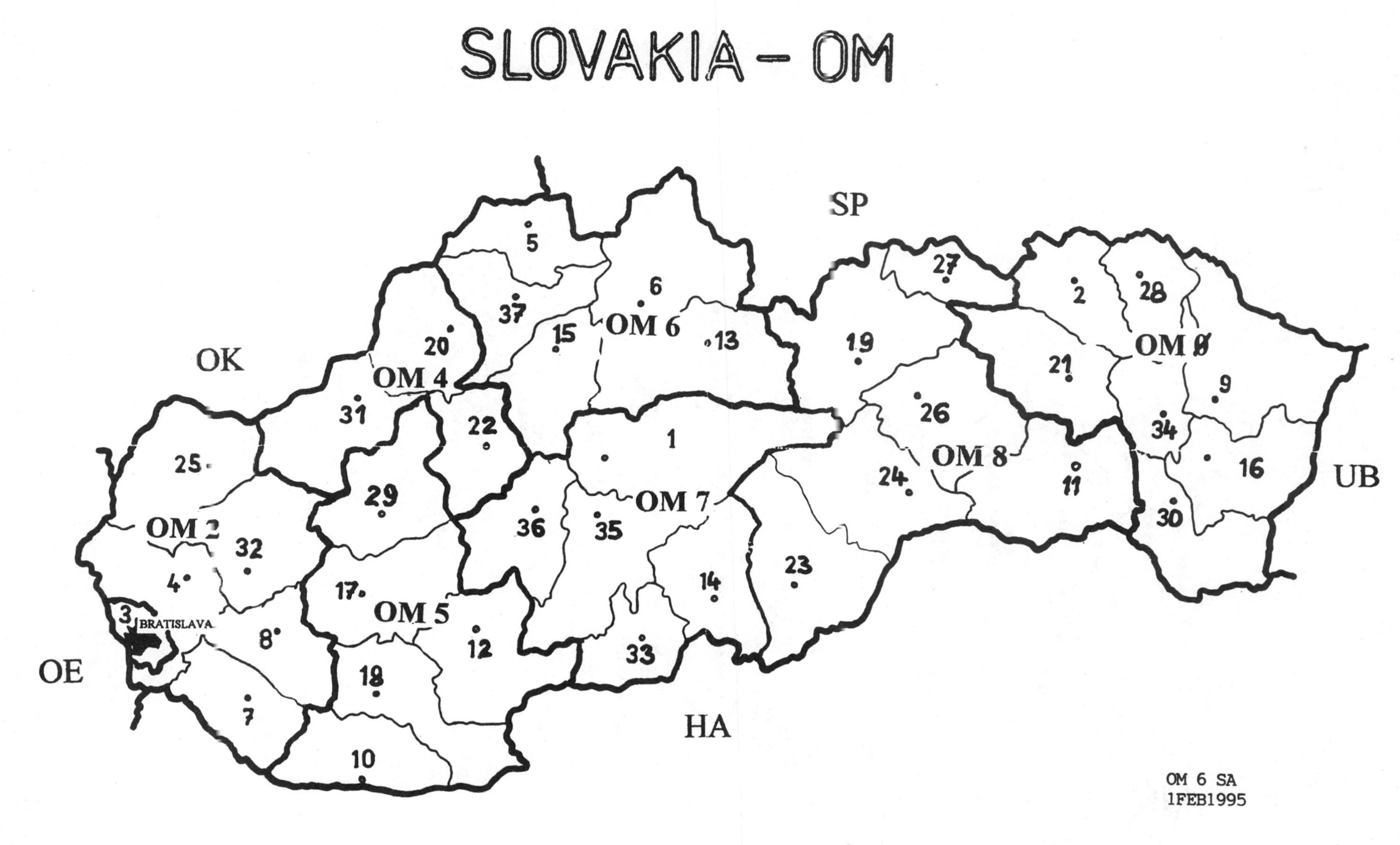

SLOVAKIA – OM
OE
OK
SP
HA
UB
BRATISLAVA
OM 2
OM 4
OM 5
OM 6
OM 7
OM 8
OM Ø
1
2
3
4
5
6
7
8
9
10
11
12
13
14
15
16
17
18
19
20
21
22
23
24
25
26
27
28
29
30
31
32
33
34
35
36
37
OM 6 SA
1FEB1995

BRITISH COUNTIES
HAM OUTLINE MAP No 10

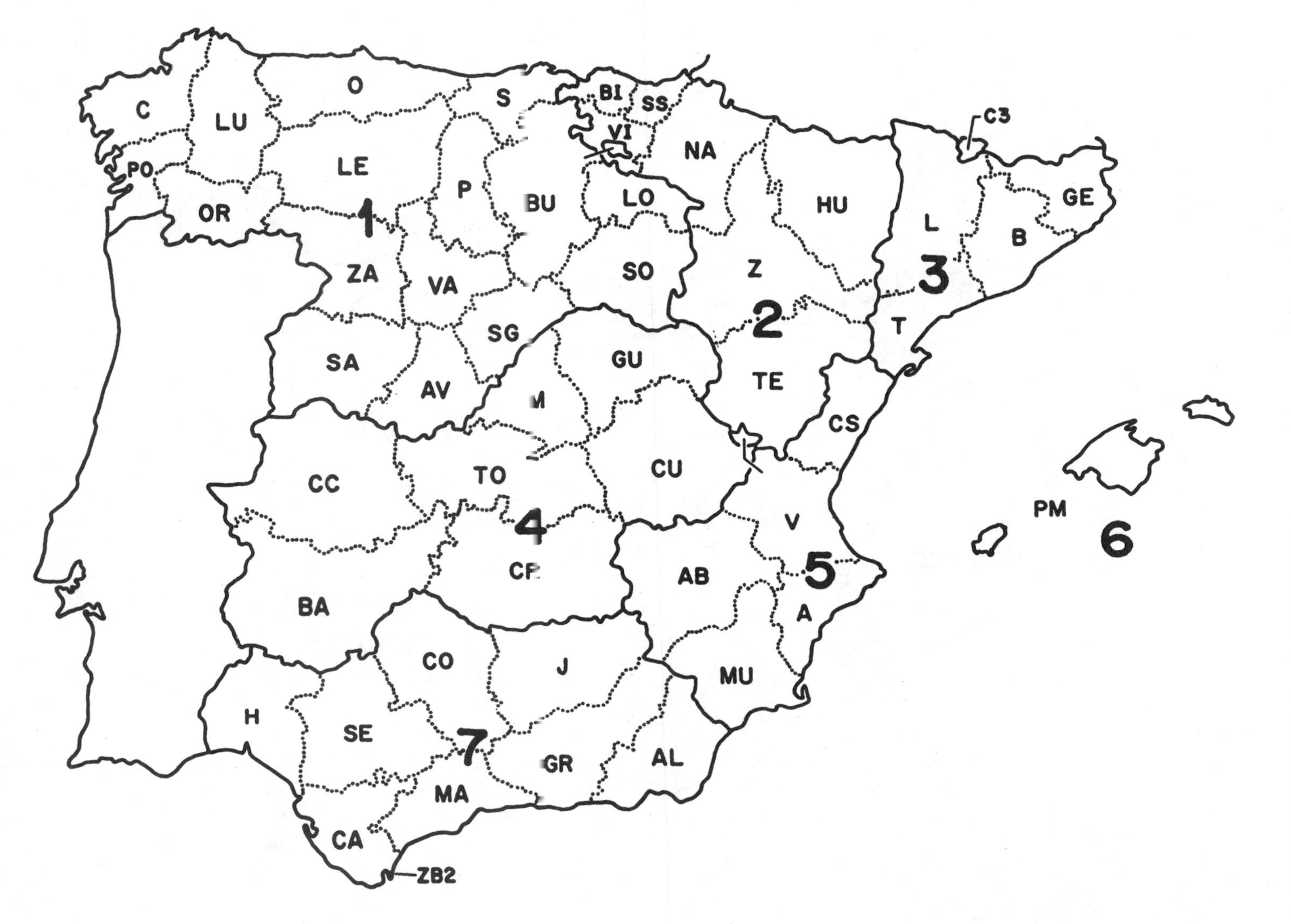

SPANISH PROVINCES
PREFIX: AM-AO EA-EH
HAM OUTLINE MAP No 41

A	5	ALICANTE
AB	5	ALBACETE
AL	7	ALMERIA
AV	1	AVILA
B	3	BARCELONA
BA	4	BARDAJOZ
BI	2	VIZCAYA
BU	1	BURGOS
C	1	LA CORUNA
CA	7	CADIZ
CC	4	CACERES
CE	9	CEUTA
CO	7	CORDOBA
CR	4	CIUDAD READ
CS	5	CASTELLON
CU	4	CUENCA
GC	8	LAS PALMAS
GE	3	GERONA
GR	7	GRANADA
GU	4	GUADALAJARA
H	7	HUELVA
HU	2	HUESCA
J	7	JAEN
L	3	LERIDA
LE	1	LEON
LO	1	LA RIOJA
LU	1	LUGO
M	4	MADRID
MA	7	MALAGA
ML	9	MELILLA
MU	5	MURCIA
NA	2	NAVARRA
O	1	ASTURIAS
OR	1	ORENSE
P	1	PALENCIA
PM	6	BALEARES
PO	1	PONTEVEDRA
S	1	CANTABRIA
SA	1	SALAMANCA
SE	7	SEVILLA
SG	1	SEGOVIA
SO	1	SORIA
SS	2	GUIPUZCOA
T	3	TARRAGONA
TE	2	TERUEL
TF	8	TENERIFE
TO	4	TOLEDO
V	5	VALENCIA
VA	1	VALLADOLID
VI	2	ALAVA
Z	2	ZARAGOZA
ZA	1	ZAMORIA

4/91 KØOST

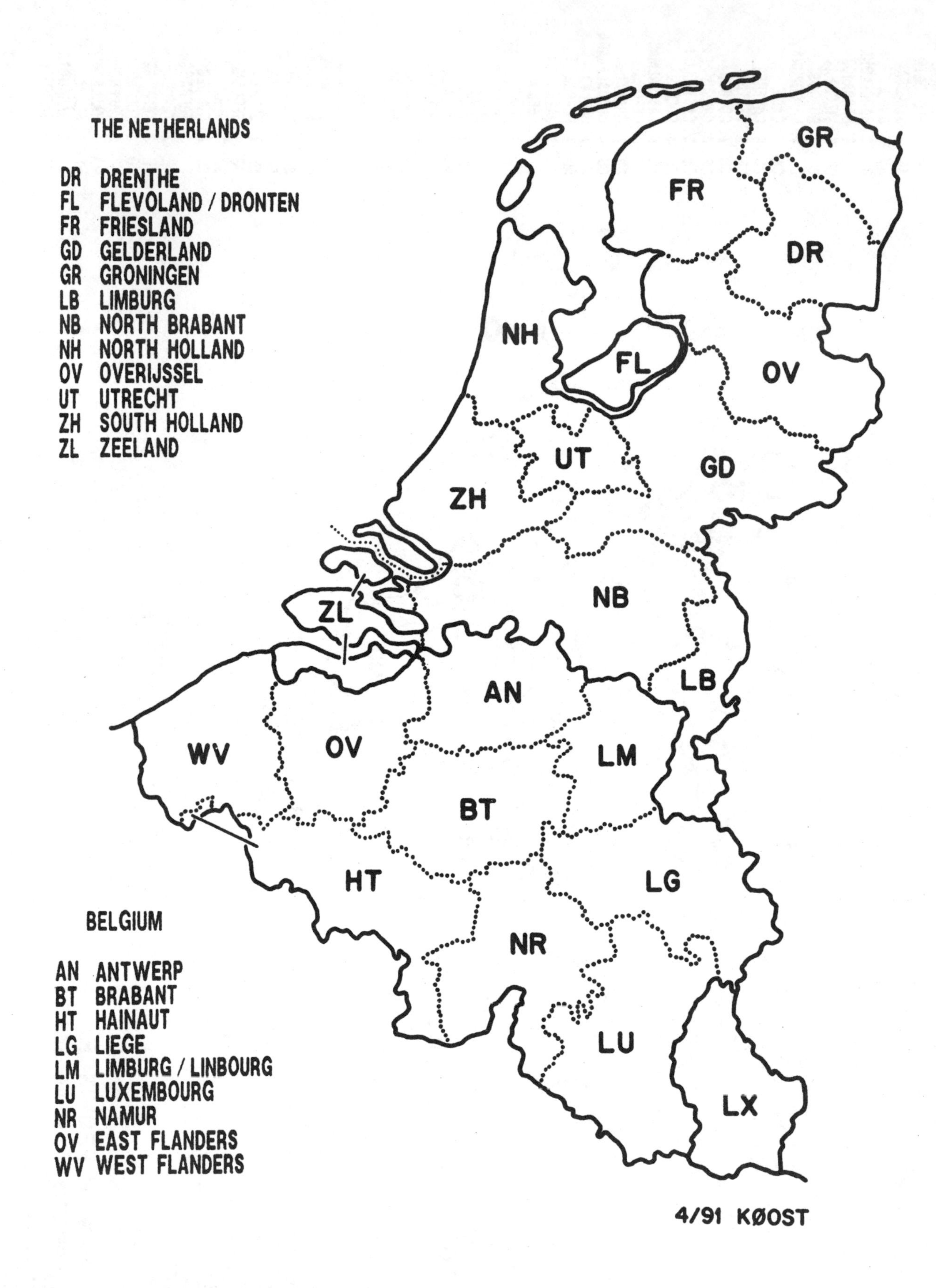

BELGIAN / DUTCH PROVINCES
PREFIX: ON-OT / PA-PI
HAM OUTLINE MAP No 42

POLISH PROVINCES
PREFIX: HF SN-SR 3Z

B = Lubuskie
C = Lodzkie
D = Dolnoslaskie
F = Pomorskie
G = Slaskie
J = Warminsko-Mazurskie
K = Podkarpackie
L = Lubelskie
M = Malopolskie
O = Podlaskie
P = Kujawsko-Pomorskie
R = Mazowieckie
S = Swietokrzyskie
U = Opolskie
W = Wielkopolskie
Z = Zachodniopomorskie

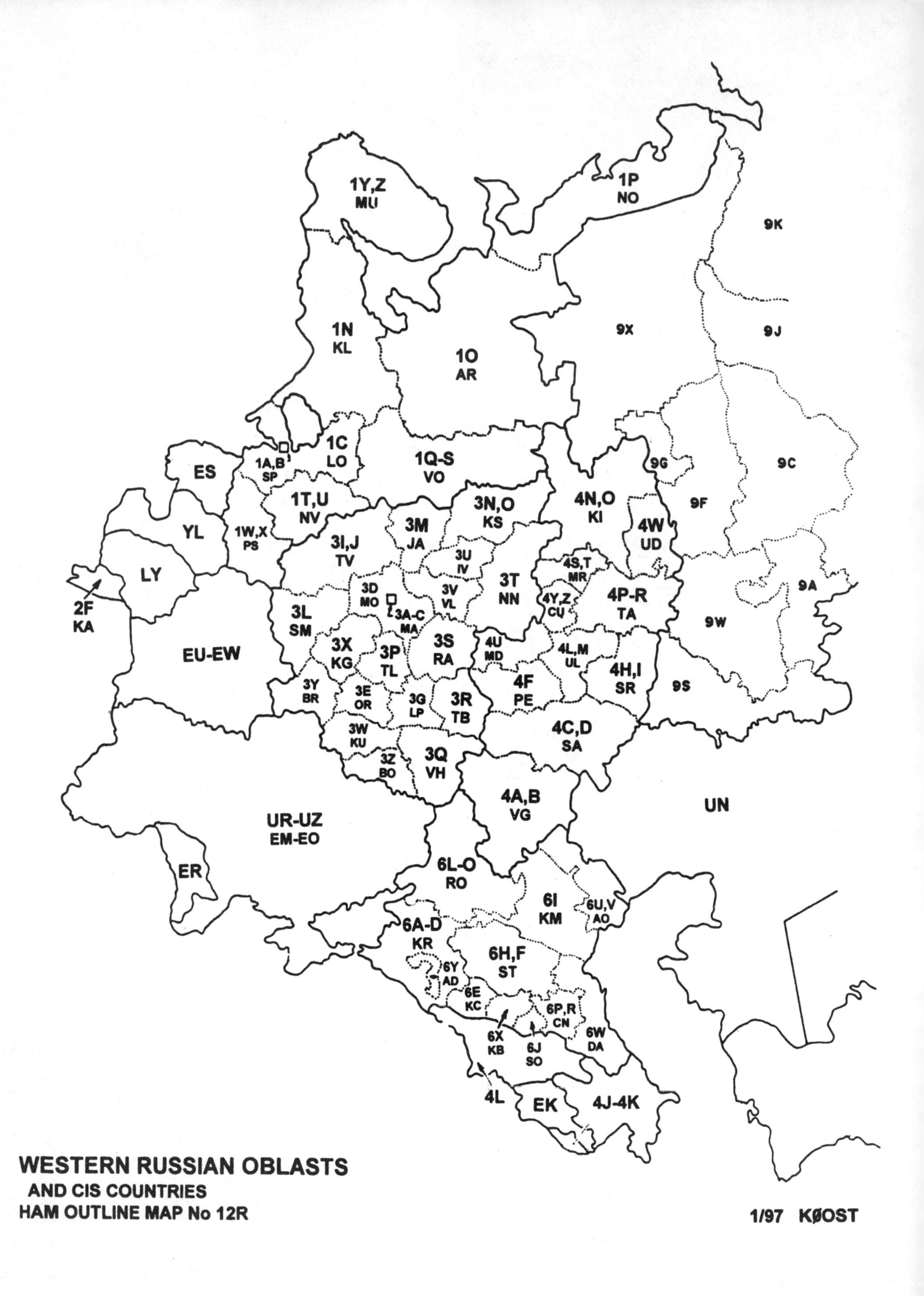
1Y,Z
MU
1P
NO
9K
1N
KL
1O
AR
9X
9J
1C
LO
1A,B
SP
ES
1Q-S
VO
9G
9C
1T,U
NV
3N,O
KS
4N,O
KI
9F
YL
1W,X
PS
3M
JA
3I,J
TV
4W
UD
3U
IV
4S,T
MR
LY
3D
MO
3V
VL
3T
NN
4Y,Z
CU
4P-R
TA
9A
2F
KA
3L
SM
3A-C
MA
9W
3X
KG
3P
TL
3S
RA
4U
MD
4L,M
UL
EU-EW
4H,I
SR
3Y
BR
3E
OR
3G
LP
3R
TB
4F
PE
9S
3W
KU
4C,D
SA
3Z
BO
3Q
VH
4A,B
VG
UN
UR-UZ
EM-EO
ER
6L-O
RO
6I
KM
6U,V
AO
6A-D
KR
6H,F
ST
6Y
AD
6E
KC
6P,R
CN
6X
KB
6J
SO
6W
DA
4L
EK
4J-4K
WESTERN RUSSIAN OBLASTS
AND CIS COUNTRIES
HAM OUTLINE MAP No 12R
1/97 KØOST

EASTERN RUSSIAN OBLASTS
FORMER UA9, Ø CALL AREAS
HAM OUTLINE MAP No 13

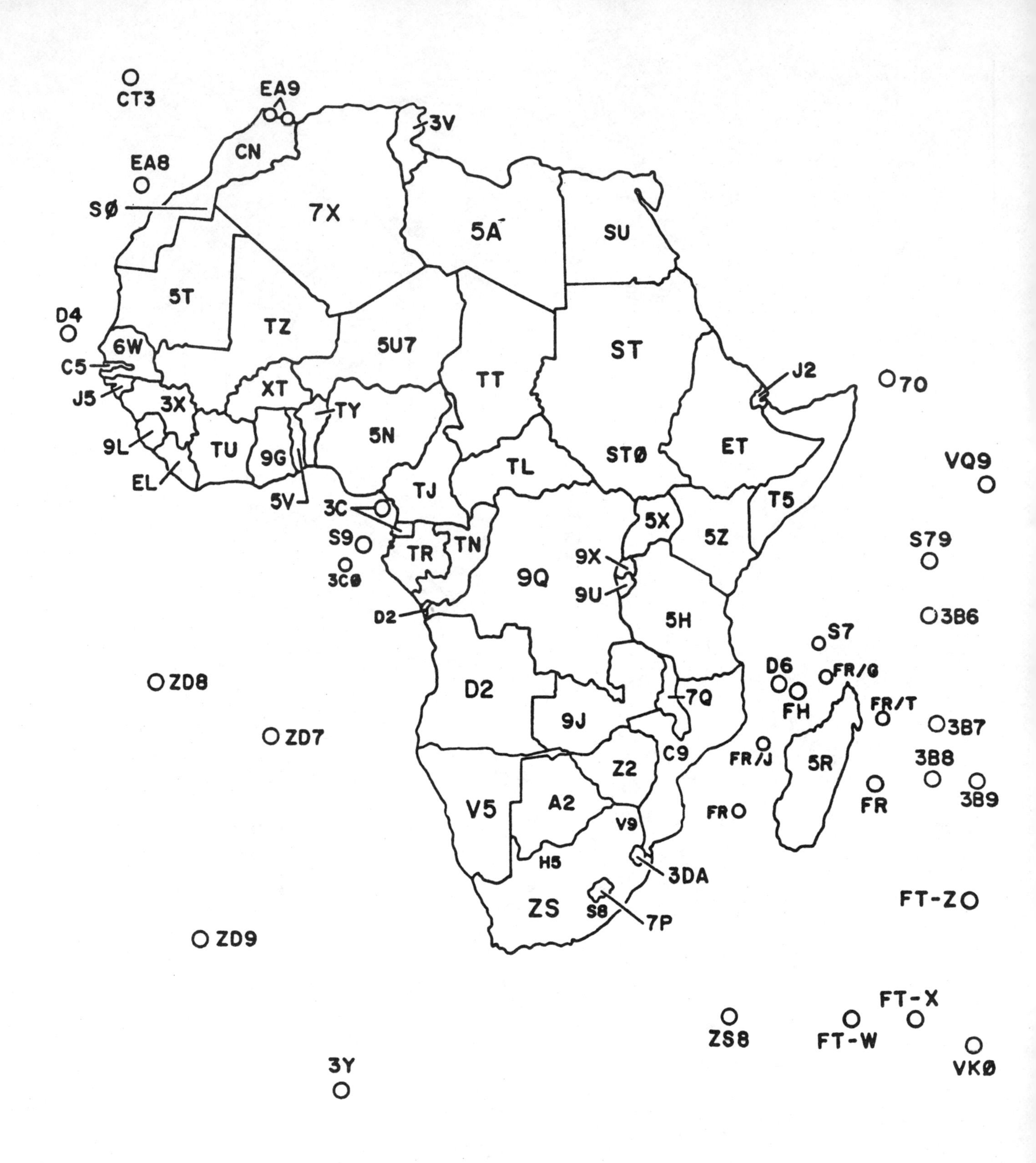

AFRICAN COUNTRIES

HAM OUTLINE MAP No 22

1/97 KØOST

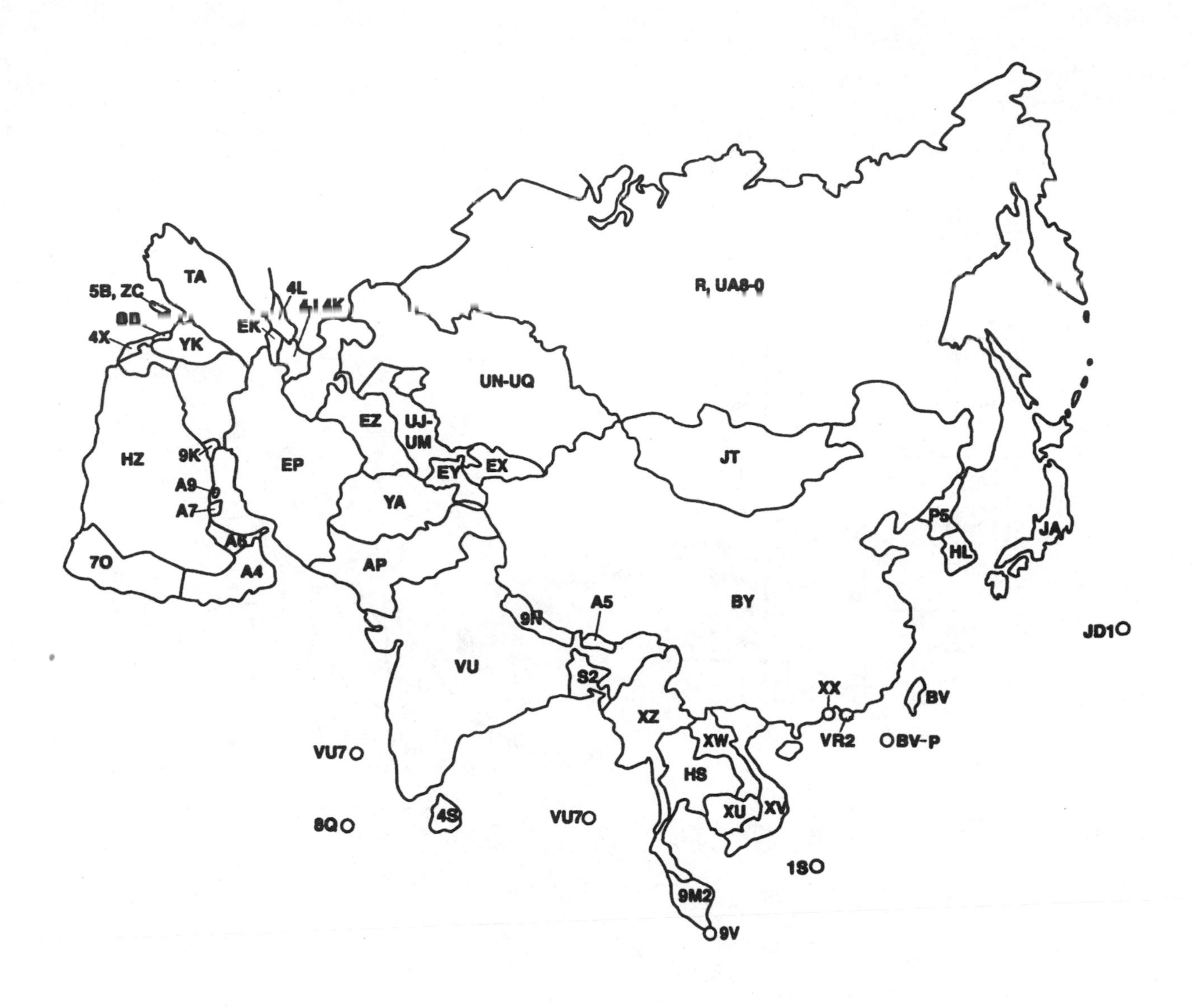

ASIAN COUNTRIES
HAM OUTLINE MAP No 1

3/00 KØOST

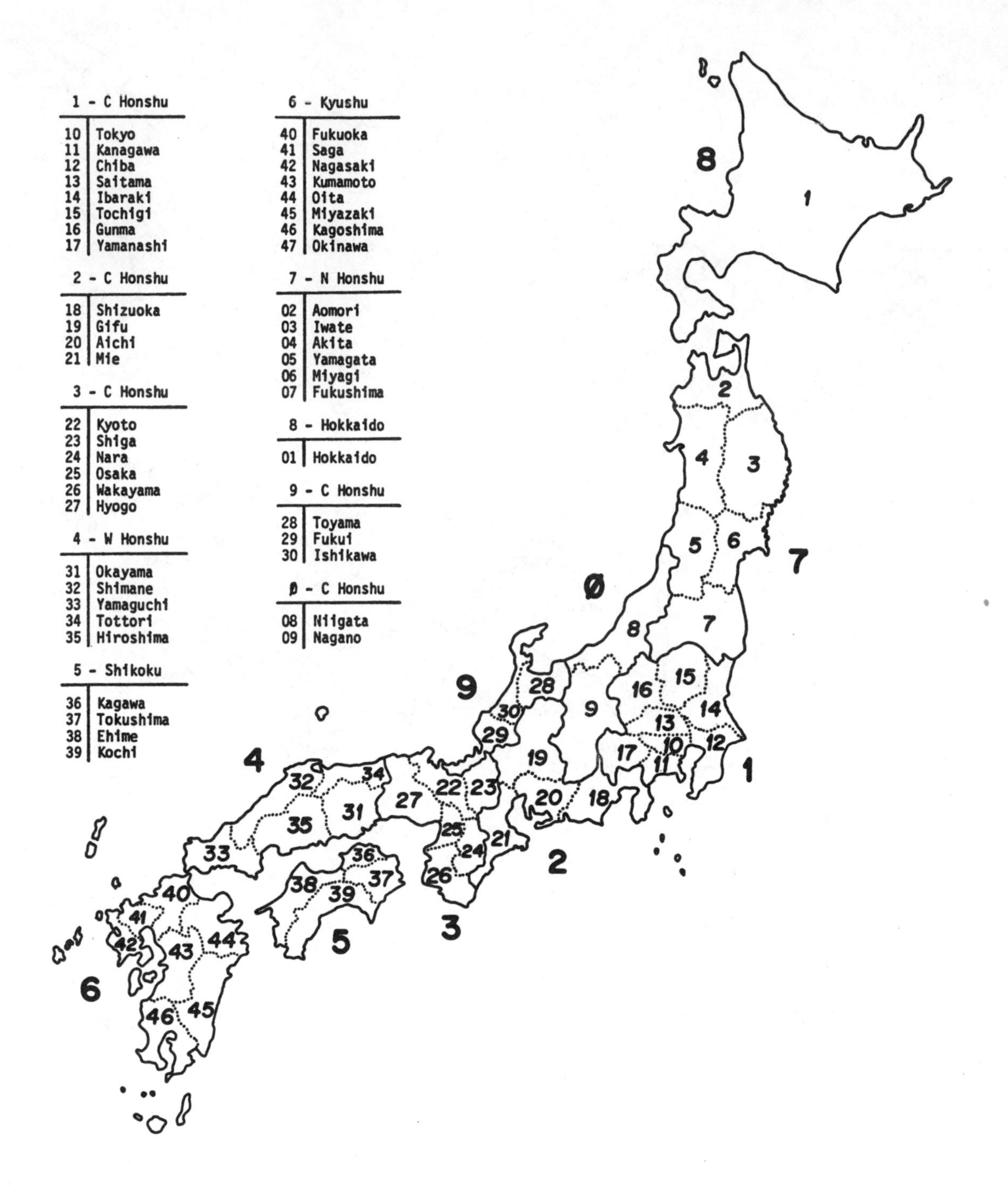

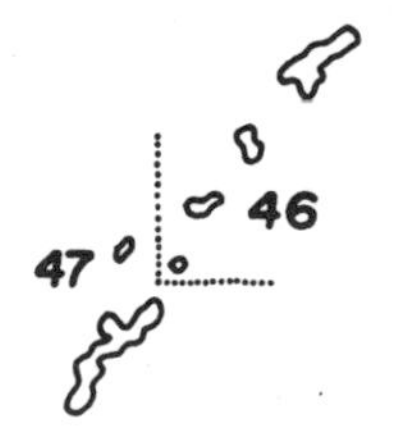

JAPANESE PREFECTURES (KENS)
PREFIX: JA-JS 7J-7N 8J-8N
HAM OUTLINE MAP No 2

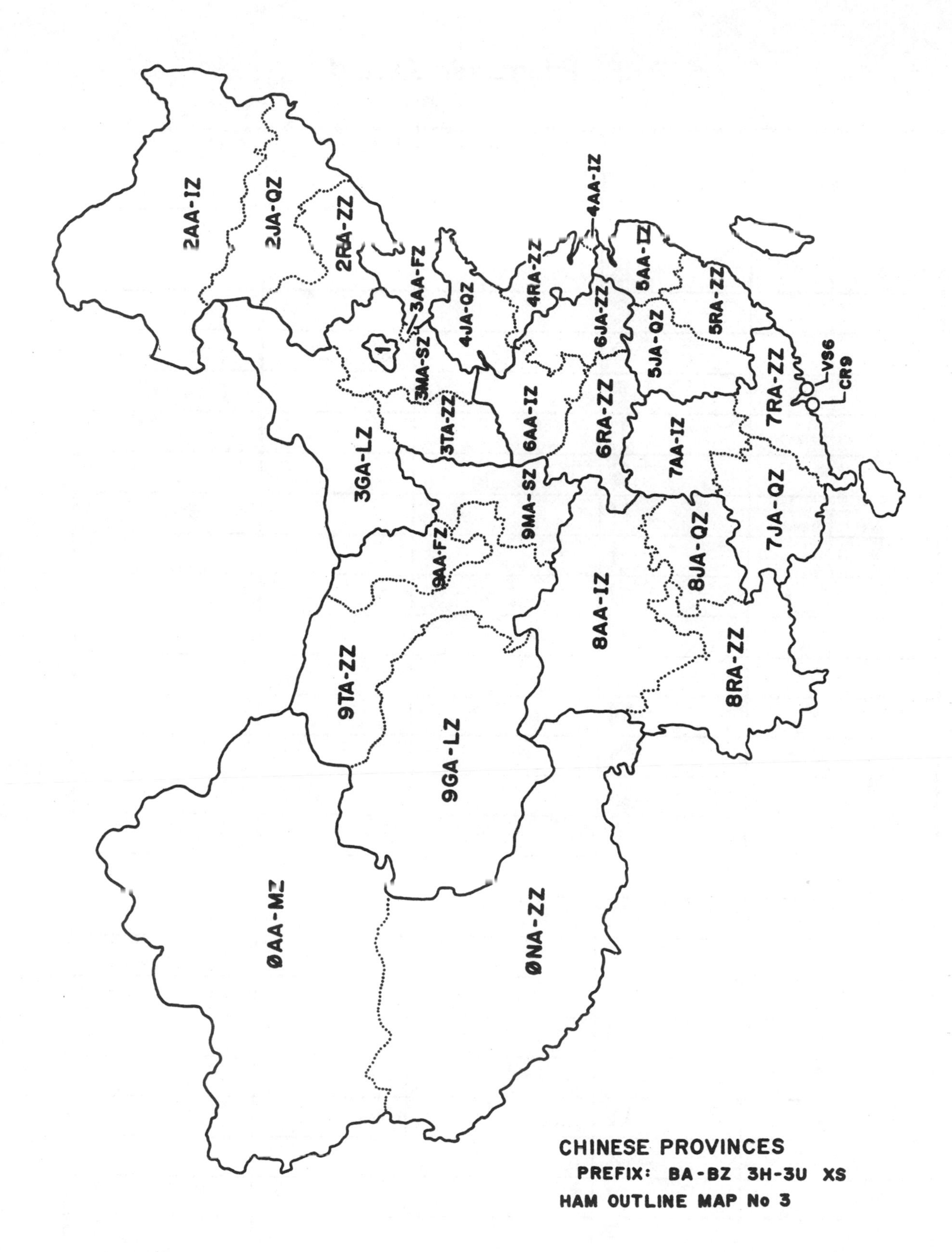

CHINESE PROVINCES
PREFIX: BA-BZ 3H-3U XS
HAM OUTLINE MAP No 3

ARRL Friendship Award

Contact Date	Callsign	First Name	Location	Other Fact
	A			
	B			
	C			
	D			
	E			
	F			
	G			
	H			
	I			
	J			
	K			
	L			
	M			
	N			
	O			
	P			
	Q			
	R			
	S			
	T			
	U			
	V			
	W			
	X			
	Y			
	Z			

Place portable indicator of callsign first, for example: HC1/K4ERO

Complete rules appear in the Operating Awards chapter of this book.

ARRL WAS APPLICATION FORM

Please print or type **CLEARLY**:
NAME, CALLSIGN: ______________________________
(Print exactly as you want it on certificate)

List any ex-calls used on any cards ______________________________

ADDRESS: ______________________________

____
(City) (State) (Zip)

FUNDS ENCLOSED

For QSL return postage
Money/SASE

$__________

Return QSL's VIA:
___ Registered
___ Certified (US Only)
___ First Class mail

For Certificate $ ________
For Endorsement $ ________

Membership ID number (from ***QST*** label): ______________
Check One: _____ Initial Application
_____ Endorsement
_____ Original WAS Number held is __________

I am applying for ONE of the following WAS Awards (each is numbered separately):
_____ Basic Award (Mixed band and/or Modes)
_____ 50 MHz _____ 144 MHz _____ 432 MHz
_____ 160 Meters _____ 222 MHz _____ RTTY
_____ Satellite _____ SSTV

_____ I am applying for the Digital Award (not numbered)
Circle One: PSK31 AMTOR PacTOR CLOVER G-TOR
Other ______________

I am applying for the following ENDORSEMENTS: (Check all that apply)
___ SSB ___ CW ___ Novice ___ QRP ___ Packet ___ EME
Single Band (Circle one): **10 12 15 17 20 40 80** (Meters)

I have read, understood, and followed all the rules of WAS (MSD-264):

______________________________ ______________
Applicant's Signature Date

HF AWARDS MANAGER VERIFICATION

I have personally inspected the confirmations with all 50 states and verify that this application is correct and true. This application is for the following SPECIALTY awards or ENDORSEMENTS:______________ (Write **NONE** if none)

______________________ ______________ H-__________ ______________
Signature Callsign (Mgr code#) Date

****************************** DIRECTIONS TO APPLICANT ******************************

1. Read WAS Rules (MSD-264) carefully.
2. Fill out both sides of this application.
3. Sort cards by state as listed on the back of this application.
4. Present application and cards to your ARRL HF Awards Manager for verification. Applications from DX stations may be certified by the Awards Manager of your IARU member-society.
5. Send application to ARRL HQ with the appropriate fee(s):
   $5 for **each** WAS certificate (the fee includes any endorsements on the same application)
   $3 per endorsement application (if applying for multiple endorsements on the SAME application, the fee remains $3)
   Example: 10 Meters, SSB and Novice. Send cards ONLY if there is no local HF Awards Manager to verify your application.
6. If mailing cards, enclose sufficient postage for return of your cards. Mail to: ARRL WAS Award, 225 Main Street, Newington, CT 06111, USA.

MSD-217(10/01)

WAS RECORD SHEET

Applicant's callsign ____________________

List any ex-calls used on any cards submitted:____

STATE	CALL	DATE	BAND	MODE
Alabama				
Alaska				
Arizona				
Arkansas				
California				
Colorado				
Connecticut				
Delaware				
Florida				
Georgia				
Hawaii				
Idaho				
Illinois				
Indiana				
Iowa				
Kansas				
Kentucky				
Louisiana				
Maine				
Maryland (D.C.)				
Massachusetts				
Michigan				
Minnesota				
Mississippi				
Missouri				
Montana				
Nebraska				
Nevada				
New Hampshire				
New Jersey				
New Mexico				
New York				
North Carolina				
North Dakota				
Ohio				
Oklahoma				
Oregon				
Pennsylvania				
Rhode Island				
South Carolina				
South Dakota				
Tennessee				
Texas				
Utah				
Vermont				
Virginia				
Washington				
West Virginia				
Wisconsin				
Wyoming				

ARRL 5BWAS AWARD APPLICATION

Please print or type **CLEARLY**: Name, Callsign:______________________________
(Print exactly as you want it on certificate)

List any ex-calls used on any cards ______________

Address:______________________________

____
(City) (State) (Zip)

For Return Postage:
$__________

Return QSL's VIA:
- ☐ Registered ☐ Certified (US Only)
- ☐ First Class mail

For Certificate and Pin:
$__________

Membership ID number (from *QST* label):

____

Is this your first WAS Award?__________

$______________enclosed for a plaque.

I have read, understood and followed all the rules of the WAS (MSD-264)

______________________________ ______________
Applicant's Signature Date

****************** HF AWARDS MANAGER VERIFICATION *****************

I have personally inspected the confirmations with all 50 states and verify that this application is correct and true.

______________________ ____________ H-__________ ____________
Signature Callsign (Mgr code#) Date

****** ****************** 5 Band WAS Rules ***************************** *****

1. The 5BWAS certificate and plaque (see below #4) will be issued for having submitted confirmations with each of the 50 United States for contacts dated January 1, 1970, or after, on five amateur bands (10, 18 and 24 MHz excluded). Phone and CW segments of a band do not count as separate bands.
2. WAS Rules (MSD-264) that do not conflict with these 5BWAS rules also apply to the 5BWAS Award.
3. There are no specialty 5 Band awards or endorsements.
4. A handsome 9 X 12 personalized walnut plaque is available for a fee of $30 US (check or money order) plus shipping ($8 in USA).
5. The fee for the 5 Band WAS certificate is $10, which includes a lapel pin.

*************************** Directions to Applicant *************************************

1. Read the 5BWAS rules above and WAS Rules (MSD-264) carefully.
2. Fill out BOTH sides of this application.
3. Sort cards by state as listed on the back of this form.
4. Present application and cards to your ARRL HF Awards Manager for verification. Applications from DX stations may be certified by the Awards Manager of your IARU member-society.
5. Send applications to ARRL HQ. Send cards ONLY if there is no local HF Awards Manager to verify your application.
6. If mailing cards, enclose the $10 fee and sufficient postage for return of your cards. Typically, 250 QSL cards weigh 30 ounces. Mail to ARRL 5BWAS Award, 225 Main Street, Newington, CT 06111 USA
7. Enclose $30 US plus shipping if you want the plaque. ($8 shipping US, foreign depending on location)

MSD-225(1/98)

5B-WAS RECORD SHEET

DIRECTIONS: Enter callsigns in the appropriate boxes by state and band. QSLs must be for contacts on or after January 1, 1970.

Applicant's callsign____________List any ex-calls used on any cards submitted:____________

STATE	80	40	20	15	10
Alabama					
Alaska					
Arizona					
Arkansas					
California					
Colorado					
Connecticut					
Delaware					
Florida					
Georgia					
Hawaii					
Idaho					
Illinois					
Indiana					
Iowa					
Kansas					
Kentucky					
Louisiana					
Maine					
Maryland (D.C.)					
Massachusetts					
Michigan					
Minnesota					
Mississippi					
Missouri					
Montana					
Nebraska					
Nevada					
New Hampshire					
New Jersey					
New Mexico					
New York					
North Carolina					
North Dakota					
Ohio					
Oklahoma					
Oregon					
Pennsylvania					
Rhode Island					
South Carolina					
South Dakota					
Tennessee					
Texas					
Utah					
Vermont					
Virginia					
Washington					
West Virginia					
Wisconsin					
Wyoming					

MSD-225 (1/98)

VHF/UHF Century Club Award Rules

1. The VHF/UHF Century Club (VUCC) is awarded for contact with a minimum number of Maidenhead 2 degrees by 1 degree grid locators per band as indicated below. Grid locators are designated by a combination of two letters and two numbers. More information on grid locators can be found in January 1983 *QST*, pp 49-51 (reprint available upon request. Send a SASE 9 × 12-inch envelope with 3 units of postage). The *ARRL World Grid Locator Atlas* and the *ARRL Grid Locator for North America* are available from the ARRL Publication Sales Department.

   (a) The VUCC certificate and endorsements are available to amateurs worldwide; however, ARRL membership is required for hams in the US, its possessions and Puerto Rico.

   (b) The minimum number of grid locators needed to initially qualify for each individual band award is as follows:

for 50 MHz, 144 MHz and Satellite -	100 Credits;
for 222 MHz and 432 MHz -	50 Credits;
for 902 MHz and 1296 MHz -	25 Credits;
for 2.3 GHz -	10 Credits;
for 3.4 GHz, 5.7 GHz, 10 GHz, 24 GHz, 47 GHz, 75 GHz, 119 GHz, 142 GHz, 241 GHz and Laser 300 GHz and above -	5 Credits

   Awards for 222 MHz and 432 MHz are designated as HALF Century.
   Awards for 902 MHz and 1296 MHz are designated as QUARTER Century.
   Awards above these are SHF.

2. Only those contacts dated January 1, 1983 or later are creditable for VUCC purposes.

3. Individual band awards are endorseable in the following increments:

(a) 50 MHz, 144 MHz and Satellite	25 Credits
(b) 222 and 432 MHz	10 Credits
(c) 902 and above	5 Credits

4. General Rules
   (a) Separate bands are considered as separate awards.
   (b) No crossband contacts are permitted, except for Satellite.
   (c) No contacts through active repeaters are permitted, except for Satellite Awards.
   (d) Contacts with aeronautical mobiles (in the air) do NOT count.
   (e) Stations who claim to operate from more than one grid locator simultaneously (such as from the intersection of 4 grid locators) must be physically present in more than one locator to give multiple locator credit with a single contact. This requires the operator to know precisely where the intersection lines are located and placing the station exactly on the boundary to meet this test. To achieve this precision work requires either current markers permanently in place, or the precision work of a professional surveyor. Operators of such stations should be prepared to provide some evidence of meeting this test if called upon to do so. Multiple QSL cards are not required. GPS readings are acceptable.

5. There are no specialty endorsements such as "CW only," etc.

6. For VUCC awards on 50 through 1296 MHz and Satellite, all contacts must be made from a location or locations within the same grid locator or locations in different grid locators no more than 50 miles apart. For SHF awards, contacts must be made from a single location, defined as within a 300-meter diameter circle.

7. Application Procedure (please follow carefully)
   (a) Confirmations (QSLs) and application forms (MSD-259 and MSD-260) must be submitted to an approved VHF Awards Manager for certification. ARRL Special Services Clubs appoint VHF Awards Managers whose names are on file at HQ. If you do not know of an awards manager in your area, HQ will give you the name of the closest manager. (Also located on the Web at: http:\\www.arrl.org/awards/vucc). Foreign VUCC

applications should be checked by the Awards Manager for their IARU Member Society in their respective country. Do not send cards to HQ, unless asked to do so.

(b) For the convenience of the Awards Manager in checking cards, applicants may indicate in pencil (pencil ONLY) the grid locator on the address side of the cards that DO NOT clearly indicate the grid locator. The applicant affirms that he/she has accurately determined the proper location from the address information given on the card by signing the affirmation statement on the application.

(c) Cards must be sorted:
(1) Alphabetically by field
(2) Numerically from 00 to 99 within that field

(d) Where it is necessary to mail cards for certification, sufficient postage for proper return of all cards and paperwork, in addition to appropriate fees as noted in #8, must be included along with a separate self-addressed mailing label. An SASE is not necessary when a certificate will be issued, since a special mailing tube is used. An 9 × 12-inch SASE with at least 3 units of postage (depending on the number of sheets to be returned) is required for all endorsements. ARRL accepts no responsibility for cards handled by mail to and from VHF Awards Managers and will not honor any claims.

(e) Enclosed with the initial VUCC certificate from HQ will be a computer printout of the original list of grid locators for which the applicant has received credit (MSD-259). When applying for endorsements, the applicant will indicate in RED on the right hand side of the page those new grid locators for which credit is sought, and submit cards for certification to an Awards Manager. A new updated computer printout will be returned with appropriate endorsement sticker(s). Thus, a current list of grid locators worked is always in the hands of the VUCC award holder, available to the VHF Awards Manager during certification, and a permanent historical record always maintained at HQ. Use of the old MSD-259 field sheet is still OK, but only computer printouts will be sent in the return.

8. FEES

(a) Beginning January 1, 1996, all VUCC applicants will be charged an initiation fee of $10.00. This fee includes one VUCC certificate and one VUCC lapel pin.

(b) Each additional VUCC certificate, whether new or replacement, is $10.00.

(c) For endorsements, a 9 × 12-inch SASE with sufficient return postage for the return of all paperwork is required. In lieu of an SASE, $2.00 ($4.00 foreign) for postage and handling may be sent. (SASE for any submission where a certificate is issued is not necessary since certificates are mailed in a protective tube.)

(d) Lapel pins from VUCCs received prior to the January 1, 1996 fee change may be purchased at $5.00 each.

9. Disqualification

(a) Altered/forged confirmations or fraudulent applications submitted may result in disqualification of the applicant from VUCC participation by action of the ARRL Awards Committee.
(b) The applicant affirms he/she has abided by all the rules of membership in the VUCC program and agrees to be bound by the decisions of the ARRL Awards Committee.

10. Decisions of the ARRL Awards Committee regarding interpretation of the rules here printed or later amended shall be final.

11. Operating Ethics: Fair play and good sportsmanship in operating are required of all VUCC members.

12. Complete information on the VUCC Program can be found on *ARRL Web* at: **http://www.arrl.org/awards/vucc/** This includes all the forms, rules and copy of the January 1983 *QST* article.

MSD-261(07/01)

ARRL VUCC AWARD APPLICATION FORM

This form is required with each submission new or endorsement

Please print or type **clearly** and use a separate application for **each** award.

Callsign: ____________________ → from Grid Locator → (   ) Ex Callsigns:______________

↑ *Required* ↑

Name: ____

Print exactly as you want to appear on certificate

Address:_______________________________________

City:________________________ State:______ Zip Code:_________

Membership ID Number:_________________ Expiration Date:__________

Check one → ☐ **Initial Application** ☐ **Endorsement**

Band: (please check only one) Use One (1) Application Per Band

☐ 50 MHz	☐ 144 MHz	☐ 222 MHz	☐ 432 MHz
☐ 902 MHz	☐ 1296 MHz	☐ 2.3 GHz	☐ 3.4 GHz
☐ 5.7 GHz	☐ 10 GHz	☐ 24 GHz	☐ 47 GHz
☐ 75 GHz	☐ 119 GHz	☐ 142 GHz	☐ 241 GHz
☐ Laser (300 GHz)		☐ Satellite	

Initial Applicants_____________Number of Grid Locators

Endorsements

_____________ + _____________ = _____________

Previous Total | Number added with this endorsement | New Total

√ "I affirm that I have observed all the VUCC rules as well as all pertinent government regulations established for Amateur Radio in my country. I agree to be bound by the ARRL Awards Committee (Decisions of the ARRL Awards Committee are final)."

√ Signature:__________________________ Callsign:___________ Date:___________

VHF Awards Manager Verification

"I have verified these contacts as set forth by the rules of the VUCC Program."

_______________ _______________ _______________ _______________

Signature | Callsign | Verification Number | Date

** **Application Directions** ***

1) Complete all fields above and read MSD-261 detailing program rules
2) Enclose award fees ($10.00 for each new award)
3) Enclose $2.00 or a 9 X 12 SASE for return postage (Endorsements only) ($4.00 foreign)
4) Sort cards alphabetically by field and numerically within the field from 00-99
5) Contact your VHF Awards Manager before sending cards to assure they are available to make arrangements for checking
6) Give Awards Manager all fees and mailing costs. They are responsible for sending paperwork to ARRL
7) **DO NOT SEND CARDS TO ARRL HEADQUARTERS**

MSD-260 (7/01)

ARRL VUCC AWARD CALLSIGN: ________________

___ 50 ___ 144 ___ 222 ___ 432 ___ 902 ___ 1296 ___ 2.3 ___ 3.4
___ 5.7 ___ 10 ___ 24 ___ 47 ___ 75 ___ 119 ___ 142 ___ 241
___ Laser (300 GHz) ___ Satellite

Directions: This sheet good for ONE field.

1. Enter your callsign and check band.
2. Enter two-letter field.
3. Enter callsign of stations worked next to appropriate square below.
4. Total grid locators for each field.

Field: ____________
(First 2 letters of locator)

Sq	Callsign	Sq	Callsign	Sq	Callsign	Sq	Callsign
00	________	25	________	50	________	75	________
01	________	26	________	51	________	76	________
02	________	27	________	52	________	77	________
03	________	28	________	53	________	78	________
04	________	29	________	54	________	79	________
05	________	30	________	55	________	80	________
06	________	31	________	56	________	81	________
07	________	32	________	57	________	82	________
08	________	33	________	58	________	83	________
09	________	34	________	59	________	84	________
10	________	35	________	60	________	85	________
11	________	36	________	61	________	86	________
12	________	37	________	62	________	87	________
13	________	38	________	63	________	88	________
14	________	39	________	64	________	89	________
15	________	40	________	65	________	90	________
16	________	41	________	66	________	91	________
17	________	42	________	67	________	92	________
18	________	43	________	68	________	93	________
19	________	44	________	69	________	94	________
20	________	45	________	70	________	95	________
21	________	46	________	71	________	96	________
22	________	47	________	72	________	97	________
23	________	48	________	73	________	98	________
24	________	49	________	74	________	99	________

TOTAL number of grid locators
worked in this field: __________

msd-259(7/01)

WAC AWARD APPLICATION

Name________________________________ Callsign________________

Mailing Address____

City/Town____________________State__________ ZIP Code________

This application is for:

- ☐ Basic certificate (mixed mode)
- ☐ CW certificate
- ☐ Phone certificate
- ☐ SSTV certificate
- ☐ RTTY certificate
- ☐ FAX certificate
- ☐ Satellite certificate
- ☐ 5-band certificate
- ☐ 6-band endorsement
- ☐ QRP endorsement (5 watts output or less)
- ☐ 1.8-MHz endorsement
- ☐ 3.5-MHz endorsement
- ☐ 50-MHz endorsement
- ☐ 144-MHz endorsement
- ☐ 430-MHz endorsement
- ☐ 1270-MHz endorsement

My 5-band certificate is dated________________

Enclosed are QSL cards from the following stations:

Enter band(s) and callsigns:

	MHz	MHz	MHz	MHz	MHz
N. America					
S. America					
Oceania					
Asia					
Europe					
Africa					

The undersigned has abided by all rules set forth for this award.

My ARRL membership does not expire until____________________

Signature______________________________Callsign__________

Date______________________________________

12/99

United States Counties

(information courtesy of CQ's Counties Award Record Book)

Alabama (67 counties)
Autauga
Balwin
Barbour
Bibb
Blount
Bullock
Butler
Calhoun
Chambers
Cherokee
Chilton
Choctaw
Clarke
Clay
Cleburne
Coffee
Colbert
Conecuh
Coosa
Covington
Crenshaw
Cullman
Dale
Dallas
Dekalb
Elmore
Escambia
Etowah
Fayette
Franklin
Geneva
Greene
Hale
Henry
Houston
Jackson
Jefferson
Lamar
Lauderdale
Lawrence
Lee
Limestone
Lowndes
Macon
Madison
Marengo
Marion
Marshall
Mobile
Monroe
Montgomery
Morgan
Perry
Pickens
Pike
Randolph
Russel
Saint Clair
Shelby
Sumter
Talladega
Tallapoosa
Tuscaloosa
Walker
Washington
Wilcox
Winston

Alaska (4 counties)
Southeastern
Northwestern
South Central
Central

Arizona (15 counties)
Apache
Cochise
Coconino
Gila
Graham
Greenlee
Lez Paz
Maricopa
Mohave
Navajo
Pima
Pinal
Santa Cruz
Yavapai
Yuma

Arkansas (75 counties)
Arkansas
Ashley
Baxter
Benton
Boone
Bradley
Calhoun
Carroll
Chicot
Clark
Clay
Cleburne
Cleveland
Columbia
Conway
Craighead
Crawford
Crittenden
Cross
Dallas
Desha
Drew
Faulkner
Franklin
Fulton
Garland
Grant
Greene
Hempstead
Hot Spring
Howard
Independence
Izard
Jackson
Jefferson
Johnson
Lafayette
Lawrence
Lee
Lincoln
Little River
Logan
Lonoke
Madison
Marion
Miller
Mississippi
Monroe
Montgomery
Nevada
Newton
Ouachita
Perry
Phillips
Pike
Poinsett
Polk
Pope
Prairie
Pulaski
Randolph
St. Francis
Saline
Scott
Searcy
Sebastian
Sevier
Sharp
Stone
Union
Van Buren
Washington
White
Woodruff
Yell

California (58 counties)
Alameda
Alpine
Amador
Butte
Calaveras
Colusa
Contra Costa
Del Norte
El Dorado
Fresno
Glenn
Humboldt
Imperial
Inyo
Kern
Kings
Lake
Lassen
Los Angeles
Madera
Marin
Mariposa
Mendocino
Merced
Modoc
Mono
Monterey
Napa
Nevada
Orange
Placer
Plumas
Riverside
Sacramento
San Benito
San Bernardino
San Diego
San Francisco
San Joaquin
San Luis Obispo
San Mateo
Santa Barbara
Santa Clara
Santa Cruz
Shasta
Sierra
Siskiyou
Solano
Sonoma
Stanislaus
Sutter
Tehama
Trinity
Tulare
Tuolumne
Ventura
Yolo
Yuba

Colorado (63 counties)
Adams
Alamosa
Arapahoe
Archuleta
Baca
Bent
Boulder
Chaffee
Cheyenne
Clear Creek
Conejos
Costilla
Crowley
Custer
Delta
Denver
Dolores
Douglas
Eagle
Elbert
El Paso
Fremont
Garfield
Gilpin
Grand
Gunnison
Hinsdale
Huerfano
Jackson
Jefferson
Kiowa
Kit Carson
Lake
La Plata
Larimer
Las Animas
Lincoln
Logan
Mesa
Mineral
Moffat
Montezuma
Montrose
Morgan
Otero
Ouray
Park
Phillips
Pitkin
Prowers
Pueblo
Rio Blanco
Rio Grande
Routt
Saguache
San Juan
San Miguel

Sedgwick
Summit
Teller
Washington
Weld
Yuma

Connecticut (8 counties)
Fairfield
Hartford
Litchfield
Middlesex
New Haven
New London
Tolland
Windham

Delaware (3 counties)
Kent
New Castle
Sussex

Florida (67 counties)
Alachua
Baker
Bay
Bradford
Brevard
Broward
Calhoun
Charlotte
Citrus
Clay
Collier
Columbia
Dade
De Soto
Dixie
Duval
Escambia
Flagler
Franklin
Gadsden
Gilchrist
Glades
Gulf
Hamilton
Hardee
Hendry
Hernando
Highlands
Hillsborough
Holmes
Indian River
Jackson
Jefferson
Lafayette
Lake
Lee
Leon
Levy
Liberty
Madison
Manatee
Marion
Martin
Monroe
Nassau
Okaloosa
Okeechobee
Orange
Osceola
Palm Beach
Pasco
Pinellas
Polk
Putnam
St. Johns
St. Lucie
Santa Rosa
Sarasota
Seminole
Sumter
Suwannee
Taylor
Union
Volusia
Wakulla
Walton
Washington

Georgia (159 counties)
Appling
Atkinson
Bacon
Baker
Baldwin
Banks
Barrow
Bartow
Ben Hill
Berrien
Bibb
Bleckley
Brantley
Brooks
Bryan
Bulloch
Burke
Butts
Calhoun
Camden
Candler
Carroll
Catoosa
Chariton
Chatham
Chattahoochee
Chattooga
Cherokee
Clarke
Clay
Clayton
Clinch
Cobb
Coffee
Colquitt
Columbia
Cook
Coweta
Crawford
Crisp
Dade
Dawson
Decatur
De Kalb
Dodge
Doolly
Dougherty
Douglas
Early
Echols
Effingham
Elbert
Emanuel
Evans
Fannin
Fayette
Floyd
Forsyth
Franklin
Fulton
Gilmer
Glascock
Glynn
Gordon
Grady
Greene
Gwinnett
Habersham
Hall
Hancock
Haralson
Harris
Hart Heard
Henry
Houston
Irwin
Jackson
Jasper
Jeff Davis
Jefferson
Jenkins
Johnson
Jones
Lamar
Lanier
Laurens
Lee
Liberty
Lincoln
Long
Lowndes
Lumpkin
McDuffie
McIntosh
Macon
Madison
Marion
Meriwether
Miller
Mitchell
Monroe
Montgomery
Morgan
Murray
Muscogee
Newton
Oconee
Oglethorpe
Paulding
Peach
Pickens
Pierce
Pike
Polk
Pulaski
Putnam
Quitman
Rabun
Randolph
Richmond
Rockdale
Schley
Screven
Seminole
Spalding
Stephens
Stewart
Sumter
Talbot
Taliaferro
Tattriall
Taylor
Telfair
Terrell
Thomas
Tift
Toombs
Towns
Treutlen
Troup
Turner
Twiggs
Union
Upson
Walker
Walton
Ware
Warren
Washington
Wayne
Webster
Wheeler
White
Whitfield
Wilcox
Wilkes
Wilkinson
Worth

Hawaii (5 counties)
Hawaii
Honolulu
Kalawao
Kauai
Maui

Idaho (44 counties)
Ada
Adams
Bannock
Bear Lake
Benewah
Bingham
Blaine
Boise
Bonner
Bonneville
Boundary
Butte
Camas
Canyon
Caribou
Cassia
Clark
Clearwater
Custer
Elmore
Franklin
Fremont
Gem
Gooding
Idaho
Jefferson
Jerome
Kootenai
Latah
Lemhi
Lewis
Lincoln
Madison
Minidoka
Nez Perce
Oneida
Owyhee
Payette
Power
Shoshone
Teton
Twin Falls
Valley
Washington

Illinois (102 counties)
Adams
Alexander
Bond
Boone
Brown
Bureau
Calhoun
Carroll
Cass
Champaign
Christian
Clark
Clay
Clinton
Coles
Cook
Crawford
Cumberland
De Kalb
De Witt
Douglas
Du Page
Edgar
Edwards
Effingham
Fayette
Ford
Franklin
Fulton
Gallatin
Greene
Grundy
Hamilton
Hancock
Hardin Henderson
Henry
Iroquois
Jackson
Jasper
Jefferson
Jersey
Jo Daviess
Johnson
Kane
Kankakee
Kendall
Knox
Lake
La Salle
Lawrence
Lee
Livingston
Logan
McDonough
McHenry
McLean
Macon
Macoupin
Madison
Marion
Marshall
Mason
Massac
Menard
Mercer
Monroe
Montgomery
Morgan
Moultrie
Ogle
Peoria
Perry
Platt
Pike
Pope
Pulaski
Putnam
Randolph
Richland
Rock Island
St. Clair
Saline
Sangamon
Schuyler
Scott
Shelby
Stark
Stephenson
Tazewell
Union
Vermilion
Wabash
Warren
Washington
Wayne
White
Whiteside
Will
Williamson
Winnebago
Woodford

Indiana (92 counties)
Adams
Allen
Bartholomew
Benton
Blackford
Boone
Brown
Carroll
Cass
Clark
Clay
Clinton
Crawford
Daviess
Dearborn
Decatur
De Kalb
Delaware
Dubois
Elkhart
Fayette
Floyd
Fountain
Franklin
Fulton
Gibson
Grant
Greene
Hamilton
Hancock
Harrison
Hendricks
Henry
Howard
Huntington
Jackson
Jasper
Jay
Jefferson
Jennings
Johnson
Knox
Kosciusko
Lagrange
Lake
La Porte
Lawrence
Madison
Marion
Marshall
Martin
Miami
Monroe
Montgomery
Morgan
Newton
Noble
Ohio
Orange
Owen
Parke
Perry
Pike
Porter
Posey
Pulaski
Putnam
Randolph
Ripley
Rush
St. Joseph
Scoff
Shelby
Spencer
Starke
Steuben
Sullivan
Switzerland
Tippecanoe
Tipton
Union
Vanderburgh
Vermillion
Vigo
Wabash
Warren
Warrick
Washington
Wayne
Wells
White
Whitley

Iowa (99 counties)
Adair
Adams
Allamakee
Appanoose
Audubon
Benton
Black Hawk
Boone
Bremer
Buchanan
Buena Vista
Butler
Calhoun
Carroll
Cass
Cedar
Cerro Gordo
Cherokee
Chickasaw
Clarke
Clay
Clayton
Clinton
Crawford
Dallas
Davis
Decatur
Delaware
Des Moines
Dickinson
Dubuque
Emmet
Fayette
Floyd
Franklin
Fremont
Greene
Grundy
Guthrie
Hamilton
Hancock
Hardin
Harrison
Henry
Howard
Humboldt
Ida
Iowa
Jackson
Jasper
Jefferson
Johnson
Jones
Keokuk
Kossuth
Lee
Linn
Louisa
Lucas
Lyon
Madison
Mahaska
Marion
Marshall
Mills
Mitchell
Monona
Monroe
Montgomery
Muscatine
O'Brien
Osceola
Page
Palo Alto
Plymouth
Pocahontas
Polk
Pottawaftamie
Poweshiek
Ringgold
Sac
Scott
Shelby
Sioux
Story
Tama
Taylor
Union
Van Buren
Wapello
Warren
Washington
Wayne
Webster
Winnebago
Winneshiek
Woodbury
Worth
Wright

ebec

oln
rd
obscot
ataquis
adahoc
erset
do
shington
(

ryland (24 counties)
gany
e Arundel
timore
timore City
vert
roline
rroll
cil
arles
rchester
ederick
arrett
arford
oward
ent
ontgomery
rince Georges
ueen Annes
t. Marys
omerset
albot
Vashington
Vicomico
Vorcester

Massachusetts (14 counties)
Barnstable
Berkshire
Bristol
Dukes
Essex
Franklin
Hampden
Hampshire
Middlesex
Nantucket
Norfolk
Plymouth
Suffolk
Worcester

Michigan (83 counties)
Alcona
Alger
Allegan
Alpena
Antrim
Arenac
Baraga
Barry
Bay
Benzie
Berrien
Branch
Calhoun
Cass
Charlevoix
Cheboygan
Chippewa
Clare
Clinton
Crawford
Delta
Dickinson
Eaton
Emmet
Genesee
Gladwin
Gogebic
Grand Traverse
Gratiot
Hillsdale
Houghton
Huron
Ingham
Ionia
Iosco
Iron
Isabella
Jackson
Kalamazoo
Kalkaska
Kent
Keweenaw
Lake
Lapeer
Leelanau
Lenawee
Livingston
Luce
Mackinac
Macomb
Manistee
Marquette
Mason
Mecosta
Menominee
Midland
Missaukee
Monroe
Montcalm
Montmorency
Muskegon
Newaygo
Oakland
Oceana
Ogemaw
Ontonagon
Osceola
Oscoda
Otsego
Ottawa
Presque Isle
Roscommon
Saginaw
St. Clair
St. Joseph
Sanilac
Schoolcraft
Shiawassee
Tuscola
Van Buren
Washtenaw
Wayne
Wexford

Minnesota (87 counties)
Aitkin
Anoka
Becker
Beltrami
Benton
Big Stone
Blue Earth
Brown
Carlton
Carver
Cass
Chippewa
Chisago
Clay
Clearwater
Cook
Cottonwood
Crow Wing
Dakota
Dodge
Douglas
Faribault
Fillmore
Freeborn
Goodhue
Grant
Hennepin
Houston
Hubbard
Isanti
Itasa
Jackson
Kanabec
Kandiyohi
Kittson
Koochiching
Lac Qui Parle
Lake
L. of the Woods
Le Sueur
Lincoln
Lyon
McLeod
Mahnomen
Marshall
Martin
Meeker
Mille Lacs
Morrison
Mower
Murray
Nicollet
Nobles
Norman
Olmsted
Otter Trail
Pennington
Pine
Pipestone
Polk
Pope
Ramsey
Red Lake
Redwood
Renville
Rice
Rock
Roseau
St. Louis
Scott
Sherburne
Sibley
Stearns
Steele
Stevens
Swift
Todd
Traverse
Wabasha
Wadena
Waseca
Washington
Watonwan
Wilkin
Winona
Wright
Yellow Medicine

Mississippi (82 counties)
Adams
Alcorn
Amite
Attala
Benton
Bolivar
Calhoun
Carroll
Chickasaw
Choctaw
Claiborne
Clarke
Clay
Coahoma
Copiah
Covington
De Soto
Forrest
Franklin
George
Greene
Grenada
Hancock
Harrison
Hinds
Holmes
Humphreys
Issaquena
Itawamba
Jackson
Jasper
Jefferson
Jefferson Davis
Jones
Kemper
Lafayatte
Lamar
Lauderdale
Lawrence
Leake
Lee
Leflore
Lincoln
Lowndes
Madison
Marion
Marshall
Monroe
Montgomery
Neshoba
Newton
Noxubee
Oktibbeha
Panola
Pearl River
Perry
Pike
Pontotoc
Prentiss
Quitman
Rankin
Scott
Sharkey
Simpson
Smith
Stone
Sunflower
Tallahatchie
Tate
Tippah

Kansas (105 counties)
Allen
Anderson
Atchison
Barber
Barton
Bourbon
Brown
Butler
Chase
Chautauqua
Cherokee
Cheyenne
Clark
Clay
Cloud
Coffey
Comanche
Cowley
Crawford
Decatur
Dickinson
Doniphan
Douglas
Edwards
Elk
Ellis
Ellsworth
Finney
Ford
Franklin
Geary
Gove
Graham
Grant
Gray
Greeley
Greenwood
Hamilton
Harper
Harvey
Haskell
Hodgeman
Jackson
Jefferson
Jewell
Johnson
Kearny
Kingman
Kiowa
Labette
Lane
Leavenworth
Lincoln
Linn
Logan
Lyon
McPherson
Marion
Marshall
Meade
Miami
Mitchell
Montgomery
Morris
Morton
Nemaha
Neosho
Ness
Norton
Osage
Osborne
Ottawa
Pawnee
Phillips
Pottawatomie
Pratt
Rawlins
Reno
Republic
Rice
Riley
Rooks
Rush
Russell
Saline
Scott
Sedgwick
Seward
Shawnee
Sheridan
Sherman
Smith
Stafford
Stanton
Stevens
Sumner
Thomas
Trego
Wabaunsee
Wallace
Washington
Wichita
Wilson
Woodson
Wyandotte

Kentucky (120 counties)
Adair
Allen
Anderson
Ballard
Barren
Bath
Bell
Boone
Bourbon
Boyd
Boyle
Bracken
Breathitt
Breckenridge
Bullitt
Butler
Caldwell
Calloway
Campbell
Carlisle
Carroll
Carter
Casey
Christian
Clark
Clay
Clinton
Crittenden
Cumberland
Daviess
Edmonson
Elliott
Estill
Fayette
Fleming
Floyd
Franklin
Fulton
Gallatin
Garrard
Grant
Graves
Grayson
Green
Greenup
Hancock
Hardin
Harlan
Harrison
Hart
Henderson
Henry
Hickman
Hopkins
Jackson
Jefferson
Jessamine
Johnson
Kenton
Knott
Knox
Larue
Laurel
Lawrence
Lee
Leslie
Letcher
Lewis
Lincoln
Livingston
Logan
Lyon
McCracken
McCreary
McLean
Madison
Magoffin
Marion
Marshall
Martin
Mason
Meade
Menifee
Mercer
Metcalfe
Monroe
Montgomery
Morgan
Muhlenberg
Nelson
Nicholas
Ohio
Oldham
Owen
Owsley
Pendleton
Perry
Pike
Powell
Pulaski
Robertson
Rockcastle
Rowan
Russell
Scott
Shelby
Simpson
Spencer
Taylor
Todd
Trigg
Trimble
Union
Warren
Washington
Wayne
Webster
Whitley
Wolfe
Woodford

Louisiana (
Acadia
Allen
Ascension
Assumption
Avoyelles
Beauregard
Bienville
Bossier
Caddo
Calcasieu
Caldwell
Cameron
Cataboula
Claiborne
Concordia
De Soto
E. Baton Rouge
East Carroll
East Feliciana
Evangeline
Franklin
Grant
Iberia
Iberville
Jackson
Jefferson
Jefferson Davis
Lafayette
Lafourche
La Salle
Lincoln
Livingston
Madison
Morehouse
Natchitoches
Orleans
Ouachita
Plaquernines
Pointe Coupee
Rapides
Red River
Richland
Sabine
St. Bernard
St. Charles
St. Helena
St. James
St. John the Baptist
St. Landry
St. Martin
St. Mary
St. Tammany
Tangipahoa
Tensas
Terrebone
Union
Vermilion
Vernon
Washington
Webster
W. Baton Rouge
West Carroll
West Feliciana
Winn

Maine (16 counties)
Androscoggin
Aroostook
Cumberland
Franklin
Hancock

Tishomingo
Tunica
Union
Walthall
Warren
Washington
Wayne
Webster
Wilkinson
Winston
Yalobusha
Yazoo

Missouri (115 counties)
Adair
Andrew
Atchison
Audrain
Barry
Barton
Bates
Benton
Bollinger
Boone
Buchanan
Butler
Caldwell
Callaway
Camden
Cape Girardeau
Carroll
Carter
Cass
Cedar
Chariton
Christian
Clark
Clay
Clinton
Cole
Cooper
Crawford
Dade
Dallas
Daviess
De Kalb
Dent
Douglas
Dunklin
Franklin
Gasconade
Gentry
Greene
Grundy
Harrison
Henry
Hickory
Holt
Howard
Howell
Iron
Jackson
Jasper
Jefferson
Johnson
Knox
Laclede
Lafayette
Lawrence
Lewis
Lincoln
Linn
Livingston
McDonald
Macon
Madison
Maries
Marion
Mercer
Miller
Mississippi
Moniteau
Monroe
Montgomery
Morgan
New Madrid
Madrid
Newton
Nodaway
Oregon
Osage
Ozark
Perniscot
Perry
Pettis
Phelps
Pike
Platte
Polk
Pulaski
Putnam
Ralls
Randolph
Ray
Reynolds
Ripley
St. Charles
St. Clair
St. Francois
St. Louis
St. Louis City
Ste. Genevieve
Saline
Schuyler
Scotland
Scott
Shannon
Shelby
Stoddard
Stone
Sullivan
Taney
Texas
Vernon
Warren
Washington
Wayne
Webster
Worth
Wright

Montana (56 counties)
Beaverhead
Big Horn
Blaine
Broadwater
Carbon
Carter
Cascade
Chouteau
Custer
Daniels
Dawson
Deer Lodge
Fallon
Fergus
Flathead
Gallatin
Garfield
Glacier
Golden Valley
Granite
Hill
Jefferson
Judith Basin
Lake
Lewis and Clark
Liberty
Lincoln
McCono
Madison
Meagher
Mineral
Missoula
Musselshell
Park
Petroleum
Phillips
Pondera
Powder River
Powell
Prairie
Ravalli
Richland
Roosevelt
Rosebud
Sanders
Sheridan
Silver Bow
Stillwater
Sweet Grass
Teton
Toole
Treasure
Valley
Wheatland
Wibaux
Yellowstone

Nebraska (93 counties)
Adams
Antelope
Arthur
Banner
Blaine
Boone
Box Butte
Boyd
Brown
Buffalo
Burt
Butler
Cass
Cedar
Chase
Cherry
Cheyenne
Clay
Colfax
Curning
Custer
Dakota
Dawes
Dawson
Deuel
Dixon
Dodge
Douglas
Dundy
Fillmore
Franklin
Frontier
Furnas
Gage
Garden
Garfield
Gosper
Grant
Greeley
Hall
Hamilton
Harlan
Hayes
Hitchcock
Holt
Hooker
Howard
Jefferson
Johnson
Kearney
Keith
Keya Paha
Kimball
Knox
Lancaster
Lincoln
Logan
Loup
McPherson
Madison
Merrick
Morrill
Nance
Nemaha
Nuckolls
Otoe
Pawnee
Perkins
Phelps
Pierce
Platte
Polk
Red Willow
Richardson
Rock
Saline
Sarpy
Saunders
Scotts Bluff
Seward
Sheridan
Sherman
Sioux
Stanton
Thayer
Thomas
Thurston
Valley
Washington
Wayne
Webster
Wheeler
York

Nevada (16 counties)
Churchill
Clark
Douglas
Elko
Esmeralcla
Eureka
Humboldt
Lander
Lincoln
Lyon
Mineral
Nye
Pershing
Storey
Washoe
White Pine

New Hampshire (10 counties)
Belknap
Carroll
Cheshire
Coos
Grafton
Hillsboro
Merrimack
Rockingham
Strafford
Sullivan

New Jersey (21 counties)
Atlantic
Bergen
Burlington
Camden
Cape May
Cumberland
Essex
Gloucester
Hudson
Hunterdon
Mercer
Middlesex
Monmouth
Morris
Ocean
Passaic
Salem
Somerset
Sussex
Union
Warren

New Mexico (33 counties)
Bernalillo
Catron
Chaves
Cibola
Colfax
Curry
De Baca
Dona Ana
Eddy
Grant
Guadalupe
Harding
Hidalgo
Lea
Lincoln
Los Alamos
Luna
McKinley
Mora
Otero
Quay
Rio Arriba
Roosevelt
Sandoval
San Juan
San Miguel
Santa Fe
Sierra
Socorro
Taos
Torrance
Union
Valencia

New York (62 counties)
Albany
Allegany
Bronx
Broome
Cattaraugus
Cayuga
Chautauqua
Chemung
Chenango
Clinton
Columbia
Cortland
Delaware
Dutchess
Erie
Essex
Franklin
Fulton
Genesee
Greene
Hamilton
Herkimer
Jefferson
Kings
Lewis
Livingston
Madison
Monroe
Montgomery
Nassau
New York
Niagara
Oneida
Onondaga
Ontario
Orange
Orleans
Oswego
Otsego
Putnam
Queens
Rensselaer
Richmond
Rockland
St. Lawrence
Saratoga
Schenectady
Schoharie
Schuyler
Seneca
Steuben
Suffolk
Sullivan
Tioga
Tompkins
Ulster
Warren
Washington
Wayne
Westchester
Wyoming
Yates

North Carolina (100 counties)
Alamance
Alexander
Alleghany
Anson
Ashe
Avery
Beaufort
Bertie
Bladen
Brunswick
Buncombe
Burke
Cabarrus
Caldwell
Camden
Carteret
Caswell
Catawba
Chatham
Cherokee
Chowan
Clay
Cleveland
Columbus
Craven
Cumberland
Currituck
Dare
Davidson
Davie
Duplin
Durham
Edgecombe
Forsyth
Franklin
Gaston
Gates
Graham
Granville
Greene
Guilford
Halifax
Harnett
Haywood
Henderson
Hertford
Hoke
Hyde
Iredell
Jackson
Johnston
Jones
Lee
Lenoir
Lincoln
McDowell
Macon
Madison
Martin
Mecklenburg
Mitchell
Montgomery
Moore
Nash
New Hanover
Northampton
Onslow
Orange
Pamlico
Pasquotank
Pender
Perquimans
Person
Pin
Polk
Randolph
Richmond
Robeson
Rockingham
Rowan
Rutherford
Sampson
Scotland
Stanly
Stokes
Surry
Swain
Transylvania
Tyrrell
Union
Vance
Wake
Warren
Washington
Watauga
Wayne
Wilkes
Wilson
Yadkin
Yancey

North Dakota (53 counties)
Adams
Barnes
Benson
Billings
Bottineau
Bowman
Burke
Burleigh
Cass
Cavalier
Dickey
Divide
Dunn
Eddy
Emmons
Foster
Golden Valley
Grand Forks
Grant
Griggs
Hettinger
Kidder
La Moure
Logan
McHenry
McIntosh
McKenzie
McLean
Mercer
Morton
Mountrail
Nelson
Oliver
Pembina
Pierce
Ramsey
Ranson
Renville
Richland
Rolette
Sargent
Sheridan
Sioux
Slope
Stark
Steele
Stutsman
Towner
Traill
Walsh
Ward
Wells
Williams

Ohio (88 counties)
Adams
Allen
Ashland
Ashtabula
Athens
Auglaize

Belmont
Brown
Butler
Carroll
Champaign
Clark
Clermont
Clinton
Columbiana
Coshocton
Crawford
Cuyahoga
Darke
Defiance
Delaware
Erie
Fairfield
Fayette
Franklin
Fulton
Gallia
Geauga
Greene
Guernsey
Hamilton
Hancock
Hardin
Harrison
Henry
Highland
Hocking
Holmes
Huron
Jackson
Jefferson
Knox
Lake
Lawrence
Licking
Logan
Lorain
Lucas
Madison
Mahoning
Marion
Medina
Meigs
Mercer
Miami
Monroe
Montgomery
Morgan
Morrow
Muskingum
Noble
Ottawa
Paulding
Perry
Pickaway
Pike
Portage
Preble
Putnam
Richland
Ross
Sandusky
Scioto
Seneca
Shelby
Stark
Summit
Trumbull
Tuscarawas
Union
Van Wert
Vinton
Warren
Washington
Wayne
Williams
Wood
Wyandot

Oklahoma (77 counties)
Adair
Alfalfa
Atoka
Beaver
Beckham
Blaine
Bryan
Caddo
Canadian
Carter
Cherokee
Choctaw
Cimarron
Cleveland
Coal
Comanche
Cotton
Craig
Creek
Custer
Delaware
Dewey
Ellis
Garfield
Garvin
Grady
Grant
Greer
Harmon
Harper
Haskell
Hughes
Jackson
Jefferson
Johnston
Kay
Kingfisher
Kiowa
Latimer
Le Fiore
Lincoln
Logan
Love
McClain
McCurtain
McIntosh
Major
Marshall
Mayes
Murray
Muskogee
Noble
Nowata
Okfuskee
Oklahoma
Okmulgee
Osage
Ottawa
Pawnee
Payne
Pittsburg
Pontotoc
Pottawatomie
Pushmataha
Roger Mills
Rogers
Seminole
Sequoyah
Stephens
Texas
Tillman
Tulsa
Wagoner
Washington
Washita
Woods
Woodward

Oregon (36 counties)
Baker
Benton
Clackamas
Clatsop
Columbia
Coos
Crook
Curry
Deschutes
Douglas
Gilliam
Grant
Harney
Hood River
Jackson
Jefferson
Josephine
Klamath
Lake
Lane
Lincoln
Linn
Malheur
Marion
Morrow
Multnomah
Polk
Sherman
Tillamook
Umatilla
Union
Wallowa
Wasco
Washington
Wheeler
Yarnhill

Pennsylvania (67 counties)
Adams
Allegheny
Armstrong
Beaver
Bedford
Berks
Blair
Bradford
Bucks
Butler
Cambria
Cameron
Carbon
Centre
Chester
Clarion
Clearfield
Clinton
Columbia
Crawford
Cumberland
Dauphin
Delaware
Elk
Erie
Fayette
Forest
Franklin
Fulton
Greene
Huntingdon
Indiana
Jefferson
Juniata
Lackawanna
Lancaster
Lawrence
Lebanon
Lehigh
Luzerne
Lycoming
McKean
Mercer
Mifflin
Monroe
Montgomery
Montour
Northampton
Northumberland
Perry
Philadelphia
Pike Potter
Schuylkill
Snyder
Somerset
Sullivan
Susquehanna
Tioga
Union
Venango
Warren
Washington
Wayne
Westmoreland
Wyoming
York

Rhode Island (5 counties)
Bristol
Kent
Newport
Providence
Washington

South Carolina (46 counties)
Abbeville
Aiken
Allendale
Anderson
Bamberg
Barnwell
Beaufort
Berkeley
Calhoun
Charleston
Cherokee
Chester
Chesterfield
Clarendon
Colleton
Darlington
Dillon
Dorchester
Edgefield
Fairfield
Florence
Georgetown
Greenville

Greenwood
Hampton
Horry
Jasper
Kershaw
Lancaster
Laurens
Lee
Lexington
McCormick
Marion
Marlboro
Newberry
Oconee
Orangeburg
Pickens
Richland
Saluda
Spartanburg
Sumter
Union
Williamsburg
York

South Dakota (66 counties)
Aurora
Beadle
Bennett
Bon Homme
Brookings
Brown
Brule
Buffalo
Butte
Campbell
Charles Mix
Clark
Clay
Codington
Corson
Custer
Davison
Day
Deuel
Dewey
Douglas
Edmunds
Fall River
Faulk
Grant
Gregory
Haakon
Hamlin
Hand
Hanson
Harding
Hughes
Hutchinson
Hyde
Jackson
Jerauld
Jones
Kingsbury
Lake
Lawrence
Lincoln
Lyman
McCook
McPherson
Marshall
Meade
Mellefte
Miner
Minnehaha
Moody
Pennington
Perkins
Potter
Roberts
Sanborn
Shannon
Spink
Stanley
Sully
Todd
Tripp
Turner
Union
Walworth
Yankton
Ziebach

Tennessee (95 counties)
Anderson
Bedford
Benton
Bledsoe
Blount
Bradley
Campbell
Cannon
Carroll
Carter
Cheatham
Chester
Claiborne
Clay
Cocks
Coffee
Crockett
Cumberland
Davidson
Decatur
DeKalb
Dickson
Dyer
Fayette
Fentress
Franklin
Gibson
Giles
Grainger
Greene
Grundy
Hamblen
Hamilton
Hancock
Hardeman
Hardin
Hawkins
Haywood
Henderson
Henry
Hickman
Houston
Humphreys
Jackson
Jefferson
Johnson
Knox
Lake
Lauderdale
Lawrence
Lewis
Lincoln
Loudon
McMinn
McNairy
Macon
Madison
Marion
Marshall
Maury
Meigs
Monroe
Montgomery
Moore
Morgan
Obion
Overton
Perry
Pickett
Polk
Putnam
Rhea
Roane
Robertson
Rutherford
Scott
Sequatchie
Sevier
Shelby
Smith
Stewart
Sullivan
Sumner
Tipton
Trousdale
Unicoi
Union
Van Buren
Warren
Washington
Wayne
White
Williamson
Wilson

Texas (254 counties)
Anderson
Andrews
Angelina
Aransas
Archer
Armstrong
Atascosa
Austin
Bailey
Bandera
Bastrop
Baylor
Bee
Bell
Bexar
Blanco
Borden
Bosque
Bowie
Brazoria
Brazos
Brewster
Briscoe
Brooks
Brown
Burleson
Burnet
Caldwell
Calhoun
Callahan
Cameron
Camp
Carson
Cass
Castro
Chambers
Cherokee
Childress
Clay
Cochran
Coke
Coleman
Collin
Collingsworth
Colorado
Comal
Comanche
Concho
Cooke
Coryell
Cottle
Crane
Crockett
Crosby
Culberson
Dallam
Dallas
Dawson
Deaf Smith
Delta
Denton
De Witt
Dickens
Dimmit
Donley
Duval
Eastland
Ector
Edwards
Ellis
El Paso
Erath
Falls
Fannin
Fayette
Fisher
Floyd
Foard
Fort Bend
Franklin
Freestone
Frio
Gaines
Galveston
Garza
Gillespie
Glasscock
Goliad
Gonzales
Gray
Grayson
Gregg
Grimes
Guadelupe
Hale
Hall
Hamilton
Hansford
Hardeman
Hardin
Harris
Harrison
Hartley
Haskell
Hays
Hemphill
Henderson
Hidalgo
Hill
Hockley
Hood

Hopkins
Houston
Howard
Hudspeth
Hunt
Hutchinson
Irion
Jack
Jackson
Jasper
Jeff Davis
Jefferson
Jim Hogg
Jim Wells
Johnson
Jones
Karnes
Kaufman
Kendall
Kenedy
Kent
Kerr
Kimble
King
Kinney
Kleberg
Knox
Lamar
Lamb
Lampasas
La Salle
Lavaca
Lee
Leon
Liberty
Limestone
Lipscomb
Live Oak
Llano
Loving
Lubbock
Lynn
McCulloch
McLennan
McMullen
Madison
Marion
Martin
Mason
Matagorda
Maverick
Medina
Menard
Midland
Milam
Mills
Mitchell
Montague
Montgomery
Moore
Morris
Motley
Nacogdoches
Navarro
Newton
Nolan
Nueces
Ochiltree
Oldham
Orange
Palo Pinto
Panola
Parker
Parmer
Pecos
Polk
Potter
Presidio
Rains
Randall
Reagan
Real
Red River
Reeves
Refugio
Roberts
Robertson
Rockwall
Runnels
Rusk
Sabine
San Augustine
San Jacinto
San Patricio
San Saba
Schleicher
Scurry
Shackelford
Shelby
Sherman
Smith
Somervell
Starr
Stephens
Sterling
Stonewall
Sutton
Swisher
Tarrant
Taylor
Terrell
Throckmorton
Titus
Tom Green
Travis
Trinity
Tyler
Upshur
Upton
Uvalde
Val Verde
Van Zandt
Victoria
Walker
Waller
Ward
Washington
Webb
Wharton
Wheeler
Wichita
Wilbarger
Willacy
Williamson
Wilson
Winkler
Wise
Wood
Yoakum
Young
Zapata
Zavala

Utah (29 counties)
Beaver
Box Elder
Cache
Carbon
Daggett
Davis
Duchesne
Emery
Garfield
Grand
Iron
Juab
Kane
Millard
Morgan
Piute
Rich
Salt Lake
San Juan
Sanpete
Sevier
Summit
Tooele
Uintah
Utah
Wasatch
Washington
Wayne
Weber

Vermont (14 counties)
Addison
Bennington
Caledonia
Chittendon
Essex
Franklin
Grand Isle
Lamoille
Orange
Orleans
Rutland
Washington
Windham
Windsor

Virginia (95 counties)
Accomack
Albemarle
Alleghany
Amelia
Amherst
Appomattox
Arlington
Augusta
Bath
Bedford
Bland
Botetourt
Brunswick
Buchanan
Buckingham
Campbell
Caroline
Carroll
Charles City
Charlotte
Chesterfield
Clarke
Craig
Culpeper
Cumberland
Dickenson
Dinwiddie
Essex
Fairfax
Fauquier
Floyd
Fluvanna
Franklin
Frederick
Giles
Gloucester
Goochland
Grayson
Greene
Greensville
Halifax
Hanover
Henrico
Henry
Highland
Isle of Wight
James City
King and Queen
King George
King William
Lancaster
Lee
Loudoun
Louisa
Lunenberg
Madison
Mathews
Mecklenburg
Middlesex
Montgomery
Nelson
New Kent
Northampton
Northumberland
Nottoway
Orange
Page
Patrick
Pittsylvania
Powhatan
Prince Edward
Prince George
Prince William
Pulaski
Rappahannock
Richmond
Roanoke
Rockbridge
Rockingham
Russell
Scott
Shenandoah
Smyth
Southampton
Spotsylvania
Stafford
Surry
Sussex
Tazewell
Warren
Washington
Westmoreland
Wise
Wythe
York

Washington (39 counties)
Adams
Asotin
Benton
Chelan
Clallam
Clark
Columbia
Cowlitz
Douglas
Ferry
Franklin
Garfield
Grant

Grays Harbor
Island
Jefferson
King
Kitsap
Kittitas
Klickitat
Lewis
Lincoln
Mason
Okanogan
Pacific
Pend Oreille
Pierce
San Juan
Skagit
Skamania
Snohomish
Spokane
Stevens
Thurston
Wahklakum
Walla Walla
Whatcom
Whitman
Yakima

West Virginia (55 counties)
Barbour
Berkeley
Boone
Braxton
Brooks
Cabell
Calhoun
Clay
Doddridge
Fayette
Gilmer
Grant
Greenbrier
Hampshire
Hancock
Hardy
Harrison
Jackson
Jefferson
Kanawha
Lewis
Lincoln
Logan
McDowell
Marion
Marshall
Mason
Mercer
Mineral
Mingo
Monongalia
Monroe
Morgan
Nicholas
Ohio
Pendleton
Pleasants
Pocahontas
Preston
Putnam
Raleigh
Randolph
Ritchie
Roane
Summers
Taylor
Tucker
Tyler
Upshur
Wayne
Webster
Wetzel
Wirt
Wood
Wyoming

Wisconsin (72 counties)
Adams
Ashland
Barron
Bayfield
Brown
Buffalo
Burnett
Calumet
Chippewa
Clark
Columbia
Crawford
Dane
Dodge
Door
Douglas
Dunn
Eau Claire
Florence
Fond du Lac
Forest
Grant
Green
Green Lake
Iowa
Iron
Jackson
Jefferson
Juneau
Kenosha
Kewaunee
La Crosse
Lafayette
Langlade
Lincoln
Manitowoc
Marathon
Marinette
Marquette
Menominee
Milwaukee
Monroe
Oconto
Oneida
Outagamie
Ozaukee
Pepin
Pierce
Polk
Portage
Price
Racine
Richland
Rock
Rusk
St. Croix
Sauk
Sawyer
Shawano
Sheboygan
Taylor
Trempealeau
Vernon
Vilas
Walworth
Washburn
Washington
Waukesha
Waupaca
Waushara
Winnebago
Wood

Wyoming (23 counties)
Albany
Big Horn
Campbell
Carbon
Converse
Crook
Fremont
Goshen
Hot Springs
Johnson
Laramie
Lincoln
Natrona
Niobrara
Park
Platte
Sheridan
Sublette
Sweetwater
Teton
Uinta
Washakie
Weston

Total US Counties = 3076

Notes: Since USA-CA started, counties that have been absorbed include Princess Ann, Norfolk and Nansemond in Virginia; Ormsby in Nevada; Washabaugh in South Dakota. Carson City, Nevada, is now considered an independent city. Richmond County in New York is now called Staten Island County. A new county has been added to New Mexico—Cibola (used to be part of Valencia County).

The following is a list of independent cities and the county for which each can be used.

Independent City	**Counts as _____ County**
[VIRGINIA]	
Alexandria	Arlington or Fairfax
Bedford	Bedford
Bristol	Washington
Buena Vista	Rockbridge
Charlottesville	Albermarle
Chesapeake	Isle of Wight
Clifton Forge	Alleghany
Colonial Heights	Chesterfield or Prince George
Covington	Alleghany
Danville	Pittsylvania
Emporia	Greensville
Fairfax	Fairfax
Falls Church	Fairfax
Fort Monroe	York
Franklin	Southampton
Fredericksburg	Spotsylvania
Galax	Carroll or Grayson
Hampton	York
Harrisonburg	Rockingham
Hopewell	Prince George
Lexington	Rockbridge
Lynchburg	Amherst or Bedford or Campbell
Manassas	Prince William
Manassas Park	Prince William
Martinsville	Henry
Newport News	York
Norfolk	Isle of Wight
Norton	Wise
Petersburg	Chesterfield or Dinwiddie or Prince George
Poquoson	York
Portsmouth	Isle of Wight
Radford	Montgomery
Richmond	Chesterfield or Henrico
Roanoke	Roanoke
Salem	Roanoke
South Boston	Halifax
Staunton	Augusta
Suffolk	Isle of Wight or Southampton
Virginia Beach	Isle of Wight
Waynesboro	Augusta
Williamsburg	James City
Winchester	Frederick
Carson City, Nevada	Douglas or Lyon or Story or Washoe
Washington, DC	Montgomery or Prince Georges, Maryland

ARRL Field Organization

The United States is divided into 15 ARRL Divisions. Every two years the ARRL full members in each of these divisions elect a director and a vice director to represent them on the League's Board of Directors. The Board determines the policies of the League, which are carried out by the Headquarters staff. A director's function is principally policymaking at the highest level, but the Board of Directors is all-powerful in the conduct of League affairs.

The 15 divisions are further broken down into 71 sections, and the ARRL full members in each section elect a Section Manager (SM). The SM is the senior elected ARRL official in the section, and in cooperation with the director, fosters and encourages all ARRL activities within the section. A breakdown of sections within each division (and counties within each splitstate section) follows:

ATLANTIC DIVISION: *Delaware, Eastern Pennsylvania* (Adams, Berks, Bradford, Bucks, Carbon, Chester, Columbia, Cumberland, Dauphin, Delaware, Juniata, Lackawanna, Lancaster, Lebanon, Lehigh, Luzerne, Lycoming, Monroe, Montgomery, Montour, Northhampton, Northumberland, Perry, Philadelphia, Pike, Schuylkill, Snyder, Sullivan, Susquehanna, Tioga, Union, Wayne, Wyoming, York); *Northern New York* (Clinton, Essex, Franklin, Fulton, Hamilton, Jefferson, Lewis, Montgomery, Schoharie, St. Lawrence); *Maryland -D.C.*; *Southern New Jersey* (Atlantic, Burlington, Camden, Cape May, Cumberland, Gloucester, Mercer, Ocean, Salem); *Western New York* (Allegany, Broome, Cattaraugus, Cayuga, Chautauqua, Chemung, Chenango, Cortland, Delaware, Erie, Genesee, Herkimer, Livingston, Madison, Monroe, Niagara, Oneida, Onondaga, Ontario, Orleans, Oswego, Otsego, Schuyler, Seneca, Steuben, Tioga, Tompkins, Wayne, Wyoming, Yates); *Western Pennsylvania* (those counties not listed under Eastern Pennsylvania).

CENTRAL DIVISION: *Illinois; Indiana; Wisconsin.*

DAKOTA DIVISION: *Minnesota; North Dakota; South Dakota.*

DELTA DIVISION: *Arkansas; Louisiana; Mississippi, Tennessee*

GREAT LAKES DIVISION; *Kentucky, Michigan, Ohio.*

HUDSON DIVISION: Eastern New York (Albany, Columbia, Dutchess, Greene, Orange, Putnam, Rensselaer, Rockland, Saratoga, Schenectady, Sullivan, Ulster, Warren, Washington, Westchester); *N.Y.C.-L.I.* (Bronx, Kings, Nassau, New York, Queens, Staten Island, Suffolk); *Northern New Jersey* (Bergen, Essex, Hudson, Hunterdon, Middlesex, Monmouth, Morris, Passaic, Somerset, Sussex, Union, Warren).

MIDWEST DIVISION: *Iowa; Kansas, Missouri, Nebraska.*

NEW ENGLAND DIVISION: *Connecticut, Maine, Eastern Massachusetts* (Barnstable, Bristol, Dukes, Essex, Middlesex, Nantucket, Norfolk, Plymouth, Suffolk); *New Hampshire; Rhode Island, Vermont, Western Massachusetts* (those counties not listed under Eastern Massachusetts).

NORTHWESTERN DIVISION: *Alaska; Idaho; Montana; Oregon; Eastern Washington* (Adams, Asotin, Benton, Chelan, Columbia, Douglas, Ferry, Franklin, Garfield, Grant, Kittitas, Klickitat, Lincoln, Okangogan, Pend Oreille, Spokane, Stevens, Walla, Walla, Whitman, Yakima); *Western Washington* (Challam, Clark Cowlitz, Grays Harbor Island, Jefferson, King, Kitsap, Lewis, Mason, Pacific, Pierce, San Juan, Skagit, Skamania, Snohomish, Thurston, Wahkiakum, Whatcom).

PACIFIC DIVISION: *East Bay* (Alameda, Contra Costa, Napa, Solano); *Nevada; Pacific* (Hawaii and U.S. possessions in the Pacific); *Sacramento Valley* (Alpine, Amador, Butte, Colusa, El Dorado, Glenn, Lassen, Modoc, Nevada, Placer, Plumas, Sacramento, Shasta, Sierra, Siskiyou, Sutter, Tehama, Trinity, Yolo, Yuba); *San Francisco*, (Del Norte, Humboldt, Lake, Marin, Mendocino, San Francisco, Sonoma); *San Joaquin Valley* (Calaveras, Fresno, Kern, Kings, Madera, Mariposa, Merced, Mono, San Joaquin, Stanislaus, Tulare, Tuolumne); *Santa Clara Valley* (Monterey, San Benito, San Mateo, Santa Clara, Santa Cruz).

ROANOKE DIVISION: *North Carolina, South Carolina; Virginia; West Virginia.*

ROCKY MOUNTAIN DIVISION: *Colorado; Utah; New Mexico, Wyoming.*

SOUTHEASTERN DIVISION: *Alabama*; *Georgia; Northern Florida* (Alachua, Baker, Bay, Bradford, Calhoun, Citrus, Clay, Columbia, Dixie, Duval, Escambia, Flagler, Franklin, Gadsden, Gilchrist, Gulf, Hamilton, Hernando, Holmes, Jackson, Jefferson, Lafayette, Lake, Leon, Levy, Liberty, Madison, Marion, Nassau, Okaloosa, Orange, Pasco, Putnam, Santa Rosa, Seminole, St. Johns, Sumter, Suwanee, Taylor, Union, Volusia, Wakulla, Walton, Washington); *Southern*

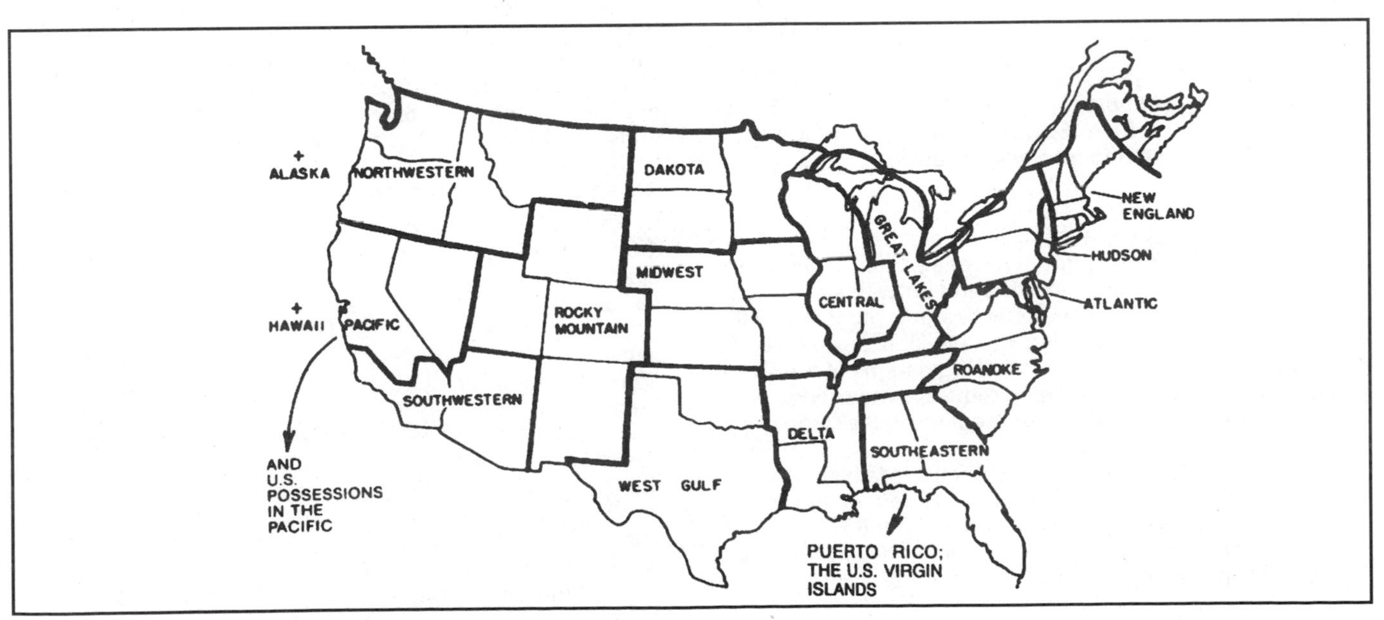

Florida (Brevard, Broward, Collier, Dade, Glades, Hendry, Indian River, Lee, Martin, Monroe, Okeechobee, Osceola, Palm Beach and St. Lucie); *Puerto Rico; U.S. Virgin Islands; West Central Florida* (Charlotte, DeSoto, Hardee, Highlands, Hillsborough, Manatee, Pinellas, Polk and Sarasota).

SOUTHWESTERN DIVISION: *Arizona; Los Angeles; Orange* (Inyo, Orange, Riverside, San Bernardino); *San Diego* (Imperial, San Diego); *Santa Barbara* (San Luis Obispo, Santa Barbara, Ventura).

WEST GULF DIVISION: *North Texas* (Anderson, Archer, Baylor, Bell, Bosque, Bowie, Brown, Camp, Cass, Cherokee, Clay, Collin, Comanche, Cooke, Coryell, Dallas, Delta, Denton, Eastland, Ellis, Erath, Falls, Fannin, Franklin, Freestone, Grayson, Gregg, Hamilton, Harrison, Henderson, Hill, Hopkins, Hunt, Jack, Johnson, Kaufman, Lamar, Lampasas, Limestone, McLennan, Marion, Mills, Montague, Morris, Nacogdoches, Navarro, Palo Pinto, Panola, Parker, Rains, Red River, Rockwall, Rusk, Shelby, Smith, Somervell, Stephens, Tarrant, Throckmorton, Titus, Upshur, Van Zandt, Wichita, Wilbarger, Wise, Wood, Young); *Oklahoma; South Texas* (Angelina, Aransas, Atacosa, Austin, Bandera, Bastrop, Bee, Bexar, Blanco, Brazoria, Brazos, Brooks, Burleson, Burnet, Caldwell, Calhoun, Cameron, Chambers, Colorado, Comal, Concho, DeWitt, Dimmitt, Duval, Edwards, Fayette, Fort Bend, Frio, Galveston, Gillespie, Goliad, Gonzales, Grimes, Guadalupe, Hardin, Harris, Hays, Hidalgo, Houston, Jackson, Jasper, Jefferson, Jim Hogg, Jim Wells, Karnes, Kendall, Kenedy, Kerr, Kimble, Kinney, Kleberg, LaSalle, Lavaca, Lee, Leon, Liberty, Live Oak, Llano, Madison, Mason, Matagorda, Maverick, McCulloch, McMullen, Medina, Menard, Milam, Montgomery, Newton, Nueces, Orange, Polk, Real, Refugio, Robertson, Sabine, San Augustine, San Jacinto, San Patricio, San Saba, Starr, Travis, Trinity, Tyle, Uvalde, Val Verde, Victoria, Walker, Waller, Washington, Webb, Wharton, Willacy, Williamson, Wilson, Zapata, Zavala); *West Texas* (Andrews, Armstrong, Bailey, Bordon, Brewster, Briscoe, Callahan, Carson, Castro, Childress, Cochran, Coke, Coleman, Collingsworth, Cottle, Crane, Crockett, Crosby, Culberson, Dallam, Dawson, Deaf Smith, Dickens, Donley, Ector, El Paso, Fischer, Floyd, Foard, Gaines, Garza, Glasscock, Gray, Hale, Hall, Hansford, Hardeman, Hartley, Haskell, Hemphill, Hockley, Howard Hudspeth, Hutchinson, Irion, Jeff Davis, Jones, Kent, King, Knox, Lamb, Lipscomb, Loving, Lubbock, Lynn, Martin, Midland, Mitchell, Moore, Motley, Nolan, Ochiltree, Oldham, Parmer, Pecos, Potter, Presidio, Randall, Reagan, Reeves, Roberts, Runnels, Schleicher, Scurry, Shackelford, Sherman, Sterling, Stonewall, Sutton, Swisher, Taylor, Terrell, Terry, Tom Green, Upton, Ward, Wheeler, Winkler, Yoakum).

RAC RADIO AMATEURS OF CANADA-CANADA; *Alberta; British Columbia; Manitoba; Maritime* (Nova Scotia, New Brunswick, Prince Edward Island); *Newfoundland/ Labrador; Ontario; Quebec; Saskatchewan.*

SECTION MANAGER

In each ARRL section there is an elected Section Manager (SM) who will have authority over the Field Organization in his or her section, and in cooperation with his Director, foster and encourage ARRL activities and programs within the section. Details regarding the election procedures for SMs are contained in "Rules and Regulations of the ARRL Field Organization," available on request to any ARRL member. Election notices are posted regularly in the Happenings section of *QST*.

Any candidate for the office of Section Manager must be a resident of the section, a licensed amateur of the Technician class or higher, and a full member of the League for a continuous term of at least two years immediately preceding receipt of a petition for nomination at Hq. If elected, he or she must maintain membership throughout the term of office.

The following is a detailed resume of the duties of the Section Manager. In discharging his responsibilities, he:

a) Recruits and appoints nine section-level assistants to serve under his general supervision and to administer the following ARRL programs in the section: emergency communications, message traffic, official observers, affiliated clubs, public information, state government liaison, technical activities and on-the-air bulletins.

b) Supervises the activities of these assistants to ensure continuing progress in accordance with overall ARRL policies and objectives.

c) Appoints qualified ARRL members in the section to volunteer positions of responsibility in support of section programs, or authorizes the respective section-level assistants to make such appointments.

d) Maintains liaison with the Division Director and makes periodic reports to him regarding the status of section activities; receives from him information and guidance pertaining to matters of mutual concern and interest; serves on the Division Cabinet and renders advice as requested by the Division Director; keeps informed on matters of policy that affect section-level programs.

e) Conducts correspondence or other communications, including personal visits to clubs, hamfests and conventions, with ARRL members and affiliated clubs in the section; either responds to their questions or concerns or refers them to the appropriate person or office in the League organization; maintains liaison with, and provides support to, representative repeater frequency coordinating bodies having jurisdiction in the section.

f) Writes, or supervises preparation of, the monthly Section News column in *QST* to encourage member participation in the ARRL program in the section.

g) Recruits new amateurs and ARRL members to foster growth of Field organization programs and the amateur service's capabilities in support of public service.

[Note: Move of permanent residence outside the section from which elected will be grounds for declaring the office vacant.]

ARRL LEADERSHIP APPOINTMENTS

Field Organization leadership appointments from the Section Manager are available to qualified ARRL full members in each section. These appointments are as follows: Assistant Section Manager, Section Emergency Coordinator, Section Traffic Manager, Official Observer Coordinator, State Government Liaison, Technical Coordinator, Affiliated Club Coordinator, Public Information Officer, Bulletin Manager, District Emergency Coordinator, Emergency Coordinator and Net Manager. Holders of such appointments may wear the League emblem pin with the distinctive deep-green background. Functions of these leadership officials are described below.

Assistant Section Manager

The ASM is an ARRL section-level official appointed by the Section Manager, in addition to the Section Manager's eight section-level assistants. An ASM may be appointed if the Section Manager believes such an appointment is desirable to meet the goals of the ARRL Field Organization in that section. Thus, the ASM is appointed at the complete discretion of the Section Manager, and serves at the pleasure of the Section Manager.

1) The ASM may serve as a general or as a specialized assistant to the Section Manager. That is, the ASM may assist the Section Manager with general leadership matters as the Section Manager's general understudy, or the ASM may be assigned to handle a specific important function not within the scope of the duties of the Section Manager's eight assistants.

2) At the Section Manager's discretion, the ASM may be

designated as the recommended successor to the incumbent Section Manager, in case the Section Manager resigns or is otherwise unable to finish the term of office.

3) The ASM should be familiar with "Guidelines for the ARRL Section Manager," which contains the fundamentals of general section management.

4) The ASM must be an ARRL full member, holding at least a Novice class license.

Section Emergency Coordinator

The SEC must hold a Technician class license or higher and is appointed by the Section Manager to take care of all matters pertaining to emergency communications and the Amateur Radio Emergency Service (ARES) on a sectionwide basis. The duties of the SEC include the following:

1) The encouragement of all groups of community amateurs to establish a local emergency organization.

2) Recommendations to the SM on all section emegency policy and planning, including the development of a section emergency communications plan.

3) Cooperation and Coordination with the Section Traffic Manager so that emergency nets and traffic nets in the section present a unified public sevice front. Cooperation and coordiantion should also be maintained with other section leadership officials as appropriate, particularly the State Governmnent Liaison and the Public Information Coordinator.

4) Recommendations of candidates for Emergency Coordinator and District Emergency Coordinator appointments (and cancellations) to the Section Manager and determinations of areas of jurisdiction of each amateur so appointed. At the SM's discretion, the SEC may be directly in charge of making (and cancelling) such appointments. In the same way, the SEC can handle the Official Emergency Station program.

5) Promotion of ARES membership drives, meetings, activities, tests, procedures, etc., at the section level.

6) Collection and consolidation of Emergency Coordinator (or District Emergency Coordinator) monthly reports and submission of monthly progress summaries to ARRL Hq.

7) Maintenance of contact with other communication services and liaison at the section level with all agencies served in the public interest, particularly in connection with state and local government, civil preparedness, Red Cross, Salvation Army and the National Weather Service. Such contact is maintained in cooperation with the State Government Liaison.

Section Traffic Manager

The STM is appointed by the Section Manager to supervise traffic handling organization at the section level–that is, of coordinating the activities of all traffic nets, both National Traffic System-affiliated and independents, so that routings within the section and connections with other nets to effect orderly and efficient traffic flow are maintained. The STM should be a person at home and familiar with traffic handling on all modes, must have at least a Technician class license, and should possess the willingness and ability to devote equal consideration and time to all section traffic matters. The duties of the STM include the following:

1) Establish, administer, and promote a traffic handling program at the section level, based on, but not restricted to, National Traffic System networks.

2) Develop and implement one or more effective training programs within the section that addresses the needs of both traditional and digital modes of traffic handling. Ensure that Net Managers place particular emphasis on the needs of amateurs new to formal network traffic handling, as well as those who receive, send, and deliver formal traff ic on a "casual" basis, via RTTY, Amtor, and Packet based message storage and bulletin board systems.

3) Cooperate and coordinate with the Section Emergency Coordinator so that traffic nets and emergency nets in the section present a unified public service front.

4) Recommend candidates for Net Managers and Official Relay Station appointments to the SM. Issue FSD-211 appointment/cancellation cards and appropriate certificates. At the SM's discretion, the STM may directly make or cancel NM and ORS appointments.

5) Ensure that all traffic nets within the section are properly and adequately staffed, with appropriate direction to Net Managers, as required, which. results in coverage of all Net Control liaison functions. Assign liaison coverage adequate to ensure that all digital bulletin boards and message storage systems within the section are polled on a daily basis, to prevent misaddressed, lingering, or duplicated radiogram-formatted message traffic.

6) Maintain familiarity with proper traffic handling and directed net procedures applicable to all normally used modes within the section.

7) Collect and prepare accurate monthly net reports and submit them to ARRL Headquarters, either directly or via the Section Manager, but in any case on or prior to the established deadlines.

Affiliated Club Coordinator

The ACC is the primary contact and resource person for each Amateur Radio club in the section, specializing in providing assistance to clubs. The ACC is appointed by, and reports to, the Section Manager. Duties and qualifications of the ACC include:

1) Volunteer a great deal of time in getting to know the Amateur Radio clubs' members and officers person to person in his or her section. Learn their needs, strengths and interests and work with them to make clubs effective resources in their communities and more enjoyable for their members.

2) Encourage affiliated clubs in the section to become more active and, if the club is already healthy and effective, to apply as a Special Service Club (SSC).

3) Supply interested clubs with SSC application forms.

4) Assist clubs in completing SSC application forms, if requested.

5) Help clubs establish workable programs to use as SSCs.

6) Approve SSC application forms and pass them to the SM.

7) Work with other section leadership officials (Section Emergency Coordinator, Public Information Coordinator, Technical Coordinator, State Government Liason, etc.) to ensure that clubs are involved in the mainstream of ARRL Field Organization activities.

8) Encourage new clubs to become ARRL affiliated.

9) Ensure that annual progress reports (updated officers, liaison mailing addresses, etc.) are forthcoming from all affiliated clubs.

10) Novice Class license; ARRL membership required.

Bulletin Manager

Rapid dissemination of information is the lifeblood of an active, progressive organization. The ARRL Official Bulletin Station network provides a vital communications link for informing the amateur community of the latest developments in Amateur Radio and the ARRL. The ARRL Bulletin Manager is responsible for recruiting and supervising a team of Official Bulletin Stations to disseminate such news and information of interest to amateurs in the section and to provide a means of getting the news and information to all OBS appointees. The bulletins should include the content of ARRL bulletins (transmitted by W1AW), but should also include items of local, section and regional interest from other sources, such as ARRL section leadership officials, as well as information provided by the Division Director.

A special effort should be made to recruit an OBS for each major repeater and packet bulletin board in the section. This is where the greatest "audience" is to be found, many of whom are not sufficiently informed about the latest news of Amateur Radio and the League. Such bulletins should be transmitted regularly, perhaps in conjunction with a repeater net or on a repeater "bulletin board" (tone-accessed recorded announcements for repeater club members).

Although the primary mission of OBS appointees is to copy ARRL bulletins directly from W1AW, in some sections the Bullein Manager may take on the responsibility of retransmitting ARRL bulletins (as well as other information) for the benefit of OBS appointees, on a regularly scheduled day, time and frequency. An agreed-upon schedule should be worked out in advance. Time is of the essence when conveying news; therefore a successful Bulletin Manager will develop ways of communicating with the OBS appointees quickly and efficiently.

Bulletin Managers should be familiar with the position description of the Official Bulletin Station, which appears later. The duties of the Bulletin Manager include the following:

1) The Bulletin Manager must have a Technician class license or higher, and maintain League membership.

2) The Bulletin Manager is appointed by the SM and is required to report regularly to the SM concerning the section's bulletin program.

3) The Bulletin Manager is responsible for recruiting (and, at the discretion of the SM, appointing) and supervising a team of Official Bulletin Stations in the Section. A special effort should be made to recruit OBSs for each major repeater and PBBS in the section.

4) The Bulletin Manager must be capable of copying ARRL bulletins directly from W1AW on the mode(s) necessary. The Bulletin Manager may, in some cases, be required to retransmit ARRL bulletins for OBS appointees who might be unable to copy them directly from W1AW.

5) The Bulletin Manager is also responsible for funneling news and information of a local, section and regional nature to OBS appointees. In so doing, the Bulletin Manager must maintain close contact with other section-level officials, and the Division Director, to maintain an organized and unified information flow within the section.

Official Observer Coordinator

The Official Observer Coordinator is an ARRL section-level leadership official appointed by the Section Manager to supervise the Official Observer program in the section. The OO Coordinator must hold a General class (or higher amateur license and be licensed as a Technician or higher for at least four years.

The Official Observer program has operated for more than half a century, and in that time, OO appointees have assisted thousands of amateurs whose signals, or operating procedures, were not in compliance with the regulations. The function of the OO is to listen for amateurs who might otherwise come to the attention of the FCC and to advise them by mail of the irregularity observed. The OO program is, in essence, for the benefit of amateurs who want to be helped. Official Observers must meet high standards of expertise and experience. It is the job of the OO Coordinator to recruit, supervise and direct the efforts of OOs in the section, and to report their activity monthly to the Section Manager and to ARRL Hq.

The OO Coordinator is a key figure in the Amateur Auxiliary to the FCC's Field Operations Bureau, the foundation of which is an enhanced OO program. Jointly created by the FCC and ARRL in response to federal enabling legislation (Public Law 97-259), the Auxiliary permits a close relationship between FCC and ARRL Field Organization volunteers in monitoring the amateur airwaves for potential rules discrepanciestviolations. (Contact your Section Manager for further details on the Amateur Auxiliary program.)

Public Information Coordinator

The ARRL Public Information Coordinator is a section-level official appointed by the Section Manager to be the section's expert on public information and public relations matters. The Public Information Coordinator is also responsible for organizing, guiding and coordinating the activities of the Public Information Officers within the section.

The Public Information Coordinator must be a full member of the ARRL and, preferably, have professional public relations or journalism experience or a significantly related background in dealing with the public media.

The purpose of public relations goes beyond column inches and minutes of air time. Those are means to an end—generally, telling a specific story about hams, ham radio or ham-related activities for a specific purpose. Goals may range from recruiting potential hams for a licensing course to improving public awareness of amateurs' service to the community. Likewise, success is measured not in column inches or air time, but in how well that story gets across and how effectively it generates the desired results.

For this reason, public relations are not conducted in a vacuum. Even the best PR is wasted without effective follow-up. To do this best, PR activities must be well-timed and wellcoordinated within the amateur community, so that clubs, Elmers, instructors and so on are prepared to deal with the interest the PR generates. Effective PICs will convey this goal-oriented perspective and aftitude to their PIOs and help them coordinate public relations efforts with others in their sections.

Recruitment of new hams and League members is an integral part of the job of every League appointee. Appointees should take advantage of every opportunity to recruit a new ham or a member to foster growth of Field Organization programs, and our abilities to serve the public.

Specific Duties of the Public Information Coordinator

1) Advises the Section Manager on building and maintaining a positive public image for Amateur Radio in the section; keeps the SM informed of all significant events which would benefit from the SM's personal involvement and reports regularly to the SM on activities.

2) Counsels the SM in dealing with the media and with government officials, particularly when representing the ARRL and/or Amateur Radio in a public forum.

3) Maintains contact with other section level League officials, particularly the Section Manager and others such as the State Government Liaison, Section Emergency Coordinator and Bulletin Manager on matters appropriate for their attention and to otherwise help to assure and promote a coordinated and cohesive ARRL Field Organization.

4) Works closely with the section Affiliated Club Coordinator and ARRL-affiliated clubs in the section to recruit and train a team of Public Information Officers (PIOs). With the approval of the Section Manager, makes PIO appointments within the section.

5) Works with the SM and other PICs in the division to:

a) develop regional training programs for PIOs and club publicity chairpersons;

b) coordinate public relations efforts for events and activities which may involve more than one section, and

c) provide input on matters before the League's Public Relations Committee for discussion or action.

6) Establishes and coordinates a section-wide Speakers Bureau to provide knowledgeable and effective speakers who are available to address community groups about Amateur Radio, and works with PIOs to promote interest among those groups.

7) Helps local PIOs to recognize and publicize newsworthy stories in their areas. Monitors news releases sent out by the PIOs for stories of broader interest and offers construc-

tive comments for possible improvement. Helps local PIOs in learning to deal with, and attempting to minimize, any negative publicity about Amateur Radio or to correct negative stories incorrectly ascribed to Amateur Radio operators.

8) Working with the PIOs, develops and maintains a comprehensive list of media outlets and contacts in the section for use in section-wide or nationwide mailings.

9) Helps local PIOs prepare emergency response PR kits containing general information on Amateur Radio and on local clubs, which may be distributed in advance to local Emergency Coordinators and District Emergency Coordinators for use in dealing with the media during emergencies.

10) Works with PIOs, SM and ARRL staff to identify and publicize League-related stories of local or regional interest, including election or appointment of ARRL leadership officials, scholarship winners/award winners, *QST* articles by local authors or local achievements noted or featured in *QST*.

11) Familiarize self with ARRL Public Service Announcements (PSAs), brochures and audio-visual materials; assists PIOs in arranging air time for PSAs; helps PIOs and speakers choose and secure appropriate brochures and audio-visual materials for events or presentations.

12) At the request of the Section Manager or Division Director, may assist in preparation of a section or division newsletter.

13) Encourages, organizes and conducts public information/public relations sessions at ARRL hamfests and conventions.

14) Works with PIOs to encourage activities that place Amateur Radio in the public eye, including demonstrations, Field Day activities, etc. and assures that sponsoring organizations are prepared to follow up on interest generated by these activities.

Most public relations activities are conducted on a local level by affiliated clubs, which generally are established community organizations. PICs should encourage clubs to make public relations a permanent part of their activities.

With the Section Manager's approval, the PIC may appoint club publicity chairpersons or other individuals recommended by affiliated clubs as PIOs. Where the responsibility cannot or will not be assumed by the club, the PIC is encouraged to seek qualified League members who are willing to accept the responsibility of PIO appointments.

Appointees should take advantage of every opportunity to recruit a new ham or member to foster growth of Field Organization programs, and our abilities to serve the public.

State Government Liaison

The State Government Liaison (SGL) shall be an amateur who is aware (at a minimum) of state legislative proposals in the normal course of events and who can watch for those proposals having the potential to affect Amateur Radio without creating a conflict of interest.

The SGL shall collect and promulgate information on state ordinances affecting Amateur Radio and work (with the assistance of other ARRL members) toward assuring that they work to the mutual benefit of society and the Amateur Radio Service.

The SGL shall guide, encourage and support ARRL members in representing the interests of the Amateur Radio Service at all levels. Accordingly, the SGL shall cooperate closely with other section-level League officials, particularly the Section Emergency Coordinator and the Public Information Coordinator.

When monitoring state legislative dockets, SGL's should watch for key words that could lead to potential items affecting Amateur Radio. Antennas (dish, microwave, towers, structures, satellite, television, lighting), mobile radio, radio receivers, radio interference, television interference, scanners, license plates, cable television, ham radio, headphones in automobiles, lightning protection, antenna radiation and biological effects of radio signals are a few of the examples of what to look for.

In those states where there is more than one section, the Section Managers whose territory does not encompass the state capital may simply defer to the SGL appointed by their counterpart in the section where the state capital is located. In this case, the SGL is expected to communicate equally with all Section Managers (and Section Emergency Coordinators and other section-level League officials). In sections where there is more than one government entity, i.e., Maryland, DC, Pacific, there may be a Liaison appointed for each entity.

Technical Coordinator

The ARRL Technical Coordinator (TC) is a section-level official appointed by the Section Manager to coordinate all technical activities within the section. The Technical Coordinator must hold a Novice class or higher amateur license. The Technical Coordinator reports to the Section Manager and is expected to maintain contact with other section-level appointees as appropriate to ensure a unified ARRL Field Organization within the section. The duties of the Technical Coordinator are as follows:

1) Supervise and coordinate the work of the section's Technical Specialists (TSs).

2) Encourage amateurs in the section to share their technical achievements with others through the pages of *QST*, and at club meetings, hamfests and conventions.

3) Promote technical advances and experimentation at VHF/UHF and with specialized modes, and work closely with enthusiasts in these fields within the section.

4) Serve as an advisor to radio clubs that sponsor training programs for obtaining amateur licenses or upgraded licenses in cooperation with the ARRL Affiliated Club Coordinator.

5) In times of emergency or disaster, function as the coordinator for establishing an array of equipment for communications use and be available to supply technical expertise to government and relief agencies to set up emergency communication networks, in cooperation with the ARRL Section Emergency Coordinator.

6) Refer amateurs in the section who need technical advice to local TSs.

7) Encourage TSs to serve on RFI and TVI committees in the section for the purpose of rendering technical assistance as needed, in cooperation with the ARRL OO Coordinator.

8) Be available to assist local technical program committees in arranging suitable programs for ARRL hamfests and conventions.

9) Convey the views of section amateurs and TSs about the technical contents of *QST* and ARRL books to ARRL HQ. Suggestions for improvements should also be called to the attention of the ARRL HQ technical staff.

10) Work with the appointed ARRL TAs (technical advisors) when called upon.

11) Be available to give technical talks at club meetings, hamfests and conventions in the section.

District Emergency Coordinator

The DEC is an ARRL full member of at least Technician class experienced in emergency communications who can assist the SEC by taking charge in the area of jurisdiction especially during an emergency. The DEC shall:

1) Coordinate the training, organization and emergency partication of Emergency Coordinators in the area of jurisdiction.

2) Make local decisions in the absence of the SEC or through coordination with the SEC concerning the allotment of available amateurs and equipment during an emergency.

3) Coordinate the interrelationship between local emergency plans and between communications networks within the area of jurisdiction.

4) Act as backup for local areas without an Emergency Coorinator and assist in maintaining contact with governmental and other agencies in the area of jurisdiction.

5) Provide direction in the routing and handling of emergency communications of either a fiormal or tactical nature.

6) Recommend EC appointments to the SEC and advise on OES appointments.

7) Coordinate the reporting and documentation of ARES activities in the area of jurisdiction.

8) Act as a model emergency communicator as evidenced by dedication to purpose, reliability and understanding of emergency communications.

9) Be fully conversant in National Traffic System routing and procedures as well as have a thorough understanding of the locale and role of all vital governmental and volunteer agencies that could be involved in an emergency.

EMERGENCY COORDINATOR

The ARRL Emergency Coordinator is a key team player in ARES on the local emergency scene. Working with the Section Emergency Coordinator, the DEC and Official Emergency Stations, the EC prepares for, and engages in management of communications needs in disasters. EC duties include:

1) Promote and enhance the activities of the Amateur Radio Emergency Service (ARES) for the benefit of the public as a voluntary, non-commercial communications service.

2) Manage and coordinate the training, organization and emergency participation of interested amateurs working in support of the communities, agencies orfunctions designated bythe Section Emergency Coordinator/Section Manager.

3) Establish viable working relationships with federal, state, county, city governmental and private agencies in the ARES jurisdictional area which need the services of ARES in emergencies. Determine what agencies are active in your area, evaluate each of their needs, and which ones you are capable of meeting, and then prioritize these agencies and needs. Discuss your planning with your Section Emergency Coordinator and then with your counterparts in each of the agencies. Ensure they are all aware of your ARES group's capabilities, and perhaps more importantly, your limitations.

4) Develop detailed local operational plans with "served" agency officials in your jurisdiction that set forth precisely what each of your expectations are during a disaster operation. Work jointly to establish protocols for mutual trust and respect. All matters involving recruitment and utilization of ARES volunteers are directed by you, in response to the needs assessed by the agency officials. Technical issues involving message format, security of message transmission, Disaster Welfare Inquiry policies, and others, should be reviewed and expounded upon in your detailed local operations plans.

5) Establish local communications networks run on a regular basis and periodically test those networks by conducting realistic drills.

6) Establish an emergency traffic plan, with Welfare Traffic inclusive, utilizing the National Traffic System as one active component for traffic handling. Establish an operational liaison with local and section nets, particularly for handling Welfare traffic in an emergency situation.

7) In times of disaster, evaluate the communications needs of the jurisdiction and respond quickly to those needs. The EC will assume authority and responsibility for emergency response and performance by ARES personnel under his jurisdiction.

8) Work with other non-ARES amateur provider-groups to establish mutual respect and understanding, and a coordination mechanism for the good of the public and Amateur Radio. The goal is to foster an efficient and effective Amateur Radio response overall.

9) Work for growth in your ARES program, making it a stronger, more valuable resource and hence able to meet more of the agencies'local needs. There are thousands of new Technicians coming into the amateur service that would make ideal additions to your ARES roster. A stronger ARES means a better ability to serve your communities in times of need and a greater sense of pride for Amateur Radio by both amateurs and the public.

10) Report regularly to the SEC, as required.

Recruitment of new hams and League members is an integral part of the job of every League appointee. Appointees should take advantage of every opportunity to recruit a new ham or member to foster growth of Field Organization programs, and our abilities to serve the public.

Net Manager

For coordinating and supervising traffic-handling activities in the section, the SM may appoint one or more Net Managers, usually on recommendation of the Section Traffic Manager. The number of NMs appointed may depend on a section's geographical size, the number of nets operating in the section, or other factors having to do with the way the section is organized. In some cases, there may be only one Net Manager in charge of the one section net, or one NM for the phone net, one for the CW net. In larger or more traffic-active sections there may be several, including NMS for the VHF net or nets, for the RTTY net, or NTS local nets not controlled by ECs. All ARRL NMs should work under the STM in a coordinated section traffic plan.

Some nets cover more than one section but operate in NTS at the section level. In this case, the Net Manager is selected by agreement among the STMs concerned and the NM appointment conferred on him by his resident SM.

NMs may conduct any testing of candidates for ORS appointment (see below) that they consider necessary before making appointment recommendations to the STM. Net Managers also have the function of requiring that all traffic handling in ARRL recognized nets is conducted in proper ARRL form.

Remember: All appointees or appointee candidates must be ARRL full members.

ARRL STATION APPOINTMENTS

Field Organization station and individual appointments from the Section Manager are available to qualified ARRL full members in each section. These appointments are as follows: Official Relay Station, Official Emergency Station, Official Bulletin Station, Public Information Officer, Official Observer and Technical Specialist. All appointees receive handsome certificates from the SM and are entitled to wear ARRL membership pins with the distinctive blue background. All appointees are required to submit regular reports to maintain appointments and to remain active in their area of specialty.

The report is the criterion of activity. An appointee who misses three consecutive monthly reports is subject to cancellation by the SM of the appropriate section leadership official, who cannot know what or how much you are doing unless you report. An appointee whose appointment is cancelled for this or other reasons must earn reinstatement by demonstrating activity and adherence to the requirements. Reinstatement of cancelled appointments, and indeed judgment of whether or not a candidate meets the requirements, is at the discretion of the SM and the section leadership.

The detailed qualifications of the six individual "station" appointments are given below. If you are interested, your SM will be glad to receive your application. Use application form FSD-187, reproduced nearby.

Official Relay Station

This is a traffic-handling appointment that is open to all

ARRL FIELD ORGANIZATION
APPLICATION FOR STATION APPOINTMENT

I, ______________________, licensee of Amateur Radio station __________, hereby submit application to the Section Manager (or designated Section-level official) of the ______________ Section, for appointment as:

Official Relay Station ____
Official Emergency Station ____
Official Observer ____ (Must hold minimum Technician Class license for 4 or more years).
Official Bulletin Station ____
Assistant Technical Coordinator ____
Public Information Assistant ____

My qualifications are:

____
(describe appropriate experience and/or interest)
____
____
____

Address: ______________________________
street city state zip

Phone number: ______________________________
work home

ARRL membership number: ______________ (from *QST* label)

I have read the description of the duties and responsibilities of the appointment, and agree to maintain current League membership, and report my station activity to designated section officials on a monthly basis.

Signed: ______________ Date: ______________

(FSD-187 7/07)

This form is available to apply for an ARRL station appointment. An online form is available at **http://www.arrl.org/FandES/field/forms/fsd187/form.html**

licenses. This appointment applies equally to all modes and all parts of the spectrum. It is for traffic handlers, regardless of how or in what part of the spectrum they do it.

The potential value of the operator who has traffic know-how to his country and community is enhanced by his ability and the readiness of his station to function in the community interest in case of emergency. Traffic awareness and experience are often the signs by which mature amateurs may be distinguished.

Traditionally, there have been considerable differences between procedures for traffic handling by CW, phone, RTTY, ASCII and other modes. Appointment requirements for ORS do not deal with these, but with factors equally applicable to all modes. The appointed ORS may confine activities to one mode or one part of the spectrum if he wishes although versatility does indeed make it possible to perform a more complete public service. The expectation is that the ORS will set the example in traffic handling, however it is done. To the degree that he is deficient in performing traffic functions by any mode, to that extent he does not meet the qualifications for the appointment. Here are the basic requirements:

1) Full ARRL membership and Novice class license or higher.

2) Code and/or voice transmission.

3) Transmission quality, by whatever mode, must be of the highest quality, both technically and operationally. For example, CW signals must be pure, chirpless and clickless, and code sending must be well spaced and properly formed. Voice transmission must be of proper modulation percentage or deviation, precisely enunciated with minimum distortion. RTTY must be clickless, proper shift, etc.

4) All ORSs are expected to follow standard ARRL operating practices (message form, ending signals, abbreviations or prowords, courtesy, etc.).

5) Regular participation in traffic activities, either freelance or ARRL-sponsored. The latter is encouraged, but not required.

6) Handle all record communications speedily and reliably and set the example in efficient operating procedures. All traffic is relayed or delivered promptly after receipt.

7) Report monthly to the STM, including a breakdown of traffic handled during the past calendar month.

Official Emergency Station

Amateur operators may be appointed as an Official Emergency Station (OES) by their Section Emergency Coordinator (SEC) or Section Manager (SM) at the recommendation of the EC, or DEC (if no EC) holding jurisdiction. The OES appointee must be an ARRL member and set high standards of emergency preparedness and operating. The OES appointee makes a deeper commitment to the ARES program in terms of functionality than does the rank-and-file ARES registrant.

The requirements and qualifications for the position include the following: Full ARRL membership; experience as an ARES registrant; regular participation in the local ARES organization including drills and tests; participation in emergency nets and actual emergency situations; regular reporting of activities.

The OES appointee is appointed to carry out specific functions and assignments designated by the appropriate EC or DEC. The OES appointee and the presiding EC or DEC, at the time of the OES appointment, will mutually develop a detailed, operational function/assignment and commitment for the new appointee. Together, they will develop a responsibility plan for the individual OES appointee that makes the best use of the individual's skills and abilities. During drills and actual emergency situations, the OES appointee will be expected to implement his/her function with professionalism and minimal supervision.

Functions assigned may include, but are not limited to, the following four major areas of responsibility:

OPERATIONS—Responsible for specific, pre-determined operational asignments during drills or actual emergency situations. Examples include: Net Control Station or Net Liaison for a specific ARES net; Manage operation of a specified ARES VHF or HF digital BBS or MBO, or point-to-point link; Operate station at a specified emergency management office, Red Cross shelter or other served agency operations point.

ADMINISTRATION—Responsible for specific, pre-determined administrative tasks as assigned in the initial appointment commitment by the presiding ARES official. Examples include: Recruitment of ARES members, liaison with Public Information Officer to coordinate public information for the media; ARES registration data base management; victim/refugee data base management; equipment inventory; training; reporting; and postevent analysis.

LIAISON—Responsible for specific, pre-determined liaison responsibilities as assigned by the presiding EC or DEC. Examples include: Maintaining contact with assigned served agencies; Maintaining liaison with specified NTS nets; Maintaining liaison with ARES officials in adjacent jurisdictions; Liaison with mutual assistance or "jump" teams.

LOGISTICS—Responsible for specific, pre-determined logistical functions as assigned. Examples include: transportation; Supplies management and procurement (food, fuel, water, etc.); Equipment maintenance and procurement-radios, computers, generators, batteries, antennas.

MANAGEMENT ASSISTANT—Responsible for serving as an assistant manager to the EC, DEC or SEC based on specific functional assignments or geographic areas of jurisdiction.

CONSULTING—Responsible for consulting to ARES officials in specific area of expertise.

OES appointees may be assigned to pre-disaster, post-disaster, and recovery functions. These functions must be specified in the OES's appointment commitment plan.

The OES appointee is expected to participate in planning meetings, and post-event evaluations. Following each drill or actual event, the EC/DEC and the OES appointee should review and update the OES assignment as required. The OES appointee must keep a detailed log of events during drills and actual events in his/her sphere of responsibility to facilitate this review.

Continuation of the appointment is at the discretion of the appointing official, based upon the OES appointee's fulfillment of the tasks he/she has agreed to perform.

Recruitment of new hams and League members is an integral part of the job of every League appointee. Appointees should take advantage of every opportunity to recruit a new ham or member to foster growth of Field Organization programs, and our abilities to serve the public.

Official Bulletin Station

Rapid dissemination of information is the lifeblood of an active, progressive organization. The ARRL Official Bulletin Station network provides a vital communications link for informing the amateur community of the latest developments in Amateur Radio and the League. ARRL bulletins, containing up-to-the minute news and information of Amateur Radio, are issued by League Hq as soon as such news breaks. These bulletins are transmitted on a regular schedule by ARRL Hq station W1AW.

The primary mission of OBS appointees is to copy these bulletins directly off the air from W1AW—on voice, CW or RTTY/ASCII—and retransmit them locally for the benefit of amateurs in the particular coverage area, many of whom may not be equipped to receive bulletins directly from W1AW.

ARRL bulletins of major importance or of wide-ranging scope are mailed from Hq to each Bulletin Manager and OBS appoin tee. However, some bulletins, such as the ARRL DX Bulletin (transmitted on Fridays UTC), are disseminated only by W1AW because of time value. Thus, it is advantageous for each OBS to copy W1AW directly. In some sections, the Bulletin Manager may assume the responsibility of copying the bulletins from W1AW; therefore, individual OBSs should be sure to meet the Bulletin Manager on a regular, agreed-upon schedule to receive the latest bulletins.

Inasmuch as W1AW operates on all bands (160-2 meters), the need for OBSs on HF has lessened somewhat in recent times. However, OBS appointments for HF operation can be conferred by the Section Manager (or the Bulletin Manager, depending on how the SM organizes the section) if the need is apparent. More important, to serve the greatest possible "audience," OBS appointees who can send ARRL bulletins over VHF repeaters are of maximum usefulness and are much in demand. If possible, an OBS who can copy bulletins directly from W1AW (or the Bulletin Manager) should be assigned to each major repeater in the section. Bulletins should be transmitted regularly, perhaps in conjunction with a VHF repeater net, on a repeater bulletin board (toneaccessed recorded announcements for repeater club members), or via a local RTTY (computer) mailbox. Duties and requirements of the OBS include the following:

1) OBS candidates must have Novice class license or higher.

2) Retransmission of ARRL bulletins must be made at least once per week to maintain appointment.

3) OBS candidates are appointed by the Section Manager (or by the Bulletin Manager, if the SM so desires) and must adhere to a schedule that is mutually agreeable, as indicated on appointment application form FSD-187.

4) OBS appointees should send a monthly activity report (such as FSD-210, under "Schedules and Net Affiliations") to the Bulletin Manager, indicating bulletin transmissions made and generally updating the Bulletin Manager to any OBS-related activities. This reporting arrangement may be modified by the Bulletin Manager as he/she sees fit.

5) As directed by the Bulletin Manager, OBSs will include in their bulletin transmissions news of local, section and regional interest.

Public Information Officer

Public Information Officers (PIOs) are appointed by and report to the ARRL section Public Information Coordinator (PIC) generally upon the recommendation of an affiliated club and with the approval of the Section Manager (SM). PIOs are usually club publicity chairpersons and must be full ARRL members. Training for PIOs should be provided regularly on a sectional or regional basis by the PIC and/or other qualified people.

Good "grass roots" public relations activities involve regular and frequent publicizing of amateur activities through local news media plus community activities; school programs; presentations to service clubs and community organizations; exhibits and demonstrations; and other efforts which create a positive public image for Amateur Radio.

Public relations are not conducted in a vacuum. Even the best PR is wasted without effective follow-up. To do this best, PR activities must be well-timed and well-coordinated within the amateur community, so that clubs, Elmers, instructors and so on are prepared to deal with the interest the PR generates.

Recruitment of new hams and League members is an integral part of the job of every League appointee. Appointees should take advantage of every opportunity to recruit a new ham or member to foster growth of Field Organization programs, and our abilities to serve the public.

Specific Duties of the Public Information Officer

1) Establishes and maintains a list of media contacts in the local area; strives to establish and maintain personal contacts with appropriate representatives of those media (editors, news directors, science reporters and so on).

2) Becomes a contact for the local media and assures that editors/reporters who need information about Amateur Radio know where to find it.

3) Works with Local Government Liaisons to establish personal contacts with local government officials where possible and explain to them, briefly and non-technically, about Amateur Radio and how it can help their communities.

4) Keeps informed of activities by local hams and identifies and publicizes those that are newsworthy or carry human interest appeal. (This is usually done through news releases or suggestions for interviews or feature stories.)

5) Attempts to deal with and minimize any negative publicity about Amateur Radio and to correct any negative stories which are incorrectly ascribed to Amateur Radio operators.

6) Generates advance publicity through the local media of scheduled activities of interest to the general public, including licensing classes, hamfests, club meetings, Field Day operations, etc.

7) Works with the section PIC to identify and publicize Leaguerelated stories of local news interest, including election and appointment of hams to leadership positions, *QST* articles by local authors or local achievements noted or featured in *QST*.

8) Maintains contact with other League officials in the local area, particularly the Emergency Coordinator and/or District Emergency Coordinator. With the PIC, helps prepare an emergency response PR kit, including general brochures on Amateur Radio and specific information about local clubs. Distributes them to ECs and DECs before an emergency occurs. During emergencies, these kits should be made available to reporters at the scene or at a command post. The PIO should help summarize Amateur Radio activity in an ongoing situation, and follow up any significant emergency communications activities with prompt reporting to media of the extent and nature of Amateur Radio involvement.

9) Assists the section PIC in recruiting hams for the section's Speakers Bureau; promotes interest among community and service organizations in finding out more about Amateur Radio through the bureau and relays requests to the PIC.

10) Helps individual hams and radio clubs to develop and promote good ideas for community projects and special events to display Amateur Radio to the public in a positive light.

11) Attends regional training sessions sponsored by section PICS.

12) Becomes familiar with ARRL Public Service Announcements (PSAs), brochures and audio-visual materials; contacts local radio and TV stations to arrange airing of Amateur Radio PSAs; secures appropriate brochures and audio-visual materials for use in conjunction with planned activities.

13) Keeps the section PIC fully informed on activities and places PIC on news release mailing list.

Official Observer

The Official Observer (OO) program has been sponsored by the League for over 50 years to help amateurs help each other. Official observer appointees have aided thousands of amateurs to maintain their transmitting equipment and operating procedures in compliance with the regulations. The object of the OO program is to notify amateurs by mail of operating/technical irregularities before they come to the attention of the FCC.

The ARRL commitment to volunteer monitoring has been greatly enhanced by the creation of the Amateur Auxiliary to the FCC Field Operations Bureau, designed to enable amateurs to play a more active and direct role in upholding the traditional high standard of conduct on the amateur bands. The OO is the foundation of the Amateur Auxiliary, carrying out the all-important day-to-day maintenance monitoring of the amateur airwaves. Following recommendation by the Section Manager, potential members of the Amateur Auxiliary are provided with training materials, and all applicants must successfully complete a written examination to be enrolled as Official Observers. For further information, please contact your SM.

The OO performs his function by listening rather than transmitting, keeping a watchful ear out for such things as frequency instability, harmonics, hum, key clicks, broad signals, distorted audio, overdeviation, out-of-band operation, etc. The OO completes his task once the notification card is sent. Reimbursement for postage expenses are provided for through the SM. The OO:

1) Must be an ARRL full member and have been a licensee of Technician class or higher for at least four years.

2) Must undergo and complete successfully the Amateur Auxiliary training and certification procedure.

3) Must report to the OO Coordinator regularly on FSD-23.

4) Maintain regular activity in sending out notices as observed.

The OO program is one of the most important functions of the League. A sincere dedication to helping our brother and sister amateurs is required for appointment. Only the "very best" are sought.

Technical Specialist

Appointed by the SM, or TC under delegated authority from the SM, the TS supports the TC in two main areas of responsibility: Radio Frequency Interference, and Technical Information. The TS must hold full ARRL membership and at least a Novice class license. TSs can specialize in certain specific technical areas, or can be generalists. Here is a list of specific job duties:

1) Serve as a technical oracle to local hams and clubs. Correspond by telephone and letter on tech topics. Refer correspondents to other sources if specific topic is outside TS's knowledge.

2) Serve as advisor in radio frequency interference issues. RFI can drive a wedge in neighbor and city relations. It will be the TS with a cool head who will resolve problems. Local hams will come to you for guidance in dealing with interference problems.

3) Speak at local clubs on popular tech topics. Let local clubs know you're available and willing.

4) Represent ARRL at technical symposiums in industry; serve on CATV advisory committees; advise municipal governments on technical matters.

5) Work with other ARRL officials and appointees when called upon for technical advice especially in emergency communications situations where technical prowess can mean the difference in getting a communications system up and running, the difference between life and death.

6) Handle other miscellaneous technically related tasks assigned by the Technical Coordinator.

Local Government Liaison

The Local Government Liaison (LGL) is primarily responsible for monitoring proposals and actions by local government bodiesand officials which may affect Amateur Radio; for working with the local PIO to alert section leadership officials and area amateurs to any such proposals or actions, and for coordinating local responses. In addition, the LGL serves as a primary contact for amateurs encountering problems dealing with local government agencies, for those who want to avoid problems and for local officials who wish to work with amateurs or simply learn more about Amateur Radio. The most effective LGL will be able to monitor local government dockets consistently, muster local, organized support quickly when necessary, and be well known in the local amateur community as the point person for local government problems. The LGL must be a Full Member of the ARRL. LGLs are appointed by and report to the Section Manager, or State Government Liaison (acting under delegated authority from the SM).

Specific Responsibilities:

1) Monitor proposals and actions of town/city councils zoning appeals boards, and any other legislative or regulatory agencies or officials below the state level whose actions can directly or indirectly affect Amateur Radio.

2) Attend meetings of those bodies when possible, to become familiar with their policies, procedures and members. Assist local amateurs in their dealings with local boards and agencies.

3) Be available to educate elected and appointed officials, formally and informally, about the value of Amateur Radio to their community.

4) Work with the PIO or PIC to inform local amateurs, the SGL and the SM of any proposals of actions which may affect Amateur Radio, and report regularly on the progress or lack thereof.

5) Work with the PIO to organize the necessary local response to any significant proposals or actions, either negative or positive, and coordinate that response.

6) Refer amateurs seeking ARRL Volunteer Counsels to HQ.

7) Register on mailing list for Planning Commission meeting agendas.

8) Work with the PIO and local clubs to build and/or maintain good relations between Amateur Radio and local officials. (For example, invite the mayor to a club dinner or council members to Field Day.)

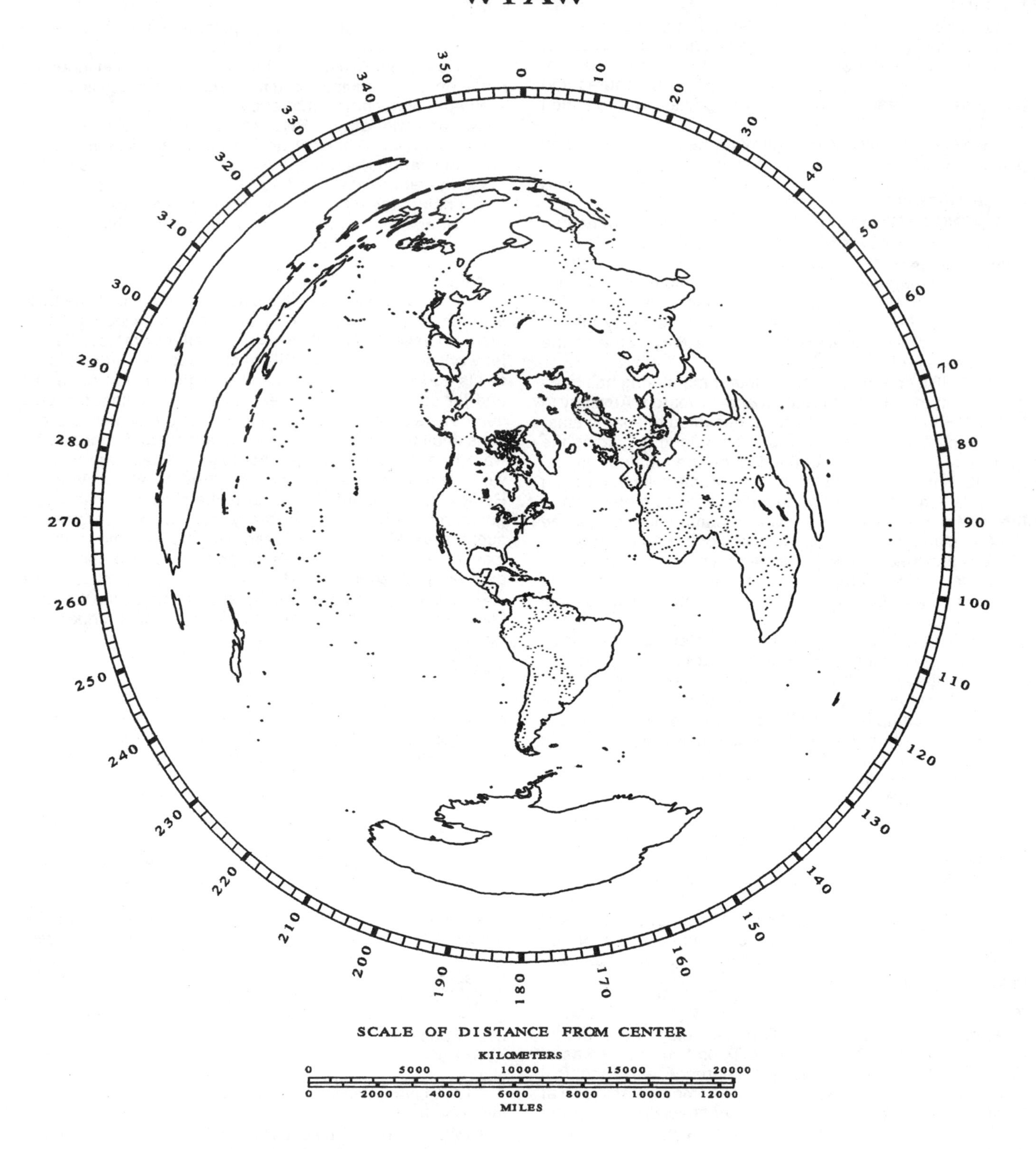
K5ZI AZIMUTHAL EQUIDISTANT MAP CENTERED ON
W1AW
0
10
20
30
40
50
60
70
80
90
100
110
120
130
140
150
160
170
180
190
200
210
220
230
240
250
260
270
280
290
300
310
320
330
340
350
SCALE OF DISTANCE FROM CENTER
KILOMETERS
0
5000
10000
15000
20000
0
2000
4000
6000
8000
10000
12000
MILES

K5ZI AZIMUTHAL EQUIDISTANT MAP CENTERED ON

Eastern USA

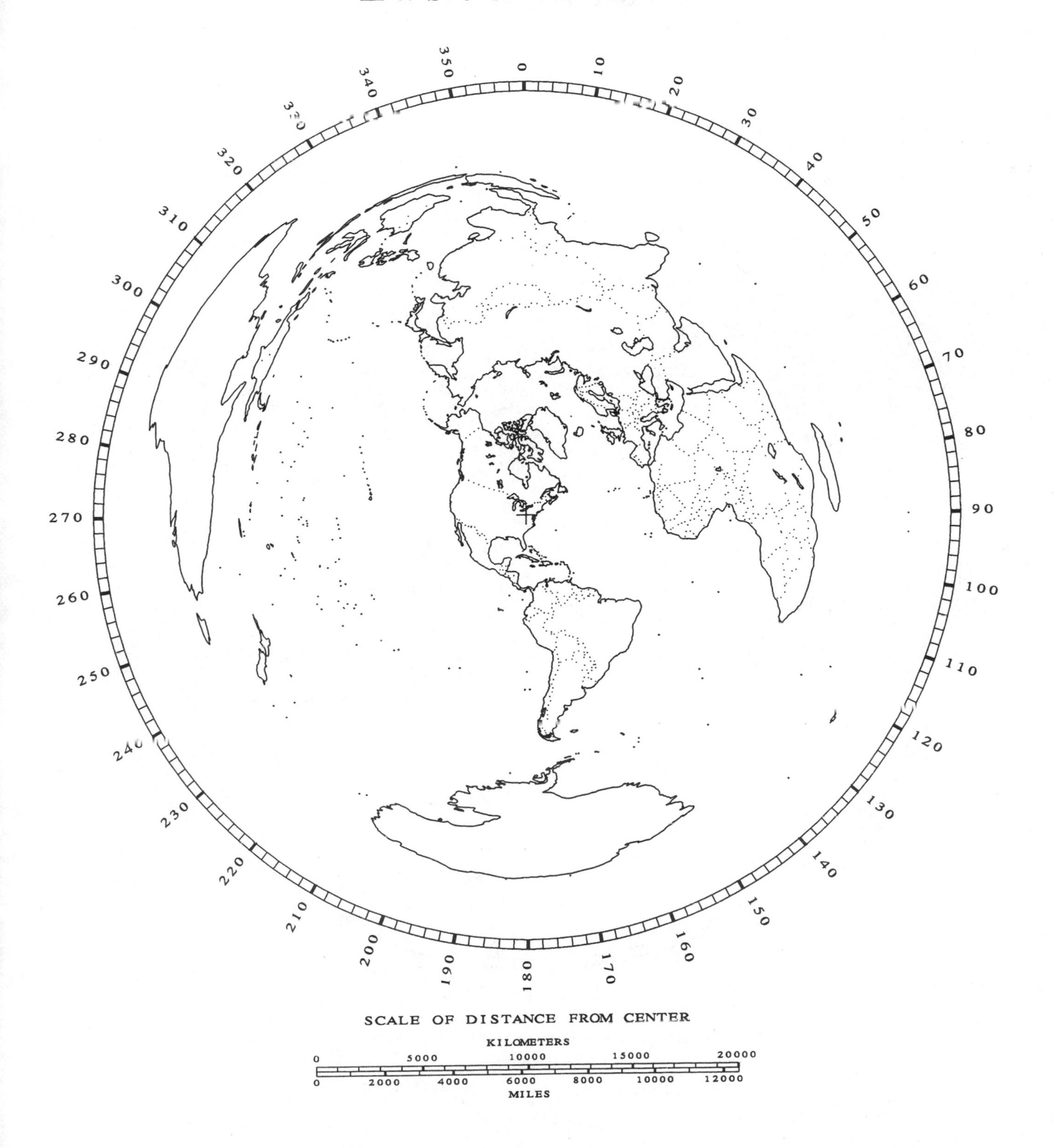

K5ZI AZIMUTHAL EQUIDISTANT MAP CENTERED ON

Central USA

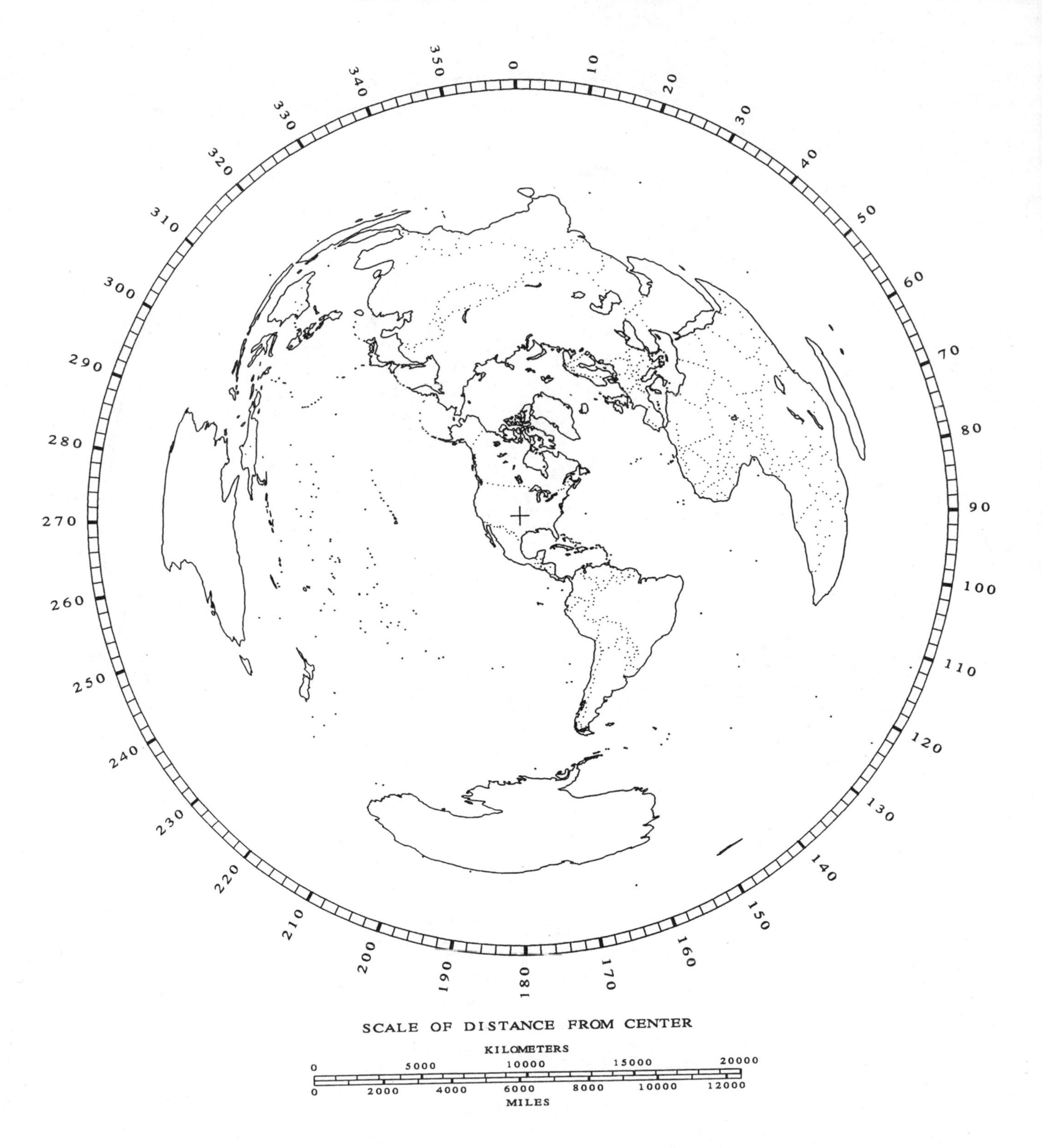

K5ZI AZIMUTHAL EQUIDISTANT MAP CENTERED ON

Western USA

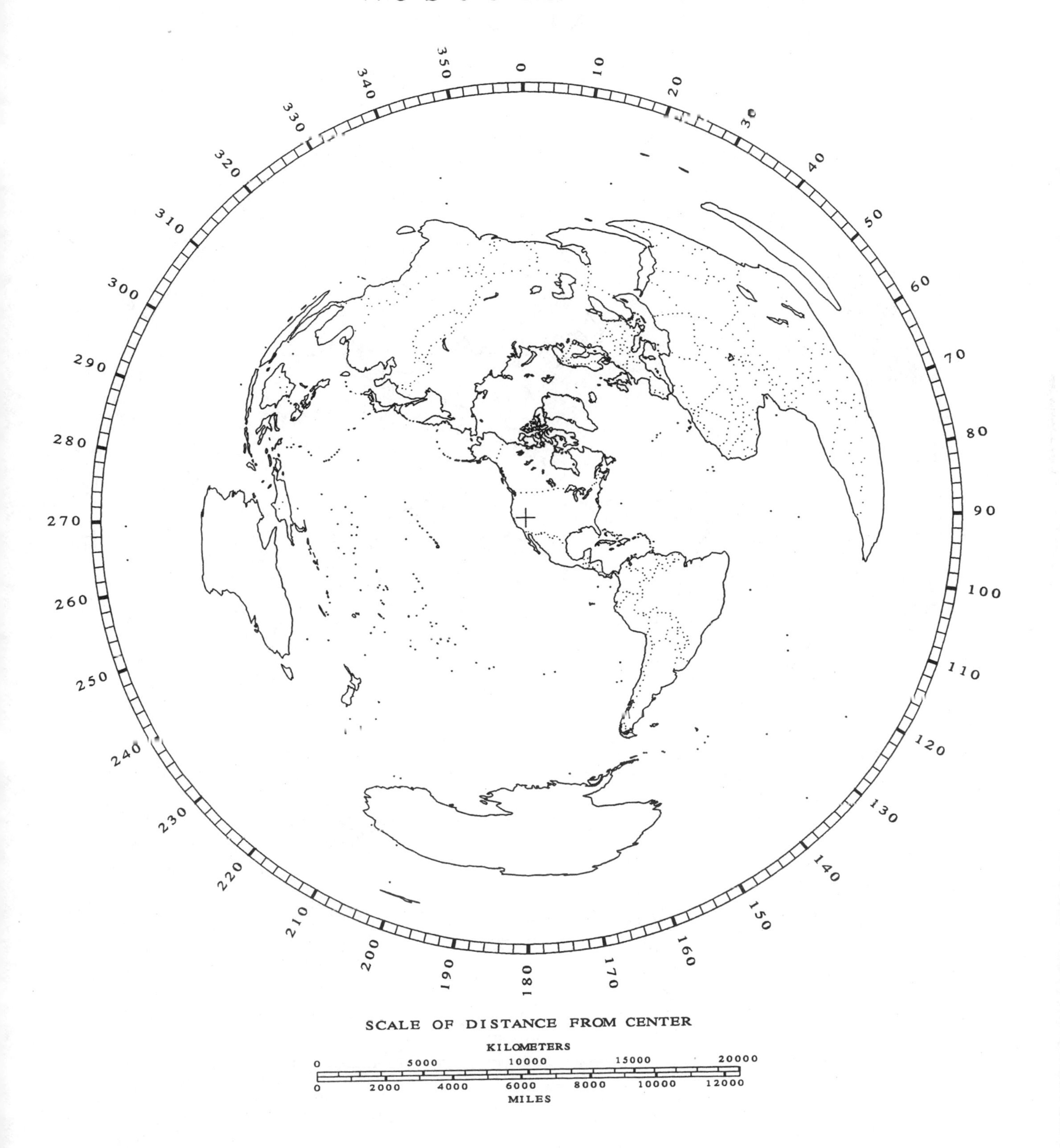

K5ZI AZIMUTHAL EQUIDISTANT MAP CENTERED ON

Alaska

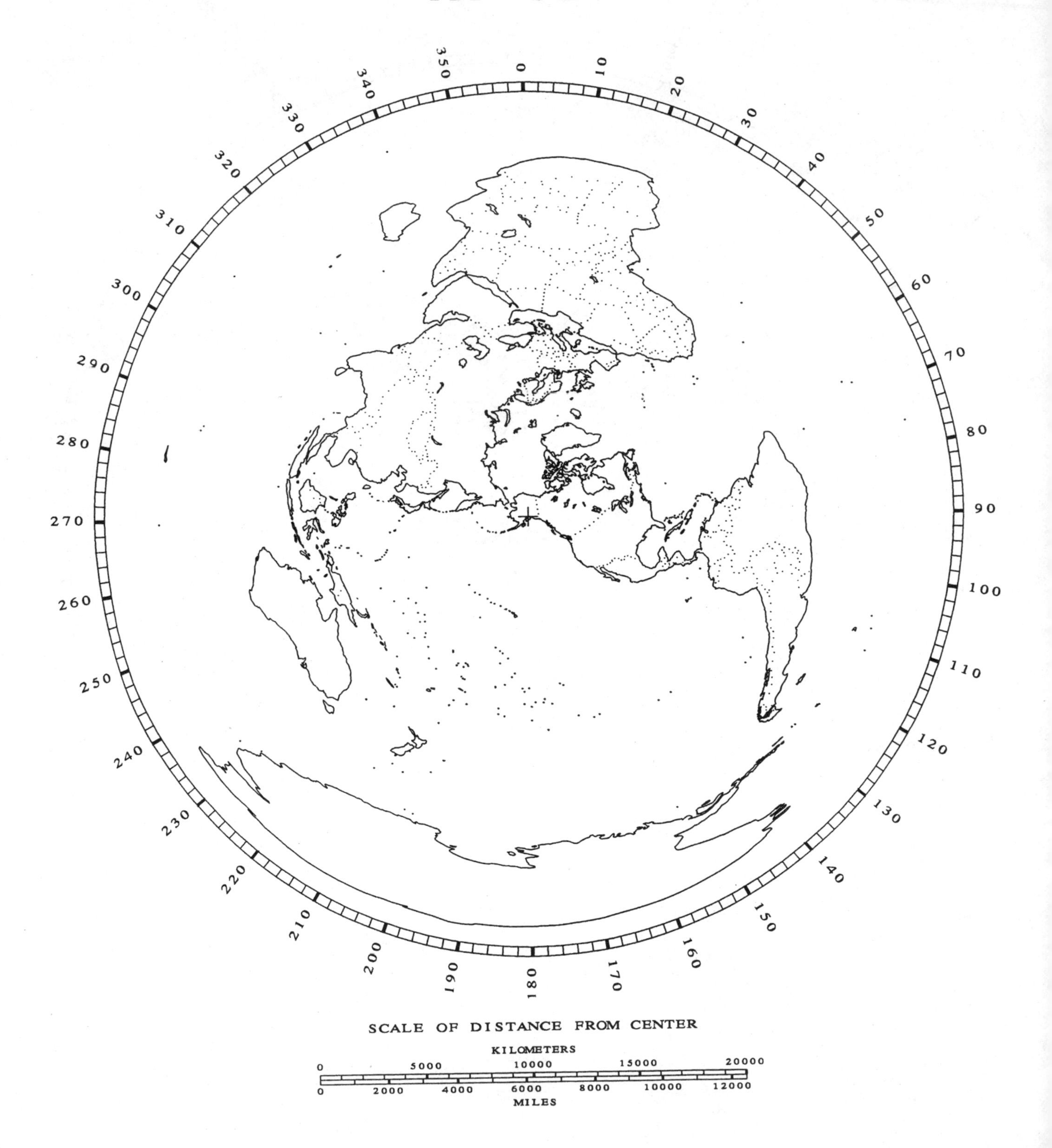

K5ZI AZIMUTHAL EQUIDISTANT MAP CENTERED ON

Hawaii

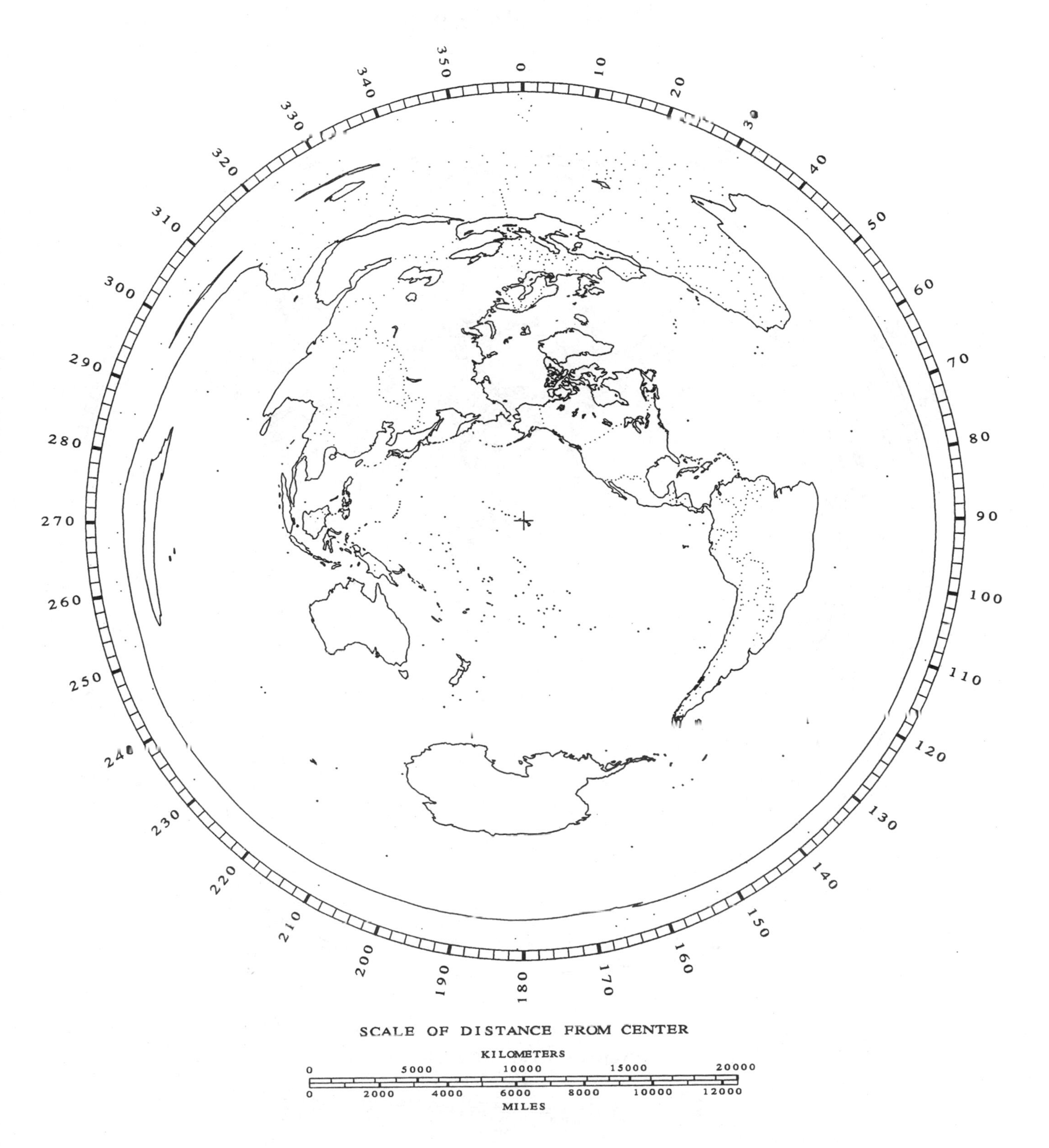

K5ZI AZIMUTHAL EQUIDISTANT MAP CENTERED ON

Caribbean

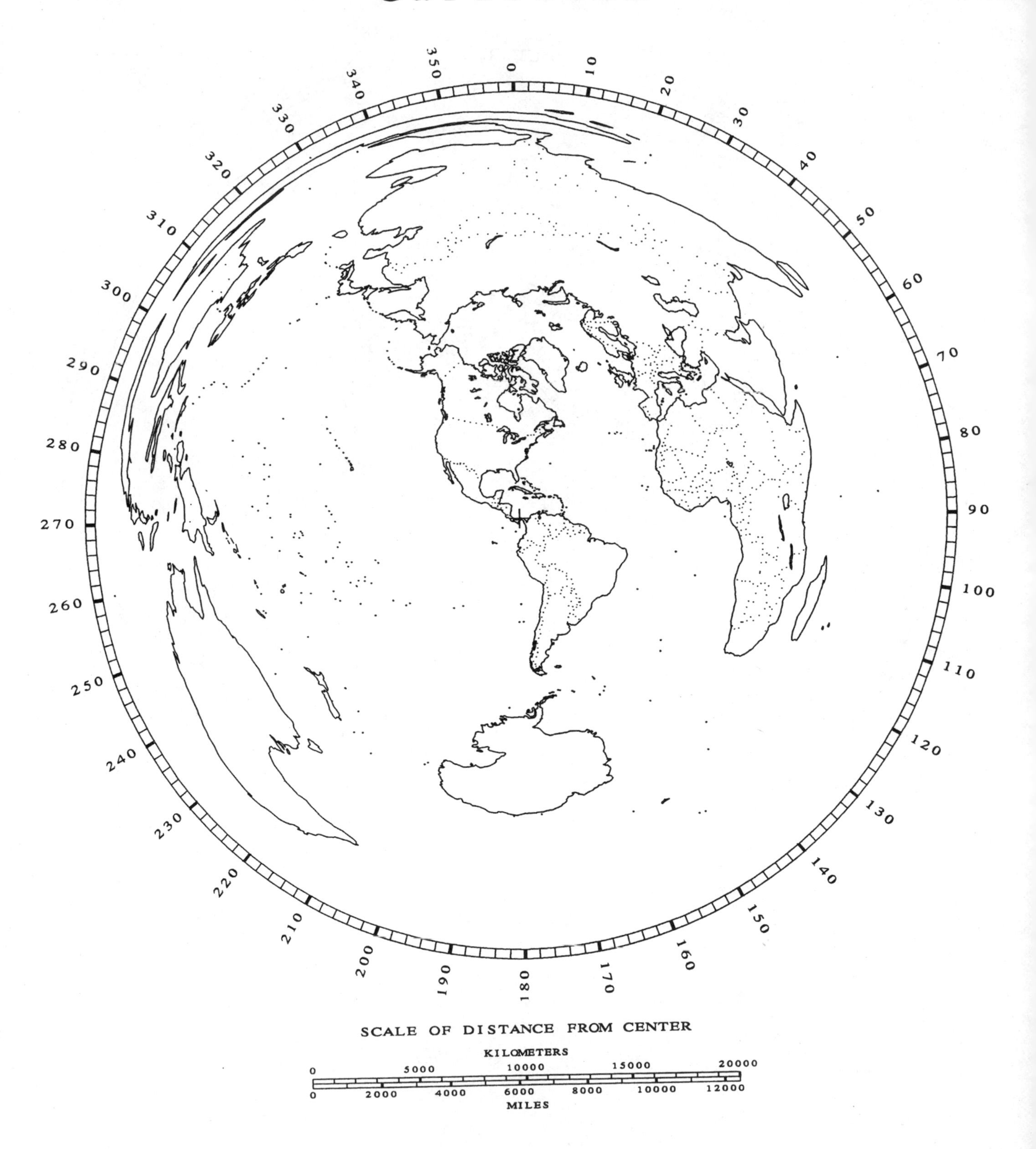

K5ZI AZIMUTHAL EQUIDISTANT MAP CENTERED ON

Eastern South America

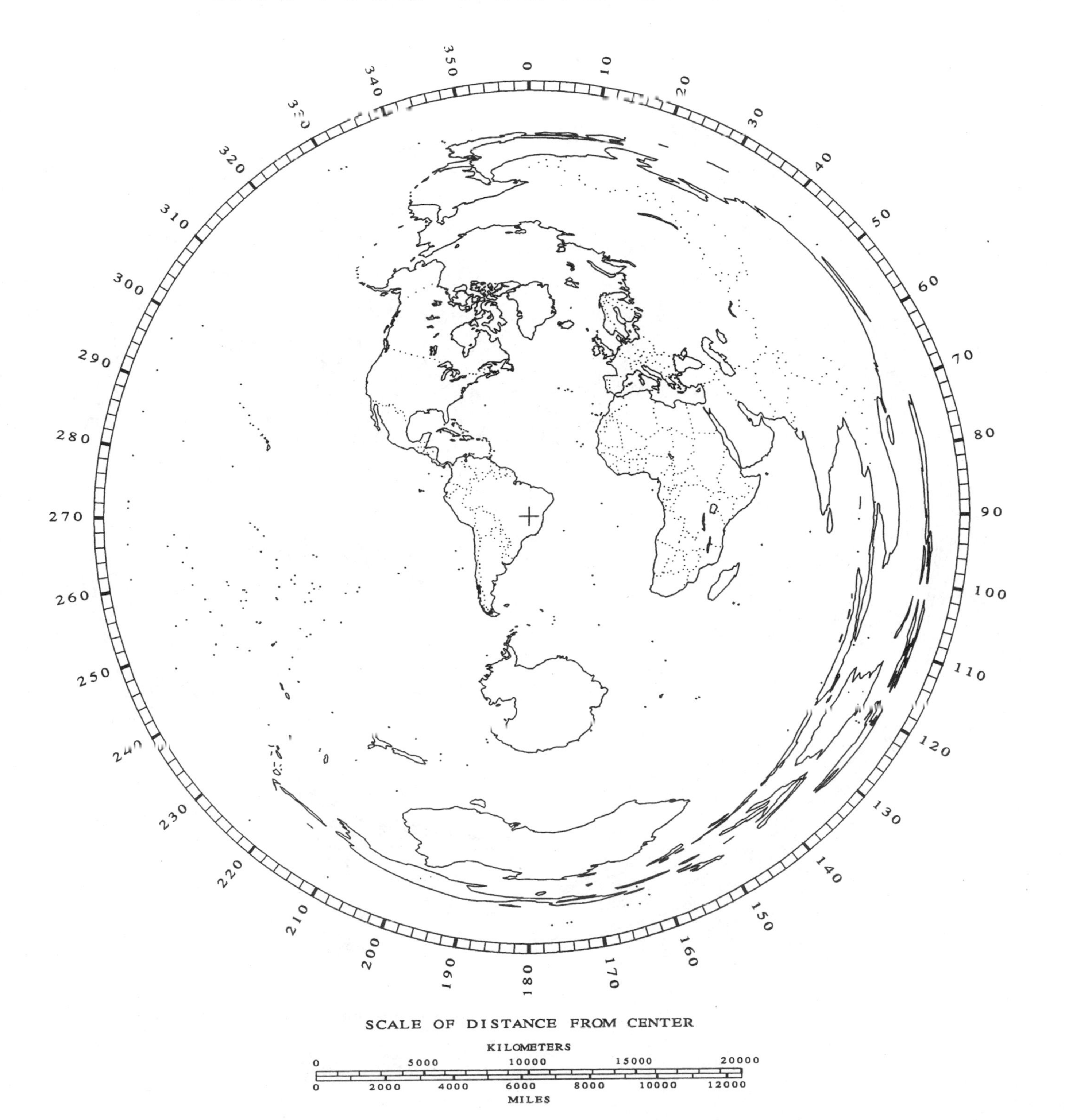

K5ZI AZIMUTHAL EQUIDISTANT MAP CENTERED ON

Southern South America

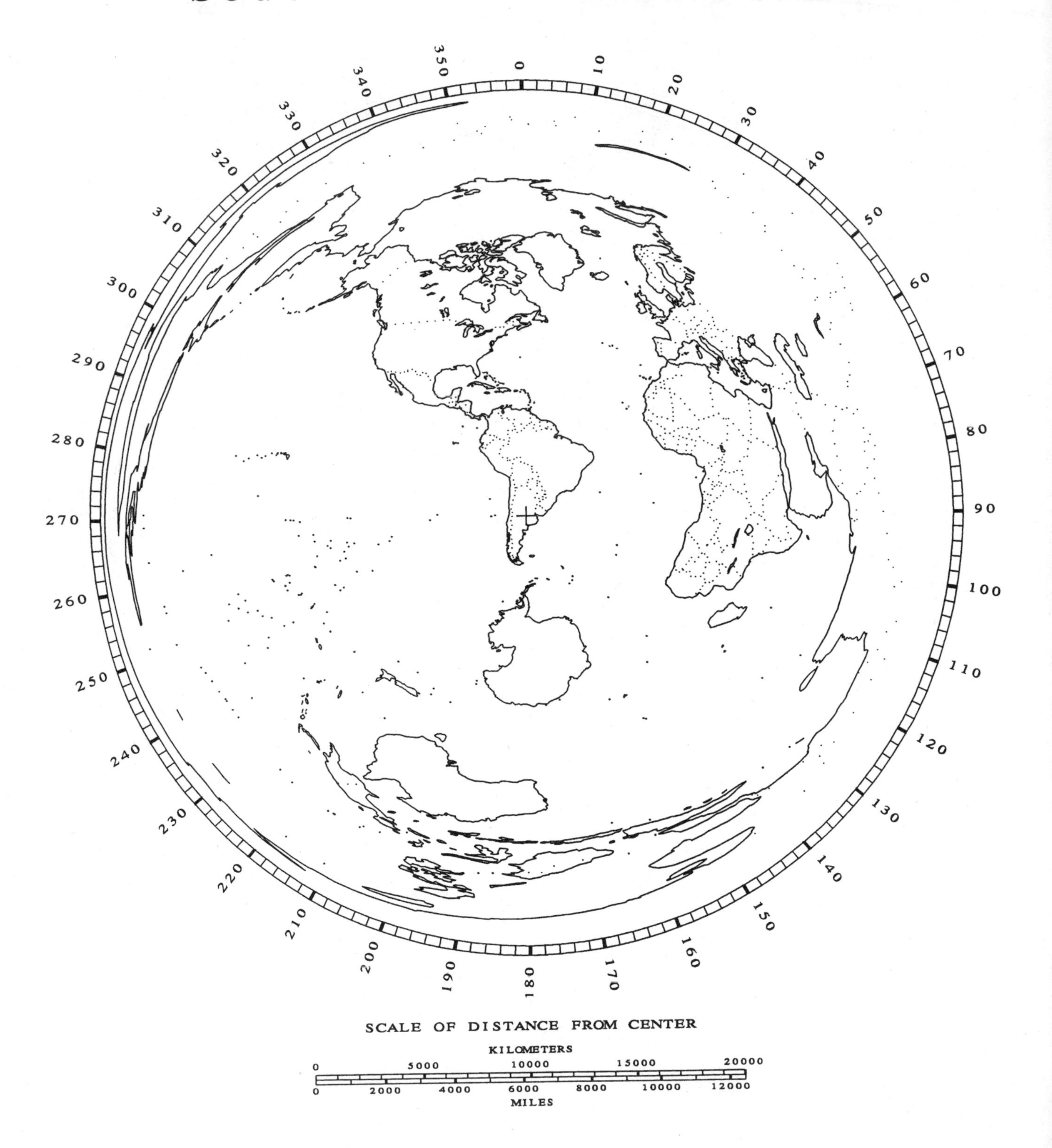

K5ZI AZIMUTHAL EQUIDISTANT MAP CENTERED ON

Antarctica

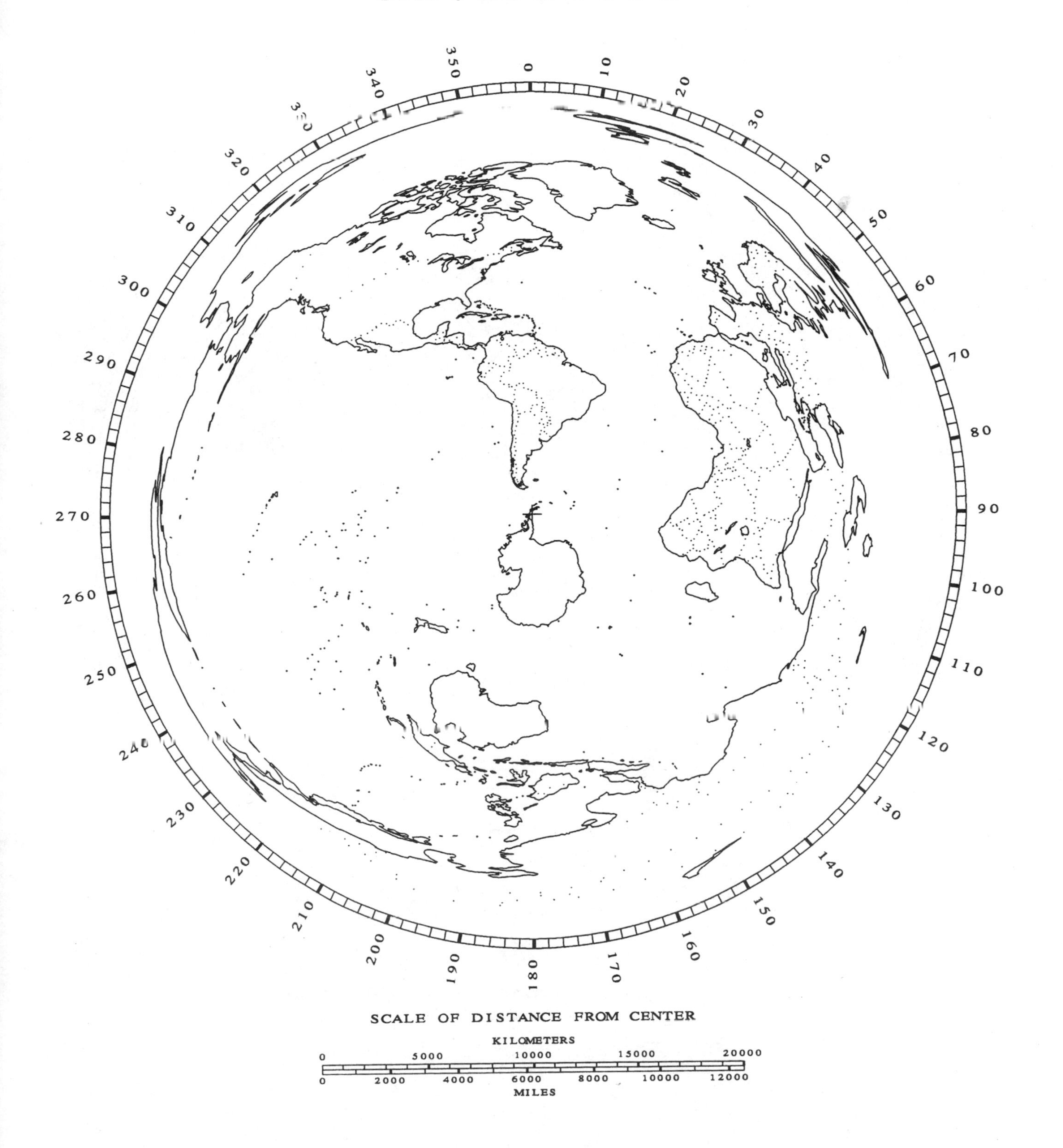

K5ZI AZIMUTHAL EQUIDISTANT MAP CENTERED ON

Western Europe

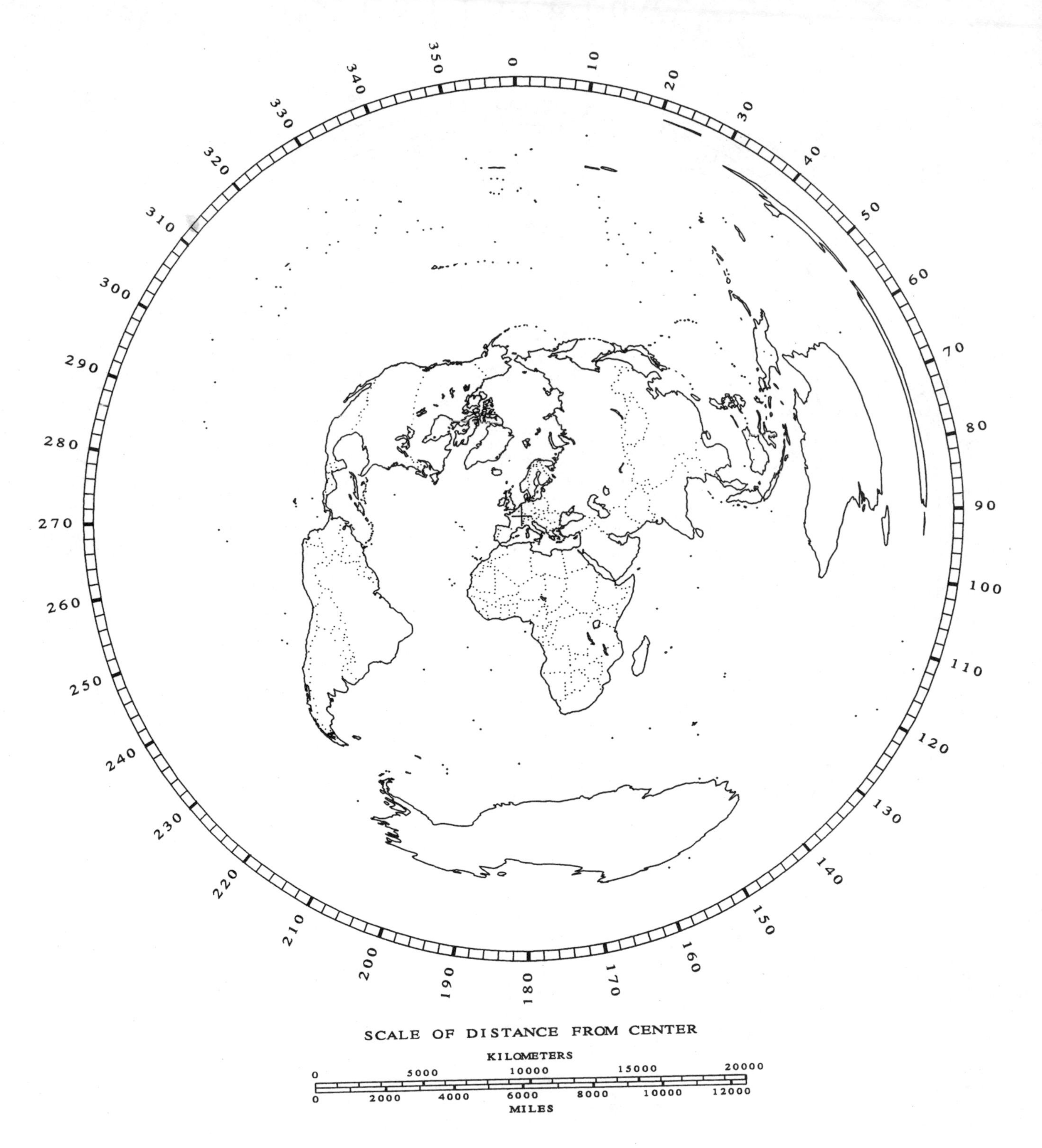

K5ZI AZIMUTHAL EQUIDISTANT MAP CENTERED ON

Eastern Europe

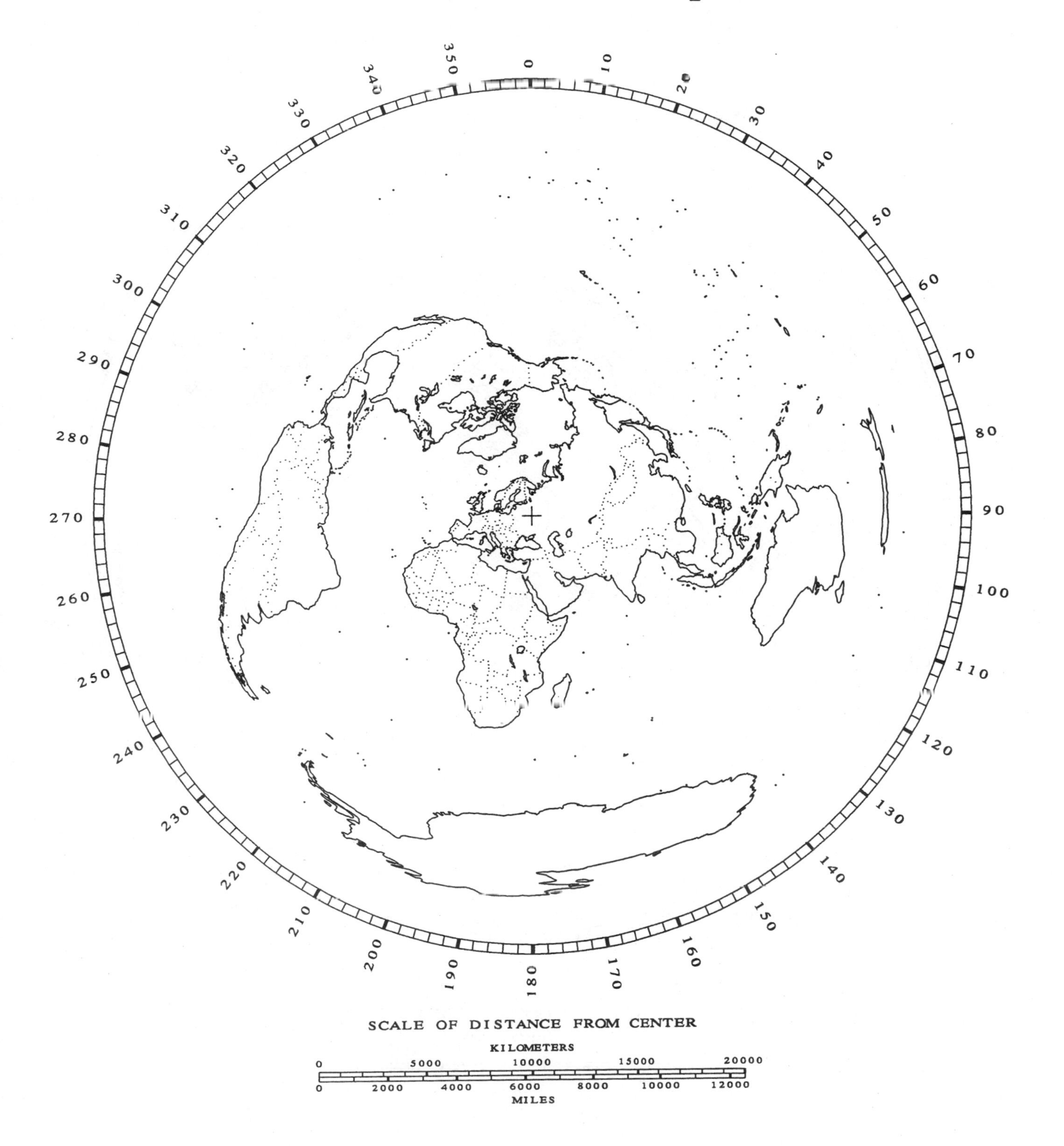

K5ZI AZIMUTHAL EQUIDISTANT MAP CENTERED ON

West Africa

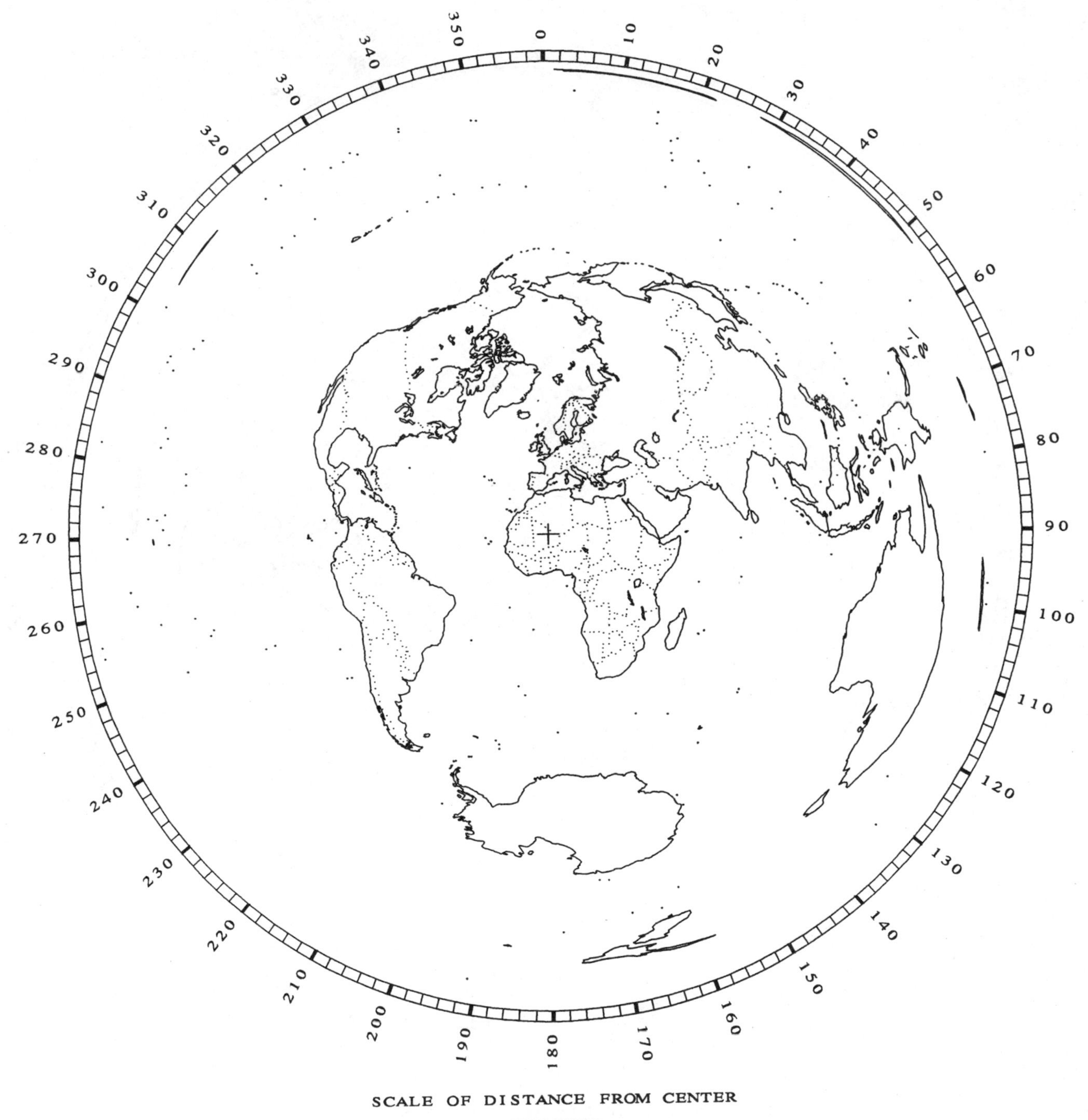

SCALE OF DISTANCE FROM CENTER

KILOMETERS

0 5000 10000 15000 20000

0 2000 4000 6000 8000 10000 12000

MILES

K5ZI AZIMUTHAL EQUIDISTANT MAP CENTERED ON

East Africa

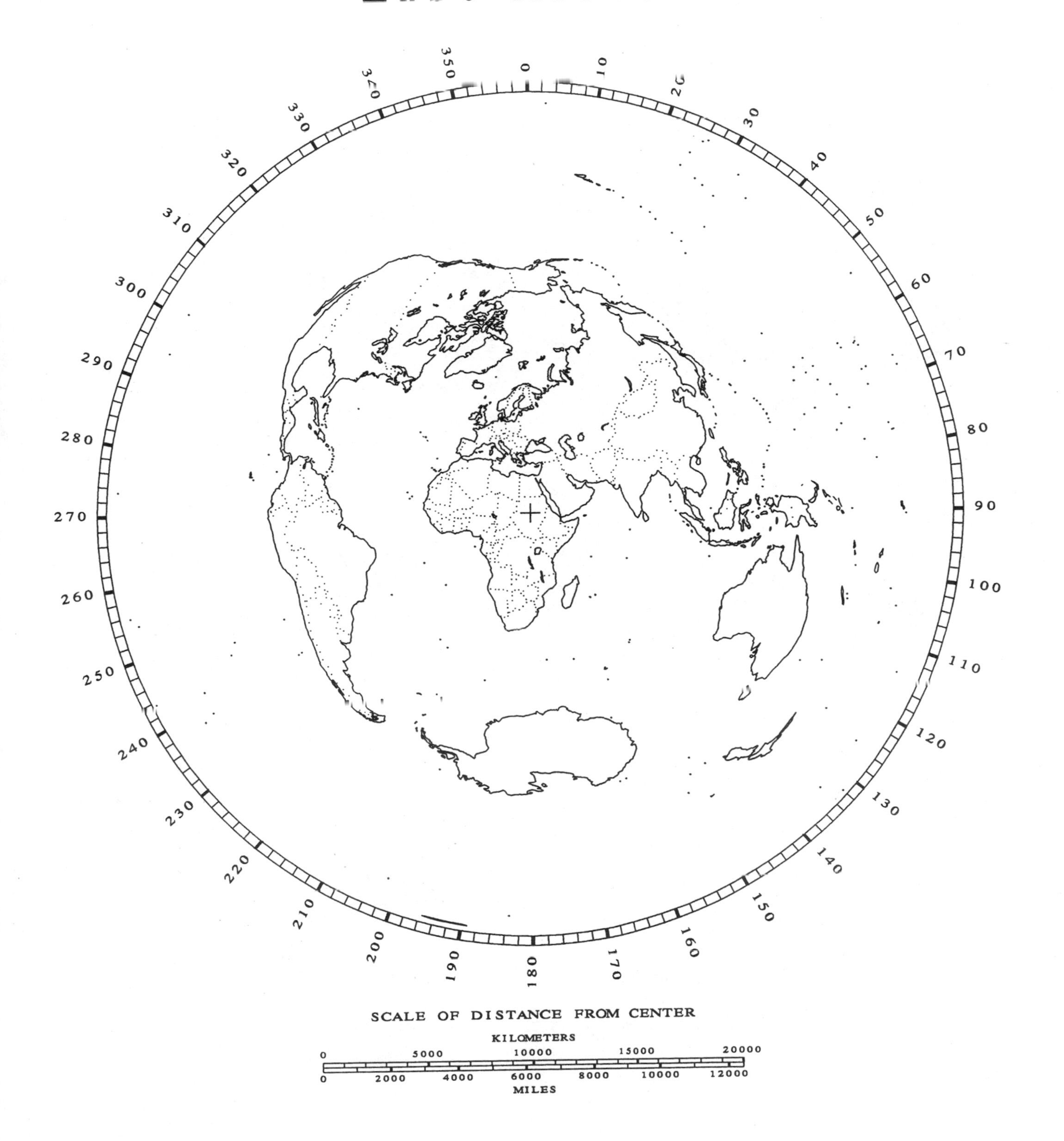

K5ZI AZIMUTHAL EQUIDISTANT MAP CENTERED ON

Southern Africa

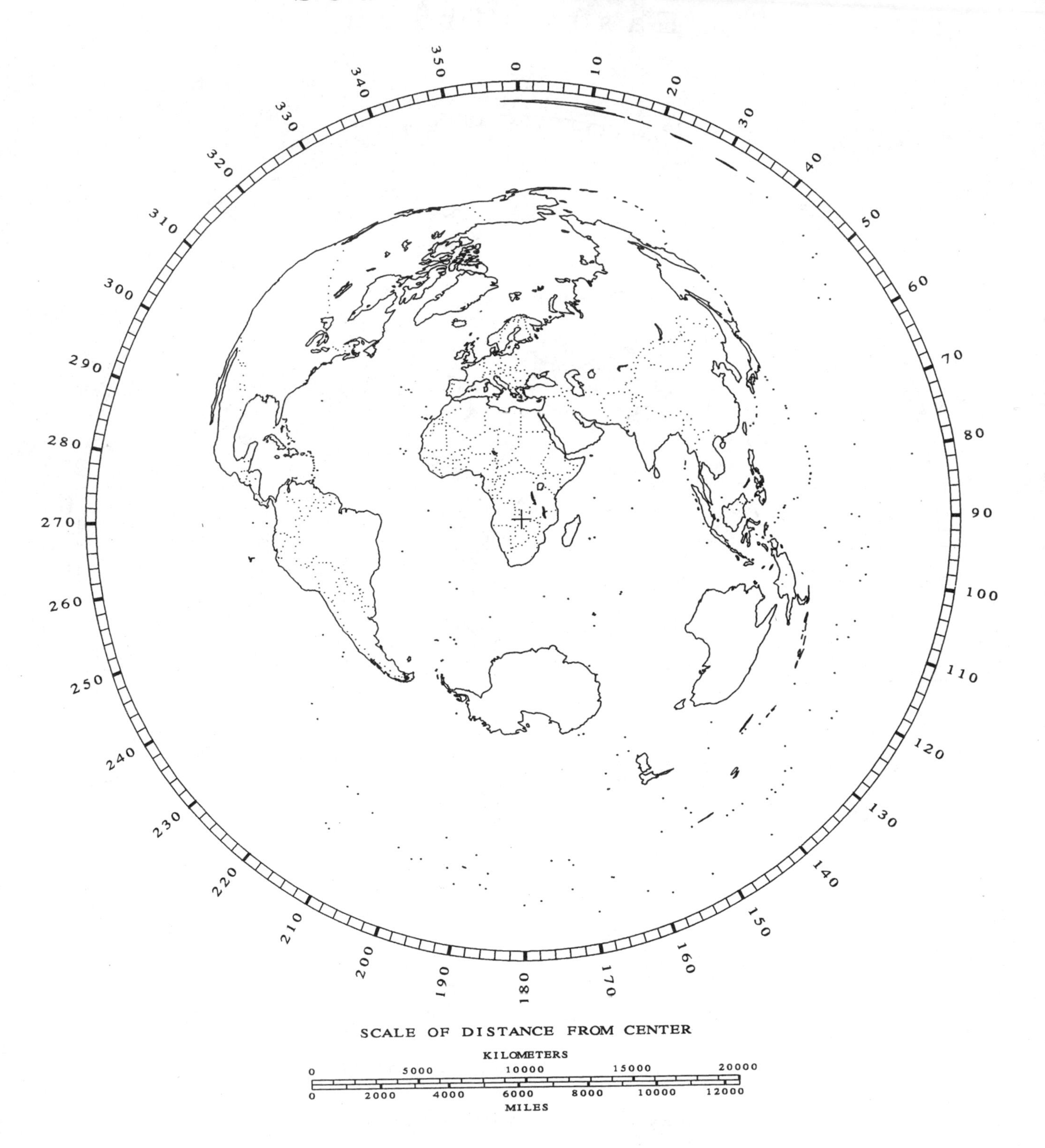

SCALE OF DISTANCE FROM CENTER

KILOMETERS

0 5000 10000 15000 20000

0 2000 4000 6000 8000 10000 12000

MILES

K5ZI AZIMUTHAL EQUIDISTANT MAP CENTERED ON

Near East

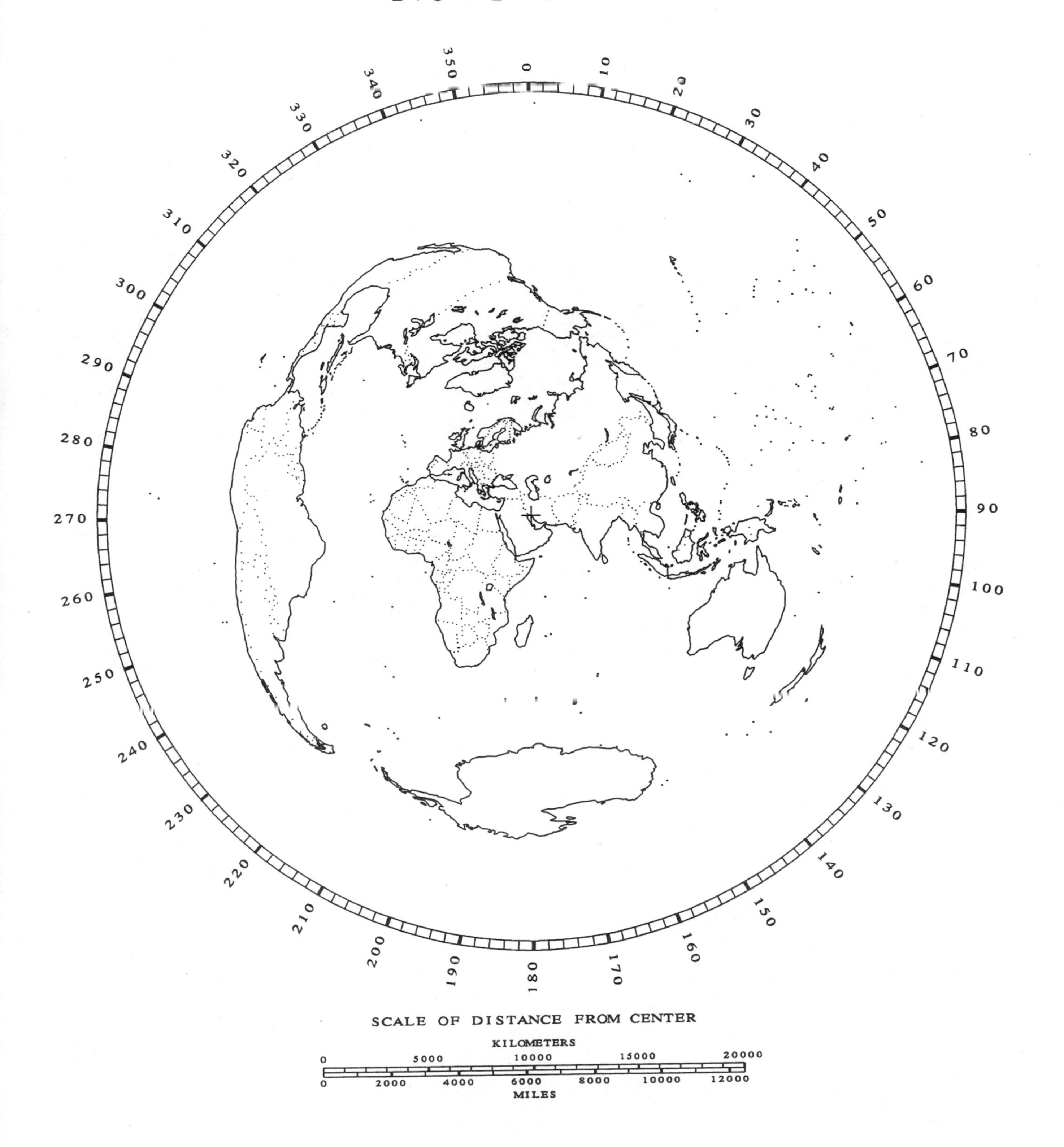

K5ZI AZIMUTHAL EQUIDISTANT MAP CENTERED ON

Southern Asia

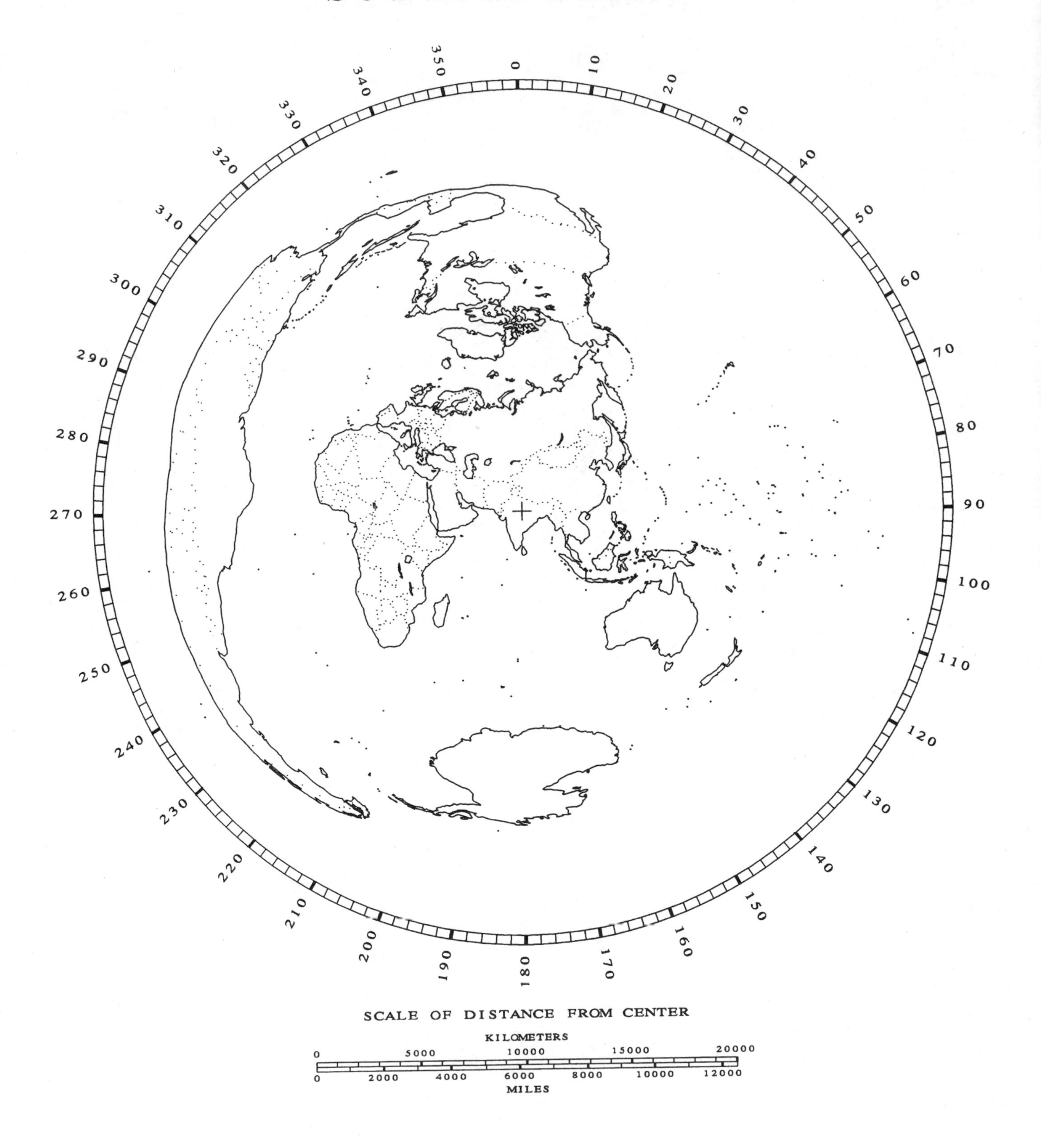

KSZI AZIMUTHAL EQUIDISTANT MAP CENTERED ON

Southeast Asia

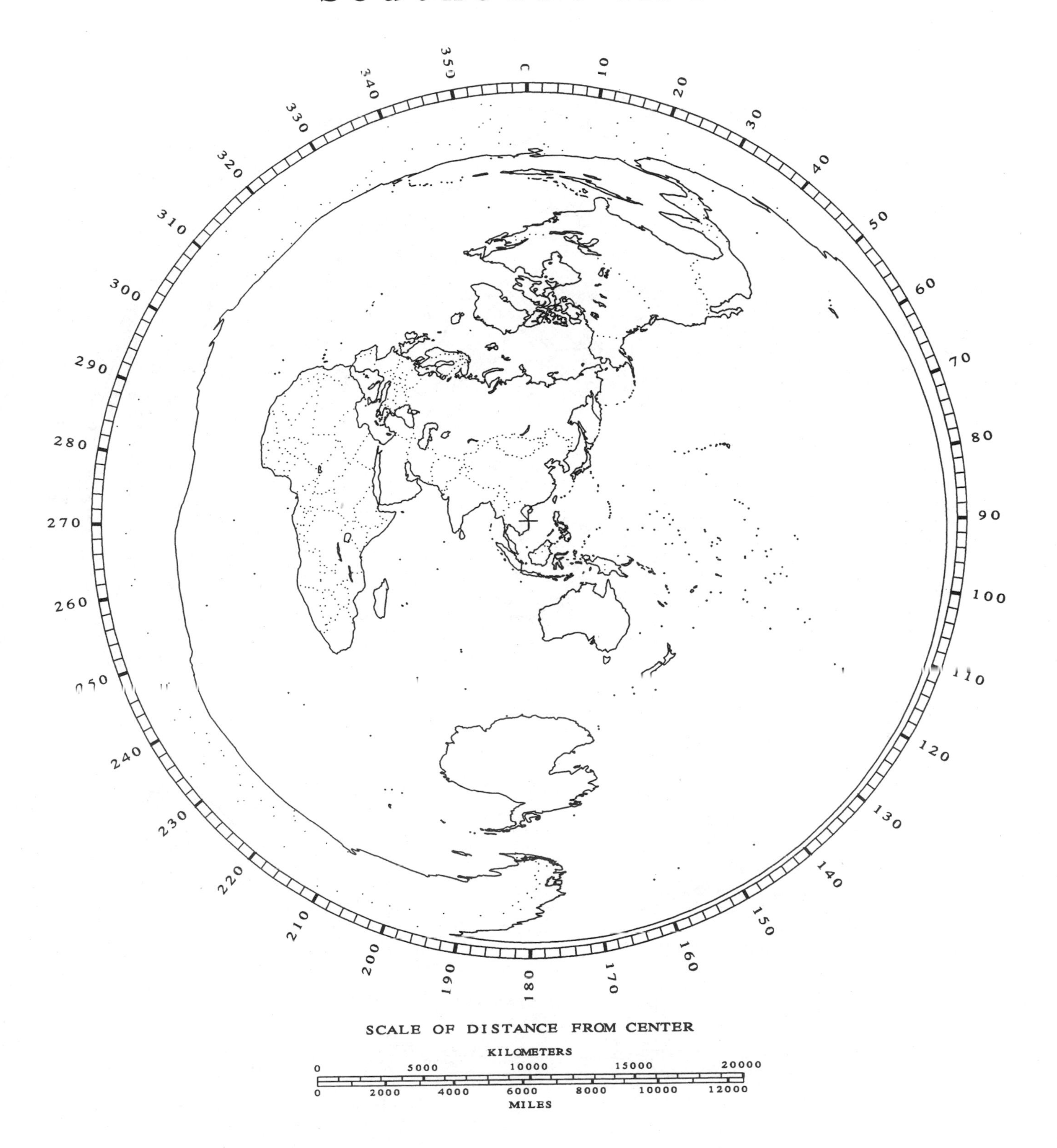

K5ZI AZIMUTHAL EQUIDISTANT MAP CENTERED ON

Far East

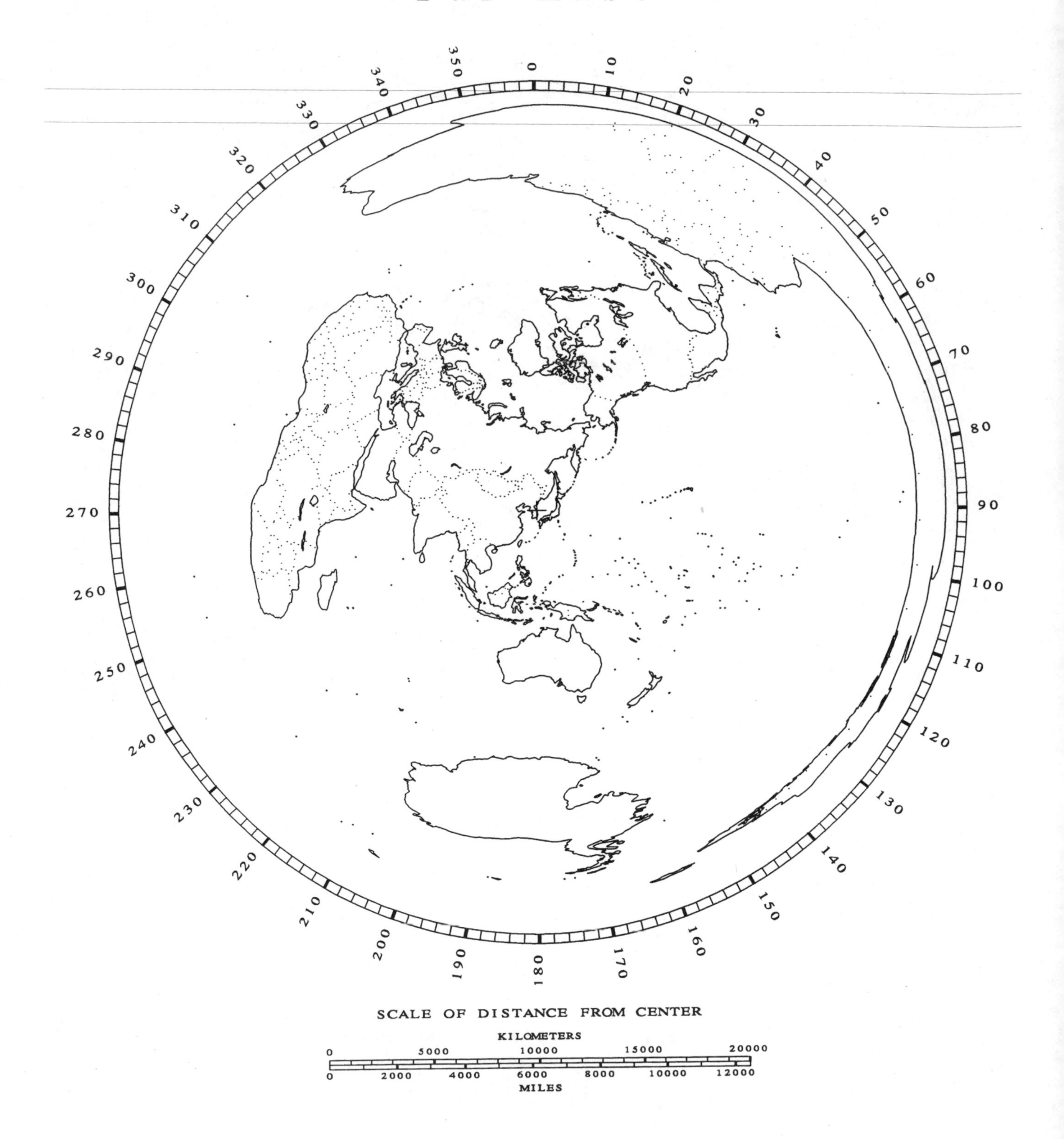

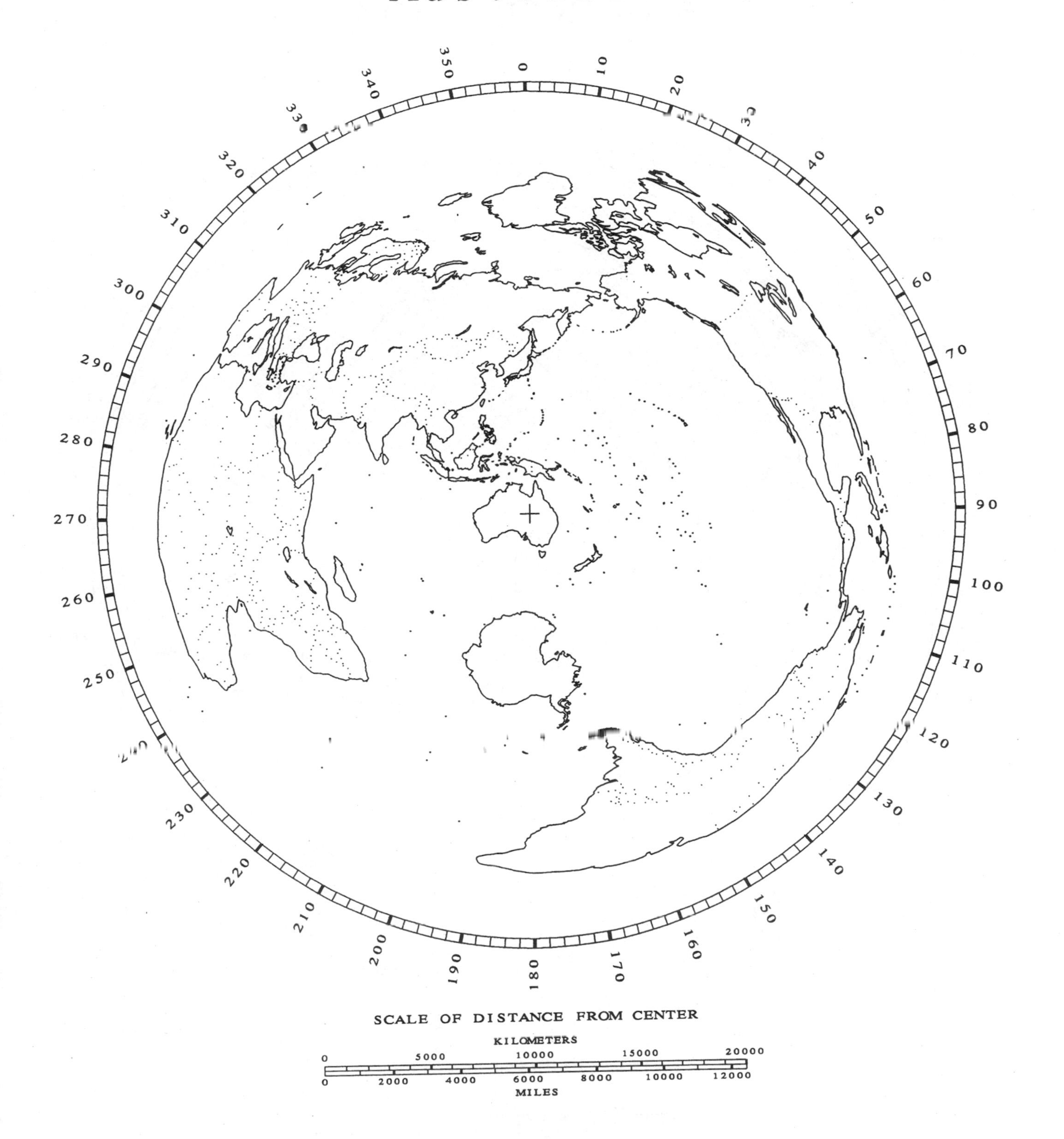
K5ZI AZIMUTHAL EQUIDISTANT MAP CENTERED ON
Australia
0
10
20
40
50
60
70
80
90
100
110
120
130
140
150
160
170
180
190
200
210
220
230
250
260
270
280
290
300
310
320
340
350
SCALE OF DISTANCE FROM CENTER
KILOMETERS
0
5000
10000
15000
20000
0
2000
4000
6000
8000
10000
12000
MILES

K5ZI AZIMUTHAL EQUIDISTANT MAP CENTERED ON

South Pacific

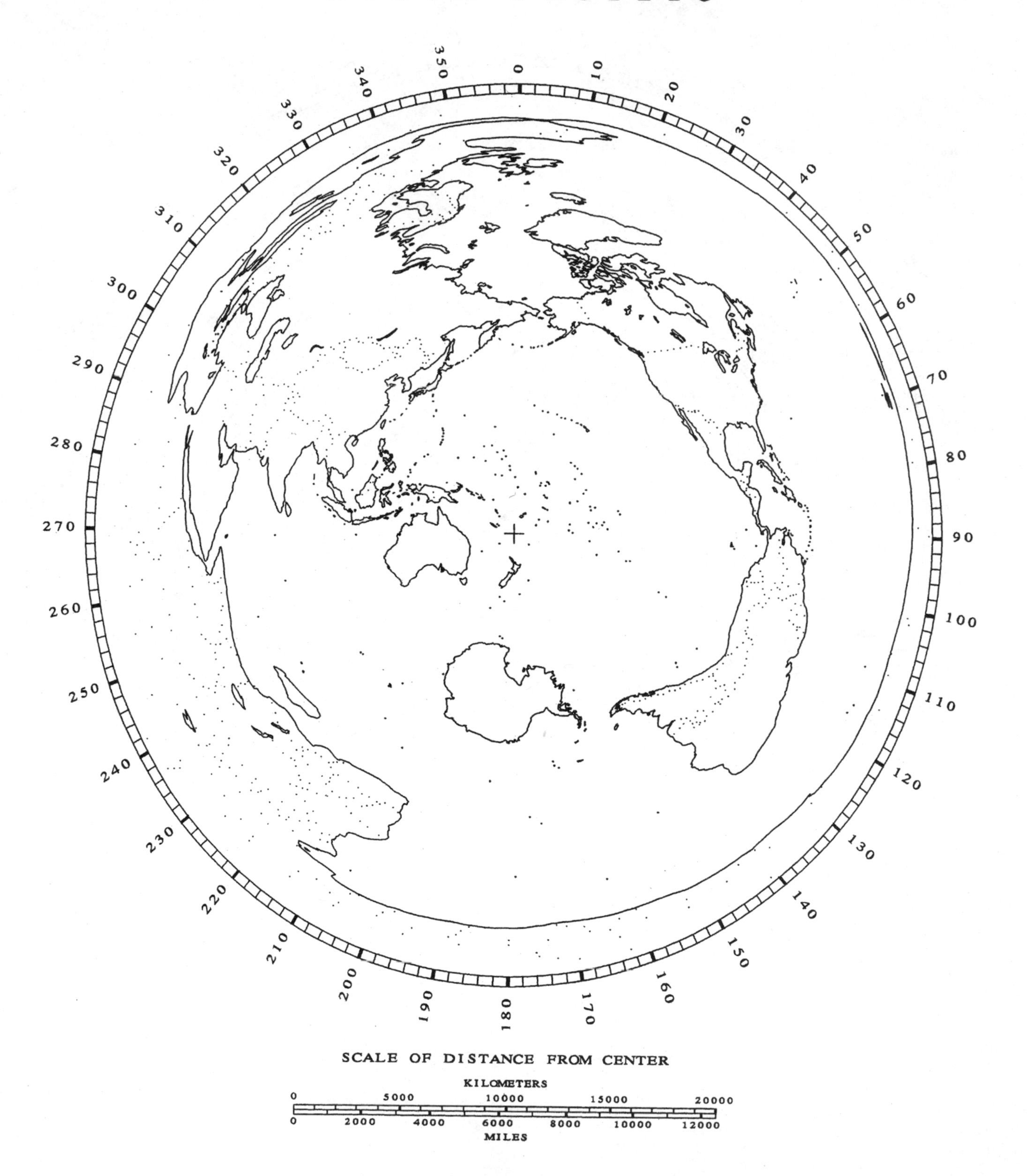

SCALE OF DISTANCE FROM CENTER

KILOMETERS

0 5000 10000 15000 20000

0 2000 4000 6000 8000 10000 12000

MILES

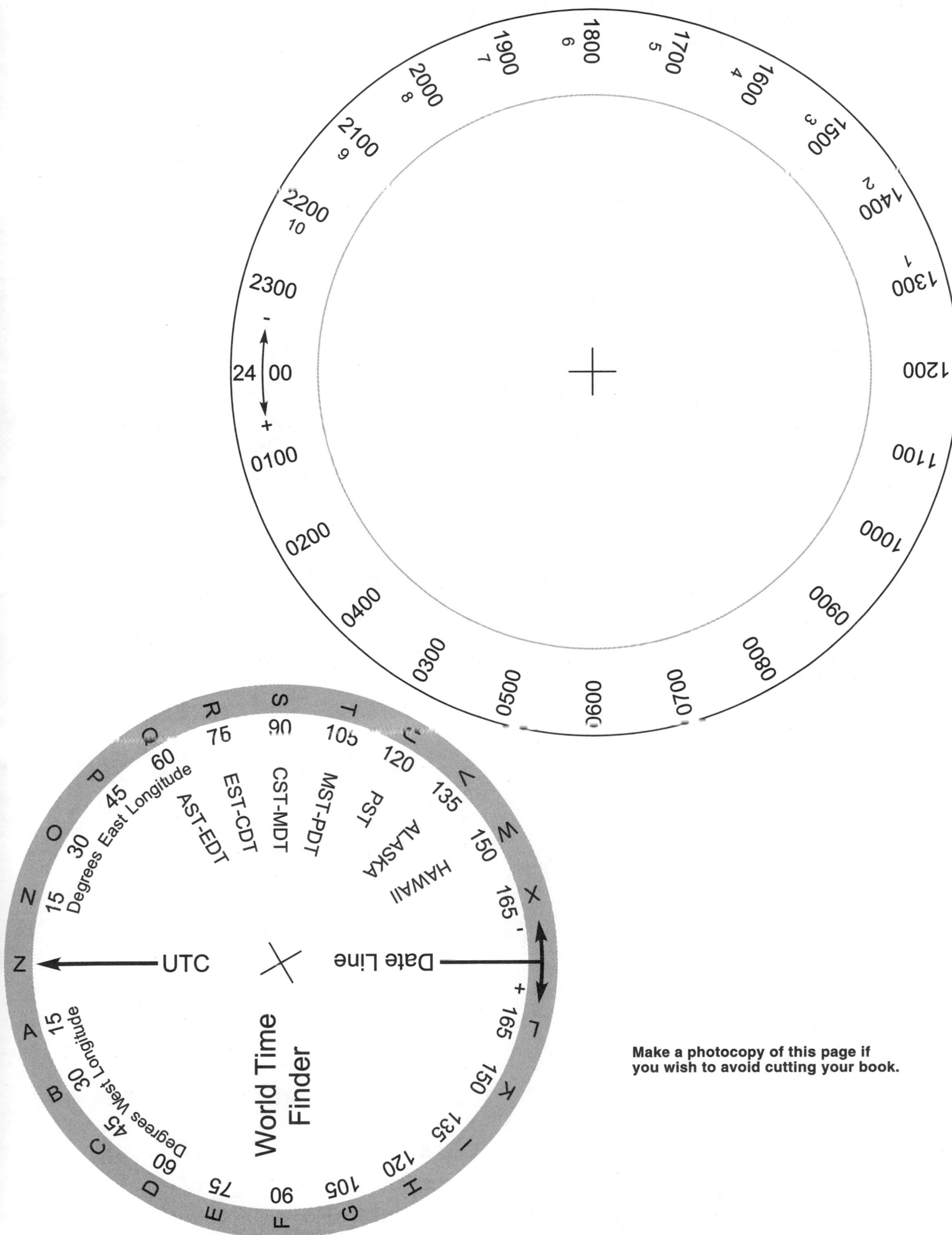

Make a photocopy of this page if you wish to avoid cutting your book.

Index

Note: HDR refers to the Ham Desktop Reference.

A

B

C

K

L

Notes

Notes

Notes

Notes

Notes

FEEDBACK

Please use this form to give us your comments on this book and what you'd like to see in future editions, or e-mail us at **pubsfdbk@arrl.org** (publications feedback). If you use e-mail, please include your name, call, e-mail address and the book title, edition and printing in the body of your message. Also indicate whether or not you are an ARRL member.

Where did you purchase this book?

☐ From ARRL directly ☐ From an ARRL dealer

Is there a dealer who carries ARRL publications within:

☐ 5 miles ☐ 15 miles ☐ 30 miles of your location? ☐ Not sure.

License class:

☐ Novice ☐ Technician ☐ Technician Plus ☐ General ☐ Advanced ☐ Amateur Extra

---

Name ______________________ ARRL member? ☐ Yes ☐ No

Call Sign ______________________

Daytime Phone ( ) ______________________ Age ______________________

Address ______________________

City, State/Province, ZIP/Postal Code ______________________

If licensed, how long? ______________________

Other hobbies ______________________

____

Occupation ______________________

For ARRL use only	OPMAN
Edition	7 8 9 10 11 12 13
Printing	2 3 4 5 6 7 8 9 10 11 12

From ______________________

____

____

Please affix postage. Post Office will not deliver without postage.

EDITOR, ARRL OPERATING MANUAL
AMERICAN RADIO RELAY LEAGUE
225 MAIN STREET
NEWINGTON CT 06111-1494

——————————————— please fold and tape ———————————————